Nature's Music

The Science of Birdsong

NATURE'S MUSIC

The Science of Birdsong

Editors

Peter Marler

Department of Neurobiology,
Physiology and Behavior
University of California
Davis California USA

Hans Slabbekoorn

Institute of Biology
Leiden University, Leiden
The Netherlands

ELSEVIER
ACADEMIC
PRESS

AMSTERDAM BOSTON HEIDELBERG LONDON NEW YORK OXFORD
PARIS SAN DIEGO SAN FRANCISCO SINGAPORE SYDNEY TOKYO

This book is printed on acid-free paper

Elsevier Academic Press
525 B Street, Suite 1900, San Diego, California 92101-4495, USA
http://www.elsevier.com

Elsevier Academic Press
84 Theobald's Road, London WC1X 8RR, UK
http://www.elsevier.com.

Library of Congress Catalog Number: 2004108979

British Library Cataloguing in Publication Data
A catalogue record for this book is available from the British Library

ISBN 0 12 473070 1
Typeset by Replica Press, India
Printed and bound in Italy

04 05 06 07 08 09 9 8 7 6 5 4 3 2 1

Contents

Preface: the Editors

PETER MARLER and HANS SLABBEKOORN

On a sad day in June 2000 Luis Felipe Baptista died unexpectedly near his home in San Francisco, California, at the age of 59. The science of birdsong lost one of its most erudite and charismatic authorities. In a prologue to this book, Baptista's long time friend and colleague Robert Bowman contributes a brief overview of some of his most illustrious accomplishments. Within months of his death still grieving colleagues decided to celebrate his remarkable career with a symposium, and ultimately with a book, on the remarkable progress that has been made by him and many other scientists around the globe in the study of birdsong.

In an enterprise spanning more than three years, the debts the editors owe to others are many. Among his many responsibilities Luis had served as Curator of Ornithology and Mammalogy at The California Academy of Sciences. The staff members of the Academy were unstinting in their support of the Baptista symposium, especially Terry Gosliner, Nina Jablonski, and Doug Bell. Sylvia Hope played a key role at many stages, including the planning and execution of the book, and Kathleen Berge was most helpful in organizing and conducting the symposium. A cadre of distinguished experts assembled on November 3 2001 for the day-long meeting. Although the speakers had been encouraged to present up-to-the-minute overviews of their area of expertise, no one was quite prepared for the extent and the richness of the new information presented. It covered all aspects of research on the vocal behavior of birds, never brought together before in the same forum. This book was the next step, after supplementing the list of authors to complete the coverage.

In an era when everyone specializes, it is all too easy for people working in one corner of the science of birdsong, say the neurobiology of vocal learning, to get out of touch with progress in other aspects of their discipline, such as the impact of the environment on the evolution of birdsong. We hope that this book will help to familiarize everyone working in the field with the remarkable progress that their colleagues have been making. Above all we hope that this book will appeal to general readers, especially to bird watchers and others who already appreciate nature's music. Luis Baptista was renowned for his skill in communicating to others, both his love of birdsong, and the excitement of scientific ornithology. In that same spirit we have strived to ensure that, wherever possible jargon has been held in check, while still maintaining the highest scientific standards, in both text and figures. The one-page 'boxes' spaced at intervals throughout the text, written specifically at the invitation of the editors, provide extra detail on some topics, and further explication of technical issues. We have solicited sound recordings of songs and other avian vocal signals wherever these could be found and assembled them in the two compact discs at the back of this book

In our editorializing it was a pleasure to interact with the sixteen biologists who joined us in writing chapters for the book. Without exception they were cooperative in dealing with our questions and suggestions, and sympathetic to our desire, wherever possible, to strike a balance between technical content and readable prose, an endeavor in which Gregory Ball and William Searcy gave some valuable help. At all stages, from preparing, editing and checking the manuscripts, organizing and formatting figures and color plates and with Judith Marler, scrutinizing proofs, Rebecca Wylie was a tower of strength. We owe special thanks to the 44 authors who made extra efforts with the boxes, creating new text and figures for the purpose and, together with the chapter authors, helping to assemble and donate sound recordings for the compact discs. Tackling the daunting task of putting the C.D.'s together would have been unthinkable without the generous help of Jack Bradbury and his staff at the Macaulay Library of Natural Sounds at the Laboratory of Ornithology, Cornell University, especially Steve Pantle, Bob Grotke and Mary Guthrie. For the color plates that Elsevier/Academic Press encouraged us to include, we are pleased to display the work of photographers who specialize in creating images of birds caught in the very act of vocalizing. They add a further dimension to our efforts to convey something of the saga of *Nature's Music*.

List of contributors

Bridget Appleby	BBC Natural History Unit, Bristol, England, UK.
Arthur P. Arnold	University of California, Los Angeles, USA
Myron C. Baker	Colorado State University, Fort Collins, USA
Gregory F. Ball	John Hopkins University, Baltimore, Maryland, USA
Robert Bowman	California Academy of Sciences, San Francisco, USA
Deborah Buitron	North Dakota State University, Fargo, USA
Clive K. Catchpole	University of London, Egham, England, UK
MarthaLeah Chaiken	Hofstra University, Hempstead, New York, USA
Claude Chappuis	Museum of Natural History, Paris, France
Sarah Collins	University of Nottingham, England, UK
Jeffrey Cynx	Vassar College, Poughkeepsie, New York, USA
Robert Dooling	University of Maryland, College Park, USA
Françoise Dowsett-Lemaire	Ganges, France
Chris Evans	MacQuarie University, Sydney, Australia
Adriana R. Ferreira	Duke University, Durham, North Carolina, USA
W. Tecumseh Fitch	University of St. Andrews, Scotland, UK
Sandra L.L. Gaunt	Ohio State University, Columbus, USA
Jean-Pierre Gautier	Universite de Rennes, Paimpont, France
Michael H. Goldstein	Franklin & Marshall College, Lancaster, Pennsylvania, USA
Franz Goller	University of Utah, Salt Lake City, USA
Thomas Hahn	University of California, Davis, USA
Paul Handford	University of Western Ontario, London, Canada
Henrike Hultsch	Freie Universität, Berlin, Germany
Darren E. Irwin	University of California, Los Angeles, USA
Jessica H. Irwin	University of California, Los Angeles, USA
Erich D. Jarvis	Duke University, Durham, North Carolina, USA
Andrew P. King	Indiana University, Bloomington, USA
Kohta I. Kobayasi	Chiba University, Chiba-city, Japan
Michel Kreutzer	Université Paris X-Nanterre, France
Donald Kroodsma	University of Massachusetts, Amherst, USA
Anthony Leonardo	Harvard University, Cambridge, Massachusetts, USA
Daniel Margoliash	University of Chicago, Illinois, USA
Peter Marler	University of California, Davis, USA
Archibald McCallum	Applied Bioacoustics, Eugene, Oregon, USA
Douglas A. Nelson	Ohio State University, Columbus, USA
Stephen Nowicki	Duke University, Durham, North Carolina, USA
Gary L. Nuechterlein	North Dakota State University, Fargo, USA

Kazuo Okanoya	Chiba University, Chiba-city, Japan
Robert B. Payne	University of Michigan, Ann Arbor, USA
Irene M. Pepperberg	Massachusetts Institute of Technology, Cambridge, USA
Jeffrey Podos	University of Massachusetts, Amherst, USA
Marilyn Ramenofsky	University of Washington, Seattle, USA
Steve Redpath	Center for Ecology and Hydrology, Banchory, Scotland, UK
Katharina Riebel	Leiden University, The Netherlands
William A. Searcy	University of Miami, Coral Gables, Florida, USA
Hans Slabbekoorn	Leiden University, The Netherlands
Jill Soha	Ohio State University, Columbus, USA
Roderick A. Suthers	Indiana University, Bloomington, USA
Carel ten Cate	Leiden University, The Netherlands
Dietmar Todt	Freie Universität, Berlin, Germany
Mari A. Tokuda	Tokyo Institute of Technology, Japan
Meredith J. West	Indiana University, Bloomington, USA
Jay Withgott	Portland, Oregon, USA
Ayako Yamaguchi	Boston University, Massachusetts, USA
Richard A. Zann	La Trobe University, Melbourne, Australia

Foreword

A tribute to the late Luis Felipe Baptista

ROBERT I. BOWMAN

This volume is a celebration of the life of Luis Felipe Baptista, and it is appropriate and timely to review some of the highlights of Dr Baptista's scientific career. Tales of his productivity are legendary, and many of his achievements were widely publicized in tributes by his friends and professional associates after his death, in speeches, by radio and television, in scientific journals, in newspapers and magazines. I knew Luis as a friend and colleague since his graduate student days at the University of California, Berkeley, commencing in the fall of 1967. He had just completed his master's degree studies at the University of San Francisco under the tutelage of the renowned ornithologist Dr Robert T. Orr, who was at that time also the Curator of Ornithology and Mammalogy at the California Academy of Sciences. It was through this first close relationship with a professional biologist that Luis was amazed to learn that some people actually made a living doing what he had been doing for pleasure since boyhood – watching birds!

As a youth raised by Portuguese-Chinese parents in Hong Kong and Macau, it was painless for him to learn to speak Portuguese, English, and Chinese. Later he picked up Spanish and German with little difficulty. He also became very familiar with the habits and voices of native birds, as well as the widely prevalent Chinese art of aviculture. Early in his youth he realized that some birds learned their songs by imitation, a major focus of his later research. Indeed his graduate research on the song of the white-crowned sparrow – an abundant bird on the Berkeley campus and at other San Francisco Bay Area localities – provided him with a plethora of thematic vocal variations. He later broadened his documentation of variations in song from California to British Columbia. He demonstrated that boundaries between dialects are not always clear-cut, and that birds living at the interface may be bilingual.

A great source of amusement and amazement for those who were privileged to hear Luis report on his white-crowned sparrow songs were his precise imitations of their dialects. He had a remarkable memory and an acute sensitivity to the many nuances of pitch and timing that individual birds displayed.

After the completion of his doctoral dissertation at Berkeley, Luis received a grant from NATO to visit Germany where he spent more than a year at the Max Planck Institüt für Verhaltensphysiologie. While stationed at Vögelwarte Radolfzell, he recorded the learned 'rain call' of the chaffinch around Lake Constance, trying to define dialect boundaries. The project was later expanded to include a transect across Germany and resulted in a remarkable documentation of geographical dialects in a bird call.

Upon his return to the United States in 1973 Luis assumed an academic appointment at Occidental College in Los Angeles where he became Professor of Biology and Curator of the Moore Laboratory collection of birdsongs. In 1980, Luis was appointed Chairman and Associate Curator of Ornithology and Mammalogy at the California Academy of Sciences in San Francisco, where he remained until his untimely death on June 10, 2000.

Without enumerating details of his more than 120 scientific publications, the majority of which concern avian vocalizations, Luis' broad and detailed knowledge of birds of the world is evident in some of the other topics he wrote about, often with co-authors, including cowbird parasitism; leaf bathing in emberizine sparrows; the use of courtship displays in taxonomy; behavioral interactions within and among animal species; handedness and its possible taxonomic significance in grassquits; taxonomic revision of the Mexican *Piculus* woodpecker complex; hybridization in *Calypte* hummingbirds; behavior and taxonomic status of Grayson's dove; photoperiodically induced ovarian growth in white-crowned sparrows; testosterone, aggression, and dominance in Gambel's white-crowned sparrows; the origin of Darwin's finches; the fine structural basis of the cuteaneous water barrier in nesting zebra finches; production and control of birdsong; the behavior, status and relationships of the endemic St Lucia black finch; inheritance and loss of the straw display in estrildid finches; field observations of some New Guinea mannikins; the role of song in the evolution of passerine diversity; seasonal changes in song behavior and song nuclei in the brain of Gambel's white-crowned sparrow; nature and its nurturing in avian vocal development; the program of reproduction and reintroduction of the dove on Socorro Island; bioacoustics as a tool in conservation studies; guidelines for the use of wild birds in research; relationships of some mannikins and waxbills in the Estrildidae; linguistics and anthropological study of Chinese birds; cognitive processes in avian vocal development; what the white-crowned sparrow can teach us about language.

Luis' breadth of knowledge of birds of the world qualified him superbly for the task of revising one of the most famous modern ornithology texts, *The Life of Birds* by J.C. Welty. At the time of his death, Luis was involved with a National Academy of Sciences project on the biology of music and its relevance to birdsong.

Much of Luis' avian research was conducted and co-authored cooperatively with colleagues and junior scientists. He shared his knowledge of birds generously, with students and the general public alike. He was a great communicator. He possessed a brilliant mind, an encyclopedic memory of facts and faces, and a warmth of personality that charmed all who visited his quarters at the California Academy of Sciences. His presence served as a magnet for visiting scientists and the lay public.

Through the years Luis' presence greatly enhanced the scientific stature of the Academy through his public lectures, foreign travel on research projects, his prolific publication record, his generous coaching of graduate students, and his concern for the survival of endangered bird species. The latter brought him close to the problem of environmental degradation and the sorrowful fate of the endemic dove of Socorro Island off Baja California. Fortunately, a few surviving individuals were to be found in the collections of mainland aviculturists. Working with Mexican biologists, Luis initiated an intensive breeding program in the hope of reintroducing the species to its ancestral homeland, once the island environment was rehabilitated. The good news is that plans were launched in collaboration now under way with Mexican authorities to reintroduce the Socorro dove to its ancestral home. The year 2003 is significant because it commemorates the 100th anniversary of the first expedition to the Revillagigedo Archipelago by the California Academy of Sciences. The sad news is that Luis will never see the fruits of his conservation efforts. But he took pride in his award from the Rolex Corporation for initiating this pioneering enterprise in the field of conservation.

The departure of Luis Baptista from the halls of science at universities, and at the California Academy of Sciences is deeply felt by his many friends and admirers around the world. This symposium volume is a deserving tribute to the legacy of his creative spirit, his voracious curiosity, and his many insights into the minds of birds. His unabashed love for scientific investigation was nourished by his faithful and supportive staff, and especially by his beloved long-term companion, Helen Horblit. I know that he would be deeply pleased and appreciative of the dramatic progress in many aspects of the science of birdsong reviewed in this volume.

Acknowledgements

CHAPTER 1

Peter Marler: I dedicate this brief historical review to my late mentor, William Homan Thorpe, the pioneer who launched the new discipline. For help in preparing the chapter I am grateful to Hans Slabbekoorn, my co-editor. In preparing figures and obtaining sound recordings, I was aided generously by Allan Baker, Jack Bradbury, Doug Nelson, Jürgen Nicolai, Steve Nowicki, Peter Slater, Bill Searcy, Jill Soha, and Rebecca Wylie. Thanks are due to the Macaulay Library of Natural Sounds at Cornell University for allowing use of sound recordings from their CD on the Diversity of Animal Sounds. For research assistance over the years I am grateful to Miwako Tamura, Mary Sue Waser, Roberta Pickert, Virginia Sherman, and especially to Susan Peters. Don Kroodsma gave generous access to material on the swamp sparrows of the Great Vly. Research was supported by the National Science Foundation and the National Institute of Mental Health.

CHAPTER 2

Sarah Collins: Thanks for comments on earlier versions of the chapter to Bill Searcy, Hans Slabbekoorn, and Peter Marler and to Selvino de Kort for discussions on the nature of birdsong. Thanks also to the students of the University of Nottingham on the Behavioural Ecology Field Course for their enthusiasm in recording birdsong and conducting playback experiments. My work was supported by a grant from the Nuffield Foundation.

CHAPTER 3

Henrike Hultsch & Dietmar Todt: We thank the editors for their help in improving the quality of this chapter, and Nicole Geberzahn for preparing most of the figures and the sound samples. The studies on nightingales reported here were supported by a fund of the German Science Foundation DFG (Az: To 13/30-1).

CHAPTER 4

Donald Kroodsma: I'm grateful that I've spent much of my adult life thinking about the diversity of bird sounds, and have key people to thank for those opportunities: my parents, Sewall Pettingill, John Wiens, Peter Marler, and my wife Melissa, who has tolerated and encouraged all of this frivolity. The National Science Foundation has provided substantial support over the years. Thanks also to Peter Marler and Hans Slabbekoorn, two gracious and patient editors, and to Luis Baptista, with whom I shared more than a quarter century of birdsong adventures.

CHAPTER 5

Peter Marler: I thank Sylvia Hope for much help in preparing and commenting on this review. Many people contributed generously to the background research, including Mike Baker, Mike Beecher, Hans-Heiner Bergmann, Jack Bradbury, Thomas Bugnyar, Nick Davies, Chris Evans, Bruce Falls, Millicent Ficken, Tom Hahn, Bill Hamilton, Berndt Heinrich, Arla Hile, Georg Klump, Walt Koenig, Indrikis Krams, Paul Mundinger, Marilyn Ramenofsky, Steve Rothstein, Hans Slabbekoorn, Georg Striedter, and Sandy Vehrencamp. I am grateful to Don Kroodsma and the Laboratory of Ornithology at Cornell University for permission to reproduce figures and sound recordings from the chapter on vocal behavior in their Home Study Course in Bird Biology.

CHAPTER 6

Hans Slabbekoorn: I thank Tom Smith of the Center for Tropical Research for moral and financial support during my studies in Cameroon, California and Colorado. Luis Baptista was a great pleasure to meet and I am grateful to him and the California Academy of Sciences for allowing me to use their acoustic laboratory. Jacintha Ellers, Peter Marler, and Katharina Riebel made helpful comments that improved my chapter. Many thanks to Thierry Aubin, Theo Beerenfenger, Frank Dorritie, Lang Elliott, Albertina Leitao, and Gary Ritchison for help in getting sound recordings. I thank Andrea Jesse for her work and pleasant collaboration on the white-crowned sparrows of the San Francisco Bay Area (Figure 1.6). R. Manin is the photographer of both pictures in Box 39. I thank Merijn de Bakker, John van Dort, Bart Houx, Caroline van Heijningen, Rob Lachlan, Helder Perreira, Martin Poot, Pien Verburg, and Paige Warren for suggestions and material about musicians inspired by birdsong. My work was supported by the Netherlands Organization for Scientific Research. NWO and the US National Science Foundation.

CHAPTER 7

Robert Dooling: I thank all of the students, postdoctoral fellows, and colleagues who have collaborated on the work contained herein over the years, especially Beth Brittan-Powell, Bernard Lohr, and Marjorie Leek who also read and commented on previous drafts of this manuscript. This work was supported by grants from the National Institute of Deafness and Communication Disorders.

CHAPTER 8

Erich D. Jarvis: For assistance with figures and/or text I thank: Art Arnold, Gregory Ball, Eliot Brenowitz, Catherine Carr, Timothy DeVoogd, Fred Gage, Adriana Ferreira, John Kirn, Lubica Kubikova, Dan Margoliash, Claudio Mello, Richard Mooney, Paul Nealen, David Perkel, Constance Scharff, Kazuhiro Wada, Martin Wild. For sound files I thank: Michael Brainard, Fernando Nottebohm, Kazuo Okanoya, Constance Scharff. In Figure 1: I am grateful to John W. Sundsten for permission to reproduce the human brain image from the Digital Anatomist Project, Dept. of Biological Structure, University of Washington at http://www9.biostr.washington.edu/da.html, and to John Kirn for illustration material.

CHAPTER 9

Roderick A. Suthers: I am deeply indebted to my students and colleagues whose names appear in the authorship of papers cited from my laboratory. Without their talents, enthusiasm and hard work many of these studies would not have been possible. I thank Masakazu Konishi for encouraging me to apply to songbirds the techniques developed for studying echolocating bats and birds. Special thanks are also due to Michel Kreutzer and Eric Vallet at the University of Paris for the invitation to collaborate in their research on canary song and to Carel ten Cate at Leiden University for the opportunity to participate in studies on dove vocalizations. The author's research was supported by grants from the U.S. National Science Foundation and the National Institutes of Health.

CHAPTER 10

Carel ten Cate: I am grateful to Mechteld Ballintijn, Gabriël Beckers, Selvino de Kort and

Hans Slabbekoorn, all of whom contributed significantly to the *Streptopelia*-project. They and other members of the Leiden Behavioural Biology group were excellent discussion partners for the ideas presented in the chapter. Selvino and Hans collected the recordings gathered on the CD. I also thank J. Jordan Price for providing the sonograms in Figure 10.1 as well as additional unpublished material. Gabriël, Selvino and, in particular, the editors of this book provided constructive and useful criticisms on earlier drafts.

CHAPTER 11

Jeffrey Podos & Stephen Nowicki: We thank the editors for the opportunity to contribute to this volume, and for their helpful suggestions on our chapter. Our laboratory groups at Duke and UMass Amherst have provided valuable feedback over the years on the ideas presented here. J.P.'s research in the Galapagos has been made possible through the kind support of the Charles Darwin Research Station and the Galapagos National Park Service. We are both grateful to the National Science Foundation for funding our research (IBN 0077891 and IBN 0347291 to JP, and IBN 9974743 and IBN 0315377to SN).

CHAPTER 12

Sandra L. L. Gaunt & D. Archibald McCallum: For discussion and comment on the manuscript for this chapter we thank Dr. Douglas A. Nelson, Director and Dr. Jill Soha, Curator of The Borror Laboratory of Bioacoustics, Museum of Biological Diversity, The Ohio State University. Tom Scott provided fruitful discussions on applying bioacoustic technology to monitoring.

CHAPTER 13

Irene V. Pepperberg: This chapter was written with support from the MIT School of Architecture and Planning and donors to *The Alex Foundation*. Research reported in this chapter was supported by donors to *The Alex Foundation* and several grants from the National Science Foundation.

CHAPTER 14

Meredith J. West, Andrew P. King, & Michael H. Goldstein: The authors' work was supported by grants from the National Science Foundation and the National Institute of Health.

Chapter 1

Science and birdsong: the good old days

PETER MARLER

INTRODUCTION

The invention of a novel analytical technique often helps to launch a new science. What microscopes were for the emergence of cell biology as a discipline, or the cathode-ray oscilloscope for neurophysiology, it was the sound spectrograph that, immediately after the Second World War, enabled the birth of the science of birdsong. There had been no lack of interest in birdsong previously, and fascinating and important discoveries were made, especially about the functions of song. But never before had researchers come together to form a coherent discipline. Until about 1950, everyone interested in birdsong had no choice but to work by ear. Only when the sound spectrograph became available was it possible, for the first time, to grapple objectively with the daunting variability of birdsong, and to specify its structure precisely. Almost immediately a multitude of new issues became accessible for scientific scrutiny and experimentation.

IN THE BEGINNING

Visible Speech, Visible Birdsongs

With remarkable rapidity sonograms became the standard method for birdsong study. Previously the only machine available for the empirical visualization of animal sounds was the same cathode-ray oscillograph that revolutionized neurophysiology. Oscillograms were useful for the study of insect sounds, but not very helpful for analyzing sounds with a more complex frequency structure, like birdsongs or, for that matter, human speech (**Fig. 1.1**). Around 1940, a group of researchers at the Bell Telephone Laboratories decided that it was time to develop methods for making the details of speech more visible and intelligible, in part because they thought it might help in teaching deaf people to learn to speak and use the telephone (Potter et al. 1947). Also it was wartime and, as with other acoustic technologies, another driving force was the need to monitor movements of ships and submarines, analyzing their far-traveling ocean-born sounds, each with its own particular signature. As a consequence, many of the details remained classified until the War was over. Soon after hostilities ended the Kay Electric Company was created as an offshoot of the Bell Laboratories with the assignment to build and market a machine for visible speech (**Fig. 1.1**). The success of the Vibralyser, as it was first called, was mixed. There were some dramatic achievements. Sound spectrograms of speech became a basic tool in linguistics, and remained so for years. As far as speech therapy was concerned, sonograms of speech proved to be difficult even for experts to read in real time, greatly limiting their value in working with the deaf. But they were perfect for the study of birdsongs.

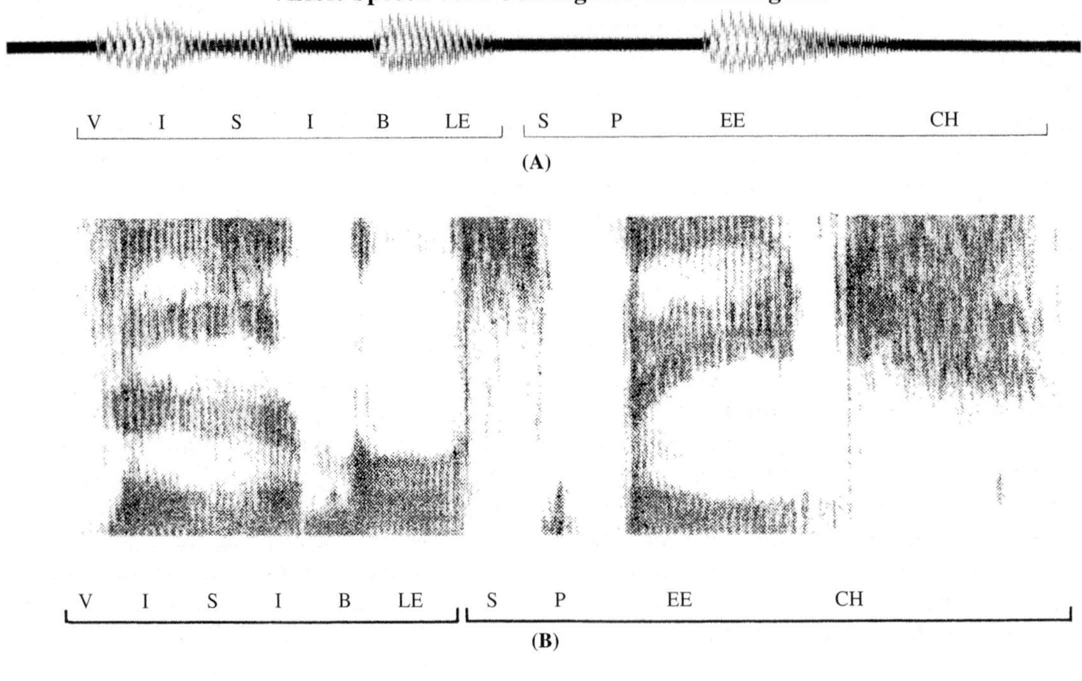

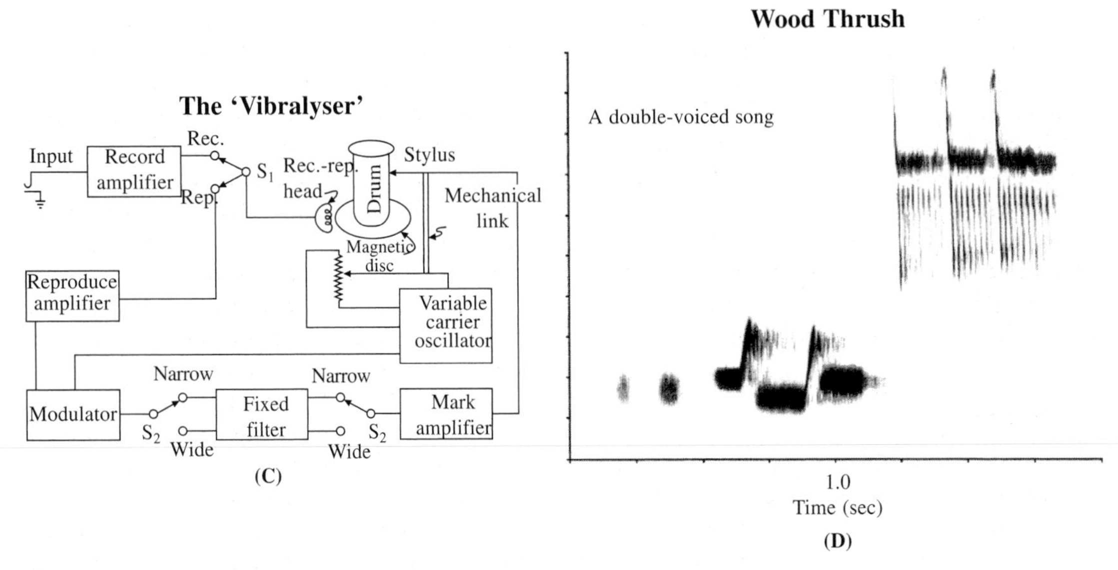

Figure 1.1 Some historical memorabilia. An oscillogram of *Visible Speech* (Potter et al. 1947) (**A**); followed by a much more informative spectrogram of the same words (**B**). The Vibralyser diagram by Borror comes from his 1953 paper with Reese (**C**). (**D**) represents a sonogram of an obviously two-voiced song of a wood thrush, from Borror and Reese (1956).

The potential of the sound spectrograph for the study of birdsong was appreciated almost immediately. During the War ornithologist and entomologist Donald Borror worked in Naval intelligence and learned of the visible speech project. When he arrived back in Ohio State, he discovered a sound spectrograph sent there for a declassified project that did not work out. He was able to arrange access, and sonograms of songs of sparrows, thrushes, and wrens soon followed (Borror 1956; Borror & Reese 1953, 1956). For the first time the extraordinary virtuosity of the avian voice was revealed in all its glorious detail. Although Borror was primarily interested in field identification and taxonomy, he proved to be a perceptive bioacoustician (Borror 1960). In some of the first sonograms of birdsong published, he showed that birds are capable of singing with two independent voices (**Fig. 1.1**), something already hinted at by Potter (1945) in comments on brown thrasher song, sonograms of which were reproduced in the original book *Visible Speech* (Potter et al. 1947).

At about the same time, Nicholas Collias, a student of Chicago ecologist W.C. Allee, and a pioneer in the study of animal behavior, collaborated with linguist Martin Joos who had worked on the sound spectrograph project, to publish a sonogram-based functional analysis of the vocal signals of chickens, describing their structure and exploring their function (Collias & Joos 1953; Collias 1960). The stage was set to launch the science of birdsong in earnest.

Studying by Ear

In an era dominated by computers, with so many elegant methods available for the analysis of complex sounds, it is hard for us now to imagine the extraordinary impact of the sound spectrograph. Until that time, the only useful descriptions we had, aside from oscillograms, were either musical transcriptions, or verbal renditions, along the lines of the 'drink-your-tea' of towhees. Evocative as they were, they hardly did justice to all the acoustic intricacies, as was immediately obvious with William Homan Thorpe's (1954) first sonograms of chaffinch song (**Fig. 1.2**).

Musical notations are another possibility. The late French composer Olivier Messiaen created some nice renditions of chaffinch song (**Fig. 1.2**), and birdsongs have engaged many composers over the centuries. The relationship between birdsong and music is an interesting one. Despite his many song-inspired compositions, Messiaen was heard to remark that birds sing "in extremely quick tempi which are absolutely impossible for our instruments" (Johnson 1975). So he often found it necessary to transcribe them to a slower tempo. The "excessively high registers" of most birds often required him "to transcribe them several octaves lower, suppressing very small pitch intervals." Long before, Barrington (1773) had made the same point. "As a bird's pitch, therefore, is higher than that of any instrument, we are consequently at a still greater loss when we attempt to mark their notes in musical characters, which we can so readily apply to such as we can distinguish with precision." Similarly, "the intervals used by birds are commonly so minute, that we cannot judge at all of them from the more gross intervals into which we divide our musical octave" (Barrington 1773, p. 266). As a consequence, even though in pieces like 'Oiseux exotiques,' and 'Quatuor pour la Fin du Temps,' Messiaen captured the essential rhythms of birds singing quite wonderfully, he often rendered particular songsters all but unrecognizable to an ornithologist. Musical transcriptions are in fact not very helpful from a scientific point of view. Sonograms do a much better job (**Boxes 1 & 2**, pp. 5–6).

As a kind of compromise, several people, especially Saunders (1935), developed schematized frequency/time diagrams that captured some of the dominant features of birdsongs (**Fig. 1.3**), and this is the method I adopted in my youthful studies of birdsong. Beginning in 1949, playing hooky from my graduate studies in botany and my duties as a plant ecologist for the Nature Conservancy, I hiked around in England, Scotland, France, and the Azores chasing chaffinches. Altogether I transcribed by ear more than five hundred

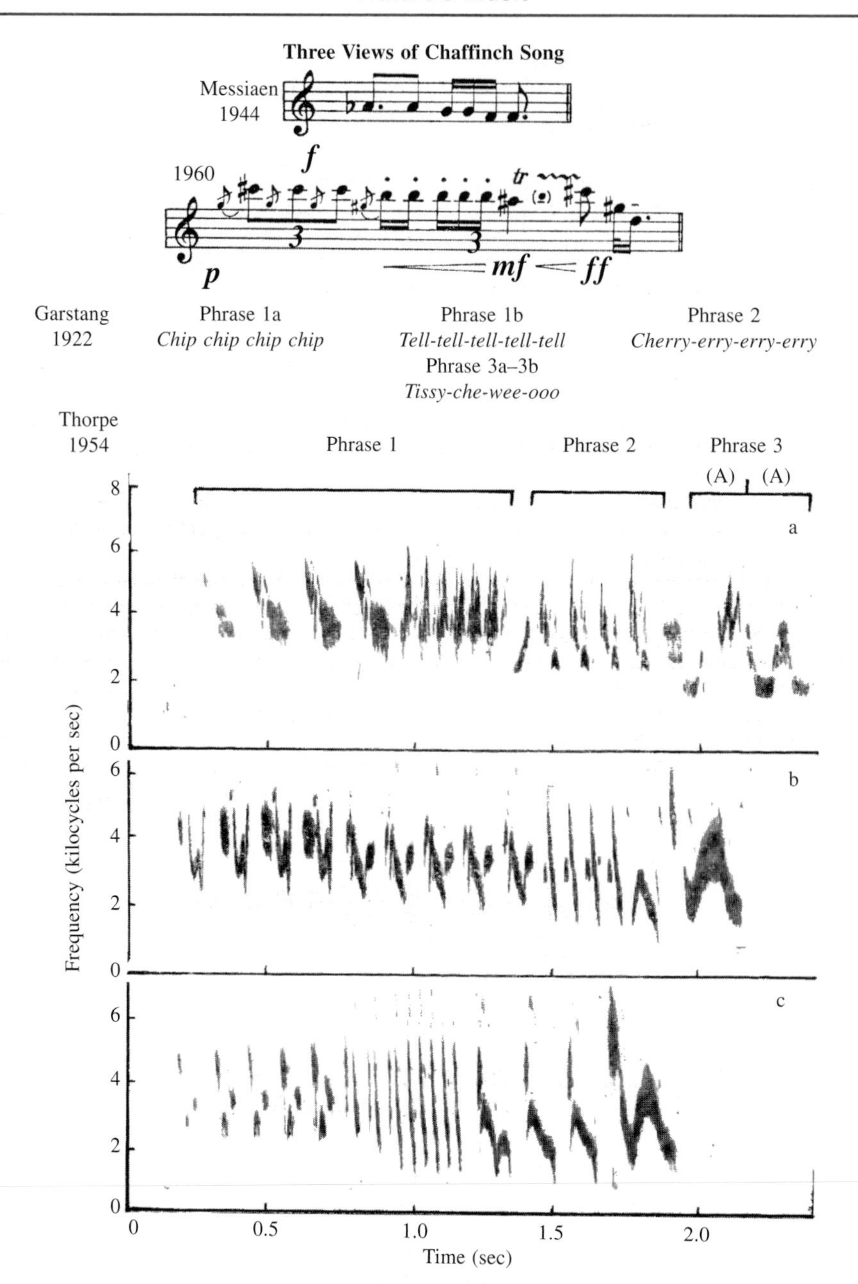

Figure 1.2 Three renditions of chaffinch song, from Garstang (1922), Messiaen (Johnson 1975) and Thorpe (1954). The musical transcriptions are from Messiaen's 'Vingt Regards sur l'Enfant-Jesus' (1944) and 'Chronochromie' (1960). Thorpe's chaffinches are three different birds.

chaffinch songs, learning much about their behavior and ecology in the process.

Looking back, my conclusions about the mosaic structure of chaffinch song dialects were not far off target. But with my primitive methods no one else could tell whether my results were believable or not, including the person who mattered most, destined to be my new boss, W.H. Thorpe at Cambridge University. Thorpe had invited me to join him in 1951 as a research

BOX 1

GRAPHIC REPRESENTATION OF SOUNDS

Three types of graphs are commonly used to visualize sounds; those below were all made from one recording of a white-crowned sparrow song. The first graph illustrates intensity fluctuations over time: the *amplitude wave form* (**A**). The x-axis represents the passing of time while the sound volume is reflected in the height of the spikes above and below this axis. Usually the y-axis indicates the relative amplitude: no sound results in no spikes and the loudest sound reaches the maximum extension possible in the graph.

Second is the *sonogram* (**B**) which includes information on the pitch or more precisely the frequency of the sound. The x-axis again represents time and the y-axis pitch, with low-frequency sound near the baseline and high-frequency sound higher up. The frequency of birdsongs usually falls between 500 Hz and 10,000 Hz or, as often indicated, 0.5 and 10 kHz. In this frequency–time graph, information on amplitude is depicted by the darkness of the gray-scale with black reflecting frequencies of the highest amplitude. This gray-scale often disappears in print, replaced by a black and white version. The gray-scale is replaced by a color-scale in many computer software programs.

Third is the *power spectrogram* (**C**), which displays frequency versus amplitude, summated for a segment of sound, or an entire song. It shows the distribution of power through the sound spectrum for a certain song, call, or note, depicting the sound energy present at each frequency.

The sonogram is the most widely used, often to measure temporal and spectral characteristics of songs. It is also used more and more in bird guides to describe songs, instead of onomatopoeic renditions or musical script. The sonographic representation of sound is objective, but the form it takes depends on the temporal and spectral resolution chosen; the same sound can appear very different if generated with different analysis bandwidths. A narrow-band analysis, with high-frequency resolution yields short, wide graphs, compressed in time (**D**); a wide-band analysis, with high temporal resolution leads to tall, narrow graphs, compressed on the frequency axis (**E**). It is technically important to state the bandwidth employed in an analysis, but generally people use similar, wide-band settings for birdsong, making sonograms more or less comparable, as in the illustrations for this book.

Hans Slabbekoorn

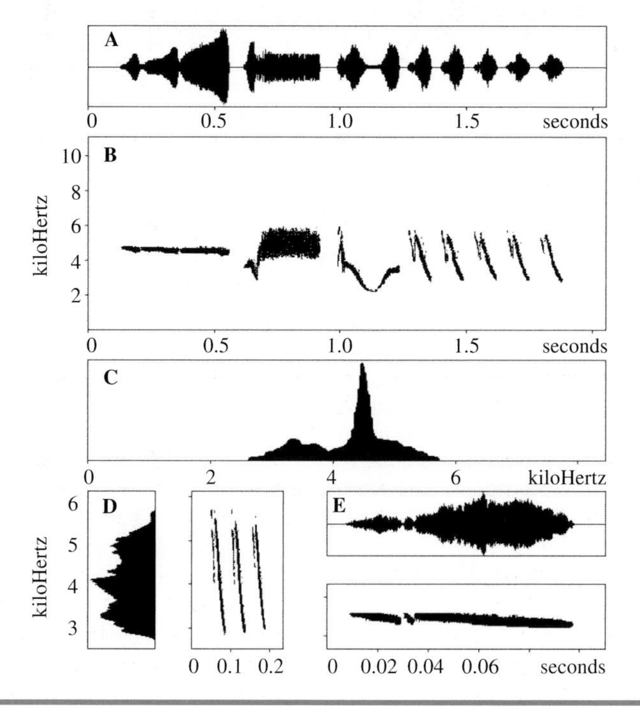

BOX 2

A ROUGH GUIDE TO READING SONOGRAMS

Birds produce an enormous variety of sounds, which can be evaluated by ear, but better by eye via sonographic representations. As a sound is played back, you can read along with a sonogram from left to right (**CD2 #3–15**). A 'note' is the usual term for the smallest sound unit in birdsong appearing in a sonogram as a continuous sound trace. A set of two or more notes repeated coherently in a 'trill' is a 'syllable,' an unrepeated cluster is a 'note complex,' and a series of note-complexes and trills is a 'song.'

Pure tone notes at a constant frequency are the simplest components of birdsongs (**A**). There are examples in the two notes of the African cuckoo (**B**), the introductory note of a white-crowned sparrow song (**C**), and the alarm 'seet' of the great tit (**D**).

The fox sparrow has a few simple notes that change gradually in frequency (**E**) mingled with other more complex notes in the territorial song (**F**). Frequency changes in such pure notes can be heard by the human ear, but the details only become clear when a recording is played at a reduced speed, bearing in mind that at a slower speed sounds are lower in pitch. We can simulate birdsong complexity by creating artificial frequency modulations at various rates (**G**) – at a slow pace, frequency upsweeps and downslurs are easily heard, but become more difficult to discern at the fast pace of the frequency modulations in yellow warbler song (**H**).

Tonal sounds often change gradually in frequency, but more discrete changes also occur, as if the bird's voice is breaking. This can lead to stereotypic note variants, with and without abrupt frequency jumps; four of the seven notes of the second song of a diederik cuckoo display a sudden increase in frequency (**I**).

Harmonics or overtones are represented in a sonogram by a typical ladder pattern. One frequency, the fundamental, appears with one or more other sound traces at frequencies that are multiples of the fundamental; harmonics of a fundamental frequency of 500 Hz occur at 1.0 kHz, 1.5 kHz, and so on (**J**). The relative amplitude of these harmonics may vary, with a strong effect on the quality of the sound, and some can even be completely missing, including the fundamental itself. Overtones are sometimes not harmonics but 'side-bands.' If a tone is frequency-modulated it can be described in two ways, as a warble or as a complex tone. Both are valid and what you see on a sonogram depends on the analyzing bandwidth. A warble appears on a sonogram with high-time resolution as a tone fluctuating in frequency. As a complex tone, on a sonogram with high-frequency resolution, you see side-bands spaced at the modulation frequency, above and below the tone. The more side-bands there are, the more emphatic the warble sounds. Amplitude modulation can also create side-bands, just one on each side of the carrier frequency. Side-bands are characteristic of some birds, such as red-winged and yellow-headed blackbirds giving their songs a buzzy, nasal tone. Harmonics are found in many bird calls and songs, such as in the zebra finch (**K**) and in the human voice.

Not all songs and calls are tonal and smooth; some sound like hissing, others are noisy and croaky. Noisy sounds lead to messy sonograms with sound diffused across the spectrum, as in some zebra finch syllables and in calls of the Eurasian jay (**L**). When birds use a double voice they produce two independent sounds at the same time, which can vary independently. The songs of the black-bellied seedcracker contain harmonically related sounds together with another independently varying note (**M**).

Hans Slabbekoorn

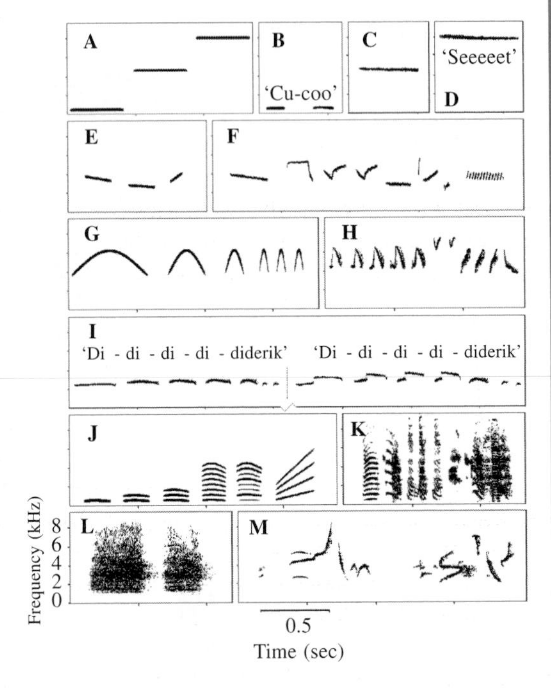

Song Sparrow: Two Songs

Musical whistle with occasional notes buzz-like

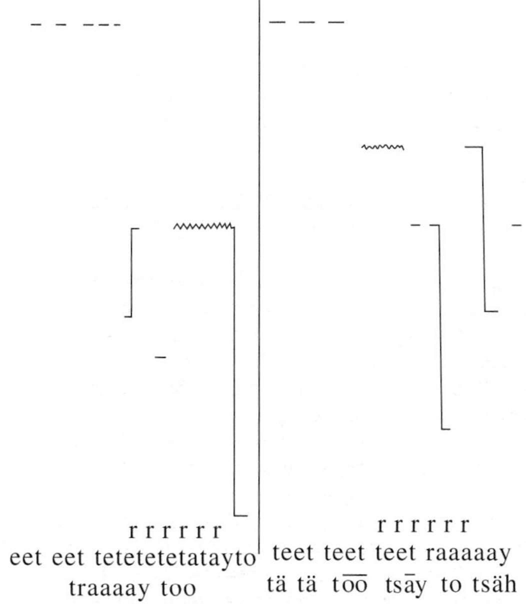

Swamp Sparrow: Three Songs

Rather sweet whistle

tweet weet weet weet
weet weet weet weet
weet weet weet

tayhayhayhayhayhayhay
hayhayhayhayhayhay
hayhay, etc.

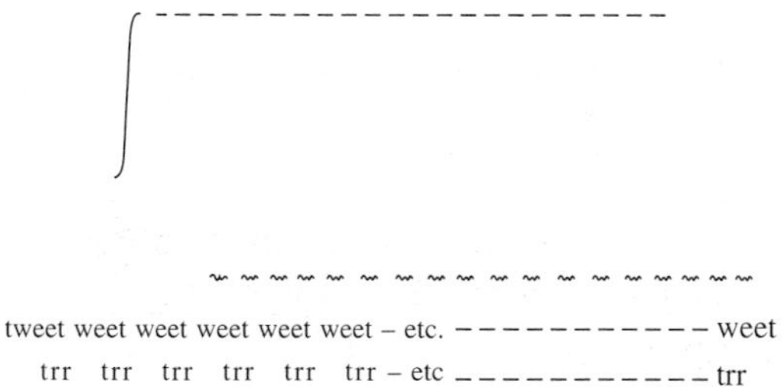

Figure 1.3 Transcriptions by Saunders (1935) of song and swamp sparrow songs.

assistant. I already had a PhD in plant ecology, so in one sense I went to Cambridge as a postdoctoral fellow. But I was a neophyte as far as animal behavior was concerned, so I became a graduate student again, this time in zoology, with Thorpe as my major professor.

SONG STUDIES IN MID-CENTURY

Thorpe's laboratory was on the top floor of the zoology department, next door to the legendary insect physiologist, Victor Wigglesworth; this was an appropriate location for Thorpe as an expert on insect behavior, and a wartime innovator in developing new biological methods for pest control. But now his mind was filled with questions about birdsong and the role of learning in its development. The breadth of his erudition was an inspiration to me, as was that of Robert Hinde, the new director of the just-founded Madingley Ornithological Field Station: both were sources of endless revelations. Above all, Thorpe had just acquired in 1950 a sound spectrograph, only the second to be imported into Great Britain; the first went to the Admiralty Research Laboratory at Teddington, presumably for use by the Royal Navy. With free access to this new machine, and the library of long playing records of bird sounds donated from the archives of the British Broadcasting Corporation (BBC), I was in heaven. We were set for the great leap forward, and never looked back.

At that time, Thorpe, like everyone else, had to make sound recordings by cutting wax discs, calling for much care and cumbersome equipment; tape recorders, including some that were at least semi-portable, arrived somewhat later. The mysteries of wow and flutter loomed large, and fluctuations in tape speed were a major problem. Among the more imaginative solutions was the heavy cast iron flywheel you screwed on to the tape transport of the famous Magnemite portable tape recorder, before lugging it off into the field. It was a while before truly portable tape recorders came on the market, driven in part by the burgeoning needs of filmmakers, a market force without which the magnificent Swiss-made Nagra recorders would never have seen the light of day.

Quality playback equipment that was portable enough for field use had to be custom made, one of a host of problems we now solve comfortably with personal computers. The elaborate interactive playback experiments of recent years (McGregor 1992; Dabelsteen & McGregor 1996; McGregor & Dabelsteen 1996) would have been unthinkable even a short time ago.

It was not until we resettled in Berkeley in 1957, that I was able to pause and gather my thoughts, and reap the full benefit of the open and generous opportunities that Thorpe had given me to sift through the treasures in the BBC birdsong recordings with the sound spectrograph. Many of the sonograms I made then only saw the light of day in 1959, in an invited chapter on animal communication I wrote for a book on Darwin, edited by a colleague from my botanical past, Peter Bell. A few are reproduced in **Figures 1.4** and **1.5**

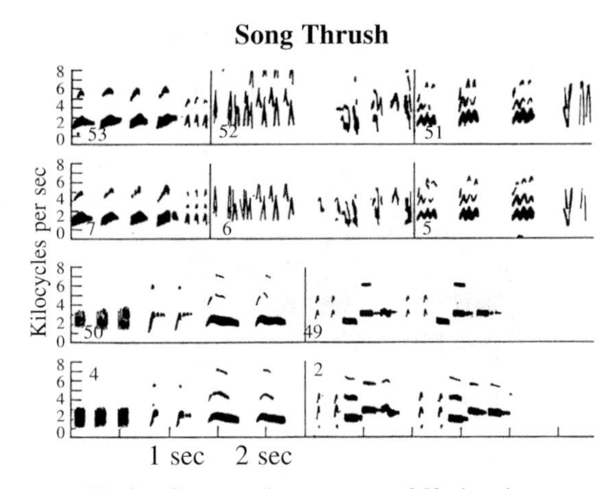

The last five songs in a sequence of 53 given by a song thrush, showing how closely they correspond to the same themes given earlier in the sequence.

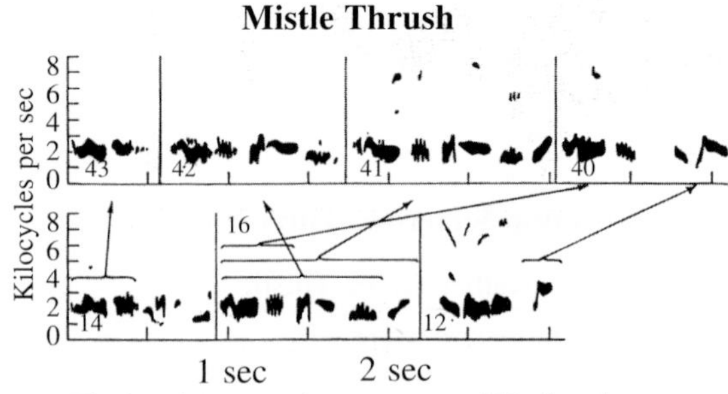

The last four songs in a sequence of 53 given by a mistle thrush compared with earlier songs in the sequence to show how new themes are produced by the recombination of old phrases.

Figure 1.4 Song thrush and mistle thrush songs from BBC archival recordings that I analyzed on Thorpe's sound spectrograph, in the mid-1950s. At that time we did not know whether birds like the song thrush have a repertoire of discrete song types, or how repertoires are constructed. From Marler 1959.

The Debut of the Chaffinch

This period in the 1950s, when Thorpe presided over the birth of the science of birdsong, was also a time of ferment in the behavioral sciences. Thorpe assembled a massive review of the learning abilities of birds in 1951, setting aside once and for all the shibboleth inherited from the era of Maier and Schneirla (1935) that birds are reflexive machines; this view went back at least to 1924, when Herrick had said that "it is everywhere recognized that birds possess highly complex instinctive endowments and that their intelligence is very limited."

Although he did not actually discover song learning, the blossoming of research on avian vocal learning, in all its dimensions, would almost certainly never have happened without Thorpe's seminal contributions. His historic papers on song learning in chaffinches, in 1954 and 1958, the first in-depth studies based firmly on sound spectrographic analyses, were crucial in launching the new discipline and they evoked enormous interest (**CD1, #3**). It was already clear that the chaffinch was an ideal subject for learning studies. Holger Poulsen in Denmark published a paper on inheritance and learning in chaffinch song which I saw in proof while I was completing my pre-Cambridge field project on song dialects. Writing in 1951, just before I arrived in Cambridge I had already concluded, somewhat prematurely, that "the geographical variation and development of dialects in the song of the chaffinch are phenotypic variations, arising and persisting because of the two processes of learning

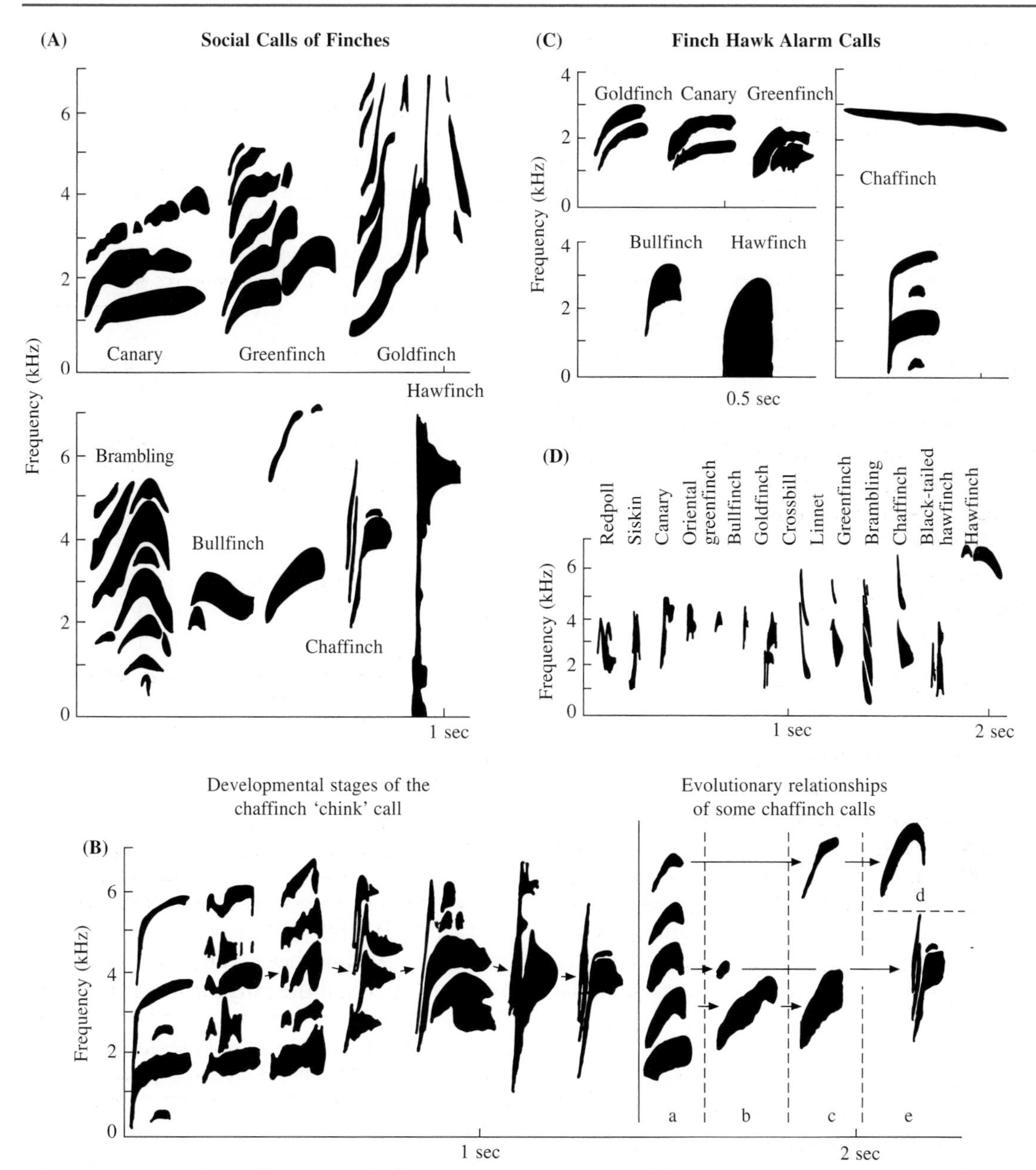

Figure 1.5 Sonograms of various finch calls I recorded in aviaries at Thorpe's Madingley Ornithological Field Station outside Cambridge in the mid-1950s. Below are the original figure legends edited from Marler (1959). (**A**) The 'social' calls of various finches, all showing affinities with the same basic pattern. Those of the chaffinch and hawfinch, also used as mobbing calls, are the most divergent, but the chaffinch 'chink' can be traced back to the same type, as shown in (**B**). On the left are developmental stages of the chaffinch 'chink' call, relating it to the basic finch type. On the right are suggested evolutionary relationships between the 'primitive' alarm call (a), used only by young chaffinches but widespread among other finches, and certain adult calls. Different harmonics may have been selected for the 'huit' call (c), the 'seee' hawk alarm (d); an unusual form of the hawk alarm call, and the 'chink' call (e). Call b shows an unusual form of the 'huit' call, which may correspond to an intermediate stage. (**C**) Hawk alarm calls of various finches, showing their essential similarity. The chaffinch only uses this call when young (below); in the adult male it is replaced by the type of call difficult to locate (above), which none of the other finches has evolved. (**D**) The 'flight' calls of some finches, showing how they tend to conform to similar patterns, with the exception of the hawfinch.

to vocalize from associates and of retaining a preference to breed in certain localities" (Poulsen 1951; Marler 1952).

Ethology on Stage

Meanwhile, the newly emerging discipline of animal behavior was in a state of high excitement. In 1950–51 Thorpe served as the Prather lecturer at Harvard, and he seized on the opportunity to spread the gospel according to European ethologists Konrad Lorenz and Niko Tinbergen; he was preparing the way for his own 1953 book. *Learning and Instinct in Animals* was a source of inspiration for the next generation. Many of the ideas had already been set forth in his paper on "The concepts of learning and their relation to those of instinct;" this was presented at a remarkable conference for the Society for Experimental Biology in 1949 in Cambridge on "Physiological mechanisms in animal behavior" that he helped to organize. Prophets gathered from various disciplines, anticipating future prospects of the behavioral and neuroethological sciences. Lorenz and Tinbergen spoke at length on the principles of ethology, aided and abetted by the Dutch pioneer Gerard Baerends and by Thorpe himself, bringing this material together for a general biological audience for the first time.

The conference proceedings amounted to a manifesto for research on proximate mechanisms in the behavioral sciences; there were contributions from such famous figures as Weiss on development, Gray and von Holst on the coordination of motor patterns, Boycott and Young on octopus brains and behavior, Lashley and his search for the engram, and many others, culminating with Thorpe's magisterial overview of learning and instinct. He mentioned song learning only in passing, along with imprinting, pleading for more study. In one sentence, Thorpe anticipated a significant contribution that ethology in general and birdsong research in particular was destined to make in behavioral science. "Where the innate powers of recognition can only carry the animal a part of the way towards its goal, the process is completed and adjusted by a tendency to learn in certain restricted times and directions (as in the tendency of a bird to learn and copy the song of its own species in preference to the song of another) so that experience completes for the individual the process commenced for him by his inherited constitution" (Thorpe 1950).

As the sound spectrograph became more widely available, its use in bioacoustics spread like wildfire. Studies of geographic variation in birdsong began to appear, first as a trickle, then as a flood, in both Canada and the USA. The role of song in territorial defense and the impact of sexual selection now became tractable research targets. The significance of local dialects came under renewed scrutiny. Sonograms provided the bedrock for the initiative that Thorpe had launched on the elusive problem of vocal learning, paving the way for study of the underlying neural mechanisms, culminating in the discovery of the song system in the avian brain more than twenty years later (Nottebohm et al. 1976).

DIALECTS

In one important step, the reality of local dialects in birdsong became clear; what had been suspected by ear became fact. Dialects exist in birdsong, on a scale so local that, for creatures as vagile as birds, it hardly seemed plausible to attribute them to genetic differences between populations. All of the early pioneers, including the leading evolutionary biologists of the day, like Julian Huxley (1942) and Ernst Mayr (1942), as well as Thorpe (1951), recognized the scientific challenge that dialects represent. The many insights that flow from immersion in the study of song dialects is evident from the career of Luis Baptista (Baptista 1975, 1990), to whom this book is dedicated. To this day dialects remain an important focus in birdsong studies; witness Kroodsma's unexpected and provocative discovery of song dialects in a sub-oscine, the three-wattled bellbird (see Chapter 4, p. 108).

The conspicuousness of the dialects in white-crowned sparrow songs in western North America

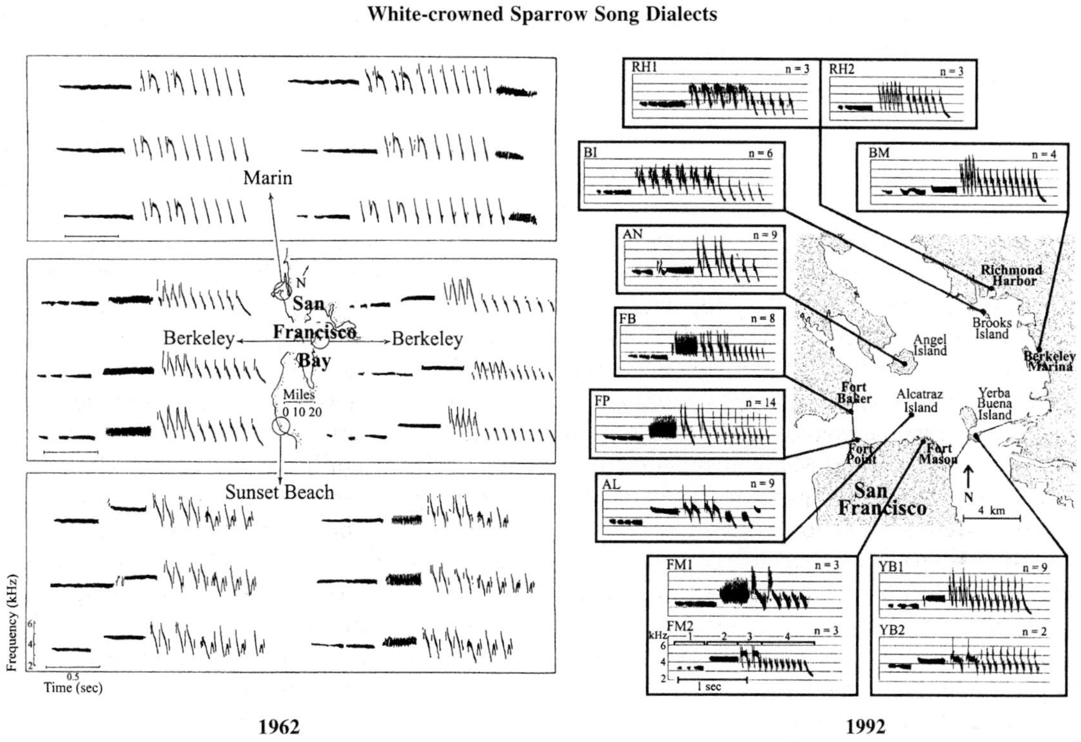

White-crowned Sparrow Song Dialects

1962

1992

Figure 1.6 Local dialects around the San Francisco Bay in white-crowned sparrow song. The figure on the left, based on Marler & Tamura (1962) was reproduced in many textbooks. On the right side are the dialects thirty years later, from Slabbekoorn et al. 2003.

(**Fig. 1.6**) was instrumental in the emergence of this species as a subject of choice in research on song variation and development, in both laboratory and field studies (Marler & Tamura 1962, 1964; Konishi 1965a; Baptista 1975; Baker & Cunningham 1985). Sonographic analyses elevated studies of song dialects of such European birds as chaffinches, corn buntings, yellowhammers, and redwings, from the birdwatcher level (Huxley 1947; Marler 1952) to the status of scientific research (Slater et al. 1984; McGregor 1983; Baker et al. 1987a; Espmark et al. 1989).

Problems of micro- and macro-geographic variation now began to come into focus, eventually providing new windows on cultural evolution and speciation (Mundinger 1982; Lynch 1996; Martens 1996; Payne 1996; Slabbekoorn & Smith 2002b). For the first time we began to gain some sense of what the entire song repertoire of a species encompasses, and of the challenges confronting a bird in recognizing its own species song and distinguishing it from the many others with which it might be confused. Given the enormous variability of birdsong this might seem to be a daunting task, and yet birdwatchers usually have no trouble telling one from another. How in fact do birds do it?

THE MYSTERY OF SPECIES UNIVERSALS

Song Recognition

There is some evidence of innate song recognition, and Thorpe was the first to suggest that it must occur. He was preoccupied with questions about song recognition from the earliest days, especially since some of his laboratory subjects seemed to resist being taught alien songs when first learning to sing (Thorpe 1961). He was greatly impressed by the instant recognizability of the songs of chaffinches that had lived for generations in completely new environments, after introduction by homesick colonials in New Zealand and South Africa (**CD1, #4**). It seemed to him that they must be able to recognize the songs they should be attending to while learning to sing. In Cambridge, his young chaffinches found his artificially created songs quite unacceptable, but they would learn re-articulated chaffinch songs. They also learned some elements of a tree-pipit song selected by Thorpe because he thought the tonal quality was reminiscent of the chaffinch (Fig. 1.7).

Thorpe tentatively concluded that there must be something distinctive about the tonal quality of songs, providing the basis for any innate 'blueprint' they might possess. He thought that young birds might be innately cued to these tonal qualities, hence their tendency to learn conspecific song. He seemed reluctant to admit the possibility that innate recognition of song might be based not on its tonal quality, but on a much more complete mental image of many or even all of the basic components of its species not necessarily all sharing the same tonal quality. As the sound spectrograph made much more comprehensive analysis possible, we could begin conducting large-scale investigations of the entire song repertoire of a species, to identify features present in all songs, and thus candidates as possible cues for species recognition. The results were fascinating, and suggested that we needed to reopen the whole question of song recognition.

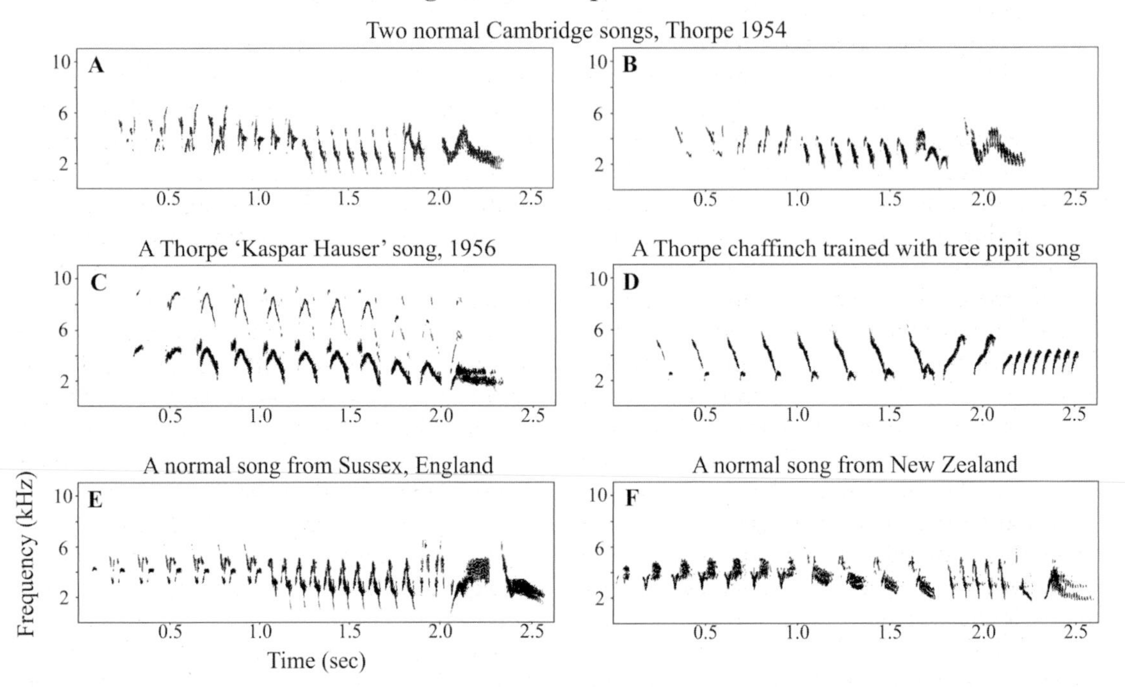

Figure 1.7 Six chaffinch songs, four of them from W.H. Thorpe's original recordings. **A** & **B** are Cambridge birds from the same local dialect, recorded by Thorpe in 1954. **C** is a bird raised in social isolation by Thorpe as a "Kaspar Hauser." **D** is a bird trained by Thorpe with a recording of a tree pipit song. **E** is a song from Sussex, England, the source of the chaffinches introduced to New Zealand between 1840 and 1860. **F** is a chaffinch song from New Zealand (**CD1 #3–4**).

We have long known that song plays a crucial role, not only in mate choice, but also in territorial defense, where same-species, same-sex birds are usually the primary targets (Howard 1920; Hinde 1956a). As methods for song playback to territorial males were perfected (Bremond 1968, 1976; Krebs 1977a), and the importance of singer location and individual identity and the ability to match songs with rivals became clear (**Box 3**, p. 49; see Chapter 4) (Brooks & Falls 1975 a, b; Falls & Brooks 1975; Falls 1978, 1982), it was inevitable that individual and species differences in song characteristics would eventually come under renewed scrutiny (Becker 1982; Falls 1992). Almost without exception, playback of songs of other species proved to be much less potent in evoking territorial response from resident males than own-species songs. But the task of establishing which particular song features were critical in own-species/other species discrimination was a daunting one, and not many people had the tenacity to grapple with it. Again, the sound spectrograph came to the rescue. Amidst the close scrutiny of all aspects of song variation, comprehensive analyses of song structure now began to emerge for a few species.

The obvious first step in using song playback to study the responses of territorial males was the selection of particular songs as stimuli. Because learned birdsongs are so incredibly variable the choice is harder than it sounds, if the choice is to be representative. Attempting to characterize the song of an entire species is a daunting undertaking, and compromises are inevitable. Archived sound recordings like those in the Borror Laboratory of Bioacoustics at Ohio State, and the Macaulay Library of Natural Sounds at Cornell University (see Chapter 12) become a treasured resource, and much time and devotion to duty is required of the researcher to gather the necessarily large volume of new material. Luis Baptista devoted much of his life to recording songs of the white-crowned sparrow. He often worked in suburbia, stopping his car in mid-traffic to record a new bird, oblivious to the resulting traffic jam. Other early enthusiasts for comprehensive sonographic song coverage included Gerhardt Thielcke and his European tree-creepers, Fernando Nottebohm's chingolos in Argentina, Peter Slater and chaffinch song, the nightingales of Henrike Hultsch and Dietmar Todt, Michel Kreutzer's cirl buntings, Peter Becker's goldcrests and firecrests, Bob Lemon's cardinals, not forgetting Don Kroodsma's life-long devotion to the voices of the wrens of the world. These long-term studies of species' song repertoires have all been invaluable in providing a solid basis for subsequent work on recognition and song ontogeny. They provided an intriguing opportunity to follow up in a new way on Thorpe's early reflections on how we should interpret the apparent contradiction between the great variability associated with the learnability of song, the clear evidence of song universals shared by all the singers of a species, and problems of song recognition.

The Indigo Bunting

Before launching his path-breaking studies of the social life of bee-eaters and birds navigating by the stars, Steven Emlen at Cornell did one of the first studies of song recognition; he quickly found the selection of playback stimuli to be a problem even with the relatively simple song of the indigo bunting (**CD1 #5**). This bird became the subject of what is still one of the most comprehensive analyses of the mechanisms underlying species and individual song recognition. Emlen addressed the problem in an interesting way, by first looking at sonograms of indigo bunting song from several parts of its range, in and around Ithaca, New York, trying to figure out what they had in common (Emlen 1971, 1972). Thompson (1970) had already conducted a similar exercise in a different part of the species range, and the comparison turned out to be fascinating and unexpected (Shiovitz & Thompson 1970; Shiovitz 1975; Thompson 1976; Margoliash et al. 1991; Baker & Boylan 1995; Payne 1996). Shiovitz and Thompson had assembled a comprehensive sonogram catalog of less than a hundred distinct syllable types, from which every one of the hundreds of songs they analyzed from Kentucky and Michigan could be assembled. Amazingly, miles away in New York, the identical catalog proved to be equally serviceable in classifying almost all of Emlen's songs (**Fig. 1.8**). He only had to add a few more types for the coverage to be complete. Since then the Thompson catalog has been the basis for every study of indigo bunting song, only updated 20 years later to 127 syllable types by Baker & Boylan (1995). The bottomline is that, over thousands of square miles, indigo bunting songs are constructed from essentially the same set of building blocks.

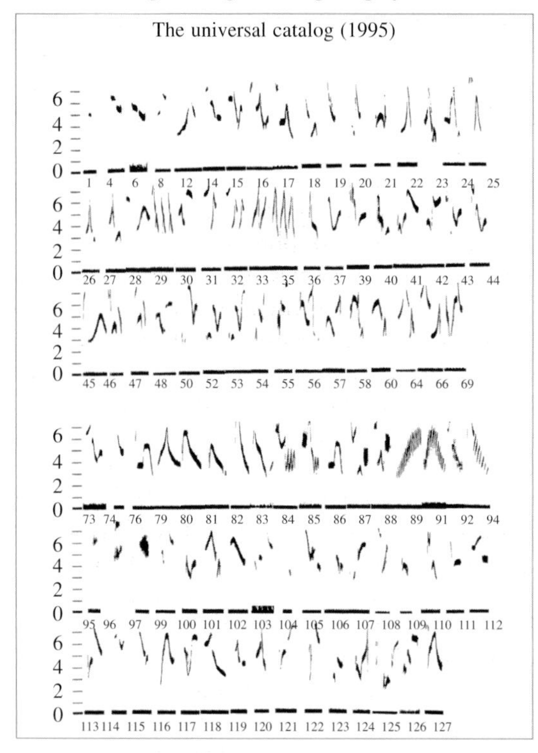

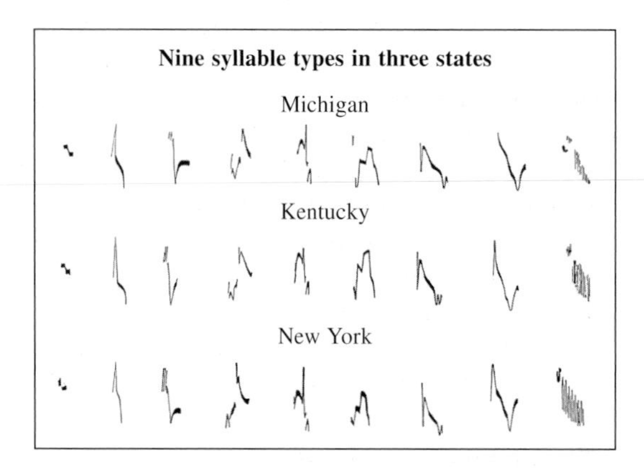

Figure 1.8 The catalog of the universal repertoire of indigo bunting song syllables, from Baker and Boylan (1996). Also shown are nine syllable types recurring in similar form in three states. From Shiovitz 1975.

Although its significance was not fully appreciated at the time, this was a very surprising discovery. Even though these are learned birdsongs, species universals are present as a dominant feature of indigo bunting song. Despite the variability that you would expect to be an inevitable consequence of the cultural transmission of song, and the copying errors that must ensue, despite the existence of dialects, and the modification of song by improvization and invention, and the great individuality of these and all other learned birdsongs, underneath it all is a surprisingly straightforward basic set of universal acoustic elements from which all indigo bunting songs are made. Indigo bunting songs are still individually distinctive, but the variability is constrained. Each one consists of an individually distinct set of sequenced syllable types and there are not only individual differences, but also local dialects based on the subset of syllables used, their sequencing, and the overall temporal structure. Much of the variability resides in the temporal structure, and it is not surprising that when Emlen (1972) experimented with artificially modified songs, he found timing and other syntactical cues to be especially important in song recognition in indigo buntings based on the relatively simple library of syllable types to which all indigo buntings conform.

A fundamental question inevitably follows: where does this shared content come from? Could it be the clue to species recognition? Emlen could not resolve this question to his satisfaction. He could not decide whether the mosaic of different syllable type collectively provides the basis for recognition, in additive fashion, or whether some, still mysterious undescribed spectral quality is responsible, a constant species-specific feature shared by all of the syllable types and perhaps by all indigo bunting vocalizations. Thinking years later about individual identification by song, Lambrechts and Dhont (1995) puzzled over this same tonal quality question. It is the same question posed by Thorpe and his chaffinches, and to this day we don't know the answer. Whatever the final solution, species universals in song will always loom large as one primary source of evidence, and it is worth checking whether there are parallels to the indigo bunting story.

The Canary

There are species universals in the songs of canaries, favored as house pets for at least 500 years, and kept as much for their voice as for their plumage (Barrington 1773). Domestic canaries have a basic set of song components consisting of about 10 syllable types or 'tours,' a much smaller set than that of the indigo bunting. The judges of song tournaments have used these tours for centuries to rank males in competitions, scoring them on the basis of their tour repertoire, and how well they are delivered. The tours are not completely stereotyped; for each tour there is a cluster of within-type variants and a male can have several variations. Thus, each singer has a repertoire of 30 or so different song syllables (Güttinger 1979, 1985). Syllable repertoires vary somewhat between strains (**Box 4**, p. 58) and fanciers are careful in selecting their breeding stock and in using prizewinners as song tutors. There is great plasticity in canary song development. Songs are learned and develop abnormally in isolation, with shrunken repertoires, especially after the elimination of auditory feedback (**Fig. 1.9**). But again there is a curious undercurrent of species-specificity to the song, posing a challenge to anyone trying to disentangle the interconnections between nature and nurture in song development (Marler & Waser 1977; Waser & Marler 1977; Marler 1997). Is there some hidden aspect of tonal quality that marks all tours as 'canary,' or is song recognition more of an additive process, based on something like a mental checklist of syllable types with which an unknown song is compared? Again, the answer is unknown.

The Swamp Sparrow

Another bird with clearly defined species universals in its song is the swamp sparrow. It has a learned song, a simple two-second trill of individually distinctive syllables. Swamp sparrow songs develop abnormally in social isolates and are highly degraded in birds deafened early in life (Marler & Sherman 1983, 1985). There are dialects, and learning early in life determines the dialect a male sings, and the dialect to which females are most responsive (Balaban 1988). Thus, swamp sparrow song satisfies the criteria for a culturally transmitted behavior. But once again, sonograms reveal that, instead of the unmanageable variability that might have been anticipated, all swamp sparrow songs can be broken down into six note types, distinguished by their duration, bandwidth, and the pattern of frequency modulation (**Fig. 1.10**). Much the same typology results from analyzing sound spectrograms by eye (Marler & Pickert 1984) and from statistical procedures such as multidimensional scaling (Clark et al. 1987; Nelson & Marler 1989).

There are rules about the order in which the note types are assembled into a syllable, varying from one locality to another. For example, a New York song syllable typically begins with a type I note and closes with a type VI note. The Minnesota dialect has an opposite rule, with song syllables typically beginning with a type VI note and ending with a type I note. Males and females will learn either dialect depending on their circumstances. Again, there is a significant degree of within-category variation, superimposed on the basic note typology (**Fig. 1.10**), and if you add this to the many ways in which note types can be combined to create a syllable, there is ample room for dialects and individuality, even in the same population. No two individuals have identical songs, and territorial males are highly responsive to individual differences as a basis for neighbor–stranger discrimination (Searcy et al.

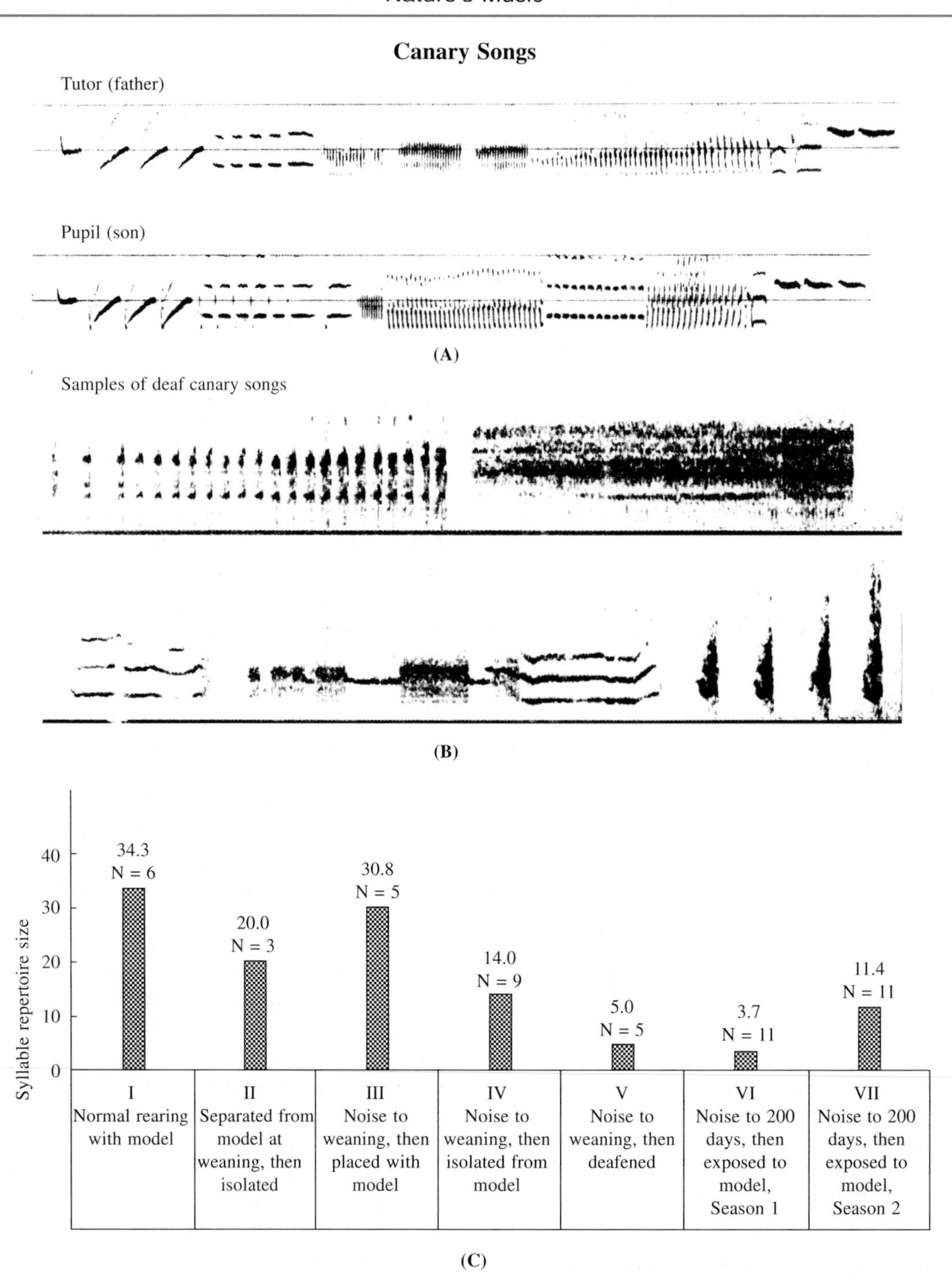

Figure 1.9 (**A**) Sonograms of part of a wasserschlager canary song, illustrating some of the syllable types. Songs of a tutor may be copied quite precisely by a young male, in this case the tutor's son. (**C**) The song syllable repertoire varies greatly with a male's history, reduced in isolated males, and shrinking to a minimum in males raised in white noise and then deafened as in **B**. From Marler et al. 1973; Marler & Waser 1977; Waser & Marler 1977.

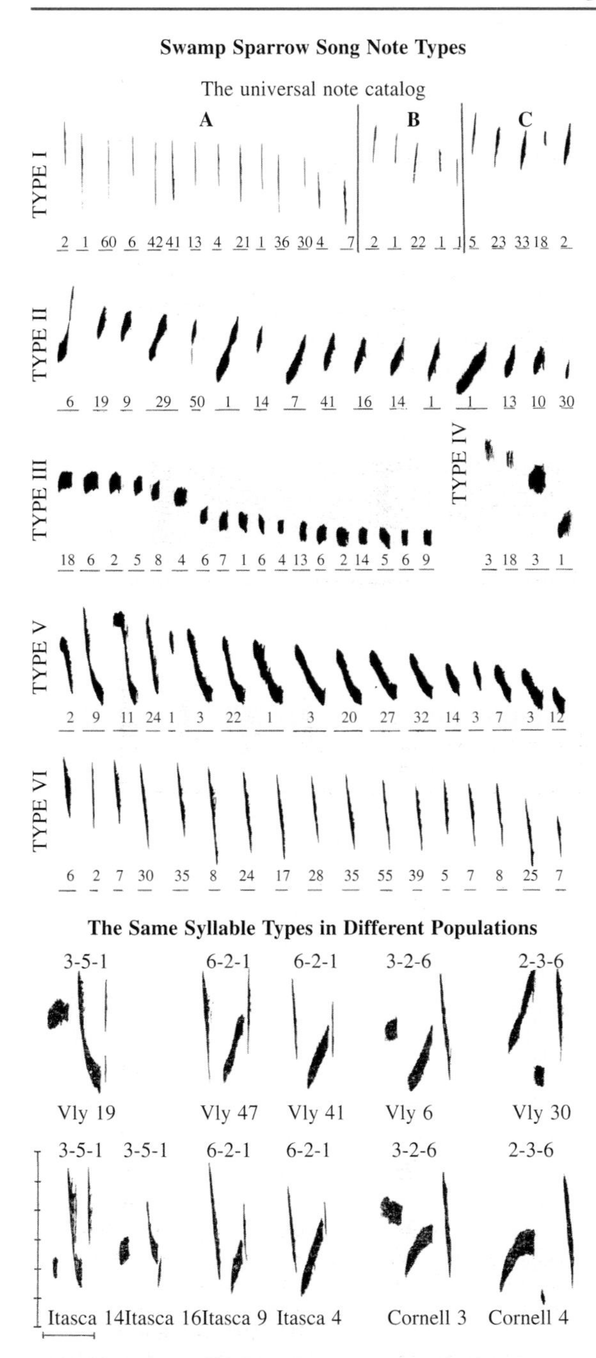

Figure 1.10 A breakdown of sonograms of 1307 swamp sparrow song note types from one population, first into 96 subtypes, and then into six major categories, defined by such features as duration, inflection, and bandwidth. Below each sonogram is the number of exemplars contained by that subtype. Below are examples of syllables shared in three populations in the Great Vly, New York; Lake Itasca, Minnesota; and locations in Maryland and New York. From Marler & Pickert 1984.

1981a, 1982); but species recognition, in both sexes, appears to be based on another set of song features, presumably species universals that are shared by all swamp sparrows (Searcy et al. 1981b, c). Of course, species-specific phonology remains a possible contributor to species song recognition, as Thorpe suspected in the chaffinch. He was especially interested in conspecific tutor recognition by young birds learning to sing. Later experiments pointed to a similar conclusion with young swamp sparrows which learn an artificial song however it is syntactically organized, as long as it is composed of natural swamp sparrow song syllables. But the same is not true of young song sparrows, which accept alien song syllables, as long as they are organized in song sparrow-like syntax (Marler & Peters 1989), showing that tonal quality is not the whole story. Responses of young white-crowned sparrows to natural and artificial songs show that they too base recognition of tutor songs of their species on a range of structural features in addition to phonology, although it too plays a role (Soha & Marler 2000, 2001). We need new methods for analyzing the fine phonological details of birdsong to carry this work forward.

Young sparrows respond innately to so many of the universal features that are present species-wide in their songs that one is tempted to entertain a theory of song learning that turns the traditional view on its head (Marler 1997). Perhaps they already have an innate library of song features in their brain, with tutor experience serving to select items from the library as a basis for song learning, each bird with its own distinct selection. This idea might help to explain why song learning is so extraordinarily rapid, as Hultsch & Todt (see Chapter 3) have demonstrated so dramatically in the European nightingale. A host of difficult experiments on the neurobiology of song learning in young birds will be needed before we can tell if this is a viable model of the vocal learning process.

The Wood Thrush and Others

Another unexpected candidate for species universals is one of the most attractive of all to our ears, the song of the wood thrush. It has three parts, and the middle one is especially influenced by learning (Whitney & Miller 1987; Whitney 1989). Improbable as it sounds, inspection of hundreds of these beautiful songs led Whitney & Miller to conclude that there is an underlying conformity; all natural middle phrases, wherever they occur in the species range, fall into about 20 rough categories, each with some degree of within-category variation. Even though these phases develop abnormally in isolation, there still seems to be a basic typology in natural song that all wood thrushes share (Lanyon 1979; Whitney & Miller 1987); it is as though each bird has some notion in its head of a complete species-wide repertoire, from which it can make its choice. Again there are species-wide song universals in these learned birdsongs, interwoven with the many other song features revealed by the sound spectrograph, varying from individual to individual and from dialect to dialect.

Species universals in birdsong pose fundamental problems that have yet to be resolved. Somehow we have to reconcile their presence with our expectations of the variability that must ensue from song learning. If experience is such a dominant factor in song development, where do these song universals come from? A simple picture of the intermingling of instinctive and learned song components, what Konrad Lorenz called 'intercalation,' is problematical, if only because most of these universals do not develop in birds raised in isolation. The gradual realization that this is an important paradox deriving from a generation of sonographic analyses of birdsong lends some credence to novel interpretations of song learning, such as the possibility that the process is 'selective' rather than 'instructive.' We usually think of birds waiting to be instructed by experience as to how they should sing. But perhaps bird brains already know much more about the song of their species than is at first evident, defined not just in terms of overall quality, but as a catalog of distinct syllable and song types. Could they already possess that knowledge, delaying expression until energized into activation by a process of 'selection?' Perhaps we should even consider the possibility that song learning results from hearing songs or syllables that match up with a pre-established set of auditory specifications in their brains (Marler 1997). We have to wait for the neurophysiologists to tell us whether this is a feasible notion. Meanwhile, species universals remain an important challenge. Physiological analysis of responsiveness to them is difficult because of the need to study the responsiveness of young birds during a very limited time-window early in life, the course of which is constantly being changed by what they hear. But this is a worthy cause, revealed by a generation of analysis by the sound spectrograph of the songs of many species; with luck, future researchers will be able to see interplay between nature and nurture at its most intricate.

SUBSONG: THE START OF A LONG JOURNEY

Once the power of sound spectrography became obvious there was a burst of enthusiasm for new approaches to sound analysis, some homegrown, some spearheaded by military needs for underwater cold war surveillance (Mulligan 1963; Hopkins et al. 1974). These became much more sophisticated with the advent of digital techniques (Clark et al. 1987; Tchernichovski et al. 2001). Each method has its own strengths and weaknesses. Baker and Logue (2003) did everyone a service by applying three different methods to the same problem, detecting population differences in a black-capped chickadee call. The approach of Tchernichovski and his colleagues was clearly the best in separating the bird's voice from background noise. In an earlier venture of his own, DuPont director and scientific ornithologist extraordinaire, Crawford Greenewalt (1968) harnessed the electronic skills of his technical colleagues to look at the fine-grained structure of birdsongs, focusing especially on the inferences that could be drawn about how the syrinx works. He showed definitively that there must be at least two independently operating sound sources in use in many birdsongs, as Borror and Reese (1953, 1956) had already concluded. But how the songbird syrinx actually achieves the feat of two-voiced singing remained mysterious until, as described in Chapter 9, Suthers developed new ways to study operations inside the syrinx of a singing bird.

Honing Singing Skills

From the earliest days, Thorpe was fascinated by the quiet 'recording' or subsong of young songbirds, so reminiscent of infant babbling and apparently unique to birds that learn to sing (Thorpe & Pilcher 1958). But beyond the idea that it was a kind of practice in singing, its significance remained obscure. Now that we understand what an extraordinarily intricate instrument the songbird syrinx is, enriched as it is by the capacity for coordinated singing with two voices, it is clear that learning how to use it is a major challenge. By hindsight, we can see why songbirds take so long in progressing from the first amateurish murmurings of subsong to the full-throated music of mature singing.

If you study sonograms of early subsong (see p. 93), it is easy to see the quavering efforts to produce pure tones, as the young bird strives, from utterance to utterance, to bring the two sides into closer coordination. Only after conquering these basic skills and learning in subsong how to guide the voice by ear, is the bird ready to embark on producing learned songs. Imitations first begin in plastic song, as the bird starts consulting the memories of songs heard previously. This retrieval process takes time to complete; captive swamp sparrows that Susan Peters and I studied took an average of 60 days to progress through to fully matured crystallized song (Marler & Peters 1982a). They proceeded in a series of stages, first stabilizing their imitations of song syllables, without much attention to order, only then recasting them at the time of crystallization into the one-phrase syntax typical of their species (**Fig. 1.11**).

Crystallization can be a quite abrupt event, apparently facilitated by a surge of testosterone from the male testes at this time (Marler et al. 1988; **Box 13**, p. 91). Singing becomes louder, and song duration, which is highly variable during the plastic song of sparrows, suddenly stabilizes. This is also the stage when sonograms reveal a narrowing of the preference for particular plastic songs, as the bird rejects many of those over-produced at earlier stages (**Fig. 1.11**). With digital analyses, we can visualize the developmental changes in greater detail than previously (Clark et al. 1987; Tchernichovski et al. 2001; **Fig. 1.12**), preparing the way for us to determine more precisely exactly how development proceeds.

When a bird has been adequately tutored, the first consequences of tutoring appear quite early in plastic song. But we have no idea how the song universals we have been discussing enter the stream of vocal development. Are they, like imitations, completely tutor-based, or does the genome begin to make its own specific contributions to the details of song development at some stage either directly or indirectly by biasing the process of imitation? What would the developmental flow look like after tutoring with a song lacking the song universals of that species? Would the pupil forgoe them, or introduce them of its own accord? The new techniques for study of early song development are opening up many new questions for us (Tchernichovski et al. 2001).

Are There Auditory 'Templates' in the Brain?

Since the 1950s and 1960s entomologists have been among the pioneers in bioacoustics. They had less need to await the sound spectrograph than ornithologists because cricket songs are readily analyzable by an oscillograph; much of this information on insect sounds was brought together by one of France's bioacoustical pioneers, Roger Guy Busnel in 1963 in an edited book. He asked me to write a chapter, but there were problems in the production of this massive tome on the *Acoustic Behavior of Animals*; it originated from a 1956 meeting at Pennsylvania State University, but it took forever to appear, and I

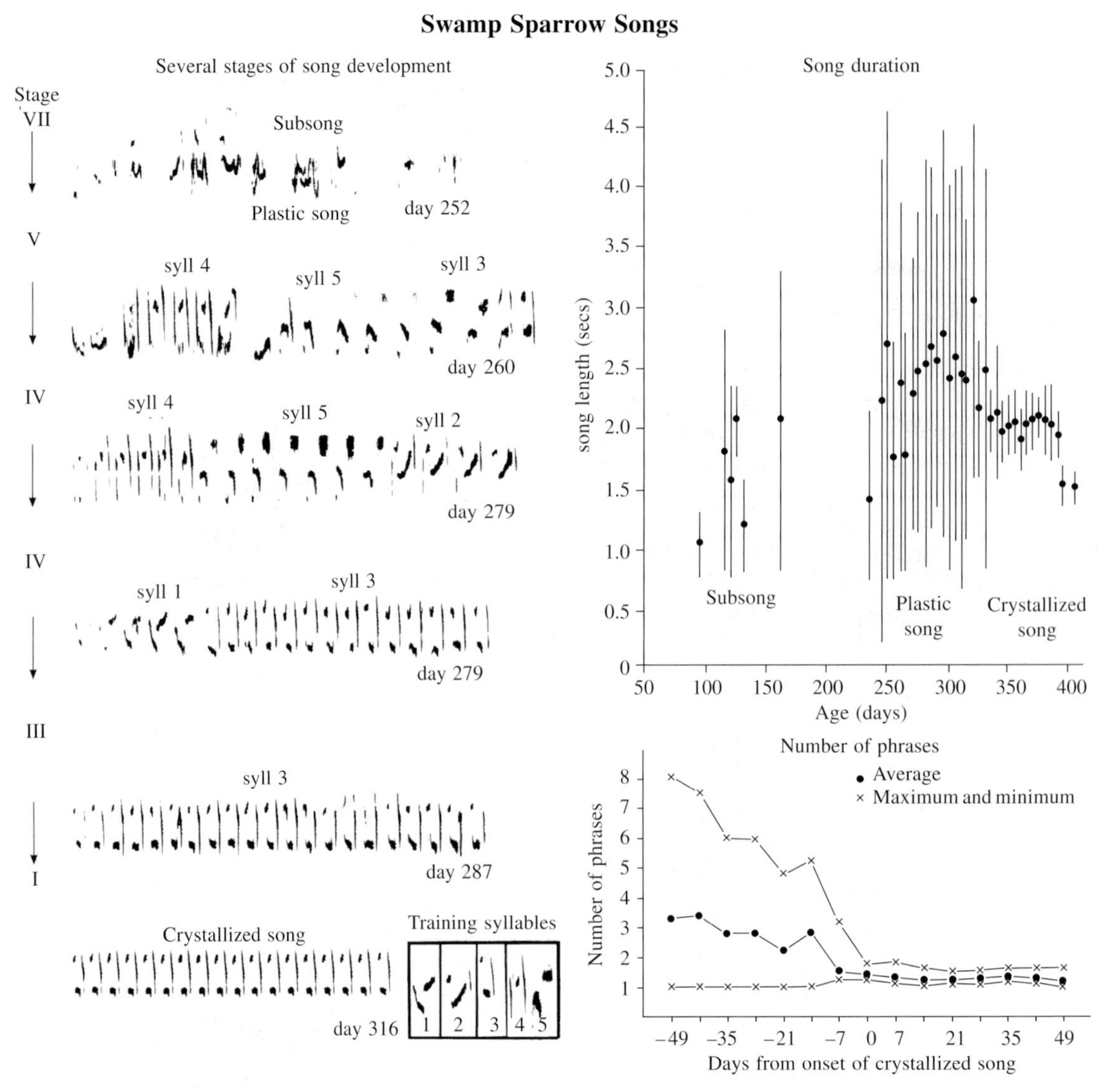

Figure 1.11 Song development in sixteen tape-tutored male swamp sparrows. On the left are some of the seven developmental stages of the same male, from day 252 subsong to crystallized song on day 316. Imitations of many training syllables appear in plastic song. As songs crystallize the number of phrases per song declines (bottom right), song duration becomes much less variable (top right), and many syllable types are discarded. From Marler & Peters 1982a & b.

began to wonder if my chapter on inheritance and learning in the development of animal vocalizations, in which I had raised the issue of auditory templates, would ever see the light of day. In it I suggested that: "the two most obvious ways in which such (genetic) control could operate are either by an inherited pattern of motor input to the sound-producing organs, or by an inherited auditory 'template' to which the animal would match the sounds produced, as a result of hearing its own voice. At present we cannot choose between these two." "It may be that inherited vocalizations are not dependent on auditory feed-back either for development or

Swamp Sparrow Song

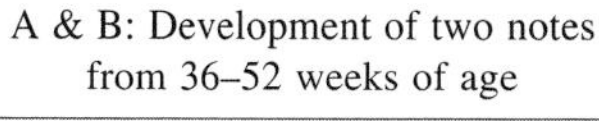

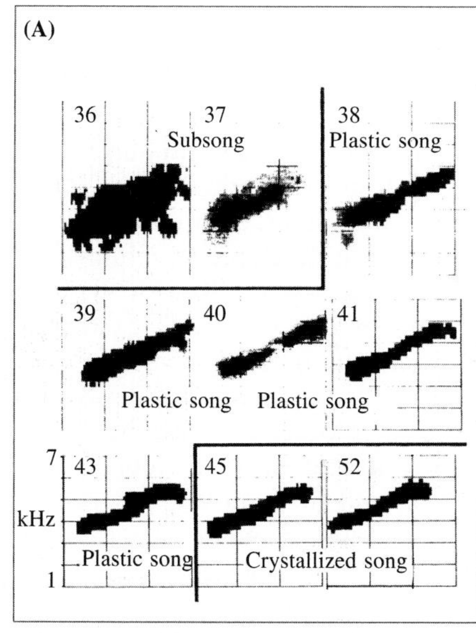

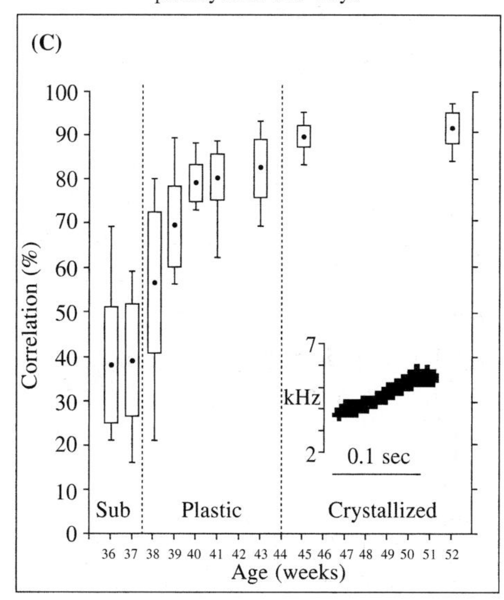

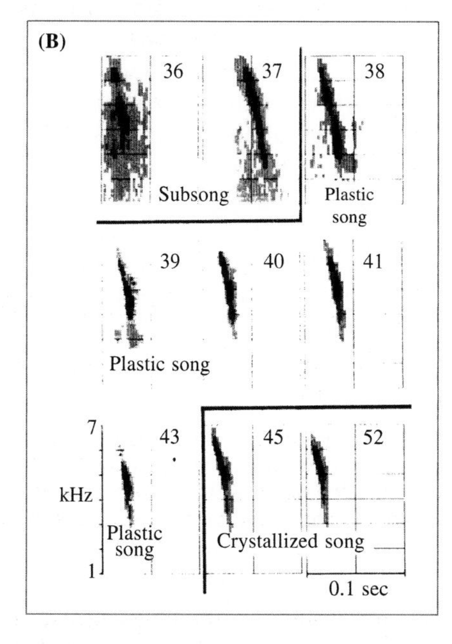

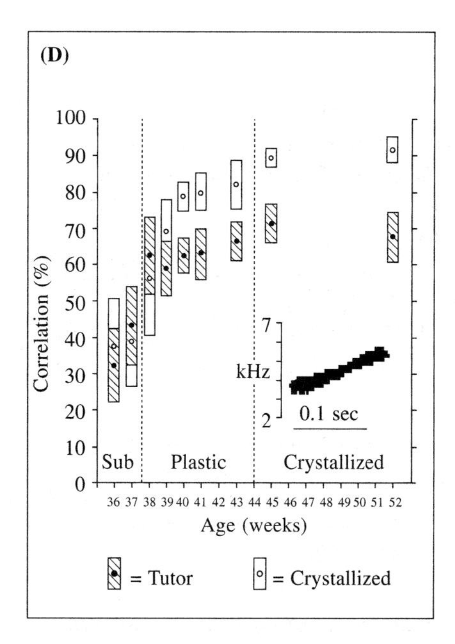

Figure 1.12 Digitized analyses of developing song notes in the swamp sparrow from subsong to crystallized song. **(A & B)** show averaged sonograms of two note types, as related to the final crystallized version of this male's song. **(C)** shows the computerized data visualized in **(A)**. In **(D)** data are plotted in relation both to the crystallized version and to this note in his tutor's song. As plastic song began, this male diverged from the tutor more and more. From Clark et al. 1987.

maintenance, while the reverse may be true in both cases for learned vocalizations" (Marler 1963, p. 233). These prognostications were more prophetic than I could have known at the time. Unwittingly, I may have been the first to use the terms 'template' and 'auditory feedback' in this context.

Auditory Feedback: Guiding the Voice By The Ear

Our understanding of the vocal learning process took a major step forward when, in a series of dramatic studies, a Fulbright scholar from Japan and graduate student in my Berkeley laboratory, Masakazu (Mark) Konishi showed that if young songbirds are deafened, their singing is radically disrupted; young birds developed a completely different song when they could not hear their own voice (Konishi 1964, 1965a, b). Juncos, robins, black-headed grosbeaks, and white-crowned sparrows all developed rather similar songs after early-deafening, amorphous and variable, rasping and buzzy in tone, almost insect-like in quality. By comparison, songs of birds with normal hearing reared in social isolation, although abnormal, were much more melodious and bird-like; they retained a significant number of the features of the normal songs of their species, especially their pure tonal quality, typical of so many birdsongs (Thorpe 1954, 1958; Lanyon 1960; Marler et al. 1962; Marler & Tamura 1964; Nowicki et al. 1989). Most of these normal song features that isolates develop were stricken from the songs of early-deafened birds.

Equally important was Konishi's demonstration that the same insect-like songs developed in birds already tutored before deafening, as long as this was done before singing had started (Konishi 1965a). It became clear that a bird's ability to hear its own voice is critical for providing access to the stored memory traces of songs learned previously. But in addition, auditory feedback provides access to innate information that a naive bird's brain encodes about how a birdsong should sound. This was the time when the concept of auditory templates gained full currency, embraced by Konishi to include the concept of acquired auditory templates, used to guide the development of learned song (Konishi 1965a).

How Much Can a Deaf Bird Accomplish?

As this work proceeded, the sound spectrograph revealed a new problem. Although the songs of most of Konishi's early-deafened birds were virtually amorphous, deaf juncos and grosbeaks did succeed in developing a minimal degree of normal song syntax (Konishi 1965b). The problem was that the age of deafening varied somewhat, and some might already have started to sing. This raised the possibility that prior song practice could influence how much song structure a deaf bird developed, as another graduate student launching his career in my laboratory, Fernando Nottebohm from Argentina (1966, 1968) then showed in experiments with chaffinches. Once chaffinch song is fully crystallized, deafening seems to have little effect, as Konishi (1965a) found with white-crowned sparrows. Deafening earlier in song development had a much more dramatic effect. It began to look as though almost all species-specific song features might be erased if auditory feedback was cut off early enough.

Another question was raised about the effects of very early experience. If early-deafened birds did succeed in developing some specific song structure, Konishi and Nottebohm pointed out that, even without any prior singing practice, the song structure of birds taken as nestlings and raised by hand "may be influenced by auditory experience before isolation, so that the resulting song pattern may not have been derived by reference to a pre-existing template" (Konishi & Nottebohm 1969, p. 46). Virginia Sherman and I decided to reopen this issue later, by focusing on two species with very different song syntax; the swamp sparrow with a simple trill, and the song sparrow with a more complex, multi-phrase song. We thought that this choice of subjects might increase our chances of establishing whether or not there are species differences in the songs of early-deafened birds.

First, we controlled for very early song experience by raising birds with normal hearing under canaries as foster parents; we found that their songs were indistinguishable from those of chicks taken from wild nests (Marler & Sherman 1983). Reassured that very early experience of conspecific song in the nest was not a factor, we scrutinized the songs of birds deafened before subsong had started, looking for contrasts between the two species; and we found some (**Fig. 1.13**). The contrasts were subtle and quantitative, and they differed in the predicted direction; there was more multipartite segmentation in the songs of deaf song sparrows than in the swamp sparrows (Marler & Sherman 1985). Thus, twenty years later, we were able to confirm Konishi's original suspicion that, buried in the noisy, amorphous songs of early-deafened birds there are indeed traces of species-specific song syntax, perhaps reflecting species differences in the pattern of respiratory flow through the syrinx, which might be less dependent on auditory feedback.

Another notable Konishi/Nottebohm finding was the reduced impact of deafening on adult

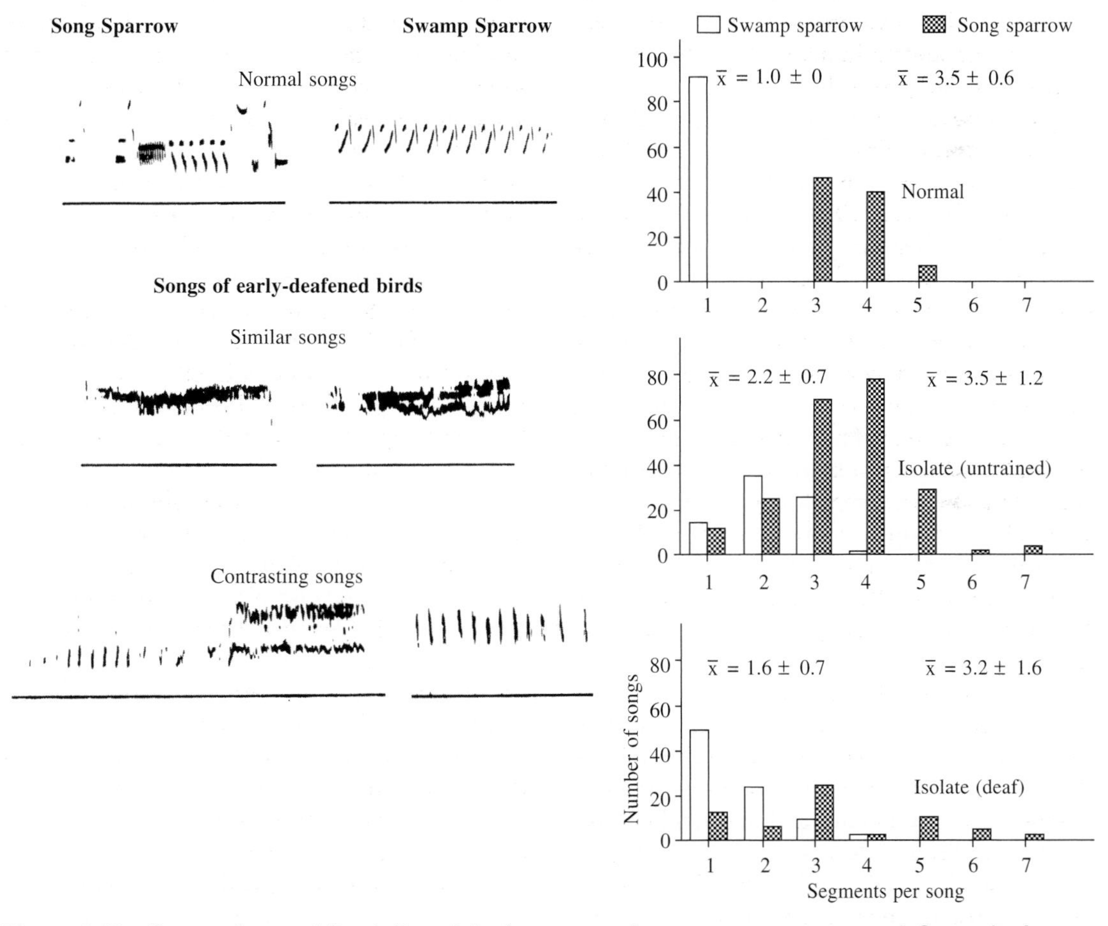

Figure 1.13 Songs of normal (top left) and deaf swamp and song sparrows compared. Some deaf songs of the two species are almost indistinguishable (middle left). Others show species differences in the degree of segmentation (bottom left), similar to those in normal songs. On the right the numbers of song segments or 'phrases' for the two species are plotted for wild birds, for isolates with their hearing intact and for deaf birds. The difference is complete in normal songs, blurred in isolates, and even less clear but still detectable, in deaf songs. After Marler & Sherman 1983.

crystallized song, subsequently a topic of much debate (see p. 33). They theorized at that time that perhaps "the motor pattern of song, once stabilized, can be maintained without any sensory monitoring," as though a motor tape takes over. Alternatively, "suppose that as a given vocal pattern is gradually stabilized, its auditory and proprioceptive feedback are compared so as to establish a reference through which the vocal output can be evaluated by either feedback. Maintenance of the pattern should then be possible by proprioception in the absence of auditory feedback" (Konishi & Nottebohm 1969, p. 44). It thus became important to determine how the structure of stabilized adult song is maintained.

The Song System: The Holy Grail

To begin to explore what role proprioception might play, Nottebohm looked at the effects on song of severing the hypoglossal nerves to the syrinx; these nerves convey not only motor impulses to the syrinx but also, he presumed, proprioceptive information from the syringeal musculature back to the brain. Almost fortuitously, he was led to a fascinating discovery. Although the right and left hypoglossal nerves are quite separate, they often make unequal contributions to operations of the syrinx, so that motor control of the syrinx is lateralized. Hypoglossectomy did disrupt song, but the effects were more severe with nerve cuts on one side than on the other. Initially it appeared that the left side played a dominant role, an uncanny parallel with left hemispheric dominance in the control of speech in our own brains (Nottebohm 1971a, b, 1972). The next question followed inexorably; does the lateralization extend up into the brain? Before this issue could even be contemplated, a detailed map of circuitry within the songbird brain was a necessity, and this became Nottebohm's next target. Using Belgian wasserschlager canaries as subjects, already under study as vocal learners in our Rockefeller laboratory (Marler et al. 1973; Marler & Waser 1977; Waser & Marler 1977), Nottebohm and his colleagues discovered a previously unsuspected, well-defined subsystem in the canary brain that is highly specialized for the control of song (Nottebohm et al. 1976, 1982; see Chapter 8).

A new page was turned in the science of birdsong. Nottebohm was a junior faculty member in my Rockefeller University laboratory at the time, so I had a ringside seat; I can vouch that this new undertaking was nothing like as easy as it sounds. In the first place, it took a good deal of courage. A neuroanatomical laboratory had to be set up, with all of the necessary techniques. Not unexpectedly, with neuroanatomical neophytes making such a radical discovery, there was skepticism about whether the results were credible. I seem to recall some difficulty at first in getting the findings published. Editorial incredulity was followed by an acute case of study section indigestion during the review of a grant application at the National Institutes of Health. Eventually, it was realized that the anatomical work was indeed sound, and that this represented a remarkable development in the study of functional vertebrate neuroanatomy.

Many discoveries followed in quick succession. I still recall the day when Fernando Nottebohm and Art Arnold, now a UCLA professor, then a graduate student in Nottebohm's laboratory, rushed into my office, with stained brain sections in hand, Art with zebra finches, Fernando with canaries. They were thrilled to the point of near incoherence by the discovery that, even with the naked eye, you could see the sex differences in the song systems of both species, flatly contradicting the current dogma that sexual dimorphism in the brains of higher vertebrates is minimal or nonexistent (Nottebohm & Arnold 1976). Shortly afterwards, Fernando was promoted to head his own laboratory at Rockefeller, as an independent researcher.

INNATENESS AND OTHER TABOOS

Some of the early excitement in the science of birdsong came from the realization that vocal learning was a perfect forum for exploring the interactions of nature and nurture. Controversies about whether genes or environments are dominant in the development of behavior reached a fevered pitch in mid-century as the impact of ethological thinking made itself felt. In laying out plans for future behavioral research, Julian Huxley (1963) and Niko Tinbergen (1963) agreed on the three main points of focus: the causation of behavior, by which they meant the underlying physiological machinery; the survival value of behavior; and its role in evolution. In the years that followed, there has been dramatic progress in all three. But Tinbergen chose to add a fourth area, the development of behavior, which was destined to become something of an orphan. Even today, we have no comprehensive theory of behavioral ontogeny. Although I believe that the science of birdsong is ultimately destined to play a key role in bringing the study of behavioral development up to par, progress to date has been slow. To understand why, I find it instructive to look back to other events taking place in the 1950s, directly provoked by the renewed interest in instinctive behavior (Marler 2004).

Lorenz, Tinbergen, and Thorpe, like Darwin, were all fully convinced of the value of the concept of instinct. As a medical student in Vienna, Lorenz was exposed to the ferment of phylogenetic theorizing in comparative anatomy and morphology, mostly based on the study of bones and skeletons. The young Lorenz was more interested in behavior, but his teachers took the view that behavior was too variable and amorphous to be amenable to the same kind of study. Encouraged by his most influential mentor, Oscar Heinroth, Lorenz became convinced that ethology could be as objective as comparative anatomy. As director of the Berlin Zoo, Heinroth was intimate with the behavior of scores of animals, many of which he raised in captivity, often by hand. Among Lorenz's zoo favorites were waterfowl, and watching them he found that many of the distinctive, species-specific displays they performed in captivity matched those in the wild, even though they had no adult mentors to instruct them. In his 1941 monograph on duck behavior, Lorenz compared in detail 20 species, all of whose displays appeared to him to develop normally in captivity. It seemed natural to regard these displays as instinctive or innate, and the concept of innateness became central to his view of behavior.

'Innate' Behavior?

Ethologists then began describing how young animals of diverse species, raised under similar conditions, consistently developed distinctive behaviors that were sufficiently stable and stereotyped to yield insights into taxonomy and phylogeny. Lorenz developed his notion of 'fixed action patterns,' building on some of Heinroth's ideas, characterizing units of behavior that could be studied in similar terms to anatomical and morphological features. This view was embraced by Tinbergen in his pivotal book on *The Study of Instinct*. It appeared in 1951, the same year I became a graduate student in Thorpe's laboratory. But over the next few years something strange happened. Tinbergen made a radical change in his position on innateness. His approach to observation and experimentation hardly changed, but his position as a theorist was quite transformed. The event he held responsible was a 1953 paper by Daniel Lehrman, a leading American animal behaviorist and authority on the intricate interactions between hormones and behavior in the reproduction of doves. He called the paper "A critique of Konrad Lorenz's theory of instinctive behavior." Lehrman's attack on the concept of innateness had a major impact, not only in ethology, but throughout animal behavior and comparative psychology. Subsequently Tinbergen became much more cautious about using the term 'innate,' as did many other leading ethologists, including Patrick Bateson, Robert Hinde, and Gerard Baerends, all influential figures. The message was not lost on students who, like me, were in the midst of launching their own careers in animal behavior.

Innateness Under Attack

For the young and old alike, Lehrman's paper was influential not only because of some cogent criticisms of ethology, but also because of his eloquence, and the tone in which it was written, which was extremely hostile. He almost ridiculed the concept of innateness, stressing the difficulty of excluding environmental influences on ontogeny, especially in the study of behavior. Lorenz was furious and, although they became reconciled later, I suspect that he never really forgave Lehrman. At times the confrontation escalated and assumed an almost religious fervor. The aftermath of these angry exchanges could be felt decades later, and I am convinced that the bitterness of the controversy inhibited many in the next generation from investigating or even acknowledging the importance of the genetic side of the developmental equation. Years later, Lehrman (1970) was adamant that "the clearest possible genetic evidence that a characteristic of an animal is genetically determined in the sense that it has been arrived at through the operation of natural selection does not settle any questions at all about the developmental processes by which the phenotypic characteristic is achieved during ontogeny," a view with which, at the time, many others seemed to agree. Thorpe was one of the few who maintained a more balanced viewpoint.

In retrospect, the Lehrman controversy generated more heat than light. Why did it continue for so long? In a symposium on the development and evolution of behavior, published thirteen years after his 1953 bombshell, Lehrman acknowledged that it had been as much a statement of faith as a scientific position. Nevertheless, after what amounted to an apology to Konrad Lorenz, he did not disavow his basic criticisms, again sprinkling salt on the wound. After admitting that "My critique does not now read to me like an analysis of a scientific problem, with an evaluation of the contribution of a particular point of view, but rather like an assault upon a theoretical point of view, the writer of which assault was not interested in pointing out what positive contributions that point of view had made," he added, in parentheses, that "this would be an appropriate point for me to remark that I do not now disagree with any of the basic ideas expressed in my critique" (Lehrman 1970, p. 22).

Of Lehrman's main points, the most crucial is his insistence that classifying behaviors into 'innate' and 'non-innate,' 'learned' or 'non-learned,' or 'acquired' and 'inherited' is counterproductive because it does not tell us anything about how behavior actually develops, and may actually distract us from trying to find out. If true, this point is of course equally valid when we define a trait such as eye color as genetically determined. But Lorenz had more philosophical issues in mind when he insisted on the value of innateness concepts, perhaps so obvious to him that he found them hard to articulate persuasively. In an attempted rebuttal, he argued that innate and learned behavior can be distinguished by the different sources of information on which their ontogeny is based (Lorenz 1965). Although not truly logical, as I will argue in a moment, his response was also vague and uncompelling, and failed to resolve the dilemma to anyone else's satisfaction.

The Canadian physiological psychologist Donald Hebb from McGill, whose 1949 book *The Organization of Behavior* is still quoted widely by neurobiologists, visited Cambridge in the 1950s; as I listened in on some of the discussions with Thorpe and Hinde, I found his take on these issues illuminating. He acknowledged the dilemma upon which Lehrman was so completely focused, and agreed that "we are no farther forward by coining a new name for instinctive behavior." Reflecting on the many problems lurking here, he opined that "we are involved here in the difficulties of the constitutional–experiential dichotomy. We must distinguish, conceptually, the constitutional factor in behavior from the experiential, but there is presumably no mammalian (or avian) behavior that is uninfluenced either by learning, or by the constitution that makes some learning easy or inevitable" (Hebb 1949, p. 166).

But interestingly, like Lorenz, Hebb still found

it heuristically valuable to distinguish between instinctive and learned behaviors, even though when we adopt a developmental approach, which is surely what Tinbergen was advocating, the instinctive/learned distinction loses its logical underpinnings (Hebb 1958). Hebb proceeded to express Lehrman's essential point, but in a more balanced way. "In distinguishing hereditary from environmental influence, it is reasonable and intelligible to say that a difference in behavior from a group norm, or between two individuals, is caused by a difference of heredity, or a difference of environment; but not that the deviant behavior is caused by heredity or environment alone" (Hebb 1953, p. 47).

The Interaction of Nature and Nurture

The need for a deeper understanding of the interplay of nature and nurture had long been acknowledged as a central problem in genetics. With characteristic lucidity, the distinguished geneticist J.B.S. Haldane (1946) pointed out that the positions of theoreticians and experimentalists on this subject are rather different. "We can only determine the differences between two different genotypes by putting each of them into a number of different environments." "The problem is, of course, exceedingly complex, but certain facts about it are simple." "We compare two pure lines of mice not only as regards color, hair form, and other characters which are little affected by nurture, but for such characters as resistance to different bacterial and virus infections, each must be tested by appropriate changes of environment." Note that Haldane is in no way presenting an anti-genetic argument. He is reminding us that an animal's genetic constitution influences how it responds to the environment, echoing a central tenet of ethological theory.

A similar point of view was expressed by some evolutionary biologists of the time, such as the Russian geneticist Schmalhausen. His book *Factors of Evolution* was written in 1943, but because of his opposition to Lysenko it was delayed by Stalin, finally published in 1947 in Russian and only in 1949 in English, as a result of the professional encouragement of Theodosius Dobzhansky at Columbia in New York. Schmalhausen takes it for granted that the environment is a necessary part of every developmental interaction. But "what matters in this interaction between organism and environment is that the morphogenetic reaction is typical of the organism under given conditions." "The organism itself determines its relationship to its environment, thus protecting itself against some influences and utilizing others. Every species profits from environment in its own way and responds to changes in environment in different ways" (Schmalhausen 1949). It was clear to him, as to Haldane (1946), that although the interplay between nature and nurture during development is ubiquitous, nurture plays a more dominant role with some traits than others, bringing us closer to what Lorenz had in mind.

In spite of this inexorable duality, people still cling tenaciously to terms like 'instinctive' and 'innate,' presumably because they serve a useful purpose as labels for behaviors in which genetic factors play an especially dominant role during development. Even Hebb (1949), ever cautious about its limited explanatory value, regarded the term 'instinctive' as heuristically useful. "This is behavior in which the motor pattern is variable but with an end result that is predictable from acknowledgment of the species, without knowing the history of the individual animal. This class of behavior must be recognized" (Hebb 1949, p. 166). If only Hebb's more temperate view had prevailed, the history of the innateness concept might have proceeded very differently. Instead, the whole issue was thrown into a turmoil from which it has taken several decades to recover. Under Thorpe's influence, the science of birdsong was one of the few approaches to animal behavior in which some creative balance was maintained. Today a new era is dawning; we can now manipulate experimentally all components of the developmental equation. The complexity is awesome, but progress is being made, and before too long the nature/nurture controversy may become a thing of the past.

Phenotypic Plasticity

Another valuable step forward is the growing recognition that the concept of 'phenotypic plasticity' is useful in the study of song development and of behavioral ontogeny in general. Entomologist and population geneticist West-Eberhard (1989, 2003) defines phenotypic plasticity as "the ability of a single genotype to produce more than one alternative form of morphology, physiological state, and/or behavior in response to environmental conditions." It serves to remind us, first, that genotypes can encode instructions for the development not just of one phenotype, but several, depending on the environment experienced; and secondly, and equally important, that genotypes often have a direct influence on which aspects of the environment are most potent in eliciting changes in patterns of gene expression, recapturing for us the essence of the Lorenzian concept of 'innate release mechanisms.' Contrary to what Lehrman implied, a discovery that an environmental change can modify the trajectory of behavioral development is not automatically ammunition for an anti-genetic argument, however illuminating it may be from an ontogenetic viewpoint.

It is not always clear why some people object to concepts of innateness with such vehemence. One reason may be the belief that whenever you invoke nature in discussions of behavioral development, there is an implied commitment to total, unequivocal predestination, and the inevitable emergence of a unimodal, stereotyped behavioral phenotype. This is simply a mistaken view. A surprising number of people still believe that notions of innateness allow no option for developmental plasticity, adaptive or otherwise. In fact, in the varied environments found in nature, the same genotype often yields not one, but many morphological and behavioral phenotypes. In some degree, phenotypic plasticity may prove to be ubiquitous, as we can illustrate briefly with a few examples.

A classic case with behavioral connotations is caste determination in social insects; the morphology and behavior of workers is tightly controlled by their environments; the presence of a queen, egg size, nutrition, and temperature all trigger different growth programs (Hölldobler & Wilson 1990). Adaptive, environmentally triggered changes in insect life-cycles are ubiquitous. Some butterflies have seasonally distinct morphs, once thought to be different species, displaying adaptive contrasts in morphology and behavior, often triggered by temperature in the tropics and photoperiod in temperate regions (Shapiro 1976). Within a species there can be genetically determined variations in the photoperiod specifications for the control of diapause, depending on latitude (Tauber et al. 1986). A wonderful case of phenotypic plasticity is the larva of the moth *Nemoria*, which mimics either a catkin or a twig. The two morphs differ radically in appearance and behavior, and larvae displaced on an oak tree actively relocate, either on catkins or twigs. The alternative growth patterns are triggered by specific aspects of diet, especially the concentration of tannins (Greene 1989). And there are many cases of apparent phenotypic plasticity in vertebrates (West-Eberhard 2003). Intraspecific variation in patterns of social organization in fish, amphibians, birds, and mammals, cued by such experiential factors as population pressure, food availability, and predation are all good candidates (Lott 1991). Some degree of genetically preordained and environmentally triggered phenotypic plasticity may be a virtually universal feature of living things (Raff & Kaufman 1983).

West-Eberhard (2003) argues persuasively that the developmental plasticity of behavior can be accommodated under the same conceptual umbrella as physiology and morphology; specific cues from the environment, physical or social, engender a change of state of the organism, modifying patterns of growth, especially in the nervous system, changing motor patterns, and inducing new patterns of responsiveness to external stimuli. Even learning, the most specialized manifestation of developmental plasticity, can be viewed as a form of cueing by particular external stimuli, interacting with genetically based sensory and motor predispositions. Perhaps this is the basis of the species universals in learned birdsongs we have been considering. The suggestion that early stages of song learning could be based on a selective process, rather than on a simple instruction, also qualifies as a possible example of phenotypic plasticity (Marler 1997).

INNATE SONGS, LEARNED SONGS?

If genomic contributions are in fact as critical as I am suggesting for understanding the development of behavior, students of song learning will have to join forces with developmental biologists and geneticists if further progress is to be made. In preparation for that eventuality it is worth dwelling yet again on our habit of labeling behaviors as instinctive or acquired, if only as an aid in the identification of behavioral preparations worthy of intensive developmental analysis. Song learning serves to illustrate some of the pitfalls.

We readily speak of songs of the eastern phoebe as innate; they develop normally in a bird raised in social isolation (Kroodsma & Konishi 1991; see Chapter 4). Song and swamp sparrow songs are learned; natural songs develop very differently from those of males raised out of hearing of their own kind (Mulligan 1966; Kroodsma 1977; Marler & Sherman 1985; **Fig. 1.14**). Because tutors are copied in detail, we do not hesitate to classify these as learned songs. However, this labeling procedure is less straightforward than it appears (Johnston 1988).

A song sparrow that can hear its own voice produces an isolate song that, although abnormal, has several song sparrow-like features (Marler & Sherman 1985). It has several parts, with a pure tonal quality, an accelerated first trill and a normal overall duration (**Table 1.1**). What do these normal features of isolate song represent? Is it useful to label them as innate? If so, how should we interpret the abnormal aspects of isolate song? Perhaps we can classify some song components as innate, and others as learned?

We get a firmer perspective on this question by a closer comparison with the swamp sparrow. As in song sparrows, isolate swamp sparrow song is simpler than normal (**Fig. 1.15**), with a less complex syllable structure, and a slower tempo. But there is another side to this coin. If we make a four-way comparison, between natural and isolate songs of both species, we find that despite the abnormalities isolate songs are nevertheless easy to tell apart.

In fact both sparrows have specific song features that are unchanged in isolate song (**Table 1.1**); if we restricted our attention solely to them, we could conclude that sparrow songs are innate. But as we know, some features of these 'learned' songs develop abnormally in isolation. Evidently some song features are less fixed than others, and less resistant to change, more subject to nurture than to nature. You might think that by focusing on features that are more readily changed, we could use potential mutability as a yardstick, in deciding if a feature is innate or learned; but it turns out that this does not work either.

The singing behavior of these sparrows differs in two particular ways that are especially interesting because they are very abnormal in isolates; consider first the number of notes per song. When we compare the two species we find that despite the generally slower tempo of isolate songs in both species, with note durations and inter-note intervals drastically lengthened, if we count the total number of notes the average is still shorter in swamp sparrows than in song sparrows, as in normal song (**Fig. 1.16**). Even more interesting is song repertoire size. A wild song sparrow has 10 or so song types, more than the three or so of a swamp sparrow. As in many birds, repertoires shrink in isolates, to about half the normal size; but despite this shrinkage, isolate song sparrow repertoires are still three times larger than those of swamp sparrows (**Fig. 1.16**). So these species differences, that we can presume are genetically based, are still displayed in song features that are developmentally labile. Should we then regard repertoire size and note number as learned or as innate? Clearly, neither conclusion is satisfactory.

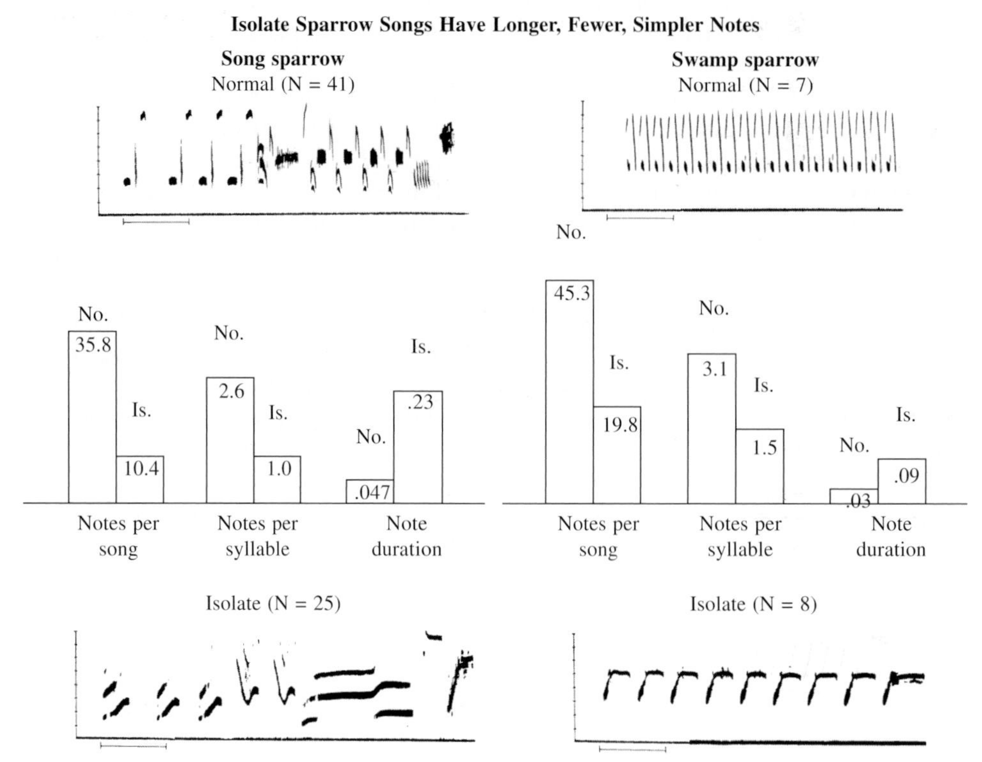

Figure 1.14 In the simplified songs of sparrows raised in isolation, both song and swamp sparrows have fewer notes per song and per syllable, and notes are longer than normal. Modified from Marler & Sherman 1983.

Table 1.1 Measures of normal and isolate songs of swamp and song sparrows compared. The symbols > and < indicate which value is larger or smaller in the species comparisons. In all cases the direction of species difference is the same in normal and isolate songs, even when the traits, like repertoire size and number of notes, are developmentally plastic. From Marler & Sherman 1985.

	Isolate Songs			Normal Songs		
	Song Sparrow (5 birds)		Swamp sparrow (5 birds)	Song sparrow (4 birds)		Swamp sparrow (7 birds)
Repertoire size	*5.0 ± 1.0*	>	*1.6 ± 0.5*	*10.3 ± 1.5*	>	*3.0*
Song duration (s)	2.898 ± 0.439	>	2.006 ± 0.292	2.573 ± 0.188	>	2.099 ± 0.135
No. of segments	3.5 ± 0.3	>	2.2 ± 0.6	3.9 ± 0.2	>	1.0 ± 0
Notes per song	*10.4 ± 2.1*	<	*19.8 ± 8.3*	*35.8 ± 8.3*	<	*45.3 ± 7.5*
Note duration (s)	0.232 ± 0.034	>	0.092 ± 0.072	0.047 ± 0.007	>	0.030 ± 0.005
Inter-note interval (s)	0.084 ± 0.025	<	0.044 ± 0.012	0.032 ± 0.005	<	0.019 ± 0.003
Number of trills	1.9 ± 0.3	>	1.7 ± 0.7	2.1 ± 0.1	>	1.0 ± 0
Total no. of trilled syllables	7.6 ± 1.9	<	17.7 ± 8.8	9.3 ± 0.6	<	14.8 ± 1.6
No. of notes in trilled syllables	1.0 ± 0	<	1.5 ± 0.7	2.6 ± 0.4	<	3.1 ± 0.7
No. of syllables in a single trill	3.6 ± 0.8	<	10.7 ± 7.0	4.2 ± 0.1	<	14.8 ± 1.6
No. of note complexes	1.5 ± 0.3	>	0.5 ± 0.3	1.8 ± 0.2	>	0
No. of notes in a note complex	2.1 ± 0.9	>	1.1 ± 0.1	9.7 ± 4.5	>	0

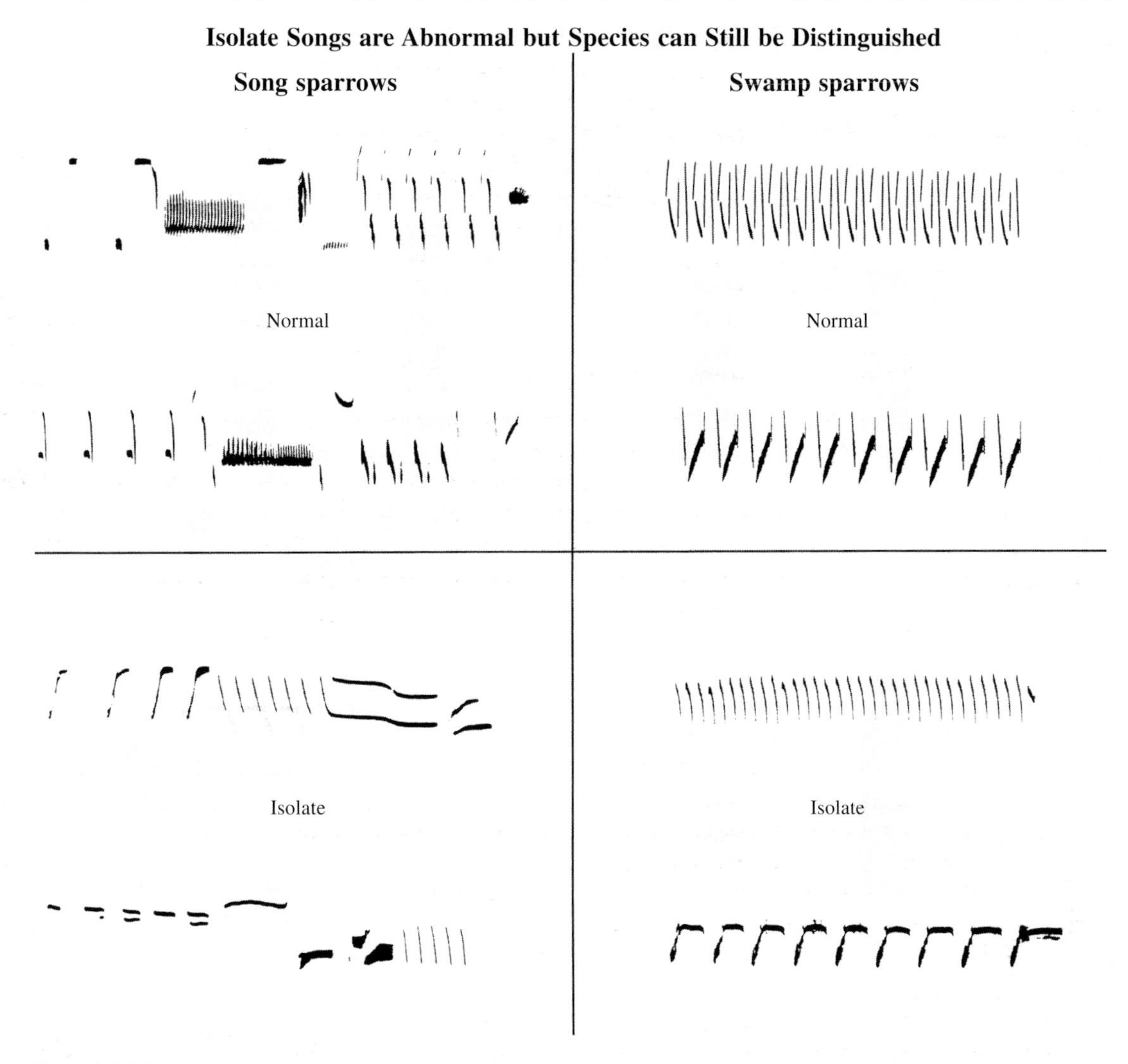

Figure 1.15 Normal songs of swamp and song sparrows, illustrating both within-species variation and between-species differences, as documented in **Table 1.1**. Below are examples of isolate songs, both simpler than normal song, but still showing clear species differences. Based on Marler & Sherman 1983.

So we find ourselves confirming what Lehrman (1953) asserted 50 years ago that in the long run classifying behaviors as 'learned' or 'innate' is not conceptually productive; however, it is not clear whether this is worth getting upset about. The habit of labeling behaviors in this fashion is so deep-rooted that we will probably never succeed in eradicating it. And it does have some value in placing behaviors on a lability continuum, with some behaviors more 'nurture-dependent,' more changeable and variable, and others more 'nature-dependent,' more stereotyped and resistant to change. But as both Hebb and Lehrman said repeatedly, while it may be valid to speak of 'differences' between individuals or populations or species, as 'innate' or 'learned', this is no more than a tentative first step in the long, arduous process of determining how these differences actually unfold in the course of ontogeny. In one sense, we can perhaps regard

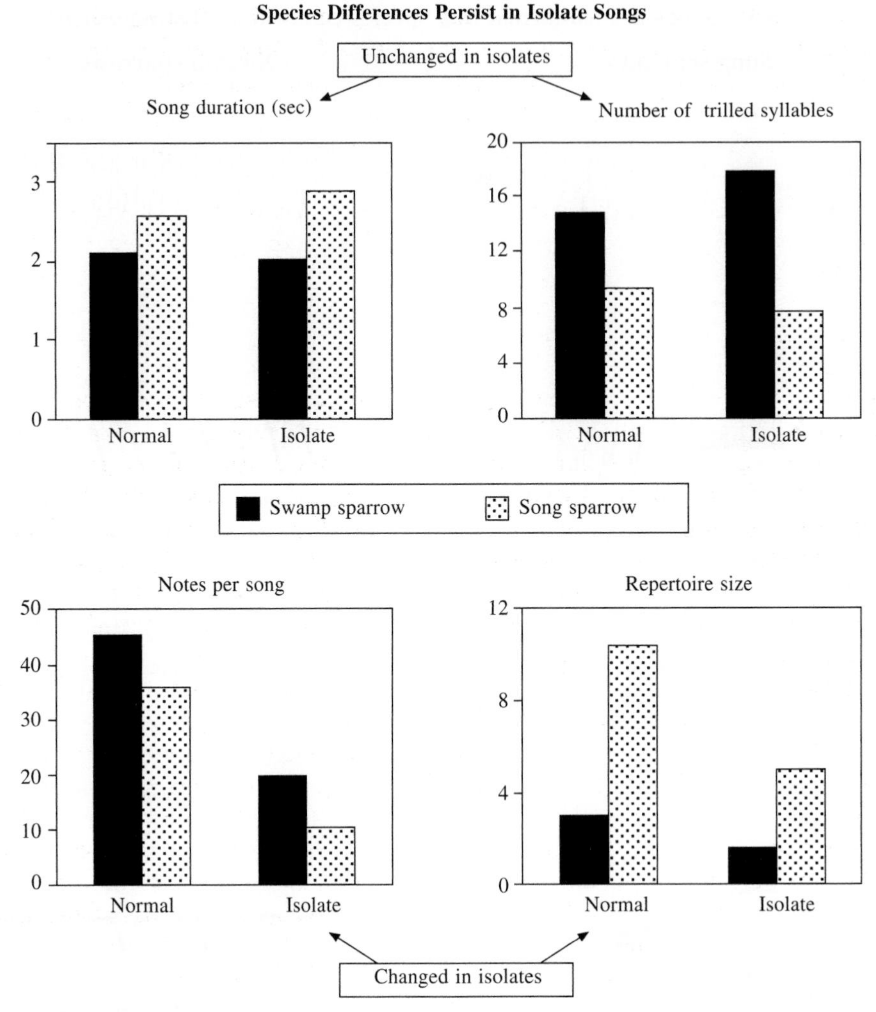

Figure 1.16 Species differences in normal song persist in songs of isolates whether the feature concerned is developmentally stable, like song duration or the number of trilled syllables (top), or whether it is developmentally plastic, like the number of notes or the song repertoire size (bottom). After Marler & Sherman 1983, with data from **Table 1.1.**

the nature/nurture controversy as having been laid to rest. But in another sense, we have so little in-depth understanding of the interactions between genotype and environment that underlie the development of behaviors like birdsong, that the enterprise of coming to understand them has hardly begun. This is our challenge for the future.

In embarking on this grand endeavor, it will behoove us to bear in mind lessons from past history. Those who resist the invocation of genetic contributions to behavioral development should be reminded that involvement of the genome need not imply a commitment to stereotyped behavior. In fact, the stereotypy itself may be deceptive (Waddington 1957); the underlying potential for flexibility could be just as great with stereotyped as with variable behaviors, but with added mechanisms that correct for the perturbations to which a developing organism must always be subject. This theme is getting much attention in birdsong studies as fully crystallized adult song turns out to be more subject to modification if you perturb the feedback controls than had ever been suspected (Nordeen & Nordeen 1992; Okanoya & Yamaguchi 1997; Woolley & Rubel 1997; Leonardo & Konishi 1999). A homeostatic approach to behavioral stereotypy may prove to be the most appropriate model to explore. To understand phenotypic plasticity we have to acknowledge that genomes evolve to cope with changing environments. The selective stakes are so high that pressures for the evolution of genome-controlled strategies for adaptive interaction with the environment, varying at the individual level, must be intense. Responsiveness to particular environmental cues—some general, some highly specific—must surely have evolved to influence the choice of alternative strategies; once again, we confront the universal duality of nature and nurture.

Behavioral scientists have led the field in analyzing and understanding the contributions of 'nurture' to ontogeny (Gottlieb 1976, 1992). The methods and concepts of modern genetics have become so sophisticated that the 'nature' side of the equation is now equally tractable. Our hope for the future lies in combining these approaches to create a unified theory of behavioral development in which the critical roles of the genome and the environment are both acknowledged to the full.

THE VERSATILITY OF BIRDSONG

In thinking about the genetics of song learning we sometimes speak of genetic 'constraints.' This may not be an inappropriate term when we are focused on species differences in song learning. One of the most remarkable characteristics of singing birds is their versatility. The potential of vocal learning sometimes seems to be almost limitless and, in some circumstances, 'constraints' may be needed to channel song learning in particular directions, according to differing species' needs. But the overarching capacity for vocal learning itself has genetic underpinnings, and these are completely pervasive in all songbirds, providing the sophisticated neural machinery that makes song learning possible. Here the term 'constraint' is less appropriate.

If we set aside species-specific limitations and take a more general view of song learning we uncover a truly astonishing range of abilities. Some birds seem able to mimic almost any imaginable sound, not necessarily in any way musical. Others create songs of such great beauty that we celebrate them as accomplished musicians. Again the Honorable Daines Barrington (1773) provides a historical perspective.

Barrington ranked the British birds he felt were the finest songsters, striving for esthetic judgments of the kind you might make in comparing paintings. Of the dozen or so top contenders, the nightingale was a comfortable winner, with the linnet and the skylark as runners-up. He scored them for mellowness of tone, sprightly notes, plaintive notes, compass, and execution. He was cautious about his judgments, admitting that "many may disagree with me about particular birds." British ornithologist Max Nicholson and sound engineer Ludwig Koch (1936) gave the blackbird and the song thrush much higher scores, perhaps because Barrington knew them primarily as cage birds; interestingly, he was familiar with the American mockingbird, regarding it as the nightingale's most formidable competitor for songster supremacy. In a footnote he says that the Indian name for the mockingbird signifies four hundred tongues.

With the advent of the science of birdsong, quite different kinds of judgments can be made about songster supremacy, based not on esthetics but on the awesome technical accomplishments of some birds. The song of the wren, for example, was low on Barrington's list; even an enthusiast like the late Reverend Edward Armstrong in Britain, an authority on birdsong, emphasized its loudness and vehemence rather than its beauty, partly because "the swift delivery makes it difficult to apprehend the exact relationship of the components" (Armstrong 1955, p. 56). Yet armed with the sound spectrograph, revealing these relationships in all their glory, Kroodsma (1981) was led to describe it as "a pinnacle of song complexity."

Every male winter wren has a learned repertoire of up to 10 different song types, and even 30 in some birds in western North America; each song is 10 seconds or so in duration, containing between 50 and 100 notes. Close inspection reveals that the notes are organized into phrases that we only register by slowing the song down (**CD1 #6**); the same phrases recur in different songs in a male's repertoire, but in different sequences (**Fig. 1.17**). It appears that when a young male learns from adults, he must break the songs down into segments and, in an extraordinary display of virtuosity, arranges them in different patterns; in this way he creates endless variety and greatly enlarges his song repertoire (Kroodsma 1981; Kroodsma & Momose 1991). The performance is startling to watch. The whole body vibrates as this song, each with its two score or more notes, audible nearly half a mile away, issues forth from this tiny bird's throat, a 'twittering flow' of sound much admired by Thoreau (Elliot 1999).

The brown-headed cowbird displays another kind of virtuosity, winning hands-down the Greenewalt sweepstakes for the range of frequencies encompassed within a single song; cowbirds span about 10 kHz. Oscines excel over all other birds in frequency range (Greenewalt 1968), an accomplishment attributable to their skill in using the different frequency tuning of their two voices to expand the vocal range (see Chapter 9). The size of some song repertoires is

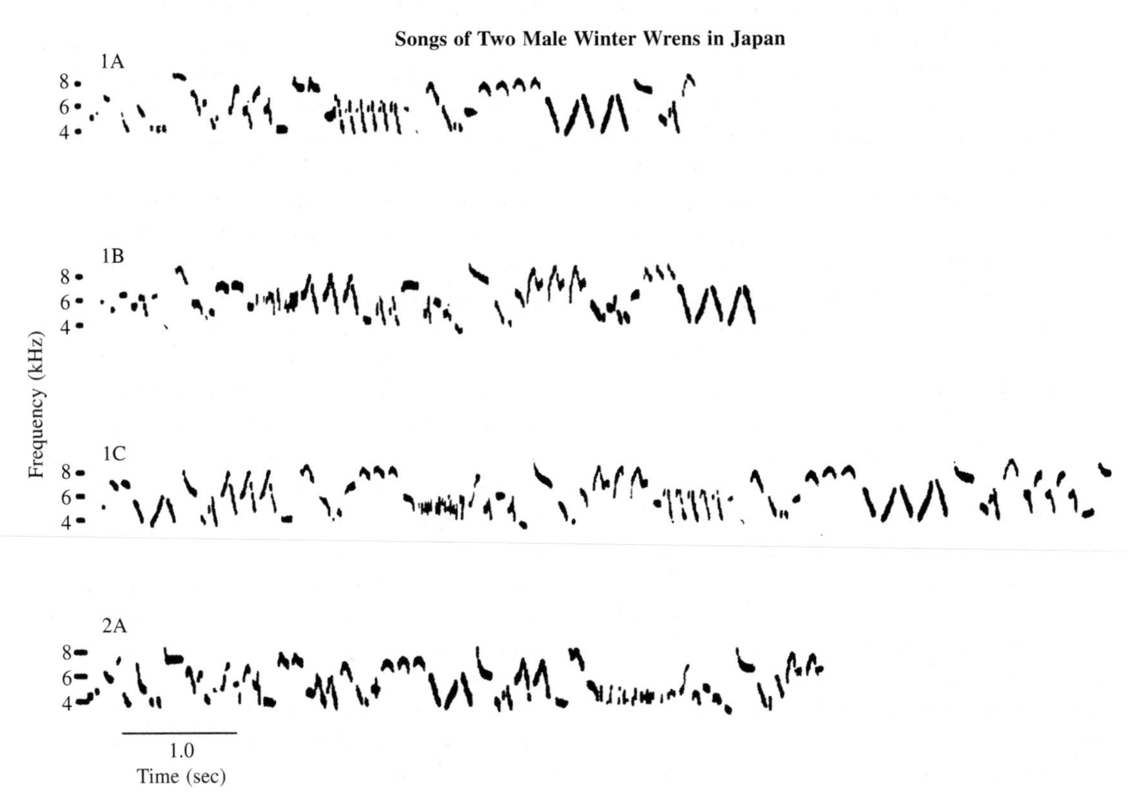

Figure 1.17 Winter wrens create novel songs by recombining phrases and stabilizing them in new sequences. Looking closely, you see that the three song types of male one (1A-C) share many phrases. Male two was a neighbor, and shared phrase types with male one (2A). From Kroodsma & Momose 1991.

another source of wonderment. Repertoires of 5–10 song types are not uncommon, and a hundred or more song types are uttered by nightingales, some wrens, and especially by members of the mockingbird family; the record holder is a brown thrasher with something over 2000 songs (Kroodsma & Parker 1977).

Another yardstick by which to measure vocal virtuosity is the range of physically distinct types of sound represented in a bird's voice. Many are quite conservative, as we have noted, restricting themselves to sounds grouped around a set of species norms; but the range of some mimics seems to be limitless. Lyrebirds and other Australian mimics make imitations, not only of scores of other birds, but of much more improbable sounds from the physical environment (Robinson & Curtis 1996). The late, much-revered Australian biologist A.J. Marshall (1954) ranked the spotted bowerbird as perhaps the finest mimic. Its own voice is nothing to write home about, but any deficiency in natural notes is remedied by the remarkably faithful reproduction (to human ears) of those of many other species, as well as the barking of dogs, the noise of cattle breaking through scrub, the sound of a maul striking a splitter's wedge, of sheep walking through fallen dead branches, and emus crashing through twanging fence-wires. A spotted bower-bird mimics a whistling eagle so faithfully that a hen and chickens fly for cover. Besides birdsong the spotted bower-bird imitates the 'whirring like' noise made by the crested bronzewing pigeon during flight, as well as wood-chopping, the crack of a stockwhip and, in fact, any other sound that it hears (Marshall 1954). A fawn-breasted bowerbird netted during a bird-banding exercise in Papua New Guinea by Ian Burrows, left lying in a bag near the nets while awaiting release, gave convincing renditions of the cries of children, and the sawing and hammering sounds of carpenters building a house (**CD1 #7**).

You might think that the calls of mammals would be far beyond the vocal compass of any bird; not so. With a little ethological trickery, a white-crowned sparrow, not normally given to mimicry, was persuaded to copy the alarm calls of ground squirrels, by embedding them in sparrow songs (Soha & Marler 2001). The best known examples of avian imitations of the sounds of mammals are of course talking birds, including canaries, parakeets, and other parrots (see Chapter 13); hill mynah birds are probably the most accomplished of all (**Fig. 1.18**; **CD1 #7**), and much ink has been spilled over just how they produce such life-like speech with a vocal tract that is structured so differently from our own (Thorpe 1959, 1961; Greenewalt 1968).

Quite aside from scientific virtuosity, for many of us the voices of birds are appreciated simply as a source of deep esthetic pleasure. In each part of the world, admirers have their own favorites (**CD1 #8**). The nightingale reigns supreme in Europe. The wood thrush and its relatives are favorites in North America, the brown-backed solitaire in Central America, the robin-chats in South Africa, the lyrebird in Australia, the tui in New Zealand, the shama in India, the list goes on (Hartshorne 1973). The musician wren of South America is one of my personal favorites (**CD1 #9**).

We are especially charmed when birds imitate our own music, something that occasionally happens even in the field. In 1956, I visited the Westphalian castle in Germany where Lorenz was awaiting the call to the new Max Planck Institute in Bavaria. We were there in part to make plans with Lorenz's student Jürgen Nicolai for a never-to-be-realized trip to the Himalayas to record finch songs; while we were there my wife and I visited the nearby home of the German ornithologist and war veteran Erwin Tretzel to hear the glorious sounds of his pet shama thrushes from India. Tretzel has described two remarkable cases of song dialects in wild birds based on sounds picked up from being around people (Tretzel 1965, 1967).

Near Erlangen in southern Germany a crested lark learned four whistles that a shepherd used to control his dog. The imitations, good enough that the dog obeyed playbacks faithfully, were acquired by other larks, coming to form a small local dialect. Then Tretzel found three European

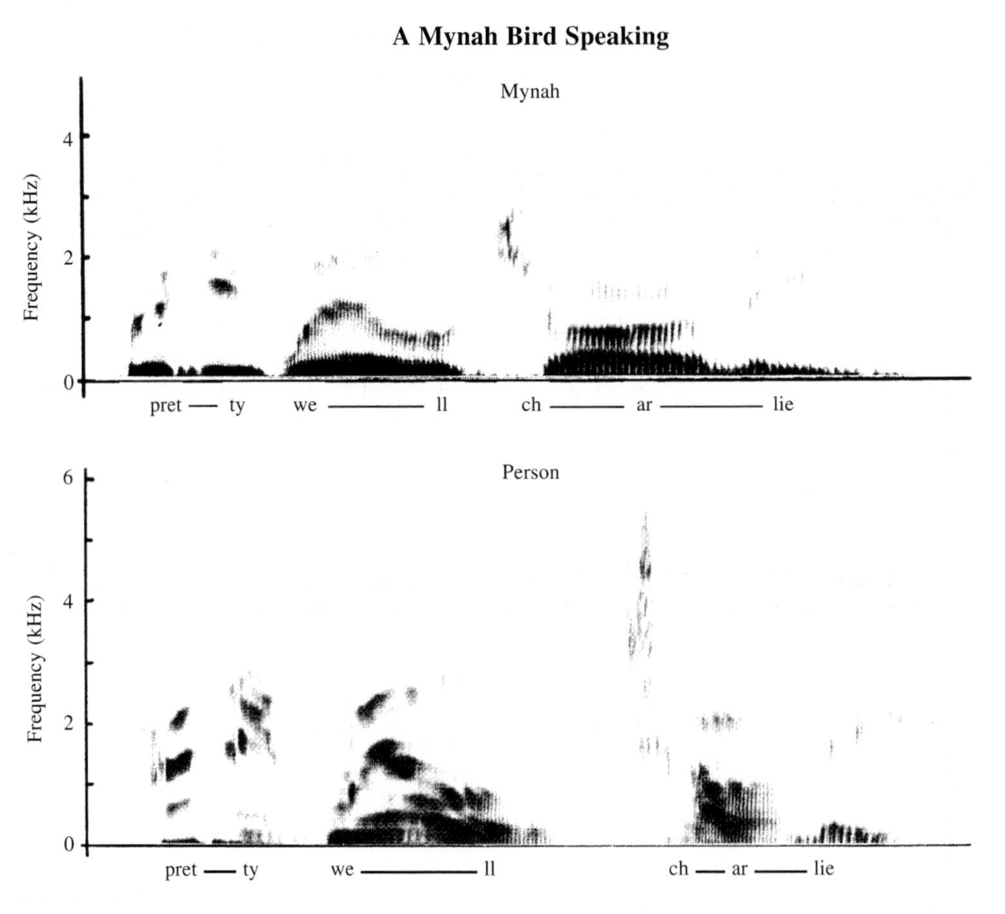

Figure 1.18 A hill mynah saying 'pretty well Charlie,' and spoken below by a person. From Greenewalt 1968.

blackbirds copying a tune that a local man whistled to his cat; he tracked them down by listening to the birds. Again a local dialect was created. Tretzel made many recordings, showing how the birds sometimes transposed and ornamented the tune. Despite great variation in the pitch of the human whistling, both the blackbirds and the larks always sang at precisely the same frequency, a tendency they shared with the many captive bullfinches, taught to imitate tunes on the flageolet, played to them by small boys for pocket money (Thorpe 1955). Evidently well-trained bullfinches fetched a good price (Barrington 1773). Jürgen Nicolai collected many examples, including a 20-second melody whistled by its keeper that was imitated by a bullfinch in its entirety (Güttinger et al. 2002; **Fig. 1.19; CD1 #10**).

The most famous musical connection is Mozart's internationally renowned pet starling whose song, according to West and King (1990), is immortalized in some of the indiosyncracies of Mozart's musical joke (K522). It took a musician who was also a devoted starling enthusiast to trace it back to its source; it makes a wonderful detective story. The link to music provides a culmination to this volume which Luis Baptista would have enriched with his own unique contributions, had he not been taken from us so prematurely. No one can quite replace him.

A Hand-reared Bullfinch Whistling a Tune

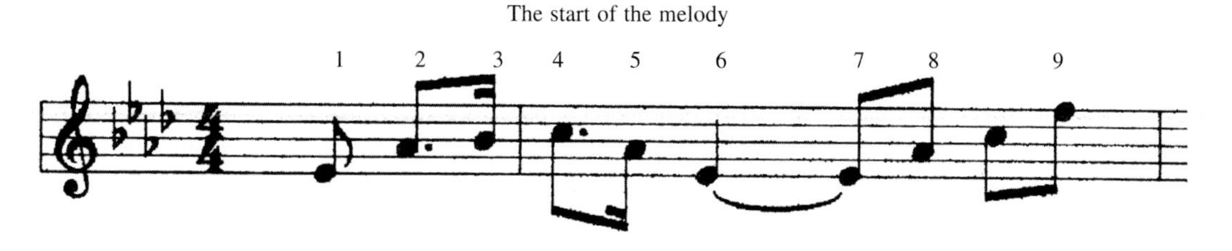

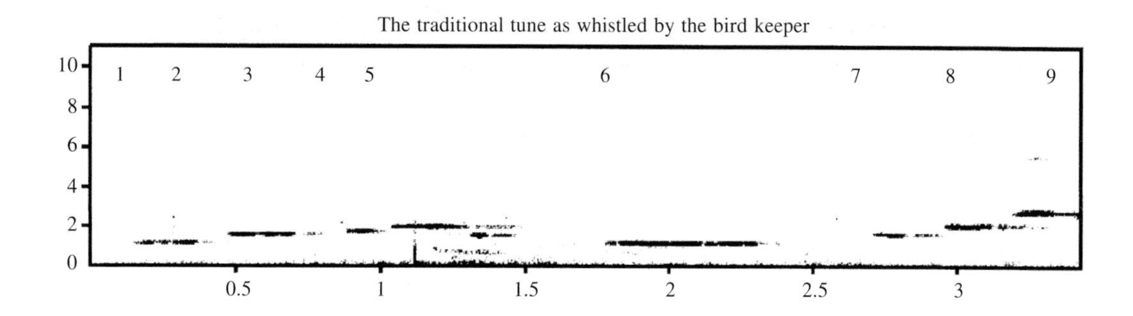

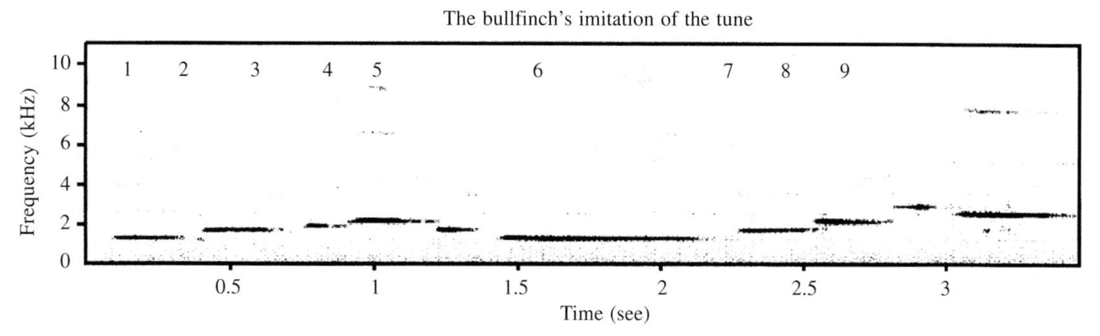

Figure 1.19 A whistled tune imitated by a male bullfinch in Germany. From Güttinger et al. 2002 (**CD1 #10**).

CONCLUSIONS

When I was a pre-1950 student, we had only subjective methods for analyzing birdsongs; perhaps we enjoyed them even more than we do today. But the verbalizations and transcripts we used, valuable for describing calls, were with a few exceptions, useless for songs. The impact of the sound spectrograph after the Second World War was revolutionary. For the first time, all of the rich complexities of birdsong were revealed for objective study. Sonographic analysis was especially productive in the hands of my mentor, W.H. Thorpe. In the 1950s his studies of song development in the chaffinch effectively launched the new discipline, which has since flourished beyond all expectations.

One of the many strengths of the science of birdsong is the tractability of avian vocal behavior for objective investigation of the ramifications of nature and nurture in behavioral development. Sound spectrography made this possible, but after half a century of yeoman's service, sonograms are now being superceded by new techniques for describing and comparing birdsongs, methods of great sophistication and power, hopefully accessing for us aspects of song structure like spectral features that are not well-suited for sonographic scrutiny. As students in the next generation dedicate their careers to furthering the science of birdsong, these new approaches will serve them well. But with every step taken, we should not forget the debt to the sound spectrograph, as we strive to integrate the new information with the vast emporium of facts and concepts derived from half a century of sonogram-based study. With the right blend of diligence, adventurousness, and good scholarship, they will succeed, both in building on the past, and ensuring that the new millennium will be even more productive than the one that has just ended.

Chapter 2

Vocal fighting and flirting: the functions of birdsong

SARAH COLLINS

INTRODUCTION

Birdsong has long been associated in our minds with mating behavior and male aggression. When males are active in defending territories and attracting mates, we think of springtime and breeding seasons. A large body of research on birdsong confirms that its two main functions are repelling rivals and attracting mates.

My aim is to review this evidence and discuss the importance of specific aspects of male song. The aspects of singing behavior that are important for each function vary between species. In a few cases, I will also outline the role that female song may play. One pervasive theme will be how the evolution of song has been driven by the twin selection pressures of female choice and male competition.

EVIDENCE THAT SONG IS IMPORTANT

Is song in fact used in male–male competition and female choice and, if so, which aspects of an individual's song make him more successful when it comes to reproduction? It is possible that individual characteristics are reflected in the song, and individuals who impress with their song may have a greater reproductive success through repelling rivals or attracting females. So a primary question is whether male song does reflect some aspect of male quality and increases the reproductive success of the singer.

Some simple experiments show that females do respond to male song, and that males can be intimidated by a rival's song. If males are removed from their territories and replaced by loudspeakers broadcasting song, the territories remain unoccupied for longer than silent control territories without broadcast song (Krebs 1976; Falls 1988; Nowicki et al. 1998a). Song evidently does repel rivals from intruding into a territory. Females, on the other hand, are more likely to be attracted to nest boxes where males have been removed and replaced with loudspeakers, rather than to silent nest boxes (Eriksson & Wallin 1986; Mountjoy & Lemon 1996; Johnson & Searcy 1996). Females also respond to song by performing copulation solicitation displays (Searcy 1992a), and tend to approach speakers playing male song in the laboratory (Clayton 1988).

If song is to serve these dual functions, both sexes will use it to assess the male, as a rival or a potential mate. However, the qualities on which males and females should base their assessment are somewhat different. Females need to find a mate who will maximize their reproductive success. We expect that factors such as the male's, age, condition, parental ability, and the quality of his territory will affect his attractiveness as a partner. Where male rivalry is concerned, competitors need to know the location of a rival male, who he is, how likely he is to attack, and his fighting ability. In some cases, the same song characteristic may give information to both males and females on aspects of the singer to which they will respond. However, in many cases, the information of interest is quite different and we would therefore expect different song parameters to be used.

MALE–MALE COMPETITION: VOCAL FIGHTING

Contests among males may be over mates, a territory, or the resources that attract females, such as a nest or feeding site. So the question is, what kind of song will best prevent a male from trying to take over a rival's nest box, territory, or female? Why should a male pay attention to the song of his rival? Why not fight and let the winner take the spoils? Of course, if a male territory owner can indicate he is such a superior fighter that, in a combat, the rival would lose, then both males may gain from avoiding a fight. Therefore, any song characteristic that suggests, either honestly or by bluffing, that a male is an excellent fighter, should be produced by the male to cause rivals to withdraw.

The outcome of fights is likely to depend on physical strength, fighting skills, and motivation to fight. Factors determining physical strength, such as size, weight, body condition, and energy reserves, become important in the assessment of rivals. Motivation to fight may vary among males, depending on what they have to gain and lose. Males who are more motivated are more willing to escalate a fight and are thus more dangerous opponents (imagine a small dog seeing off a larger one by not giving up, constantly yapping and snapping). Males who have a breeding female on their territory may be more motivated to fight; a male trying to take over a territory may find it easier to move on and try his luck at a new site rather than risk a fight with a highly motivated opponent.

However, what is to prevent a male from singing a song that indicates he is a superior fighter, or highly motivated to fight when in fact he is not? If it is possible for inferior males to signal that they are superior or highly motivated, then an intruder should take account of the possibility of deception. Why retreat when you may in fact be stronger than your rival? This conflict is at the heart of the evolution of the signals that occur between rival males. Territory owners should do all they can to repel rivals, but rivals should only withdraw if they can determine that the signal is a true indicator of superior fighting ability or of willingness to escalate the intensity of a fight.

There are several possible solutions to the signaling problem. Signals that are costly to produce are likely to be honest (Zahavi 1975, 1977) and the cost itself ensures their honesty. We may assume that an inferior male simply cannot produce a signal as costly as the one that a superior male is able to produce. There may be production costs such as the energy needed to utter a particular signal, and males with higher energy reserves may be able to generate a 'stronger' signal than those with lower reserves. This type of cost would lead to a gradual increase in the intensity of a particular song parameter in relation to a male's condition. If body condition affects fighting ability, as is quite likely, then a song reflecting this characteristic will also indicate a competitor's strength.

Alternatively, costs could be imposed by other individuals. A good analogy may be the way humans use aggressive shouting to intimidate rivals. Aggressive shouting will often cause rivals to withdraw, but it also runs the risk that a particularly strong or aggressive individual, rather than being intimidated, will respond physically. If your signal was all bluff you may get injured in the ensuing fight. Therefore, it is better not to signal that you will act aggressively unless you can follow it up if challenged. Replace shouting with singing in a particular manner and you have a good idea of what may happen in territorial disputes in birds. The cost of cheating, by pretending to be stronger or more aggressive than you are, will be related to the level of probing, or testing, by rivals. Bluffing may occur at some level, and its frequency will be related to the likelihood and cost of being probed.

In this scenario, the signal itself need not be costly to produce, but should indicate the likelihood of attacking, level of aggression, or motivation to fight. This kind of signal is known as a 'conventional signal' (Guilford & Dawkins 1995), so called because the specific form of the signal is arbitrary and a matter of convention. The definition of a conventional signal is that "the signal is more or less arbitrarily related to the message, many signals can carry the same message" – costs are target receiver dependent, so signals can be cheap to produce" (Guilford & Dawkins 1995). As long as everyone understands what different arbitrary signals mean, the system works. With conventional signals there are more likely to be different signal categories rather than gradual increases in the intensity of the same signal, as is likely to be the case for a physically costly signal.

Some signals are affected by physical restrictions and are thus, potentially, indicators of male quality. For example, larger males can produce deeper frequency sounds due to their larger vocal apparatus; a smaller male simply cannot cheat by producing a lower frequency sound. If larger males are better fighters, as is quite likely, then it makes sense to withdraw if you hear a low frequency sound indicating a male is larger than yourself. But this connection is not foolproof. Sometimes physical restrictions can be overcome. In some species the trachea has become elongated so that it is no longer proportional to body size. The frequency of the song, or call, now indicates trachea size and not body size (Fitch 1999). Smaller males can cheat. It would be interesting to see whether the elongated trachea of a large male is still longer that that of a small male. When a cat increases its apparent size by raising its fur on end, a large cat will still look bigger than a small one.

Testing the Role of Song in Male Rivalry

Three main methods have been used to address questions about the role of male song in rivalry. The first is observational. Birds are observed and recorded during aggressive encounters, and their song characteristics are compared to nonaggressive interactions. For example, male barn swallows emphasize rattles in the song during aggressive encounters, indicating that rattles are a relevant component of competitive interactions between males (Galeotti et al. 1997). In another approach, song traits are correlated with measures of male quality that might be important to other males, such as size or age, or with consequences of success in rivalry, such as territory size or quality. In male barn swallows, males who produce lower frequency rattles are in good condition (Galeotti et al. 1997), a useful fact for a rival to know. It seems likely that rattle frequency is important in male rivalry situations.

A second method is territorial song playback. A speaker is placed in the subject male's territory, usually near the edge, and a song is played back to the territory owner. Usually two versions of a song, or two song types, are played sequentially with a pause between the two. The response of the territorial male to each song is measured; indicators of a strong response are an increased singing rate, flying close to the speaker, and increased calling. It is assumed that a male territory owner will respond more strongly to the song that would be more effective in male rivalry situations.

However, there is a problem with this method. You wish to test how the characteristics of the resident male's song function in male rivalry situations, but what you are in fact testing is how songs of intruders or rivals are perceived, recognized and responded to by territorial males (Searcy & Nowicki 2000). The focus is assumed to be on intruder deterrence, but a different question is actually being asked – how worried does a territory owner get in response to different intruder songs? But the important question in male–male competition is usually – how good are different songs at deterring intruders? In general this may not be a problem, because the same song characteristics may be important to both intruders and resident males. However, you may simulate a very powerful intruder (i.e. a male with a song more typical of a territory owner) and the territory owner who hears that playback may withdraw rather than attempt to fight back. This could result in a situation where very strong stimuli actually *decrease* the territorial response compared to an 'average' strength stimuli (**Fig. 2.1A**). If you were testing the effectiveness of a song as a keep-out signal by playing a territorial male's song to an intruder, then there would be a perfect negative relationship between song stimulus strength and intruder response (**Fig. 2.1B**). Male willow warblers produce a particular syllable type in agonistic encounters (Type A), especially associated with imminent attack (Jarvi et al. 1980). Playback of type A songs to territory owners, simulating a very aggressive intruder, results in withdrawal of the territory holder.

For this reason a third method, the speaker replacement experiment, is perhaps a better way to study how effective different songs are in repelling rivals from a territory. Here a male is removed from his territory or nest box and is replaced by a speaker broadcasting particular song stimuli. We already know that territories broadcasting song remain unoccupied longer than control territories. This type of experiment has been conducted with thrush nightingales, great tits, white-throated sparrows, and song sparrows (Searcy & Nowicki 2000). Unfortunately, this approach is sometimes impractical as intrusions into a territory are infrequent and difficult to observe. In addition, the simulated playback 'singer' is static rather than singing from song posts around the territory. For this reason territorial playback experiments are still the most common method of addressing questions on the functions of territorial song, and how song variants differ in their effectiveness at intimidating rivals. Although care must be taken in the interpretation of the results, it is relatively easy to perform, and does indicate whether males are able to perceive, recognize, and respond to differences between songs. It is thus valuable in identifying the song parameters that are involved in male competition.

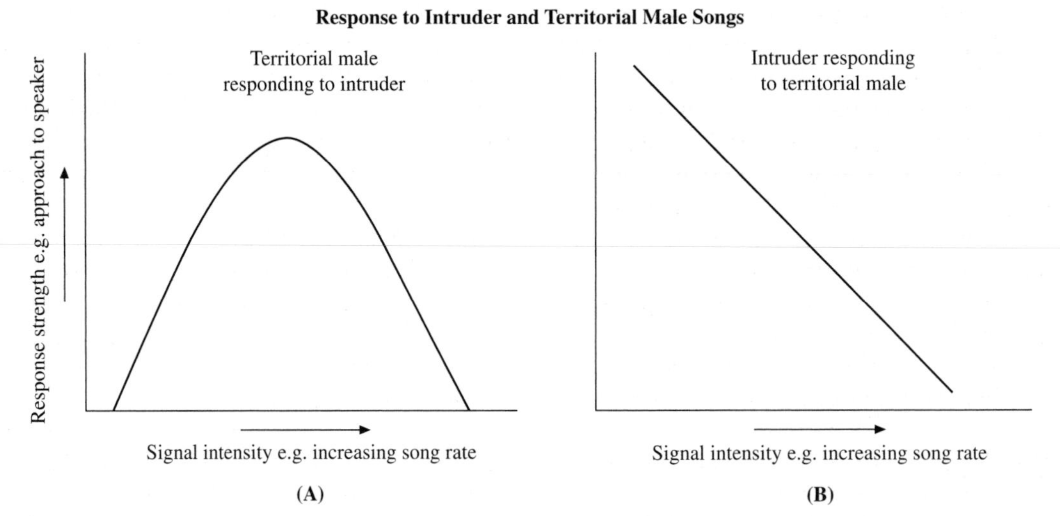

Figure 2.1 The response strength of a male to playback of male songs differing in signal intensity; **(A)** response of a territorial male to intruder song stimuli; **(B)** an intruder responding to territorial male song stimuli. In **A**, the response strength eventually drops with increasing signal intensity.

Simple Changes in Song Characteristics

The simplest way to compare the quality of conspecific individuals is through differences in very basic song characteristics, such as frequency or amplitude changes, song production rate, or the inclusion of particular notes.

One interesting idea is that lower frequency sounds may be indicators of relaxation, because tenseness in the sound production muscles causes an increase in frequency (Zahavi 1982). A male who appears relaxed in the face of danger may be seen as a greater threat; to anthropomorphize, the male appears very confident of winning the fight. However, it is more likely that lower frequencies indicate a larger body size (Beani & Dessi-Fulghieri 1995), although the evidence for this is contradictory. As already mentioned, in some species the relationship between the size of the body and the vocal apparatus may be decoupled by tracheal elongation (Fitch 1999). Relationships between body size and low frequency have been found across species (Ryan & Brenowitz 1985), but are not always found within a species, although males generally produce lower frequency sounds than females (Ballintijn & ten Cate 1997). In chickadees, playback of male songs with low whistle frequency results in retreating by territory owners (Shackelton & Ratcliffe 1994), suggesting that a low frequency may indicate a more able competitor. The relationship is reversed in roosters; dominant males produce higher frequency sounds (Leonard & Horn 1995). Lambrechts (1996) suggested that high frequency sounds are more energetically expensive to produce than lower frequencies. The reason given is that for most species, including humans, sounds in the bottom quarter of the vocal frequency range are easier to produce than those in the upper frequency range. In summary, although there is evidence that simple changes in the sound frequency of songs may be involved in signaling competitive ability, relationships between frequency and male competitive ability are not always clear.

In the barn swallow (**CD1 #11**), Galeotti et al. (1997) found that the rattle produced at the end of the song was longer in individuals with higher testosterone levels, a likely indicator of greater aggressiveness (**Fig. 2.2**). In addition, the frequency of the rattle was lower in males in good body condition; also, in competitive situations, the rattles in the song are emphasized by males, suggesting that these signals may be important in male–male competition (**Fig. 2.3**). Differences between individuals in the structure of the rattles can convey important information to the receiver, such as the probability of escalation (longer rattles = more likely to escalate), and the physical quality of the singer (lower frequency rattles = good condition). Rattle characteristics are related to male reproductive success, indicating that the greater competitive ability of males with long, low frequency rattles improves the male's breeding success. The beginning of the song, a complex series of notes, appears to be important in female choice (Møller et al. 1998).

Similarly, a specific song element is important in territory defense in the water pipit (Rehsteiner et al. 1998). Male songs contain the 'snarr' (**Fig. 2.4**), a rasping element with a broad frequency range, very similar to the barn swallow rattle (**CD1 #11**). Male water pipits with more 'snarr' notes had territories that overlapped less with neighbors, were heavier and were mated more often than males with fewer snarrs. However, breeding success, measured by the number of chicks successfully raised, was not related to the 'snarr' score. The 'snarr' is easily locatable (Dooling 1982a; Wiley & Richards 1982), so a male may be more apparent at his territory boundaries, perhaps with an aversive effect on potential intruders. In addition, the heavier weight of males with higher 'snarr' scores means they are likely to be more formidable opponents. As in the barn swallow a harsh, potentially energetically expensive note is used to signal competitive ability. In both cases differences in the signal are related to differences in male attributes, and this information is potentially useful to rivals selecting their subsequent behavior. We have good reason to believe that harsh loud elements signal male condition, and thus fighting ability, because they are energetically expensive to produce (Obwerger & Goller 2001).

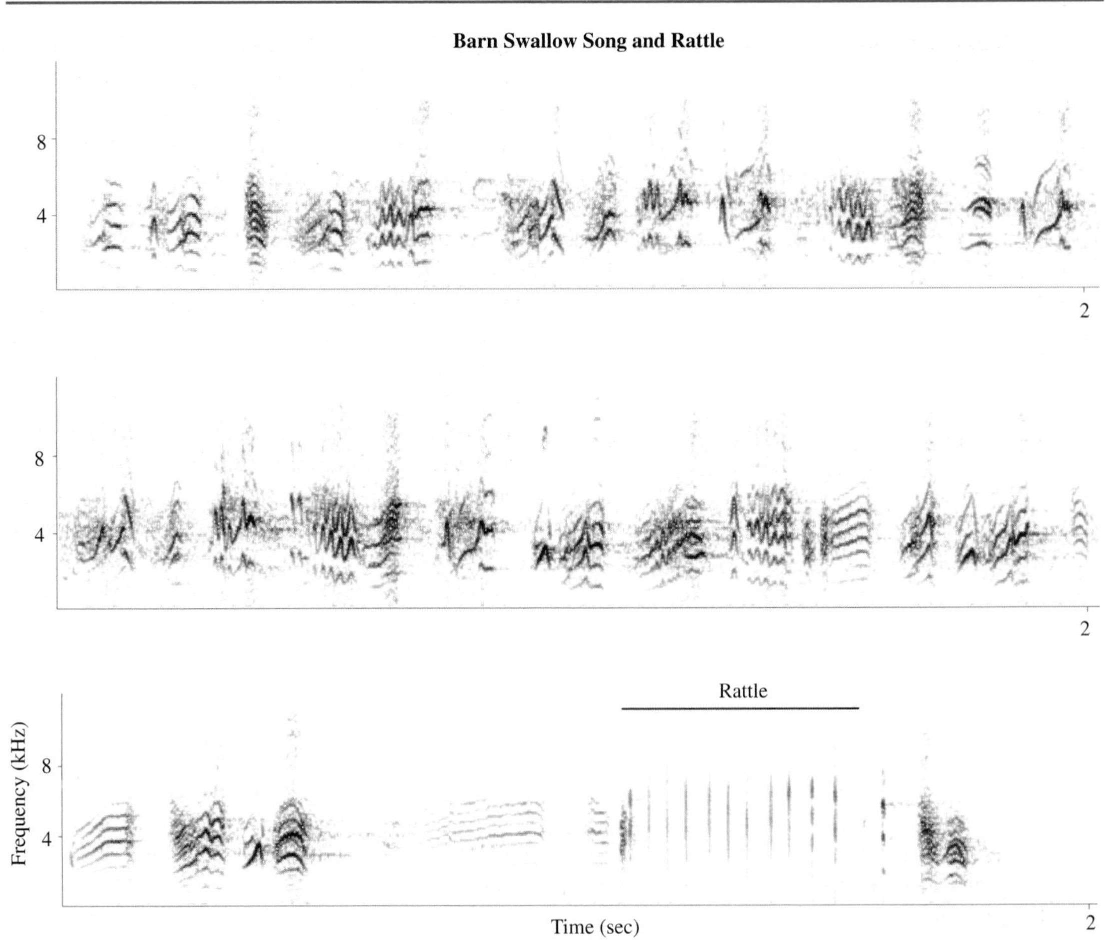

Figure 2.2 Barn swallow males emphasize the rattles in the song during competitive encounters with other males. The rattle is longer in individuals with higher testosterone levels and the frequency of the rattle is lower in males in good body condition. Only one rattle is illustrated **(CD1 #11)**.

Increases in song intensity require more energy and thus are readily explained as indicators of male condition (Obwerger & Goller 2001). In blackbirds differences in the potential for aggression are signaled by increasing the intensity of the song, although the structure remains the same (**CD1 #12**). The low intensity song is slower, has longer motifs, shorter twitters and is often somewhat quieter than high intensity song (Dabelsteen & Pedersen 1990; **Fig. 2.5**). Males change from low intensity to high intensity song when rivals start singing from outside their territories. The change seems to indicate a territory owner's increasing motivation to fight. Males who try to bluff may risk becoming exhausted and thus disadvantaged in any subsequent fight.

One well-studied species is the collared dove, (ten Cate et al. 2002). Males defend a territory for most of the year and coo from song posts around their territories. An important function of the coo is communication between males. The elements within a coo may consist of a constant frequency or, in adult males, may contain a frequency modulation (a jump in frequency; **Fig. 2.6; CD1 #13**). Individuals differ in the number of coo elements which are modulated, the percentage of modulated coos overall, and the peak frequency of coo elements are both correlated with weight (heavier males produce

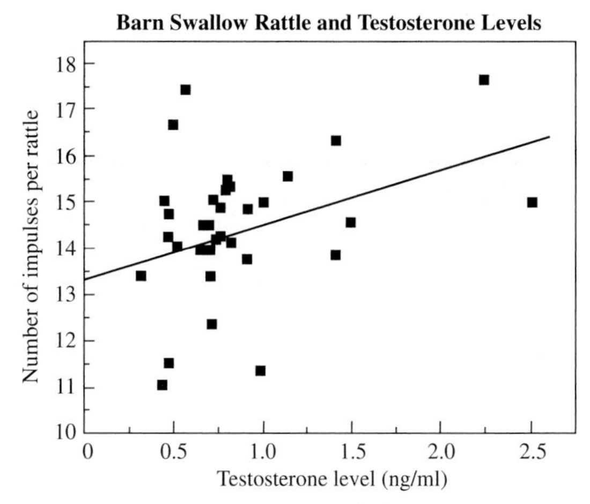

Figure 2.3 The relationship between plasma testosterone levels and the number of impulses per rattle in barn swallow song. From Galeotti et al. 1997.

higher frequency elements, more of which are modulated). Territorial males respond more strongly to playbacks of modulated coos than to nonmodulated coos. The increase in response is due specifically to the modulation, i.e. the *change* in frequency, not the fact that the overall frequency is higher. Ten Cate et al. (2002) suggest that their production by larger males may indicate that modulations are costly; the frequency jump may be difficult to produce except by males in good condition (Lambrechts 1996). An alternative is that there may be a predation cost due to the locatability of signals with simple frequency modulations (Dooling 1982a; Wiley & Richards 1982), especially in open environments. Fitter males that can easily escape from predators, or males willing to take a higher risk may be more likely to modulate their coos.

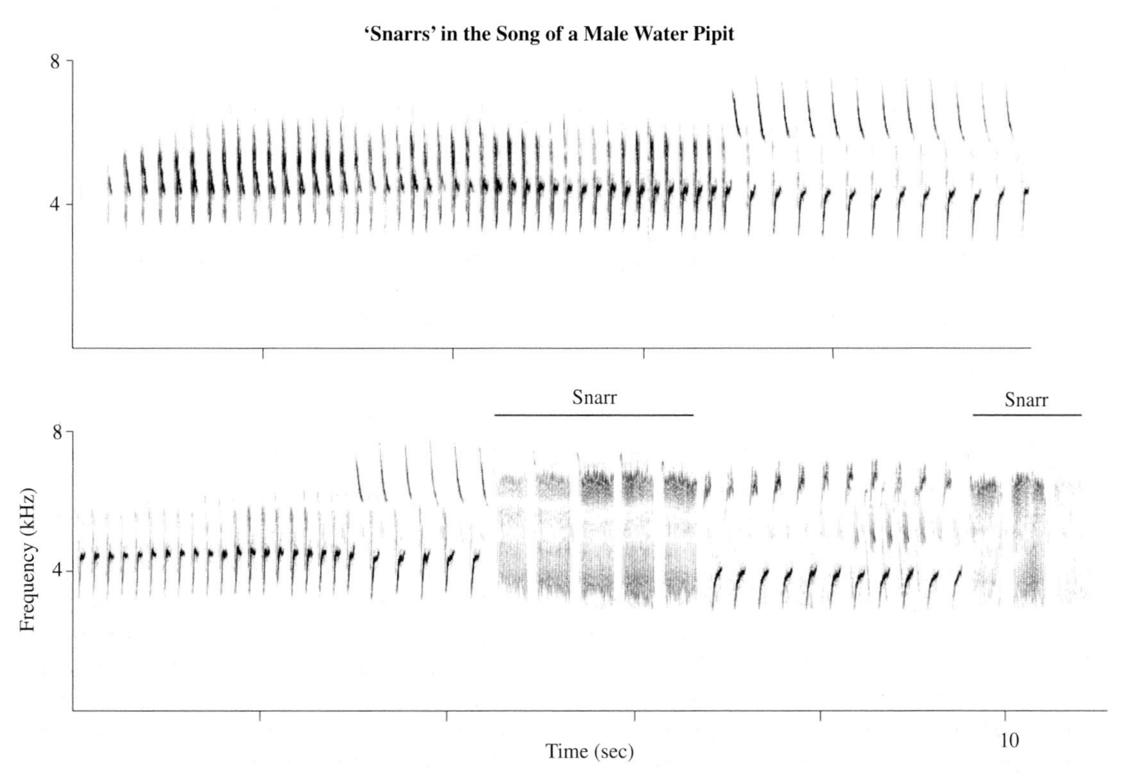

Figure 2.4 Water pipit songs, the bottom one with a 'snarr' note (**CD1 #11**).

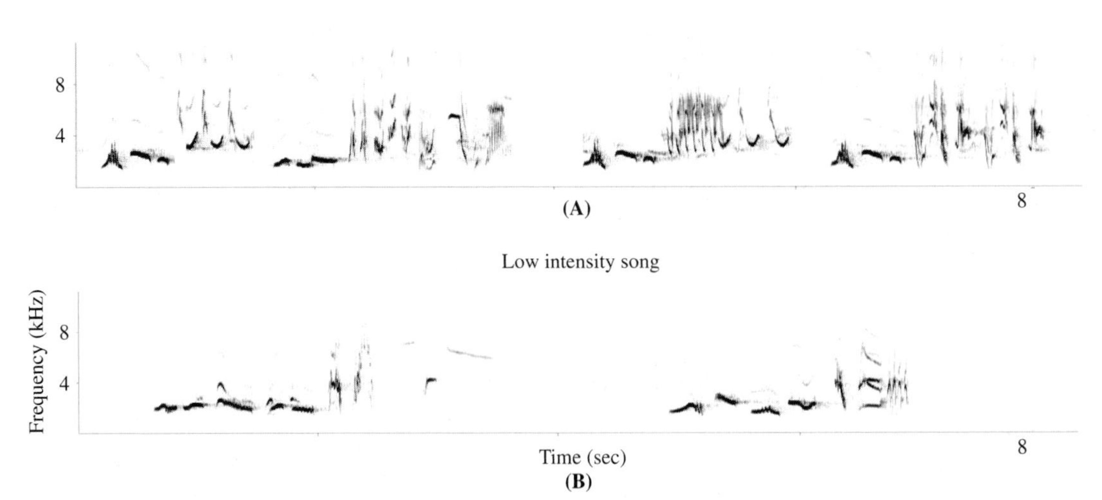

Figure 2.5 European blackbird songs. **(A)** High intensity song used in aggressive situations; **(B)** low intensity song, which is slower, has longer motifs, shorter twitters and is often somewhat quieter than high intensity song. From Dabelsteen & Pedersen 1990 **(CD1 #12)**.

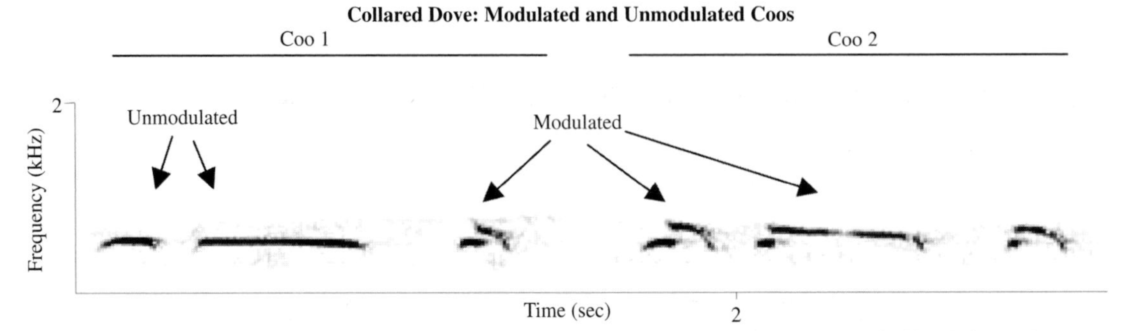

Figure 2.6 Male collared doves produce modulated and unmodulated coo elements. The left coo has two unmodulated elements and a third modulated one. The right coo consists of three modulated elements **(CD1 #13)**. From Slabbekoorn & ten Cate 1997.

Song Repertoires and Complexity

There are two types of repertoires: (1) a *song repertoire,* where a male sings several different song types but individual song types do not vary much; (2) a *syllable repertoire,* where a number of syllables are recombined to produce different songs (**Fig.** 2.7). In the case of song repertoires a male may produce one version of his song several times before switching to a new type (eventual variation), or he may switch types after every song (immediate variation). Repertoires have often been studied in the context of female choice, but less frequently when it comes to male–male competition. However, in great tits and red-winged blackbirds, speaker replacement experiments show that intruders are less likely to intrude into territories where males have larger repertoires (Krebs et al. 1978; Yasukawa 1981a).

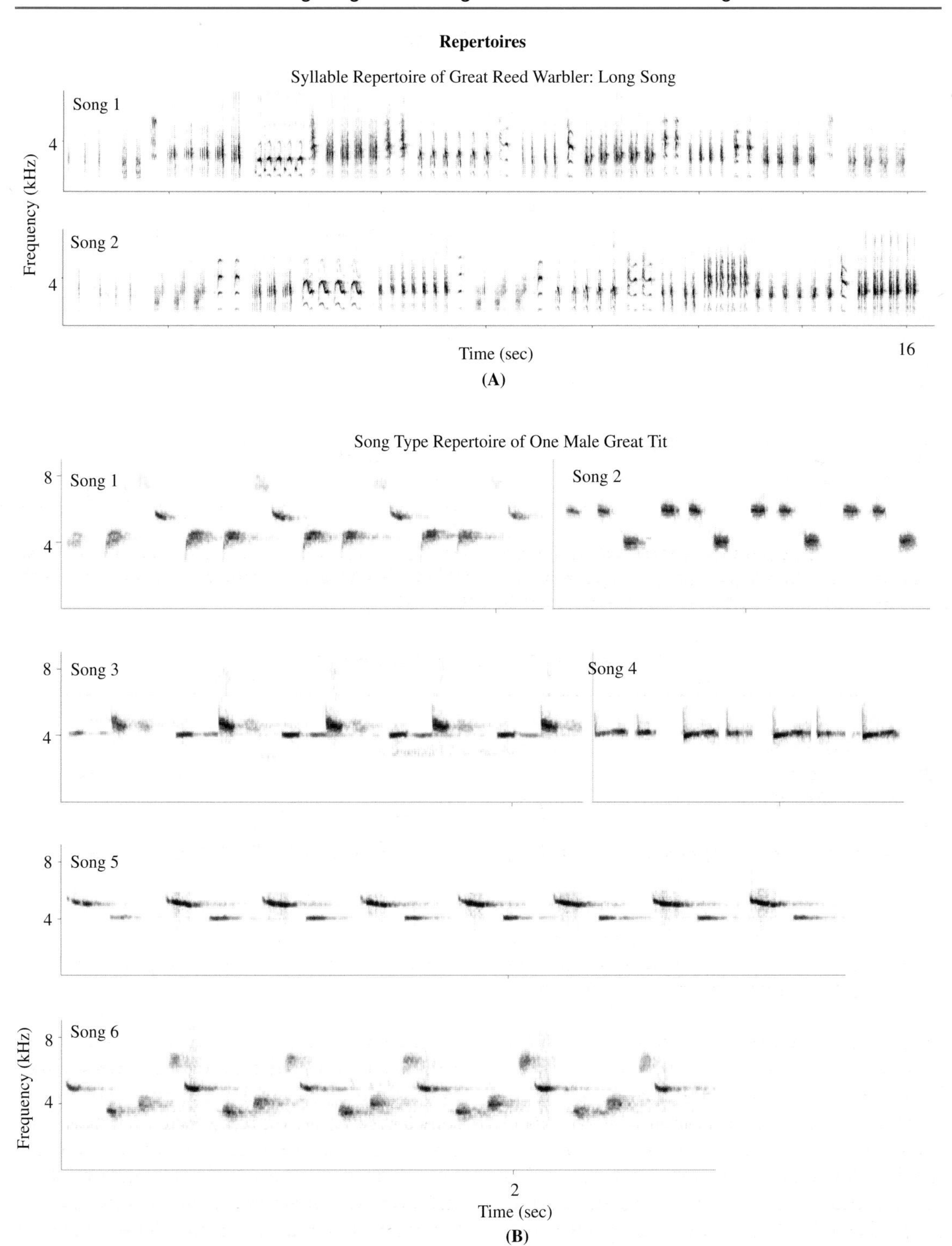

Figure 2.7 Examples of the two main types of song repertoire, **(A)** a syllable repertoire: a great reed warbler long song; **(B)** a song type repertoire: a great tit (**CD1 #14**).

The success of males with larger song repertoires could occur because intruders habituate to song types (Searcy et al. 1994). If a male switches between songs in his repertoire, the response of the receiver increases in strength after each switch. Thus, a repertoire may be a way of maintaining a response. The magnitude of the recovery of the aggressive response depends upon the similarity of the following song type, and similar song types, even from different individuals, appear to be classed together by the receiver. Interestingly, males treat two different song types from the same individual's repertoire as if they were from different individuals, at least initially. It was suggested that repertoires evolved so that males could fool rivals into believing that there are multiple opponents, although this suggestion was subsequently disproved (Krebs 1977a; Yasukawa 1981a).

In European starlings a complex song, with a large syllable repertoire, is a more effective deterrent to intruders at the nest box (Mountjoy & Lemon 1991), and there is a correlation between repertoire size and the probability of winning encounters (Eens 1997). In a study of song sparrows, song repertoire size correlated with territory tenure and reproductive success (Hiebert et al. 1989). Males with large repertoires acquired a territory more quickly and survived for longer. Why these males with larger repertoires survive longer and are more successful in defending a territory is not known, but the increased survivorship implies that they are in better condition. In some species males learn new songs or syllables every year resulting in an increased repertoire size with age, as in starlings and sedge warblers. It is thus possible that males with larger repertoires are more successful at maintaining territories, not because they have more complex song, but simply because they are older, more experienced, and in better condition, as indicated by their increased survivorship.

Male Louisiana waterthrushes respond to increased aggression from rivals with more complex songs than those used in routine territorial advertisement (Smith & Smith 1996). However, great reed warblers (**CD1 #14**), and reed warblers, produce shorter, less complex songs in encounters with other males (Catchpole 1983), as do some other species. Clearly, species differ in whether less or more complex songs are used in aggressive encounters, indicating that the role of repertoire size in male competition is not a simple one.

Although a correlation between repertoire size and reproductive success has been found in several species, this is often due to female choice, not male–male competition. It is more likely that female choice drives the evolution of repertoires, which have then been co-opted in some species, for use in competition between males. The reasons why repertoires may be a good indicator of male fitness will therefore be discussed in more detail later when addressing female choice (see p. 46).

Song Matching

Song matching occurs when one bird responds to another by singing either the same song (song type matching), or by singing a song that is shared by the two birds but is not being sung at that moment (repertoire matching; **Box 3**, p. 49). Song matching interactions have been studied both by conventional and by interactive playbacks (McGregor et al. 1992), when the experimenter responds to the territory owner with either matching or nonmatching songs. This technique has greatly facilitated the study of dyadic interactions between males, and the effects of both song matching and song overlapping.

It is likely that song matching of both types is a conventional signal (Vehrencamp 2001; see p. 41). Either matching or nonmatching could indicate a potential aggressive response. Both individuals know their own willingness or ability to fight but, unless they are familiar, not that of their opponent. A cheat would always signal that he is willing to fight so as to scare off rivals. However, some rivals will not be intimidated and these are likely to be the most aggressive. Therefore, the cheat will suffer a cost if his bluff is called by an aggressive opponent. The best strategy is for the signal given to correlate

BOX 3

BROADCAST SONG REPELS MALES AND ATTRACTS FEMALES: REPERTOIRE EFFECTS

The defense of territory is thought to be the principal function of song in communication between males. The most direct test of this function is a 'speaker replacement experiment,' in which males are removed from their territories and replaced with loudspeakers, from which the song of the species under study is broadcast. Intrusion by other males onto speaker-occupied territories is monitored and compared to intrusion onto control territories, either left vacant after the owners have been removed or occupied by loudspeakers playing a control sound such as white noise. This design was pioneered by Göransson et al. (1974) with thrush nightingales, and used subsequently with great tits (Krebs 1977b; Krebs et al. 1978), red-winged blackbirds (Yasukawa 1981a, b), white-throated sparrows (Falls 1988), and song sparrows (Nowicki et al. 1998b). In all cases, intrusion was delayed or less frequent on the territories defended by song than in control territories. Yasukawa's study on red-winged blackbirds is particularly important because he demonstrated not only that song keeps intruders out but that a repertoire of song types is more effective in repelling intruders than is a single song type (see below).

A similar experimental design has been important in demonstrating one of the principal functions of song in male–female communication – the attraction of females to the male and his territory. Here song is broadcast from vacant areas, and visitation rates of females are compared with areas where either no stimulus or a neutral stimulus is broadcast. This design has been applied mainly to hole-nesting species; nestboxes provide convenient focal points for observation of female visits. It was pioneered by Eriksson and Wallin (1986) with pied and collared flycatchers, and then applied to European starlings by Mountjoy and Lemon (1991), to house wrens by Johnson and Searcy (1996), and to hoopoes by Martin-Vivaldi et al. (2000). In all cases, more unmated females were attracted by song playback than by controls, though whether repertoires are more attractive to females has yet to be directly investigated.

William A. Searcy & Stephen Nowicki

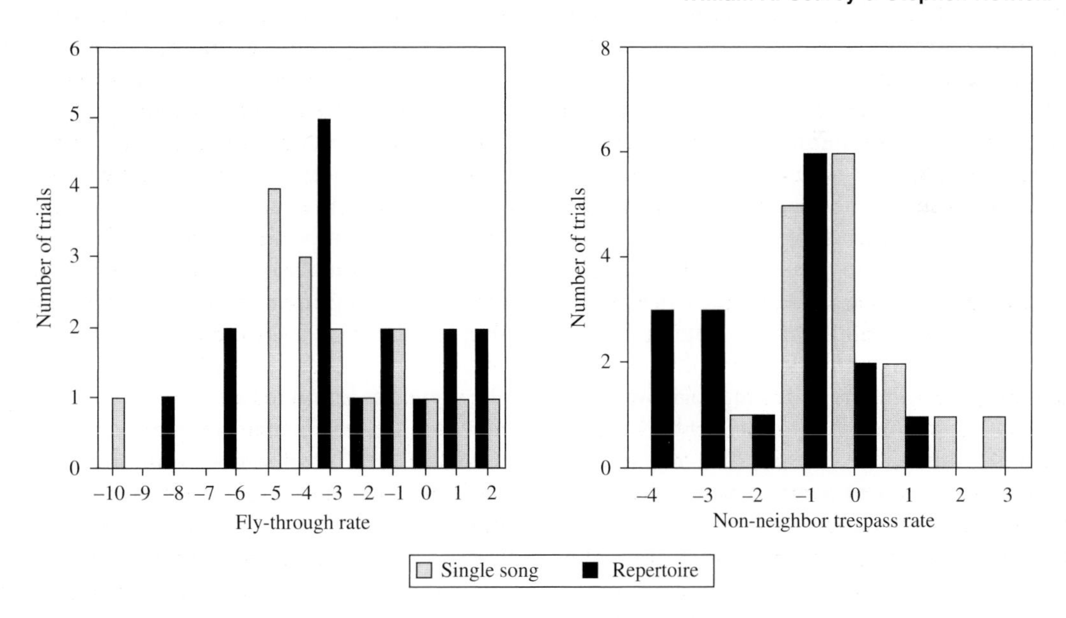

Fly-through rates and non-neighbor trespass rates of red-winged blackbirds were reduced in speaker-occupied territories (Yasukawa 1981a, b). Rates are shown relative to controls (experimental rate minus control rate); negative values indicate a decrease during playback. Playback of both single song types and repertoires of eight song types reduced fly-through rates, whereas only repertoire playbacks reduced non-neighbor trespass rates.

reasonably well with an individual's probability of retaliating. Individuals pay attention to whether rivals match or not, because they gain useful information from the signal. In general, when contestants are equal in ability, fighting is more likely as the outcome cannot be decided by recourse to the signal alone. If conventional signals are indeed arbitrary, there should be species in which both matching and nonmatching is the aggressive signal. So far, only in one out of ten or so studies, on the bobolink, is there evidence that a nonmatching signal is used (Capp & Searcy 1991).

Male song sparrows may respond by repertoire matching in singing interactions with neighbors. It is assumed that repertoire matching indicates to the neighbor that they are recognized, or that the responder is indeed the neighbor (Beecher et al. 1996). It does not appear to be a very aggressive signal, but is more aggressive than singing a nonmatching song. Interestingly, male song sparrows who do not share many songs with a neighbor, and therefore cannot engage in repertoire matching, suffer more aggressive encounters with neighbors (Wilson et al. 2000; Wilson & Vehrencamp 2001). Thus, repertoire sharers are able to communicate more effectively than nonsong sharers who must resort to direct confrontation (Vehrencamp 2001). Song sparrows converge on songs of neighbors after dispersal, perhaps so that they can perform repertoire matching with neighbors (Beecher et al. 1996). Males responded by singing a song shared with the neighbor in 87% of playbacks. To match repertoire the male song sparrow needs to know beforehand what songs they both share. Therefore, in the case of repertoire matching, a shared history is a necessity, but not for song type matching. Neighbors may be responded to in a less threatening way because they are less of a threat than floaters–the 'dear enemy' effect (Falls 1982). Unlike floaters, neighbors already have a territory, although of course some neighbors may be looking to expand. Neighbors are more of a threat at the start of the breeding season when territories are being established.

Song type matching is associated with escalated encounters in early season territory establishment between neighbors, and at all times of the season with intruders, and is associated with subsequent approach in several species including chaffinches (Hinde 1958), great tits (McGregor et al. 1992), cardinals (Lemon 1974), and song sparrows (Nielsen & Vehrencamp 1995). Territorial males tend to match song type in response to unknown individuals and this appears to be a signal of potential escalation (Krebs et al. 1981). The wood thrush avoids matching intruder playback stimuli (Whitney & Miller 1983), perhaps because the stimulus is too strong. Song type matching is clearly a signal of aggressive intention in the song sparrow. Males that match song types follow up with more aggressive responses and they respond more aggressively to signals that match their own song type (Vehrencamp 2001).

Territorial male great tits perform song type matching duels with rivals, and matching indicates that a direct attack may follow (**CD1 #14**). However, because the degree of similarity of the 'intruder' song to those of the territory owner is important in determining whether a song is type-matched or not, neighbors are song-matched more often than strangers (Falls et al. 1982). This is unexpected, as matching is thought to indicate potential escalation and usually fights are less common between neighbors. However, neighbors are more easily matched by chance as well as by design, and if the effect of overall song similarity is controlled for, strangers are matched more often than neighbors (**Fig. 2.8**). The same was found to be true in corn buntings. However, western meadowlarks match strangers' songs, but not those of neighbors (Falls 1985), without any need to control for similarity (**Fig. 2.9**).

It has been argued that song matching is a conventional signal, and does not reflect male characteristics related to fighting ability or motivation. However, there is a counterargument: matching may actually indicate something about male quality; more specifically, the number of songs you share with your neighbor may indicate

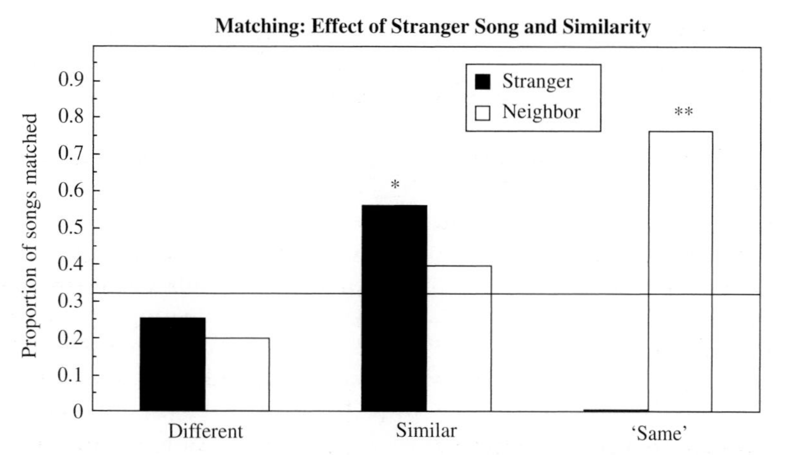

Figure 2.8 The proportion of songs matched in response to playback stimuli differing in the degree of similarity and type. Great tits are more likely to match songs to strangers when similarity of song types is taken into account (from Falls et al. 1982). *P < 0.05, **P < 0.01, significance above the line showing the chance level of matching.

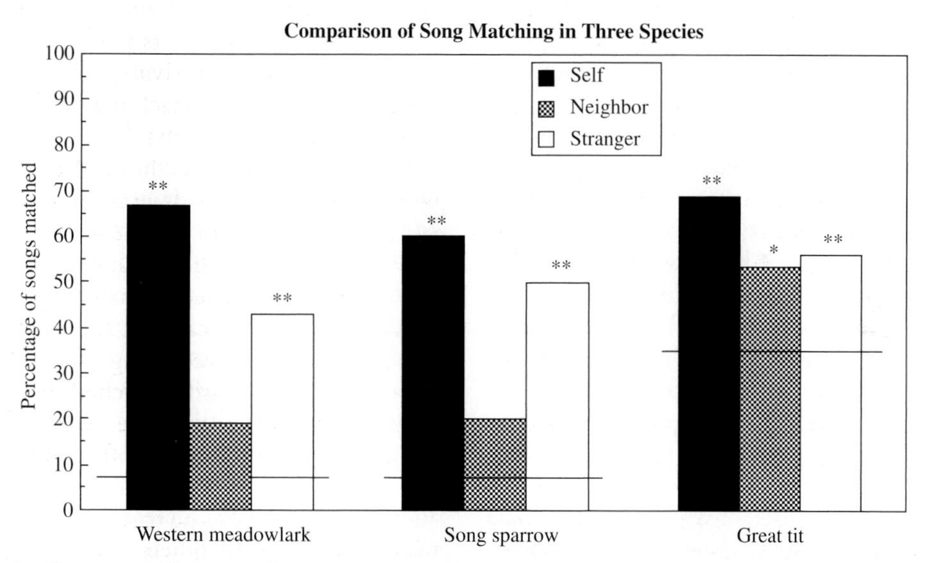

Figure 2.9 Song matching rates in response to self, neighbor, and stranger songs for three species. The lines represent chance levels of matching. *P < 0.05, **P < 0.01 – significance above chance. The data are for meadowlarks (Falls 1985), song sparrows (Stoddard et al. 1992), and great tits (Falls 1982).

quality. You are more likely to be able to match to neighbors or strangers if you have a larger repertoire. Larger repertoires are associated with increased age, and perhaps condition. Extensive song sharing with neighbors may also indicate you did not disperse far from your song tutor and natal area. Perhaps males who have to travel further to obtain a territory are of lower quality, or are less successful due to their lack of experience in that habitat. By this argument, the degree of song sharing with a neighbor, and thus the ability to song match, is related to dispersal distance and repertoire size. If a male song sparrow does not share many songs with his neighbors he suffers the consequence of having to fight more often with them (Vehrencamp 2001). In indigo buntings, matching has consequences for reproductive success. Males that match neighbors are more successful breeders, and individuals that match are those that have remained in their natal area (Payne 1982). Thus, whether song matching is a conventional signal or not is not clear.

Overlapping Songs

Birds can time their singing to overlap with that of another bird. Such overlapping appears to be a signal of readiness to escalate contests (Dabelsteen et al. 1996, 1997; Hultsch & Todt 1982). As with song matching, overlapping will depend to some extent on the behavior of both overlapper and the singer being overlapped. Overlapping, extensively studied in nightingales (see Chapter 3), occurs particularly in disputes over territories (Hultsch & Todt 1982); it appears to be aversive as males either avoid posts on which their song is overlapped (Todt 1981) or adjust their singing to avoid being overlapped (Hultsch & Todt 1982).

Nightingales also respond more strongly to an individual they have heard overlapping other males, compared to an individual they have heard being overlapped (**Fig. 2.10**). This implies that eavesdroppers use information gained from listening to dyadic interactions and that overlappers are apparently judged to be a greater threat (Naguib & Todt 1997). Birds also use their own direct experience with an individual to determine their interactive strategy. If a male generally overlaps them, they are more likely to respond strongly even when the individual is not overlapping in that particular bout. Individual nightingales have a consistent tendency to be overlappers or nonoverlappers (Naguib 1999).

Overlapping is also an aggressive signal in great tits (Dabelsteen et al. 1996), European robins (Brindley 1991; Dabelsteen et al. 1997), and blackbirds (Wolfgramm & Todt 1982). Overlapping is likely to be a conventional signal; it is hard to see how overlapping is costly to the overlapper in terms of energy output. However, an overlapper may reduce the amount of information that the rival can signal, thus reducing its efficiency (see Chapter 3).

Song Switching and Singing Rate

Birds with a song type repertoire often appear to use the rate of change to a new song type as a graded indicator of potential aggressiveness. Switching among song types has been argued to reduce habituation (Falls & D'Agincourt 1982), deceive intruders (Krebs 1977a), or reduce exhaustion of the vocal muscles (Lambrechts & Dhondt 1988). Some species use high switching rate as the default nonaggressive signal and reduce switching in the context of aggressive encounters, whereas in other species males increase their switching rate in aggressive encounters (Vehrencamp 2000). Which behavior occurs probably depends upon its normal, non-aggressive singing style. Species that switch song types relatively often may be more likely to reduce switching rate in territorial encounters. Species that switch less often may be more likely to increase switching rate in aggressive encounters. For example, in the song sparrow an increase in switching rate is associated with subsequent aggressive approaches, and receivers approach a switching stimulus more aggressively (Stoddard et al. 1988). The same is true for cardinals, meadowlarks, and Carolina wrens (Vehrencamp 2000). However, in red-winged blackbirds, banded wrens, and dunnocks the switching rate is decreased in territorial defense situations. In addition, dunnocks increase switching when males interact with females (Langmore 1997; **Fig. 2.11**).

Switching at a high rate does not appear to have a direct cost, although there is a potential cost of retaliation, namely, a stronger approach to individuals that switch at a high rate (Nielsen & Vehrencamp 1995), at least in some species. However, in other species the retaliation cost is in response to a lower switching rate. Thus, there is no consistent rule about whether high or low switching rates are aggressive.

An increase in the rate of singing, with more

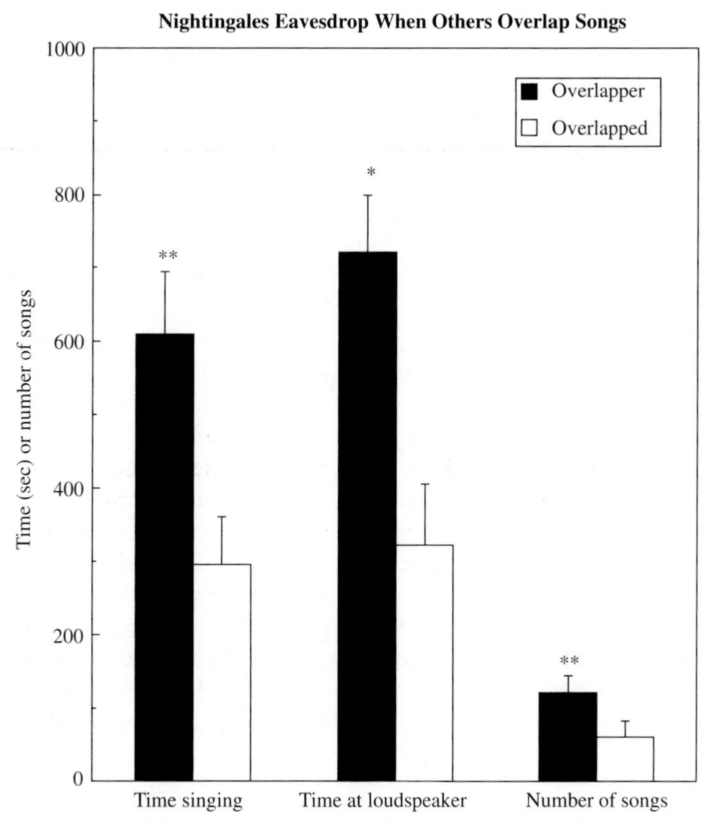

Figure 2.10 Nightingale males respond more strongly to a male they have heard overlapping songs, than to a male they have heard being overlapped. The data show means (+SE) of response measures of males in a two-speaker playback experiment.$^{*}P < 0.05$, $^{**}P < 0.01$. From Naguib & Todt 1997.

songs produced or songs produced at a higher rate, is more likely than the rate of switching to be energetically costly, and an increase in song rate is observed in aggressive interactions in some species. In the song sparrow (Kramer et al. 1985) males increase their singing rate in response to playback of intruder stimuli. There are other birds in whom song rate increases in aggressive interactions, such as the great tit (Weary et al. 1988). Song rate changes are most likely to relate to the increase in arousal and aggressive motivation as an intruder approaches. Although song rate increases in the song sparrow during interactions with other males, other work suggests that matching repertoires and switching rate are used by this species for territorial defense. Given that song rate in song sparrows does not appear to predict the outcome of fights (Bower 1999), it more likely to be a byproduct of increased excitation rather than an important aggressive signal in this species. In general, an increase in song rate appears to be a signal used less by males to assess rivals than by females to assess potential mates.

We thus find that the diversity of the song parameters that are important in territorial interactions is remarkable (**Table 2.1**). In some species song type appears to be important, in others frequency modulations; in yet other species aggressive responses are determined by complex dyadic interactions sometimes involving the use of repertoires. These differences are likely to be due to ecological and social differences between species. In species with repertoires where there

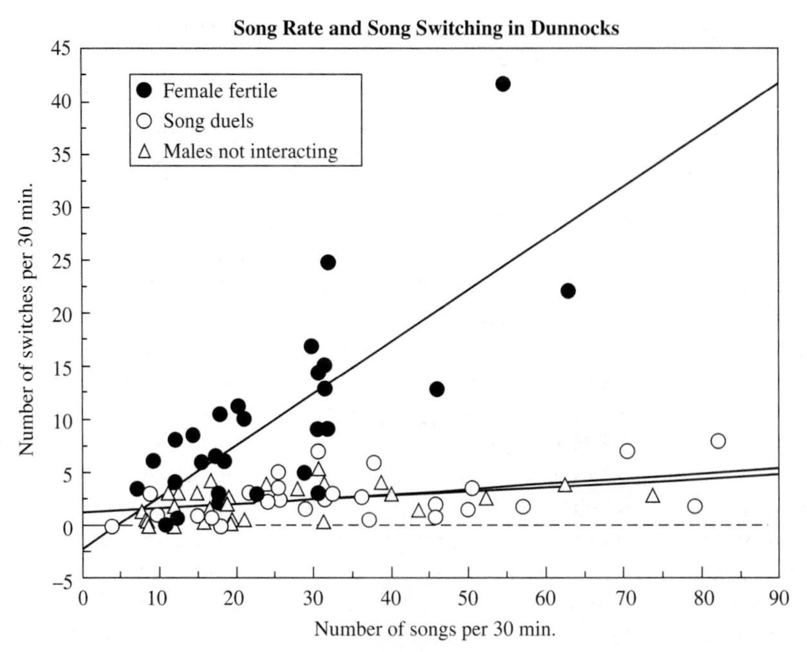

Figure 2.11 The numbers of songs uttered and the rate of song type switching by male dunnocks vary with the situation. ● During the females' fertile period. O When males engage in song duels. Δ When males are not interacting with other males. From Langmore 1997.

Table 2.1 Relationship between song characteristics and male–male competition. ↑ indicates an increase and ↓ indicates a decrease in a song characteristic with increasing male fighting ability, motivation, or attribute that is likely to be related to male fighting ability.

Song characteristic	Relationship between song characteristics and presumed male fighting ability, physical attribute, or motivation
Frequency	partridge ↓; chickadee ↑; barn swallow ↓; rooster ↑; collared dove ↑
Intensity	blackbird ↑; barn swallow ↑
Repertoire	great tit ↑; red-winged blackbird ↑; starling ↑; song sparrow ↑; Louisiana water thrush ↑; great reed warbler ↓; reed warbler ↓
Matching	bobolink ↓; chaffinch ↑; great tit ↑; cardinal ↑; song sparrow ↑; wood thrush ↑; corn bunting ↑; western meadowlark ↑; indigo bunting ↑
Overlapping	nightingale ↑; great tit ↑; robin ↑; blackbird ↑
Switching	song sparrow ↑; cardinal ↑; meadowlark ↑; Carolina wren ↑; yellow warbler ↑; chestnut-sided warbler ↑; redstart ↑; red-winged blackbird ↓; dunnock ↓; banded wren ↓
Song rate	great tit ↑; song sparrow ↑; red-winged blackbird ↑; dunnock↑; banded wren ↓

is song sharing between neighbors you are likely to find repertoire matching. In species with very large repertoires song matching may be difficult, but overlapping or switching is certainly possible. Switching is, of course, only possible if a species possesses a song repertoire. Simple song characteristics such as frequency changes or increases in intensity or rate may be used more often in species that lack repertoires or have simple songs. Birds that can be seen singing from song posts may be subject to different evolutionary pressures than those who sing from cover, and are less subject to predation or retaliation by opponents. However, in the best-studied species it appears that multiple signals are used in male competition. Male song sparrows use a variety of signals to indicate the likelihood of attack; the same is true for great tits, which appear to use matching repertoires, song rate, and overlapping as signals in aggressive encounters. Whether the same is true for less well-studied species remains to be seen.

Different species may use different song parameters in male–male interactions, and the same parameter may mean different things. In some species, males produce simpler songs in territorial interactions while others increase song complexity (Smith & Smith 1996; Catchpole 1983). Clearly, the meaning of song complexity and song repertoire size vary with the species and under different circumstances. So, although we think we can make sensible predictions about which parameters indicate competitive ability or aggressive potential in males, the fact that species differ in the combination of signals used to interact with rivals suggest that our predictions are not always correct. These are among the many fascinating challenges that confront us in functional studies of birdsong.

MATE CHOICE: VOCAL FLIRTING

In many birds females choose mates on the basis of differences in song (Searcy 1992a). What are the benefits a female might obtain from choosing a particular male over another? If the choice is active rather than random, the time spent assessing and energy used traveling between potential mates will impose some cost. The usual assumption is that females should choose to mate with a male that allows her to have more surviving offspring, by providing 'direct' benefits. For example, a male in good condition may feed young at a higher rate, increasing the female's reproductive success. Females who are mated to a male on a territory with good food resources are likely to have heavier, healthier offspring. However, the idea that females may choose a male, when receiving nothing more than sperm from him, was controversial for many years. It was accepted that some males could be more fertile than others, in which case a female should choose the most fertile mate, with a potential direct benefit of higher fecundity for her. But why should females choose between males if none of them will provide care for the offspring, a territory from which she can obtain food for the offspring, or a well-hidden nest site?

The answer is that a female can gain 'indirect' benefits from mating with certain males; although she gains no material benefit from them, her offspring may do so. Some males may have 'good genes' which are passed on to the offspring who are then more likely to survive and reproduce. For example, good genes could increase resistance to common parasites (Hamilton & Zuk 1982). Another potential indirect benefit first spelled out in Fisher's runaway hypothesis is sometimes known as the 'sexy son' theory (Fisher 1930). If a female is attracted to a particular male, perhaps because he has an especially attractive color patch, then her daughters may inherit her preference bias and her sons may have brighter patches. Having the preferred coloration, they are more likely to be chosen. A female choosing a preferred male has sons who are also preferred, and daughters who will have preferred sons. Choosy females have sexier sons, and so more grandchildren. The idea that a female may have a bias towards a male with a particular ornament, because of a bias in her perceptual or recognition system, is known as 'sensory bias' (Ryan & Rand 1993). In the case of song we can imagine that a male with a song that is close in frequency to an alarm call, or a male whose song contains a note in the most sensitive part of the hearing range may be more stimulating and thus preferred over other males. A 'sensory bias' may initiate the Fisher runaway process. In the case of sensory bias and the Fisher runaway process, there may be no difference in the fitness, condition, or quality of preferred and non-preferred males. When favored males pass on either good genes or direct benefits, we expect to be able to measure some difference, such as body mass, parasite load or territory quality, between preferred and nonpreferred males.

The division between direct and indirect benefits is not always so simple; for example, a male that is free of parasites will not infect the female (direct benefit), but he may also have superior resistance genes that are passed on to the offspring (indirect benefit). A high quality territory is a direct benefit, but males with 'good genes' may be more likely to obtain such a territory through defeating rivals. The fact that the male has a good territory indicates that he is likely to have more food and will also be in better condition. So, in addition to access to more resources on the territory, the female will also have a high quality mate in good condition. Song may also indicate more subtle differences in a male's suitability as a mate. For example, are the male and female genetically compatible, within the same species and population? Here, both the male and female benefit from mating with the right species so there is no conflict in male and female strategies. We shall see how some of these factors may play a role in making decisions regarding mate choice.

How can different song characteristics indicate that a male is of high quality in terms of providing direct or indirect benefits to the female; which song parameters might females use to choose among males differing in the benefits they may provide? It is clear that females sometimes choose males on the basis of differences in plumage characteristics that reflect mate quality. For example, female widow birds prefer males with longer tails, and female house finches prefer males with a deeper red coloration (Andersson 1989; Hill 1990, 1991). Females often appear to prefer males with more exaggerated ornaments; could the same be true for song? We can presume that degree of exaggeration is a useful choice criterion, because enhanced ornaments tend to be more expensive to produce. Therefore, males who are in good condition, of a higher quality, or have good territories are able to 'afford' larger or more conspicuous ornaments (Zahavi 1975; Grafen 1991).

There are song characteristics that can serve as costly ornaments, serving to differentiate between males of differing quality. Females will be selected to ignore, or devalue, any signal where a low quality male could fake the signal of a high quality male; a costly signal is one way of ensuring that males are honest. Costly signals are likely to occur when females are choosing for direct benefits or good genes. Lower costs are likely to be found in the case of choice for compatible males, or preferences due to sensory biases. Conventional signals are less likely to be important for female choice. This is not to say that a female may not choose a male whose dominance is maintained through the use of conventional signaling. The male may even be chosen on the basis of the conventional signal. The difference is that the female herself does not impose the receiver-dependent costs, but simply observes the outcome of his competitive interactions with other males (West et al. 1981).

Studies on female preference involve a number of techniques, generally quite different from those applied to the study of male responses to song (Searcy 1992a). In the field or in aviaries, different songs may be played from nest boxes to influence which is occupied first by females. Females may be implanted with estradiol, which makes them more sexually receptive. They are then played songs from different males or different song types, and the number of copulation solicitation displays to each is counted as a measure of preference for that song. Field observations can reveal which males are mated first, provided that there is control for the effects of other variables such as territory quality (Searcy 1992a). Finally, a few studies have used operant conditioning techniques; females press a button or sit on a particular perch to hear a song, and the song she chooses to hear more often is taken as a measure of preference (Collins 1999; Riebel & Slater 1998a). All preference studies strive to relate differences in song to variation in male characteristics. For example, if females prefer males with high song rates, do they then obtain a male with a good territory, with good genes, or both?

Specific Song Structure

An interesting study on female song preferences in canaries showed that males can possess what are called 'sexy syllables'(**Box 4**, p. 58). Female domestic canaries prefer songs that they had been exposed to when young but, in one particular strain, songs of a certain type were preferred whether they had been heard earlier or not; even isolated females preferred these 'sexy' song types (Vallet & Kreutzer 1995; Kreutzer et al. 1996; Nagle & Kreutzer 1997). The essential components are notes with abrupt frequency falls and short inter-note intervals. The preference may be due to a sensory bias (Ryan & Rand 1993); perhaps that note fits precisely the acoustic sensitivity of females. However, this note also requires very precise coordination of both the left and right syrinx for production, and could be an indicator of male quality (see Chapter 9). On the other hand, the female preference could be a carry-over from the wild ancestors. Canary song has been highly modified during domestication. New note types have undoubtedly been added by selective breeding and tutoring, and females may not have evolved responsiveness to them. An obvious question is: are 'sexy syllables' part of the original note repertoire of wild canaries, perhaps becoming less frequent after domestication and selective breeding for song quality? 'Sexy syllables' are not all that melodious to the ears of canary fanciers (**Box 39**, p. 298).

Preference for high quality singing skills has been found recently in the swamp sparrow, in whom females favor certain kinds of trilled notes (Ballentine et al. 2004). There is a trade-off between the rate of a trill and the frequency range of the component notes. Males that produce fast trills are constrained to a narrow frequency bandwidth, so that broad frequency trills are produced at a low rate. However, some males are nearer the performance limits than others, indicating a high quality vocal performance ability. Females give more copulation solicitation displays to such a song. This shows that females use a vocal performance indicator, relating to the trill–bandwidth trade-off, as a choice criterion; whether this indicates differences in male quality is not yet clear.

Female brown-headed cowbirds display a preference for a specific song structure that reflects male dominance, and they prefer the song produced by isolated males (West et al. 1981). It appears that differences in song are determined by the outcome of male interactions. Only dominant males are able to maintain production of potent songs, because males punish rivals producing potent songs, which have a different frequency pattern from nonpotent songs. The fact that isolated males have potent songs is thus easily explained; isolated males are never attacked. This song type is, therefore, a good indicator of the number of fights lost and won, and females may be using it to choose a dominant male. This species is a brood parasite, and the male only provides his genes to the female, but a female may benefit from mating with a male with higher quality genes. A dominant male will not help protect the chicks directly, but dominance may be related to genetic quality, inherited in turn by the offspring. Or perhaps offspring of dominant males are more likely to become dominant themselves and are thus preferred by females because they will have sexy potent songs.

Although it is not yet clear what benefits, if any, females obtain from choosing males with specific note types, this relatively new area of study brings together studies on sound production and mate choice, and is likely to generate more exciting findings in the near future.

Song Repertoires

Measures of song repertoire size are based on the number of different syllables that constitute a song, or the number of different song types. Either way, there are many indications that females prefer to mate with males who have larger repertoires, or more complex songs (e.g. great tits, McGregor et al. 1981; Lambrechts & Dhondt 1986; great reed warbler, Catchpole 1986; song sparrow, Hiebert et al. 1989; aquatic warblers, Catchpole & Leisler 1996), although direct experimental proof is still lacking (see Chapter 4, p. 127). Furthermore, what benefits females obtain is unclear. The most common correlate of repertoire size is male age. Males of some species make yearly additions to their

BOX 4

FEMALE CANARIES ARE ESPECIALLY RESPONSIVE TO 'SEXY' SONGS

In most songbirds it is unusual to find a song phrase composed of a string of repeated syllables that have both a high repetition rate and a wide frequency range (Podos 1996). It may be that motor constraints on the syrinx and its neural controls impose limits, because producing such a song phrase requires unusual abilities or more energy (Suthers 1999a; see Chapter 9). Depending on the strain, male domestic canaries mostly use their left syrinx during song, but they use their right syrinx to expand the range into higher frequencies. These aspects of song production are controlled by the nervous system, and the lateralization of production of the different song parameters to the right and left sides extends from the periphery up into the brain (Hallé et al. 2003). The repertoire size of male common domestic canaries varies greatly, some singing less than ten different syllables and others more than thirty. Thus, some males can be regarded as virtuosos.

Experiments show that females prefer song with a structure that challenges the motor constraints on rate and frequency range; they favor song syllables with a high repetition rate and a wide frequency bandwidth (Vallet & Kreutzer 1995; **CD2 #16**). The song structures are potent in evoking responses in the brain (Del Negro et al. 2000), and the actual mating preferences do not arise through learning (Nagle & Kreutzer 1997). Both trill rate and frequency bandwidth add to the higher response strength, and supernormal, artificially created combinations of the two characteristics beyond what naturally occurs, lead to the highest copulation solicitation display rates (Draganoiu et al. 2002). Further study is needed to distinguish whether the 'sexy' song phrases elicit more interest from the female because they are an honest fitness signal, or because they exploit a female sensory bias.

Although there is an innate component, the preferences of female canaries are also influenced by the songs that a female experiences when she is young (Nagle & Kreutzer 1997; Depraz et al. 2000); songs are effective in inducing copulation solicitation displays, but they do not stimulate other reproductive activities such as nest building or egg laying (Leboucher et al. 1998). For activities such as these, Kroodsma (1976) demonstrated that song diversity is more important than monotony. Also, female canaries are able to associate the male who is their mate with his particular song and do not confuse it with the song of another familiar male (Béguin et al. 1998). Thus, a female uses the different parameters of male song in various ways, depending on the phase of her reproductive cycle and her prior song experience, both as a young bird, and as an adult.

Michel Kreutzer

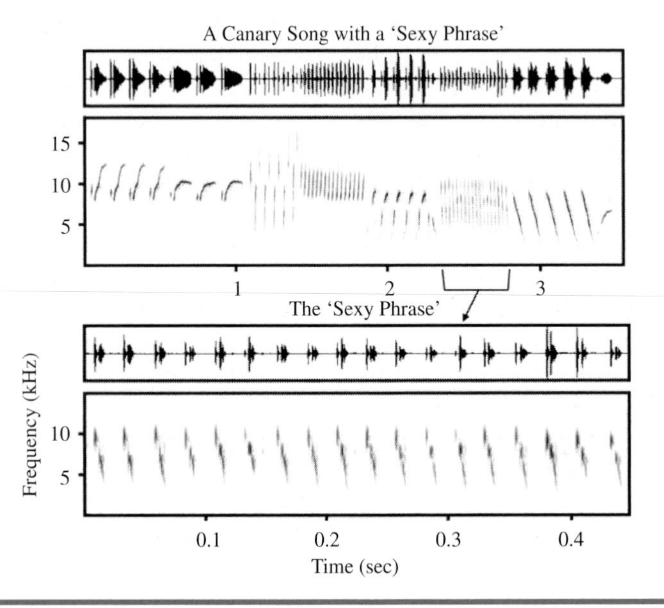

repertoire; therefore, the older the male the more songs he has learned. In the song sparrow, older males have larger repertoires (Hiebert et al. 1989), but the correlation is produced not by adult learning but by the attrition of males with smaller repertoires, who are more likely to die. A similar situation has been found in the willow warbler (Gil et al. 2001). In these cases repertoire size indicates the probability of a male surviving.

Older male pied flycatchers are in better condition and have more complex songs (Lampe & Espmark 1994). However, males with larger repertoires are less likely to survive than males with smaller repertoires from a particular year. There is still a correlation of repertoire size with age, because males learn new songs between breeding seasons but, within an age cohort, males with larger repertoires are less likely to survive. This reduced survivorship is in contrast to great tits (McGregor et al. 1981; Lambrechts & Dhondt 1986) and song sparrows (Hiebert et al. 1989), as already mentioned, where males with larger repertoires are more likely to survive. Lampe & Espmark speculate that testosterone may drive the increase in repertoire size across years. If this is so, males with larger repertoires may suffer the immune cost of increased levels of testosterone (Møller et al. 2000). Female pied flycatchers still prefer males with larger repertoires (Lampe & Sætre 1995). In aviary experiments, females preferred to build nests with males with more complex songs. Sætre et al. (1995) have shown that older males feed the young better than younger, less experienced males. So, female pied flycatchers choosing a male with a large repertoire obtain a male in good condition with greater breeding experience, but with a reduced probability of survival to the next year.

Older male starlings also have larger repertoires (**Fig. 2.12**), and males with larger repertoires are preferred as mates both in the field and in aviary experiments (Eens et al. 1991; **CD1 #15**). Other than a nest hole, male starlings do not defend a territory, so females gain no food resources. Exploring the role of repertoire size, Mountjoy and Lemon (1991) found that more complex songs deterred male starlings from entering nest boxes, and that females were attracted to nest boxes defended by males with large repertoires (Mountjoy & Lemon 1996). Eens (1997) found that females prefer larger repertoires. Is repertoire size a signal to males or to females? Eens et al. (1993) found that males sang more songs and included more song types when presented with a female than with a male. Thus it appears that, in starlings, repertoire size functions as an attractor for females, but is also important in male competition.

Female great tits prefer males with three to five song types, rather than those with only one or two (**Figs. 2.7B & 2.13**, Baker et al. 1986). Great tits with larger repertoires survive better and, in general, are more successful breeders. This may be because they have better territories, but assessment of song is probably quicker and easier than a comprehensive assessment of territory quality. Why male great tits with larger repertoires obtain better territories is unclear, although speaker replacement experiments suggest that males with larger repertoires are better able to defend territories (Krebs 1977a). Here again, male competition and female choice are not easily separated. Indeed, in some species the correlation between repertoire size and male mating success disappears when territory quality is taken into account, as in red-winged blackbirds (Yasukawa et al. 1980). In this species, females appear to be assessing territory quality and choosing better territories directly, even though males on these territories do indeed have larger repertoires. However, female great tits do seem to base their choice on repertoire size, which may indicate both male and territory quality, rather than territory quality alone.

The sedge warbler is particularly well studied. In a field study, Catchpole (1980) showed that males with larger syllable repertoires attracted females earlier in the breeding season; also females display more to recordings of larger repertoires (Catchpole et al. 1984). Buchanan and Catchpole (1997) found that repertoire size is one of several cues used in mate choice, and that variability in the trait is important. Repertoire size varied from year to year and was always correlated with mating

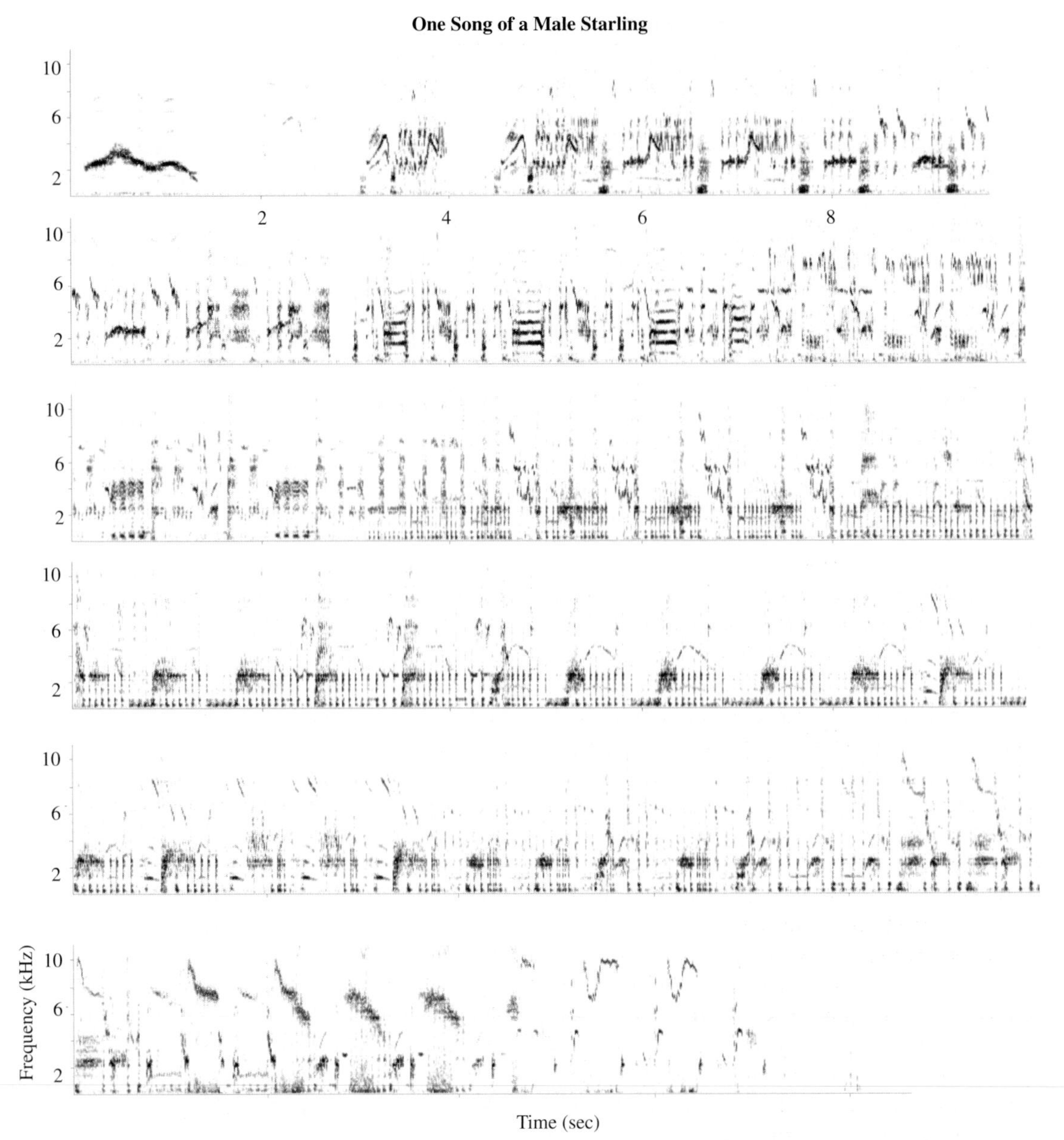

Figure 2.12 A male starling song containing all of the basic syllable types and some heterospecific imitations. The male was singing in a nest box, while a prospecting female was in the vicinity. From Eens 1997. (**CD1 #15**).

success, even when controlling for territory quality. As in other species, older males have larger repertoires. The most striking finding is that males infected with parasites have smaller repertoires (Buchanan et al. 1999). Therefore, females obtain parasite-free, older males by choosing males with larger repertoires. In a related species, the great reed warbler, repertoire size was correlated with the number of extra pair copulations. Female great reed warblers appear to base their choice of mate on repertoire size *and* territory quality. However, females who are

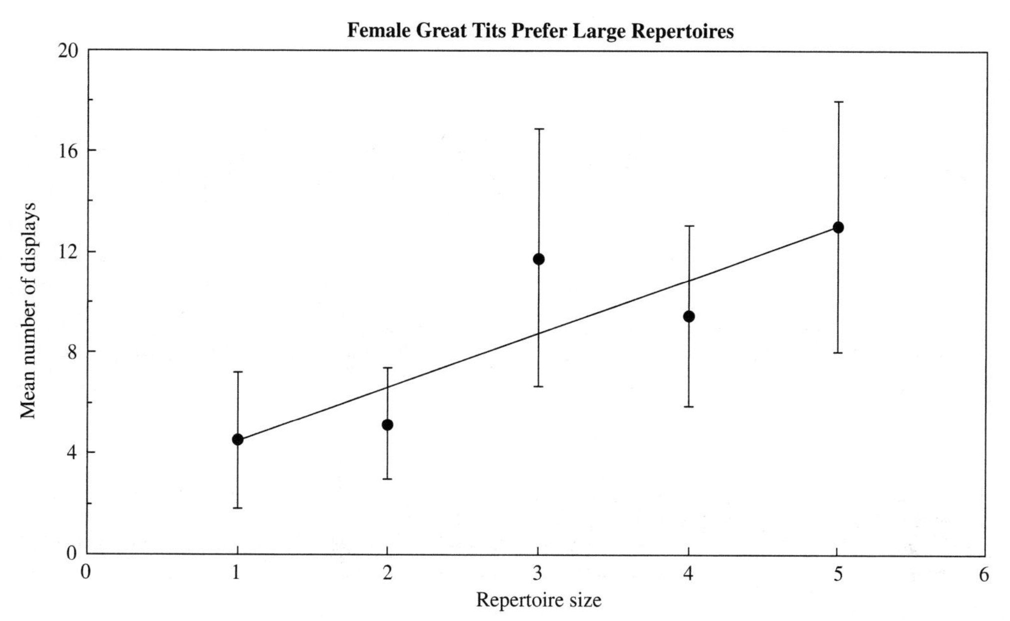

Figure 2.13 Mean number of copulation solicitation displays given by 11 female great tits as a function of male song repertoire size. One to two song types elicited a significantly smaller response than repertoires of three to five song types. From Baker et al. 1986.

paired choose to have extra pair copulations with males whose repertoires are larger than those of their partner (Hasselquist et al. 1996). The relative post-fledging survival of the young is correlated with the genetic father's repertoire size rather than the foster father's (**Figs. 2.14 & 2.7A**). Females thus obtain some kind of indirect genetic benefit from choosing to mate with males with larger repertoires (Hasselquist et al. 1996).

So far, it appears that males with larger repertoires are generally older, have fewer parasites, survive better, and have offspring that survive better. Thus, repertoire size does seem to indicate male quality, but how can it do this? It is not obvious why singing many different songs should be costly. However, if a male is parasitized or in poor condition he sings fewer songs, consistent with the idea that there is a cost. The larger repertoires of older males is usually assumed to reflect the time it takes to learn more songs. But in some species the number of songs a male sings is fixed in the first year or so; they do not learn more songs as they age, but males with smaller song repertoires die, thus producing the observed correlation between age and repertoire size.

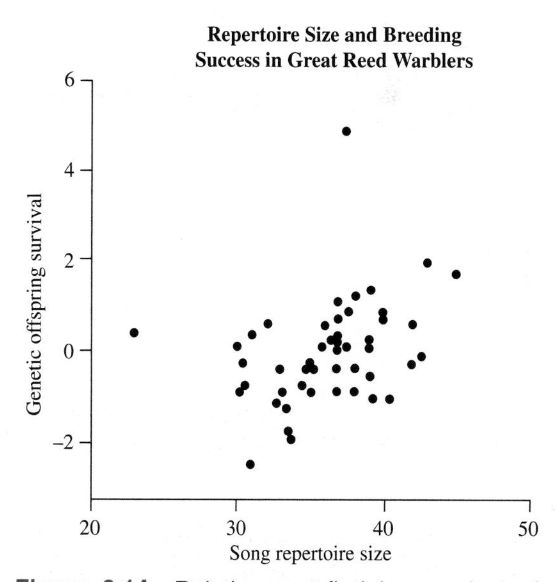

Figure 2.14 Relative post-fledging survival of genetic offspring correlates with the standardized male song repertoire size in great reed warblers. From Hasselquist et al. 1996.

It has been suggested that there is a cost to learning a large number of songs in terms of the brain space needed to store them. The assumption is that males with larger repertoires have a larger area in the brain dedicated to song learning (the higher vocal center – HVC). Thus, repertoire size indicates a male's ability to bear the 'space costs' of having a large HVC. However, does repertoire size depend on HVC size and is it really costly? In general, species with larger repertoires have larger song system nuclei in their brain; and, in some species, individuals with more complex songs have larger song nuclei (see Chapter 8). Sedge warbler males with larger repertoires and more complex, longer songs have larger HVCs (Airey et al. 2000). However, although females appear to prefer males with larger repertoires (Buchanan & Catchpole 1997), as mentioned above, there is no difference between paired and unpaired males in their HVC size (Airey et al. 2000); thus female choice for repertoires is not selecting males with larger HVCs. Possibly there is a cost to maintenance of the song center (Gil & Gahr 2002). In the canary, some song system nuclei shrink outside the breeding season (see Chapter 8), suggesting that there may be a cost to its maintenance when not needed. However, it is hard to see why maintaining this area of the brain would increase energetic costs compared to all other active areas. An alternative explanation is that the cost may be developmental (Nowicki et al. 1998a). Birds who suffer early nutritional stress develop smaller song repertoires perhaps because they are deficient in the resources required to develop the song area of the brain, lack the energy to practice singing, or are too weak to pay attention to their song tutors. A nutritionally stressed bird would learn fewer songs as a youngster and, in those species where there is a fixed learning period, the male would have fewer songs for the rest of his life. It is known that defects in early nutrition can affect adult survival (Birkhead et al. 1999), so females should avoid males who have suffered physically when young. Nowicki et al. (2002a) recently tested whether females discriminate against males who learn their songs poorly, as is likely to happen if they have been exposed to nutritional stress. Experienced females preferred males who include a relatively high proportion of learned versus improvised material in their songs, and who copy the learned notes accurately, i.e. are good learners. This demonstrates that variation in learning abilities among males matters to females when choosing a mate. Females could use learning ability as an indicator of quality, assuming it indicates a history of developmental stress (Nowicki et al. 2002a; **Box 5**, p. 63).

Another possibility is that the cost is not energetic but a result of decreased immunocompetence. As mentioned for pied flycatchers (Lampe & Epsmark 1994), repertoires may be controlled to some extent by the male hormone testosterone. Testosterone has a suppressing effect on the immune system, so that a bird with higher levels has a less efficient immune system (Møller et al. 2000). That may be why, in the sedge warbler, males who are parasitized have smaller repertoires. Male testosterone levels decrease as the immune response necessary to fight off the parasites gets under way. The cost of a larger repertoire may thus lie in the hormones needed to develop it, not in learning or singing. Males will differ in their fitness, in the form of disease resistance, and in energy reserves, and thus in the immune system depression that they can bear (Gil & Gahr 2002). A fit male will have higher testosterone levels and still laugh off an infection, whereas a male in poor condition will need to put more of his energy into fighting it off. In either case testosterone levels, and also repertoires, are likely to decrease when a male is infected, but by a smaller degree in fitter males. The early nutrition and the immunosupressant hypotheses both presume that males with a larger repertoire are fitter and thus indicate the benefit a female receives from mating with males with larger repertoires. Males with a large repertoire have been healthy during development and are better able to bear the immunosupressant costs of testosterone. As yet, these ideas are still speculative, but are being addressed by current work (**Box 6**, p. 64).

BOX 5

THE QUALITY OF SONG LEARNING AFFECTS FEMALE PREFERENCES

It is well known that male song influences mating preferences in female songbirds, and many aspects of song are learned by imitation from other males of the same species. Females are more responsive to normal, learned songs than to those developed by a male in social isolation, and are completely unresponsive to the very abnormal songs of deaf males (Searcy et al. 1985; Searcy & Marler 1987; **B; CD2 #17**). But the differences between the songs of wild birds, isolates, and deaf birds are gross, involving many different features. What about the finer details of songs that all qualify as normal for the species, and yet vary greatly from bird to bird? Do they influence female responsiveness? In the case of the learned details of song dialects, we know that females base their preferences on those same traits that are affected by learning (Baker et al. 1987b; Balaban 1988). Therefore, we speculated that how well a male learns his songs ought to affect female preferences; in other words females should prefer well learned over poorly learned songs. In the first direct test of this proposition, song sparrows were taken soon after hatching from a population in Hartstown, Pennsylvania, raised in the laboratory, and tape-tutored with songs recorded from the same population (Nowicki et al. 2002). Songs of the hand-reared males recorded at one year were divided into well-learned and poorly learned categories, based on quantitative comparisons with their tutor songs, drawn from the local population (**A**). We then presented these songs to adult female song sparrows from the same population, and measured the numbers of copulation solicitation displays they performed. First, we compared female responses to a set of 10 songs that had both a high proportion of copied notes (mean = 98%), and high note copying accuracy (mean spectrogram cross-correlation [SCC] = 0.69), and another set with a low proportion of copied notes (mean = 29%), and low note copy accuracy (mean SCC = 0.54). The females showed a strong, preferential response to the songs with high proportions of well-copied notes. Then we took a pool of songs, all with a high proportion of copied notes, and chose two subsets, one with high note copy accuracy (mean SCC = 0.71), and the other with low note copy accuracy (mean SCC = 0.59). Females responded more strongly to the songs with the more accurately copied notes, though the preference was weaker than in the first experiment. Evidently, as predicted, female song sparrows prefer well-learned to poorly learned songs.

William Searcy & Stephen Nowicki

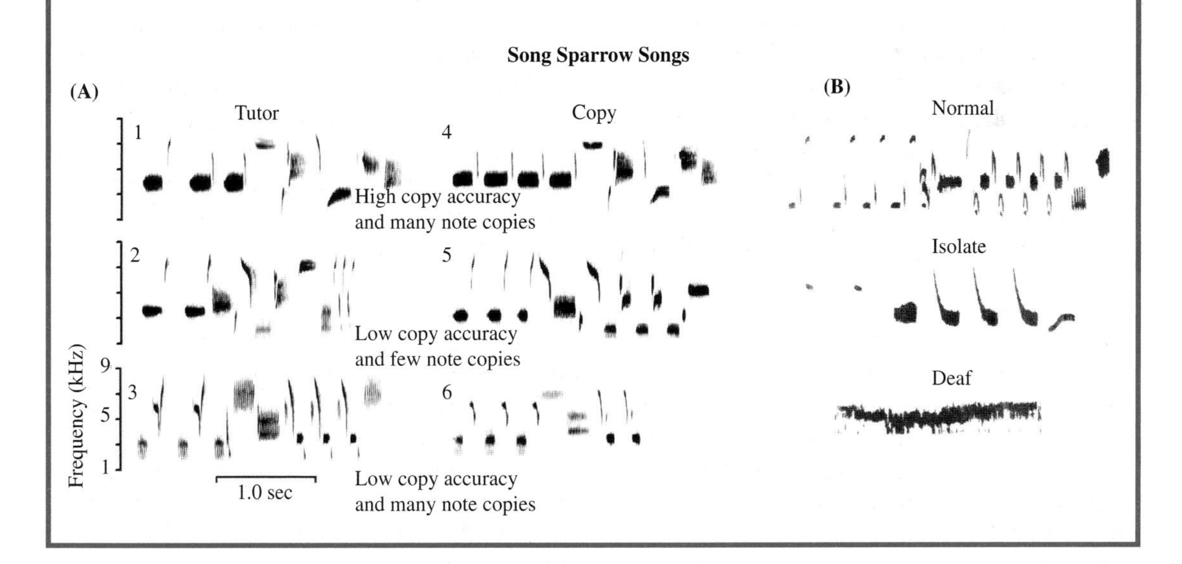

BOX 6

THE HOOTING OF TAWNY OWLS: STRUCTURE AND FUNCTION

The nocturnal tawny owl is highly vocal, giving a variety of calls, the most distinctive of which is the 'to-woo' hoot used to announce ownership of their territory (**CD2 #18**). Tawny owls have individually distinctive hoots (**A**) and, with the aid of a computer, hoots of different birds can be distinguished, using the length of different call components with an accuracy of 99% (Galeotti & Pavan 1991; Appleby & Redpath 1997a). Moreover, their calls are consistent from year to year, so it is possible to monitor owl populations by comparing sonograms between years (Appleby 1995). The structure of owl calls was found to differ between populations (Appleby & Redpath 1997a), but neighbors had dissimilar calls, suggesting that they do not learn from each other. Owl calls did not appear to vary in relation to habitat (Appleby & Redpath 1997b) but, in an Italian study, owls living in woodland seemed to have lower pitched hoots than those living in farmland (Galeotti 1998). Playbacks showed that tawny owls recognize the calls of their neighbors, and exhibit a swifter and more aggressive response to stranger calls than to local territory holders, presumably because strangers represent more of a threat to their territory (Galleoti & Pavan 1993). Males and females often cooperate in defending a territory against an intruder. In playback experiments, females were more likely to respond to female than male calls in their territories, while males showed no distinction, except that they were more likely to respond aggressively to female calls if they had already bred successfully with their mate (Appleby et al. 1999). So, in addition to the territory, males seem to defend a reproductively valuable partner. There is also some evidence that hoots contain other information that may be useful to mates and competitors. Larger males were found to have lower frequency hoots (Appleby & Redpath 1997b). Furthermore, owls with more blood parasites responded more slowly to a challenge and they gave calls in which the highest frequency was lower, and the range of frequencies was smaller (Redpath et al. 2001; **B**). Thus, there is potential for individual owls to assess male parasite load from the speed of response and the structure of the call.

Steve Redpath & Bridget Appleby

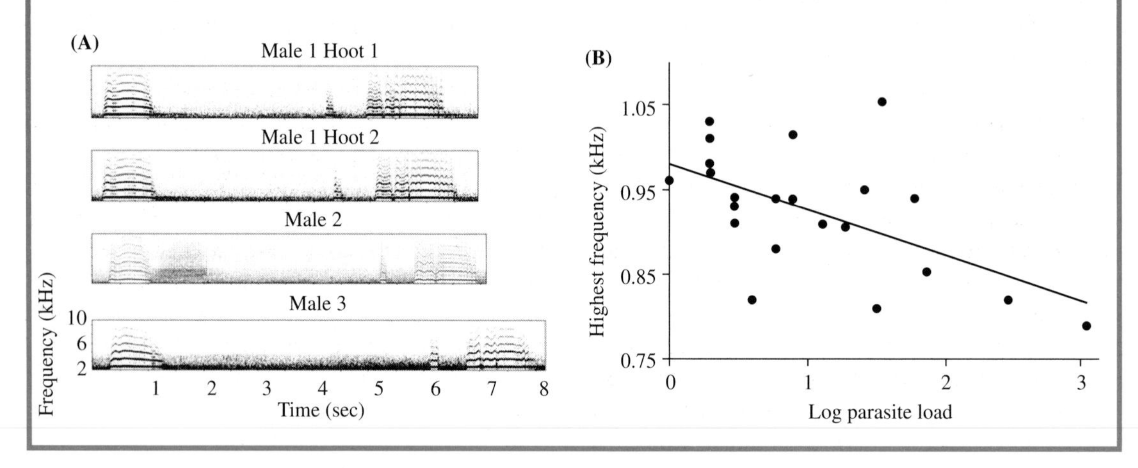

As a final point, there is always the possibility that repertoires are preferred by females as a result of sensory bias (Ryan & Rand 1993), and that the benefits they obtain are an incidental byproduct. Searcy et al. (1994) showed that male response to song increases when the song switches to another type in red-winged blackbirds. He hypothesized that the new song type results in dishabituation. Females show the same effect (Searcy 1988); thus female habituation to a stereotyped song may set the stage for the evolution of song repertoires. Females do not habituate so quickly to males who have more song types or a more complex song. Therefore, an apparent preference for the complexity of larger repertoires is observed. To test for a general bias across all species for decreased habituation to repertoires, we must show female preference for repertoires in a species where there are no repertoires. The evidence so far is somewhat contradictory (Searcy & Marler 1984). The results using species with song repertoires supported the dishabituation hypothesis. Female field sparrows and white-throated sparrows showed no preference for an artificially constructed repertoire, something neither species possesses naturally (Searcy & Marler 1981). Female common grackles do show a preference for a combined song of four males (a repertoire) over a single male (Searcy 1992b). However, in this species the preference may be an ancestral trait, rather than evidence of a general bias, as males in all other closely related species have repertoires (Gray & Hagelin 1996).

Zebra finches have no close relatives with repertoires and, although there is some variation in syllable structure within a male zebra finch song, they do not have a true repertoire (Helekar et al. 2000). In a preference experiment females were trained to press buttons to receive a song (Collins 1999; **CD1 #14**). One button resulted in the playing of an artificial 'repertoire' and one a stereotyped song (**Fig. 2.15A**). All showed a preference for the repertoire (**Fig. 2.15B**). Thus, in a species where males do not possess a repertoire, nevertheless, females show a preference for multiple song types. Repertoires may have evolved to prevent habituation and, through the cost of learning a repertoire, high quality males may incidentally have larger repertoires. Of course, the question then remains – why do some species lack repertoires and why are some females not biased towards them (Gray & Hagelin 1996)? Perhaps individual identification is more important in some species, or the relative cost of developing a repertoire varies. A short lifespan, a short breeding season, or high predation risks may also affect whether repertoires are favored.

Song Familiarity

In a few species it appears that female choice for song is affected by familiarity. The function of the preference could be mate recognition. Female song sparrows prefer a neighbor's song more than a stranger's songs (**Fig. 2.16**), and give the strongest response to the song of their mates (O'Loghlen & Beecher 1999). In general, females tend to respond more to any song that is similar to that of their mate, but this is obviously not the same mechanism that leads them to a particular mate in the first instance. Preference for familiarity, in songs that are not those of the mate, has also been observed in previously unmated females, indicating that preference for a familiar song is not just due to mate recognition. Female zebra finches prefer songs that are similar to those of their father (Clayton 1990a), or that they heard frequently when young (Riebel 2000; **Box 7**, p. 67). Female brown-headed cowbirds prefer the song of males from a culture to which they have been exposed extensively (Freeberg et al. 1999). So, familiarity can influence female preference for a mate, but the benefits to the female and the function of the preference are not always clear. The most plausible explanation is that the preference for familiarity is a consequence of the way that females imprint on species' song characteristics. Females use song to choose a male of their own species, and they learn some species' song characteristics by imprinting on songs heard when they were young. Songs more similar to those heard when young fit the species-specific template better than

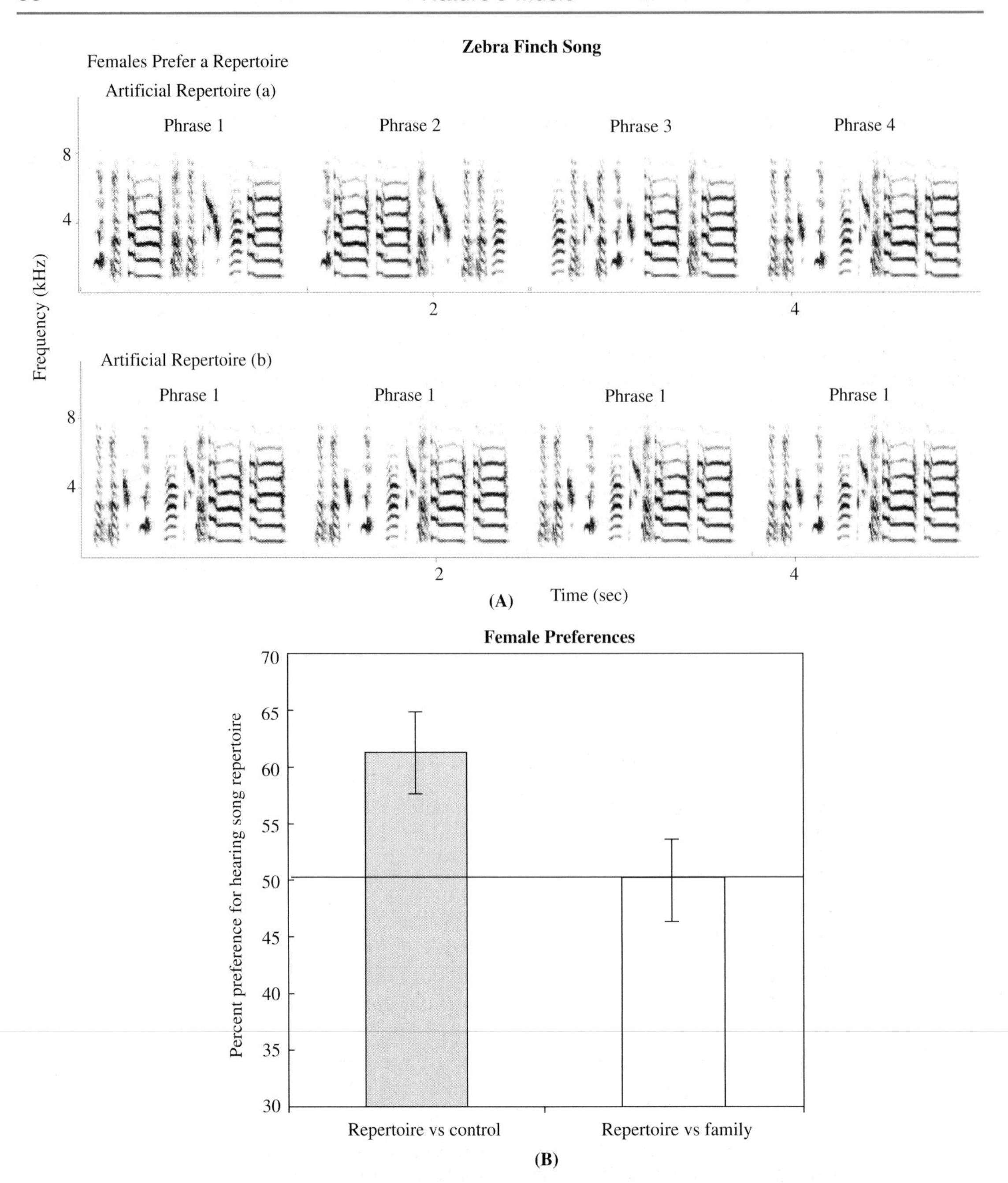

Figure 2.15 **(A)** Two sets of song stimuli given to female zebra finches – an artificial, varied repertoire (a) and a repeated, stereotyped song (b), as a control (**CD1 #16**). Females preferred listening to the repertoire stimulus. **(B)** In a repertoire versus family experiment, the 'family' stimulus consisted of four very similar songs from different individuals within a family. Here, there was no preference. From Collins 1999.

BOX 7

LEARNING TO SING, LEARNING TO LISTEN: DEVELOPMENT OF SONG PERCEPTION

Given the importance of learned song in mate attraction and stimulation, it would be surprising if potential receivers did not engage in some perceptual learning, with so much of the fine detail of adult song culturally rather than genetically inherited. After cross-fostering between subspecies, female zebra finches preferred the songs of their foster parents, not their genetic fathers (Clayton 1990a). Focusing in on within-population variation of zebra finch song, and sidestepping the effects of social interaction with the singer (Riebel 2000), I limited young females' song exposure to tape tutoring between days 35–65 post-hatching, within the sensitive phase for male song learning. After sexual maturity at 3–4 months of age, song preference was tested by teaching them to peck red buttons for song playback. In three tests (see figure, solid symbols), they consistently chose the taped song they heard early in life over an unfamiliar song, strongly suggesting that preferences for within-population variants of song are culturally inherited. In striking contrast, females raised without any exposure to adult male song (see figure, open symbols) behaved inconsistently, changing preferences between repeated tests. Early exposure to song might turn out to be as crucial for the development of perceptual competence in the receiver as it is in vocal production by the sender. Interestingly, when we tested males and females in the same context with the same songs, preferences for tutor songs were equally strong, although of course only males learned to sing them (Riebel et al. 2002). This supports the idea that early perceptual learning is independent of learning for production: only some learn to sing, but all are likely to learn to listen.

Katharina Riebel

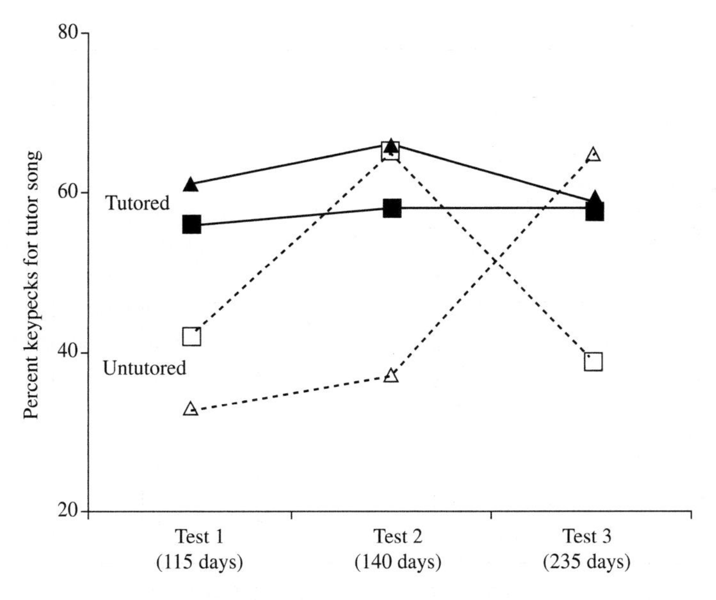

After tutoring, two females consistently preferred the tutor song over an unfamiliar song in three tests at different ages (solid symbols). Two untutored females showed no consistent preference (open symbols). The tutored females had each heard a different song and were given the other's tutor song as an unfamiliar song. The same song combinations were used with the untutored females. Data from Riebel 2000.

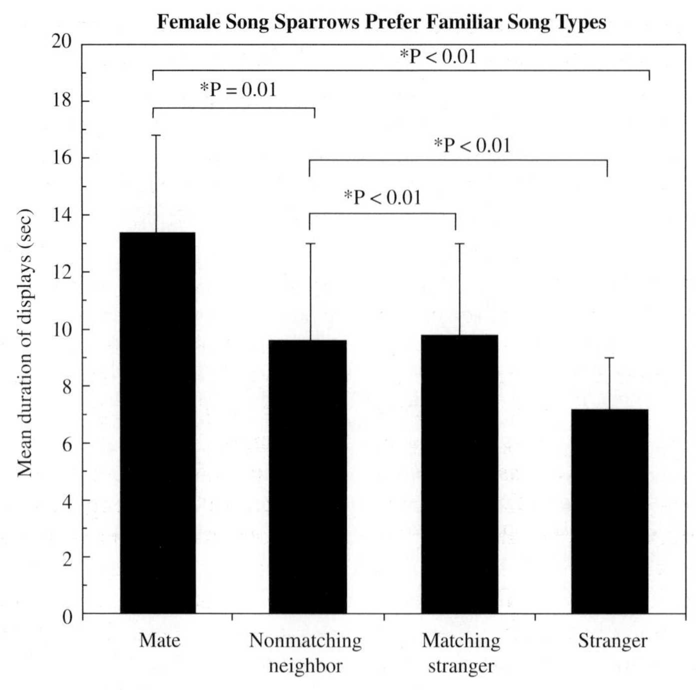

Figure 2.16 Song preferences of female song sparrows. The figure shows the mean duration of solicitation displays (+SE) during playback of 8 repetitions of a song in one of four categories; mate, nonmatching neighbor, matching stranger, and stranger. *Wilcoxon signed ranks test. From O'Loghlen & Beecher 1999.

unfamiliar songs (ten Cate & Vos 1999). Another explanation is that preference for familiar, or local song results from females using similarity to local songs as a measure of the learning ability of the male (Searcy et al. 2002). As already mentioned, song learning ability may reflect the degree of developmental stress experienced and thus, potentially, male quality (Nowicki et al.1998a).

A preference for a familiar song is known to be vital to breeding success in the village indigobird, which is a brood parasite. Male indigobirds learn their song in part from the host (Payne 1973a; Payne & Payne 1994) and therefore have a similar song (**Box 41**, p. 315). Females prefer male indigobirds who sing the same song as their foster parent species. There are some differences between the host and parasite songs, however, and female indigobirds appear to prefer songs of males of their own species rather than the actual host's song. Given that indigobirds have a number of specializations that are specific to a particular host species, it is important for a female to find a male from a lineage that has the same host preference.

Local dialects are the most widely discussed aspect of female preference for familiar song types. Some bird species have the equivalent of regional accents and if females preferred familiar song they would tend to mate with males that had the same dialect as the one where they spent their youth (Cunningham & Baker 1983). Balaban (1988) showed a preference for local males in swamp sparrows, and Searcy et al. (1997, 2002) found a preference for local males in song sparrows. Female great tits prefer local males (Baker et al. 1987a). It has been proposed that, because of local genetic differences between dialect populations, it is advantageous to mate with males from the same population. Therefore, females should prefer males with a dialect resembling that of their own population to be sure of obtaining the best male. The outcome would be subpopulations of locally adapted birds.

Whether there are locally adapted populations maintained by females' song preferences has been an area of intense discussion, with sometimes conflicting results being found by different researchers (Petrinovich & Baptista 1984; Baptista 1985; Baker & Cunningham 1985; Chilton & Lein 1996). The most extensive studies are on *Zonotrichia* sparrows (see Chapter 3; **Box 12**, p. 89). As they disperse, female white-crowned sparrows will encounter males both from their own population and from the neighboring population with a different dialect. If females prefer to mate with males from their own population they should choose to mate with males from the same dialect population as their natal one, using song as a cue. Work by Cunningham and Baker (1985) suggests that females do prefer males that sing the dialect of their natal population. However, it appears that females learn some of their song preferences after dispersal and, if they disperse to the area where the neighboring dialect is more prevalent, may prefer males of a different dialect to that of their natal population. In addition, some studies have suggested that males also learn or modify their song after dispersion, in which case the dialect would not necessarily reflect the characteristics of the natal population (Baptista & Morton 1988). However, this is evidence against the local adaptation hypothesis, not against a preference for local song.

Another approach is to induce a female to sing by implanting her with testosterone, the assumption being that the song she sings is that of her natal population. The local adaptation hypothesis predicts that the dialects of the male and female in a pair should generally be the same (Tomback & Baker 1984; Baptista & Morton 1982). Tomback & Baker (1984) found an association between the dialects of mates, but Baptista & Morton (1982), and Petrinovich & Baptista (1984) found that captive females sang songs unlike those of their mates. Preferences can also be studied in the laboratory, by implanting females with oestradiol and playing back songs from different dialects. Baker et al. (1981, 1987b) found that natal dialect songs were preferred, and received more copulation solicitation displays. But Chilton et al. (1990) found no preference for natal dialects. Chilton & Lein (1996) suggested that, in another subspecies, female white-crowned sparrows did not base their choice on dialectical variation in male song, and that it is unlikely that mate choice decisions promote genetic isolation. They did find an association between the songs of mates, but they suggest that females learn the song of their first mate. Females did not respond more strongly in the laboratory to playback of songs of the local type, suggesting that song type was not important although, given that the females were paired, perhaps they were less likely to respond to male song. It should be noted that Chilton and Lein were working with a migratory subspecies, and Baker with a subspecies that is resident the year round and has differently structured song dialects, perhaps an important distinction (**Box 12**, p. 89). Data of Bensch et al. (1998) suggest that female great reed warblers choosing local males would obtain a more fit male, in better condition. This could be because lower quality males are forced to disperse further, philopatric males may be locally adapted, or may have better local experience. Similar arguments were advanced in the discussion of song sharing; local variants of song may be important in both female choice and male competition for similar reasons.

To summarize, female preference for familiar song is often observed, and may be a byproduct of the mate or species' recognition mechanisms (Nelson 1989a), or a way of finding high quality males (Searcy et al. 2002; Bensch et al. 1998), or it may be due to preference for locally adapted mates (Baker & Cunningham 1985).

The Rate of Singing

Song rate, usually defined by the overall time spent singing, or singing within a defined time period, rather than by the speed of note production, is an obvious candidate for a costly signal. The evidence regarding the energy costs of singing is somewhat contradictory. In the

Carolina wren (Gaunt 1987) and the sage grouse, oxygen consumption has been shown to increase with increased song rate. In sage grouse, vigorous males expend twice as much energy per day as males that do not display (Vehrencamp et al. 1989). However, a recent study has shown that metabolic rates increase by only a multiple of 1.05–1.38 during singing (Obwerger & Goller 2001), suggesting that singing for longer is not costly. However, there are other indications that singing may be costly. Males sing more when provided with extra food (Alatalo et al. 1990) and when the temperature is higher (Gottlander 1987; **Fig. 2.17**). Male barn swallows decrease their song rate when they have a high ectoparasite load (Møller 1991). The song rate of pied flycatchers is correlated with food availability and temperature. This suggests that when males have more energy to spare they sing more, so it is possible that birds in better condition will sing more. Of course, some of the cost of singing may result from the incompatibility of singing and feeding, and from singing taking up time and energy that could be spent in other activities.

In one of the first studies to show that singing rate might be important for reproductive success, Payne & Payne (1977) showed in village indigobirds that males with higher song rates had a higher reproductive success. Because this bird is a brood parasite, there are no benefits associated with a territory, from the mate or parental care. The increased reproductive success is likely to be due to the higher genetic quality of males with a higher song rate. Female indigobirds obtain only indirect benefits from males, but song rate can also indicate that a male can provide direct benefits. In stonechats, song rate was correlated with participation in parental care, in terms of feeding nestlings and defending them from predators, indicating that song can indicate behavioral differences between males (Greig-Smith 1982a). Although no mate choice study was conducted, it appears that song rate would be an efficient parameter for female stonechats to use. Other studies have shown not only that song rate correlates with reproductive success or behavior, but that females prefer males who sing at a higher rate, in willow warblers (Rædesater et al. 1987), white-crowned sparrows (Wasserman & Cigliano 1991), and starlings (Eens et al. 1991). Females have good reason to prefer males with a high song rate, which can indicate territory quality (Gottlander 1987), male condition or ability (Houtman 1990; Beani &

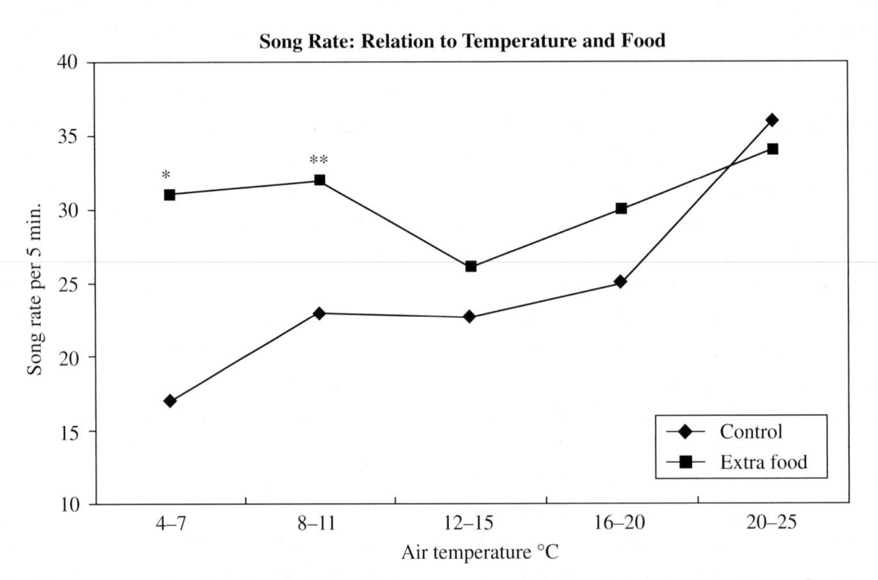

Figure 2.17 The song rate of male pied flycatchers varies with food and temperature. Some differences between the groups are significant *P < 0.05, **P < 0.001. From Gottlander 1987.

Dessi-Fulghieri 1995), or male genetic quality (Houtman 1990; Møller et al. 1998).

Female pied flycatchers prefer males with a high song rate and thus obtain a mate with a good territory (Gottlander 1987; Alatalo et al. 1990). Song rate was not correlated with male physical characteristics in pied flycatchers (Gottlander 1987), which suggests that song rate is not related to intrinsic male characteristics but to territory quality. The assumption is that in better territories a male will generally be well fed, or able to find food more easily, and thus can sing at a higher rate. Zebra finch males provide parental care but no territory. In laboratory experiments, males with higher song rates were preferred by females (Collins et al. 1994) and these males had heavier offspring, indicating a higher genetic quality; song rate is also heritable (Houtman 1990). All males had *ad libitum* food so differences in song rate are unlikely to be food related; female zebra finches probably obtain mates with good genes and in good condition, by using song rate to differentiate between males. Sons of females preferring males with high song rates actually get double benefits; not only are they heavier, and thus presumably more likely to survive, but they also inherit their father's high song rate. If they have a high song rate they are more likely to be chosen by females, i.e. they are 'sexy' (Fisher 1930).

A study on a nonsong bird, the partridge, showed that females prefer males that produced the 'rusty gate' call at a high rate (Beani & Dessi-Fulghieri 1995). This call is affected by the levels of testosterone (Fusani et al. 1994), and increased vigilance postures are observed in these males. Presumably, more vigilant males will protect the young and female more effectively. Male display rate is also important in the sage grouse. The probability that a visiting female will mate with a male is related to his display rate (Gibson 1996). Males display in aggregations called 'leks', which females attend over several days before choosing a mate, so they can determine long-term display rate differences between males (**Fig. 2.18**). Presumably, these males have increased energy reserves or a higher genetic quality. Barn swallow

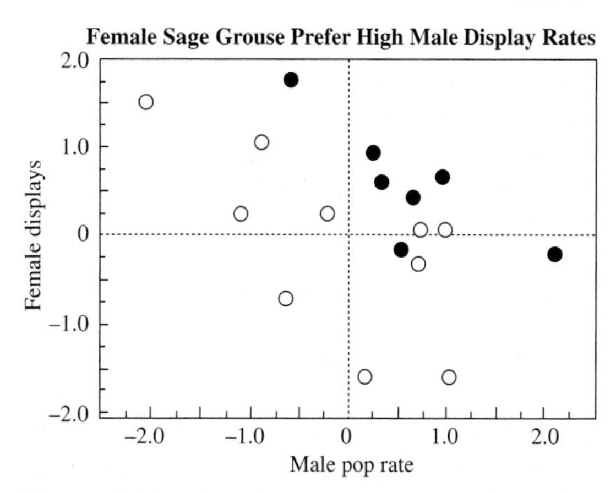

Figure 2.18 Female sage grouse prefer males with a high display rate and a short inter-pop interval. ● chosen; O not chosen. From Gibson 1996.

females choose males with high song rates for extra pair copulations (Møller et al. 1998). Again, they only obtain sperm from the male, so are presumably choosing for intrinsic male qualities.

An important question is the reliability of song rate as a measure of male quality. The song rate of male Ipswich sparrows is affected by food provisioning and by temperature, as in other species. Therefore, in warmer weather and in mild years, the song rate provides less useful information as all males can sing at a high rate. In colder, harsher years, only males who are in good condition can sing at a high rate, so song rate becomes a good choice criterion. Indeed, females seem to rely on song rate only in harsher years (Reid & Weatherhead 1990).

In all species studied, if females show a preference at all, they favor males with a higher song rate. It is easy to see why females should use this parameter to choose between males. Males singing more are likely to have greater energy reserves, which indicates to the female that she will obtain a male with a better territory, in good condition, with good genes, who will be a good parent, or some combination of these qualities.

So we find that the role of song in female choice is better understood than its role in male competition. In general, there are fewer contradictions, as can be seen in **Table 2.2**.

Table 2.2 Relationship between song characteristics and female choice. ↑ indicates that an increase in the song characteristic in question is preferred by females and ↓ indicates that a decrease is preferred.

Song characteristic	Relationship to female choice or physical attribute
Frequency	partridge ↓
Performance	occurs in canaries, swamp sparrows, and brown-headed cowbirds
Repertoire	great tit ↑; red-winged blackbird ↑; European starling ↑; song sparrow ↑; aquatic warbler ↑; great reed warbler ↑; willow warbler ↑; pied flycatcher ↑; sedge warbler ↑
Familiar dialect	white-crowned sparrow ↑; zebra finch ↑; brown-headed cowbird ↑; village indigobird ↑; swamp sparrow ↑; great reed warbler ↑; great tit ↑; song sparrows ↑
Song rate	Carolina wren ↑; barn swallow ↑; sage grouse ↑; pied flycatcher ↑; village indigobird ↑; stonechat ↑; willow warbler ↑; white-crowned sparrow ↑; zebra finch ↑; Ipswich sparrow ↑; partridge ↑ (of one note only)

All species where females choose on the basis of male repertoire size prefer males with larger repertoires. Also, all females that pay attention to song rate prefer males that sing at a higher rate. When it comes to female preference for particular note types, too few species have been studied to draw any general conclusions. Female preference for song types has generated some controversy, especially in the white-crowned sparrow. However, the controversy is perhaps over the importance of dialects for promoting population isolation and local adaptation, rather than over the question of whether there is preference for local song types. Preference for local songs has been found in a number of species; in others, no preference has been found but, so far, no preference for foreign dialects has ever been shown. Perhaps preference for local dialects depends upon processes such as the habitat characteristics that facilitate transmission of different song frequencies, or adult movement patterns (Slabbekoorn & Smith 2002b).

There are still unanswered questions about preference for repertoires, and whether repertoires indicate individual male fitness. Although age is a known predictor of repertoire size in many species, it is not the case for all. Repertoire size, as a way for females to choose older males, is a factor only in some species. Studies showing that parasite load correlates with repertoire size are an important step towards answering this question. Male song rate, perhaps the best understood song parameter with respect to female choice, is almost certainly a signal of adequate energy reserves. It indicates individual condition or territory quality, or both, and females have a good reason to prefer males who sing more.

ONE SINGER: MULTIPLE FUNCTIONS

A song typically repels rivals and attracts females. There are three ways for a song to perform both functions.

(1) The same song characteristics could be used to indicate the singer's ability to perform both functions; for example, a high song rate could attract females and repel rivals. In great tits, red-winged blackbirds, and starlings repertoire size appears to be used by both males and females to assess singers. As shown in **Tables 2.1** and **2.2**, in a few species males and females use the same song characteristics for assessment.

(2) Different aspects of singing behavior could be effective in interactions with males and females, even when the same song types are sung to both sexes. Yellow warblers, chestnut-sided warblers, and American redstarts repeat one song type when signaling to potential mates, and sing a series of song types, including that sung to females, when signaling to potential rivals (Spector 1992; Kroodsma et al. 1989; Weary et al. 1994). Great tit males match rivals to deter competitors, while the females use song repertoire size to choose a mate. The songs sung in each

case are the same, but different parameters are used as the response criterion.

(3) Either (i) different song types, or (ii) different parts of the same song are effective in interactions with males and females (see p. 57).

The challenge posed by the dual function of song can be addressed in several ways, depending on the combination of evolutionary pressures to which a particular species is exposed, acting sometimes in concert and sometimes in conflict (**Box 8**, p. 74).

Song Types

The use of different song types, or different song parts, for interactions with males or females is perhaps the most interesting solution to the dual function challenge. Given that male rivals and potential mates are interested in different qualities of the singer, it is not difficult to understand why this has occurred. In some cases songs sung to males and females differ in complexity as well as type. The great reed warbler uses long songs for mate attraction and short songs for territory defense (Catchpole 1983). The aquatic warbler produces complex songs (known as C songs) to attract females, and shorter songs of one or two phrases to interact with rival males (Catchpole & Leisler 1996). Male dusky warblers sing an individually specific stereotyped song (S song), or an extremely complex, variable song (V song); the S song is used to guard the territory and interact with neighbors, and is produced frequently when the female is fertile, presumably to guard against potential extra-pair copulations (Forstmeier & Balsby 2002). Females choose a mate based on the V song and prefer those that are more complex.

In a few species different parts of the same song may perform different functions. In chaffinch song, the end flourish seems to be more important in mate attraction (Riebel & Slater 1998a), and the trill in male–male competition (Leitao & Riebel in press). Females prefer relatively longer flourishes and males respond more strongly to relatively long trills.

The song of the blackcap (**CD1 #17**) also consists of two parts, an initial warble and a terminal whistle (**Fig. 2.19**). There are indications from unpublished work of mine that the whistle is important in territorial conflicts and the warble in interactions with females. The whistle is louder and easily locatable; the warble is complex, quieter, hard to locate, and inaudible at a distance, and becomes shorter as the breeding season progresses, when there is no longer a need to attract a female. Thus the warble and whistle may perform different functions. In water pipits and barn swallows certain notes address a particular function. The 'snarr' in water pipits and the rattle in barn swallows both appear to have competitive functions (see p. 43). The rest of the song of both species is involved in attracting females. These examples give a flavor of the various ways in which males find efficient solutions to solve the problem of the dual functions of song.

Evolution Through Sexual Selection

Sexual selection has affected the evolution of song characteristics in all species. It has driven the evolution of some very complex patterns of singing behavior, such as interactive singing, in which the details provide information about the singer. This is thus a level above a simple sing–receive–react sequence of behaviors. Song learning itself may have evolved to allow males to sing different song types that are more stimulating to females and more intimidating to rivals. Sexual selection may also have affected processes of speciation; female preference for particular song types may have driven speciation in groups such as the indigobirds.

The same set of evolutionary pressures, attracting a mate and repelling a rival, result in different outcomes from species to species. For example, in the appraisal of individual differences by zebra finches, song rate has evolved as the main intraspecific song parameter used in female mate choice, whereas in the great reed warbler repertoire size is the most important choice

BOX 8

SPECIES AND MATE RECOGNITION AMONG CALLING GREBES

Western grebes and Clark's grebes are sibling species that often breed in mixed colonies. They both use special advertising calls to attract potential mates as the first step in pair formation (**CD2 #19–20**). This is followed by an elaborate courtship sequence that includes the striking 'rushing-on-water' and 'weed dance' displays. Displays of the two species are nearly identical, and male western grebes often engage male Clark's grebes in joint 'rushing' or 'barge-trilling' displays to attract the attention of females. In fact, until the mid-1980s, Clark's grebe was considered a color-phase of the western grebe. Storer (1965), however, reported strong assortative mating within the two types. In addition, spectrograms revealed that the advertising calls of Clark's grebes lack the distinctive 10–20 ms gap in calls of western grebes (Nuechterlein 1981; **A**). These calls are clearly involved in choosing mates rather than territorial defense, since birds court and form pairs outside the nesting colonies. Playbacks from floating blinds to actively courting birds revealed that males of mixed species populations readily distinguish the two call types based on whether a gap in the call is present. Early in the season, courting males of both types only answer and approach female calls of their own type. However, when an artificial 10–15 ms gap was spliced into the same Clark's female calls that they had been ignoring, western grebe males began answering and approaching as to western grebe female calls. Hybridization between the two species does occur, and follow-up experiments suggest an explanation. As the courting season progresses, late-courting western grebe males are much more likely to answer and approach playback calls from Clark's grebe females than earlier in the season. Hybrid pairings are most common late in the season, and are unlikely to be the result of mistaken species identity. Rather, individuals of both sexes may become less choosy as the pool of available mates decreases (Nuechterlein & Buitron 1998).

Although grebe advertising calls are not complex, they convey a wealth of additional information (**B**). Advertising calls of females are higher and longer than the calls of males, and playbacks show that unpaired males easily distinguish this difference. Playbacks to males that are courting in a crowded area demonstrated that they are able to instantly use the individual characteristics of an unpaired female's advertising call to answer her selectively and repeatedly while approaching from a distance. After pair formation, both sexes use the individual characteristics of their mate's advertising call to locate each other when separated. Paired birds are less vocal, and usually only answer playback calls from their own mate. Calls are given in bouts of 1–7 similar calls. When communicating with their mates, grebes tend to use shorter call-bouts (1–2 calls) than unpaired birds (3–7 calls). In playbacks with varying bout lengths, courting males were less likely to answer a given female call when it was presented to them in short rather than long bouts. Thus, variations in the structure of their rather simple advertising calls enable grebes to identify the species, sex, pairing status, and individual identity of courting birds. This may allow grebes to reserve their more spectacular and energetically demanding displays for demonstrating other qualities such as health and vigor.

Gary L. Nuechterlein & Deborah Buitron

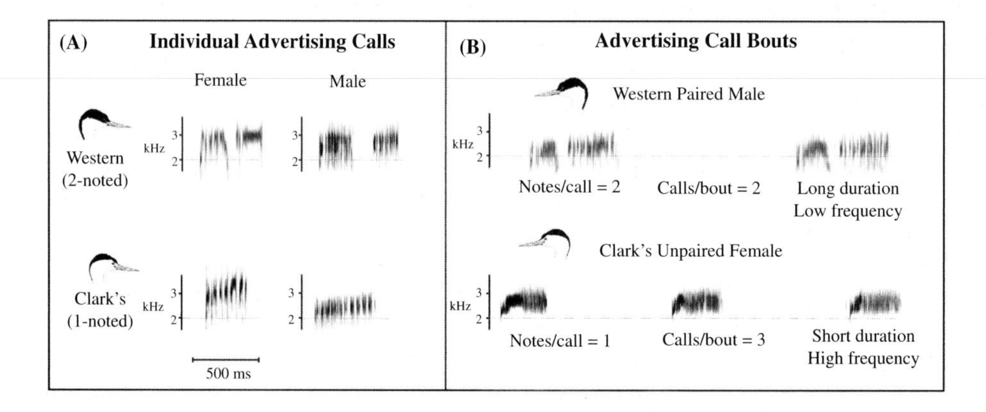

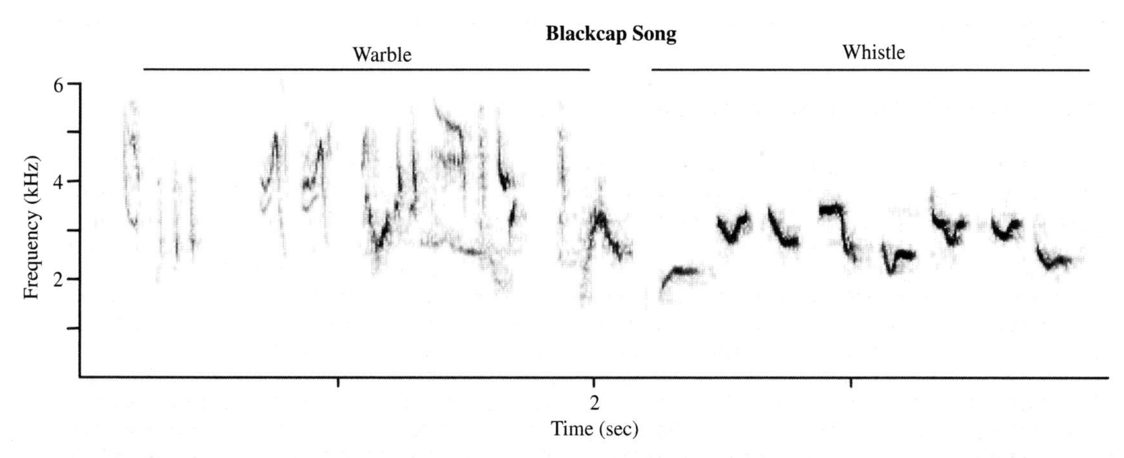

Figure 2.19 A song of a male blackcap showing the whistle and warble sections. The warble is quiet, whereas the whistle is loud and easy to locate (**CD1 #17**).

criterion. The structure, complexity, and temporal arrangement of songs is also affected by other aspects of behavior and ecology (Slabbekoorn & Smith 2002b). For example, species in which males can attract more than one mate tend to have more complex syllable repertoires (Read & Weary 1992). The great disparity between the most successful and least successful males in polygamous species adds to the pressure to attract females by performing virtuoso songs; the selection pressure for complexity is stronger than in monogamous species, in which, because the male is involved in rearing the chicks, females are more likely to base mate choice on signals indicating that the male will be a good parent (Catchpole 2000). Interestingly, song type repertoires (as opposed to syllable repertoires) are larger in species where males provide more parental care (Read & Weary 1992). Whether a species is migratory or not is another important factor. Catchpole (1980) suggested that migratory species have more elaborate songs because they are under pressure to pair early, defend a territory forcefully, and breed successfully in a short time, so that sexual selection is more intense. We need to study more species before we can determine more accurately just how selection pressures affect song patterns across species (Read & Weary 1992).

FEMALE SONG AND DUETTING

Generally speaking, males sing and females do not, because it is usually a male responsibility to attract a female rather than vice versa. Also, males are generally the sex that holds territories and competes with rivals for access to females. Of course, this assumes that the only functions of song are to defend territories or attract mates. However, in a number of birds females sing songs much like males (Langmore 2000). Female song has been found in at least 40 species; it is thought to be more common in tropical species, which are less well studied, so the true number of species in which there are female singers may be much higher. Females are generally harder to observe singing and, in a number of sexually monomorphic species, singing birds assumed to be male were subsequently found to be female. There are singing females in superb fairy wrens (Cooney & Cockburn 1995; **CD1 #18**), white-crowned sparrows (Baptista et al. 1993), blue-breasted waxbills (Collins pers. obs.), European robins (Lack 1965), long-tailed manakins (Trainer et al. 2002), dusky antbirds, and starlings, to name but a few. Langmore (1998, 2000) has outlined the situations in which females have been found to sing.

Functions of Female Song

Female song may have several possible functions. It may attract a partner or a male for extra-pair copulation, induce copulation with a mate, or aid in retaining a mate by maintaining the pair bond. Alternatively, females often have as strong a reason as males to defend their territory from intruders, who may compete with her for food for herself and her offspring, try to take over nest sites, or engage in extra-pair copulations with her mate. The potential functions of female defense differ from those of males only in that a territory is not obtained specifically to attract a mate.

One of the best known examples of a female singer is the European robin; females sing in the winter and defend their own territories. The song functions just as male songs usually do in other species (Lack 1965), to defend a territory in order to have sole access to the resources. Superb fairy-wren females sing to defend their territory, perhaps because males are always sneaking off for extra-pair copulations (Cooney & Cockburn 1995). However, the song of the female fairy-wren may encourage other males to visit her territory so she can engage in extra-pair copulations herself (**Fig. 2.20**). Female song functions in territory defense in several other species where males are frequently absent from the territory, as in fairy-wrens, or where food resources are scarce in the winter and females defend winter territories, as in robins and mockingbirds (Lack 1965). Female Australian magpies in communal groups sing to defend a territory against threats from other colonies (Brown & Farabaugh 1991; **CD1 #8**). Interestingly, in this species, the syllables of the female's song are more complex than those of the male.

In several polygynous species, female song is used in aggressive interactions between females as a form of mate defense. Red-winged blackbird females produce simple 'teer' songs in the context of female–female aggression (Yasukawa & Searcy 1982). The more females a male obtains in his 'harem' the fewer resources there will be for each, so it pays females to repel rivals.

The females of all three species of cordon-bleu finch sing occasionally (Goodwin 1982; Gahr & Guttinger 1986; Collins personal observation; **CD1 #19–20**). Males and females defend a small area around the nest site. The song appears to be less complex than that of the male (Collins personal observation; **Fig. 2.21**). The females seem to sing mostly before egg laying in the breeding season, but can sing at other

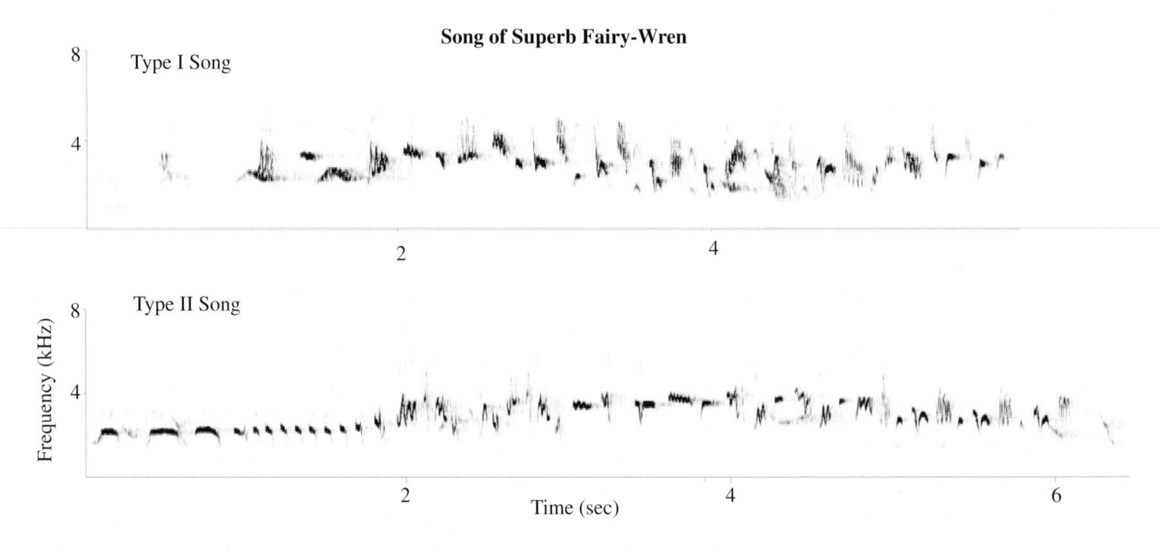

Figure 2.20 Superb fairy-wren territorial songs. Type I song uttered by both males and females; type II song used only by males **(CD1 #18)**.

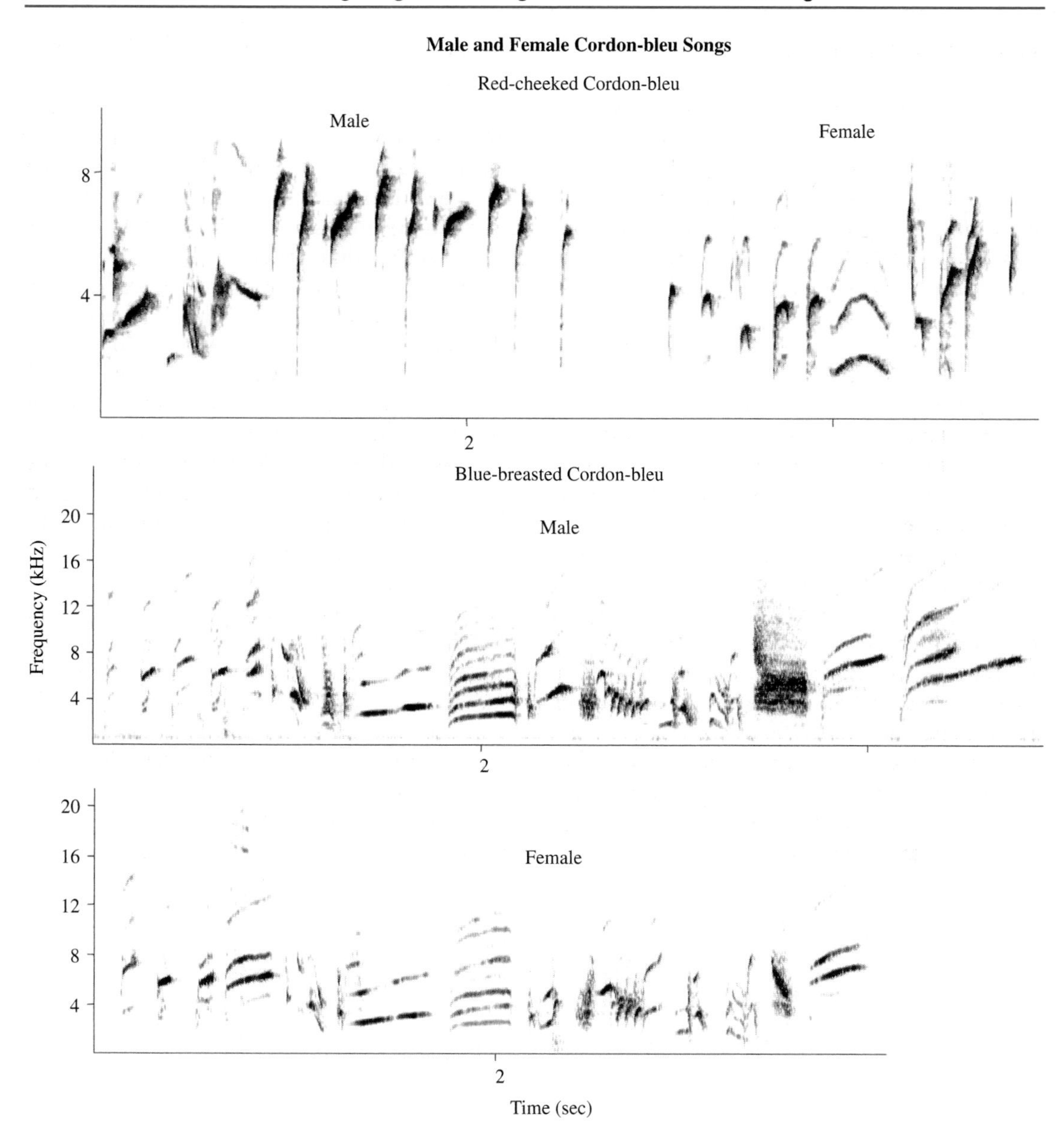

Figure 2.21 Male and female songs from the red-cheeked (**CD1 #19**) and the blue-breasted cordon-bleu (**CD1 #20**).

times of the year. The function is probably to maintain the pair bond or synchronize breeding (Gahr & Guttinger 1986). In slate-colored boubous (Sonnenschein & Reyer 1983) female song has been shown to synchronize breeding.

The only species for which there is evidence that female song functions in mate attraction is the alpine accentor (Langmore et al. 1996). Accentors breed in groups of up to four males and four females. Males feed the chicks of females they have mated with, so it is to a female's advantage to mate with all males in the group. When females are fertile there is intense competition through song to attract a male for copulation; playback of female song attracts males. The same is possibly true for the dusky antbird, in which females sing after losing a mate (Morton 1996a). Langmore (2000) suggests that this function of female song may be important in tropical species with a territory that is defended the year round; widowed females may need to attract a new mate without undue risk of losing their territory.

Duetting – Song in Synchrony

So far, I have described cases where males and females sing in bursts, or bouts, independently of each other. The song is directed at intruders, rivals, mates or potential mates, but is, essentially, done alone. There is another style, involving duetting, in which the songs of partners are interdependent.

Duetting occurs "when members of a mated pair sing in combination with one another, either synchronously or alternately" (Langmore 1998). The duet may function in territorial defense. It may be more effective than solitary defense, perhaps because each member of the pair can defend against intrusion by a same-sex intruder, as in dusky antbirds (Morton 1996a), or the Polynesian megapode (Goth et al. 1999). In other species, such as bay wrens, the duet may function in mate guarding (Levin 1996 a, b), but with different roles for male and female song. Levin found that the female part of the duet functioned to repel female intruders and the male part to guard his mate from extra-pair copulations.

There is one case of duetting involving two males. In the long-tailed manakin, pairs of males duet to attract a female, usually with a lead male and a follower. Male pairs may stay together for years and, over time, come to match each other's song in terms of frequency (Trainer et al. 2002). Eventually, the subordinate male may take over as the lead male.

More work is needed on female song and duetting. The function of female song has been studied in only a few cases. Duetting has received more attention, but again the function is not always clear. Generally female song and duetting appear to function similarly to male song; repelling rivals, defending resources, and attracting a mate. There could be a contrast between female and male song in the synchronization of breeding. We do not know what kind of female songs are more successful at attracting mates or repelling rivals. As for males, songs that are more complex, more energetically expensive, and more salient may be the most effective. As yet there has been little work on song learning by females. Do females learn from males, females, or their mate? Slate-colored boubou females learned their song from females and males learned from males (Wickler & Sonnenschein 1989), but whether this pattern recurs is not known. In general, the same kind of selection pressures appear to be acting on female song as upon male song.

CONCLUSIONS

The study of the functions of birdsong is one of the most active areas of research in behavioral ecology. Song is clearly important for both competition and attracting a mate. The need to advertise your quality in such a way as to convince receivers that they should mate with you or leave you alone has driven the evolution of song in a number of ways. Individuals within a species differ in the types of song they sing, and potential rivals and mates pay attention to those differences and act accordingly. The differences between individuals in song relate to differences in aggressiveness, male quality, and the resources made available. By using song to judge individual rivals and mates, birds avoid fights they would lose and obtain better mates. The specific parameters that signal quality differ between species, and song types and singing behavior may differ depending upon whether the song is addressed primarily to rivals or mates.

One common outcome of the need for singers to advertise is that songs become more complicated and versatile, and perhaps more costly to produce and thus more cheat-proof as signals. This will be true for interactions with

Chaffinch Wothe

Winter wren Elliott

Indigo bunting Elliott

Bullfinch Wothe

Plate II

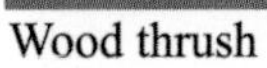
Wood thrush Elliott

Crested lark van Rossum

Song sparrow Elliott

Hedge accentor Wothe

males and females. However, there is still argument about why song repertoires appear to be most important to females for choosing a mate in one species whereas, in another, song rate is more significant.

To achieve a deeper understanding of the patterns and processes involved in the evolution of song, we need to integrate the study of sexual selection with the neurobiology of song learning, mechanisms of sound production, and the use of comparative studies to reconstruct the evolutionary history of song. I predict that some of the most exciting research to come will focus on female preferences for specific note types, and on the physiological and morphological factors affecting repertoire size. We have far to go before we fully understand all the nuances of birdsong as a way of flirting and fighting.

Chapter 3

Learning to sing

HENRIKE HULTSCH AND DIETMAR TODT

INTRODUCTION

The development of birdsong is among the most intriguing mysteries of nature. It is not always easy to decide what is more appealing to our ears: the elaborate melodies of an adult bird or the soft rambling of a youngster intoning long cadences of endlessly varying sounds. The early stages of song development remind us of the playful babbling of a human infant who obviously enjoys vocalizing without caring about the presence or absence of a companion. Research in both fields, song development and language acquisition, has shown that such apparently joyful and play-like activities are an expedient means for eventually mastering the orchestration of the voice, as well as other significant skills.

Above all, students of behavioral development appreciate the process of learning to sing as a chance to explore all of the intrinsic mechanisms that underlie the long journey from the first tentative experiments to eventual vocal expertise. For two reasons, this opportunity is almost unique in the animal kingdom. First, young birds, like newborn humans, are vocally active from an early age, permitting us to identify and measure their abilities as they progress. Second, birds, like us, develop their vocalizations guided by information that they extract from sounds heard early in life and then memorize, often for many weeks or months. Vocal learning has only been recorded in a few taxa of birds, and is extremely rare in other animals, including mammals, aside from cetaceans and some bats.

The development of singing in birds also ranks as a prime biological model for understanding how nature and nurture interact, and how they jointly affect the growth of behavioral competence. Research on this subject profits from contributions from many different disciplines, including neurobiology and genetics, as well as behavioral biology. There are further advantages. Behavioral studies offer the prospect of uncovering aspects of song development at three different levels. Besides an investigation of so-called developmental trajectories, including rules that describe the many transformations of a juvenile's vocal utterances, there are two kinds of reference patterns to which each vocalization can be compared; the stimulus patterns that a bird has experienced earlier, and the target songs that he finally performs as an adult.

Learning to sing takes some time. Usually it begins when a young bird leaves the nest, and then continues for several months. It comes to an end only shortly before the bird establishes his own territory and flags its turf by singing. Although some species also learn new songs later in life, the first year of song development is crucial in all songbirds. Their many accomplishments can be subdivided into two major stages: the auditory phase of song memorization, and the motor phase of song development.

Here we treat the various aspects of learning to sing in some detail, keeping as a framework typical biological concepts that are often used to explain the behavioral accomplishments of organisms, either as a consequence of evolutionary adaptation, or as mechanisms to solve a particular problem successfully. To give an example: young birds are often exposed to a great variety of different voices. How do they avoid memorizing 'wrong stimuli,' while managing to pick up and later develop the typical song patterns of their species? To address such questions, we will follow the time course of an individual's vocal development. First, we consider some problems that arise in the first few months of life, during a bird's natal summer, and then we discuss strategies that allow him to cope with issues that arise later. Our approach will be selective, focusing especially on results from recent birdsong research, without attempting to be comprehensive.

Most of what we know about learning and development of song comes from a small handful of bird species. They include three North American birds, the song sparrow, the swamp sparrow, and the white-crowned sparrow; one famous European bird, the common nightingale; and finally two domesticated birds, the zebra finch, originating from Australia, but today raised in many laboratories around the world, and the canary. These songbirds differ in a number of significant characteristics. The size of song repertoires, for instance, may range from one to some hundreds of songs. Species comparisons allow us to address questions about the nature of the memory mechanisms underlying these differences. Some species are sedentary, whereas others are migratory, providing another approach to the strategies that birds use in adapting vocally to new neighborhoods. In dealing with such diverse objectives, we will refer to representative species, including the chaffinch, the starling, and the marsh wren.

CHALLENGES IN A YOUNG BIRD'S WORLD: HOW TO PROCEED FROM HEARING TO MEMORIZING

We all agree that learning provides opportunities to do something new. However, some prerequisites must be met to ensure that learning is useful and free of risk. The mechanisms of associative learning are a good example. In associative learning, both animals and humans extract information about the regularities of the external world by using stimuli from the environment to predict important events (Pavlovian conditioning), or by using their own actions to achieve desirable consequences or prevent undesirable ones (operant conditioning). Thus, one reason why learning is so widespread and obviously successful is because it allows animals to adapt to a changing world in a dynamic way by evaluating the causal structure of their environment. Learning to sing, however, is a rather different accomplishment. Imagine a young bird equipped with a well-developed auditory system who, after leaving his nest, is exposed to a growing number of different acoustical stimuli. How does he decide which of these stimuli are worth memorizing and, later, which are suitable for reproducing as imitations? There are several behavioral adaptations and intrinsic mechanisms that help to cope with such problems.

Predispositions About What to Learn: Advantages of Being Selective

The risk that a fledgling's song learning is misled by nonspecies-typical auditory stimuli is minimized by a mechanism called pattern selectivity. Properties of this mechanism were first studied by Thorpe (1958) and later elaborated to a theoretical framework known as the "auditory template model of song development" by Konishi (1965a) and Marler (1976). Thorpe, a pioneer of the experimental research on song learning in birds, found that young chaffinches preferentially acquire conspecific song models. Based on a number of subsequent studies including some other species (Konishi 1965a; Konishi & Nottebohm 1969), Marler postulated an inherent predisposition to listen to and memorize species-typical sound patterns selectively. This concept implies that even birds without prior experience are able to recognize such patterns as relevant and to distinguish them from the song patterns

BOX 9

ACOUSTIC CUES FOR SPECIES-SELECTIVE SONG LEARNING

Songbirds with widely different songs share a common mechanism of song learning by cultural transmission. Young birds acquire song models from adult conspecifics and subsequently imitate them with varying degrees of fidelity. Two factors, social cues and genetic predisposition, interact to ensure that, in most species, of all the sounds heard during the sensitive phase, young songbirds learn the song of their species. Laboratory experiments show that a live singing bird is a potent tutor: young songbirds raised by adults of another species often learn the song of their foster parents. But when tape-tutored with an assortment of songs in social isolation, without exposure to live tutors, those birds that can learn from tapes will selectively learn the song of their own species. What acoustic cues direct this selective learning? This question can be addressed in laboratory tutor experiments using manipulated model songs.

The first such experiments, on song sparrows and swamp sparrows (Marler & Peters 1977, 1988b), showed that, like the acoustic properties of the species-typical songs themselves (**A; CD2 #21–22**), the cues that guide selective song learning by young birds vary across species. Swamp sparrows, whose song consists of a single trill, imitated only species-typical notes regardless of the syntax, or temporal pattern, in which these were presented. Song sparrows, whose song is more complex, imitated only species-typical notes from single trill songs, but imitated both song sparrow and swamp sparrow notes when these were presented in more complex, song sparrow-like syntax. Thus swamp sparrows rely on note structure as the primary acoustic cue for selective song learning, while song sparrows can use both note structure and the overall temporal pattern of song.

An acoustic cue for selective song learning has also been identified in the white-crowned sparrow (**CD2 #23–25**). This species' song consists of one or two pure-tone whistles followed by two to four phrases whose structure and sequence varies across subspecies and dialects. Because the introductory whistle is universal, it can provide a reliable cue for species-specific song learning. To test whether the whistle does in fact guide song learning, I tutored young white-crowned sparrows with three song types: (**B1**) normal conspecific, own-species song, but with the whistle removed, (**B2**) heterospecific, other-species song with an introductory whistle added, and (**B3**) corresponding heterospecific songs without added whistles (Soha & Marler 2000). The young birds preferentially learned heterospecific material with introductory whistles over conspecific material without whistles, confirming that the whistle does contribute to selective song learning in this species. One male white-crowned sparrow learned a 'song' made up of ground squirrel alarm calls with a whistle added (**B4**). In general, features of a species' song which are both universal and distinctive from those of sympatric species' songs are most likely to serve as acoustic cues for selective song learning.

Jill Soha

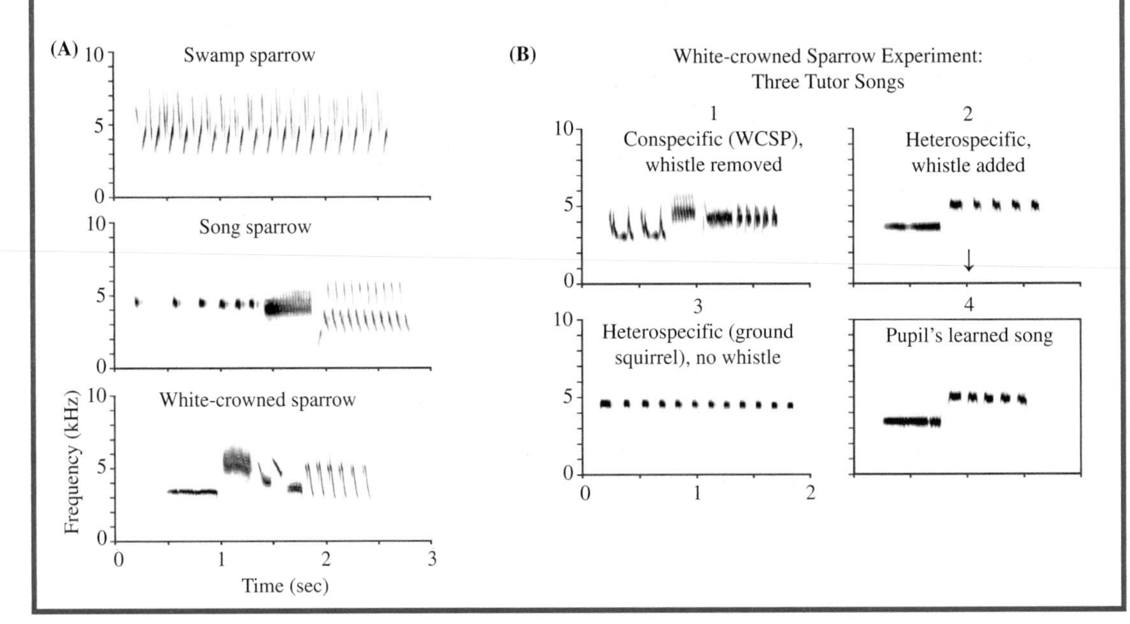

of other species. Experiments supplementing this insight have been conducted in young swamp sparrows that were presented with a choice design. Swamp sparrows were allowed to hear songs of their own species and also of two other species, the closely related song sparrow and, for a control, the domestic canary (Dooling & Searcy 1980). The study showed that only species-typical songs were processed in such a way as to elicit clear physiological responses, in this case a change in heart rate. Since that time, more and more evidence has accumulated for the crucial role that such predispositions play as guidelines escorting a bird through the first steps of song learning (Marler & Peters 1980; **Box 9**, p. 82).

Sensitive Phases: The Time for Easy Learning

Listening and responding to the singing of conspecifics is not a year-round activity, but has a clear seasonal distribution. There is also an age-related pattern in a bird's readiness to memorize and then to develop its song vocally. With some exceptions, the first year is the most important time for song learning in a bird's life. Within this year, there is usually an early and limited period of time when birds are especially receptive to song stimuli and when they memorize them most readily. This period is called the 'sensitive phase.'

The discovery that perceptual aspects of song

BOX 10

STARLING SONG: AN OPEN-ENDED LEARNER

European starlings are eccentrics in the birdsong literature, thanks to the unique structural and developmental features of their songs, among the most complex known. They sound like a random assortment of whistles, mews, clicks, and screeches, with the occasional sneeze or car alarm. But behind this exuberant cacophony, all starling songs conform to a common set of rules (Eens 1997). A single song can last up to a minute. It consists of a succession of motifs, also termed phrases, lasting about 0.5–1.5 s. Each motif tends to be a variation or a repetition of the one before. Each adult male has a unique repertoire of about 60–80 motifs, if one counts all the variants. The song usually begins with long low-pitched whistles separated by pauses. These are followed by a sequence of variable motifs, often incorporating imitations of other species or mechanical sounds. Next come a series of rapid click-like beats that sound like a speeded-up machine gun. The song concludes with loud, high frequency shrieks (see **Fig. 2.12** p. 60), often accompanied by enthusiastic wing waving. Both males and females sing, but female songs are less frequent and less structured.

Perfecting these songs is the work of a lifetime, or at least of several years. Starlings are true open-ended learners–that is, they not only alter their motif repertoires from one year to the next (Eens 1997), but they also memorize and reproduce motifs they hear for the first time as adults. This was established by exposing young birds to a changing succession of tape recordings or live tutors in a controlled acoustic environment, and monitoring their singing over the course of two years (Chaiken et al. 1994). In contrast with canaries, which develop new repertoires every spring, starlings can apparently memorize and produce new songs at any time of year (Böhner et al. 1990). They can even acquire new motifs after spending their first year in acoustic isolation (Chaiken & Böhner unpublished).

Developmental studies raise intriguing questions about the roles of nature and nurture in song acquisition. Starlings, like most other songbirds, develop highly abnormal songs when raised in social and acoustic isolation (Chaiken et al. 1993). It was initially assumed that any species-typical features that were lacking in the songs of isolates must normally be acquired by imitation. However, young starlings seem to develop a good approximation of normal song under any conditions other than total individual isolation, a situation that is reminiscent of what was found some years ago in Arizona juncos (Marler 1967). More or less normal song was developed by starlings that were either group-reared and exposed to recordings of nightingale song (Böhner & Todt 1996), reared in pairs with no exposure to adult song (Chaiken et al. 1997), or even reared with a devocalized companion (Chaiken 2000). It appears that individual birds possess the capacity to develop more or less normal song, but they need the social stimulation that companions provide to realize their full potential.

MarthaLeah Chaiken

acquisition are linked to a sensitive phase had great impact on the progress of learning research, and drew attention to similarities between sexual imprinting and song learning. Although these two kinds of learning serve rather different functions, the formation of a social preference in one case, and the acquisition of behavior for communication in the other, they share a number of common mechanisms that make them apparently distinct from other types of learning such as Pavlovian and operant conditioning (Immelmann & Suomi 1981; Slater et al. 1988; ten Cate 1989; Slater et al. 1993; ten Cate et al. 1993). For example, both sexual imprinting and song learning are biased towards species-typical signals, both need relatively few exposures to stimuli (Bischof 1997), and both use highly specialized neural circuitry to acquire and store stimulus representations.

Inquiries into the circumstances in which song learning occurs also led to improvements in the design of learning experiments in hand-raised birds, which in turn revealed further facets of the song learning process (Konishi 1985; Marler 1987). Today we know, for example, that the sensitive phase of song acquisition is a kind of time window that facilitates stimulus memorization during a species-typical age period and restricts the ability to learn new songs later on. The window typically opens around fledging, when appropriate song stimuli are perceived, and closes after a limited amount of time. The closing of sensitive periods is not merely due to a clock-like mechanism that constrains learning after reaching a certain age. Rather, the evidence suggests that the accumulation of a certain amount of experience with the song patterns to be learned plays a crucial role. This has been shown by experiments where males were prevented from hearing song stimuli early in life. In several species, such deprivation consistently resulted in a prolonged ability to acquire songs (Eales 1987; Nottebohm 1989; Slater et al. 1993; Nelson 1997). Flexibility in the timing, especially with regard to the closure of sensitive phases, is a biologically significant characteristic of the song learning process, allowing birds to cope with some of the variation and uncertainty that exists in their social and ecological environments. In addition to these general properties, sensitive phases show considerable variation across species, as correlates of their particular ecology (**Fig. 3.1**).

The timing of sensitive phases provides a further mechanism designed to help a young bird to avoid learning the wrong song. There is coordination between the timing of sensitive phases and the time when adult songsters of the bird's own species are vocally active. In functional terms, it is expedient to specify a time to listen and memorize song models for the development of the bird's own song repertoire.

Depending on the properties of the sensitive phases for song learning, songbirds can be subdivided into two different groups: closed-ended learners that are age-limited and open-ended learners. The first group is well studied and encompasses song sparrows, swamp sparrows, white-crowned sparrows, chaffinches, and zebra finches. For the second group, in contrast, detailed studies are available only for a few species, especially starlings (Eens et al. 1992; Chaiken et al. 1994; **Box 10**, p. 83), nightingales (Hultsch 1991a; Todt & Geberzahn 2003), and canaries (Marler & Waser 1977; Nottebohm & Nottebohm 1978; Nottebohm et al. 1986); even with apparently age-limited species, there are sometimes signs of at least the potential for adult song plasticity (**Box 11**, p. 86). Besides investigations conducted in the laboratory, a number of field studies suggest that supplementary learning may be rather widespread among oscine birds. The debate about limitations on when songs can be learned is an interesting one, as we will review later (see p. 98).

Social Influences: Who to Attend to

In addition to a predisposition to learn preferentially song patterns with a particular species-typical acoustic morphology, social experience can have a significant impact on the memorization of songs (Baptista & Petrinovich 1984, 1987; Baptista 1996). For instance, indigo

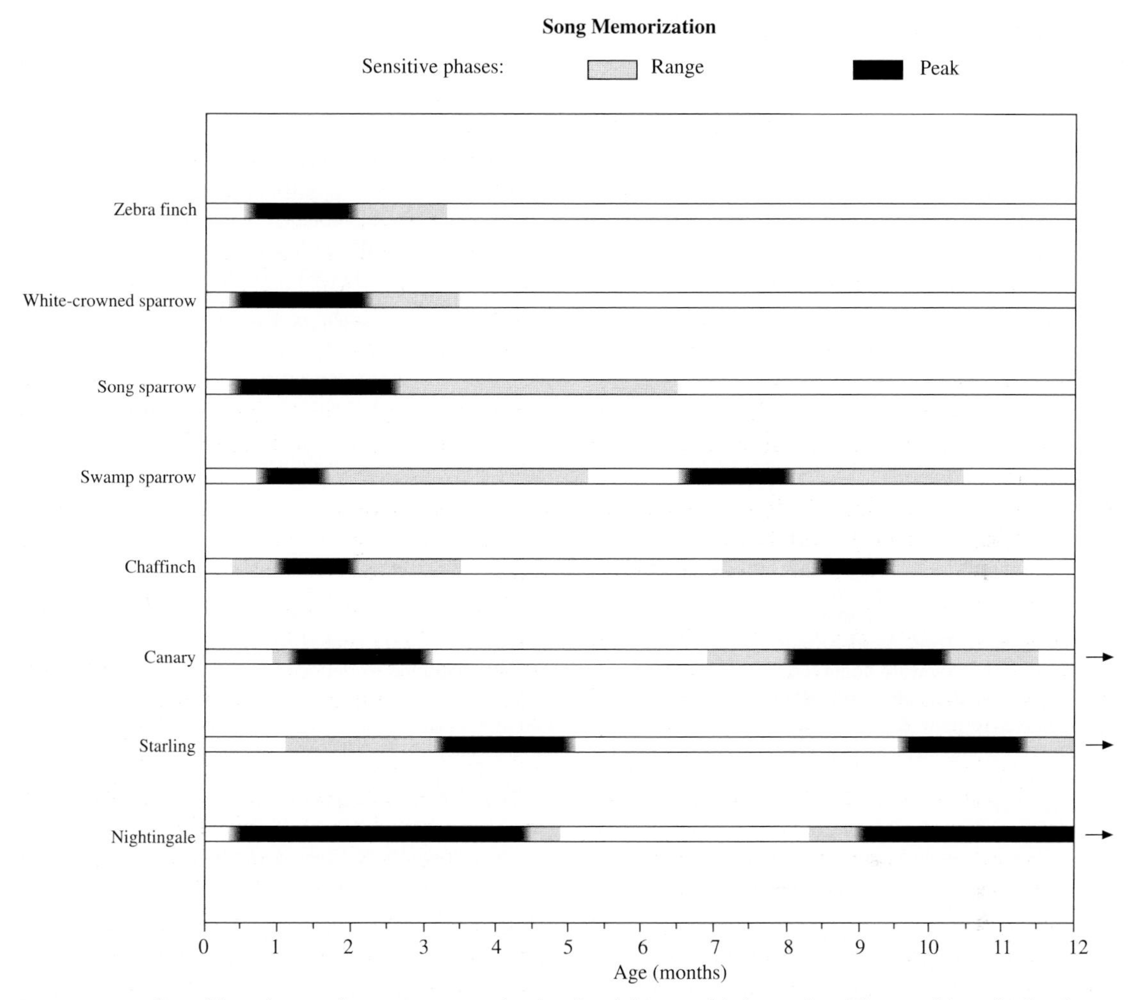

Figure 3.1 Sensitive phases for song memorization in eight songbird species. Those with a single phase are given first. References (top–down): (1) Böhner 1990; Zann 1997; (2) Marler 1970a; Petrinovich & Baptista 1987; Nelson et al. 1995 (3) Marler & Peters 1987; Nordby et al. 2001; (4) Dooling & Searcy 1980; Marler & Peters 1988a; (5) Thorpe 1958; Slater & Ince 1982; (6) Marler & Waser 1977; Nottebohm & Nottebohm 1978; (7) Chaiken et al. 1994; Böhner & Todt 1996; (8) Hultsch & Kopp 1989; Todt & Böhner 1994.

buntings and zebra finches can acquire their songs from the father, flock members, or from territorial neighbors (Payne 1981; Böhner 1983; Clayton 1987; Williams 1990). Such social selectivity is affected by prior experience with individual conspecifics, or individuals of other species that are accepted as tutors by mimics (Chaiken et al. 1993; Todt & Böhner 1994). Factors contributing to these effects include parental care by a potential tutor, aggressiveness, or just visual and acoustic contact (Todt et al. 1979; Kroodsma & Pickert 1980; Payne 1981; Böhner 1983; Clayton 1987; Eales 1987; Williams 1990). Prominent among the prerequisites of tutor acceptance is the age of the pupil, whereas some biological properties of the tutor seem to be less important. Young bullfinches (Nicolai 1959), nightingales (Hultsch & Kopp 1989) and starlings (Böhner & Todt

BOX 11

HOW 'CLOSED' ARE CLOSE-ENDED LEARNERS?

Songbirds are traditionally categorized as open- or close-ended learners. Open-ended learners, such as canaries and starlings, continue to learn new songs throughout life. Close-ended learners, such as song sparrows and zebra finches, appear to learn song once and then repeat these songs in a stereotyped or crystallized manner for the rest of their lives. Notice that this close-ended learning can be regarded in two ways, a limited time in which to learn song, and a rigid motor program. The limited time period in which to learn song may be analogous to our own critical period for learning language. But when we humans speak, we do so with exquisite non-stereotypy. Given an increase in background noise, we can alter our voice's amplitude (the Lombard Effect). We rely on accurate feedback. When we hear our own speech slightly delayed we often stop, hesitate, and stutter. One traditional view of close-ended singers has been almost the opposite, following on classical ethological notions. Once launched, song production is a consummatory action. As with sneezing or swallowing, the act, once begun, is almost impossible to stop. As one researcher put it, "when all aspects of song, including its duration, frequency range, and the form of the component sounds, become fixed, the song is said to be crystallized" (Konishi 1989).

The past decade or so has produced evidence that whatever is close-ended in some songbirds may be more than a little ajar. The first experiment addressing this issue showed that zebra finches would sometimes interrupt song when disturbed by a strobe light (Cynx 1990), but the birds almost always completed individual syllables first. This suggests that syllables are units of song, a finding replicated in at least four other avian species: nightingales (Riebel & Todt 1997), pigeons (ten Cate & Ballintijn 1996), and song and swamp sparrows (Cynx unpublished). Then, Nordeen and Nordeen (1992) reported that adult zebra finch's songs degraded weeks to months after they were deafened. Williams & McKibben (1992) also reported evidence of song plasticity, showing that zebra finches, after undergoing peripheral nerve cuts to the syrinx, produced reconfigured song elements as the nerves re-grew. Evidently there is some sort of auditory feedback loop between song production and perception; adult zebra finches actively listen to song and make adjustments as they produce it. Zebra finches and budgerigars not only monitor their song production, but can immediately regulate the amplitude of song as background noise varies (Cynx et al. 1998; Manabe et al. 1998). Leonard & Konishi (1999) showed that, confronted with chronic delayed auditory feedback, zebra finches would display disrupted song after a few weeks. We then showed that in zebra finches, like humans, there were both immediate and transitory effects (Cynx & von Rad 2001). Finally, zebra finches will also alter the amplitude of their songs and calls based on the social context (Cynx & Gell 2004).

It is clear that adult song is a highly dynamic and closely monitored act. In these regards, it has all the characteristics of the way humans speak. However, so far as is known, close-ended learners cannot learn new song elements from a model. A demonstration to the contrary would leave close-ended learning not only ajar, but decidedly open. The Williams and McKibben results hint at this possibility. Similarly treated birds deprived of a brain circuit known to be essential to song learning fail to show the reconfiguration of song elements (Williams & Mehta 1999). If a way can be found to encourage adults of these species to learn new songs from a model, it would have important implications, suggesting that even close-ended learners retain the mechanisms for song learning throughout life.

Jeffrey Cynx

1996) readily learn songs even from a human, provided that hand-raising has begun before they open their eyes.

The significance of social factors varies remarkably among oscine birds. This has been shown by exposing young birds either to a live tutor or to playbacks of tape-recorded song models presented to birds in acoustic isolation. In some species, such as song sparrows and swamp sparrows, tape tutoring can be as effective as live tutoring (Baptista & Petrinovich 1984; Marler & Peters 1987, 1988a). In others, such as canaries and starlings, the amount of song material acquired is much smaller when birds are kept isolated and hear only tape recordings (Marler & Waser 1977; Chaiken et al. 1993). Finally, there are species, such as tree-creepers of the genus *Certhia*, which do not seem to learn at all from tape recordings (Thielcke 1970a).

Birds differ in whether they impose more complex social demands which, in turn, may change with age or experience. Shifts in social selectivity have been documented for white-crowned sparrows, starlings, and nightingales. White-crowned sparrows, for example, will learn readily from tape until 50 days of age (Marler 1970a), but thereafter will only accept live tutors as song models (Baptista & Petrinovich 1986). In contrast, four-month-old starlings learn better from live tutors than from tape recordings alone, and tape tutoring is more effective at 12 months than at four months of age (Chaiken et al. 1993). A similar change can be observed in nightingales. During the first part of their sensitive phase, from day 13 to day 40 post hatching, nightingales learn only if their familiar, live tutor is present during a playback presentation; but later on, and especially after having experienced typical nightingale song, they also memorize species-typical songs heard when they are alone (Todt et al. 1979; Todt & Böhner 1994). One reason for such differences between white-crowned sparrows and nightingales may be the clear differences in repertoire size, imposing weaker or stronger constraints on the selectivity of learning. Individual white-crowned sparrows rarely sing more than one song as adults (Baptista 1975), while the normal repertoire of a nightingale consists of up to 200 different song types (Hultsch 1980). In this species, it might be too restrictive to limit learning to one or two familiar tutors if a bird is still to develop a large repertoire. Nevertheless, social selectivity early in life may serve to refine global innate preferences for the type of songs which a male will acquire later on and subsequently attend to as an adult.

The range of variation in the prerequisites for song learning, as studied under laboratory conditions, prompted detailed research into the mechanisms of socially selective learning. These studies, in which the quality, amount, or timing of social stimulation was monitored or manipulated, suggest that the concept of social selectivity should be revisited. It is perhaps best conceived of as a graded phenomenon, having the effect of learning 'more' or 'better' from a social exposure without necessarily initiating learning (Böhner 1983; Baptista & Petrinovich 1984; Eales 1985; Clayton 1987; Williams 1990; Chaiken et al. 1993; Slater et al. 1993; Todt & Böhner 1994). Moreover, the wide range of contextual variables found to influence selective learning, such as parental care, aggression or proximity of the tutor, leads us to question whether it is based on a truly social mechanism designed, for example, to tag song stimuli as significant or nonsignificant depending on the tutor's presence or behavior. Rather, learning could ensue because of the activation of perceptual mechanisms that, in turn, make young birds more aroused or more attentive to the tutor's vocalizations. In an attempt to pin down factors associated with selective learning, Hultsch et al. (1999b) showed that young nightingales preferentially acquire songs that were paired with a stroboscope that flashed synchronously with the playback of particular songs (**Fig. 3.2**). The fact that birds acquired more songs and copied them better with this regime (in the presence of both a social/human tutor and strobe), than in the control situation (a social/human tutor with no strobe), provides corroborative evidence that the manifestation of social selectivity in song learning may indeed be mediated or at least influenced by basic psychological mechanisms such as attention or arousal.

Growing Up in a Natural Environment: When Many Factors Matter

Most of what is currently known about the mechanisms that help young birds to acquire their species-typical songs came from studies conducted under the standardized conditions of a laboratory. Animals are adapted to particular biological environments, with all of their complexities and uncertainties, and the relevance of laboratory-based findings to song development under completely natural conditions is not always clear (Beecher 1996). On the other hand, in a given bird's natural environment there are many different factors, often difficult to tell apart and

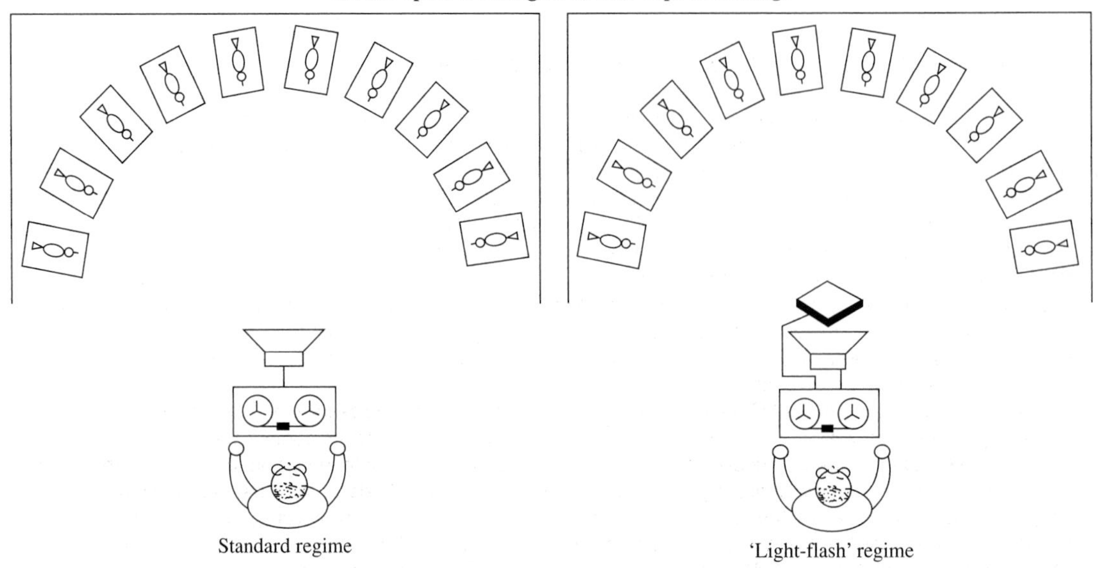

Figure 3.2 Sketch of two tutoring sessions. Left: birds hear particular songs from a tape player connected to a loudspeaker. Right: birds hear songs presented with stroboscopic light-flashes. Birds such as nightingales, starlings, and song thrushes learn such 'strobed' songs, provided the human present is their normal caretaker and has hand-reared them.

frequently overlapping in time, which can affect learning accomplishments. Laboratory studies make it possible to investigate them separately, but with an obvious trade-off. An expedient way to deal with this problem is to re-examine conclusions drawn from laboratory research in the light of subsequent studies in the field.

Some examples illustrate this approach. There are several subspecies of white-crowned sparrows that differ in their ecology and especially in their migration habits (Nelson et al. 1996, 2001; Nelson 2000a;). They also differ in how songs vary geographically (**Box 12**, p. 89). In sedentary populations the songs developed by neighboring males are very similar; but in migratory populations the songs of territory neighbors were no more similar to each other than to the songs of non-neighbors. It was inferred that the distinction in the annual cycle between sedentary and migratory habits somehow determines the pattern of vocal learning among territory neighbors. However, the most important difference between subspecies was found, not in the birds' sensitive phase for song memorization, but in their sensorimotor development, during the stage of selective song attrition, a topic we return to later (see p. 93; **Box 14**, p. 94).

Another example was recently documented for cross-species imitations in the nightingale and its twin species, the thrush nightingale. In some parts of Europe, where the two species occur sympatrically, thrush nightingales readily imitate nightingale songs (Mundry 2000). Nightingales, on the other hand, have never been found to copy the other species' songs. Under laboratory conditions, however, they can be persuaded to imitate thrush nightingale songs quite accurately, provided they hear the songs from a tutor who has previously presented them with species-typical songs (Hultsch et al. 1999a). Corroborative findings have also been reported for bullfinches. In the field, male bullfinches produce a very simple song. But when trained

BOX 12

VARIATION IN SONG LEARNING AMONG WHITE-CROWNED SPARROWS

Luis Baptista described the white-crowned sparrow of North America as "the white rat of ornithology." His frequent collaborator, Barbara DeWolfe, deserves much of the credit for stimulating six decades of research on variation in its reproductive physiology and behavior (Blanchard 1941). Ornithologists recognize five subspecies that differ in their geographic range and appearance, as well as their song, ranging from Arctic–alpine habitats to the mild conditions of coastal chaparral in California (**CD2 #26–30**). DeWolfe took advantage of this geographic variation to apply what biologists term the 'comparative approach' to study how the physiological preparations for breeding differ between four subspecies, one that is sedentary in its habits, the Nuttall's white-crowned sparrow; another that is a short-distance migrant, the Puget Sound white-crowned sparrow; and a third, a long distance migrant, the Gambel's white-crowned sparrow. A similar approach has been taken in studies of song development in these three subspecies and a fourth, the mountain white-crowned sparrow, another short-distance migrant. We lack equivalent data on the eastern subspecies.

With the striking exception of Gambel's sparrow, males in these subspecies form dialects; the songs of birds in one local area are similar, and differ from those of more distant birds. What is responsible for this contrast? By combining experiments in which nestlings of each subspecies are brought into the laboratory, hand-reared, and tutored with tape-recorded sounds, together with field observations of birds in their natural habitats, several intriguing differences between the subspecies have emerged. The sedentary Nuttall's sparrow, which lives year round on the coast of California, has a much longer sensitive phase for song memorization than any of the migratory subspecies (Nelson et al. 1995; Nelson 1999). Perhaps because young males do not disperse far from where they learn their songs during this extended period, the songs of territory neighbors are much more similar to one another than is the case in the migratory subspecies (Nelson et al. 2001). The short-distance migrants, the mountain and Puget Sound sparrows, also form vocal dialects, but the degree of resemblance between the songs of territory neighbors is less than in the Nuttall's sparrow. Male Puget Sound sparrows, and possibly the mountain white-crown as well, are much more likely to learn several dialects early in life, overproduce song when they arrive on the breeding grounds, and choose the best match available to them based on what their territory neighbors sing (Nelson 2000c). Because migratory males disperse farther than their sedentary cousins, thereby rendering the future site of breeding somewhat unpredictable, learning several dialects may provide males with the flexibility to conform to a wider range of dialects when a territory vacancy arises. In the migratory Gambel's sparrow, on the other hand, the lack of dialects and the low degree of song sharing among territory neighbors seems to be a consequence of its late and short breeding season and low site fidelity. By the time males arrive on their far northern breeding grounds, the song overproduction phase has already passed, and males lack the ability to choose a song that best matches their neighbors' song, as their relatives do breeding at lower latitudes. Studies such as these suggest that the process of song learning is flexible, and can respond adaptively to variation in local ecological conditions.

Douglas A. Nelson

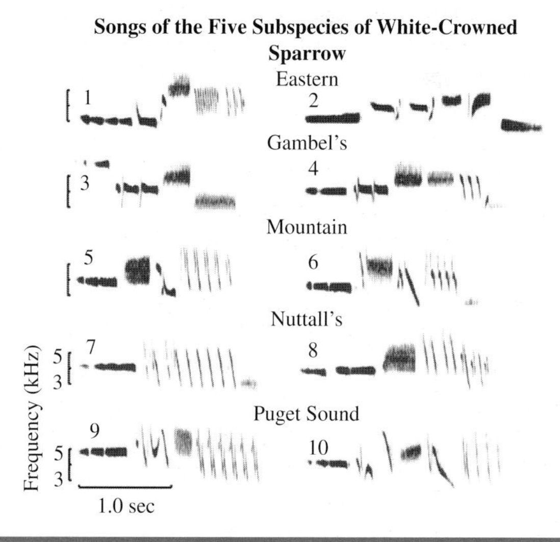

by a social tutor who presents them with more complex learning stimuli than those from their biological father, the birds will readily memorize and later perform song patterns that, instead of being three to five notes long, contain series of more than 30 elements (Nicolai 1959). They are so skillful at this that they were kept as pets in the past, and trained to whistle folksongs (see Chapter 1, p. 37; **CD1 #10**). There were even scores published, to be played on the flageolet for teaching to bullfinches (Thorpe 1955). Findings like these demonstrate the extent to which natural behavior can be radically transformed by changing the circumstances in which a young songbird grows up.

PROPERTIES OF SENSORIMOTOR LEARNING: HOW TO ACHIEVE VOCAL EXPERTISE

The ontogenetic development of singing shows a number of characteristic traits that are widespread across oscine birds (Marler 1991). For example, in the typical songbird, the early phase of auditory learning usually precedes the phase of vocal learning by an interval of several weeks. Such observations have led to the notion of song acquisition as a two-stage learning process, one to attain an auditory representation of the songs that are heard, and one to attain a motor program translating the memorized sound patterns into behavior. During the song memorization phase, while listening to song, birds may themselves vocalize not only during food begging, but also with the soft and variable sounds of subsong. Although subsong may be stimulated by hearing song, its acoustic structure seems not to be specifically related to that of the stimuli. The next phase of vocal development ushers in a new stage in a bird's life, often covering a span of several weeks or months. Vocal development can be subdivided into three prominent stages: subsong, plastic song, and crystallized full song (Marler & Peters 1981, 1982a; **Fig. 3.3**).

Subsong: The Riddle of Amorphous Vocalizations

Early in life, all songbirds perform coherent sequences of special vocalizations from which mature song emerges only gradually, with many changes in timing and structure. First to emerge are soft, rambling sounds which are classified as 'subsong' (**Fig. 3.4; CD1 #21**). In structural terms, subsong is never the same from moment to moment, and thus rather difficult to analyze. This may explain why detailed studies on subsong are lacking to date. New digital techniques promise to help with this problem (Tchernichovski et al. 2000). The great variability of subsong invites comparison with the playful activities of young mammals, or the vocalizations of human babies before they start babbling. For students of birdsong the occurrence of subsong remains a puzzling matter. The most plausible explanation is that it serves to train the vocal apparatus and thus to improve sensorimotor control, perfecting the art of two-voiced sound production (see Chapter 9). Subsong gradually merges into plastic song, the next phase of song ontogeny (**Figs. 3.5, 3.6, 3.7; CD1 #22–24**)

Plastic Song: Overproduction and Rules for Getting the Song into Shape

We know that plastic song has started when rehearsal of previously memorized song patterns begins, and birds shift gradually to more stereotyped vocalizations. As the phonetic morphology of plastic song is elaborated, the first precursors of acquired imitations emerge. Increasing numbers of syllables and phrases can be discerned, recurring in a similar form, making it possible to arrange them in categories and to make measurements. The progress in song quality can be documented with measures such as pattern stereotypy, vocal amplitude, and repertoire size (Marler & Peters 1981; Podos et al. 1999; Brumm & Hultsch 2001; Hughes et al. 2002).

The study of developmental trajectories is a particularly promising approach (Clark et al. 1987; Tchernichovski et al. 2001). For example,

BOX 13

THE ROLE OF TESTOSTERONE IN SONG LEARNING

Male birdsong is generally regarded as a typical secondary sexual characteristic, under the control of steroid hormones from the gonads. It is associated with testis growth and regression. Male song usually waxes and wanes seasonally with the fluctuation in testosterone levels in the blood. There are testosterone receptors in the syrinx. Song decreases after adult castration, and testosterone treatment reinstates it. The administration of testosterone induces song in females, and augments it in intact males. In the field, the testosterone profiles of males in spring correlates well with seasonal song development, although when they peak there seems to be a reduction in vocal plasticity (Nottebohm 1993). Testosterone has long been viewed as a major factor in song acquisition as well as production, presumed to act either directly on the brain, or after aromatization into estrogen within the brain. The dogma was so well entrenched that when the first report was published of young male zebra finches (9 males) learning to sing normally after castration, it was greeted with skepticism. Despite his careful precautions, suspicions lingered that Arnold (1975) might have left fragments of the testis behind. Then Kroodsma (1986) reported a similar result with marsh wrens (3 males) that was again met with a degree of skepticism. Only when ethologists and endocrinologists joined forces, and confirmed that testosterone was absent from the blood of male castrates, were critics finally convinced (Marler et al. 1988). Their subjects were song and swamp sparrows, castrated at 3–4 weeks of age (17 males) and then tutored with a changing program of tape-recorded songs. Almost all birds learned from the tutor tapes (**A1** & **A2**). They came into song about a month later than normal (**B1** & **B2**), passing through subsong, then plastic song. They progressed far enough to tell that learning had taken place, and had occurred at the normal time. Rather than crystallizing song, however, they persisted with plastic song, and then regressed (**A3** & **B3**), much as reported earlier by Arnold and Kroodsma for castrated zebra finches and marsh wrens. Five castrated swamp sparrows were then treated with testosterone. Within 1–3 weeks they all produced normal crystallized song. It seems that testosterone is not required for song memorization and the early stages of production, but is needed for song crystallization. Hormone assays confirmed that all but three of the castrations were complete. Unexpectedly, estradiol was found in the blood of both castrated and intact male sparrows, suggesting that it might have a role in song learning. The source of the estradiol is not known, but there are indications that it may originate within the brain itself (Schlinger & Arnold 1991).

Peter Marler

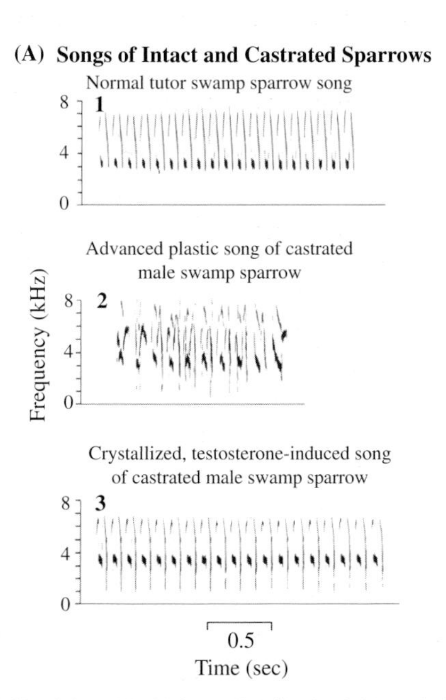

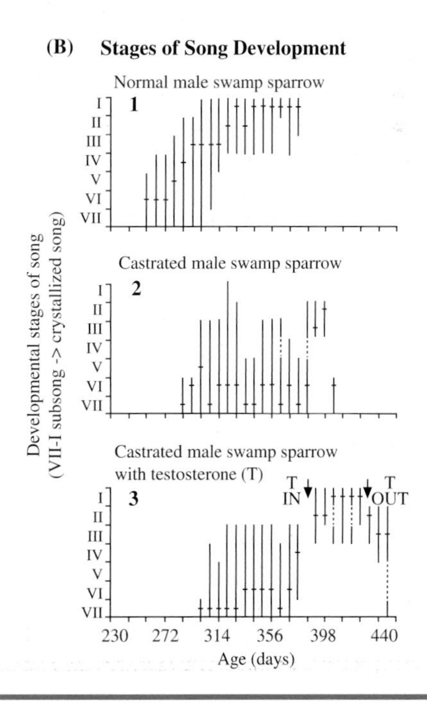

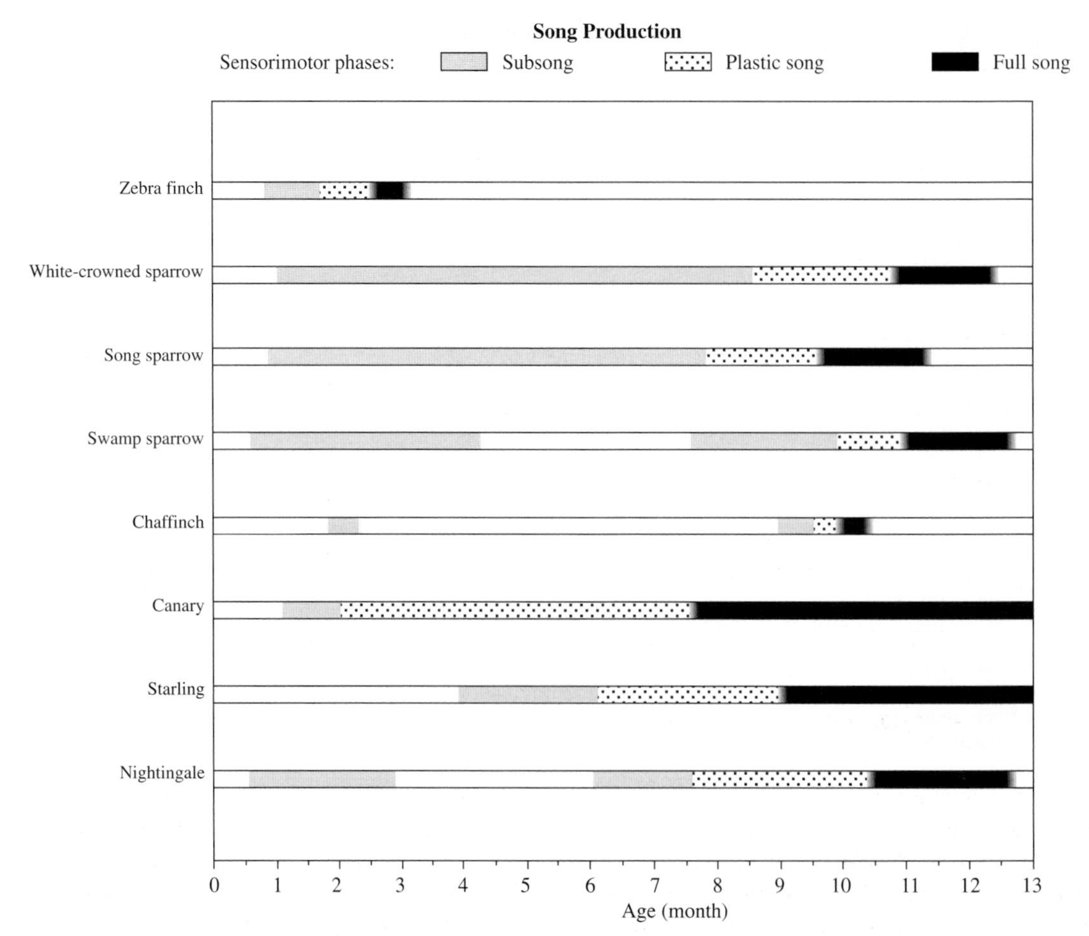

Figure 3.3 Sensorimotor phases of eight songbird species. References (top–down): (1) Zann 1997; (2) Nelson et al. 1995; (3) Marler & Peters 1982a; (4) Marler & Peters 1982b; (5) Nottebohm 1971a; (6) Nottebohm 1985; (7) Chaiken et al. 1993; (8) Hultsch 1985.

it has been shown that in some birds, note and syllable structure take on adult form ahead of song syntax. Thus, for some time after achieving the stereotyped production of song constituents, birds may still exhibit plasticity in arranging these constituents into sequences. They may sing incomplete songs, with some song constituents missing, or a typical succession of song phrases may be inverted. Another finding is the coupling of different trajectories during ontogeny, as with changes in vocal amplitude and pattern quality. As they proceed through plastic song, birds usually sing not only better but also louder. Exploring this relationship further, Brumm and Hultsch (2001) recently found something remarkable in young nightingales: besides long-term developmental coupling between loudness and song stereotypy, there are also clearly correlated changes in song quality and volume on a short-term basis. Even within a given performance that contains song material of different pattern quality, the correlation with loudness remains significant.

Song Crystallization: Rapid Progression to Adult Song

Good things take time, but important things may be a matter of moments. This is the case with the timing of song development, if we compare the long period of premature vocal

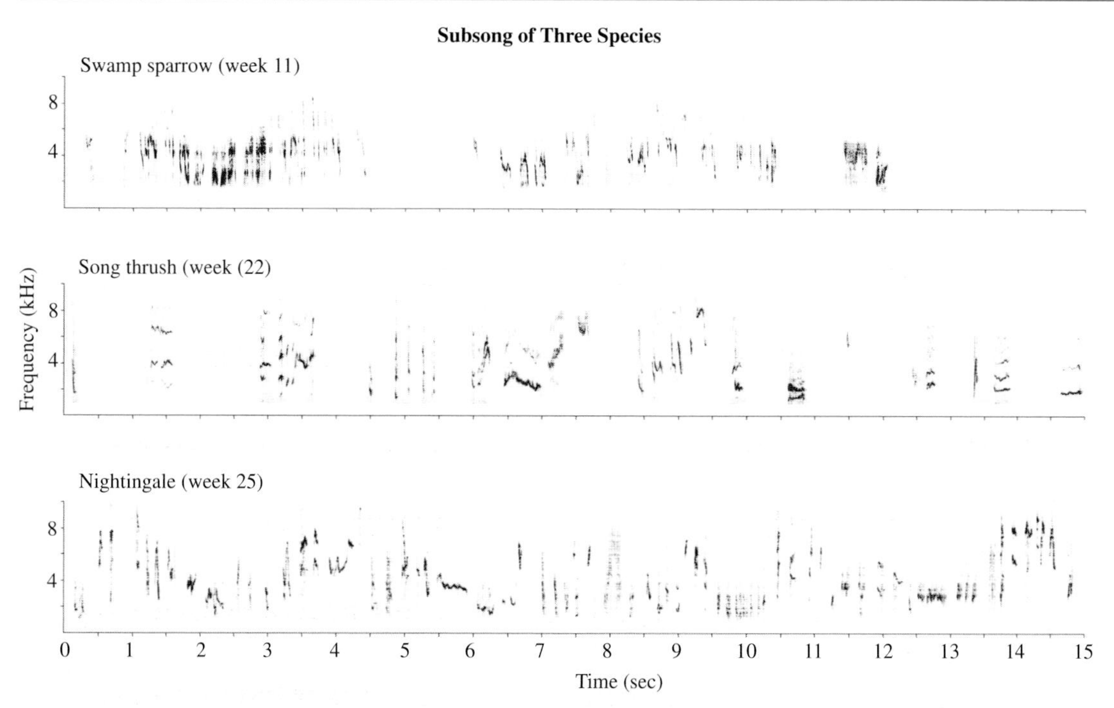

Figure 3.4 Sonograms of subsong of a swamp sparrow, a song thrush, and a nightingale (**CD1 #21**).

activity to the relatively short period in which birds advance to the stage of full adult song. During this phase, called 'song crystallization,' distinguished by a rather rapid progression from plastic song to mature singing, we see several basic improvements (apparently hastened by an increase in testosterone production; **Box 13**, p. 91). First, the diverse, highly variable types of song, so typical of the preceding stages of development, come to an end, and the previously plastic structure of vocalizations gets 'frozen.' Second, the time structure of singing often changes. Most birds shift from producing more or less continuous song sequences to a temporally segmented performance. This is done by the regular insertion of silent intervals between successive song patterns. Finally, birds cull from their originally vast range of diverse sound patterns a limited number, assembled into songs which are typical for the species. The extent of this attrition varies. It is less drastic in species like nightingales, with large song repertoires, than in birds with smaller repertoires like song sparrows, swamp sparrows, or white-crowned sparrows. In nightingales it averages about 8%, whereas it can go as high as 80% in sparrows. To some degree, such pattern attrition is a typical step in process of pattern selection in all of these species (Marler & Peters 1982b, 1987; Nelson 1992; Todt & Hultsch 1996, 1999).

In studies on various North American sparrows, including the field sparrow, the attrition process has been investigated in more detail, giving rise to the theory of 'action-based learning' (Marler & Nelson 1993; Nelson & Marler 1994). The core idea is that social experience at the time of crystallization, as the male establishes his first breeding territory, determines which songs a bird will select and produce in his adult repertoire (**Box 14**, p. 94). In the field, such social experience will be provided by interactions like counter-singing with neighboring adults. As a result of action-based learning, a juvenile bird will retain those songs in his own repertoire that match the species-typical structure and even the particular song types of his neighborhood. Thus, some vocal

BOX 14

INSTRUCTION AND SELECTION IN BIRDSONG LEARNING

A common definition of learning is any "behavioral change effected by experience" (Immelmann & Beer 1989). If we take this definition as a starting point, the study of song development in birds reveals some of the diverse ways that experience can influence behavior. One way to characterize the effect of experience is to ask whether it is instructive or selective (Nelson & Marler 1994). For many years, students of birdsong thought that learning by instruction played a dominant if not exclusive role in song development. For example, a naive young bird imitating the song of its tutor during the sensitive phase for learning appears to be a prime example of instruction, similar to the process by which college students learn from a lecturer. In the past two decades, ethologists and neurobiologists have begun to examine the importance of learning by selection, both in birdsong learning and other domains (Cziko 1995). In learning by selection- or 'action-based' learning (Marler & Nelson 1993), experience acts on a pre-existing repertoire of behaviors and its underlying mechanisms, leading some to be performed in preference to others. The repertoire could be acquired previously by instruction, or genetically pre-encoded, or a combination of both. In some respects, learning by selection resembles the operant learning studied by psychologists in the laboratory, but the concept of classical reinforcers like food, central to classical learning theory, appears to be irrelevant.

Although there were early suggestions in the observations of Margaret Morse Nice on song sparrows (Nice 1943), the first evidence for a role of learning by selection came from a detailed study of the swamp sparrow (Marler & Peters 1982b). This was followed by an observational field study of the field sparrow (Nelson 1992). Adult male field sparrows sing a single 'simple' song type. About 40% of young males, possibly yearlings, return to the breeding grounds in early spring singing two or more 'simple' song types. Singing song types in excess of what is retained in the final adult repertoire is termed 'overproduction.' Over a period of days to weeks, the young males eventually drop the extra song types, and continue to sing the one type that best matches the type sung by their most actively singing neighbor, as shown. Matched counter-singing between territory neighbors may provide the selective force that determines which song type is kept, and which is dropped from the repertoire. A similar selection process has been described in the white-crowned sparrow (Nelson 2000c), and the concept could even be relevant to understanding brain mechanisms for early song memorization (Marler 1997). The idea that many song features may be pre-encoded in the brain, but unexpressed until they are experienced, could help to explain how the species-specificity of song is maintained.

Douglas A. Nelson

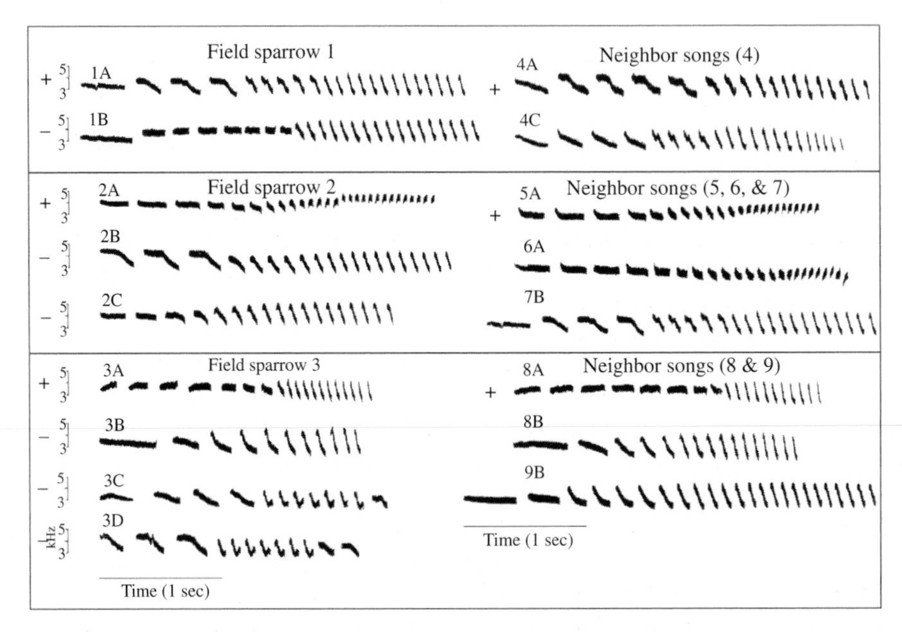

Early in the season, three male field sparrows (1–3), on the left, had song repertoires of two, three, and four song types, respectively. Letters designate the same type within panels. After a week or two each was reduced to a single song type (+) that matched one song of a close neighbor. Songs of six neighbors (4–9) are shown on the right. Cross-correlation analyses showed that the discarded songs (-) had the highest dissimilarity scores when compared with neighbors' songs. From Nelson 1992.

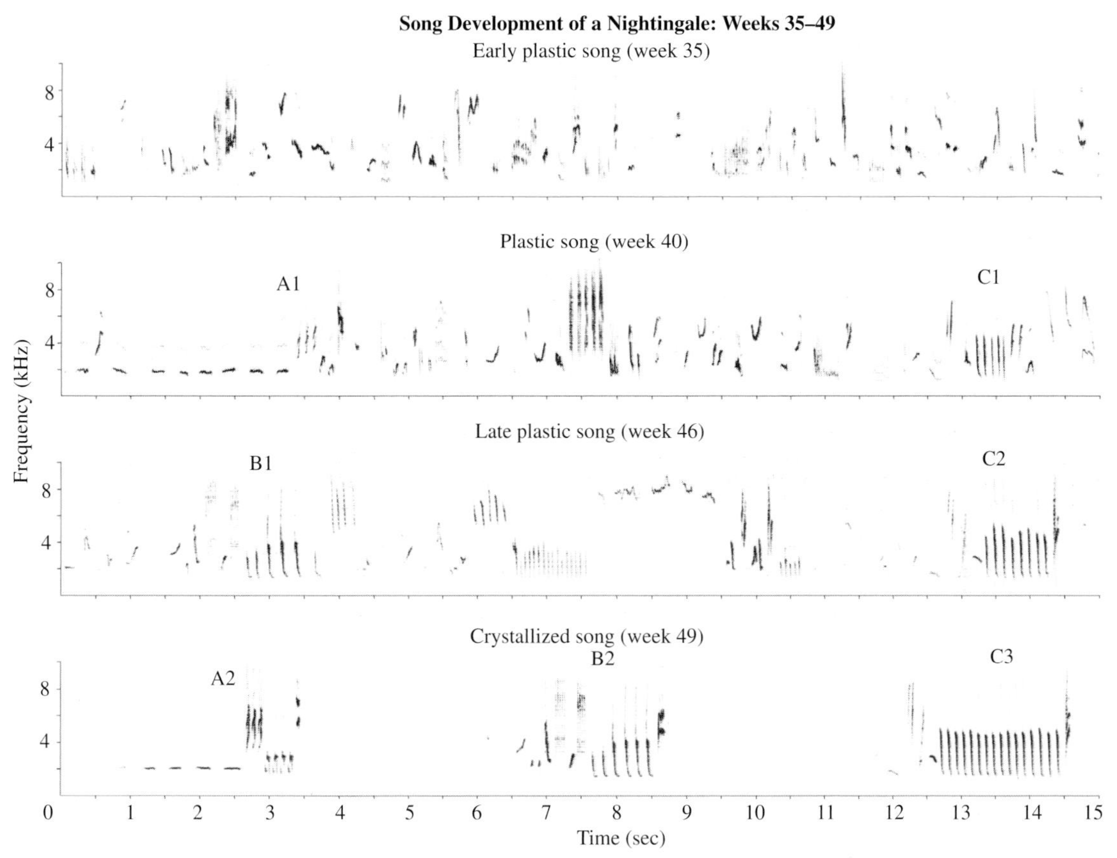

Figure 3.5 Song development of a nightingale (week 35 to week 49). At the bottom are three crystallized imitations (**A2, B2,** & **C3**) with precursors shown in the second and third row (**CD1 #22**).

material that is not incorporated in his actual repertoire may nevertheless be retained in memory, as a kind of 'dormant' pattern reservoir that can get reactivated later in life, as during social interaction with new neighbors (see p. 95), as suggested by both laboratory and field studies (Marler & Peters 1982b, 1988a; Nelson 1992; Geberzahn & Hultsch 2003). Corroborative evidence was recently documented by Geberzahn et al. (2002), who exposed nightingales, at an age of 11 months, to playbacks of songs with which the birds had previously been tutored, but which they had not imitated, either entirely or in part, during their vocal development. Exposure to these songs instantly elicited song matching, triggering the selective production of imitations that had not previously been uttered. This finding suggests that the nightingales had indeed memorized these previously unsung song patterns.

Inquiry into Mechanisms of Song Development

We are still grappling with many questions about the mechanisms that underlie song development (Székely et al. 1996; Tchernichovski et al. 2001; Brainard & Doupe 2002). A seemingly simple issue concerns how birds proceed when they are uttering their first and often fragmented precursors of song imitations. Do they 'hit' on them by chance, or do they have more efficient strategies of memory retrieval than this? We will come back to these and other related questions (see p. 103).

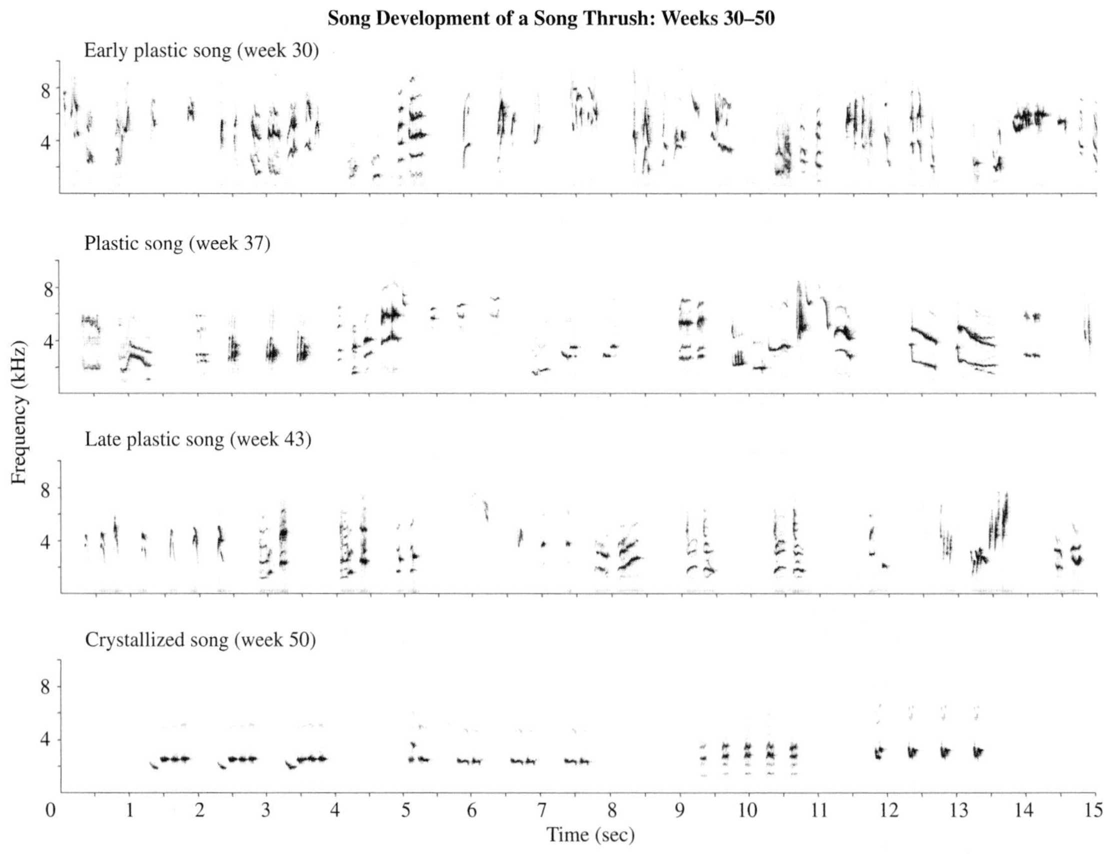

Figure 3.6 Song development of a song thrush (week 30 to week 50; **CD1 #23**).

Another question about song development is whether and how far the sensorimotor phase can be modified by environmental or physiological variables, and what the consequences may be for singing later in life. Does the relatively long duration of the ontogenetic process have any functional implication for the resulting singing behaviors (Whaling et al. 1995). White-crowned sparrows were given testosterone implants at different ages, thus inducing an artificially accelerated song ontogeny. By this procedure, the length of the pre-imitation period after tutoring, and also the duration of their sensorimotor phase of song development, was reduced by several months. The structure of crystallized songs was clearly abnormal after such accelerated development, and similar to the songs of individuals raised in acoustic isolation. This finding suggests that the storage phase may not be a time of passive retention, but may involve intrinsic processes of active consolidation and/or maturation of song material.

FUNCTIONAL ASPECTS OF SONG LEARNING: HOW TO DEVELOP AN APPROPRIATE REPERTOIRE

Singing serves several functions, elaborated in other chapters of this book. However, some aspects also merit a brief review here, such as how song learning and vocal development provide birds with signal repertoires that allow them to engage in complex communicative exchanges with each other (**Figs. 3.8 & 3.9; CD1 #25**).

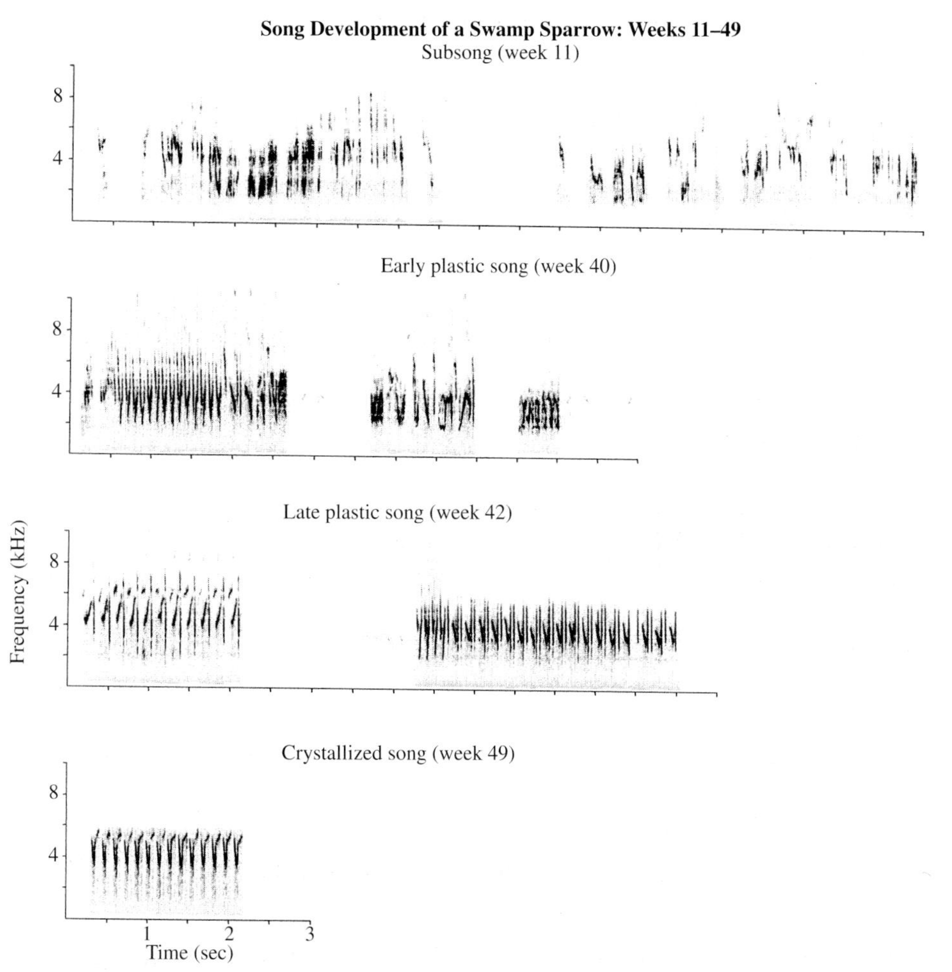

Figure 3.7 Song development of a swamp sparrow (week 11 to week 49; **CD1 #24**).

Ways to Prepare for Vocal Interaction: The Advantages of Song Sharing

Many birds use their songs to interact vocally with their neighbors. Such communication includes mutual interchanges that display both temporal and pattern specificity. The most widespread category of pattern-specific response is vocal matching, which requires that neighbors use shared parts of their signal repertoires, such as particular song types (Todt & Naguib 2000). How is song sharing among neighbors achieved? Song-sharing or repertoire-sharing will be an obvious result of learning as juveniles, if the young males do not disperse far from their natal areas. If these individuals experienced the same set of songs, they will not only imitate from the repertoires of the same population of adult birds that acted as tutors, but also share some of their imitations. However, when juveniles are not philopatric and disperse to distant breeding grounds, repertoire sharing is no longer a straightforward consequence of early learning.

Given the benefits of sharing songs among neighbors Nordby, Campbell, and Beecher (2001) postulated that the ability to learn new songs would not be confined to a bird's natal summer, but would continue until the next breeding season. To test their idea, the authors examined whether song sparrows that were first exposed

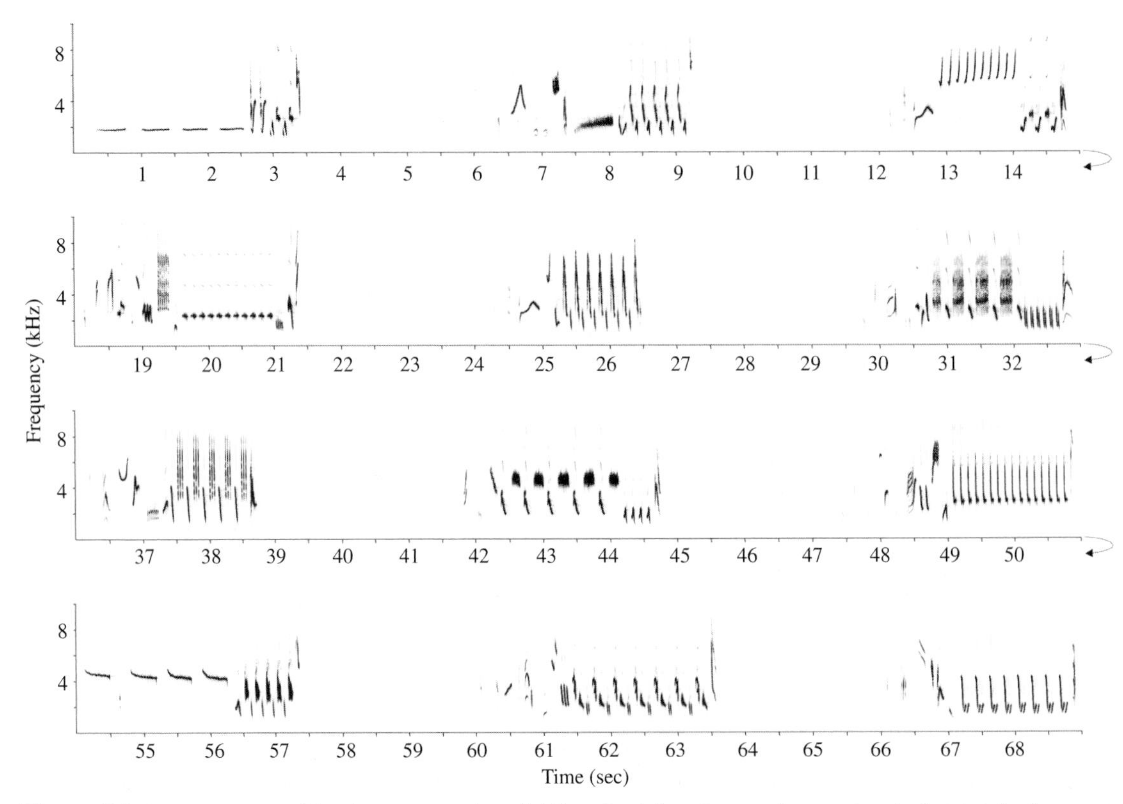

Figure 3.8 A sequence of twelve songs of a nightingale, taken from a longer bout of solo-singing (**CD1 #25**).

to songs at an age of 30–90 days would acquire new songs presented by other tutors at an age of 140–330 days. The results were positive, i.e. most males imitated songs heard during the late tutoring. However, since songbirds may not need to experience a model song very often or over an extended period of time (Hultsch & Todt 1989a), the findings do not permit the conclusion that the ability of these birds to learn new songs lasted much longer than day 140.

A number of field studies have demonstrated that neighboring males of several species make mutual adjustments to the composition of their song repertoires. As a consequence of such findings, it was suggested that these males can learn new songs later in life. In the laboratory such supplementary learning has been documented, as in the starling (Chaiken et al. 1994) and the nightingale (Hultsch 1991a, 1993a, b). However, most investigators in the field failed to exclude subjects that had previously experienced the apparently novel songs already in their natal summer, and thus may have learned but just not vocalized them. Based on results of song learning experiments with various species, including white-crowned sparrows, swamp sparrows, song sparrows, and field sparrows, the field observations of this later learning have been reinterpreted; the suggestion is that the seemingly novel songs could have been developed from song material that was acquired during the natal year, but was then discarded during crystallization (Marler & Peters 1981, 1982b, c, 1987, 1988a; Marler 1990a; Nelson 1992; Nelson & Marler 1994; Hough et al. 2000). If correct, this interpretation would place a rather different burden on the underlying brain mechanisms; proximate mechanisms for late song repertoire

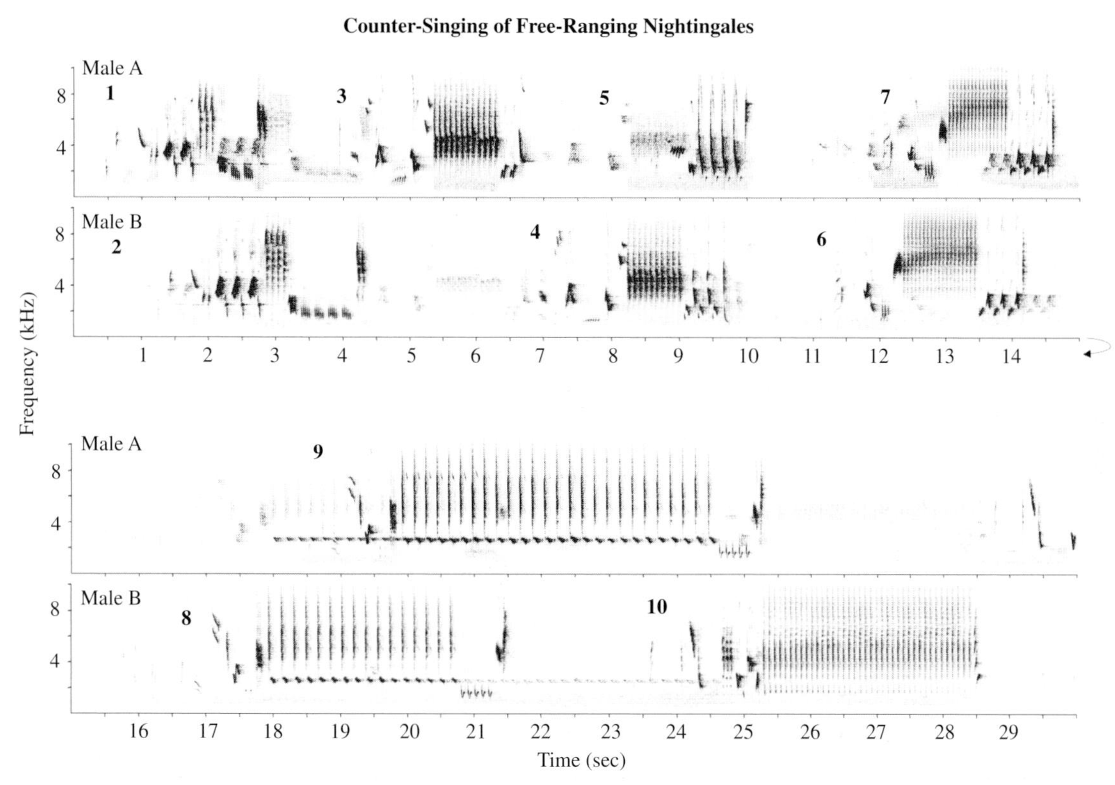

Figure 3.9 Matched counter-singing by two neighboring nightingales, males A and B. Songs 2 and 4 of male B are matching replies to songs 1 and 3 of male A. Then the birds switch roles. Songs 7 and 9 of male A show matching replies to male B's songs 6 and 8. Note that besides song matching and role switching, the timing of responses is also meaningful. The predominance of overlapping replies indicates an agonistic interaction (Hultsch & Todt 1982; see Chapter 2, p. 52; **CD1 #25**).

adjustment merit further investigation. By whatever means it is achieved, it is clear that matching songs with rivals at the location of a young male's first territory exert a major influence on his song development (see Chapter 4).

Skills for Vocal Interaction: The Role of Shared Singing Programs

As adults, the majority of oscine birds divide up their singing into songs of a few seconds' duration, which then become the most significant units of vocal interaction. These songs can be employed in various forms of interactions, among which vocal matching is one of the most conspicuous, but there are others. For example, there is variation in the interval after which one bird replies to another. The response latencies of birds vary in length, and result in a temporal overlap of songs if latencies are short. As first shown for free-ranging nightingales (Hultsch & Todt 1982) and now confirmed for other species (Todt & Naguib 2000), the overlap signals a kind of vocal threat. This makes an effective assertive response available for neighbors who do not share parts of their song repertoires. Such time-specific responses seem simple, but they require considerable skill, both for instantly retrieving songs from memory and for the production of specific songs at very short notice.

In bird species that develop vocal repertoires, males have other ways of interacting with each other. The nature of these responses usually depends on the singing style of the species, and the intrinsic program underlying their song sequencing. Birds that frequently repeat a given type of song before switching to another type, singing with so-called 'eventual variety,' sometimes coordinate their switching (Falls & Krebs 1975; Kramer & Lemon 1983). Birds that, in contrast, do not sing in bouts of one type, but perform their songs with 'immediate variety' (see Chapters 2 & 4) may abandon their normal song sequencing and switch to a song that allows them to change the order, and sequentially 'overtake' the singing sequence of a neighbor. Such complex vocal responses have been found in European blackbirds (Todt 1970, 1981), nightingales (Todt 1971; Hultsch 1980), and in marsh wrens (Verner 1976). In nightingales these complex modes of responding are based on their ability to memorize and learn not only single song types, but also the sequencing of tutored songs (Hultsch et al. 1999a), as described in more detail later (see p. 102).

An Emphasis on Individuality: Song Recombinations and Inventions

The species-specific predispositions brought to song learning ensure that species recognition is not a problem in oscine birds. But how do birds achieve another socially relevant accomplishment, the ability to recognize individuals? There are several possibilities. Individual differences sufficient to permit personal identification may emerge as side-effects of development. In addition, the individuality of a songster can be emphasized by introducing into the process of song development a drive to create song improvisations. Such improvisations occur either as novel recombinations of learned song material or as completely new song inventions (Marler & Peters 1982a, c; Hultsch 1993a, b; Hughes et al. 2002), sometimes so comprehensive and pervasive that virtually entire new songs may be created (Kroodsma et al. 1999b; see Chapter 4).

Both song recombinations and inventions are widespread, and are especially frequent in species that develop large vocal repertoires (Catchpole & Slater 1995). In this case, they represent a strategy for enlarging the repertoire of a given songster, and the introduction of individuality may be a side-effect. There is evidence that nightingales, who during their first year of life normally show a capacity for learning about eighty different song types, may sometimes compensate for deficits in this repertoire by song improvisation, as may happen if they missed the chance to acquire enough songs during their natal summer (Röll 2002). Studies of their song performance revealed that birds treat song recombinations and inventions rather differently. With recombinations birds working through their repertoire sing the new songs in close sequential juxtaposition to the imitations of the songs from which they took the original material. In the second case, however, birds perform the new inventions together, associated in sequentially coherent bouts and segregated from bouts of songs based completely on vocal imitation (Hultsch et al. 1998). This contrast supports the view that the development of inventions is a special accomplishment in the creation of song. In her recent investigation of pattern stereotypy in nightingales, Hughes showed that song inventions are affected both by species-specific predispositions, as well by learning processes (Hughes et al. 2002). However, which kind of learning plays the major role here remains a matter for further research.

Cycles of Song Ontogeny, and Aging

The stability and persistence of song patterns varies remarkably across species. Whereas some birds do not modify their patterns for years, individuals of other species change their patterns, sometimes quite radically. Starlings (Eens et al. 1992) and canaries (Nottebohm & Nottebohm 1978; Nottebohm et al. 1986), for example, modify their song repertoires from year to year. An even more striking case has been documented for indigo village birds (Payne 1981). Here, males may share several different songs which they change jointly on a yearly basis. As a consequence, in spite of repertoire modifications, they still have shared songs in their repertoires.

Less spectacular but more widespread are seasonal changes in the song. Most oscine species do not sing all year round. When they come into song each breeding season, they may modify both the amount and the structure of their vocalizations, even during a given year. Singing at the start of the season is often reminiscent of the early stages of a bird's song development in his first year of life. Given that Nottebohm and his colleagues (Nottebohm et al. 1986) uncovered a striking seasonal neuronal turnover in the nuclei of the song control system (see Chapter 8), age-related changes in song production are not surprising. From this point of view it is perhaps more remarkable that individuals of many species, in which this seasonal, neuronal turnover still occurs, retain the same songs without change for their entire lives. More neurophysiological and behavioral data are needed before we fully understand the effects of cyclic song system ontogeny and aging on singing behavior.

COMPARATIVE ASPECTS OF LEARNING TO SING

The accomplishments of birds that engage in vocal learning invite comparisons with the achievements of other organisms. However, aside from cetaceans (Janik & Slater 1997; McCowan & Reiss 1997) and bats (Boughman 1996), there are no valid mammalian comparisons other than humans. Similarly among birds, apart from oscine songbirds, only a few other taxa, namely hummingbirds (Baptista & Schuchmann 1990) and parrots (Paton et al. 1981; Pepperberg 1993), are known to acquire their vocal signals by imitation, though Kroodsma (see Chapter 4) makes a persuasive case that certain suboscines, specifically bellbirds, also learn to sing. With the exception of oscine birds and parrots (see Chapter 13), in-depth studies of this issue are rare.

There are some fascinating parallels with the development of speech. Some typical accomplishments of humans, like language acquisition and item learning, are comparable to certain aspects of song learning in birds. Such comparisons have a long history (Marler 1970b; Kuhl 1989), and were recently greatly expanded by Doupe and Kuhl (1999). Certain parallels have been highlighted as crucial to the development of both birdsong and human speech. First, both behaviors have to be learned to achieve the normal signal repertoire of the species. Second, such learning relies on the auditory perception, memorization, and imitation of sound patterns. In both, perception precedes the production of vocal material. Third, acquisition is best accomplished early in life, during sensitive periods, and is apparently guided by specific predispositions. Finally, vocal expertise is successfully reached only after progression through particular stages of development, in which vocal practice plays an essential role (see p. 90). As a general framework, these three parallels help us in dealing with some more specific issues.

How to Select One From Many Items: Evidence for Specificity

During the early phase of development, in which patterned sounds are memorized, young birds and human infants are confronted with a rather similar problem: instead of hearing a single auditory stimulus, they are exposed to a medley of many vocalizations. Children need to look out for cues that help them to parse the speech flow produced by their caretakers into segments. Once this is accomplished, it obviously helps them to identify and store information about particularly frequent segments more efficiently, especially words and combinations of words (Juszyk et al. 1992). It is interesting to ask whether young birds apply a similar strategy.

Experiments to approach this question were conducted in nightingales, a species that, like humans, memorizes and develops a large repertoire of different vocal patterns. To identify the strategies they use, our studies on their song acquisition were usually based on the following experimental design.

To compose a learning program, we select

particular songs from our catalog of master song types and record them successively on tape to form a string of songs. In a standard master string, each successive song in the sequence is of a different song type. Likewise, each of the different sequences to which a young bird is exposed during the tutoring periods consists of a unique set of song types. We thus label a particular tutoring regime by the particular song sequence which birds hear in that situation (Hultsch et al. 1984). The acquisition success of the tutored males, assessed by how many songs they imitate, then allows us to see whether a particular tutoring regime influences their singing. Additional checks are provided by making audiovisual recordings, which allow us, for example, to measure a bird's motility during a given tutoring experiment (Müller-Bröse & Todt 1991). With this design we have identified a number of conditions in which nightingales not only imitate single song patterns accurately (**Fig. 3.5**), but also readily memorize and produce long song sequences.

Song acquisition is always affected in some degree by how a bird is tutored although, in some respects, birds can be remarkably unaffected by change in the training regime. For example, while for songs experienced only five times the acquisition success of nightingales is low (approximately 30%), birds imitate around 75% of song types heard 15 times. On the other hand, more frequent exposure than this does not significantly improve acquisition success. Also, the number of song types in a sequence can be increased considerably, say from 20 to 60 song types, without any need to raise the frequency of exposure (**Fig. 3.10**). The ease with which birds cope with such an increase in the number of songs to be acquired (Hultsch & Todt 1989a) does not jibe with predictions from classical learning theory. Traditionally, exposure frequency needs to be increased in proportion to the number

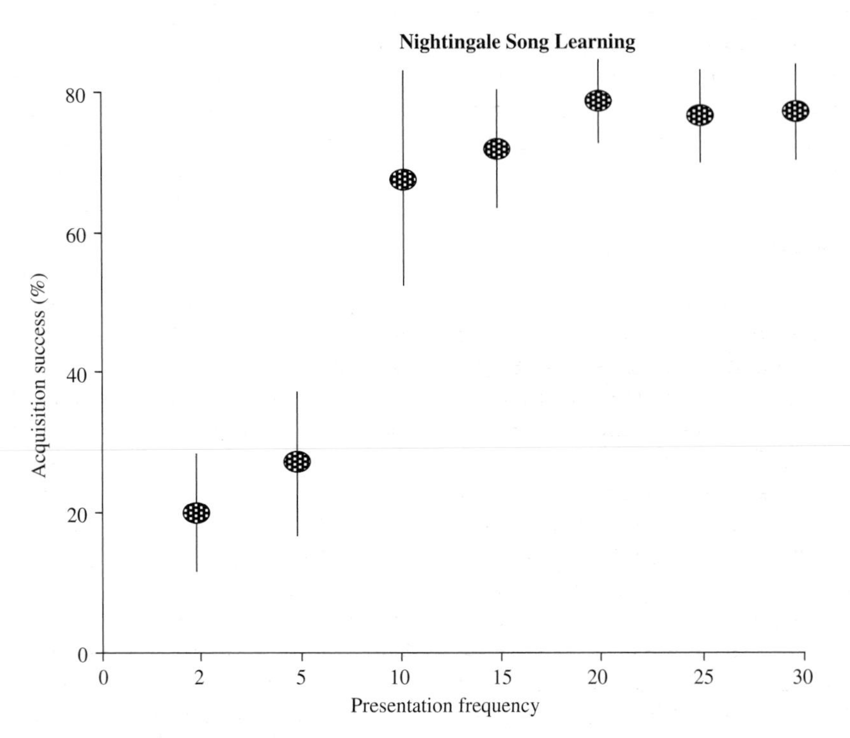

Figure 3.10 The relationship between the presentation frequency of tutor songs and acquisition success, represented as the percentage of model songs a bird imitates.

of stimuli to be acquired (Crowder 1976). These findings are relevant to the issue of song acquisition, considered as a kind of special process or template learning (Marler 1976). Any inquiry into the nature of the memory mechanisms underlying song learning has to take into account the fact that specific evolutionary adaptations are involved in the process.

One such adaptation was recently revealed by experiments testing how nightingales deal with exposure to song sequences after erasing the silent intersong intervals and other cues that might help birds to recognize each song as a separate sequential unit. The imitations they developed were each uttered separately, showing that the birds did not mind the lack of such cues in the least; they were still able to determine boundaries between successive tutor songs. Thus, in contrast with human infants, whose parsing of the speech stream is based on experience with their mother tongue, young nightingales may have some innate mental concept of how their song patterns are segmented (Hultsch et al. 1999a). Clearly, nightingales differ from humans in this regard. But what about learning accomplishments at a higher level of sequential organization, such as the imitation of a long sequence of songs?

Memorizing Serial Strings: Splitting Versus Chunking

When humans are presented with a serial learning task they usually handle it by a strategy called 'chunking'. Given a string of different items to memorize, such as words or sentences, they split the string into 'chunks' approximately four units long, which turns out to be optimal for processing in short-term memory (Bower 1970; Simon 1974; Cowan 2001). Interestingly, nightingales also perform well in extracting and memorizing information about the serial succession of song types experienced during training, and they achieve it by a quite similar maneuver (Hultsch & Todt, 1989a, b, c, 1992, 1996; Todt & Hultsch 1998). For example, when birds are exposed to a long sequence of, say, 50 songs, they first split the sequence into segments of about three to five successive songs and then store information about each segment or subset of songs separately. As with humans, the evidence suggests that these processes reflect properties of short-term memory, with the songs composing a segment treated as a 'chunk' or single unit. The consequences become apparent during the birds' own song production when they also deliver imitations derived from a given sequential segment in a similar manner, as a 'package' of song types. This effect has been termed 'package formation'.

To begin with, it remained open whether the formation of song packages really reflects short-term memory constraints, during auditory song memorization, or whether 'packaging' emerges later, during the motor phase of song development. There is now clear evidence for the first interpretation. In one study young nightingales were exposed to a set of 12 different songs that was presented 15 times, enough for them to memorize each song readily (see p. 102). Instead of the usual experimental design, however, the serial succession of song types was altered from one tutoring exposure to the next. The results were remarkable: first, as adults, all birds formed normal packages. Second and even more astounding, their packages were made up of imitations of songs which the birds had heard as a sequence only during the first tutoring session. Evidently, package formation can occur during a single exposure and the very first experience seems to play a key role (Hultsch 1992; Hultsch & Todt 1996).

Although the formation of song packages has not yet been studied in other songbirds, it seems that it also plays a role in other avian species. This can be inferred from studies of singing in the field, which report that series of three to five sequentially associated song types are indeed widespread in birds with a song repertoire (Todt 1970; Lemon & Chatfield 1971). It will be worthwhile to investigate the role of chunking in song acquisition in a wider range of species.

There are further parallels between humans and birds. For instance, if we are exposed to long strings of different items often enough, we will eventually be able to remember and reproduce the entire sequence, by rearranging the packages in the right order. Marsh wrens (Kroodsma 1979) and nightingales (Hultsch & Todt 1992) behave very similarly. If nightingales are allowed to hear a song string fifty or a hundred times, they imitate the entire sequence of tutored songs in proper serial succession. The imitation of serial order first emerges within packages, and only later, with more frequent exposure, comes to reflect the complete succession of packages. It appears that, with more opportunities to listen to a song string, order information gradually gets consolidated, suggesting a truly hierarchical process, progressively incorporating higher and higher levels of song organization (see p. 105).

Retrieval From a Large Memory: Hierarchical Representations

Rapid and faithful memorization of significant items can be a real advantage, but it is only one side of the coin. For efficient use of the learned material it has to be supplemented by a system for organizing memories, which is also optimally structured for retrieval. In humans, we know that the more information one has to keep in mind, the more important this issue becomes. One optimizing principle is to organize mental representations of learned items in a hierarchical manner, a concept that we had in mind in asking how nightingales organize the memory system for their large song repertoire.

In one experiment we presented young birds with several different strings, each holding more than 10 songs. We only allowed them to listen to these strings 10–20 times and, as a result, the birds did not imitate the serial order of the tutor songs. Nevertheless, we found that they had clearly memorized the set of tutor songs, distinguished as having been acquired in the same learning context. The birds revealed this to us by singing packages developed from the tutored song sequence acquired jointly in the same learning context as one large association group that was sequentially separated from song

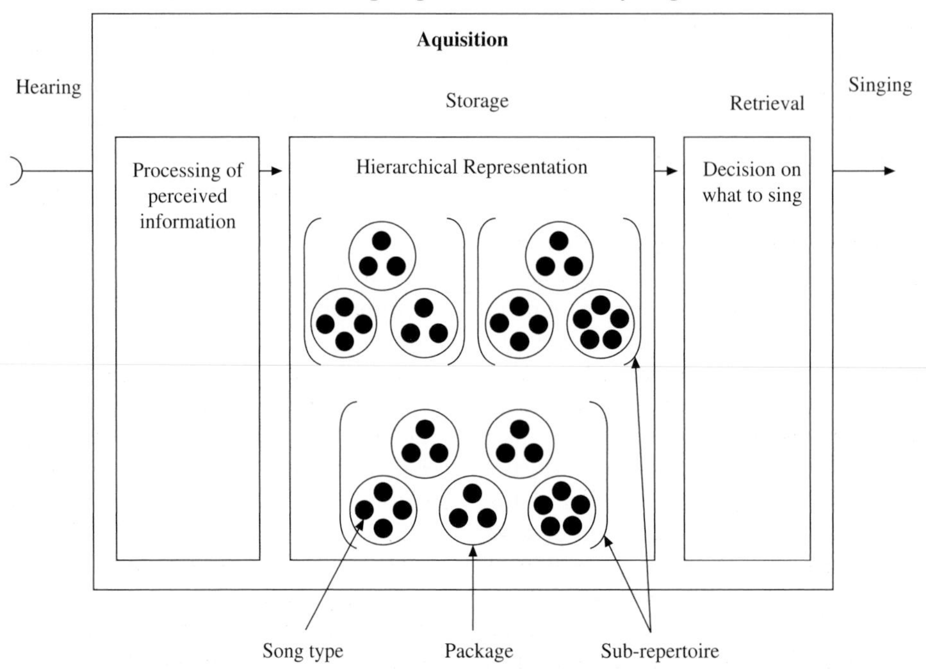

Figure 3.11 A diagram of the mechanisms we postulate to explain the relationships between hearing, memorization, sensorimotor learning, memory retrieval, and singing.

packages experienced in another learning context. This achievement, which we originally attributed to a 'context effect' (Hultsch & Todt 1989b), suggested that the internal representation of learned song material is organized in a hierarchical manner. In addition, it provided a mechanistic account of a phenomenon earlier reported from field studies by Hultsch (1985). In free-ranging nightingales, she found that the large song repertoires of these birds, up to 200 song types per male, can be viewed as an aggregation of sub-repertoires, each holding a limited number of songs. A sub-repertoire was characterized by strong sequential associations among 'members,' with much weaker relationships between members of different sub-repertoires.

This suggestion of a hierarchical organization of song memories led us to a further, more general parallel between the learning of humans and birds (**Fig. 3.11**). If the first step in this 'memory hierarchy' is placed at the level of songs, the next involves the packages, composed of a few songs, and at a still higher level are the larger association groups or sub-repertoires, composed of packages. Both song order and sub-repertoires reflect the serial organization of input during auditory learning and it follows that they are exposure induced. As such, they are to be distinguished from the package groups, which can be characterized as self-imposed associations. A crucial question about the formation of exposure-induced song associations is whether they simply reflect the memorization of stimulus chains, or whether they are linked to more abstract cognitive abilities, such as the formation of perceptual or conceptual categories. For the sub-repertoires, for instance, such categorical concepts might be 'songs heard from a particular individual,' 'at a particular location or time,' or 'songs specified by a particular quality.' Preliminary evidence suggests that birds are indeed using such cues, and experiments to systematically examine this matter are under way.

Some of the rules accounting for a hierarchical organization of song memories can be discerned already during the ontogeny of singing. Developmental trajectories of the three hierarchy levels which, in bottom up order, are songs, packages, and sub-repertoires, unfold at different times and reflect those levels of order in the singing that develops. When singing becomes functional as a territorial display in adults, a repertoire that is hierarchically prestructured is a candidate mechanism for facilitating song retrieval in situations demanding rapid vocal responses. In making decisions about 'what to sing next' it is expedient if the bird does not have to do a 'serial' search from the whole pool of developed song types. The number of decision steps and the decision time are both reduced by using a search routine that can focus on particular subsets of patterns (Todt et al. 2001).

Thus, a hierarchically organized format for storing representations of song has clear adaptive value in birds who, like the nightingale, have to manage large repertoires. In their vocal interactions these versatile songsters respond to each other using sophisticated rules that are based on pattern-specific and time-specific relationships between the mutually exchanged songs. For example, during 'rapid matching,' a male has to identify a neighbor's song and immediately select and retrieve a song of the same type from his own repertoire, doing so with a latency of approximately one second (Hultsch & Todt 1982; Wolffgramm & Todt 1982; see Chapter 2, p. 48).

How to Develop Optimal Units of Interaction: Songs Versus Sentences

There are many different kinds of communication. However, in an ideal interactive case, when two individuals are contributing equally to the signal exchange, one can expect that the signaling routines will not be arbitrary, but will follow certain rules or conventions. The significance of such rules has been documented for both the verbal and nonverbal dialogues of humans (Burgoon & Saine 1978), and for the vocal duels of songbirds (Todt & Naguib 2000). The suggestion was also made that, in ideal interactions, signal patterns would be used which strike a compromise between two opposing prerequisites: the signals should be long enough to convey a given message but, at the same time, not so long as to delay a potential reply. Some exceptions aside, most human sentences and most birdsongs that are used in an interactional context are only a few seconds long, and are thus candidates for optimal units of vocal interaction (Hultsch et al. 1999a). With this shared property of songs and sentences in mind, we can examine some of their other features and contrast some of them from formalistic and developmental points of view.

From a formal perspective, most sentences are composed of several constituents; words or phrases that follow each other according to syntactical rules. For example, a sentence often begins with a word that is particularly frequent in a given language. Similarly, most birds develop songs with certain syntactical rules. The species-typical composition of most songs requires particular types of notes, syllables, motifs, and phrases, occurring at specific positions in a given type of song. Although such structural features are more rigidly fixed than those of sentences and thus are not acceptable as genuine lexical syntax, changes of note positions are meaningful in both cases (Todt 1974). Finally, many species, however versatile their singing style, may nevertheless start successive songs with the same type of introductory note (Todt & Hultsch 1999). In other words, from a strictly formal perspective, there are certain similarities between sentences and songs.

There are also similarities from a developmental perspective. At the age of about three months human infants produce only vowel-like sounds. Then, at about seven months, they begin their 'canonical babbling' by incorporating consonants into their vocal repertoire. During this period, when they produce word-like syllables, children rehearse repeatedly and have already begun to imitate the prosodic features of their particular mother tongue. By about twelve months, infants usually start to use so-called 'one-word sentences,' and at some later time these develop into sentences composed of two or more words (Doupe & Kuhl 1999). This progression can be compared to the succession of stages in the development of singing in birds, including their subsong, plastic song, and eventually their use of crystallized full song (see p. 92).

Finally, our findings on mechanisms for song memorization add to this comparative framework. The formation of song type packages shows striking similarities to the chunking of information in human serial learning (Bower 1970; Simon 1974). In humans, chunking is related to cognitive processes, so that a chunk can be defined as a functionally meaningful unit, whether it be a simple phrase (a 'morpheme') or a sentence, or more. In song acquisition in birds, on the other hand, the equivalent to a chunk is a song, and a song can also be regarded as a meaningful unit. This can be inferred from the way songs are used as units of interaction, as during vocal communication between neighbors. Also, there are good reasons for thinking that such interactions reflect underlying cognitive accomplishments, as for instance when a bird's singing performance is organized according to categorical cues acquired from the context in which song memorization occurred. Such contexts may be defined temporally, spatially, or socially. The exposure-induced song type associations developed by nightingales, the sub-repertoires, represent such an accomplishment at a higher level of behavioral organization. Finally, some birds, such as warblers, learn the situations in which to use their songs (Kroodsma 1988; Spector et al. 1989). But the particular cues on which such categorization is based, and the mechanisms by which a categorical representation is achieved, remain to be explored. This is one of many issues that need to be addressed in future research.

CONCLUSIONS

The study of song development and learning to sing is fascinating for many reasons. Mechanisms have been uncovered that are basic to learning in general, including predispositions that influence decisions about what to learn; sensitive phases, when the potential for developmental plasticity peaks; and social factors that motivate learning selectively, all of which help a young bird to cope with the challenges of learning to sing. These mechanisms have two main effects: they allow for selection among potential learning stimuli and, at the same time, they facilitate pattern memorization. Birdsong studies have also revealed many significant properties of sensorimotor learning. They demonstrate the value of a step-wise progression through successive stages, as when birds proceed through the different stages of vocal development, subsong, plastic song, and song crystallization, before they achieve the final stage of adult singing. And, in many cases, song development still continues, after birds have finished their first year of life and become sexually mature.

Learning to sing is also fascinatingly diverse, as becomes clear when it is considered in a functional and comparative framework. It provides a male bird with a repertoire that serves both mate attraction and effective territory defense. From a territorial viewpoint, it paves the way for specific vocal interactions among males who have had the chance to learn similar songs beforehand and thus share parts of their vocal repertoires. Finally, when discussed from a comparative perspective, the potential of studies of song learning is unique: in the animal kingdom, vocal learning is a rare achievement that merits comparison with essentially human accomplishments such as language development (Doupe & Kuhl 1999). Specific differences aside, there are several parallels in the learning mechanisms involved. This is evidenced for example in the ways in which a human and a particularly accomplished avian singer like a nightingale cope with the management of huge arrays of songs, and their myriads of constituents, as we deal with our prodigious word repertoires. By using these shared properties as a frame of reference, learning to sing can be exploited as an excellent biological model for research on memory and its underlying neural mechanisms. There is no more ideal subject for study of the evolutionary strategies of problem solving, in a class of organisms with behavior that is in many ways as flexible as that of mammals.

Chapter 4

The diversity and plasticity of birdsong

DON KROODSMA

INTRODUCTION

Somewhere, always, the sun is rising, and somewhere, always, the birds are singing. As spring and summer move between northern and southern hemispheres, so, too, does this singing planet pour forth, like a giant player piano, in the north, then the south, and back again, as it has now for the 150 million years since the first birds appeared.

Ten thousand species strong now, their voices and styles are as diverse as they are delightful. Some species learn their songs, just as we humans learn to speak, but others seem to leave nothing to chance, encoding the details of songs in nucleotide sequences in the DNA. Of those that learn, some do so only early in life, some throughout life; some from fathers, some from eventual neighbors after dispersing from home. Some species sing in dialects, others don't. Especially in temperate zones, males do most of the singing, but females in the tropics are often as accomplished as their mates. Some songs are proclaimed from the treetops, others whispered in the bushes; some ramble for minutes on end, others are offered in just a split second. Some birds have thousands of songs, some only one, and some even none. Some are pleasing to our ears, and some not.

It is this diversity in ten thousand voices from our planet earth that I celebrate in this chapter. How these species differ from each other is the first step to appreciating them, of course, but those *how* questions then give way to *why* questions, the questions that become the focus of this chapter. Why do some learn and others not? Among those species that learn, why dialects in some and not others? When, where, and from whom does a young bird learn its songs? Why such impressive vocabularies in some species, so diminished in others? And why do most limit their learning to songs of their own species, but some mimic those of others? It is these *why* questions that intrigue us as we try to understand the individual voices that contribute to our singing planet.

TO LEARN OR NOT

As a young bird matures, it must become competent in using its vocal repertoire. How it becomes competent varies considerably among species.

At one end of the spectrum are those species in which no imitative learning[1] seems to occur. Among certain New World suboscine flycatchers, for example, each individual has encoded in its

[1] I offer an early caveat here for the critical reader: the 'correct' use of words that I use in this section (e.g., learn, imitate, etc.) are themselves hotly debated, and I discuss these issues briefly beginning on p. 114.

DNA all of the instructions needed to produce the details of a perfectly normal song. The eastern phoebe is perhaps best studied (Kroodsma 1985, 1989; Kroodsma & Konishi 1991). Each male of this species has two different songs, the 'fee-bee' and the 'fee-b-be-bee' (**CD1 #26**), and regardless of the auditory experience of young birds, the fine details of these songs and how they are used develop normally (**Fig. 4.1**). Even when the young bird is deaf, and cannot hear itself sing, it still produces normal songs, thus verifying that, as the young bird practices, it is not learning to match its song to either an imitated or an inborn (or 'innate') model that is stored in the brain. Some others too, such as *Myiarchus* and *Empidonax* flycatchers, and other suboscine relatives most likely develop their songs without imitating others of their species (Lanyon 1978; Kroodsma 1984).

In striking contrast to the vocal development in these suboscine flycatchers is the vocal learning that has apparently evolved independently in at least three different lineages: certain hummingbirds (Snow 1968; Wiley 1971; Baptista & Schuchmann 1990; Gaunt et al. 1994), the parrots (Farabaugh & Dooling 1996; Pepperberg 1999; Bradbury et al. 2001; Wright & Wilkinson 2001), and, of course, the songbirds, the scientific study of which date back to Barrington (1773, p. 249), who presented to the Royal Society his experiments and observations on "a subject that hath never before been scientifically treated." Learning in these species seems to involve two stages: the young bird must first memorize the sound to produce, and then it must practice singing, when it learns to match what it sings with the memory of that sound (Nottebohm 1999; see Chapter 3). Learning has perhaps evolved in other less well-studied groups, too, and may include a musk duck and a grouse (Sparling 1979; McCracken 1999).

Why Did Imitative Song Learning Evolve?

One of the most frequent questions asked among those who study birdsong is why some species learn their songs and others do not (Nottebohm 1972; Baptista & Kroodsma 2001). Proposed explanations for the evolution of vocal learning range from protecting the inner ear during loud vocalizing (Nottebohm 1991), to an arms race over 'ranging,' the use of sounds to estimate

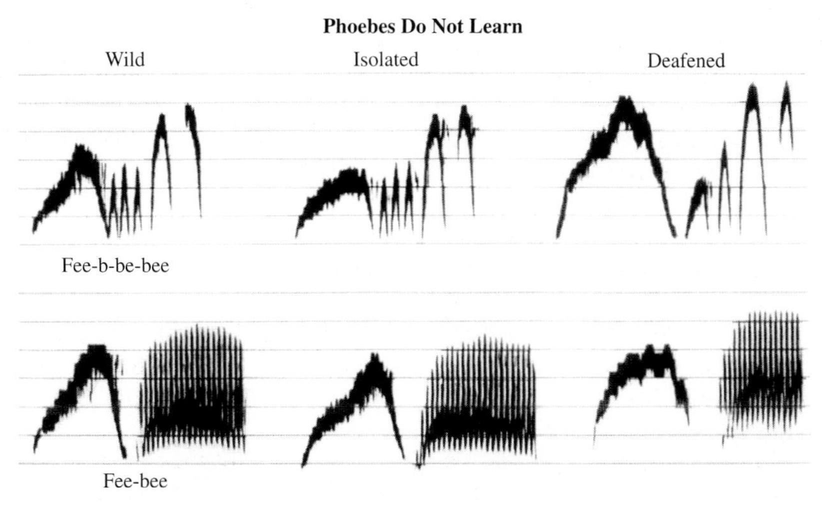

Figure 4.1 Eastern phoebes are flycatchers, and do not learn their songs. Two normal song forms of wild phoebes (left), are the 'fee-bee.' and the 'fee-b-be-bee' (**CD1 #26**). These two songs develop normally in captive birds reared in isolation (middle) and even when they have been deafened (right). Thus a young phoebe does not have to imitate or learn from another adult to memorize the proper song, and doesn't have to 'learn' from himself as he listens to his own song. From Kroodsma & Konishi 1991.

distance of conspecifics (Morton 1996b; see Chapter 6), to a context in which strong sexual selection in polygynous mating systems may have favored learning (Aoki 1989), but no proposed explanations are widely accepted. This question is especially difficult to answer among songbirds and parrots, the two groups that together comprise about half of all bird species, because learning presumably evolved among the ancestors of these lineages, close to 100 million years ago, and we cannot know the ancestral social circumstances that selected for imitating sounds.

Perhaps the best hope for understanding why some birds learn is to examine far more recent evolutionary events, events in which we can better identify the social circumstances that may have led to the origins or even the loss of vocal learning. If the musk duck learns its sounds (McCracken 1999), for example, what is it about the social system of this particular duck that might have favored vocal learning?

Vocal learning has also evolved relatively recently among the Cotinginae (**Fig. 4.2**), a suboscine group that diverged from the New World flycatchers, the Tyranninae, perhaps 35 or so million years ago (Sibley & Ahlquist 1990; Sibley & Monroe 1990). This learning was first suspected by Barbara and David Snow in the 1970s (Snow 1977), and has now been confirmed. In Central America, for example, songs of male three-wattled bellbirds differ strikingly from Nicaragua through Costa Rica to Panama, and

Figure 4.2 A male three-wattled bellbird, singing from its display perch. The bellbird is a close relative of flycatchers, thought to be 'non-learners.' Yet male bellbirds learn their songs by imitating others in the population.

three 'dialects' have been described. The dialects are identified by their unique sounds, best described as harsh 'quacks' (southern half of Costa Rica and northern Panama, referred to as the 'Panamanian dialect'); more 'musical' 'bonks' and loud 'whistles,' the whistles delivered on a single pitch (northern half of Costa Rica, referred to as the 'Monteverde dialect'); and an odd assortment of frequency-modulated whistles (Nicaraguan dialect). At Monteverde, in central Costa Rica, many young males are bilingual, having learned the full repertoire of songs for both the Monteverde and Panamanian dialect (**Fig. 4.3; CD1 #27**), but as a young male matures, a process taking six to seven years, he typically abandons songs of one of the dialects and perfects those of the other.

Not only does a young male bellbird learn the songs of a given dialect, but an adult must continually 'relearn' these songs throughout his life. Within each dialect, songs change gradually over time, as has been documented especially well for the Monteverde dialect in Costa Rica, and from one year to the next the songs of each adult change to keep pace with the overall population change. The frequency of the loud 'whistle' in this dialect has declined since the 1970s, for example, from roughly 5,500 Hz to 3,500 Hz, an average drop of 60 to 70 Hz/year (**Fig. 4.4**)

Confirming this learning process once and for all is a young, captive-reared bare-throated bellbird in Brazil. This young male was initially housed with two female Chopi blackbirds, and he learned the sounds typical of the blackbird (Fandino & Kroodsma unpublished).

And what is the social system of the bellbird species best studied, i.e., in what context might this learning have evolved? Male three-wattled bellbirds display from dispersed leks, exactly the kind of situation that Aoki (1989) proposed would be critical for the evolution of imitative learning.

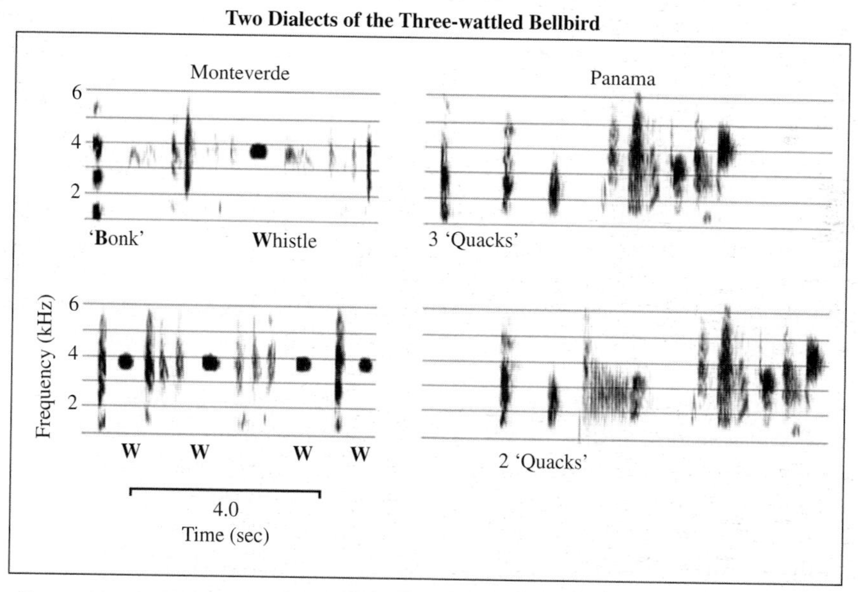

Figure 4.3 Two song dialects of the three-wattled bellbird **(CD1 #27)**. The Monteverde dialect occurs in the northern half of Costa Rica, and the two most commonly used songs are illustrated here. Songs contain loud 'bonks' and 'whistles' together with softer 'swishing' sounds; the '**B**' is beneath the loud bonk, the '**W**' beneath the loud whistles. The Panamanian dialect occurs largely in the Talamanca mountain range of southern Costa Rica and northern Panama, and is distinguished by its harsh 'quacks' and unique swishing sounds. One song begins with three harsh quacks (top), the other with only two (bottom). In these sonograms, the softer swishing sounds have been amplified to make them more visible.

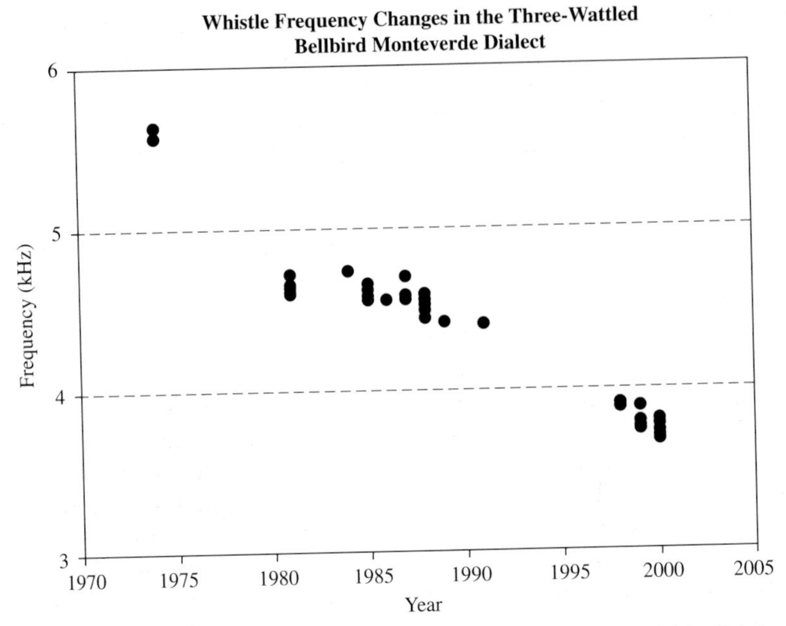

Figure 4.4 Not only do young bellbirds learn their appropriate dialect but adult bellbirds must continually relearn their songs, as songs within a population change gradually from one year to the next. Illustrated is the continued decline of the 'whistle' in the songs of the Monteverde dialect, just one of many changes that have occurred in the songs of this dialect since the 1970s. (Unpublished data collected by Barbara & David Snow, Jill Trainer, David Nutter, Debra Hamilton, George Powell, Julio Sánchez, David Stemple, Don Kroodsma, and others.)

Furthermore, the long-lived males within a dialect apparently all know and listen to each other, as they modify their songs together over time, suggesting a highly competitive framework among familiar rivals in which singing occurs.

Perhaps it is relevant that learning in several other groups has also been documented in non-monogamous mating systems, thus further supporting the idea that imitative learning evolved when males competed strongly for mates, resulting in some males being highly successful and others not (Aoki 1989). Among hummingbirds, for example, it seems that learning has been documented only in lekking or non-monogamous species (Snow 1968; Wiley 1971; Baptista & Schuchmann 1990; Gaunt et al. 1994); note, however, that those are the species in which learning would be most easily documented, too, because males sing a lot and one could easily compare the songs from one lek or location to another. What is needed among hummingbirds is a rigorous, comparative survey, to identify species and lineages in which learning does and does not occur, coupled with study of the life histories in which these developmental styles occur. Among other groups, suggestive evidence for learning has also been garnered for the lekking long-tailed manakin (Trainer et al. 2002). Furthermore, the mating system of humpback whales (Payne & McVay 1971; Tyack 1983; Cerchio et al. 2001) is also non-monogamous, perhaps much like that of the three-wattled bellbird; in both the whale and the bellbird, males apparently compete with their songs and continually relearn them from each other within the 'dialect,' resulting in marked changes in song over time within each generation (Noad et al. 2000).

Why Imitative Learning Might Be Lost

The conditions under which imitative learning evolved might also be revealed by appreciating the conditions under which it is lost in a lineage, or becomes less important or less apparent. Consider, for example, the marsh wren and its relatives. The effects of vocal imitation are clearly evident in all but one lineage of this genus: they are clear in both eastern and western lineages of the marsh wren in North America (Verner 1976; Kroodsma & Canady 1985), the Apolinar's wren in Colombia (Caycedo unpublished), the Merida wren in Venezuela (Kroodsma et al. 2001b), and populations of the sedge wren in Central and South America (Costa Rica, Brazil, Falkland Islands; Kroodsma et al. 1999c, 2002). In all these lineages, neighboring males have nearly identical songs, and distant males have different songs, revealing that songs are imitated within local dialects; furthermore, this imitative learning has been extensively documented in captive-reared marsh wrens (Kroodsma & Canady 1985). The one oddball is the North American population of the sedge wren (Kroodsma et al. 1999b); captive-reared males largely improvise their songs, and neighboring territorial males have unique song repertoires, each male 'making up' a hundred or more different songs (**CD1 #28**).

What seems to be unique to these North American sedge wrens is a semi-nomadic life style. Males banded early in the season at one breeding location typically leave there mid-season, apparently going elsewhere to breed, as they arrive at lower latitudes during July and August, breeding there, too (**Fig. 4.5**). Out of 300 males banded on their breeding territories, only one returned the following year (Johnson unpublished). In all other *Cistothorus* wren groups, birds tend to be resident or highly site-faithful. Sedge wrens in Argentina (Gabelli unpublished), in the Falkland Islands, and in

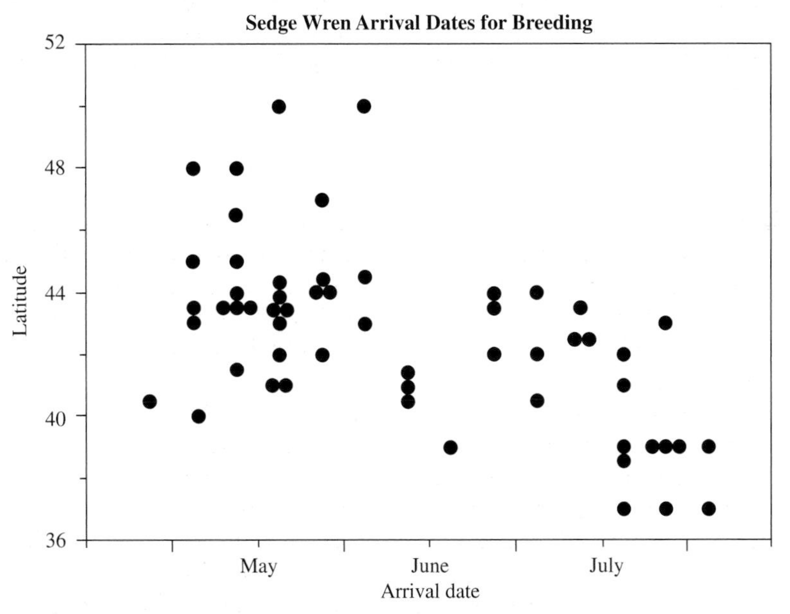

Figure 4.5 The unique, breeding-season dispersal patterns of sedge wrens in North America. Like most migratory songbirds, sedge wrens arrive at lower latitudes first and higher latitudes later, as during the month of May. During June and July, however, they seem to leave their first breeding destination, choosing another place to breed on the way back south, arriving in the more southern States of Iowa, Nebraska, Illinois, Indiana, Ohio, and Kentucky during June and July, sometimes even early August. This semi-nomadic behavior is believed to have contributed to the sedge wren's unique style of song development, in which it improvises rather than imitates songs. From Kroodsma et al. 1999b.

Costa Rica seem to remain on their territories throughout the year, as do Apolinar's and Merida wrens in the Andes. In North America, many populations of the western marsh wren are also resident the year around. Migrant marsh wrens in eastern North America often return to the same marsh year after year, too (Tintle 1982; Kroodsma unpublished), though some populations of the marsh wren in the Great Plains are less site-faithful (Leonard & Picman 1988).

Why, then, might the importance of vocal imitation have been reduced among the North American sedge wrens? Sexual selection must be strong in their polygynous mating system (Crawford 1977), perhaps helping to account for the song repertoires of up to 300 different songs in the repertoire of a single male (Kroodsma et al. 1999b). Given that a local community of singing males and their accompanying females persists only a few weeks, just long enough to raise one family, males have little opportunity to imitate each other's songs. If they were to imitate each other in order to play the complex, matched counter-singing games that western marsh wrens play (Verner 1976; see p. 123), then they'd have the perhaps impossible feat of relearning their repertoire of a few hundred songs about two or more times each year. The alternative, of somehow maintaining an identical, large imitated repertoire over large geographic spaces, seems impossible, as local dialects inevitably seem to develop when repertoires exceed five or so song types (Kroodsma 1999). Under these semi-nomadic circumstances, it seems that selection for imitation is reduced or lost, and instead males improvise, using some shared rules for generating species-typical songs.

The bellbirds and wrens seem to tell the same story about the evolution of vocal learning. In bellbirds, vocal imitation has evolved fairly recently, and occurs now in a polygynous, lekking mating system, in which competing, long-lived males are highly familiar with each other. In the North American sedge wrens, vocal imitation has been lost fairly recently, apparently in a polygynous mating system in which relationships are fleeting and competing males are highly unfamiliar with each other.

The Meanings of Words: Learn, Imitate, Innate, Inborn . . .

One final thought on word choice seems important at this point. The meaning of words such as 'innate' and 'inborn' and 'learn' and 'imitate' are often unclear, and the ambiguities can cloud thinking (Johnston 1988; see Chapter 1). When discussing the flycatchers above, I readily slip into 'shorthand' and say that 'no learning' occurs, but I'd be more accurate if I consistently wrote that "the details within the frequency–time envelope of the song seem to develop normally without exposure to adult song; the instructions on how to produce those details must somehow be encoded in the nucleotide sequences of the DNA." In contrast, if a bird 'imitates,' it copies the details of a song from another bird, and that imitation is believed to occur in two stages.

Technically, a bird could still learn but not imitate, although, and here's the confusing part, in common usage 'learning' and 'imitating' are used interchangeably. The sedge wrens, for example, don't seem to imitate, but they might learn something from songs that they hear in their environment, by improvising on them, for example; they might then learn from themselves by auditory feedback as they practice, as they match their song output to what they somehow 'know' to be a 'proper' song for a sedge wren, so that learning but not imitation is required in order for normal songs to develop. I should add here, too, that 'to learn a song' can have yet another meaning. A bird can learn to recognize a song, such as that of a neighbor, but never attempt to reproduce it, just as we humans learn to recognize many sounds in our environment.

What is important to realize is that neither the sedge wren nor the eastern phoebe has imitated songs, but how they acquire their songs is markedly different from each other. In a North American sedge wren, the precise details of his hundreds of songs are not encoded in the genes

– rather, it seems more appropriate to think that some general instructions are encoded there, instructions that enable the bird to produce songs within a range of characteristics suitable for the species. This developmental process is very different from what the non-imitating flycatchers do, and it seems that no songbird encodes the details of its songs genetically as does a flycatcher like the eastern phoebe.

DIALECTS, AND OTHER FORMS OF GEOGRAPHIC VARIATION

Song dialects are one of the consequences of imitative learning. Birds in a given neighborhood, or more extended area, have songs more like each other than birds in distant neighborhoods. Dialects happen because birds tend to breed and therefore to sing in the same local area where they learned their songs. These dialects are a striking feature of white-crowned sparrow song in coastal California (Marler & Tamura 1962; Baptista 1975), as sharp boundaries often occur between adjacent dialects, and not surprisingly these dialects have been the focus of numerous studies (Baker & Cunningham 1985; Kroodsma et al. 1985; Baptista 1999).

Species Differences in How Songs Vary Over Space – A Brief Survey

How birdsong varies over geographic space differs markedly among species. In birds that do not learn their songs, the same songs tend to occur over large geographic expanses, presumably because the genes that define the songs occur over those same expanses (e.g. flycatchers, Lanyon 1978; shorebirds, Miller 1996). When songs do differ from place to place, especially if they vary discontinuously, they provide a first clue that the genotypes of the birds in the populations differ too, and these vocally unique populations may then warrant species status, as with suboscine antbirds (Isler et al. 1997), owls (Robbins & Stiles 1999), and suboscine flycatchers (Stein 1963; Sedgwick 2001). By definition, birds that do not learn their songs cannot have 'dialects,' as that term is usually restricted to geographic differences that arise due to learning.

Among birds that learn songs, dialects have been routinely documented in a host of species, such as neotropical hummingbirds (Gaunt et al. 1994) and parrots (Wright 1996), as well as numerous songbirds, including the sunbirds of Africa (Grimes 1974; Payne 1978), the European wren (Catchpole & Rowell 1993), redwings (Bjerke & Bjerke 1981), corn buntings (McGregor 1980; McGregor & Thompson 1988), house finches in North America (Mundinger 1975), saddlebacks in New Zealand (Jenkins 1978), and, indeed, in almost every songbird that has been studied carefully (Krebs & Kroodsma 1980).

Why Dialects Form

How and perhaps why these dialects form is best illustrated with the following examples. When a first-year male indigo bunting returns from migration, he often learns the songs of an adult neighbor where he settles, thus resulting in several neighboring males with nearly identical songs (Payne 1996). Such 'micro-dialects' also occur in the chipping sparrow, for much the same reason. A young male learns his song from one adult male at his breeding location, either during his hatching summer, like the song sparrow, or after migration the following spring (Liu 2001); the most typical result is a dialect of just two birds, the tutor and his pupil (**Fig. 4.6; CD1 #29**). A similar pattern is revealed in the well-studied, sedentary song sparrow in western Washington. A young male during his hatching summer copies up to 10 entire songs from individual males in a small neighborhood where he will eventually establish his territory; he preferentially learns songs that are shared among his neighbors, too, as if to maximize the probability that he will share songs with those neighbors that survive to his first and subsequent breeding seasons (Beecher 1996; Beecher et al. 1996, 2000a; Nordby et al. 1999). Sedentary song sparrows of southern California appear similar in behavior (Wilson et

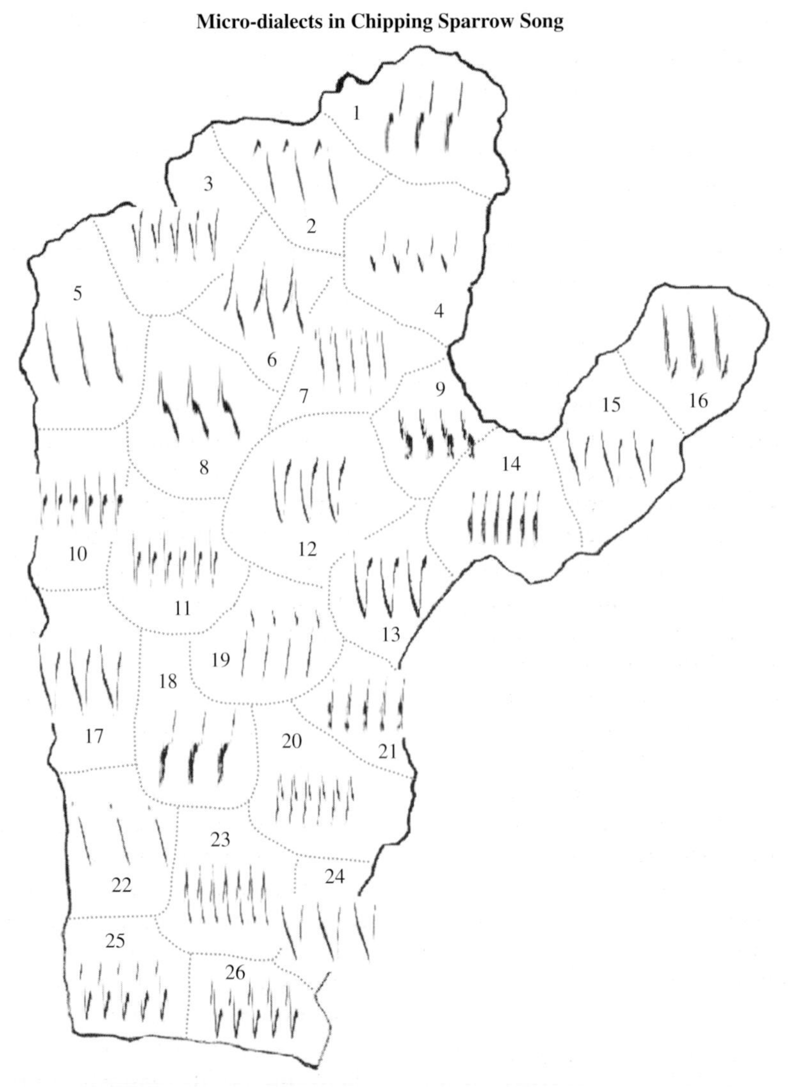

Figure 4.6 Micro-song-dialects in the chipping sparrow. Amid the great diversity of song types used by chipping sparrows, micro-dialects are created when a young male, after dispersal, rejects whatever he had learned from his father and learns the song of one particular male at his future breeding location. Illustrated here are partial sonograms of the songs of 26 males on their 1997 breeding territories at the Quabbin Cemetery in western Massachusetts. Male 16 was the father of male 13, who learned his song from male 12. Male 8 was the father of male 25, who learned his song from male 26. Both male 12 and 26 were also present in 1996, the previous year, so the juvenile could have learned his song during either his hatching year or the following spring. These micro-dialects are disrupted when adults disperse to other territories in subsequent years, e.g. males 17 and 24 were neighbors in 1996. From Liu 2001. (Reprinted with permission from *Handbook of Birds of the World,* Vol. 6, p. 23.)

al. 2000), though the more migratory eastern populations typically share fewer songs with neighbors, perhaps as a result of improvising more songs (Searcy & Nowicki 1999), which in turn may be a consequence of being less sedentary or more migratory.

Why dialects occur is also revealed by differences among closely related species or even among populations of the same species. Among the white-crowned and other *Zonotrichia* sparrows, for example, the resident, coastal white-crowns of California are the ones with the striking dialects (Marler & Tamura 1962; Baptista 1975), not the highly migratory birds in Alaska (DeWolfe et al. 1974; Nelson 1999) or the highly migratory white-throated sparrow of eastern North America (Lemon & Harris 1974). Among eastern and spotted towhees, it is in the resident populations that neighboring males share songs, not in the migratory populations (Ewert & Kroodsma 1994). Among black-capped chickadees, the simple 'fee-bee-ee' song occurs from the Atlantic to the Pacific, with no local dialects, whereas on Massachusetts' offshore islands there are dialects only a few kilometers across; the migratory and irruptive nature of mainland chickadees may prevent formation of dialects, whereas dialects seem to be the default condition in more highly resident songbirds, such as with the chickadees on the islands (Kroodsma et al. 1999a). And in the marsh wren and its relatives, too, stable communities of known adults seem to favor vocal learning and dialects (Kroodsma 1999). The more frequent occurrence of local dialects in resident populations compared with migrant populations may be a common pattern among songbirds (Handley & Nelson in review).

What Song Dialects Tell Us About Group Memberships

These examples of dialects may provide hints as to the conditions under which they form, but to truly understand this question we need to know more. Specifically, we need to know the 'sphere of influence' of the songs, and whom the singer 'expects' to influence with his song. Some paruline warblers, I believe, provide just the clues that we need, because males of many species have two functionally distinct song categories (Ficken & Ficken 1967; Morse 1970a; Lein 1978; Kroodsma 1981; Spector 1992; Lemon et al. 1994; Staicer 1996a; Bolsinger 2000), and the relative selective forces on the two songs can then be better disentangled.

A prime example is the chestnut-sided warbler (Byers 1995, 1996a, b; **Fig. 4.7; CD1 #30**). The songs that males use during contests with other males are shared only with immediate neighbors, forming micro-dialects; these songs must be learned at the site where a young male will breed, and the immediate local neighborhood of territorial males is the sphere of influence for these songs. In contrast, the songs that males use to attract females are highly stereotyped, and do not occur in dialects, instead occurring throughout the geographic range of the species; these songs are thus highly conserved, both temporally and geographically, as a male apparently strives to attract a female that might immigrate from any geographic origin. For these two song systems, then, the sphere of expected influence for a song matches the pattern of geographic variation, local songs used to establish relations among immediate territorial males, widespread songs used to attract females.

This line of thinking is reinforced by the *Cistothorus* wrens (see p. 113). Resident populations develop dialects, and the local songs in resident populations address local birds. But no geographic variation occurs in songs of the North American sedge wren; the improvised songs of this wren are capable of addressing any sedge wren, no matter what its origins or where it has traveled throughout the species' breeding range. The expected sphere of influence of the songs is again reflected in their geographic distributions.

Characteristics of songs convey membership in various groups, from the entire membership of a species to members of a more local population (Hopp et al. 2001). Because learning local songs takes some time, a male who sings local songs declares honestly his tenure within the community, and his songs might therefore be used by others to interpret whether he belongs or not (Feekes 1977). Females might use the song as a cue, for example, to know how long a male has been a resident of the local population (Rothstein & Fleischer 1987).

Vocalizations could vary geographically for

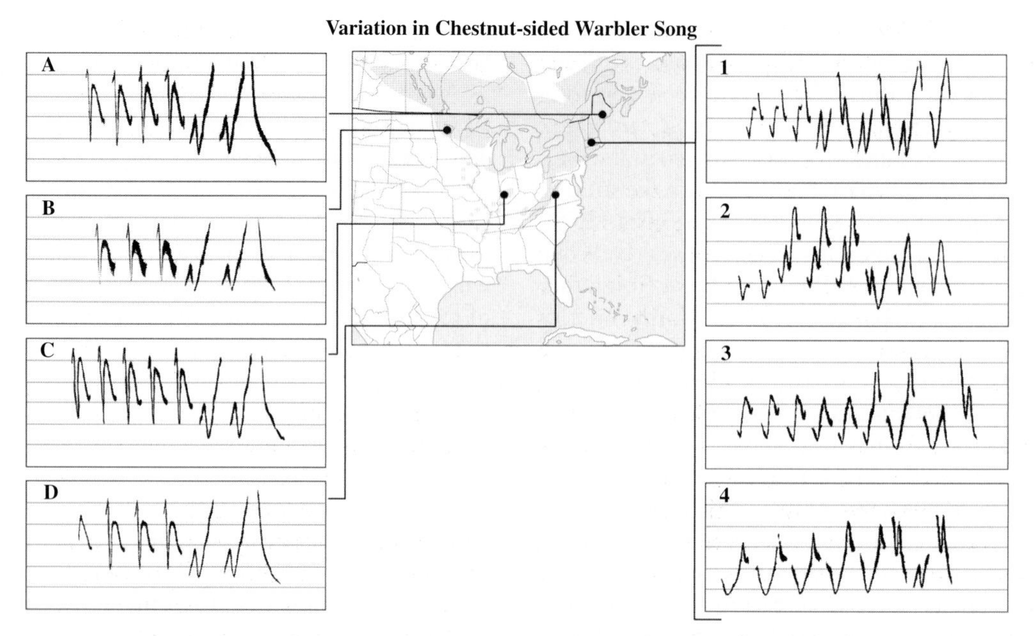

Figure 4.7 Geographic variation in the two song categories of chestnut-sided warblers, illustrating how the expected sphere of influence for a song affects how that song varies from place to place. Songs used to attract females are highly conservative, and occur throughout the geographic range of the species (left), thus enabling a male to attract any female to breed with him. Essentially the same song occurs in Maine **(A)**, Minnesota **(B)**, Indiana **(C)**, and Virginia **(D)**. In contrast, songs used during interactions with territorial males are highly localized, occurring in micro-dialects (right); songs **1–4** illustrate the variety of songs found in several neighboring dialects in the Berkshire Hills of western Massachusetts. Songs used to attract females can be learned from any adult at any time during the first year and at any place, but songs used in male-male interactions must be learned on or near the territory where the male will breed. (Reprinted with permission from *Handbook of Birds of the World*, Vol. 6, p. 36).

other reasons, too, of course. One such reason is the habitat in which the song is used, as habitats can determine which types of sound transmit best (Hansen 1979; Wiley 1991; see Chapter 6). As habitats vary across the geographic range of a species, the characteristics of songs that transmit best might vary, too (Nottebohm 1969; Bowman 1979; Gish & Morton 1981; Handford & Lougheed 1991; Lougheed & Handford 1992; Slabbekoorn & Smith 2002b).

From this discussion it is clear that, in most species in which vocal learning occurs, the micro-distribution of learned songs or song components reflects the social relations among birds, not the genetic structure of the populations (Payne & Payne 1997; Wright & Wilkinson 2001; Slabbekoorn & Smith 2002b; Ellers & Slabbekoorn 2003). To a large extent, birds seem to move freely about, choosing where to breed, and then learning the fine structure of songs in that breeding neighborhood, or somehow adapting their singing to their particular lifestyle, although some less obvious features of the songs might still betray the origins of the singer (Slabbekoorn & Smith 2002b).

But there remains the possibility, in the highly dialectal white-crowned sparrow populations of California, for example, that birds are not entirely free to move and that song dialect boundaries may inhibit dispersal. Perhaps there is a direct "relationship between song 'dialects' and the genetic constitution of populations . . . if young birds . . . are attracted to breed in areas where they hear the song type which they learned in their youth" (Marler & Tamura 1962, p. 375). This idea was the focus of spirited debates during the 1970s and 1980s (Petrinovich et al. 1981; Baker & Cunningham 1985), and was never really settled. The debate lingered because of the absence of good field data on how young white-crowns react when they disperse from their parents' territory and encounter dialect boundaries; in the absence of good field data, inferences on timing and distance of dispersal had to be drawn from learning experiments in the laboratory. And in the laboratory, it was routinely shown that young birds learn best during their first 50 days of life, before dispersal was believed to occur. Because dialects and boundaries between them are fairly stable, it was therefore reasoned that dialect boundaries must inhibit dispersal, or else dialects could not be maintained.

Others have argued that the learning experiments in the laboratory are highly artificial. A young male may be able to learn best before he disperses, for example, but that doesn't mean that he can't or doesn't modify his singing behavior, or learn a new song after crossing a dialect boundary. The potential for learning a song after dispersal has been demonstrated by using live birds instead of loudspeakers to tutor young birds – the sensitive phase can often be prolonged then, when young birds are given a more stimulating environment (see p. 84).

The most recent data suggest that, in mountain white-crowns, birds in different dialects may be slightly different genetically (MacDougall-Shackleton & MacDougall-Shackleton 2001), but even these intriguing data don't quite address the important point. It is perhaps almost inevitable that recently formed dialects will be genetically different to some extent, because it is believed that new dialects often form by a small number of birds establishing territories in new habitat (Baker & Thompson 1985). By this 'founder effect' then, birds in different dialects might differ genetically, just because the founding birds for each dialect were not entirely representative of the source population (**Box 15**, p. 120). The big question remains unanswered, I believe, as to whether an existing dialect boundary influences the dispersal of free-living birds, thus helping to isolate birds in different dialects and, in the process, hastening the formation of new species (Baptista & Trail 1992).

WHEN, WHERE, AND FROM WHOM TO LEARN

These three factors are inextricably intertwined, as illustrated by the life history of a typical songbird. Initially, a young bird is cared for by its parents, and typically the male parent sings during this time, often while feeding the young. During the third week of life, a young male is capable of memorizing songs that he hears from the father or his neighbors. By three to five weeks of age, the young bird becomes independent, and usually disperses from the natal territory, now being exposed to songs from other birds at other locations. Some time late during its first summer, or perhaps the next spring after migration, the young bird establishes a territory, and sings in doing so. The big question, then, is this: when, where, and from whom did he acquire the songs that he now sings? Surprisingly, relatively few good answers exist to that question, because a satisfactory answer requires that young birds be tailed in nature, and that can be a daunting task.

Some hints to answers come from numerous laboratory studies, which reveal that learning among many songbirds occurs most easily early in life, during what is called the 'sensitive period' (see Chapter 3, p. 83). A young white-crowned sparrow, for example, learns most readily during its first 50 days of life (Marler 1970a). Zebra finches learn best during the first two to three months of life (Immelmann 1969; Slater et al. 1988), and a marsh wren's peak of sensitivity is roughly between day 20 and 60 (Kroodsma 1978).

Although laboratory studies like these provide a wealth of information about the learning abilities of young birds, these studies cannot provide the answers we need about the process

BOX 15

BIRDSONG ON ISLANDS: THE SINGING HONEYEATER

The legendary research expeditions of Alfred Russell Wallace in the Malay Archipelago and Charles Darwin's voyage on the H.M.S. Beagle stand as unparalleled examples of scientific insights gained from the study of animals living on islands. Island faunas continue to intrigue zoologists today, and have been the focus of a number of studies over the past 50 years. Of particular interest is the effect of an island's isolation on birdsongs and calls. One common finding is that the range of sounds an island bird expresses is often much smaller than the repertoire of the same species on the adjacent mainland. The effect of isolation on the vocal repertoire can be seen in the singing honeyeater, a species found on some islands in the Indian Ocean adjacent to Western Australia (Baker 1996; Baker et al. 2001). Rottnest Island is 20 km offshore from the coastal city of Perth, and singing honeyeaters are abundant in Perth as well as on Rottnest. The island became isolated from the mainland about 6,000 years ago as the sea level rose. The singing honeyeater population on the island may derive from a rare colonizing event sometime since the island was formed. Evidence indicates that the 20 km distance over water is a serious obstacle for this species, so new colonists rarely attempt the crossing. Comparisons of the repertoire of songs and their constituent note composition between the mainland and Rottnest show a large reduction of vocal diversity on the island. The Rottnest Island population has about six different note types from which all the birds construct their songs; the mainland Perth population has more than three times as many. One interpretation of this pattern is that a very small number of singing honeyeaters colonized Rottnest sometime ago, perhaps in the distant past, and established a breeding population having only a small sample of the notes present in their source population on the mainland. A playback study (Baker 1994) revealed that the island population did not recognize songs from the mainland population, and vice versa, suggesting that the song is an effective territorial signal only in the bird's own population. Apparently the small set of notes present in the colonist's songs formed the available models from which young birds learned, and a tradition based on this small vocabulary has continued over generations, maintained by cultural transmission. This example of a founder effect seems to explain the features of the vocal signals of a number of bird species on various islands around the world. Well known to geneticists, the founder effect also explains the often-observed loss of genetic diversity in island populations.

Myron C. Baker

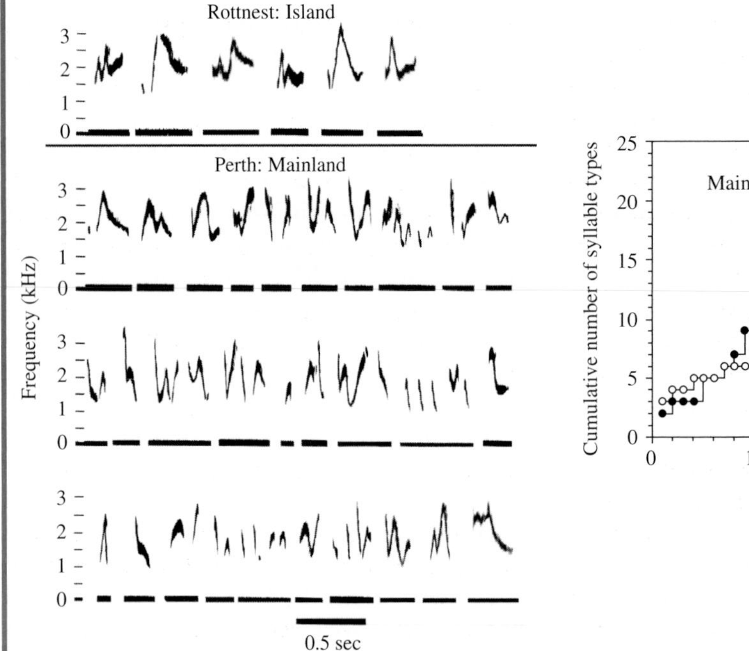

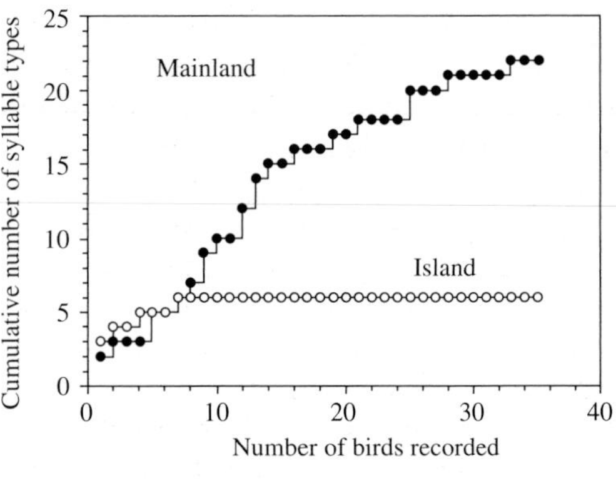

of learning in nature. The primary reason is that the laboratory provides a highly artificial condition, devoid of the rich, natural social environment in which a young bird learns in nature (Baptista & Petrinovich 1984; Pepperberg 1985; see Chapter 13). Use of live tutors has often proven more effective than use of loudspeakers to train young birds, for example, often extending the sensitive period for learning a song (Kroodsma 1978; Baptista & Gaunt 1997a; Payne & Payne 1997), though there is some disagreement about this (Nelson 1997).

So we must turn to field studies to learn when, where, and from whom. One field study does show nicely that a young male of two species of Darwin's finches tends to imitate and sing the song of his male parent (Grant & Grant 1996a). Most other studies show that, even though a young male is fully capable of learning from his singing father, he abandons his father's song for songs encountered after departing from the natal territory.

The well-studied zebra finch, for example, memorizes the song of the father before leaving home. The song that the young male eventually sings, however, depends on the social experiences that he encounters over the next month or so (Slater et al. 1988; Zann 1990, 1997).

Among other species, too, a young bird leaves home and learns the songs of males at the location where he will breed. Two early examples of this pattern were Bewick's wrens (Kroodsma 1974) in Oregon, USA (**CD1 #31**) and saddlebacks in New Zealand (Jenkins 1978). A young white-crowned sparrow can also disperse, often across a dialect boundary, and then at the new location learn a song unlike his father's song (Baker & Mewaldt 1978; Petrinovich et al. 1981; De Wolfe et al. 1989; Bell et al. 1998; Baptista 1999). Young male indigo buntings learn songs of adult neighbors after their first migration (Payne 1996). A young male chestnut-sided warbler also learns after migration, when he acquires the male–male interaction songs from adults in the neighborhood where he will defend his own territory and breed (Byers & Kroodsma 1992; Byers 1996a). His other song, the male–female interaction song, could be learned from anyone, anywhere, and at any time before he establishes his territory (Kroodsma & Byers 1998). Young brown-headed cowbirds disperse to non-natal dialect areas, too, learning the songs and flight whistles at the new location (O'Loghlen & Rothstein 1995). A young chipping sparrow departs home and, either late in the hatching summer or the following spring, acquires the song of one male at his future breeding location (Liu 2001). And young song sparrows of western Washington imitate the details of songs from adults after dispersing to the future breeding territory (Beecher 1996; Nordby et al. 2001).

Indeed, given how routinely song dialects are documented among songbirds, there must be, in the vast majority of species, a premium on having songs that are nearly identical to one's breeding neighbors. Although the singing father may provide an important learning experience for a young bird during the earliest part of his sensitive period, the father's songs are somehow lost, or written over, as the juvenile acquires songs that will be needed at his own breeding location. Early memories, such as those of the father's song, may be retained in some fashion, as revealed by white-crowned sparrows that may practice these old memories again each spring (Hough et al. 2000).

And, although an early sensitive period seems common among birds, so that few if any new songs are acquired after the first breeding season, birds of some other species do appear to routinely learn new songs throughout life. Male domesticated common canaries continue to add new components to their song each year (Nottebohm et al. 1986), as do northern mockingbirds (Derrickson 1987) and the ubiquitous European starling (Eens et al. 1992; Eens 1997; Hausberger 1997).

Exactly how these new components are added to the repertoire may not always be clear, but two possibilities exist. Some components may have been learned anew in the year that they appeared, or some may simply be recalled from previous exposure, perhaps early in life, during the process of selectively forgetting certain songs in the repertoire, termed 'selective attrition,' (Marler & Peters 1982b). Both processes undoubtedly occur in nature, and the ability to learn later in life probably occurs more often than suspected. In field sparrows, for example, how a yearling male acquires his songs during his first breeding season has been interpreted as involving only selective attrition (Nelson 1992), but field and chipping sparrows are also capable of learning new songs at that age (Liu & Kroodsma 1999; Liu 2001). Songs in individuals of other species change gradually over time, indicating that adult birds are continually relearning their songs, as in yellow-rumped caciques (Trainer 1989) and three-wattled bellbirds (Kroodsma et al. unpublished).

REPERTOIRE SIZE

Rivaling the study of song dialects by avian bioacousticians has been the study of song repertoire size (see Chapter 2). As with song dialects, repertoire size among songbirds varies greatly. Some species, such as the cedar waxwing, have no vocalization that we would consider a song (Witmer et al. 1997); nor do most corvids sing in the usual sense (Thompson 1982; Brown & Farabaugh 1997). A male Henslow's sparrow has a pathetically simple song, often said to be "one of the poorest of vocal efforts of any bird" (Peterson 1947; **CD1 #32**); a male chipping sparrow also has just one song, consisting of a lengthy series of a single repeated unit (Borror 1959). Male white-crowned sparrows typically have just one song, consisting of several different phrases, though males in dialect contact zones may be bilingual, singing the song of each dialect (Baptista 1975). Larger still is the repertoire of a song sparrow, with 10 songs or more (Mulligan 1966; Searcy & Nowicki 1999); a western marsh wren, with 100 to 200 songs (Kroodsma & Verner 1997); a sedge wren, with 300–400 songs (Kroodsma et al. 1999b); and a brown thrasher, with well over 1000 songs (Boughey & Thompson 1981; **CD1 #32**).

How Repertoires are Used

How birds use their repertoires also varies (Hartshorne 1956). A bird with only one song in its repertoire repeats that one over and over, though undoubtedly with meaningful variation from one song to the next (Ratcliffe & Weisman 1985; Searcy & Nowicki 1999; Searcy et al. 2000). Some species, such as the New World paruline warblers, have two categories of songs, and they use them in different contexts, one primarily intra-sexual, the other primarily inter-sexual (Ficken & Ficken 1967; Morse 1970a; Lein 1978; Lemon et al. 1985; Spector 1992; Byers 1996b; Staicer 1996b). Other species, those with a 'repertoire' of multipurpose songs, have a choice. Often, with a relatively small repertoire, the male sings with 'eventual variety,' producing one song type several times before switching to another, as in the song sparrow (Nice 1943). Species with larger song repertoires often sing with 'immediate variety' (Hartshorne 1956), such that successive songs are different and more sharply contrasted with each other, as with the marsh wren (Verner 1976; **Fig. 4.8; CD1 #33**). Birds that sing with immediate variety, like the northern mockingbird (Derrickson 1988), often sing rapidly or continuously, as if eager to display their song repertoire.

Among species that have these multipurpose songs, what 'good' is a repertoire of songs? One possible answer is that males can convey their motivational status by how often they switch to a new song type; the more switches, the more intense the motivation, whether the context be intra- or inter-sexual (Kroodsma & Verner 1978; Kramer & Lemon 1983; Searcy & Yasukawa 1990). Indeed, evolution of large song repertoires in some species may have occurred under ecological or social circumstances in which competition for resources was continually high, as in dense populations, where competing males interact frequently and intensely (Kroodsma 1999).

If neighboring males learn their song repertoires from each other, additional information can be encoded in a counter-singing

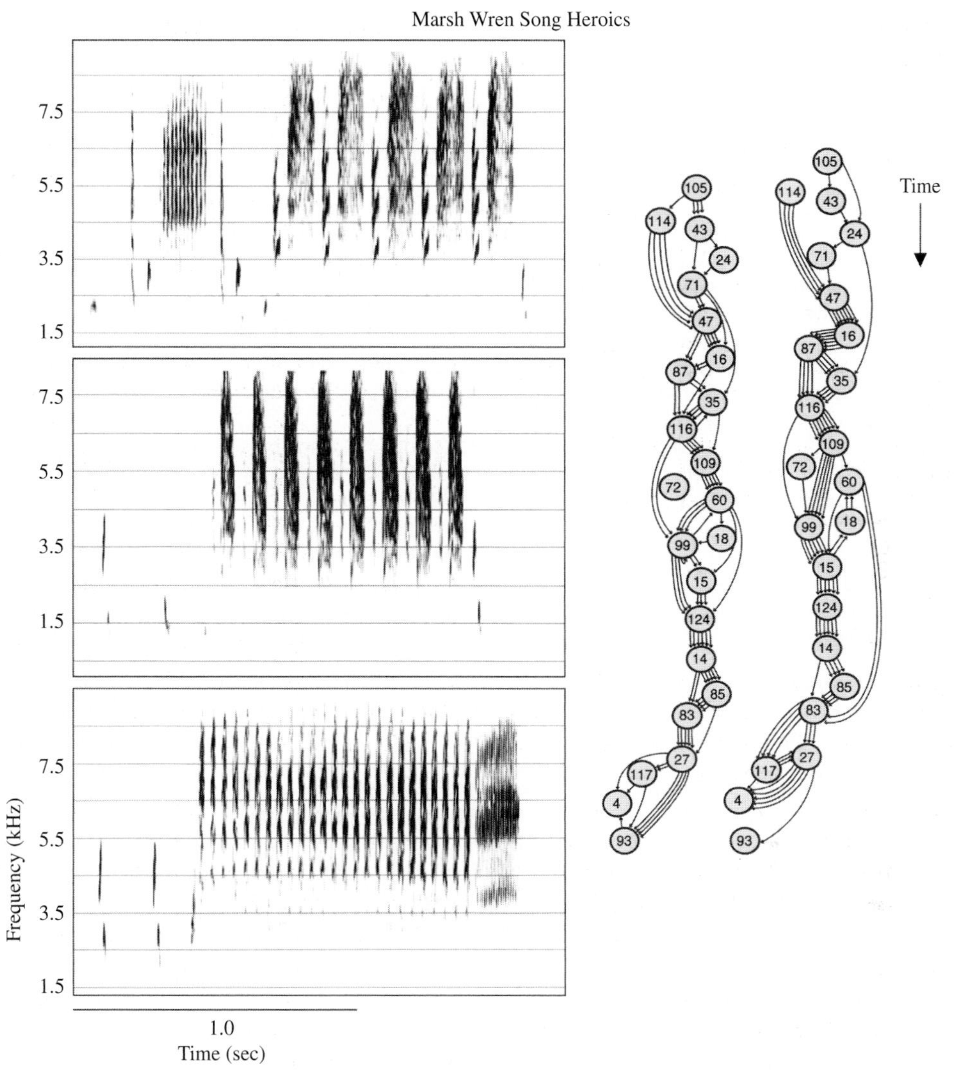

Figure 4.8 Male western marsh wrens have song duels with their large repertoires. Neighboring males learn songs and song sequences from each other, and when they counter-sing, they can reply with the same type, advance to the next one expected in the sequence, or ignore the neighbor by replying with neither of those songs. Who leads and who follows in these exchanges may reveal dominant and subordinate birds, respectively. On the left are three sample songs from one male's repertoire of more than 100 song types. On the right are flow diagrams of the singing of two males at Turnbull National Wildlife Refuge, Washington; the two males shared many song types and sang them in much the same sequences. Each circled number is a different song type in the series, and each arrow indicates that the male sang that sequence once; multiple arrows between two song types show favorite sequences. After Verner 1976. (Reprinted with permission from *Handbook of Birds of the World*, Vol. 6, p. 34.)

exchange. When replying to a singing neighbor, a male has several choices. He can reply to his neighbor with the same song that his neighbor just sang, in 'matched counter-singing,' or, he might choose a different song, as if ignoring the neighbor. If songs occur in predictable sequences, the male might also choose to advance to the next song in the sequence, thereby leading the

BOX 16

VOCAL MATCHING IS A POTENT SOCIAL SIGNAL

Hinde (1958) was the first to note that birds are especially aroused when they hear their own song. His caged male chaffinches were highly responsive to own-song playback, and the study was especially interesting because the songs of some males were abnormal: they had been raised in social isolation. Despite the abnormality, their recorded song was more potent than that of a normal chaffinch. Evidently a rival who sounds like you is especially provocative. More than twenty years passed before neurophysiologists discovered neurons in the songbird brain that are stimulated by song, most acutely by playback of the birds' own song (B.O.S.), even in birds singing abnormally (Margoliash 1983, 1986; Sutter & Margoliash 1994). One interpretation was that these neurons might be part of a learned song 'template,' to which the young male then matches his own voice, a notion that still has some currency. But then it was shown that much of the B.O.S. phenomenon actually arises *after* a male's song development is completed, and not before (Doupe 1997). Another possibility is that this is a mechanism to sensitize birds to songs like their own.

If we assume that they possess a mechanism to calibrate others' songs by self-reference, how might birds make use of it? Is song matching aggressive or affiliative? Song matching by adult territorial males during counter-singing seems to have definite aggressive connotations, especially when one song overlaps another (Todt 1981; Vehrencamp 2001). But with nonoverlapping song matching, there are also hints of other, subtler functions for vocal matching. The intricacies of vocal duels in marsh wrens, giving long series of matched songs with one leading and one following, suggest a role in establishing and reinforcing male dominance (Kroodsma 1979). There may even be circumstances in which song matching is more affiliative, serving "as a vocal greeting that confirms established social relationships" (Todt 1981). The flight call matching that occurs between mates in many finches is surely more affiliative than agonistic, and its use in winter flocks suggests that it is related to group cohesion rather than disruption. Field playback studies on orange-fronted conures in Costa Rica suggest that they not only make on-line adjustments to improve matching of the 'chee' loud contact call during an interaction, presumably affiliatively, but may seek a *mismatch* in some cases, if an aggressive signal seems more appropriate (Bradbury 2003; Vehrencamp et al. 2003).

Research on the details of song and call matching in all of these contexts deserves much more attention. They may also provide insights into how song perceptions are organized. We describe songs by the type they represent to us, but there is significant variation within a song type (Searcy et al. 1999). How similar must 'matched' songs be? Is this variation significant to the birds themselves? Does within-type variation have any significance in communication, or is it a matter of habituation? Are there degrees of song mismatching, with extreme cases becoming more alien than others, as studies of male responses to close and distant dialects sometimes suggest (Baker et al. 1987b; Balaban 1988; Slabbekoorn & Smith 2002b)?

Matched singing also affects song development. As a young male sparrow closes out the period for song memorization, and starts to actually sing, he produces more songs than he will need for a mature repertoire. As he establishes a territory, he discards some songs, and concentrates on those that are a best match with those of his territorial opponent (Nelson 1992); presumably he finds them more effective in staking out a claim. For whatever reason, there is a drive for young males to match a rival's song types, as well as they can.

There is undoubtedly something special about songs and calls of others that match your own, but the precise function of vocal matching remains unclear. Sometimes it is aggressive, sometimes affiliative. Could it be that the primary attribute of a matched song or call is that it is guaranteed to catch a listener's attention? This would make it initially neutral, but inherently potent in any learning situation. Perhaps placement on an aggressive/affiliative axis is a result of experience, with different placements associated with, say, combat, a greeting ceremony, or courtship feeding? These would each result in an acquired connotation specific to the particular song or call employed. At some stage, neurobiologists should be able to clarify the situation by tracking vocal matching circuits in more detail, and determining whether there are connections with parts of the brain involved in emotion and arousal.

Peter Marler

neighbor rather than following him (**Fig. 4.8**). Still another choice is to respond to the neighbor not with the identical song that he just sang, but rather with another song that is shared between them, illustrating 'repertoire matching' rather than 'song type matching'. These types of interactions have been documented in a variety of songbirds, such as northern cardinals (Lemon 1968), yellowhammers (Hansen 1981), common nightingales (Todt 1981; Hultsch & Todt 1986; Todt & Hultsch 1996; Todt & Naguib 2000), western meadowlarks (Falls 1985), song sparrows (Beecher et al. 2000a; Burt et al. 2001), and marsh wrens (Verner 1976). These types of male–male exchanges could provide information to listening females about the relative health or desirability or dominance of males (Kroodsma 1979), and could have been an important force in the evolution of large song repertoires in some species.

Song Repertoires and Sexual Selection

Increasingly, a large repertoire is viewed as a sexually selected trait; the larger the better, presuming that the size of the repertoire may honestly convey a male's condition to a potential mate (Searcy & Yasukawa 1996; Nowicki et al. 1998a; Airey et al. 2000; Catchpole 2000; **Boxes 16 & 17**, pp. 124 & 126). Supporting this notion are correlations between repertoire size and reproductive success in nature: a male's repertoire size is correlated with his harem size in the red-winged blackbird (Yasukawa et al. 1980), for example, as it is in the great reed-warbler (Catchpole 1986). In the sedge warbler, the repertoire size of the male parent is correlated with the hatching date of his offspring (Buchanan et al. 1999). An exciting correlation was found by Hasselquist et al. (1996), who showed that the song repertoire size of a great reed-warbler was correlated with his suitability as an extra-pair partner and with the survival of his offspring. More recently, working on the same species, Nowicki et al. (2000) concluded that the amount of stress a juvenile warbler experiences is indirectly related to his repertoire size as an adult, with repertoire size thus providing an honest indicator of health. Laboratory data are frequently cited as consistent with these ideas of repertoire size as a sexually selected trait, as a number of studies have concluded that females respond more strongly to larger than to smaller song repertoires (Searcy & Yasukawa 1996) although questions have been raised about the experimental design (Kroodsma 1990).

Although a multitude of studies are 'consistent with' the idea that song repertoires are a sexually selected trait, no study, to my knowledge, explicitly tests this idea. No study shows that females actually use repertoire size in making mating decisions. Furthermore, alternative explanations are available for most, if not all, of the correlational studies and playback studies reported above. The laboratory studies typically suffer from lack of adequate replication of repertoire stimuli used in the playbacks to females (McGregor et al. 1992; Kroodsma et al. 2001a), and as a result can, at most, show that females are sensitive to variation in male song, but not to repertoire size in particular. In the field, too, it may be that repertoire sizes in many of these studies are poorly and perhaps inadequately documented, as for the European sedge warbler (Wilson 2001).

Besides the drawbacks of poor experimental designs, these laboratory studies also suffer from an interpretational problem. In nature, we know that females of many, and perhaps most 'monogamous' species mate not only with their own social partner but also with other nearby males, so that offspring in a nest have multiple fathers (see Chapter 2). As a female makes her mating decisions, she no doubt listens to multiple males interacting in complex ways, each male singing his full repertoire of songs. Showing in the laboratory that a female is more responsive to a larger than to a smaller repertoire may simply be simulating her more natural listening experience in nature. For me, a laboratory study on common grackles (Searcy 1992b) best illustrates this interpretational problem. These birds nest in noisy colonies, with many birds singing simultaneously, though each male has

BOX 17

SEXUAL SELECTION AND COMPLEX SONG: THE SEDGE WARBLER

The sedge warbler has 'an acoustic peacock's tail' of a song (**CD2 #31**). It is one of the longest and most complicated of all birdsongs. Unlike most birds, the male sedge warbler has neither a repetitive, stereotyped song, nor a repertoire of song types. It uses the average repertoire size of about 70 syllable types to construct songs up to a minute long with up to 50 syllables (see below). Males improvise extensively from their repertoire, so no two songs are ever quite the same. Like the tail of the peacock, these complex male signals have evolved through the process of sexual selection by female choice. In nature, males with larger syllable repertoires always attract females before their rivals with less complex songs (Catchpole 1980). Similarly in the laboratory, females were played synthetic songs through speakers; the more repertoire size increased, the more females gave sexual displays (Catchpole et al. 1984).

But why should female sedge warblers favor males who sing more complex songs? Sexual selection theory assumes that by doing so females are really picking males who are also superior in other ways. For example, such males may be stronger, healthier, and better help-mates for the female. Male sedge warblers with larger repertoires are indeed better fathers, working harder to feed their young (Buchanan & Catchpole 2000). They are also healthier and less likely to be infected with diseases such as malaria (Buchanan et al. 1999). In a long-term study of the closely related great reed warbler, Hasselquist et al. (1996) found that young of males with larger repertoires were more likely to survive to breeding. The accumulated evidence suggests that large repertoire size indicates to females that males are in better condition, and will provide superior genes for offspring viability.

How has song complexity become a reliable signal of male quality? Sexual selection theory suggests that singing a more complex song has additional costs which only the best males can afford to pay. Neurobiologists have long known that song production and learning are controlled by elaborate pathways in the brain. Perhaps more complex songs need more brain space, with a costly investment in additional neurons? We have now shown in the sedge warbler, that brains of males with larger repertoires do indeed have a larger HVC, the main song control nucleus (Airey et al. 2000). In males raised in acoustic isolation, HVC develops to the same size as in sibling controls raised with exposure to song (Leitner et al. 2002). This suggests that the neural basis for song has a strong genetic component and may have some heritable characteristics. The isolated males also developed quite normal songs, and had repertoires as large or larger than their sibling controls. The drive to improvise seems so strong that, even without exposure to song, complexity flourished rather than diminished. Clearly, the song of the sedge warbler still has much to teach us about how complex sound signals and their underlying brain mechanisms have evolved together.

Clive K. Catchpole

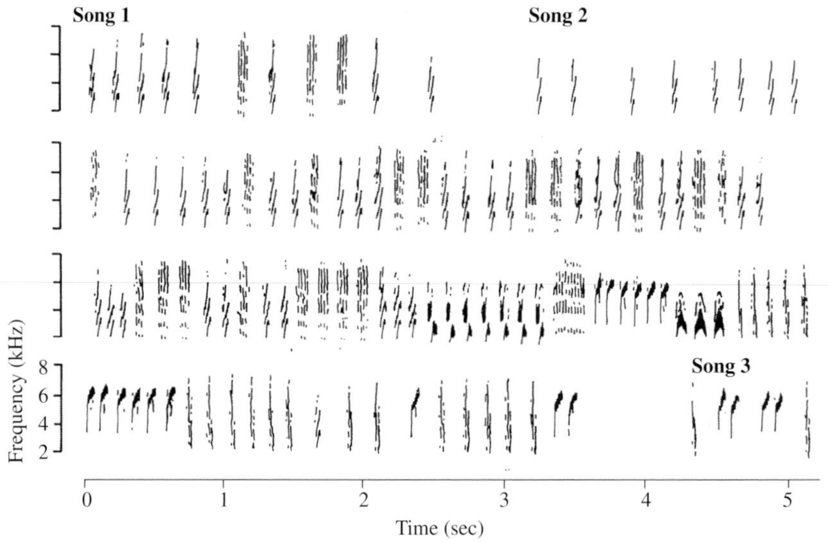

One complete sedge warbler song (2) and parts of two others (1 & 3). From Catchpole 1976.

Canary Alvarez-Buylla

Great reed warbler Wothe

Western grebe Nuechterlein

White-crowned sparrow Nelson

Nightingale Schaap

Field sparrow Elliott

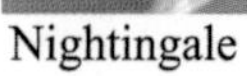

Sedge wren Elliott

Eastern phoebe Elliott

only one particular song that he sings; in the laboratory, a female is more responsive to four songs than to one. To me, a simple explanation for these kinds of results is that female grackles prefer to nest in colonies where they hear many singing males rather than all alone with a single male; that, after all, is what they do. It seems far more contorted and less parsimonious to invoke a 'sexual-selection-drives-large-song-repertoires' interpretation, such that each female actually prefers a larger repertoire in her mate than sexual selection has yet been able to achieve.

The correlational data from the field, as alluring as they are, still fall short. Take the great reed warbler story, for example, for which perhaps the best data exist. If a female is to have offspring with an extra-pair partner, she seems to consistently choose a male who has a larger song repertoire size than does her social partner (Hasselquist et al. 1996). Even though a convincing 10 out of 10 females mated outside the pair bond with a neighbor who had a larger repertoire than the mate, there are no data showing that females actually used the male songs in making their choices. Females could have been using any number of cues, for example, such as the number of songs the neighbor shared with the mate or the local population (Beecher et al. 2000b), the rate of singing (Alatalo et al. 1990), or even some non-vocal cue, such as territory quality, especially if, as happens in most species, older birds tend to have better territories and repertoire size increases with age, as occurs in some warblers of this same genus (Wilson 2001).

In spite of these alternative explanations (Mountjoy & Leger 2001), the idea of repertoire size as an honest indicator of male quality has great appeal and seems almost universally accepted. Nowicki et al. (2000, p. 2419), for example, 'statistically explain' just a little over one percent (1.6 %) of the variation in repertoire size by correlating it with a measure of developmental stress, and then conclude that they have "evidence for song learning (of repertoire size) as an indicator mechanism in mate choice." This particular study begins with the statement that "females of many songbird species show a preference for mating with males that have larger song repertoires" (Nowicki et al. 2000, p. 2419) and then smoothly slides into a discussion of mate choice, without the needed evidence that females actually choose a mate on the basis of repertoire size (or even song, for that matter); so appealing is the notion of mate choice based on repertoire size that it is essentially assumed to be the case (**Box 17, p. 126**).

What is missing from all of these studies is direct experimental data that females, under natural conditions, choose mates or extra-pair matings based on repertoire size. Until repertoire size can be manipulated and assigned to different males at random, a not impossible task under semi-natural, controlled conditions (Brenowitz et al. 1995), a true test of repertoire size as a sexually selected trait cannot be done (Kroodsma & Byers 1991).

And, of course, even if such selection for mate choice based on repertoire size does exist, it is certainly not universal among songbirds, and perhaps even rare, because so many species have small repertoires or no repertoire at all. Furthermore, selection seems, in some species, to constrain males to a small repertoire even though they are capable of more. Young sparrows practice singing more song types than the ones they eventually use (Marler & Peters 1982b), for example, and some males at dialect boundaries are bilingual, having a larger repertoire than the average male elsewhere (Baptista 1975). Also, black-capped chickadees typically have a single song that is varied in frequency (Horn et al. 1992; **CD1 #34**); in the laboratory, however, birds achieve song repertoires up to four different types, just as repertoires form in certain isolated populations of this species (Kroodsma et al. 1999a). Why repertoires are so constrained in some chickadee populations is unknown, though social factors involving female choice of mates are a likely suspect (Otter et al. 1998), the female perhaps demanding simplicity or uniformity among males.

Countering this widely accepted maxim that "larger repertoire sizes impress females," we find example after example. Why does a male chestnut-sided warbler, with three or even four different mate-attraction songs in his repertoire, sing only one of them most of the time, or why does he have more male–male interaction songs in his repertoire than male–female interaction songs (Byers 1995)? Or why does a sedge wren with 300 to 400 different songs take days to reveal them, as if he didn't care whether anybody knew how many songs he was capable of singing (Kroodsma et al. 2002)? And what is even more impressive about western marsh wrens than their large repertoires of over 100 song types is the fact that there's so little variation in repertoire size among neighbors, as if the premium is on having identical songs with which to interact, not on having more songs than one's neighbor (Verner 1976; **Fig. 4.8**). In song sparrows, too, the better predictor of success, as measured by territory tenure, is not repertoire size, but rather the number of songs shared with immediate neighbors (Beecher et al. 2000b).

Clearly, repertoire size, although an exciting feature of singing in many songbirds, needs to be examined as just one of many potentially 'honest' features of male singing that might be used to impress others (Gil & Gahr 2002). And focusing on tests of the role of repertoire size alone, in isolation from other plausible hypotheses, leads to all of the hazards of 'scientific' inquiry warned of by Chamberlin back in 1898 (reprinted for the 'modern' scientist in 1965).

MIMICRY

Most species, like the white-crowned sparrow, learn songs only of their own species. The white-crowns are selective in what they choose to learn, it had been postulated, because they have an inborn template that guides them in choosing appropriate songs for imitation (Marler 1976; Konishi 1985; Nelson & Marler 1993). Supporting this idea is recent evidence that the introductory whistles in the typical song may play a key role in guiding song development (Soha & Marler 2000), and that preferences may even exist for the songs of one's own subspecies (Nelson 2000b). This selectivity is not absolute, because under the right circumstances it can be overridden, as when a social tutor of another species is provided in the laboratory (Baptista & Morton 1981; Baptista & Petrinovich 1986).

Under some circumstances in nature, too, this natural tendency to learn only conspecific song is overridden. So rare are these events that they are often called to our attention. Examples of such odd singers in nature include a chestnut-sided warbler singing an indigo bunting song (Payne et al. 1984), a Lincoln sparrow singing a white-crowned sparrow song (Baptista et al. 1981), and house wrens singing Bewick's wren songs (Kroodsma 1973).

Many occurrences of interspecific mimicry occur between close relatives or potential competitors, often in zones of secondary contact. In the Great Plains of North America, for example, sibling species of meadowlarks, grosbeaks, buntings, and orioles routinely learn each other's songs (Lanyon 1957; Baker & Boylan 1999); in Europe, too, such species pairs often learn each other's songs (Helb et al. 1985). Such interspecific vocal learning may occur in highly aggressive contexts (Baptista & Catchpole 1989), as would be expected in zones of sympatry where members of sibling species defend territories against each other. This song convergence in sympatry seems common, and divergence uncommon (Doutrelant & Lambrechts 2001).

Mimicry of call notes has been described for several species, too, as some species incorporate not the songs but only the calls of other species into their own song, including white-eyed vireos and lesser and Lawrence's goldfinches (Adkisson & Conner 1978; Remsen et al. 1982). Mimicry of calls of other species, when used in the presence of a predator, might help rally other birds to mob the predator, as may occur in the thick-billed euphonia (Morton 1976; Remsen 1976) and the phainopepla (Chu 2001a).

Why Does a Mockingbird Mock?

Even more celebrated are the renowned mimics that occur throughout the world (Baylis 1982):

BOX 18

TRACING MIGRATORY PATHWAYS BY VOCAL MIMICRY: THE MARSH WARBLER

Vocal mimicry has interested and puzzled ornithologists for ages (Baylis 1982; Baptista & Catchpole 1989; Hausberger et al. 1991). It occurs in many groups, is not restricted to particular habitats or geographical areas, or particular mating systems such as monogamy or polygyny. However, some tentative generalizations seem to be possible: mimicking species tend to be continuous singers with large repertoires and are mostly year-round residents. Relatively few migrate over long distances, one exception being the marsh warbler, which is a remarkable mimic. It breeds in Europe and winters in south-east Africa (Dowsett-Lemaire 1979a). Imitations of 212 different species have been reported. They learn their repertoires on both breeding and wintering grounds, and their route through north-east Africa can be traced by their imitations of some African birds with highly localized distributions (**CD2 #32–34**), such as the Boran cisticola (**A**), imitated by virtually all marsh warblers (e.g. 19 of 20 full repertoires taped in Belgium). North-east African bird sounds imitated include calls of the vinaceous dove (**B**), and the song of the brubru shrike (**C**) heard over a wider geographic range. Altogether 99 European and 113 African species have been identified. Individual repertoires contain hundreds of motifs belonging to 80 or more species, with about a fifth of entire songs still unidentified, perhaps consisting of imitations of unfamiliar species. Marsh warblers appear to be unselective imitators; the most recurrent motifs belong to the most noisy and widespread species. Imitations are usually brief, often less than one second; European and African bird sounds can be recombined into complex motifs. The learning process is completed by the time first-year birds leave their winter quarters (Dowsett-Lemaire 1981a), and the repertoire remains unchanged for the rest of their life span. It takes up to 35–45 minutes of continuous singing for the full repertoire to be heard. The function of such imitative prowess is unclear: there is no relationship between the size of repertoires and mating success (Dowsett-Lemaire 1981b). The mating process is often a very quick affair, with little chance for the female to evaluate the extent of a potential mate's repertoire. An unmated male usually stops singing almost entirely once a female enters his territory and at most gives brief snatches of song while following her around in her search for nest sites. Another unusual aspect of marsh warbler vocal behavior, is the 'social singing' of mated males during incubation (Dowsett-Lemaire 1979b): males share incubation with females and when off-duty, in fair weather, they may engage in bouts of peaceful singing with their neighbors, congregating in little trios or quartets at the edge of their territories. Some males with isolated territories are so keen to participate that they will cross inhospitable country to join a singing chorus. 'Social singing' appears to have no function other than play, and is cancelled in bad weather when more time must be spent looking for food.

Françoise Dowsett-Lemaire

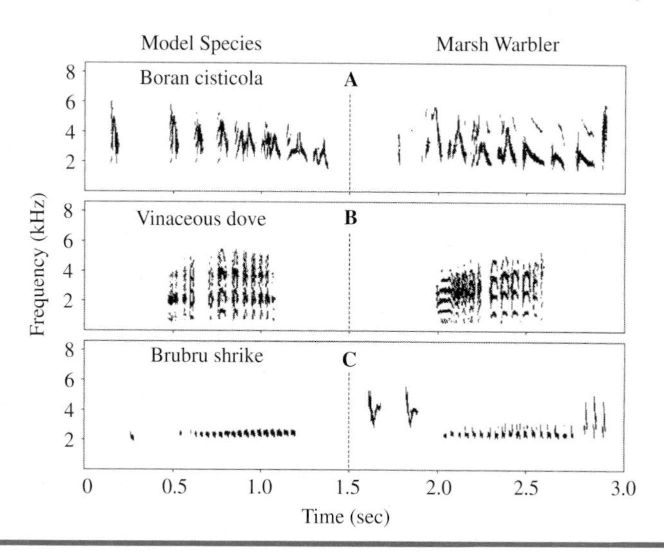

the neotropical Lawrence's thrush (Hardy & Parker 1997), the nearctic northern mockingbird (Borror 1964), the European starling (West et al. 1983; West & King 1990; **CD1 #15**), the Australian lyrebirds (Chisholm 1946; Robinson 1974, 1975; **CD1 #35**), and other species including African birds, too (Harcus 1977). Mimicry by the marsh warbler is especially fascinating, because it mimics bird sounds both at European breeding locations and at African over-wintering sites (Dowsett-Lemaire 1979a; **Box 18**, p. 129); the mimicked calls of the phainopepla also track its travels (Chu 2001a).

A special case of interspecific song learning occurs among the viduine finches of Africa (Nicolai 1964; Payne 1973b; Payne et al. 1998). These finches are brood parasites, and each viduine depends on a specific host waxbill species to rear its young. Male viduines learn the songs of their waxbill hosts and use those songs during courtship (**Box 41**, p. 315). In areas where several sibling viduine species occur sympatrically, the mimicked song of the host may be an important cue for a female to identify males of her own species.

For these diverse forms of mimicry, no single functional explanation can suffice, and several seem plausible. Learning songs of a closely related species may enable an individual to defend a territory against all competitors, whether of the same or different species. Using calls of other species in the presence of a predator could attract additional mobbers. And the host song mimicry by viduine finches makes good sense, given the reproductive habits of these brood parasites.

But why, to repeat one of the most repeated questions about birdsong, "why does a mockingbird mock" (**CD1 #36**)? Could it be a cheap and easy way to develop a large repertoire (Nottebohm 1972) of over 100 songs (Derrickson 1987)? Initially, that seems like a reasonable explanation, but in the context of what the mockingbird's close relatives do, one would more likely conclude that the mimicry *inhibits* the development of a large repertoire. The gray catbird and brown thrasher acquire song repertoires two to ten times the size of the mockingbird's (Boughey & Thompson 1981; Kroodsma et al. 1997), but they do so largely by improvising, or making their songs up, not by mimicking other birds.

Perhaps the mockingbird's mimicry functions interspecifically in some way, the aggressive mockingbird mimicking sounds of other species to help maintain more exclusive rights to resources (Hartshorne 1973; Baylis 1982)? But why, then, the more subtle form of mimicry in the songs of the vireo and goldfinches mentioned above? As Baylis (1982, p. 80) concluded over two decades ago, "avian vocal mimicry is a phenomenon that has raised many questions. Very few have been definitively answered."

CONCLUSIONS

In this chapter, I have sampled only a small portion of the extraordinarily diverse singing behaviors among the roughly 10,000 extant birds. In perhaps half of all these birds, including suboscine flycatchers, the precise details of songs are somehow encoded in the genetic material; in contrast, parrots, hummingbirds, and songbirds have independently evolved the ability to imitate vocalizations, leading to a far greater diversity of sounds and vocal repertoires (but see Chapter 8). New evidence shows that taxonomic boundaries between non-learning and learning taxa are blurred, however, as some close relatives of the flycatchers, the bellbirds, also imitate. The recent evolutionary origin of vocal imitation among bellbirds and the recent loss of imitative singing by a songbird, the sedge wren, suggest that strong competition for mates among familiar males in non-monogamous mating systems could have been the evolutionary impetus for imitation among birds.

Among species that learn their songs, dialects are almost universal. The distribution of song characteristics in these dialects, or the lack of dialects, informs us, in part, about the memberships of groups, and about which birds a male can best influence with his songs. In maintaining these dialects, young birds typically begin imitating songs from a singing parent or a nearby neighbor, but then replace those songs with others that are learned after the young bird leaves home and settles elsewhere, as he learns the nuances of songs from particular birds at that new location. One way or another, the most powerful influence determining the song that a young male eventually crystallizes is provided by the males with whom he competes as he establishes his first breeding territory.

Song repertoires vary greatly among birds, from none to one to dozens to hundreds to thousands, the larger repertoires occurring in species that have acquired the ability to learn, such as songbirds. The notion that large repertoires are used to impress females is appealing, but as yet lacks definitive experimental evidence. And why some birds limit their learning to songs of their own species and others mimic remains a mystery. "Why does a mockingbird mock?" remains an unanswered question.

Chapter 5

Bird calls: a cornucopia for communication

PETER MARLER

INTRODUCTION

The contribution of bird calls to nature's music is minimal. We all know that songs display avian vocal tract virtuosity at its finest. As an intense focus of processes of sexual selection, involvement with reproduction and territoriality has led to a wide range of acoustic adaptations that calls can hardly match. My aim in this chapter is to show that from a functional point of view, the calls of birds actually range far more widely than songs, approaching in diversity and complexity the calls of primates. They provide us with some of the finest illustrations from the animal kingdom of adaptations of signal structure to function (Hauser 1996; Bradbury & Vehrencamp 1998). Some bird calls have semantic content, meaningful to others, and rich in information. They are clearly not the mainstay of the musical side of nature's music, but calls are a goldmine of insights into animal semantics, and many other aspects of vocal communication.

First, we must consider how to distinguish calls from songs. Although there is no sharp boundary, we focus here on cases where the distinction is unambiguous. On a structural level, songs are usually longer and more complex acoustically, involving a variety of different notes and syllables, ordered in statistically reliable sequences; calls are often short, monosyllabic, with simple frequency patterning, often delivered in what often appears to be a disorderly fashion. Functionally, whereas songs play a somewhat restricted role, in territory establishment and maintenance, and mate attraction, the functions of calls include not only reproduction, but also predator alarm, the announcement and exchange of food, and the maintenance of social proximity and group composition and integration. Calls are more deeply involved than song with immediate issues of life and death.

HOW MANY CALLS DOES A BIRD HAVE?

Call Repertoires

All birds have a repertoire of calls, sometimes quite small, sometimes unmanageably large. Relatively few have been intensively studied, and entire repertoires have not often been thoroughly documented. You have to be intimate with the entire behavior of a species, in all seasons and circumstances, throughout the lifecycle, to give a reliable estimate of call repertoire size; and assembling such a catalog is by no means straightforward. There are always problems about where boundaries between call categories should be drawn. Arguments about splitting and lumping arise constantly as you try to determine how many calls a species possesses. Given the uncertainties, estimates are bound to vary. Then further questions arise. Do all species' members have access to the entire repertoire, or are some calls restricted by sex, age, or social status? Does call usage vary with the season? We can broach some of these questions by first considering two well-studied birds – the chaffinch, a small bird of the European woodlands, and the common farmyard domestic chicken.

Chaffinch Calls

Chaffinches are present in gardens and woodlands year round. They are most conspicuous in the breeding season, when male chaffinches out-voice females (Marler 1956; Poulsen 1958; Bergmann & Helb 1982; Bergmann 1993). Males not only sing, but also have a lot of different kinds of calls, more than females. In winter, the sexual contrast disappears; the call repertoire of non-breeding male chaffinches shrinks from eight call types, plus song, down to two (**Fig. 5.1**). At this time of the year, both sexes have just two calls – the flight 'tupe' and the familiar 'chink,' which serves as a separation signal, and doubles as an alarm signal (**CD1 #37**). Five reproductive

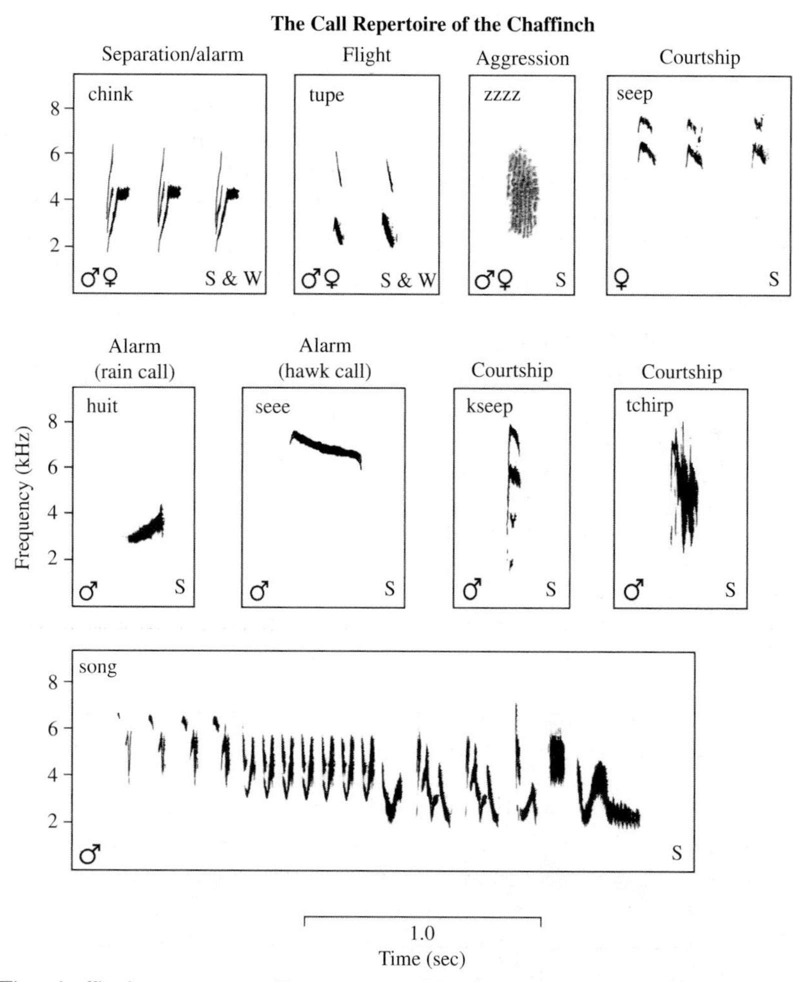

Figure 5.1 The chaffinch, a common European woodland bird, has eight basic calls. Two – a flight call and a separation/alarm call – are used in summer (S) and winter (W) by both sexes (**CD1 #37**). Subsong is not shown. This 1993 catalog for German chaffinches (Bergmann 1993) is identical to the earlier ones for British and Scandinavian chaffinches (Marler 1956; Poulsen 1958), except that the 'zzzz' aggressive call was not recorded by Bergmann.

calls are completely absent from the male's winter repertoire. One of these male breeding season signals is concerned with aggression, two with courtship, and two are specialized alarm calls; one is for aerial predators, and the other is the so-called 'Regenruf' or 'rain call,' thought among other things to forecast bad weather. Females employ only one of these five male breeding season signals, the rarely heard aggressive 'zzzz,' used by both sexes in the reproductive season especially when they attack other birds near the nest (Poulsen 1958). One sexual call is uniquely female, announcing readiness to mate. This adds up to an adult chaffinch repertoire of eight call types, only three of which are given by both sexes. The call types are distinct and discrete, but each one displays some degree of within-category variation, whether it be in structure, loudness, pattern of delivery, or the number of repetitions (Marler 1956), all with potential meaning to the birds themselves.

The chaffinch call repertoire provides a potentially powerful set of communicative signals. When we hear them, we find some are immediately identifiable, standing as potential cues for species recognition. Some sound quite similar to those of other birds, such as the 'seet' hawk-alarm call, also used by other woodland birds. The 'zzzz' of an aggressive chaffinch closely resembles an equivalent song sparrow call (see p. 165). Certain chaffinch calls are quite stereotyped, such as the 'tupe' flight call and others, like the separation/alarm 'chink' call vary a lot, as might be expected with multi-function signals. Most calls are geographically uniform, but there are very obvious local dialects in the chaffinch rain call (**Figs. 5.2 & 5.3**). As we have seen, the chaffinch call repertoire is sexually dimorphic. This is common and, in some birds, such as the ovenbird, it is even more extreme, with almost no overlap between male and female usage (Lein 1980). The physiological basis and development of such differences in call usage between the sexes is almost unstudied.

Chicken Calls

The chicken is a galliform, along with 235 other species. They include many important game birds, and the ancestors of such domesticated species as pheasants, turkey, quail, as well as the domestic chicken. Phylogenetically, they are about as remote from songbirds as they could be (see p. 265). Most galliforms are ground-dwellers, rarely going up in trees except to roost or to flee from predators. There is a certain uniformity to their behavior. Almost all are highly social. Domestic chickens are a case in point. Left to themselves they assume a pattern of social organization much like that of the ancestral junglefowl. Roosters

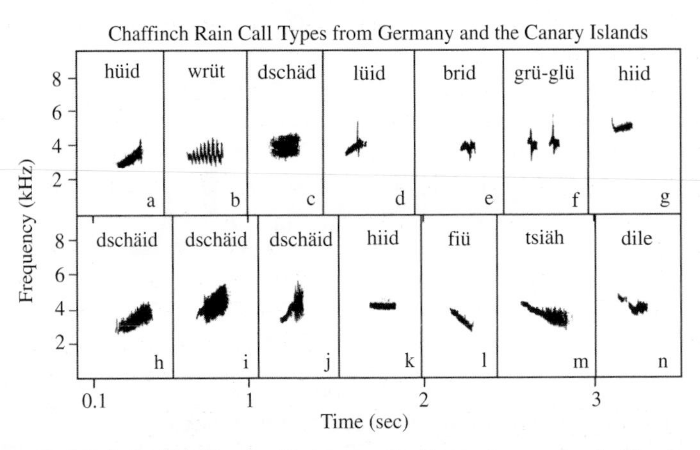

Figure 5.2 There are local dialects in the 'huit' alarm call of breeding male chaffinches, also known as the 'rain call' or 'Regenruf.' Shown with Bergmann's renditions in German are examples from Germany (a–k) and from the Canary Island subspecies (*Fringilla coelebs tintillon*). From Bergmann 1993. (**CD1 #37**).

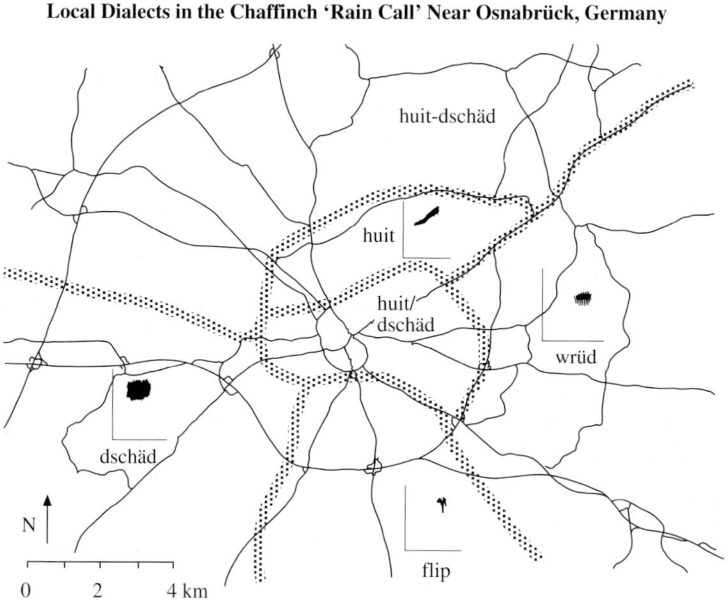

Local Dialects in the Chaffinch 'Rain Call' Near Osnabrück, Germany

Figure 5.3 A map of six chaffinch rain call dialects near Osnabrück, Germany. From Bergmann 1993.

set up a harem of several females and defend it against all comers. They are territorial in the breeding season, with a small harem; in winter they revert to a looser social organization, with dominance hierarchies and overlapping home ranges (McBride et al. 1969).

There are endless contrasts between galliforms and passerine songbirds. Although learning appears to play little or no role in the development of galliform vocalizations, their repertoires are as rich and varied as any known to us. Depending on which observer is doing the counting, adult chickens and junglefowl have a repertoire of up to about eighteen call types, plus crowing as their analog of song (Collias & Joos 1953; Konishi 1963; Collias 1987). The two sexes differ in call usage in somewhat the same way as in chaffinches. Among their alarm calls, for example, the familiar, loud clucking of chickens used as a ground-alarm call is given by both sexes in all seasons (**Fig. 5.4**; **CD1 #38**). But the long drawn-out alarm call for aerial predators that you hear mainly in the breeding season is virtually confined to males, with an interesting exception; females use it briefly when they are broody with recently hatched young. While they are in this distinctive physiological state, presumably under the influence of the hormone prolactin, the vocal behavior of hens changes; they now give aerial alarm calls to warn their chicks, and food calls to attract them to food. At other times, both of these calls are male prerogatives, muted after castration, and reinstated with the male hormone testosterone (Gyger et al. 1988) – offering, by the way, a perfect opportunity for detailed study of the physiology and neurobiology of call production.

Food calls are prominent in galliforms. Roosters use them extensively, during parental behavior, and whenever they are courting females (**Fig. 5.4**). Although acoustic details vary, chickens share food calls and many types of alarm and social signals with pheasants and quail, and their relatives. All galliforms seem to have large call repertoires. Bobwhite, chukar partridges, and California quail are typical. They have from fourteen to nineteen distinct calls used in maintaining contact, preparing to roost, warning of danger, and in sexual and aggressive interactions (Stokes 1961; Williams 1969; Johnsgard 1973).

How do chickens come to have twice as many calls at their disposal as chaffinches? Finches and

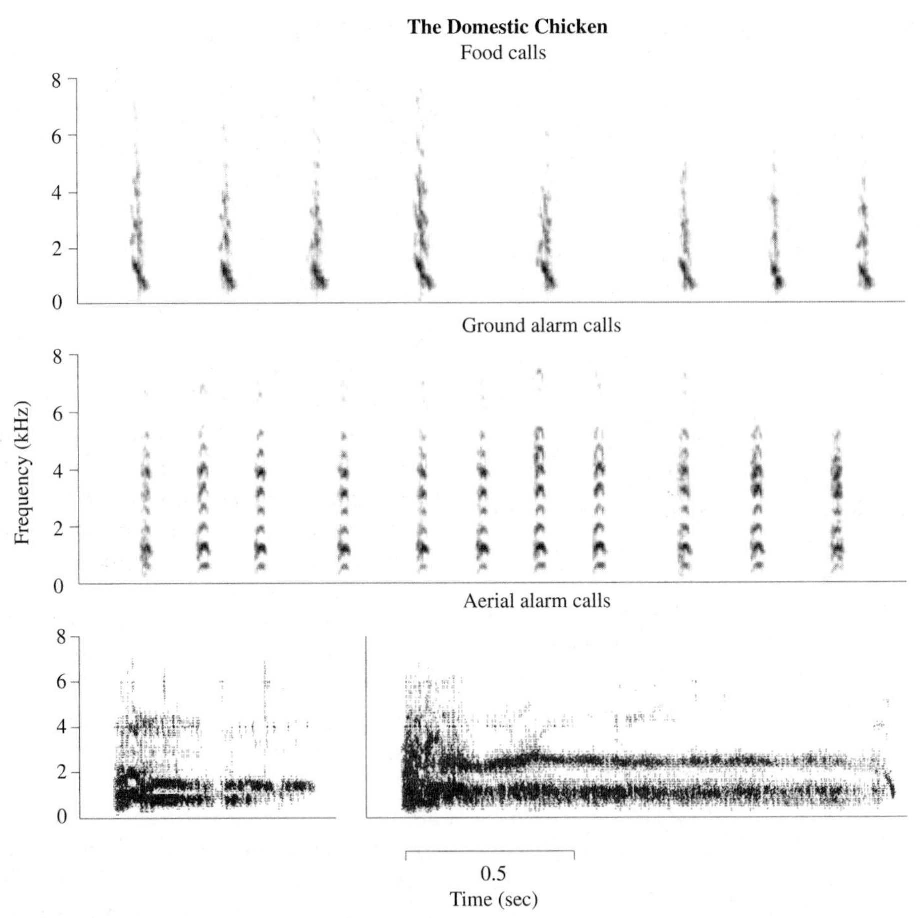

Figure 5.4 The domestic chicken's food call, used by both sexes, is a soft, slow, low-pitched 'took'. For comparison, the louder, faster, more strident 'buck,' used by both sexes as a ground alarm call is shown in the middle, and two aerial predator alarm calls below. From Evans et al. 1993; Evans & Evans 1999. (**CD1 #38**).

fowl differ in so many respects that we can only speculate about how the differences in their call repertoires arose. Some appear to be reflections of contrasting lifestyles. Galliforms are highly social, usually terrestrial, and relatively sedentary, living in structured flocks. Chaffinch society is very different. Unlike the orderly groupings of galliforms, chaffinches form loose, nomadic flocks that are fluid in composition; there is a transformation into an almost solitary state in the breeding season. Reproductive pairs of chaffinches are widely dispersed, and they vigorously exclude others of the same sex from their territory. The chaffinch's general lifestyle is shared with many temperate zone birds, though in some breeding is semi-colonial, as in wild canaries and other cardueline finches. The chaffinch call repertoire is fairly typical of many songbirds. Snow (1958) estimates that adult European blackbirds have seven basic call types. Estimates for two tropical species, the redvented bulbul and the oriental magpie robin are similar (Kumar and Bhatt 2000, 2001). Some repertoires are larger. In Europe, von Haartman and Löhrl (1950) list twelve call types for adult pied and collared flycatchers, most of them used by both sexes. There are similar estimates for the great tit and the black-capped chickadee (Gompertz 1961; Hailman & Ficken 1996; Ficken et al. 1978), and that of the Carolina chickadee seems to be somewhat larger (Smith

1972). This is about twice the size of the call repertoire of the New World dusky flycatcher (Pereyra 1998). In Costa Rica, the long-tailed manakin, remarkable for its complex lek behavior, has another large call repertoire, estimated at thirteen call types, mostly used by males (Trainer & McDonald 1993). If we broaden the search to include more distant relatives and tropical species, it becomes clear that some have substantially different call systems. The collared jay appears to have an exceptionally large call repertoire (Hardy 1967). Corvids, like Steller's jays (Hope 1980), including all the crows, ravens, jays, and magpies, tend to have quiet, inconspicuous songs with no obvious territorial function, but large repertoires of calls with many modulations and intergradations. The rich call systems of ravens and other corvids may relate to their complex social organizations, with cooperative breeding common and exclusive pair territoriality rare. The complexities of parrot social systems may vie with corvids, with equally complicated call repertoires.

It seems likely that many parrot calls are learned, leading to the thought that the capacity for vocal learning, either directly or indirectly, augments the size of the call repertoire. In fact the available evidence points in the opposite direction. We have noted the large call repertoires of many galliform birds, which do not engage in vocal learning. The same is true, as far as we know, of the pale-winged trumpeter, a forest bird of the Amazon basin. Yet the call repertoire of this non-passerine relative of the cranes and rails is estimated at at least 11 discrete call types, plus a so-called 'tremolo song' (Seddon et al. 2002). Except for roosting in trees, trumpeters, like most galliforms, are terrestrial, living in small stable groups. Their call repertoire includes contact and feeding calls, several alarm calls, and two separation calls, as rich a repertoire as any known from passerine birds or parrots. Whatever the benefits of vocal learning, it doesn't look as though augmentation of the call repertoire is one of them, at least in songbirds. Parrots, however, may turn out, with further study, to be a different story.

Parrot Call Repertoires

The fission–fusion pattern of social organization, which is beginning to emerge from field studies of parrots, adds another dimension. Their social groupings are highly fluid, and yet with a quite orderly, though elaborate, underlying organization, based on the mated pair. Their large call repertoires are on a par with those of galliforms, estimated at 10–15 call types for a sampling of New World parrots (Bradbury 2003; **CD1 #39**). Not only is the repertoire large, but there is also a great deal of within-type variation, at least some of which must depend on individual experience for development (**Fig. 5.5**). Call duetting in paired parrots seems to be widespread. There are indicators of call dialects at the level of several social groupings, including flocks, night roosts, and mated pairs, in addition to broader scale geographic variation (Saunders 1983; Wright 1996, 1997; Wanker & Fischer 2001; Wright & Dorin 2001; Bradbury 2003). Call-type matching seems to have progressed to a higher art form in some parrots, with hints of an ability to vary call structure from moment to moment, depending on whom a bird is communicating with at the time (Bradbury 2003; Vehrencamp et al. 2003), and call individuality is widespread. With such complexities, a simple call typology can hardly capture the potential of these parrot calls for communicative complexities, although it can still serve to highlight issues of interest. Alarm calls, for example, seem to be relatively rare in parrots, and yet they dominate the call repertoires of many passerine and galliform birds. The diverse contexts in which birds use alarm signals and other call types form the framework for what follows, beginning with antipredator behavior, and the many bird calls that, in one way or another, signify danger.

ALARM CALLS: THE HAZARDS OF PREDATION

Aside from species at the very top of the food chain, virtually all birds are subject to some kind

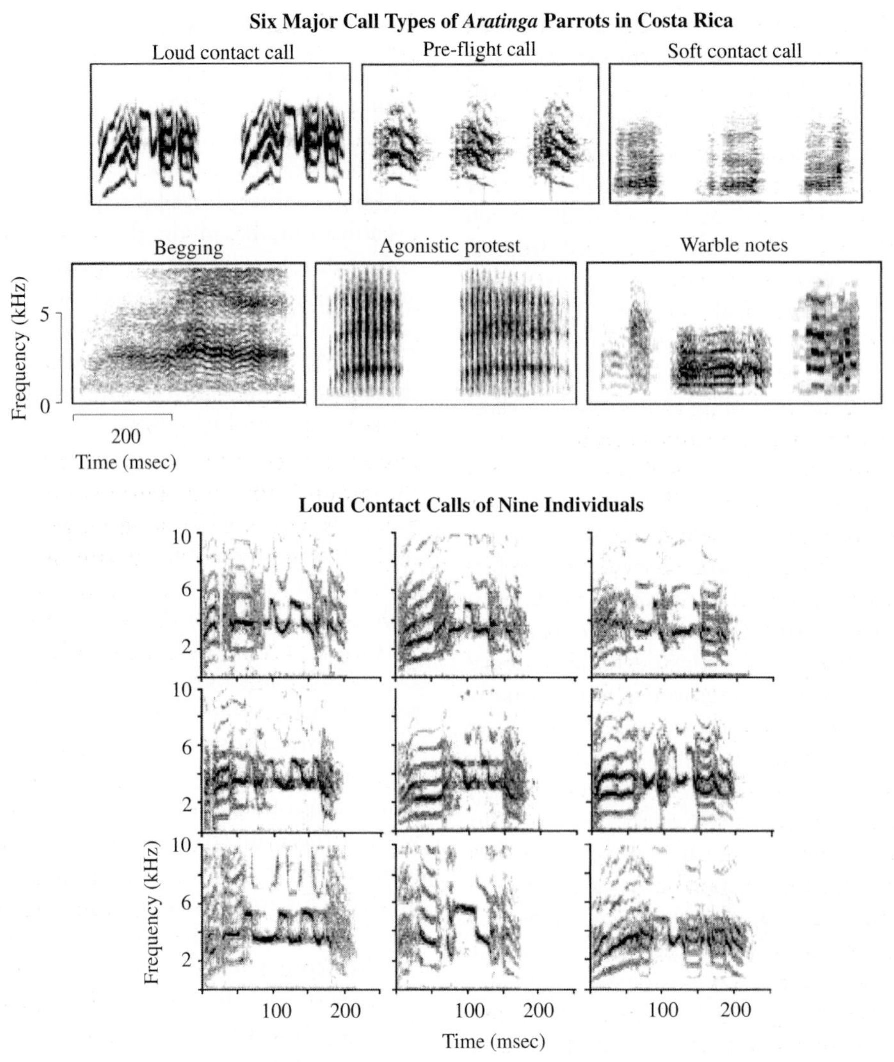

Figure 5.5 Six of the main call types of Aratinga parrots in Costa Rica (top; **CD1 #39**). Below are loud contact calls of nine orange-fronted conures showing individual differences. From Bradbury 2003.

of predation, whether as adults, juveniles, or as eggs. They are predated by snakes, weasels, cats, coyotes, and other mammals, and of course other birds. If a sudden predatory threat is detected nearby, the most logical response might seem to be to dash for the nearest cover, to freeze, and above all, to keep quiet. This is indeed an accurate description of a few birds, such as very young plovers, gulls, and galliforms, and of others, but for one striking fact. Birds that keep silent, whatever the danger, are very much in a minority. It is a mark of avian sociality that almost all birds possess alarm calls as key components in their suite of antipredator responses.

There Are Many Types of Alarm Calls

The selection pressures shaping alarm calling are potent and diverse. Risks for the caller make alarm calls costly. The type of call given in a particular situation obviously varies with the degree of danger and the vulnerability of the caller and its companions, depending in turn on the predator. When it is smaller than you, it may be best to attack. It is better to flee from a larger predator, or to investigate more closely if the identity of the predator is in doubt. The hunting habits and prey preferences are of course crucial. The most appropriate escape strategy will differ according to whether the source of danger is far away, or close by and approaching rapidly; whether it is in the air overhead, or perched and poised for attack. The environmental context is important as well. Is cover available? What are the caller's social circumstances? Is it alone or in a group? Are the birds nearby strangers or family? Not surprisingly, the literature abounds with accounts of alarm calls of many types, including alerting calls, attack and defense calls, hawk and mobbing calls, predator attraction and pursuit-deterrent calls, distress and on-guard calls, and distraction calls that divert a predator's attention towards others (Klump & Shalter 1984). We will focus on some of the better-studied examples, chosen for the general design features that they illustrate.

The relationships between vocal structure and function can become unmanageably complicated if calls serve multiple functions, as some do, but alarm calls nevertheless provide some of the best illustrations we have of their interdependence. Some alarm calls are highly variable in structure and delivery, and it is not uncommon for them to double as contact or separation calls. Or they may vary according to other nuances of the caller's physical or social situation. With some alarm calls there appears to be a high premium on individual and species identity, but others are anonymous, sometimes so similar across species that they provide a common code that enables birds to move and feed as one multi-species flock. It is especially in this situation, when several species have dedicated a similar call to the same, overriding function, that there is a good prospect of uncovering the linkages between structure and function. Among the many factors that must be borne in mind as we try to tease apart the contributions of different acoustic features to alarm call design, one is, simply, how far away can it be heard?

Who is Listening? Active Space and Alarm Calls

The active space of a call is the area around a signaler in which a call can be detected. This is a more complex concept than it appears (see Chapter 6). First, we have to take account of what an actual listener can hear, reminding us that the active space of a call varies, depending on who is listening. Sometimes a call is addressed not only to companions, but to others, not necessarily members of the same species; the predator that triggers a call may itself be an addressee. Then there are the physical circumstances of the caller – is it positioned on the ground, limiting transmission, or up on a perch, or best of all for long-range broadcasting, high in the air? Is the environment quiet or noisy? Is the vegetation dense or open? Temperature gradients and humidity have dramatic effects on sound transmission, changing with season and time of day, all modifying the active space of a call (Wiley & Richards 1982; Larom et al. 1997b; see Chapter 6).

The acoustic structure of the call itself is of course critical. Generally speaking, the lower the pitch, the further a sound travels, except for two factors that can intervene; background noise can mask a particular band of frequencies; also, at ground level, low frequencies can be cancelled out by reflections from the ground; both are relevant if the active space of a signal is to be estimated with any precision (see Chapter 6).

Detecting, Recognizing, and Locating Alarm Calls

Another important factor to be borne in mind as we consider alarm call design is the distinction between detection and recognition (Bradbury & Vehrencamp 1998). We are all familiar with the experience of hearing a sound, and yet failing to identify it (see Chapter 7). A sound must be perceived with some clarity if we are to determine its precise nature, and discriminate between it and another call type (Wiley 1994). The greater the number of acoustic subtleties involved in the recognition process, the harder the task will be, especially with a complex call that is soft, and heard under marginal conditions. Quick, accurate recognition, important in high-stake situations such as alarm calling, may be easier with simple, distinctive calls. On the other hand, in more relaxed circumstances, at close range, call simplicity is a less urgent requirement. In this situation, subtler and complex acoustic vehicles of meaning can be employed and reliably perceived; the conveyance of individual identity becomes feasible, for example. We need more psychoacoustical studies of the acoustic characteristics that facilitate or hinder call recognition under the various conditions that birds encounter in nature.

What about the localizability of alarm calls? Given adequate hearing, reasonable conditions of quiet and audibility, and enough repetitions, any sound is potentially localizable; but it is a familiar fact of life that some sounds are easier to locate than others. Students of audition have long known that broadband sounds, containing a wide range of frequencies are easier to locate than narrowband pure tones. Sudden breaks in a call provide location cues based on differences in the presence of transient frequencies, and arrival time at the left and right ears is most easily compared if the call is repetitive and its onset is abrupt. These are all factors that contribute to alarm call design, as exemplified by 'hawk alarm calls,' given by small birds when there is a flying predator overhead.

Hawk Alarm Calls

For most bird calls, the concept of an ideal type hardly arises, especially when signals with multiple functions are so common. But for signaling in very dangerous situations, say with a hunting hawk overhead, the concept has some relevance. Suppose an alarm call is to be used by a bird in mortal danger, and assume that individuality and even species-specificity can be sacrificed in the interests of functional efficiency. Let's assume further that there is a premium on concealment, both by withholding cues that might make caller location easier, and by limiting the call's active space so that, with luck, it will be inaudible to the predator that triggered it. Can we imagine a 'perfect call' being designed to fulfill these requirements?

Such a 'perfect call' would be a narrowband pure tone, pitched high enough so that it does not travel far, thus limiting active space. It would begin and end smoothly, to minimize other obvious directional cues. The call should be used sparingly because repetitions make caller localization easier. These are all characteristics of 'seet hawk calls' triggered in many small birds by a bird of prey in flight. Such calls serve to communicate alarm to companions while at the same time minimizing risk to the caller. They are pitched high, in a range where hawks cannot hear very well (Klump et al. 1986). 'Seet' hawk alarm calls, as they are known onomatopoeically, also address the possibility that, in a dangerous situation, quick and unambiguous call recognition, even when only faintly heard, is a benefit. These are among the simplest of all bird sounds; they are anonymous, with a similar structure shared by different species; as a consequence of this similarity communication between prey species is facilitated, as many observers have noted (**Fig. 5.6**). Many small birds do indeed come close to achieving the perfect alarm call; they limit individual and species distinctiveness, restricting active space, and make a commitment to one highly specialized function (**CD1 #40**).

Not all birds conform to simple design rules for alarm calls. Hawk calls of Steller's jays are similar to those of the smaller passerines in some respects, but they are very loud and of longer duration, similar to the calls of hawks themselves. Jays give these calls most frequently when a hawk is sighted overhead (Hope 1980). Because jays are moderately large and aggressive birds, they may be less constrained to make their hawk calls cryptic. In addition, this loud hawk call ranges a longer distance while remaining relatively hard to localize compared to jay mobbing calls, discussed later. The combination is suited to warning of a fast moving, overhead predator.

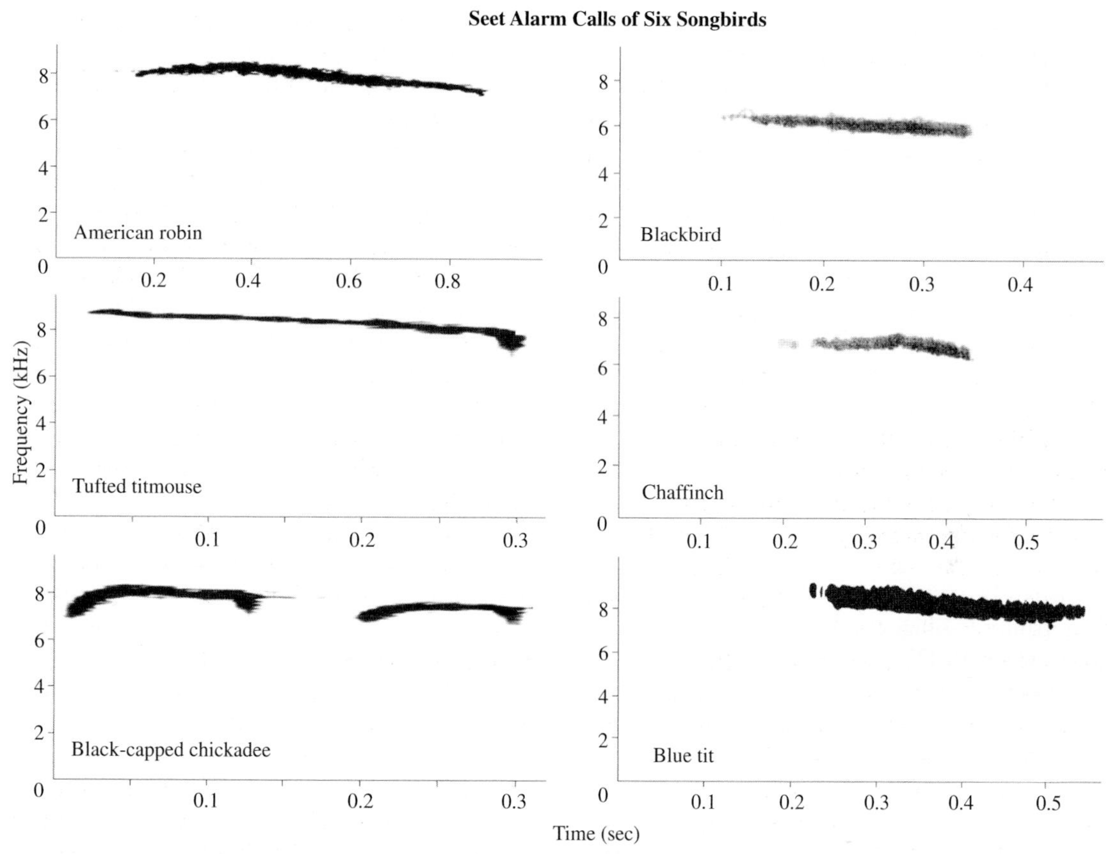

Figure 5.6 Hawk alarm calls, as used by six species of songbirds, three North American (left) and three European. All are long, drawn-out 'seet' calls, very high pitched, given sparingly, with a narrow bandwidth, hard to locate, and a limited active space. From Marler 1955; Klump & Curio 1983; Kroodsma 2000. (**CD1 #40**).

Active space requirements will vary with such things as caller body size, the social system, habitat, background noise, and vulnerability to predators of various types (see Chapter 6). The aerial predator alarm call of chickens has some features of the ideal type, but is lower pitched and ranges further (see p. 179). The alarm calls typically given by Australian passerines to flying predators are, like seet calls, narrowband, relatively high-pitched sounds, but they are shorter, louder, and are often more repetitive than hawk calls of North American and European species, verbalized more often as 'chits' than as 'seets' (Jurisevic & Sanderson 1994). We are reminded that what is ideal for one species may be less so for another, even a close relative, if it has a different lifestyle. Sometimes two phylogenetically distant species have alarm calls that are structurally more similar than two closely related ones if the former share a similar environment and the latter do not.

Mobbing Calls

A predator that is stationary or lurking on the ground presents a somewhat less urgent threat than a hawk flying overhead, poised to strike. Many birds will mob a ground predator, rather than seeking cover and hiding, using a very different alarm call. Mobbing is the typical response to hawks and owls perched nearby, nest predators such as jays and crows, and to mammalian predators as well. So-called ground predator or mobbing calls are virtually antithetical in structure to the hawk overhead call, and they differ in many aspects of function as well. Mobbing birds behave, not cryptically, but noisily and conspicuously, with constant movement and raucous calls. Bolder species such as Brewer's blackbirds and jays will dive at the predator, pecking and chasing as it tries to evade them. In contrast to the soft, shrill, infrequently uttered hawk call, mobbing calls are typically loud, harsh, or ringing in tone, given in long, repetitive sequences (**Fig. 5.7**; **CD1 #41**). Typical examples are the 'chink' of chaffinches (Marler 1956), the 'churring' of great tits (Gompertz 1961), the 'pitt' of the pied flycatcher (von Haartmann & Löhrl 1950), the 'chit' of fairy-wrens (Rowley & Russell 1997), the 'kik-kik' of village weaverbirds (Collias 1963), the 'kecker' of guinea fowl (Maier 1982), and the 'wah' call of Steller's jay (Hope 1980).

The many contrasts between 'hawk' calls and 'mobbing' calls are well illustrated by the great tit. Bringing together a wealth of data on call structure, hearing abilities, and background noise, Klump and Shalter (1984) compared the functional properties of two great tit signals, the 'seet' hawk call, and the 'churr' mobbing call. They calculated that the great tit's major predator, the sparrow hawk, cannot hear the 'seet' beyond about ten meters; a 'churring' great tit can be heard by a predator three or four times further away. This fact alone suggests that some alarm signals are aimed as much to the cause of the alarm as to the caller's companions (Rooke & Knight 1977). One aspect of mobbing call function is 'pursuit deterrence,' the idea that calling informs the predator that it has been detected, and encouraging it to go and hunt elsewhere, while inducing others to join in the effort to persuade the predator to depart (Curio 1971).

During mobbing, other birds often approach and join the chorus (Curio 1971). Playbacks of the chaffinch 'chink' in a German woodland elicited the approach of 30 species, including not only chaffinches but also great and blue tits, European robins, and blackbirds (Fleuster 1973). It is interesting to note that the 'hawk calls' that these same four birds share with the chaffinch, also elicit cross-species responses (Marler 1955). However, whereas their hawk calls are acoustically almost indistinguishable, their mobbing calls are distinct. While they are all wideband, loud, repetitive, and easily localized, the similarity ends there. Evidently the constraints on the physical structure of mobbing calls are more relaxed than those for hawk calls, as broad surveys of Australian and North American birds suggest (Jurisevic & Sanderson 1994; Ficken & Popp 1996). Also, mobbing calls often serve additional functions that may favor different acoustic features, besides being far-reaching and localizable. The chaffinch 'chink,' for example, functions not only as an alarm call, but also as a separation signal, a function requiring a degree of specific distinctiveness that may be less critical when mobbing a predator.

Distress Calls

Different again from hawk and mobbing calls is the 'screech' given when a bird is held in the grip of a predator; this is the so-called 'distress call,' another highly specialized signal. Distress calls are easy to locate. They are often piercingly loud, probably among the loudest of all bird calls. Birds of the same and other species are attracted from a considerable distance, sometimes including other predators. A bird of prey may attack the captor, sometimes distracting it sufficiently that the victim escapes unharmed, only to be caught by the newcomer (Perrone 1980; Inglis et al. 1982; Högstedt 1983; Klump & Shalter 1984). Hunters and birdwatchers use simulated distress calls as a lure, to bring birds out of cover. This call is especially characteristic of older juveniles when captured. The duration is in the middle range as calls go, up to half a second, and distress calls are repeated several times with a brief pause between. Their tone is harsh, sometimes little more than a blast of noise, but with the frequency adjusted for long-range

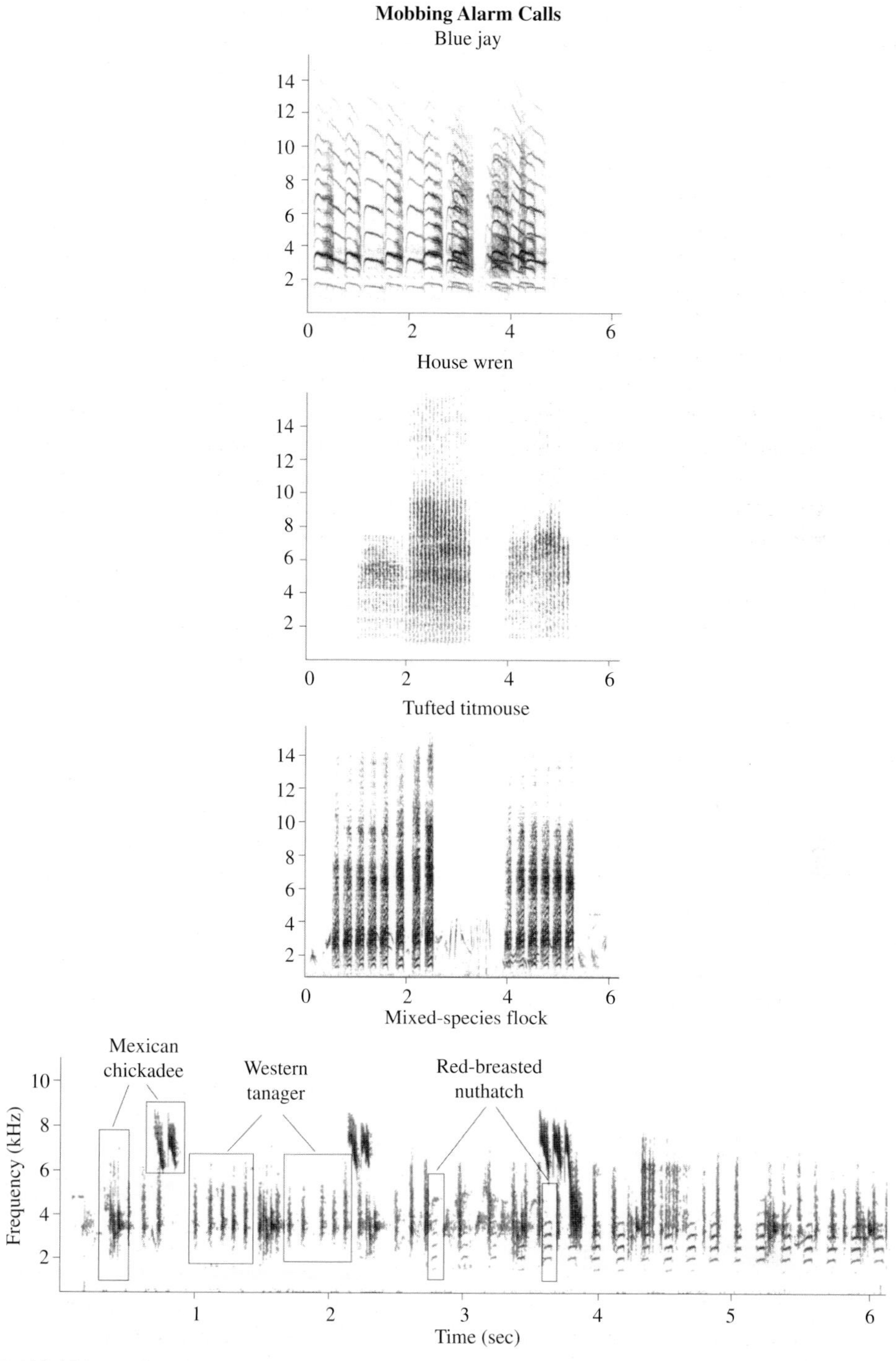

Figure 5.7 Mobbing calls, given in response to a predator like a fox or an owl (**CD1 #41**). These are loud, short and repetitive calls with a wide bandwidth. They can be heard at a considerable distance, and they attract others, often more than one species. At the bottom are three species giving mobbing calls together. These calls share obvious acoustic features, and are all easy to locate, but they are by no means indistinguishable, in contrast to the 'seet' calls in **Fig. 5.6**, which are hard to distinguish. From Kroodsma 2000.

transmission (Mathevon et al. 1997). Their grating tone is sometimes created by the multiple, dissonant frequencies that result from rapid frequency and amplitude modulation (Marler 1969). The most common verbal description of distress calls is that they are scream-like, a good description of the distress calls of sparrows.

Like willow tits (Haftorn 1999), swamp and song sparrows often give distress calls as you remove them from a mist net (Stefanski & Falls 1972a, b). The two sparrows call very similarly, with rasping buzzes that sweep from three to six kilohertz, lasting about a fifth of a second (**Fig. 5.8; CD1 #42**). The harsh tone is created by rapid modulations imposed on two bands of energy only a few cycles apart, creating the irregular beats that contribute to the noisy quality. With up to twenty calls repeated twice a second, the result is a startling noise, audible at considerable distances and guaranteed to capture the attention of everyone in the neighborhood. Distress calls are given by late nestlings and fledglings and by adults of both sexes. Tape recordings evoke strong responses from wild birds, peaking during egg laying, and when fledglings and nestlings are present. There is emphatic cross-species responsiveness, with song and swamp sparrows each reacting almost as strongly to the other's call as to their own (Stefanski & Falls 1972b). Evidently this is not a call designed to function in a strictly private, species-specific manner, as is the case, for example, with their songs (Searcy 1983; Searcy & Andersson 1986).

The distress calls of scores of other birds are very similar, including woodpeckers, junglefowl, quail and partridges, guineafowl, gulls, parrots, birds of prey, and innumerable passerines (Sumner 1935; Stokes 1961, 1967; Collias 1963; Boudreau 1968; Williams 1969; Stefanski & Falls 1972a, b; Johnsgard 1973; Greig-Smith 1982b; Inglis et al. 1982; Maier 1982; Collias 1987; Bremond & Aubin 1990; Koenig et al. 1991; Conover 1994; Jurisevic & Sanderson 1998; Venuto et al. 2001).

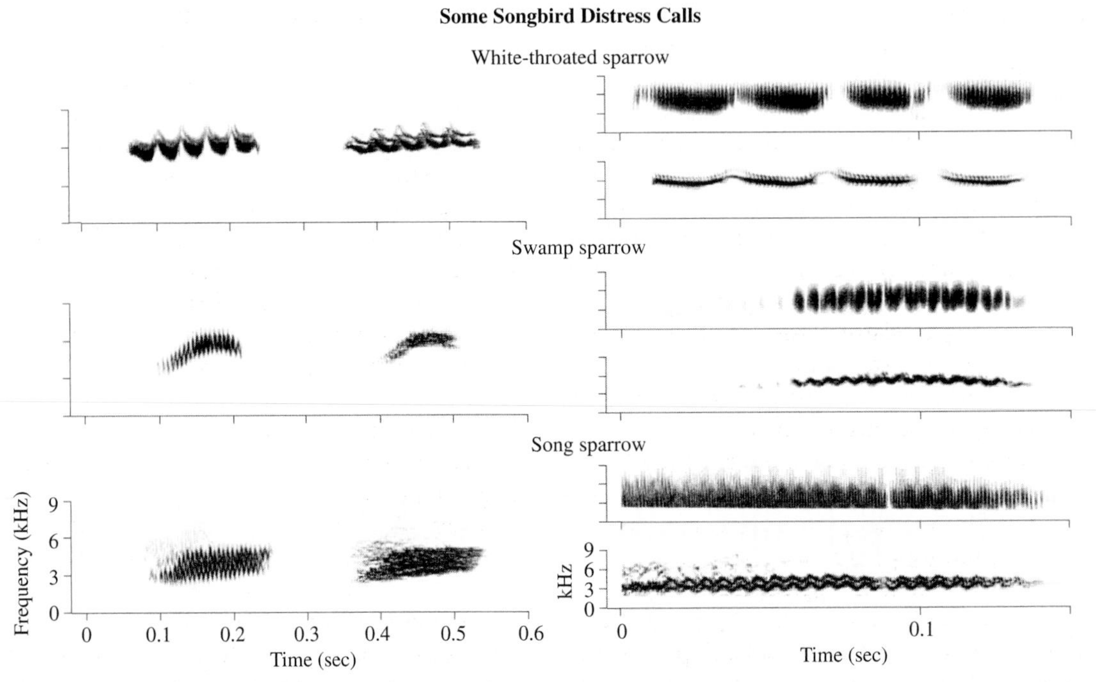

Figure 5.8 Distress calls of three adult sparrows being taken from a mist net. All are long, loud, repeated calls with a rough, rasping quality, characteristic of the distress calls of many birds. From Stefanski & Falls 1972b. (**CD1 #42**)

This is another call whose physical structure appears to be rather narrowly constrained by the functions it serves, although it must be said that the precise adaptive significance of distress calling is sometimes unclear; much ink has been spilt over competing hypotheses (Stefanski & Falls 1972a; Högstedt 1983; Klump & Shalter 1984; Conover 1994; Wise et al. 1999). Among the most compelling is the predator attraction hypothesis, including the notion of prey pirating by another predator, and the possibility of the victim escaping in the confusion. The predator itself can become so preoccupied with retaining its prey that it may itself be vulnerable to predation. A careful study of the North American acorn woodpecker revealed that, despite the close kinship groups in which these birds live, their distress call (**CD1 #43**) is not attractive to other woodpeckers, a finding that seems to favor the predator attraction view (Koenig et al. 1991). In two anecdotes Koenig describes confrontations between an observer and, in one case, a bobcat, lured in by a researcher high up in a tree holding a distress-calling bird in a bag; on another occasion, a grey fox approached repeatedly when a woodpecker gave distress calls (Koenig et al. 1991). Such anecdotes are valuable, given the rarity with which actual predator attacks are observed.

Unlike the acorn woodpecker, some birds do attract their own species with distress calls. The birds that are drawn in make themselves conspicuous, darting in close and even dive-bombing the predator. This behavior favors a 'call-for-help' interpretation of distress calling. It is especially common in parents who sometimes succeed in distracting the predator away from their young even as they endanger themselves, perhaps a case of pursuit attraction behavior? Drawing the attention of all members of the bird community to the predator's presence, in the same way that mobbing calls do (Perrone 1980; Conover & Perrito 1981; Högstedt 1983; Klump & Shalter 1984; Chu 2001b), may reduce prospects of future hunting success and encourage the predator to give up and go elsewhere. The precise adaptive significance of distress calling undoubtedly varies between species according to differences in ecology, social organization, the phase of the lifecycle, and nesting habits, such as hole-nesting. We have to remind ourselves that the relative uniformity of distress call structure among unrelated birds does not necessarily imply complete consistency of function. To be meaningful, interpretations of function must obviously take account of the particular situations in which studies are conducted and the state of the participants, considerations that will in turn guide the researcher's choice of methods for analysis.

Other Ways of Signaling Alarm

Other alarm calls do not fit neatly into one functional or structural class. Some serve an alerting function, and have a rather generalized physical structure that may blend with other calls, such as mobbing or flight signals. The colony-breeding, red-winged blackbird is thought to have as many as seven alarm calls, including alerting signals, some with functions that are still rather ill-defined; the suggestion has been made that simply switching call types may serve to alert the colony (Beletsky et al. 1986; Searcy & Yasukawa 1995). The structure of such calls may be somewhat arbitrary, although as we gain a better understanding of how a call type with multiple functions evolves, we may find even subtler links between structure and function.

Different again are the hissing calls produced by some hole-nesting birds when they are cornered by a predator (von Haartman 1957; Gompertz 1967; Thielcke 1970b). These 'defensive calls' are remarkably convincing imitations of a snake, certainly effective enough to give humans pause. Here, the physical structure of such calls is obviously by no means arbitrary.

For very vocal birds, just becoming silent can communicate alarm. Listening to the hubbub of noise from a colony of red-winged blackbirds you can sometimes hear the moving shadow of silence that precedes a Cooper's hawk as it quarters the colony looking for a meal. In its wake, the blackbirds soon launch into a 'cheer' that serves as their hawk alarm call (Orians & Christman 1968; Searcy & Yasukawa 1995). Conversely, the resumption of a chorus of contact calling after an alarm can function as an 'all clear' signal. Even other species may respond, as when a downy woodpecker, responsive as it is to the alarm calls of chickadees and titmice, becomes less vigilant once the titmice switch to contact calls (Sullivan 1984).

Alarm calls with multiple functions are probably more widespread than we think. The Steller's jay of North America has a loud, particularly noticeable alerting call, 'wek,' given when an intruder approaches on the ground or for other sudden alarming events. It is a brief burst of short, loud, abrupt, harsh notes with a rising and falling chevron structure, a structure noted by Morton (1975) in the alerting calls of many species of birds and mammals. Creatures nearby jays, other birds foraging, and even squirrels – respond by dashing for cover and becoming silent. The wek also serves as a short-range flight call, a distance contact signal, and an agonistic call used in close range interactions between two jays (Brown 1964; Hope 1980). There is an all-clear wek, after which animals that have sheltered will then creep out and resume their activities. How can the same call work in so many different ways? According to the context, weks may be slow or fast, few or many. The apparent pitch of the call varies, although there is no easily discerned difference in acoustic form of the alerting and all clear calls. The contrasting response to the same call may be keyed by the prior alarm, which would then set the context for interpreting the next burst of weks as an all clear signal. Conversely, why use the same call in so many different contexts? It may be that different aspects of call structure are useful in each separate function, or there may be wek sub-types, each used in a different situation, waiting to be discovered when we study them more closely.

CONTACT AND SEPARATION CALLS: KEEPING IN TOUCH

When birds move together, especially when they are in dense cover, whether it be long grass, brush, or the forest canopy, they often use contact calls to keep in touch. These are typically brief, soft sounds often covering a range of frequencies, audible only at close range, murmured repeatedly as group members move around. The group may be a flock, a family, or simply a mated pair. The sounds are often 'peeps,' 'ticks,' or 'clucks,' sometimes noisy in quality, but more often narrow band tonal sounds, perhaps sweeping across a range of frequencies.

To anyone familiar with small Eurasian or North American birds, the following comments of Rowley and Russell (1997) on the contact calls of Australian fairy-wrens and grass wrens will sound familiar (p. 74). These "narrow-band, high-frequency calls, of short duration … are not long-range calls, and are likely to be audible only over short distances, and are thus less likely to betray a bird's location to a searching predator." This description would be valid for the contact calls of a host of other birds. Not all are high pitched; the soft 'ut ut' contact signal of the California quail, given by adults of both sexes throughout the year, is a low-frequency call (**Fig. 5.9; CD1 #44**). Similar soft, repetitive contact calls recur in many galliforms, including junglefowl, chickens, and quail (Ellis & Stokes 1966; McBride et al. 1969; Williams 1969; Collias 1987). Again, these often-repeated calls are designed to be audible only at close quarters. When birds get out of earshot, contact calling often gets louder and more frequent, eliciting counter-calling by other group members until all are reunited.

An alternative is to switch to a 'separation call,' though a similar function may be served by a variant of the contact call, louder and given at a higher rate (**Fig. 5.9**). The crested tit uses a soft 'tic' for contact and a louder 'hack' for separation, with intergradations between them (**CD1 #45**). Field experiments show that the longer-range separation signal of crested tits places the caller at greater risk from sparrow hawk predation, which is no doubt why the tits are more circumspect about using it, compared with the contact call (Krams 2001). Separation calls can be quite distinct and individualized. New World parrots have a 'soft contact call,' repeated often, with or without a reply, while moving in vegetation. Their 'loud contact or separation call,' given in flight and while perched, used to establish a two-way vocal connection with a specific bird, such as the mate, is more highly structured and individually distinctive (Bradbury 2003). California quail have several separation calls, depending on who is separated from whom, and the distance of separation (**Fig. 5.9**). Bobwhite quail have at least three forms of the separation call, used by both sexes (Stokes 1967). Two of them – one loud, one soft – are used when members of a mated pair become separated, the third when birds have gathered at the roost; this is a loud, penetrating call, given in the evening and before morning dispersal, thought to be addressed at least in part to neighboring coveys, who are, perhaps, not attracted but repelled. As Stokes points out, the birds of a single covey can be attracted to the separation calls of members of their group, which they recognize, but repelled by calls from another covey.

Sometimes calls used as separation signals are borrowed from another context, such as flight or alarm. Many finches lack a call dedicated primarily to maintaining contact, but they often use forms of the flight call, differing in tempo and loudness, for both contact and separation. Crossbills that drift apart while foraging for pinecones in dense forest give a burst of loud flight calls, attracting companions and eliciting replies (see p. 152). Chaffinches separated from the winter flock give the 'chink' call, which also has other functions, including predator mobbing (Marler 1956). A single bird separated from the winter flock is very likely to give a 'chink' call

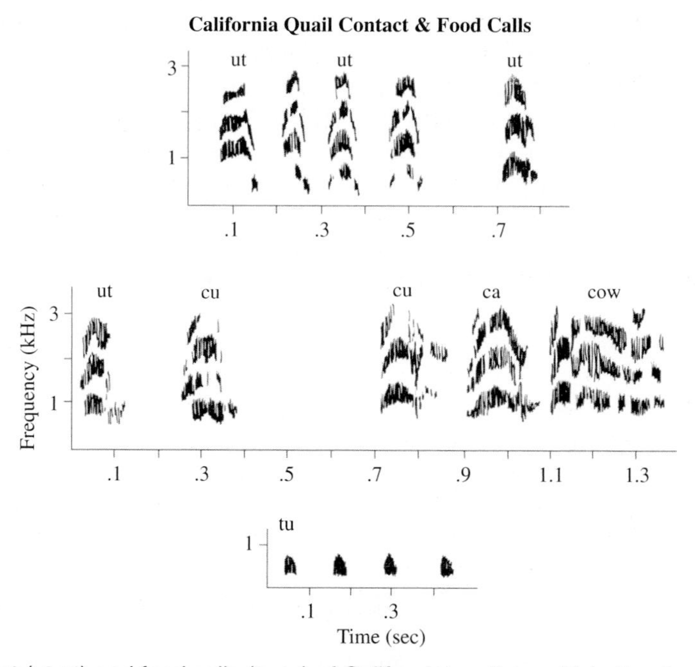

Figure 5.9 Contact (ut-ut) and food calls (tu-tu) of California quail. Loud 'ut-ut' calls are used as a short-range separation call. Over a longer range both sexes use the 'cu-ca-cow' as a separation call, in and out of the breeding season. From Williams 1969. (**CD1 #44**)

on alighting, and to seek contact. When a group alights, it has less need to make contact, and its members are much less likely to call (**Fig. 5.10**).

Flight Calls: Taking Wing Together

The ability to fly brings not only benefits, but also complications, of a social nature. A bird that suddenly takes flight can cause others to panic, and some may lose contact with their companions. This is less likely if a bird signals its intentions beforehand, and many birds have a flight call that accompanies preflight displays. Orange-fronted conures have a loud and harsh 'peach' preflight call (Bradbury 2003; Vehrencamp et al. 2003). A chaffinch foraging with others in open ground will hop into a bush and start to give the 'tupe' that is its preflight signal (**Fig. 5.1**). Others may then begin flight calling too, swelling into a chorus that spreads through the flock until they all take off. If no one else joins in, the caller becomes silent again and resumes foraging. This pattern of behavior recurs in the life of many birds, from geese to goldfinches; it serves to initiate and synchronize movement of groups as large as a flock of hundreds and as small as a pair. There may be further refinements. The tree sparrow in Europe is said to have at least two flight calls, one before and during take-off, and another in flight, and the pattern of use may even forecast whether the flight will be long or short (Berck 1961; Deckert 1962). Over smaller distances – from shrub to shrub – soft, repetitive, contact calls can serve a similar function, a familiar sound to Californians, as flocks of the very sociable bush tit move through the chaparral (Miller 1921).

Some migratory birds move at night and, as they do so, their flight calls are often quite conspicuous. On a quiet night, you can hear them with the unaided ear, and even better if you aim a microphone set in a parabolic reflector up into the sky (**Fig. 5.11; CD1 #46**). Members of the thrush family, especially the wood thrush, are particularly vocal at night and species can readily be distinguished (Tyler 1916; Browne 1953; Graber & Cochran 1959, 1960). Some birds, such as the bobolink, use the same flight call day and night, but there are hints that others use a different call at night. Some are distinctive not only as to species, but also individually, and it is tempting to speculate that this facilitates spacing between flock members at night, especially since they seem to spread out more than during the day (Hamilton 1962). Birds in captivity vocalize frequently during nocturnal migratory restlessness, presumably behaving as they would when actually flying in the wild (Hamilton 1962; **Box 19**, p. 150).

Some birds continue to utter calls while flying; cardueline finches give a burst with each upsweep of their roller coaster flight pattern. The calls of

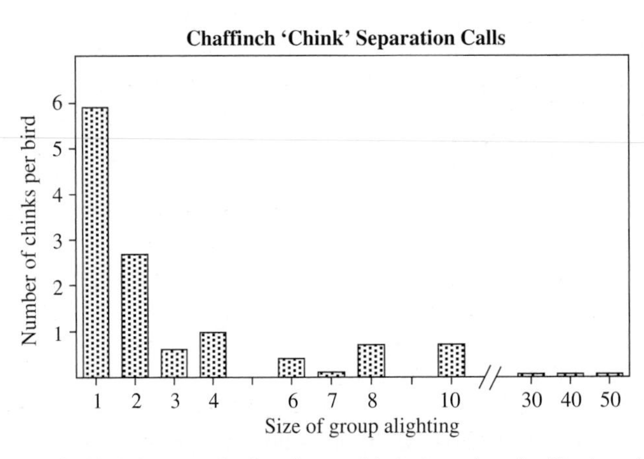

Figure 5.10 The number of chink 'separation' calls per bird given by chaffinches in winter after alighting varies with the group size. From Marler 1956; Bergmann 1993.

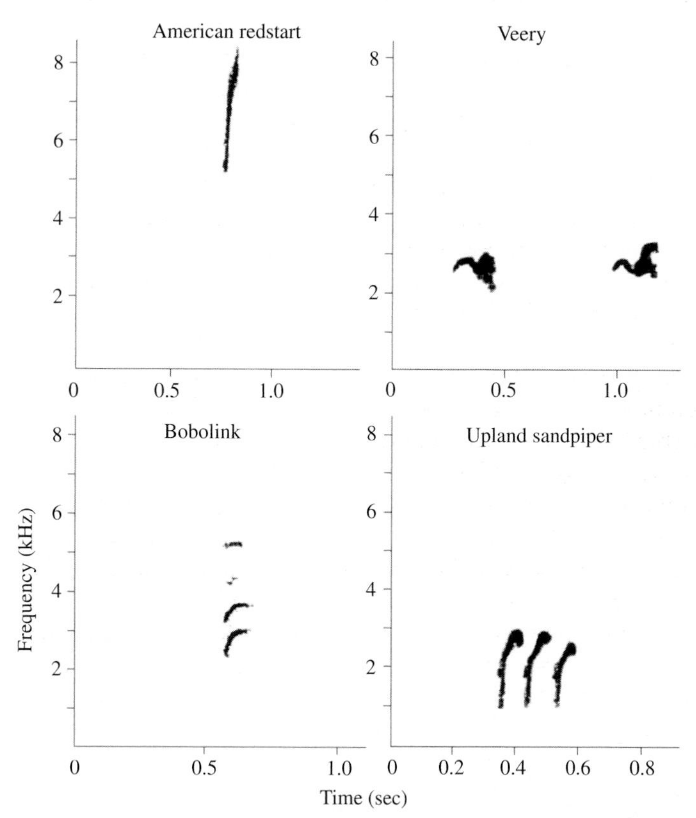

Figure 5.11 Distinctive flight calls recorded from four bird species migrating after dark. On a quiet night they can be heard with the naked ear and the caller can often be identified. From Kroodsma 2000. **(CD1 #46)**

this group, which includes siskins, crossbills, twites, and other members of the canary subfamily, have another important characteristic. Their flight calls are acoustic markers of group membership and personal identity (**Fig. 5.12**). Observing nesting American goldfinches at close quarters, Mundinger (1970) noted that incubating females, which are periodically fed in the nest by the male with a cropful of regurgitated food, began to give begging calls before the male arrived. Playbacks showed that each female recognizes her own mate's flight call as he approaches the loose colony in which these birds breed. Subtle details of call structure are shared by mated birds, developed after the pair bond is established. In the closely related European siskin, the details of flight call structure are labile throughout life, converging on a pattern shared within the social unit that prevails at that season – the breeding pair in summer, the flock in winter. As with song (see Chapter 3, p. **97**), the ability to match call type and accent with companions appears to be a potent factor in social integration (**Box 20**, p. 152). It seems that many cardueline finches learn their flight calls, perhaps all of them (Mundinger 1970; Marler & Mundinger 1975; Samson 1978; Mundinger 1979; Adkisson 1981).

Most remarkable of all is the red crossbill story. What was originally thought to be a single, highly variable species, with the peculiar crossed bills designed for stripping pine nuts from cones with amazing speed and efficiency, turns out to be a cluster of at least nine 'morphs,' each favoring

BOX 19

FLIGHT CALLS IN A NOCTURNAL MIGRANT

For migratory birds that fly thousands of kilometers between breeding and wintering grounds annually, communication among individuals in a migrating cohort may be of vital importance. Nocturnal migratory songbirds tend to fly in loose flocks, unlike the cohesive formations of waterfowl, seabirds, and shorebirds (Kerlinger 1995). In the latter, calls must surely aid in the coordination of flock movements during both day and night; the function for flight calls given by songbirds during migration is less apparent, though they are probably associated with some form of social contact during flight. We have identified a nocturnal 'pseet' call, emitted by free-living Gambel's white-crowned sparrows during fall and spring migration. Birds deliver this call at the time of nocturnal departure from low-lying shrubbery and while flying overhead. These calls appear to have a maximum frequency of 10 kHz, which exceeds that of the daytime 'pseet' calls that have a maximum frequency of 9 kHz. The higher-frequency calls may indicate a specific function for nocturnal flight in comparison with its daytime use during foraging and roosting. White-crowned sparrows held in outdoor aviaries give 'pseet' calls while expressing nocturnal restlessness, a behavior that is specifically associated with migratory flight activity. We recorded the occurrence of these calls both during the fall migration stage and in the wintering stage when migratory activity had ceased. Calling rates during the fall migratory stage begin to increase by 2200 hours, peak around 0130 hours, and decline after 0530 hours, coinciding with migratory restlessness recorded in individually-housed birds held in registration cages (Ramenofsky et al. 2003), and diverges dramatically from calling patterns in the wintering stage. Thus, the nocturnal flight calls of white-crowned sparrows appear to be closely associated with migratory activity. We speculate that they play a role in identifying and locating other individual migrants flying in the dark and within acoustic range.

Mari A. Tokuda & Marilyn Ramenofsky

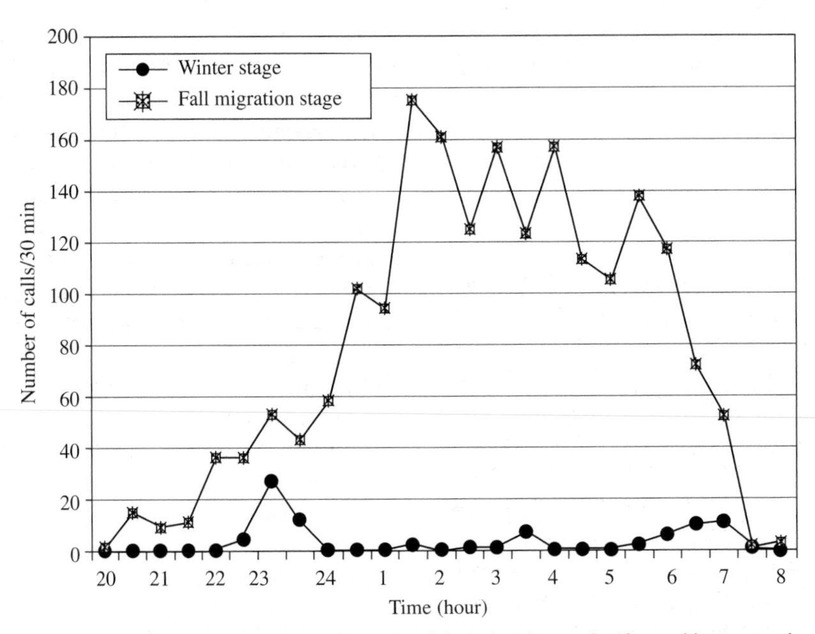

Numbers of 'pseet' calls recorded throughout the night and early morning from white-crowned sparrows held in an outdoor aviary during fall migration (October 2002), and the non-migration period in winter (January 2003).

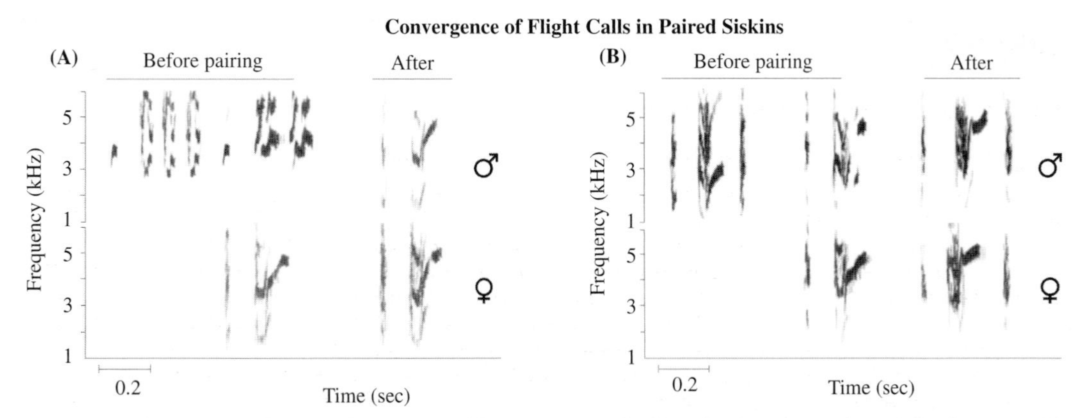

Figure 5.12 Distinctive flight calls are used by some cardueline finches for pair and flock recognition. Paired birds modify their calls to match, as shown here in a pair of European siskins **(B)** and a mixed pair consisting of a male pine siskin and a female European siskin **(A)**. From Mundinger 1970.

its own subset of conifers (Groth 1993a, 2001). These morphs are distinguished by bill size and structure, and each has distinctive, apparently learned flight calls (**Fig. 5.13**), a signal that may help to maintain the integrity of the nomadic flocks for which this bird is famous (**Box 20**, p. 152). More study of this fascinating cardueline phenomenon of flight call learning would pay handsome ethological and ecological dividends.

ROOSTING CALLS: A PLACE FOR THE NIGHT

Aside from food and water, a sheltered roosting place is perhaps a bird's most valuable resource, especially when predation is a significant risk. Even the most terrestrial of birds roost off the ground if they can, and the competition for safe refuges can be intense. Many seek safety in numbers, and vocal signals often play a role in the group assembly process, whether a huge flock, or a family, with parents inducing young to join them at night. White-crowned and golden-crowned sparrows 'pink' as they assemble in the evening (Hill & Lein 1985). Budgerigars have a soft call they give at the roost as they settle down for the night (Wyndham 1980). The roosting calls of huge, dense flocks of starlings, blackbirds, grackles, and crows are a familiar sound, sometimes in unexpected places like airports and shopping malls. Such roosts are thought to give some species access to a system of 'information centers,' facilitating communication about concentrations of food as birds leave the roost to forage (Ward & Zahavi 1973). Though little studied, roosting calls are employed by chickens, junglefowl, and other galliforms. As night falls, a broody hen will give a long, drawn-out 'purr' to attract chicks after she has settled in a roosting site (Collias & Joos 1953). When bobwhite quail have gathered at the roost and before they disperse in the morning, they give the loud penetrating contact call already mentioned; it is thought to be addressed at least in part to neighboring coveys, serving to keep roosts separated (Stokes 1967). Territory-holding European blackbirds are especially vocal when they go to roost, chasing away others who try to join them. Snow (1958) interprets this evening chink calling as "an assertion of ownership at a time when territories are habitually invaded by strange birds."

CALLING TO FOOD

Vocalizations that signal the discovery of food and, even more interestingly, announce the caller's readiness to share it, are fascinating for many reasons (Hauser 1996). They were once regarded as illustrations of altruistic behavior, a serious

BOX 20

FLIGHT CALLS AND GROUP IDENTITY IN RED CROSSBILLS

Red crossbills have long perplexed ornithologists. Renowned for their nomadic opportunistic breeding, with timing dictated less by the seasons than by the erratic availability of conifer seeds, their activities seem to fly in the face of dogma regarding migration and reproduction of temperate zone birds (Griscom 1937). They displayed baffling variation in plumage, body size, and bill morphology until an intelligible pattern finally began to emerge when loxia fans started to look at the birds' vocal behavior. Monson and Phillips (1981) first hinted that crossbills might, "speak a number of different 'languages' undecipherable to our ears. . . ." We know now that crossbills *do* 'speak different languages,' in the form of distinct flight call 'types,' and that one need not be a crossbill to decipher them. This discovery has brought order out of the previous systematic chaos (Groth 1988, 1993a), and has fascinating implications for the role of the cardueline habit of call learning in the culture, ecology, and evolution of these birds.

Flight calls are prominent in the vocal repertoire of crossbills. They are short (about 50 msec) frequency-modulated chirps used in a variety of circumstances, most notably during solitary or group flight, but often also when perched. They are effective in long-distance communication: loud flight calls produced by a perched individual or group, or by a researcher's caged 'call bird' on the ground, can induce free-flying crossbills to approach from hundreds of meters away. The calls also appear to be important in coordinating group activities. Birds foraging in conifers often perform a collective 'call crescendo' immediately prior to taking synchronous flight, evidently helping them to reach a consensus to depart.

In North America, there are at least eight different forms of the flight call, with each bird restricted to a single type (Groth 1988, 1993a). The call types differ in their fine acoustic structure, some subtly, some quite strikingly (**A**); **CD2 #35–37**. For example, call types 1 and 2 differ only slightly, in frequency range, rate of the downward frequency sweep, and whether or not they contain a small initial upward modulation, but they sound very much alike. On the other hand, types 2 and 4 are dramatically different, one being a smooth, mellow-sounding downward-sweeping note, and the other ending with strong emphasis on a sharp-sounding frequency rise. Most call types are sufficiently distinctive that a distant passing group reveals unambiguously its call type identity, both to researchers and to other crossbills.

Birds of a particular call type consistently fit into only one of the four general body size classes, and tend to gravitate to particular types of conifers that best suit their dimensions and morphology (Benkman 1993). It is likely that as conspicuous and reliable advertisements of birds' morphology and habitat preferences, flight calls help individuals to locate appropriate habitats, mates and groups. Learning appears to play a role in their development. They are changed by cross-fostering with parents having a different call type (**B**, from Groth 1993a), and adults probably converge in their flight call signatures when they form pairs (Groth 1993b), though evidence suggests that this may be more of a matter of adjustments within a call type than of changing call types entirely during adulthood. Many fascinating questions about call development cry out for investigation. Are any of the other type-specific crossbill calls, such as alarm calls, learned besides the flight call? Is each crossbill 'morph' predisposed to learn calls of its own kind, or can they all be learned with equal facility? Do social factors play a strong role in the learning process, as one might suspect? Do these and other carduelines prefer to associate with birds whose calls match their own; is there any animosity towards birds that sound different? What are the ecological consequences of flight call matching? One thing is clear: the crossbill call repertoire provides a uniquely rich area for study of the role of culturally transmitted non-song vocalizations in the social behavior, ecology, and evolution of birds.

Thomas Hahn

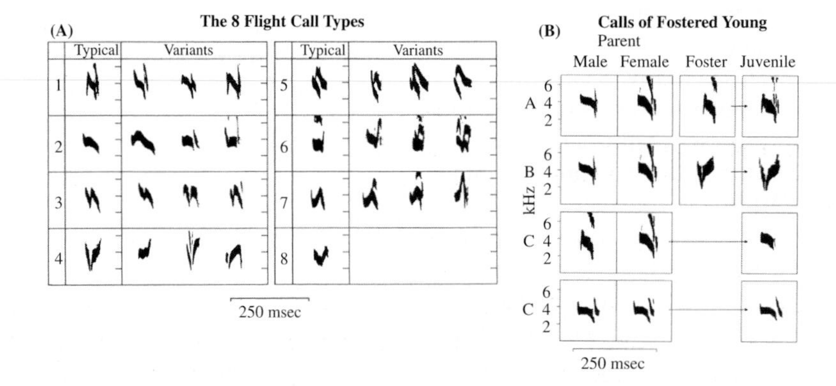

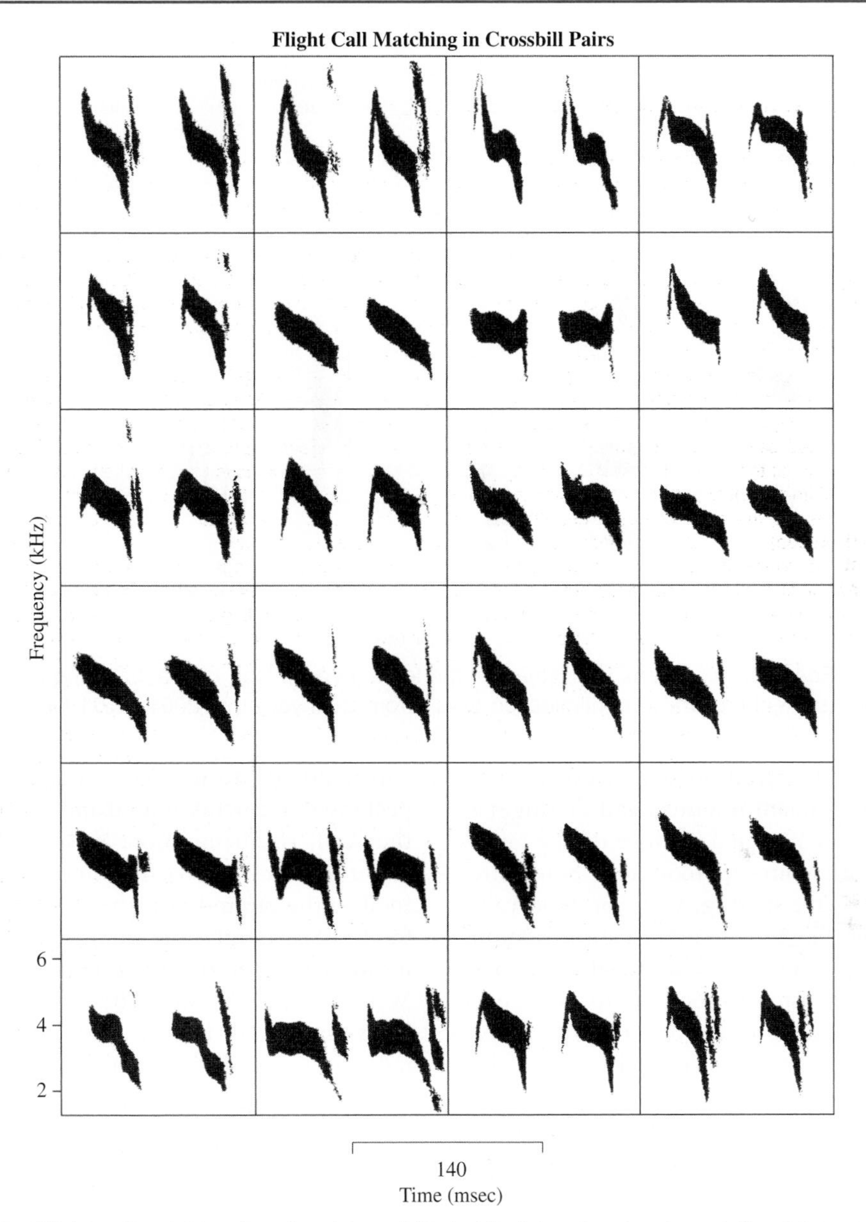

Figure 5.13 Flight calls of 24 pairs of red crossbills in Virginia, showing how calls of the same general type can vary in subtle ways so that calls of paired birds are matched. For each pair males are on the left. From Groth 1993b.

challenge for philosophers and biologists trying to understand the evolution of selfless behavior, before Hamilton (1964) developed his concept of inclusive fitness. Food calls provide a potentially valuable opportunity to explore what 'readiness to share' actually means. Is the sharing passive, as seems to be true of the food 'yells' of ravens in winter that invite others to come and partake of a deer carcass? The ravens beg from each other in the process, and attract still more by their begging calls (Heinrich 1989; Heinrich et al. 1993; Bugnyar et al. 2001; **Fig. 5.14; CD1 #47**).

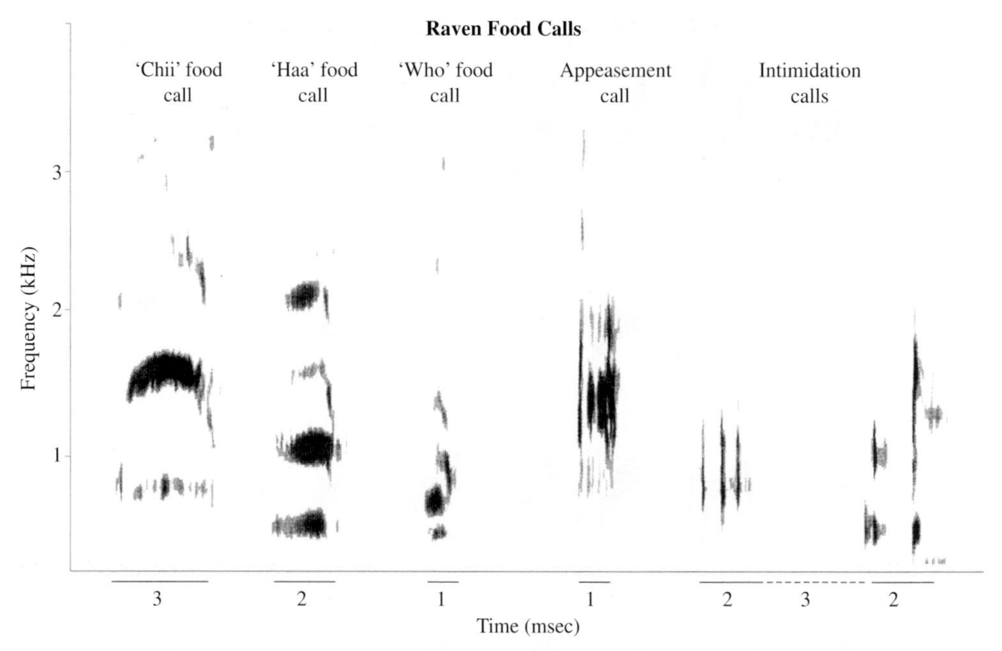

Figure 5.14 Food calls of ravens in Austria. On the left a juvenile 'chii' call, a 'haa' and a 'who'. On the right an appeasement call and two intimidation calls. From Bugnyar et al. 2001. (**CD1 #47**)

Here the food being advertised – a carcass – is a major, overabundant resource, and sharing may involve less of a personal sacrifice on the caller's part than with a bite-size food item. Some birds do display active sharing, with the food caller overtly proffering food to others, and abstaining from eating it himself. This is what happens when a rooster utters its food call to a hen; once she approaches, he offers her a choice insect. Again, there seem to be mutual benefits. The male establishes a sexual bond for future reproduction, and the female acquires food and a potential mate, but still the caller presumably incurs some cost. It is not easy to balance the cost/benefit equation.

Galliform birds provide us with the richest array of food calling behaviors known in any animal group. The 'tu tu' food call of California quail is typical. It can be triggered by presenting mealworms and, whether from a live bird or a loudspeaker, it elicits approach and feeding (**Fig. 5.9**). It is given more by males than females (Williams 1969) and accompanies the tidbitting display used in male courtship by California quail and many galliforms. Males make exaggerated pecking movements at food and other objects as they call, vocalizing more as the female comes closer; females approach and peck, either at the food or the ground near the rooster's beak. The food call is also given by adults to chicks, which respond by running up and taking the food. Williams (1969) describes the behavior of California quail, but he could just as well be talking about bobwhite quail, chukar partridges, ring-necked pheasants, or junglefowl and chickens (Williams et al. 1968; Stokes & Williams 1971, 1972).

In what is by far the most comprehensive documentation of avian food calls we possess, a brief paper by Stokes & Williams (1972) outlines the behavior of some fifty galliforms, including sonograms of calls of more than 20 species. In every case there is a single, soft note type, apparently readily localizable, and always repeated from two to eight times per second (**Fig. 5.15**). Larger birds, such as pheasants, have longer notes, given at a slower rate than smaller birds like partridges and quail. It might be significant that

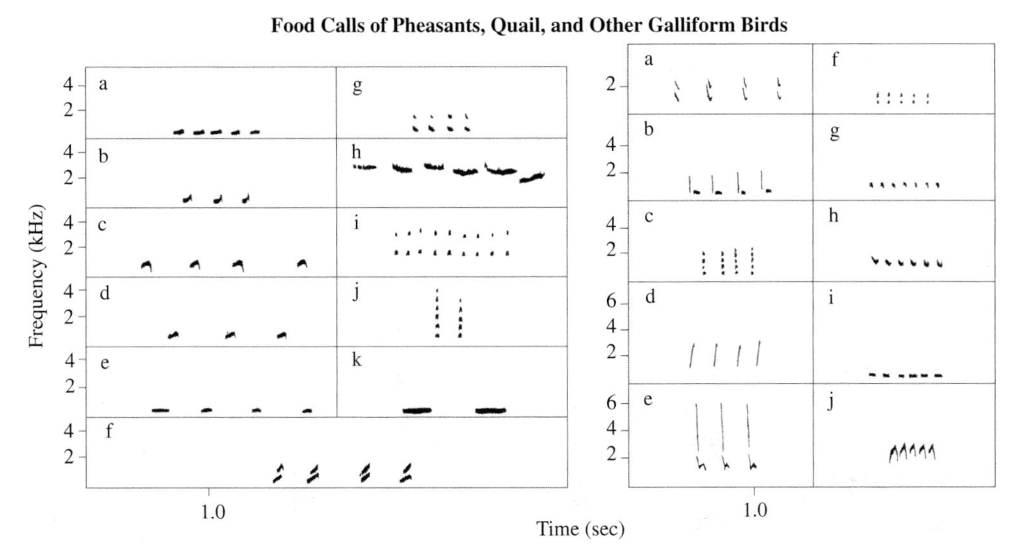

Figure 5.15 Food calls of a variety of galliform birds, eleven pheasants on the left, and ten quail and partridges on the right. In every case there is a string of two or more soft calls. From Stokes & Williams 1972.

pheasants, which are often polygamous, food call only while courting, whereas in quail and partridges, more often monogamous, both sexes food call the year round. Perhaps food calling helps to maintain the pair bond. It certainly plays a key role in courtship in the domestic chicken.

Although food sharing is sometimes passive, in galliforms the food is often actively presented. Stokes (1971) describes here a broody red junglefowl hen feeding her chicks. Sometimes, with very young chicks, she picked up the food, "held it above or in front of the chick, and waited until it took the morsel from her beak." Later, "when she located food, she sometimes carried the morsel a few steps to the chick and let it take the food from her beak or dropped it in front of the chick. More commonly she gave a food call, which alerted the chicks and brought them on the run to her. By this time she usually dropped the morsel... without picking it up." Later still, "her food call became louder, faster, and longer. In addition she sometimes dabbled at the food with exaggerated up and down head movements, two or three times per second. Chicks responded to these auditory and visual signals when as far as 10 m from the hen. The hen stopped signaling as soon as the chicks arrived." When a cock was present and found something edible "he gave the same food call as the hen and the same exaggerated head movements, at which the chicks and hen ran to him." Stokes continues:"I have seen a cock freeze with his beak a few cm above the ground until the chick came and took the morsel, but more commonly the hen ran up to him. As she came to within 0.5 m of him he usually turned away from her, and she then in turn repeated the call over the food until her chicks arrived" (Stokes 1971, pp. 22–24).

Now Stokes describes male junglefowl, using the food call during courtship. "The display movements and call were virtually identical to those both cock and hen used while feeding chicks." But the cock "usually picked up a great variety of items: the tiniest of morsels, whole peanut shells, palm nuts, wood chips, leaves and leafy twigs up to 0.5 m long. Hence the feeding was sometimes symbolic...If a hen did not run to the cock immediately, he usually intensified his calling and motions. At times he carried the item a few steps towards the hen, especially if he

BOX 21

CALLING OTHERS TO FOOD

Many birds produce distinctive calls when they discover food. This behavior is widespread in galliforms (Williams et al. 1968; Collias 1987; Marler et al. 1986a, b; Evans & Marler 1994) and has also recently been described in ravens (Bugnyar 2001). Companions typically respond by approaching rapidly, and may provide functional benefits through enhanced anti-predator vigilance (Elgar 1986), or defense of a rich and ephemeral food source (Heinrich 1988). In other species, calling males attract conspecific females, which are then courted (Marler et al. 1986a).

The food calls of male chickens are brief, pulsatile sounds (**A**) that are ideal for exploring the cognitive processes involved in communication. With food, isolated males call at an appreciable rate, but this is much increased when a hen is present, particularly if she is unfamiliar; calling is completely suppressed by a rival male (Marler et al. 1986b). Social context acts specifically upon signaling behavior. The presence or absence of a companion does not change performance on a key-pecking task that is reinforced by food, showing that the audience effect is not simply a form of social facilitation (Evans & Marler 1994). These experiments also revealed that food calling is dependent upon the presence of food, items and cues reliably associated with food and is dissociated in time from courtship display (Evans & Marler 1994).

Studies of male behavior clearly suggest that food calls may signal feeding opportunities. Two playback experiments were designed to test whether hens are sensitive to this predictive relationship. The first compared food calls with ground alarm calls, which have similar acoustic characteristics; the second compared food calls with contact calls, which are also produced during affiliative social interactions. Hens responded to recorded food calls in a very distinctive way: they moved about, pausing repeatedly to fixate on the ground in front of them (**B**). There was no such increase in substrate-searching behavior when ground alarm calls or contact calls were played back (Evans & Evans 1999). Hens were thus reacting specifically to the likely availability of food, rather than to a general property of repeated broadband sounds (McConnell 1991), or cues suggesting the presence of a non-aggressive companion (Evans & Evans 1999).

Roosters sometimes produce deceptive food calls in the absence of edible objects (Gyger & Marler 1988). If these signals are not inherently honest, as in the case of condition-dependent displays (Zahavi 1975), then how is the observed level of reliability maintained? Energetic costs are unlikely to be the answer because crowing, a louder and more sustained signal, causes only a trivial increase in metabolic rate (Horn et al. 1995). The costs of deceptive alarm calling and food calling are likely instead to be social. Vocal recognition is an essential prerequisite for detection of deceptive signalers. Spectrogram cross-correlation analyses showed that individual identity accounts for a substantial proportion of variation in food call structure. Playback experiments, using a habituation/dishabituation paradigm, confirmed vocal discrimination by hens. When a different male's call was presented following habituation, the response evoked was significantly greater than when hens heard a new call from the same male, showing that food calls are individually distinctive. The next step was to establish whether hens can track individual differences in male reliability. This is a more challenging task because it requires them to construct a 'lookup table' linking individual vocal characteristics with the probability of experiencing a particular class of environmental event. Hens heard a series of food calls from two males, one honest (each playback was followed by food) and the other dishonest. Hens learned this relationship in three tests and had consistently different responses thereafter. They also remembered differences in male reliability for at least 24 hours and generalized successfully to completely novel calls of those males (Evans & Evans unpublished). Male behavior thus seems to reflect an ability of hens to discriminate against dishonest callers, showing that social factors can indeed constrain signal reliability.

Chris Evans

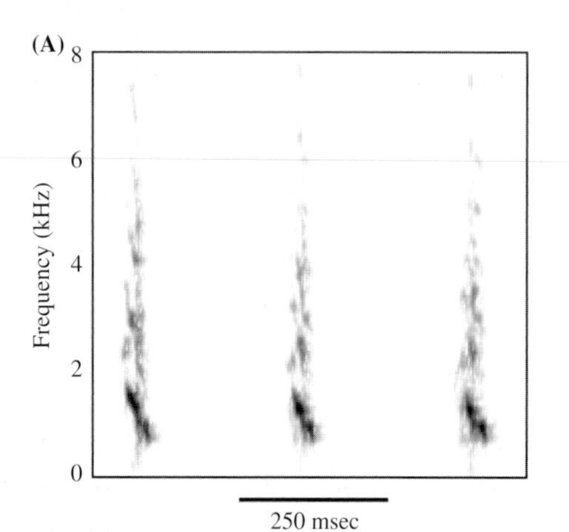

found it partly or well-hidden, but only rarely would he go right up to her. Occasionally he let the hen take food from his beak. He would repeatedly pick up the food and drop it, dabble with it on the ground, or more rarely freeze with his beak over it. A cock might continue such behavior for well over a minute, and then if no hen arrived, eat the morsel". Stokes points out that "the sight of a mealworm must be a strong stimulus for the cock to eat it, but he rarely did so while a hen was within about 10 m of him. Exceptions were when a cock had been dominated recently by another cock standing nearby, after some general disturbance in the vicinity, or when the hen and chicks were actively feeding where food was plentiful" (Stokes 1971, pp. 24–25).

These lengthy quotations serve to reinforce the case, developed later, that food calls symbolize food, and as a reminder of the fundamental importance of careful, perceptive observations of natural behavior, providing the basis for experiments that followed Stokes' work. Laboratory studies of bantam chickens, close relatives of junglefowl, confirm that roosters do indeed have a distinctive call for food, and call more to the choicest foods (**Box 21**, p. 156). A hen is more likely to approach a calling male than a silent one, and approaches more quickly when a hidden, calling rooster has a choice food than when he has a less favored item and calls more slowly (Marler et al. 1986a).

Audience Effects and Deceptive Food Calling

As already hinted at by Stokes, a bantam rooster with food is more likely to vocalize if there is both food and a hen close by, but less likely to call if there is food and a male nearby (Marler et al. 1986b; Evans & Marler 1994). This is an example of an *audience effect*, when a bird presented with a stimulus for calling, such as food, calls more if a potential audience is present than when it is alone (Gyger et al. 1986; Marler et al. 1991). A hen is an especially potent food call audience, appropriate for what is, after all, an act of courtship. Her presence not only encourages a male with food to call more, but may even induce him to call with inedible objects such as a twig, an act of deception to which the female often responds. She does so unknowingly some of the time, because the deception is most common when the hen is too far away to see whether the rooster actually has food (Gyger & Marler 1988). Rooster food calls are individually distinctive, and hens distinguish between males who have behaved deceptively in the past, and those whose food calling was followed by access to food.

Interestingly, although ubiquitous in galliforms, food calling is less common in passerine birds. The raven's food 'yell' (Heinrich 1989), the house sparrow's 'chirp' (Elgar 1986), and the 'squeak' of cliff swallows (Brown et al. 1991) are all well-documented cases of food calling, and they all result in assemblages of birds at a bonanza of food. There are intriguing indications that one note type in the syntactically complex 'chickadee' call of titmice, the C-note, may induce Carolina chickadees to approach a food tray (Smith 1972; Hailman & Ficken 1986; Freeberg & Lucas 2002). The more C-notes there are, the stronger the approach. But with all of these food-calling passerines, there appears to be no active food sharing, and it is not completely clear whether the calls are specialized for communicating about food or whether they are more generalized aggregation signals. The contact call of tree swallows is used by adults coming to the nest with food to elicit begging by the young, another kind of 'food calling' (Leonard et al. 1997). Some of these swallow calls may be further examples of signals with multiple functions, achieving specificity by the context in which they are used.

BEGGING CALLS: ASKING FOR FOOD

The donation of food in response to begging calls is a crucial event in the lifecycle of many birds. To function efficiently, begging calls need to be recognized, and localized, as well as elicit feeding. Courtship feeding is common and we know that adults of some species give begging calls that are reminiscent of those of their young (**Fig. 5.16**). Aside from the great tit (Hinde 1952; Gompertz 1961; Thielcke 1976), adult begging calls have been little studied, although female food begging during courtship is not uncommon. Calling for food by young has received much more attention. In birds raised in a nest, the proffering of food to offspring by parents and other close relatives is common, if not universal. Nidicolous (nest-inhabiting) birds usually remain in the nest to be fed and brooded there for 10–30 days or more, depending on the species. Nidifugous (nest-fleeing) birds, on the other hand, are incubated for a longer period, hatch at a later stage of development, and leave the nest very soon after hatching. Begging calls are less common in nidifugous birds, but their role in life can be just as important, as we will see.

In nidicolous birds, begging behavior forms part of the suite of responses with which nestlings greet a parent when it comes to feed or brood them. After the young leave the nest, nidicolous birds typically remain completely dependent on their parents for some time after fledging but their behavior when they depart from the nest changes radically. Young chaffinches leave the nest about twelve days after hatching, and do not return to it. Instead, they disperse, roost separately, and continue to be fed, only achieving full independence at about a month of age. In this post-nestling phase, the dynamics of parent–young interactions are quite different from the nestling period; this is a point sometimes overlooked by theorists concerned with the

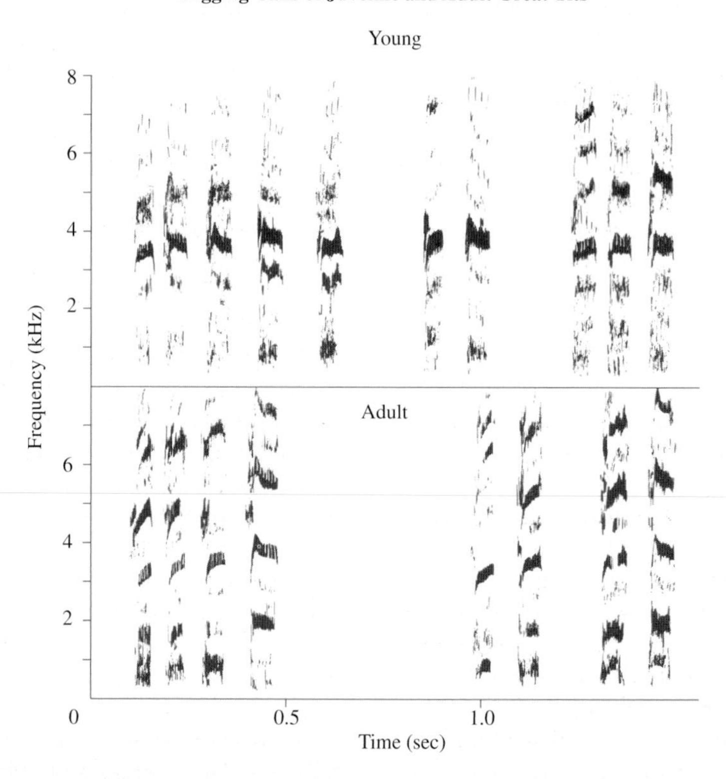

Figure 5.16 Begging calls of great tits. Nestling calls are above, and lower-pitched calls of an adult female begging from a male are below. From Thielcke 1976.

economics of the investment that parents make in raising their young, most of whom focus attention on the more accessible nestling phase. It is clear from the changed behavior of young songbirds as they leave the nest that the demands placed on their vocal signals are very different.

Starting about ten days after hatching, the soft 'peep' of nestling chaffinches is joined by a shorter, louder 'cheep,' becoming the dominant begging call when young fledge a couple of days later. This call is appreciably more complex, as the young bird hones its skill in coordinating two-voiced sound production (Wilkinson 1980; see Chapter 9; **Fig. 5.17**). An equally radical change in begging calls, around the time of fledging, is heard in many songbirds, such as indigo and lazuli buntings (Thompson & Rice 1970; Thompson 1976), and red crossbills (Groth 1993a). The changeover starts around fourteen days in young tree-creepers (Thielcke 1965), which leave the nest hole two days later. Their begging calls change from a short soft 'sip' to a much louder, longer call, more effective in attracting parents to them with food (**Fig. 5.18**). The begging calls of nestling budgerigars' morphs into the first loud contact call as fledglings mature (Brittan-Powell et al. 1997; Bradbury 2003).

Do Parents and Young Recognize Each Other?

The tree-creeper fledgling call is relatively stereotyped, whereas in bank swallows, which fledge at about eighteen days, something more complex takes place (Beecher et al. 1981a, b; Loesche et al. 1991). A quite different 'signature' begging call emerges in young bank swallows a couple of days before leaving the nest (**Fig. 5.19**). In the crowded colony in which these hole-nesting birds breed, there is constant intermingling of families. The threat of confusion is averted by the parents' remarkable ability to recognize their own young on the basis of this call, which is quite polymorphic (**CD1 #48**). Interestingly, it is less stereotyped in social than in solitary swallow species (**Fig. 5.19**). Bank swallow parents can be induced to adopt cross-fostered young, as long as the switch is made before about fifteen days of age. In another colonial songbird, the piñon jay, newly fledged young gather together in a crèche, and adults and young must recognize each other in the crowd at feeding time (Balda & Balda 1978; McArthur 1982). Vocal recognition of group membership also occurs in adults of the social Mexican jay (Hopp et al. 2001). It may be that similar transitions from simple 'cheeps' to more complex fledgling calls occur in other colonial songbirds, such as Australian grassfinches (Güttinger & Nicolai 1973; Zann 1975), and it could be that there too parents recognize their young.

Well-documented cases of parent–young recognition by voice are infrequent enough in songbirds that we take special note, and wonder if something distinctive about their way of life brings forth the adaptations of call structure required. In colonial seabirds, on the other hand, parent–young recognition has been found in every species in which it has been sought (Beer 1970). Seabirds that nest in dense colonies on level ground provide many examples, including gannets (White & White 1970; White et al. 1970), gulls (Beer 1969; Evans 1970; Miller & Emlen 1975), terns (Lashley 1915; Davies & Carrick 1962; Buckley & Buckley 1972; Busse & Busse 1977), and penguins (Thompson & Emlen 1968; Jouventin 1982; Seddon & van Heezik 1993; see Chapter 6). In addition to the numerous cases where parents recognize the begging calls of their young, sometimes chicks mirror this behavior, and recognize their parents' voices.

Most extraordinary of all are murres and other auks that nest on ledges on precipitous cliffs overlooking the sea, studied intensively on islands off the coast of Norway (Tschanz 1965, 1968; Ingold 1973). Chicks often become separated from their parents, both on the nesting ledge and after they jump off the ledge into the ocean. Once separated, they use vocal signals to reestablish contact. Razorbill chicks have four

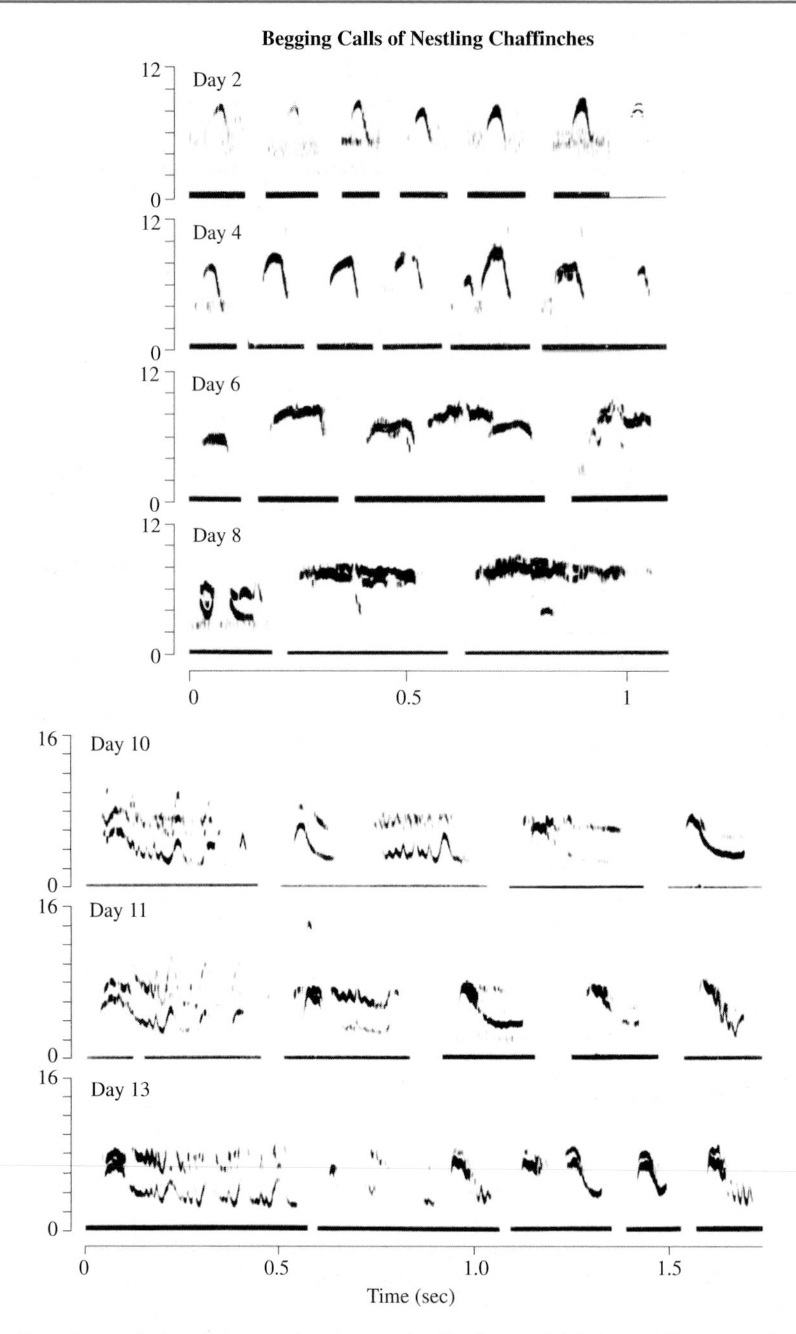

Figure 5.17 Begging call development in young chaffinches, which leave the nest at about 12 days of age. Calls at 2–8 days are shown above. Calls at 10–13 days are below, with long begging calls on the left, 'cheeps' on the right, and intergrades between. Poorly coordinated double voicing can be seen, especially in the cheeps. From Wilkinson 1980.

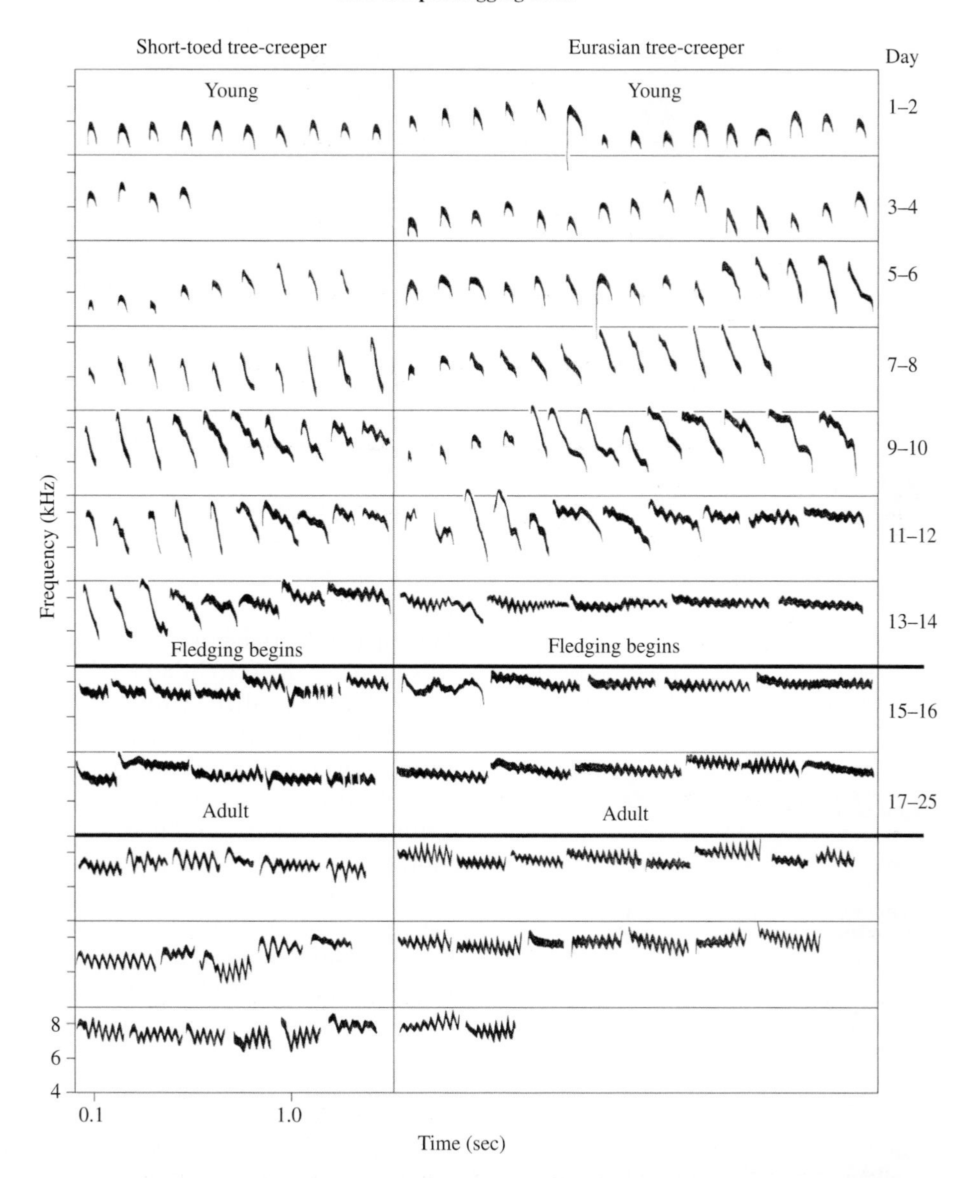

Figure 5.18 Begging calls of nestlings and fledglings of two European tree-creepers, from 1 to 25 days of age. They leave the nest at 16–17 days and call structure begins to change to the adult form a couple of days earlier. For comparison, some adult begging calls are shown at the bottom. The short-toed tree-creeper is on the left, and the Eurasian tree-creeper is on the right. From Thielcke 1965.

calls, the loudest and most resonant being the 'leap' call. By ten days of age, parents have learned to recognize the leap calls of their own young. The voices of both chicks and adults are individually distinctive. As in some gulls, so in murres, chicks learn to recognize their parents by voice. Astonishingly, they begin to memorize their parents' calls even before hatching. As the

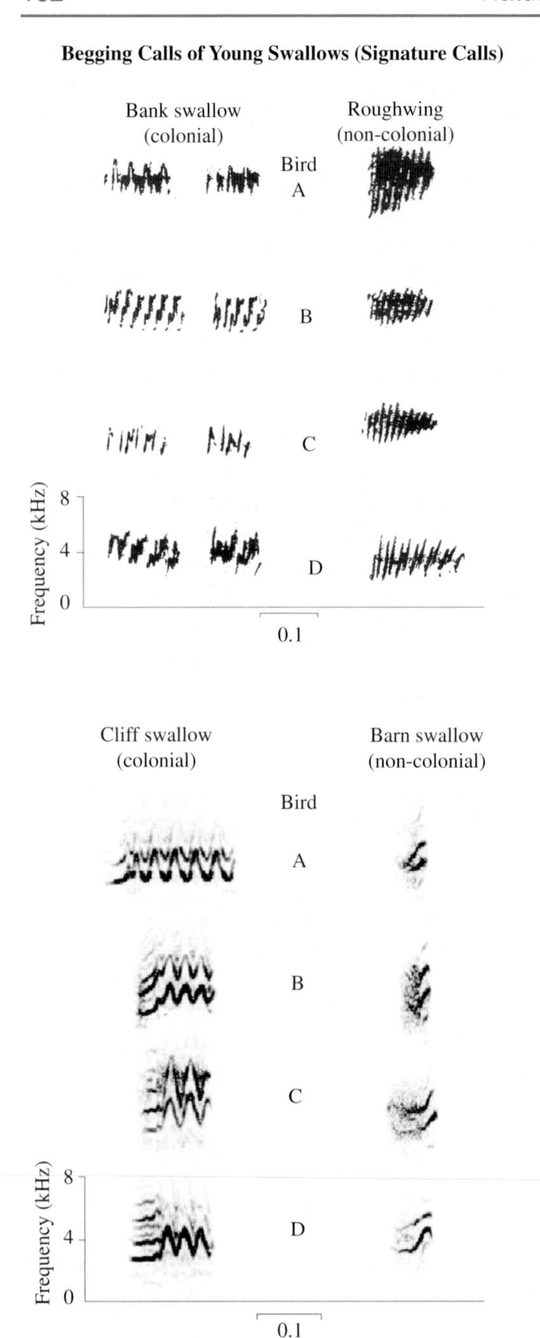

Figure 5.19 A comparison of the variability of the begging or 'signature' calls of four swallow species. Two are colonial (left) and two are non-colonial. Individual differences are more striking in the colonial species. From Beecher 1990. (**CD1 #48–49**)

eggs start to pip, parents attend closely, and give food calls, to which the chicks respond. The chicks learn rapidly, and are able to distinguish their parents' calls from others even before they leave the egg.

These are highly precocial birds and their brains and sense organs are fully functional and active up to four days before they are free from the egg. Similarly, young of other nidifugous birds, including chicks and ducklings, vocalize and respond to sounds in the days before hatching. Exchanges of calls can hasten efforts to break out of the egg, helping to synchronize hatching, a considerable benefit in birds that must keep up with a parent after leaving the nest to survive (Vince 1964; Gottlieb 1971; Baptista & Kroodsma 2001).

What Do Begging Calls Accomplish?

The apparently straightforward view that begging calls induce parents to bring food proves to be a deceptive simplification (Bradbury & Vehrencamp 1998). The rate of begging does relate in a general way to how hungry the nestling is. At least over the short term, an increase in calling – whether natural or experimentally contrived – results in more food being brought to the nest (von Haartman 1953; Rydén & Bengtsson 1980; Bengtsson & Rydén 1983; Smith & Montgomerie 1991; Redondo & Castro 1992; Mondloch 1995; Searcy & Yasukawa 1995). Over the long term, however, there is no guarantee that the increase will be maintained, or that a higher growth rate will ensue (Stamps et al. 1989). We have long known that the relationship between benefits to the young and investment on the part of parents is complex, and that what is most favorable to offspring is not necessarily in the best interests of their parents (Trivers 1974; Godfray 1991, 1995; Zahavi & Zahavi 1997). There is always the possibility of cheating, by reaping benefits without paying costs. The obvious antidote to cheating is the costliness of begging behavior but, although logical and plausible, this is hard to prove. The study of cheating and the obstacles to cheating requires reliable, quantitative metrics for gauging both costs and benefits, and these are not easy to come by (Godfray 1995).

The threat of predation may place a limit on the conspicuousness of begging behavior (Haskell 1994), an argument with support from a ten-year field study of begging calls in a group of Arizona bird species: begging calls were quietest and most high pitched, and thus most cryptic, in those birds most subject to nest predation (Briskie et al. 1999). There is a general tendency for nestling begging calls to be easily locatable, but also to be soft and high pitched (Redondo & Arias de Reyna 1988a). Aside from these studies, little progress has been made in relating begging call structure to the functions served. Begging calls can be extraordinarily diverse, even in close relatives (Popp & Ficken 1991), and it is likely that fledgling calls will prove to be even more varied. Of course, at this stage of a bird's development, many aspects of juvenile call structure may be in developmental transition, on the path to adulthood.

Cuckoo Begging Calls are Informative

Aside from selection pressure for adequate locatability and crypticity, could it be that most of the details of begging call structure are irrelevant at this stage of the lifecycle? Clear evidence to the contrary comes from a surprising source–cuckoos. When a nest parasite lays an egg in another bird's nest, it exploits the likelihood that the foster parents will raise it as one of their own. With nidicolous birds, this requires the young parasite to evict the host's young, or at least to beg for food as effectively as its nest mates. If, the details of begging behavior of the host young, including such things as call structure, are irrelevant to the host parents, then the requirements imposed on the fostered chick will be relatively undemanding. If on the other hand, the details of call structure are important to the host parents, the stage is set for the evolution of mimicry. An abundance of evidence confirms that various forms of host mimicry are crucial to the success of nest parasitism (Chance 1940; Davies 2000). Does mimicry extend to begging calls?

Because the nestling cuckoo evicts all host young from the nest and then grows to several times their size, it has a double problem. It not only has to elicit feeding by the host, but requires as much provisioning as an entire brood of the foster parents' nestlings. Remarkably, it accomplishes this by mimicking one aspect of the host's begging behavior, the rate of calling (**Fig. 5.20; CD1 #50**). A cuckoo chick calls very rapidly, with what appears to be a chorus of rapid 'si-si-si' calls, sounding like an entire brood of host chicks (Davies et al. 1998). That this high calling rate is critical was shown in elegant fashion by supplementing the begging calls of a single thrush nestling with tape recordings of cuckoo calls, with the result that the feeding rate was brought up to what a healthy cuckoo requires.

The common cuckoo's begging call is a relatively simple sound, not necessarily matching the fine details of the calls of the wide range of hosts that it parasitizes. It is therefore of special interest that some more specific brood parasites seem to match the begging calls of their hosts closely. In New Zealand, the long-tailed cuckoo and the shining cuckoo exploit different hosts (**Fig. 5.21**). The two sets of host nestlings call very differently, a rapid 'trill' in some cases, and separate 'peeps' in others, and these contrasts are apparently matched closely by nestlings of the two cuckoo species that exploit them as nest parasites (McLean & Waas 1987). Begging calls of the South African diederik cuckoo are said to match those of two different hosts–cape sparrows, and bishop birds – exploited in different parts of the cuckoo's range (Reed 1968). These and other cases of apparent begging call mimicry (Mundy 1973; Fry 1974; Redondo & Arias de Reyna 1988b) demonstrate that the detailed acoustic structure of begging calls can indeed be behaviorally significant in communication between parents and young – another fertile area for further study.

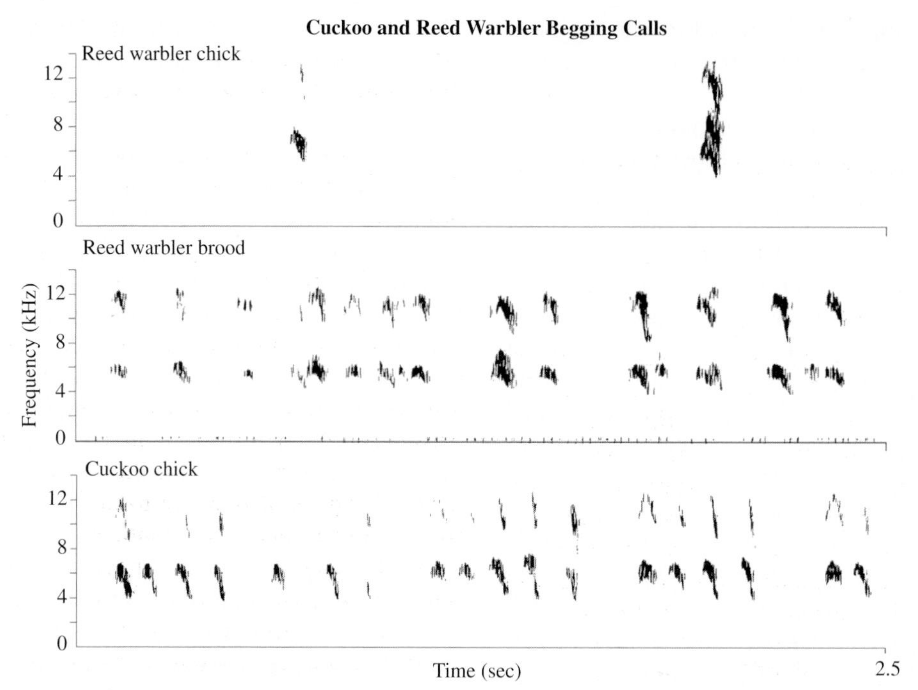

Figure 5.20 Begging calls of nestlings of a common cuckoo (bottom; **CD1 #50**), a single reed warbler (top), and a brood of reed warblers (middle). The cuckoo calls as rapidly as an entire host brood. From Davies 2000.

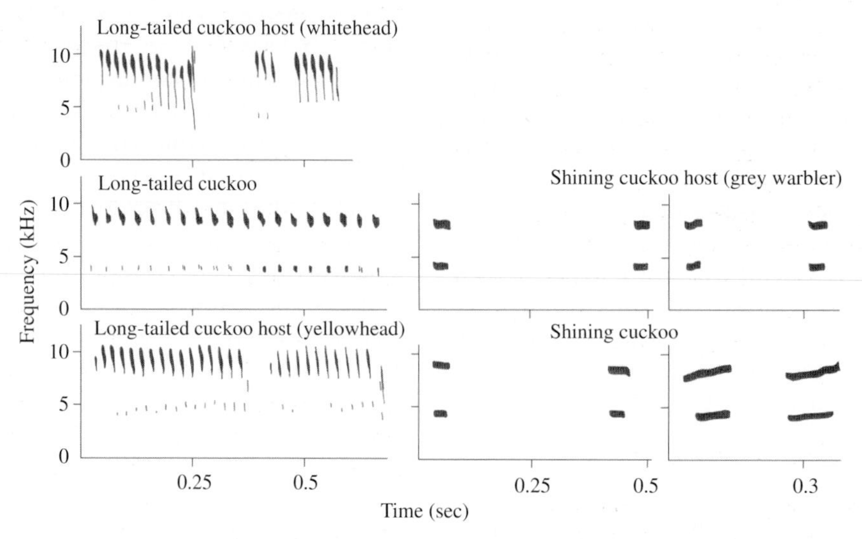

Figure 5.21 Begging calls of young long-tailed and shining cuckoos and some of their warbler hosts in New Zealand. Whiteheads (top) and yellowheads (bottom) are long-tailed cuckoo hosts, and grey warblers are shining cuckoo hosts, shown while resting and actively begging. From McLean & Waas 1987.

AGONISTIC CALLS: SIGNALS FOR AGGRESSION

When birds engage in combat or confront each other with aggressive displays at close quarters, they often vocalize. Versions of song are sometimes heard in such circumstances in passerine birds in the breeding season. The singing is usually soft and often fragmentary, as during the 'puff-sing-wave' with which a breeding male song sparrow drives off a territorial intruder (Nice 1943). However, song sparrows of both sexes also have a distinctive 'zhee' aggressive call, which turns out to be a very basic item in their call repertoire. It is used throughout the year, and it may accompany threats directed at other species, as well as in aggressive interactions with other song sparrows. In the breeding season, female song sparrows are also prone to give the 'zhee' in response to aggressive male courtship, especially when pounced upon by males other than the mate. So the 'zhee' accompanies song sparrow threat behavior in a wide range of circumstances.

Chaffinches of both sexes have a 'zzzz' aggressive call (Poulsen 1958), very similar to that of the song sparrow (Poulsen 1958). However, it is used much less often in the chaffinch and easily overlooked (Marler 1956; Bergmann 1993), except for one situation. When chaffinches are disturbed by placing stuffed birds near the nest, the 'zzzz' call is often provoked as the mount is attacked or threatened, along with the 'head-forward' aggressive display (Poulsen 1958). It is from this situation that the one sonogram we have of this call was made (**Fig. 5.1**). Some New World parrots have a 'squawk' that they use during aggressive interactions (Bradbury 2003).

Aggressive Calls and Body Size

The relatively low pitch of the drawn-out 'zzzz' used as a threat signal illustrates another link between call structure and function. Morton (1977) lists some 28 bird species, from a range of taxa that give a harsh, rasping or growling call in similar agonistic situations. There is an intriguing resemblance to the distress call (see p. 144), suggesting that perhaps this too should be regarded as at least partly aggressive in nature. Although there are exceptions, aggressive calls of animals in general tend to be harsh and low pitched, consisting of single, relatively long notes (Collias 1960). To the extent that placement in the lower register of an animal's pitch range is a correlate of body size, this has potential significance when calls are used to resolve aggressive conflicts. Body size is usually a good predictor of success in combat, and it also imposes a general limit on the minimal pitch of vocal signals, if they are to be loud enough to impact the behavior of others. Honesty is implied, because of the supposedly inevitable physical limitation imposed by body size, but there are exceptions. Some birds possess an elongated trachea, forming loops or coils in the chest that serve to exaggerate a bird's apparent size (Fitch 1999). The effect is to shift downwards areas of frequency emphasis, or 'formants' (**Box 43**, p. 326), a change that whooping cranes, for example, perceive and respond to (Fitch & Kelley 2000). But although pitch is not necessarily an 'honest' acoustic feature, low frequencies are nevertheless widely represented in agonistic calls.

Motivational–Structural Rules

The possibility of general relationships between the tonal quality of calls and their use in hostile and non-hostile interactions gave rise to the hypothesis of motivational–structural rules (Morton 1977, 1982), a relationship first noted by Darwin (1871). The concept focuses especially on the two distinct axes of close-range agonistic interactions, represented by winners and losers. The most secure aspect of the motivational-rules hypothesis is the frequent association of low pitched, rasping calls with aggressive display, a connection that embraces not only birds, but amphibians and mammals as well (Eisenberg

1974; Morton 1977; Hauser 1993). An exchange of threats may escalate into full-scale, rough-and-tumble combat, and even death. As an alternative to escalation, one animal may acknowledge defeat and either flee or display submissive or friendly behavior. The motivational-rules hypothesis proposes that prospective winners will tend to use harsh, wide-band, low-frequency sounds, and prospective losers become increasingly likely to favor relatively high-pitched, tonal sounds. Mixed, transitional calls are of course likely, as individuals pass through the range of motivational states that arise in the course of agonistic interactions.

There is a plausible physiological basis for anticipating that losers of a fight will tend to have higher-pitched calls, at least statistically speaking, because of the muscular tension in the vocal and other musculature associated with defeat, submission, and fearfulness. If we throw the net wide enough, to encompass not only non-hostile agonistic calls, but also alarm calls in general, a general relationship does seem to emerge; the sounds used tend to fall in the upper range of a species' register and to be tonal rather than harsh. It is suggested that calls that rise and then fall in pitch, pitched in the midrange, may indicate ambivalence or indecisiveness – an idea worth pursuing.

Thus, according to the motivational-rules hypothesis, animals may follow some emotion-based rules in how they vocalize, applying them generally, across call-type boundaries. Although they do not seem to be universally applicable, the supporting evidence is widespread enough to make it worthwhile to explore whether animals detect and respond to these and other emotional nuances of call structure separately, perhaps processing them in different parts of the brain than other acoustic features – another topic for future research.

CALLING ACROSS SPECIES BOUNDARIES

Birds often live with many species in close quarters, and vocalizations are bound to be heard, not only by the caller's own species, but by others as well (Morse 1970b). Do the others respond, or just turn a deaf ear? It may be a relatively straightforward matter to simply ignore them if alien calls are very distinctive. On the other hand, some call types are so similar across species that we ourselves have to scrutinize them closely to identify the caller. Are these similarities accidental, or could there be selective forces encouraging them? Might strong species distinctiveness be functionally disadvantageous for some signals (Marler 1957)? Unlike vocalizations involved directly or indirectly with reproductive isolation, where species privacy is an issue, distinctiveness may be a drawback in keeping an open channel between species.

A predator with general tastes surely suffers if the alarm calls of one bird alert other potential prey, as is often reported by observers of mixed species flocks. The many different species that scramble to join a predator-mobbing scene indicate the occurrence of cross-species responsiveness to the alarm calls that are employed. Similarly, with distress calls, the threatening hisses and growls that are used communicate readily between species. The facility with which they do so is obvious, engaging predators as well as other prey. We have to conclude that, in addition to interspecies call similarities that are direct convergent adaptations to physical requirements, such as localizability and active space, the facilitation or hindrance of interspecific communication is also a factor that cannot be ignored (Stefanski & Falls 1972a, b; Latimer 1977; Gaddis 1980; Maier et al. 1983; Ficken 2000; **Fig. 5.22**).

Interspecific interaction is hard to document but, in a cleverly designed study, Krams & Krams (2002) obtained some convincing evidence. They presented a stuffed owl to chaffinches living in two different communities. Both were species-rich, but one consisted of resident species, the other of migrant species only recently arrived. Tests were made in two periods, one early in the breeding season, the other a week later. When mobbing occurred, the pattern of recruitment

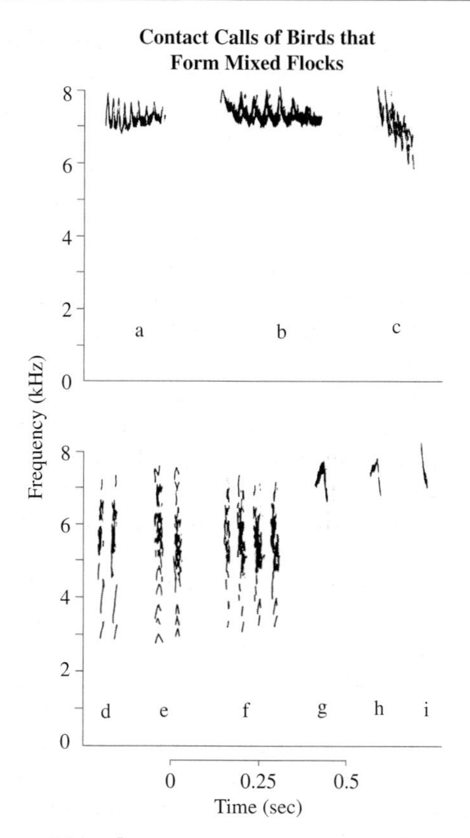

Figure 5.22 Contact calls of birds that flock together are sometimes similar, as in the 'zeet' of golden-crowned kinglets (a), the 'buzz' of brown creepers (b), and the A-note of Mexican chickadees (c). At the bottom left are Mexican chickadee C-notes (d), ruby-crowned kinglet 'chedit' notes (e), and bridled titmice C-notes (f). On the right are contact calls of boreal chickadees (g), golden-crowned kinglets (h), and brown creepers (i). The structural similarity of these three groups of birds, which do form mixed flocks, may facilitate interspecific communication. From Ficken 2000.

of other species was similar in the two communities. However, the intensity of chaffinch mobbing behavior was very different. In the 'resident' community, early and late tests yielded similar data – strong mobbing. In the 'migrant' community however, while late tests gave normal results, in the early tests chaffinch mobbing was weak or nonexistent. The authors speculate that the remarkably low early season scores for chaffinch mobbing of the migrant, mixed species community may reflect the relative anonymity of membership in this newly formed association. They argue that the lack of mutual familiarity may lead chaffinches to hold back from the risks of predator mobbing until community members of whatever species became better acquainted individually, and perhaps more able to keep book on 'tit-for-tat' interactions across species boundaries – an idea worth pursuing.

Mixed Flocks in the Tropics

Among the most remarkable multi-species assemblages are the complex mixed species flocks that, although common everywhere, are especially widespread and well organized among insectivorous birds of the humid tropics. Their membership often includes up to twenty species, sometimes more (Davis 1946; Short 1961; Moynihan 1962; McClure 1967; MacDonald & Henderson 1977; Greig-Smith 1978; Morse 1980; Powell 1980; Wiley 1980; Hutto 1987). Membership can be consistent over long periods of time. There is often a stable home range, superimposed on the territories of its members, with a systematic changing of the guard as the flock moves through its domain (Munn & Terbourgh 1979).

Certain species are especially likely to provide a core around which others group themselves, and it has been speculated that one criterion for fulfilling the role of nuclear species is a repertoire of alarm calls, together with a complex of other antipredator adaptations. Another asset is well-developed sentinel behavior that appears to benefit other species as well as the sentinel's own (Munn 1984; Gaddis 1987).

In the Peruvian rainforest, up to fifty species are recorded participating in mixed flocks, and two sentinel species appear to be especially important, the bluish-slate antshrike, and the white-winged shrike-tanager (Munn 1984). They are usually found at flock center, and are always the first to give alarm calls when a bird-eating hawk attacks. This must be a valuable service to other flock members, all of who respond. There is even intriguing evidence of sentinels giving a false alarm when vying with other flock members for a choice insect that the flock has flushed. In this apparent case of deceptive behavior, when a sentinel alarm calls in the absence of a predator, other flock members pause momentarily in their pursuit of the insect prey, long enough for the mendicant to pirate the choice food item from them (Munn 1986). Fascinating cross-species interactions like these deserve much more of our attention in the future.

Bird Call Dialects

For years it has been conventional dogma that bird calls are innate. This judgment is based on the common observation that, when birds are raised away from their own species, their songs are frequently abnormal, but their calls are not. But we don't have many detailed studies of call development. With the discovery of developmental plasticity in several bird calls this situation is likely to change. It seems that calls are modified to some degree by learning in more birds than we think.

Call dialects provide several promising leads. Some fifty years ago, the demonstration that local dialects in song are a scientific reality helped to focus researchers on cases where dialects seemed to imply song learning, and to convince the research community that vocal learning was a worthy target for intensive investigation. It is now clear that local dialects exist in some bird calls. An early European study demonstrated dialects in the 'rain call' of the chaffinch (Sick 1939), confirmed later in other parts of Germany (Detert & Bergmann 1984; Baptista 1990; Bergmann 1993; **Fig. 5.3**). The possibility of developmental plasticity is indicated by the hand-reared chaffinches from Scotland that learned the unusual rain calls of Corsican birds when tutored by them (Riebel & Slater 1998b). The chaffinch 'chink' is to some degree developmentally plastic (Marler 1956), and its structure varies in different chaffinch populations, notably in the Canary Islands and the Azores (Marler & Boatman 1951). Local dialects in the flight call of the brown-headed cowbird (**Fig. 5.23; CD1 #51**) have been thoroughly documented (Rothstein et al. 1986; Rothstein & Fleischer 1987). The extraordinary plasticity of the flight calls of cardueline finches has already been noted. The evening grosbenk is another example (Sewall et al. 2001). A nine-year-old siskin was still able to modify its flight call to match those of new companions (Mundinger 1979; **Fig. 5.12**), and hand-reared Norwegian twites clearly learned their flight calls from each other (Marler & Mundinger 1975; **Fig. 5.24**).

Learning appears to play a role in the development of red crossbill flight calls, which change after young are transferred to foster – parents having a different call dialect (Groth 1993b; **Fig. 5.13**). Many fascinating questions about crossbills await investigation. Are any other crossbill calls learned besides the flight call? Is this a case of open-ended learning, or is there just one sensitive period early in life? Is each crossbill 'morph' (see p. 152) predisposed to learn calls of its own kind, or can all dialects be learned with equal facility? Do social factors play a role in the learning process, as one might suspect? Do these and other carduelines prefer to associate with birds whose calls match their own; is there any animosity towards birds that sound different? What are the ecological consequences for crossbills of their flight call-matching behavior? What is the precise functional significance of call dialects in crossbills and all other carduelines?

Certain chickadee calls are subject to modification by learning. There are local dialects in their 'gargle,' – a surprisingly complex aggressive vocalization (Ficken & Weise 1984; Ficken et al. 1987; Ficken & Popp 1995; Miyasato & Baker 1999; Baker et al. 2000; **Fig. 5.25**). The 'chickadee' call is another amazingly complex vocalization (Ficken & Popp 1992; Hailman & Griswold 1996). When black-capped chickadees are kept together, they tend to converge on a similar shared pattern (Mammen & Nowicki 1981; Nowicki 1989; Hughes et al. 1998), which then provides the basis for vocal recognition of group membership (Nowicki 1983).

The contact call of the zebra finch, a favored

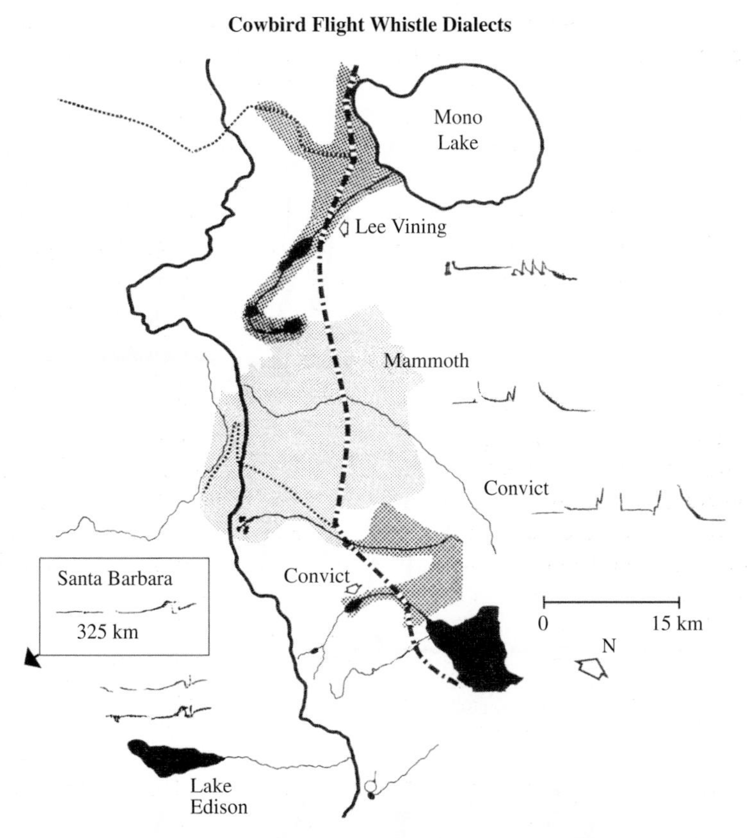

Figure 5.23 A map of flight whistle dialects of the brown-headed cowbird in the eastern Sierra Nevada, California. Sonograms are sketched for some of the main call types. Calls around Lake Edison are very similar to those in Santa Barbara, more than 300 km to the southwest. From Rothstein & Fleischer 1987. (**CD1 #51**)

subject for neurobiological investigation, is another example of call learning. The process is especially interesting because it differs in males and females; only the male call is learned. We have information on the underlying circuitry for call learning in the zebra finch forebrain, which turns out to be shared with song learning, but only in the male (Simpson & Vicario 1990).

The parrot family is another likely place to look for learned calls (see Chapter 13), and supporting evidence is emerging from field studies. There are clear dialects in the flight calls of Australian ring-necked parrots (Baker 2000), and in the contact calls of yellow-naped parrots and orange-chinned parakeets in Costa Rica (Wright 1996; Bradbury et al. 2001). Cross-fostered cockatoos will learn the contact calls of the foster parent (Rowley & Chapman 1986; **Fig. 5.26**). There are numerous demonstrations of group convergence in details of budgerigar contact calls when they are placed together (Farabaugh et al. 1994; Bartlett & Slater 1999; Hile et al. 2000; Hile & Striedter 2000). Convergence occurs in both sexes, although males do so more rapidly (**Fig.** 5.27). Especially exciting are indications that orange-fronted conures may make subtle adjustments as they respond to playbacks, to increase contact call matching, perhaps as an affiliative response (Vehrencamp et al. 2003). Conversely, change in call structure to *decrease* the degree of call matching may have aggressive connotations.

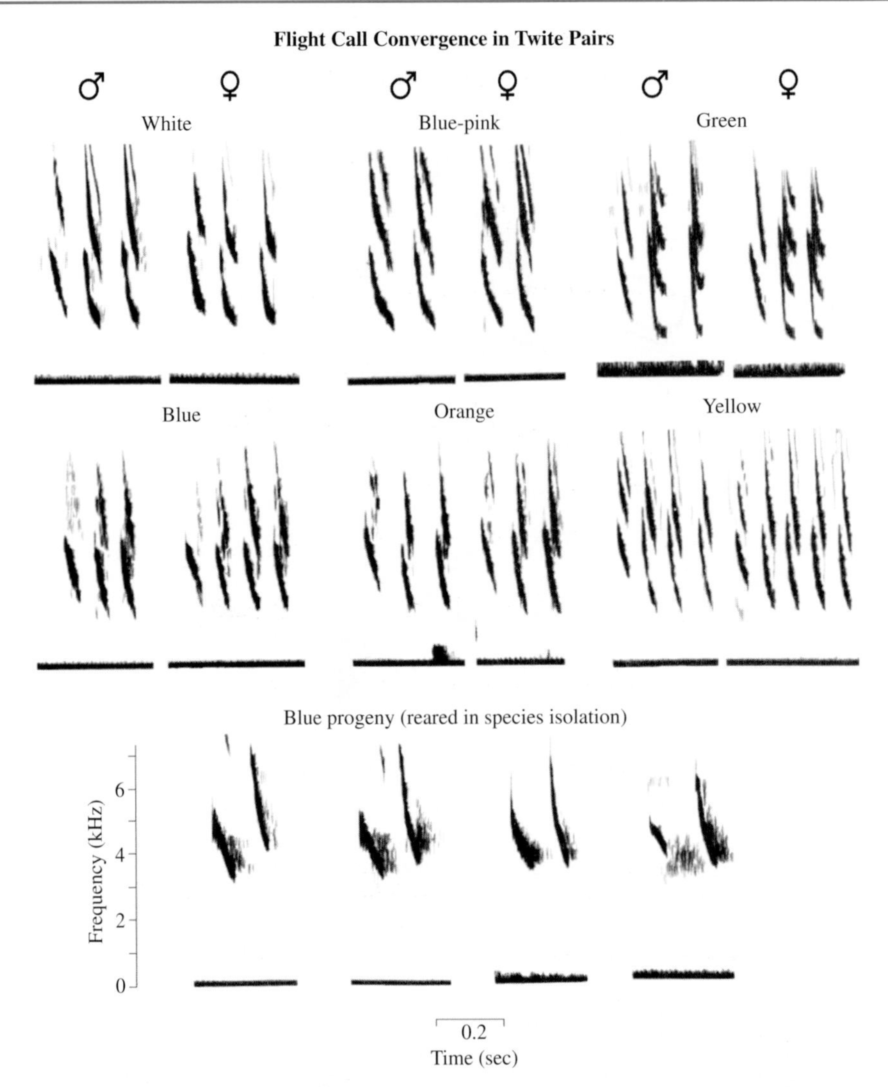

Figure 5.24 Flight calls of the twite, a Norwegian cardueline finch, are distinctive from pair to pair. Breeding season flight calls of six females and their mates are shown at the top. Below are flight calls of four adult birds, raised as a group in species isolation from the nestling phase, that were offspring of the blue pair (bottom left). They presumably learned from each other. From Marler & Mundinger 1975.

Contact calls of budgerigars and other parrots provide an excellent opportunity to explore the nature and neuroanatomical basis of the learning processes underlying call plasticity (Striedter 1994). Comparisons between the parrot family and songbirds provide an invaluable window, not only on the necessary mechanisms – behavioral and physiological – that make vocal learning possible (see Chapter 8), but also on its functional consequences.

VOCAL SEMANTICS: WHAT DO BIRD CALLS MEAN?

If, as most people believe, bird calls are involuntary, arising not from cogitation but as impulsive displays of emotion, the question of

Black-capped Chickadee Gargle Calls

One bird's gargle call repertoire

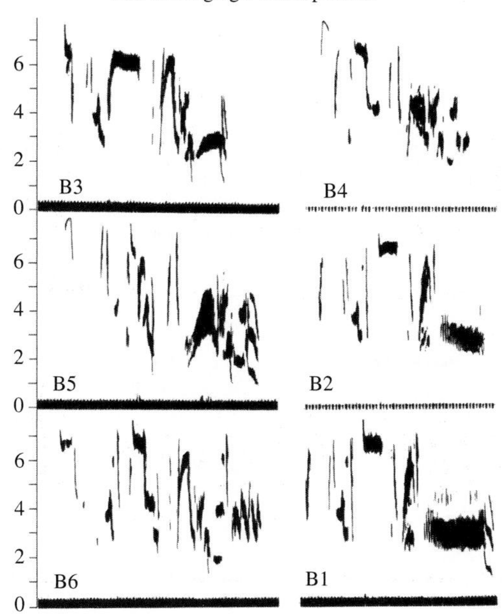

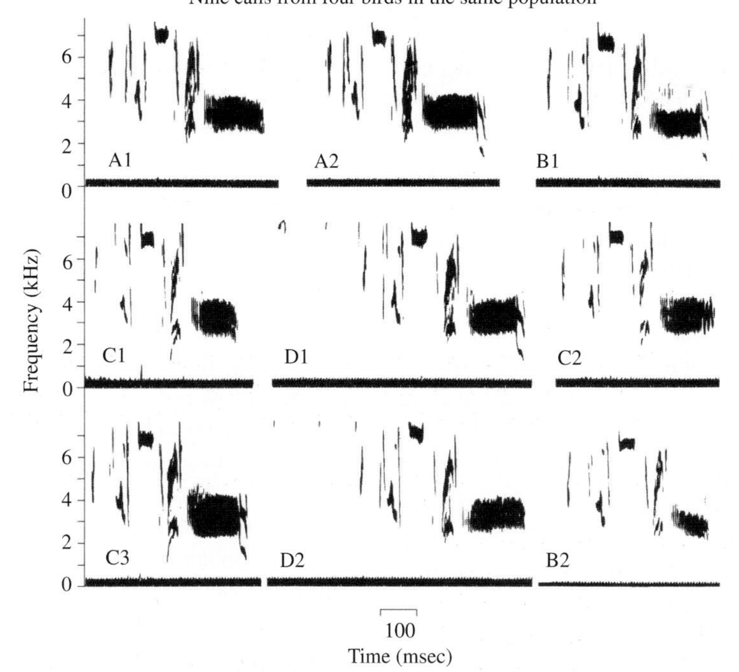

Figure 5.25 Black-capped chickadees have a repertoire of six or seven distinct gargle calls. At the top are six from the same Colorado bird. Birds living in one area share gargle call types. Below are nine similar calls, sampled from the repertoires of four neighboring birds. From Baker et al. 2000.

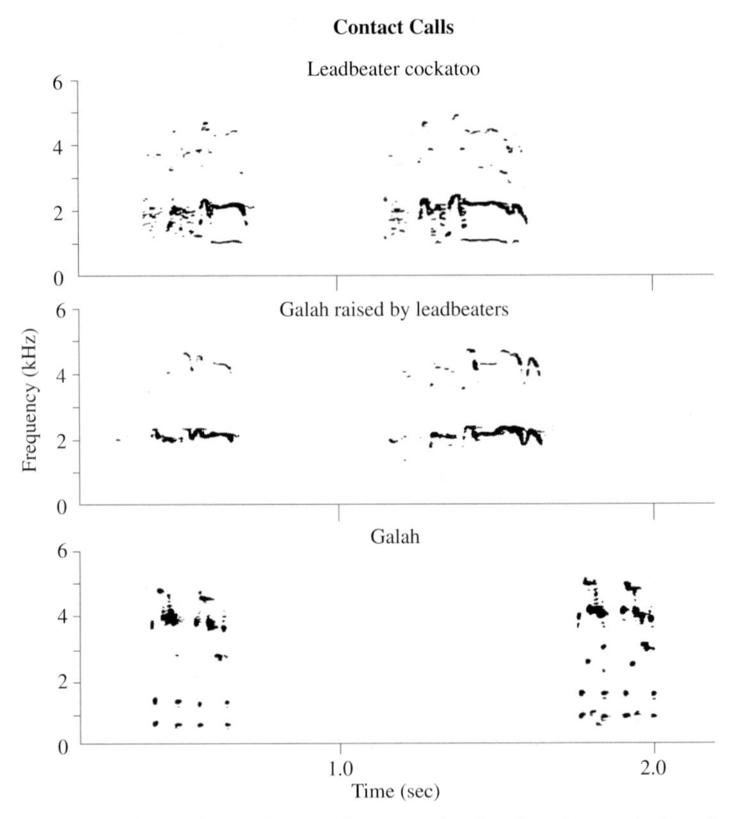

Figure 5.26 Contact calls of two Australian cockatoos, the leadbeater cockatoo (top) and the galah (bottom) and a galah raised by leadbeater foster parents. From Rowley & Chapman 1986.

meaning might seem to be moot. We ourselves do not usually think of emotional displays as suffused with meaning, at least to the same degree as language (Marler 1998). But emotion-based displays can of course convey a lot of information. When displays are direct, involuntary reflections of the physiological state of the signaler, and in so far as variations in a caller's motivational state are significant to others, such impulsive displays can play an important role in social communication, as we ourselves are very much aware, even though they are not meaningful signals in the conventional sense. Some physiological states, such as irritability or aggressiveness, have obvious relevance to others, if they choose to take notice. It behooves a combatant to note with care signs of increased aggressiveness or of fearfulness in an opponent. Such signs of impending behavioral change can have direct and immediate significance. Useful information is signaled to companions when chaffinch flight calls reveal a growing urge to take flight. It is of potential value to parental investment when increasingly hungry nestlings give more begging calls.

There is, however an important caveat here. We should not think of emotion-based signals as representing anything in the outside world, in a symbolic sense. If anything, they represent themselves, as one aspect of a complex physiological state that others can receive information about and respond to, by taking flight themselves, or bringing food. When a bird externalizes its physiological preparations for changing its behavior, this provides a means for others to make appropriate adjustments, by joining in flight and synchronizing group behavior, or averting starvation of their young.

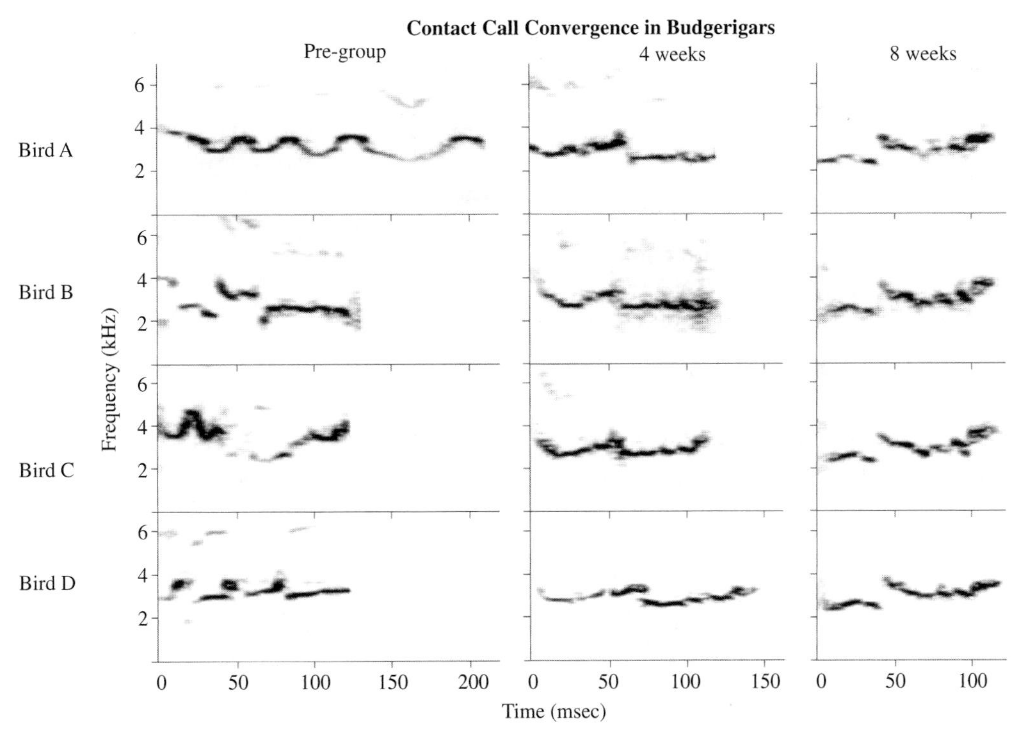

Figure 5.27 The contact calls of four adult budgerigars converged after they were placed together. From Hile & Striedter 2000.

But it would be illogical to regard a call, whose meaning is primarily a reflection of an endogenous change in the signaler's condition, as a symbol.

We take a symbol to be a sign that refers to something else, of which it is not itself a part. When I am hungry my stomach rumbles, but this is a sign of hunger, and doesn't symbolize anything. The rumbles bear what is called an 'iconic' relationship to the state of hunger; by definition, symbols are 'non-iconic.' The situation is fundamentally different with calls whose meaning is dependent, not just on the caller's physiological condition at the moment, but on something in the caller's world, directly responsible for initiating call production, something in other words that the call symbolizes (Hauser 1996).

Beyond philosophical considerations, why is this an important issue? For one thing, it has implications for the kind of brain mechanisms involved. If we wish to specify the demands placed on the bird's brain during call production, a symbolic mode of operation requires us to add an additional set of operations, acting prior to call production, to cope with processing and interpreting sensory information about whatever is being represented or symbolized. The nature and urgency of the particular triggering event in the environment must also be assessed before action is taken, whether this is a potential predator, or a choice morsel of food. We find ourselves considering the bird's brain in terms of cognition as well as physiology. Thus studies of the meaning of bird calls have implications for both cognition and neuroscience, suggesting where we should look in the brain for the underlying mechanisms. And it is not uninteresting that some bird calls do seem to function as symbols, bearing an abstract relationship to their external referents, representing a degree of behavioral and cognitive sophistication beyond what is usually expected in animal communication. We do not claim that such bird calls are semantic in precisely the same way as a human word, but that there may nevertheless be some parallels.

Do Bird Calls Symbolize Anything?

The most obvious candidates for symbolic functioning in birds are alarm and food calls. As already pointed out, both provide the tactical advantage that the potential symbolic referents – food and predators, – are accessible and susceptible to experimentation. When we hear food calls, we can assume that a bird has just encountered food, at least when operating in an honest mode. We can test experimentally whether the relationship is a dependent one or not. The more nutritious and novel the food, the more calls are given. The more a rooster food calls, the faster a hen will approach (Marler et al. 1986a). Urgency is another issue that contributes to the information content of a call.

When a female approaches a food-calling rooster, sexual activity is likely to ensue. For this reason you could argue, incorrectly as it turns out, that it is more appropriate to think of food calls as courtship signals, with only an incidental association with food. There are several reasons for thinking that this hypothesis is incorrect, and for arguing instead that the link to food is direct and primary. If a rooster's access to a hen is controlled experimentally, so that his introduction to her is timed separately from his access to food, there is always a surge in calling when food is presented (Evans & Marler 1994). A hen responds to food calls by looking down at the ground as though expecting food (Evans & Evans 1999), and the same call is given by broody hens to their chicks, inducing them to feed, with no hint of sexual behavior. As we have already argued, the case that food calls symbolize food is a strong one.

There is an apparent contradiction, however, when 'food' calls are uttered by a rooster as he presents to a hen, not food, but some inedible object such as a twig, or a leaf. Closer study suggests that this is not an error, but a case of deception. The hen does respond to these 'non-food' calls as though they still in some sense represent food to her. He is most likely to call with something inedible when his female is some distance away, more than two meters distant, where she cannot tell for sure whether he actually has food or not (Gyger & Marler 1988). Also, he is less likely to engage in prevarication if the hen he is calling to is familiar than when she is a newcomer, and thus a more attractive target of solicitation and a stronger stimulus for prevarication (Marler et al. 1986b). The notion of deception may only make sense if we regard the rooster as using food calls in an intentional way, although we may need to consider a distinction others have made between functional deception and intentional deception (Hauser 1996). The susceptibility of food calling to modulation by the presence or absence of an appropriate audience is consistent with the existence of an intent to influence the behavior of others, an interesting insight into the behavior of a creature that we do not usually think of as the acme of intelligence.

How about alarm calls? Are they also used in a semantic function? The first question to address is whether a recorded alarm call, played in the absence of predators, still elicits predator-specific responses. The answer is unequivocally positive. We have already noted the contrasting responses of a titmouse to, on the one hand, a distant soaring hawk and, on the other, to a perched owl. The flying hawk evokes fleeing, crouching and freezing; the owl evokes approach and mobbing. The same two contrasting response modes are evoked by playback of the two call types a rooster gives to aerial and ground predators (Evans et al. 1993; Marler & Evans 1997). The birds react as though the calls stand as non-iconic symbols of these two predator classes. The classes are large and ill defined, but the distinction is still valid and functionally important.

Functional Reference and Cognition

Even if we conclude that food and alarm calls do function in a symbolic fashion, thus fulfilling

Marsh warbler — Wothe

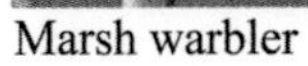

Chipping sparrow — Elliott

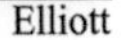

Jungle fowl — Andersen

Chestnut-sided warbler — Elliott

Blue Jay Read

Black-capped chickadee Elliott

Chaffinch nestling Berkhoudt

Red crossbill Jacobs

the criteria for referentiality, it still behooves us to be cautious about what kind of cognitive processes can be inferred. This is the reason the term 'functional reference' was coined (Marler et al. 1992; Hauser 1996), as a way to study possible cases of symbolic behavior while refraining from commitments about what is going on in an animal's mind. Admittedly, there are signs of complex cognition in avian food and alarm calling. We may even detect hints of intentionality, as when birds use alarm calls dishonestly, seemingly with the intent to deceive. As noted earlier, sentinels in mixed flocks sometimes use alarm calls deceptively to pirate food from other flock members (Munn 1984). In great tits, a subordinate ousted from a food table by a dominant sometimes gives a 'seet' alarm call with no predator present, regaining access to the food in the confusion (Møller 1988). There are audience effects on bird calling, consistent with intentional use. Aerial alarm call use increases in the presence of an audience, even though its presence does not seem to affect other signs of the caller's state of alarm (Gyger et al. 1986; 1987; Karakashian et al. 1988; Marler et al. 1991), as though alarm calling can be controlled somewhat independently of other responses to danger. Alarmed though a bird may be by a predator, alarm calls can still be withheld if there is no companion nearby; conversely, alarm calls can be given when there is no predator, in cases of deception. However although intentionality may play some role, we should not leap to the conclusion that this implies complex cognition; we may underestimate the potential complexity of reflexive mechanisms.

Nevertheless, studies of food and alarm calls make a strong case that they function semantically. An obvious question follows: what do they actually symbolize to others? There is an immediate problem. Our window on call meaning is provided by how others respond. But how do we distinguish between meaning as an imperative command to act in a certain way, and meaning as a label for something in the outside world? With animals as subjects, these alternatives are hard to separate. Then, if we entertain the naming hypothesis, there is the further problem of deciding what exactly is being labeled. Does an alarm call label something very specific, such as a particular predator species, a class of predators, or something much more general, such as a certain style and degree of danger? We know that for aerial alarm calls the first view must be incorrect, because the same hawk shape elicits a different alarm call when it is seen as a distant, moving silhouette, and at close quarters and ready to strike (Klump & Curio 1983). Perhaps meanings change with context? The referential specificity of alarm calls must surely vary adaptively from species to species, general in some, narrowly proscribed in others. These are among the many issues that must be addressed before we can gain a full understanding of avian semantics (Evans 1997, 2002). Above all, the issue of intentionality in animal communication remains as challenging as ever (Hauser et al. 1999).

Indexing: Another Factor in Call Meaning

While some bird calls display functional referentiality, others, like generalized alarm and aggressive calls, have more the quality of emotional displays. Philosophically, if not personally, we usually think of emotion-based signals like our own facial expressions and intonations of speech as rather depauperate in information content. It seems natural enough for us to downgrade the communicative status of signals that are emotional or motivational in nature, by comparison with referential signals, if only because symbolic behavior appears to involve a greater degree of cognitive abstraction than a 'simple' emotional display. And it is obvious that referential signals have far greater potential to convey precise information.

But there are important caveats here. With careful study, we find that communication by emotional displays can be very complex, especially when prevarication is involved (Ekman 1982; Scherer 1985; Fridlund 1994). Furthermore, if a bird couples a call with some kind of indexing behavior, such as head-pointing or gaze direction, a certain object or point in space or a particular group member can be precisely specified: the combination adds significantly to the communicative potential of emotion-based signals. Although the cognitive underpinnings of purely emotional signals are more limited, from a strictly functional point of view, their 'here-and-now' nature may have advantages. Furthermore, if a call that connotes fearfulness is given by a bird looking at another individual, this may provide a valuable clue about the intended recipient. It must be admitted that little is known about how accurately one bird can extrapolate the gaze direction of another into space. With modern techniques, experimental data on questions like this should now be more accessible.

Years ago, Chance (1962) noted the potential value of studies of the patterning of social attentiveness within animal groups, now being followed up by primate researchers (Perrett et al. 1984; Perrett & Emery 1994; Emery 2003). It would be fascinating to complement these primate investigations with parallel studies on the attention structure within groupings of social birds, within which calls are being used to communicate.

CONCLUSIONS

What have we learned from this walk through the world of bird calls? We have made selections from a huge literature, often more in the nature of accounts of natural history than the products of experimentation under controlled conditions. But many contemporary descriptive studies of natural behavior have high standards, both for precision and reproducibility, and most of the findings reported here will probably stand the test of time.

We have presented a catalog of some of the functions that bird calls serve, and their great importance in the life of birds is clear, especially in communication about issues that are vital for individual survival. Of the dozen or so functional classes we have considered, almost half are alarm calls, communicating about dangers of one kind or another. Some of the dangers are mild, and others are deadly serious. Interestingly, it is among alarm calls that many sexually dimorphic calls are to be found. It is not uncommon for them to be restricted to the breeding season, often as male prerogatives.

Call repertoire usage often changes seasonally, more richly exploited when birds are breeding. The increase is especially evident in males, whose call repertoires are often larger than those of females. In a sense, the vocal behavior of many male breeding birds is as ornamented as their appearance. Of the links to reproduction, some are direct, with calls used in courtship and mating, or as alternatives to song. The connection is just as potent, but less direct, when the use of male alarm calls peaks at the time of incubation and with the presence of young. The functions of calls are often complicated by use in several distinct circumstances. Within a given call category there are often sub-types that differ with the context in tonal quality, loudness, or the pattern of delivery.

Aside from the potentially lethal hazards of predation, calls are involved in many aspects of social life. With creatures as mobile as birds, group cohesion is a major issue, hence the widespread importance of contact, separation, and flight calls. While the formation of groups is sometimes a casual affair, in many species it is highly orchestrated, especially when flocks have a consistent membership of known individuals. There is often a need for calls that carry acoustic markers of individual identity. Contact and separation calls identify birds as members of a social group, whether it be a mobile flock, a local resident population, or a breeding pair. A remarkable number of bird calls are not only individually distinctive, but also different from one population to the next, known to be discriminated by the birds themselves. The number of well-documented cases of local dialects in bird calls is growing, especially striking in the flight calls of cardueline finches. It is increasingly evident that the supposedly strict innateness of bird calls is a myth.

No part of the avian call repertoire is more deserving of intensive study than calls that facilitate the sharing of food. They are important for the survival of adults and their offspring, and they also provide a fascinating window on some of the cognitive processes underlying avian call production.

Perhaps because bird calls are an especially intense focus for natural selection, their acoustic structure is often highly adapted to their function. Selection to maximize active space is reflected in quite radical adjustments of acoustic structure and calling behavior in signals like distress calls and, of course, song. But for every such case there are others with functional requirements that dictate a limited active space; 'seet' alarms and many contact calls are obvious examples. Adaptations to facilitate or hinder call localization are frequent, and alarm calls in particular have been documented in enough birds that the rules relating structure to function are beginning to emerge, though it must be admitted that there are still more hypotheses than facts about how birds actually use their alarm calls in nature.

Whenever there are selection pressures that favor species specificity in a bird call, there must sometimes be a degree of conflict with the predation-related specializations of call structure we have described. Specifically-distinctive signals set the stage for what we might term conspecific privilege, the particular facilitation of communication between members of the same species. Male song is the most obvious example, but some calls also qualify. A degree of species specificity is permissible when constraints imposed by other functional adaptations are not too severe, as often appears to be the case with flight calls, and with separation and mobbing calls. But when other functional requirements become demanding there may be no choice but to sacrifice species specificity, as in the hawk alarm calls of many birds. With calls like these, used by cohabiting bird species, selective advantages must surely accrue from the facilitation of cross-species communication, something that appears to be common in nature, and deserves more study. The mixed flocks for which tropical insectivorous birds are justly famous, will be an especially promising source of new data and concepts.

It is probably true that compared with calls, songs are more variable, and more often learned, but the list of exceptions is multiplying. One reason call learning has been overlooked is the relative simplicity of call structure that makes learned nuances less obvious. It may turn out that, although there could be as many learned calls as learned songs, the amount of learned variation that each kind of call displays is structurally limited, constrained by their simpler acoustic morphology. This structural simplicity may result in part, as with song, from the crowding of many calls into the acoustic space that they occupy in a species and in a community. Under such circumstances, a simpler structure may be favored; extreme variability must hinder quick, reliable identification of calls to some degree, but this is speculation. What we need are more facts, more species studied, chosen carefully with particular theoretical issues in mind, and more experimental testing of explanatory hypotheses. Tests need to be conducted both in the field and in the controlled conditions of the laboratory. With more data there will be almost no limit to the plausible speculations psychological, cognitive, ethological, and ecological, that bird calls bring to mind. For the adventurous, there are even hints of some elementary analogies with language. What more could one ask?

Chapter 6

Singing in the wild: the ecology of birdsong

HANS SLABBEKOORN

INTRODUCTION

The majority of animals communicate through some combination of visual, chemical, and acoustic signals. Visual signals are maybe the most obvious, broadcasting their message through patterns, colors, and behavioral displays. Everybody is familiar with the colored eye-spots in peacock tails, signaling a male's quality and used to seduce females, or the high-jumping gazelle, signaling to predators in pursuit that it will not be an easy catch. Chemical signals are used in territorial scent-marking behavior when coyotes defecate at prominent locations in their territory and domestic dogs urinate at street corners. However, many animals, especially the avian kind, rely heavily on acoustic signals. Given a choice of modality, why is it that so many birds use sounds to communicate? Why are vocalizations so often used for functions crucial to survival and reproduction, such as establishing territories or attracting mates?

The answer is to be found in a number of features that make sounds different from other signals. Sound can be transmitted over long distances, and audibilities of over a hundred meters are no exception. Elephants are even able to communicate over several kilometers with very low-pitched sounds, and whales exploit the transmission advantages of the ocean to communicate over hundreds of kilometers. Also, sound is, at least in some degree, multi-directional. A call uttered at one location may be heard by other individuals in any direction: facing the caller, behind or beside it, above or below it. Similarly, a receiver can hear sound coming from any direction and does not necessarily need to face the sound source or pay directional attention to detect it. A third feature with obvious advantages is that sound can be used in daylight as well as in complete darkness. And acoustic signals can penetrate vegetation and are less affected by obstacles in the landscape that may totally obscure visual signals.

These characteristics, signaling distance, omni-directionality, and penetration of the natural environment, not only explain why under certain conditions animals favor acoustic signals over visual or chemical ones, but also explain what type of sound is most appropriate for signaling purposes. Individuals that we record today are the offspring of those birds that did well in the past and were favored by selection over other birds that used other kinds of sounds. The selection pressures on acoustic signals vary with species, social context, and with the habitat and by investigating these factors, we may be able to understand why birds sing as they do: why some signals are low or high-pitched, loud or faint, fast or slow, simple or complex.

In this chapter the primary focus is the habitat as a source of selection pressures on sound signals. Despite its advantages for communication over long distances and in low-visibility habitats, sound transmission from sender to receiver in natural environments is by no means without problems. Sound attenuates and degrades with distance, in particular when there are obstacles in the landscape that hinder penetration. Turbulence in the air, due to wind or temperature gradients, causes irregular amplitude fluctuations that may distort a song. Many reflective surfaces, such as trunks, branches, leaves, and water and ground surfaces cause echoes that may interfere with signal perception. Furthermore, in a perfect world there would be no other sounds to confuse the receiver, but in real life the air is full of sounds day and night, produced by neighboring birds and other animals, and by abiotic factors such as wind or rain. Ambient noise in the background interferes with signal detection and recognition. Thus, in dealing with the ecology of birdsong, we have to consider how sound changes when radiating from sender to receiver as a function of habitat characteristics, and to what extent sound perception is affected by ambient noise that varies with habitat and the animals that live there.

SOUND TRANSMISSION: GETTING THE MESSAGE ACROSS

Signal Attenuation

Transmission through the environment affects birdsong, when traveling from sender to receiver, by attenuation and degradation (Wiley & Richards 1982; Forrest 1994). First, as already noted, the energy present in a sound is not beamed to any great degree, but tends to spread in all directions. The result of this omnidirectional sound transmission is a rapid loss of energy through the effect of spherical spreading. As a consequence, sound attenuates at 6 dB per doubling of distance. This does not mean that birdsong is equally loud in all directions away from the sender because of the orientation of the bird while singing (Witkin 1977; Larsen & Dabelsteen 1990). For example, male sage grouse produce a whistle during their strut display which is highly directional with most sound energy projected sideways (Dantzker et al. 1999). This is why they never face a female when they try to attract her with an acoustic display. Song generally sounds loudest in front of the bird, because that is where the bill is pointing. High-frequency sounds in particular may be beamed in one direction, resulting in a longer transmission distance. For low frequencies, it is more difficult to restrict the beam of radiating sound energy, especially with a small beak; you need a large horn to make speech directional through a megaphone.

Sound attenuation also results from the absorption of energy by the atmosphere and by obstacles in the environment. Its extent varies with frequency and depends on the air temperature and humidity. In general, higher frequencies attenuate faster than lower frequencies, and attenuation by absorption is greater as temperature increases, and reduced with increasing humidity (Wiley & Richards 1982). The variation in attenuation over a day or between seasons is determined by the effects of both temperature and humidity, which can amplify or counteract each other. When the air temperature rises in the morning, humidity often drops, and both lead to higher sound absorption rates. But when temperature rises in concert with humidity, as before a spring thunderstorm, sound transmission characteristics may not change markedly. The combined impact of temperature and humidity varies considerably among different habitat types in different climates. In hot, dry desert climates, attenuation rates are particularly high, especially for high-frequency sounds.

Obstacles such as trees, branches, and leaves also have an impact on sound transmission (Aylor 1971; Marten & Marler 1977; Marten et al. 1977; Martens 1980). Sound striking the surface of such obstacles is absorbed and reflected in different directions, leading to additional loss of sound energy (often referred to as 'excess attenuation'). If we consider the direct pathway from sender to receiver, sound may be reflected out of this pathway, and then bounced back into it from the ground, resulting in the so-called 'ground effect.' The reflected sound waves lead to attenuation at some frequencies and amplification at others (**Fig. 6.1**) because of differences in phase that lead to wave interference (Marten & Marler 1977; Embleton 1996). In addition to solid obstacles, reflections also result from atmospheric turbulence caused by wind, temperature variation, or air compression around mountains or buildings.

Signal Degradation

Degradation refers to any change in spectral, temporal, and structural characteristics occurring between sender and receiver, all of which might be important for signal recognition. The scattering of sound waves through vegetation gives rise to degradation in several ways. The wavelength of a sound in relation to the dimensions of an obstacle determines how much a sound is reflected. Low-frequency sounds have long wavelengths and are hardly affected by small obstacles such as leaves. High-frequency sounds have shorter wavelengths and are affected by smaller obstacles during transmission. As a consequence, attenuation and degradation due to reflections are both frequency-dependent: low frequencies transmit more easily through vegetation than high frequencies. The result is that birdsongs are degraded with distance in a predictable way, as though they have been passed through a low-pass filter (**Fig. 6.2**).

Echoes are sound reflections that arrive at a receiver with a delay due to a longer transmission pathway. Reflection from a wall causes a relatively discrete echo separated from the direct sound by a silent interval (**Fig. 6.1**). Scattering from vegetation or atmospheric turbulence leads to multiple echoes. In nature, the many extra pathways can lead to a range of delays, resulting in a sound 'tail' (**Fig. 6.1**). Echoes are often referred to as 'reverberations' and may result in serious signal degradation: intervals between song notes that are originally silent may be filled with sound, the original length of notes can become blurred, and sound tails may overlap in time

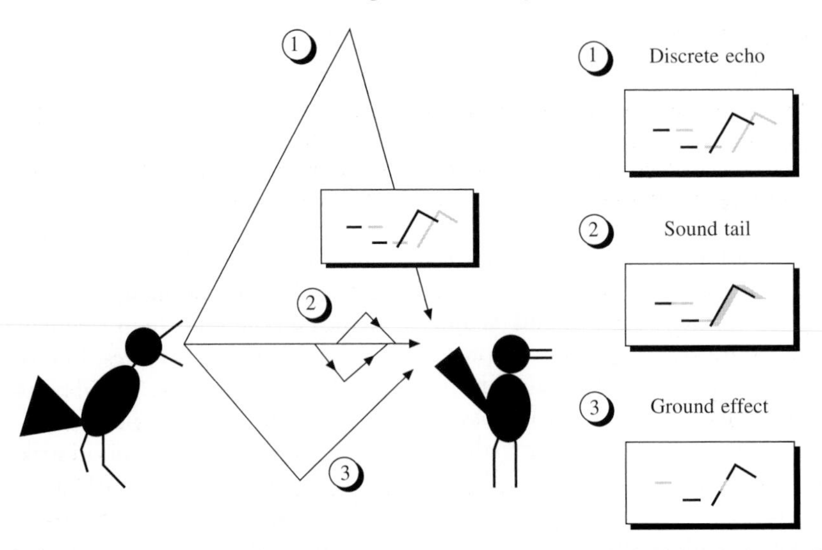

Figure 6.1 Three ways in which sound reflections cause signal degradation. (1) A clear reflection with a long delay leads to a discrete echo. (2) Many small short delay reflections, as from vegetation, lead to a 'sound tail.' Constant frequency notes have longer tails than frequency-modulated notes. (3) A strong reflection from the ground can cancel out some frequencies and amplify others because of phase differences between the direct and the reflected sound wave.

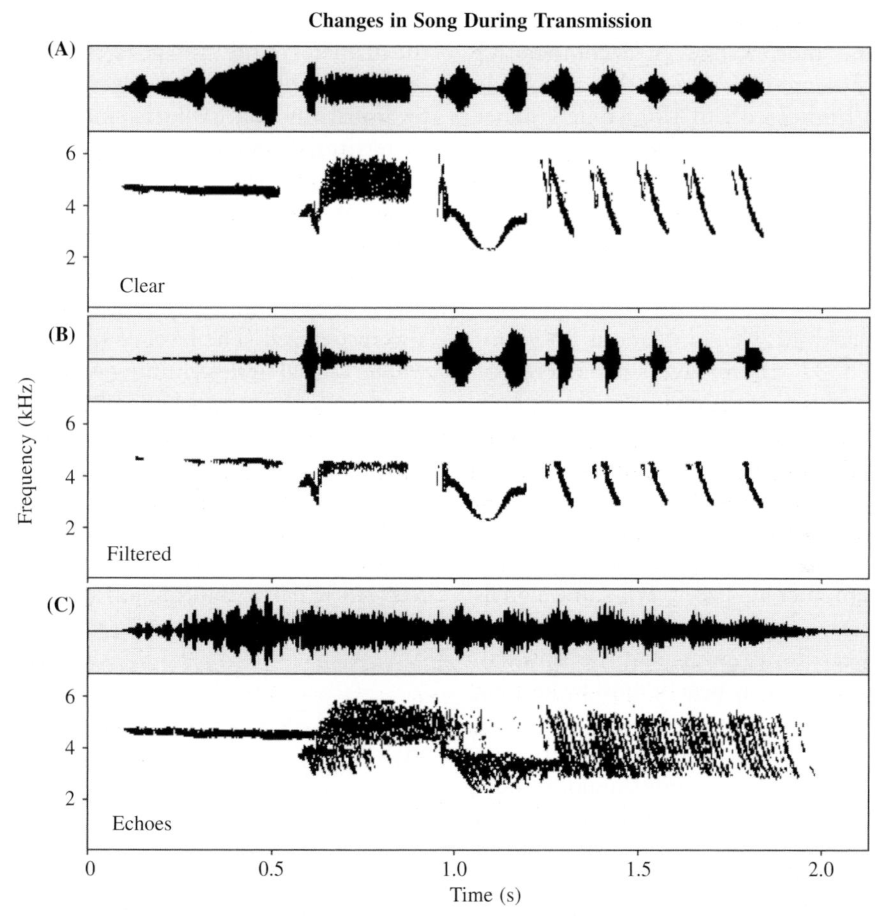

Figure 6.2 One clear and two artificially degraded songs of a white-crowned sparrow (**CD1 #52**). **(A)** An original close-range, clean recording. **(B)** High-frequency attenuation by low-pass filtering. **(C)** Multiple reverberations or 'echoes.' Both the sonograms and the amplitude wave patterns are shown. Thus, the song that a bird hears in nature is often rather different from the original.

with subsequent notes, masking some of their acoustic features (**Fig. 6.2**). In addition, reverberations often come in very irregular patterns, because of the heterogeneous distribution of obstacles in the environment and variation in their configuration in relation to the position of different senders and receivers. The irregular arrival of echoes leads to erratic amplitude fluctuations in song notes and their tails (also visible in **Fig. 6.2C**), changing the original amplitude pattern. These are all obstacles that birds must deal with as they communicate with sound under natural conditions.

Active Space of a Bird's Song

Signal attenuation and degradation have important consequences for the 'active space' of bird sounds. Active space refers to the area around the sound source over which the signal remains detectable and recognizable for potential receivers (Brenowitz 1982a). Other factors besides attenuation and degradation play a role in determining this distance: loudness of the song, the hearing sensitivity of potential receivers (see Chapter 7), and the level and spectral characteristics of the ambient noise. The maximal volume of a song is somewhat constrained by body size and morphology. In a comparative survey of 17 songbird species, the amplitude level ranged from 74 dB to 100 dB at 1 meter from the bird's beak (Brackenbury 1979). There was a weak relationship with body size: larger birds have louder songs. However, there were inconsistencies; for example, the tiny winter wren, weighing about 10 grams, produced very loud songs averaging 90 dB, whereas the much larger European blackbird, weighing about 96 grams only reached 87 dB on average. There are obviously selection pressures other than those related to body size that cause variation in song output level across different species.

Acoustic signals can be expected to evolve so that their active space is related to the optimal distance for accomplishing their communicatory function (Lemon et al. 1981). For some signals this may be 'the further the better', for others one or two territories away may be optimal. Territorial songbirds may benefit most by keeping away direct competitors without challenging males that are not a direct threat to the territory. This implies that territorial birds should be more likely to optimize than to maximize transmission distance. As a result, selection on song output level should have led to a correlation between song amplitude and territory size. Indeed, in a comparative survey on more than 30 bird species, larger birds produced louder songs. But independent of size, species with a larger territory size also had a louder song (Calder 1990), although this study did not correct for phylogenetic trends, which must be considered to draw firm conclusions from the data (see Chapter 10). Another study suggests that males produce songs just loud enough to serve their territorial function (Brenowitz 1982a). The active space of red-winged blackbird song was estimated to be 160–170 meters. The maximum distance across an individual's territory in the population studied was 60–80 meters, so that a male's song reached throughout its own territory and that of its immediate neighbor, irrespective of perch location. If a bird was perched at the edge of his own territory, the song could just reach receivers two territories away. An audible distance of twice the diameter of the average territory for a species seems a likely result of selection related to a territorial function, a rule that may apply generally to territorial animals. The long calls of some African rainforest monkeys also have an active space of twice the diameter of their average-sized home range (Brown 1989).

Important for the concept of active space is that a song should not only be detected, but also recognized. The level of signal degradation will be crucial in determining whether a receiver can decode the message encoded in the acoustic characteristics of a detected song. Studies of the recognition of degraded signals via playbacks in the natural environment have shown that, while signal degradation during transmission can be detrimental, it can also be beneficial, enabling receivers to derive information about the location of the singer.

COMMUNICATION WITH DEGRADATION: INSIGHTS FROM PLAYBACKS

Signal Degradation: Burden or Benefit?

Whether by frequency-dependent attenuation, reverberation, or by irregular amplitude fluctuations, signal degradation obviously interferes with a researcher's ability to recognize patterns on sonograms, but it also affects recognition by the birds themselves. Several bird species are known to respond differently to clear song and to degraded song played back in the field at the same amplitude level. On average, one would expect a weaker response if degradation hinders detection or recognition. This is indeed found to be true, both for species living in relatively closed, densely foliated habitat such as great tits (McGregor & Krebs 1984), and those living in more open, windy areas such as western meadowlarks (McGregor & Falls 1984). However, this does not necessarily imply a lack of detection or recognition: the birds may still have recognized the song, but they may have inferred from the degradation that the singer was too far away to merit a strong response (McGregor & Falls 1984; McGregor & Krebs 1984). Other playback experiments, altering song characteristics that would either degrade quickly under natural conditions (e.g. high frequencies) or that remain relatively unaffected during transmission (e.g. low frequencies) indicate that this may indeed be the case.

Although it is widely accepted that degradation during transmission limits the information that can be conveyed by complex signals, the sensitivity of males to song playback indicates that some birds are well adapted to recognize degraded song. There is often sufficient redundancy in characteristic song parameters to allow recognition of degraded song (Brémond 1978; Dabelsteen & Pedersen 1992). Brenowitz (1982b) conducted experiments with red-winged blackbirds, using complete and partial songs. A typical redwing song consists of one to six short introductory notes, followed by a long trill, the whole song lasting about one second. The trill is the loudest part and has a wide frequency range, with most energy between 2–6 kHz. However, after transmission over more than 100 meters, the only part of this frequency range still audible is roughly 2.5–4.5 kHz. The lowest frequencies are weaker to start with and the higher frequencies are lost through frequency-dependent attenuation. Playbacks reveal that the lower part of the trill (up to 4 kHz) is sufficient to elicit a response, but the higher part of the trill (above 4 kHz) by itself is not. Thus, enough features are retained in a normally degraded song for other redwings to recognize and respond to it. Retention of potency by those song features that are relatively resistant to degradation has been demonstrated in birds from a variety of taxa including songbirds, waders, and gulls (Stefanski & Falls 1972a; Robisson 1987; Brémond & Aubin 1990; Mathevon & Aubin 2001).

All studies mentioned so far focus on the territorial response of males, but song is also addressed to females, and degradation may again play a role. A laboratory study on brown-headed cowbirds tested the impact of song degradation on its potency in eliciting a copulatory posture in females (King et al. 1981). Cowbird song is composed of three phrases: a series of relatively low-amplitude notes, followed by a brief note and a final whistled phrase, both much louder and higher in frequency. The response strength decreased dramatically with the degree of degradation, indicating that it is important for the male to be close to the female. The lower audibility of the first component and overall reduction in signal-to-noise ratio are the main factors that cause distant cowbird songs to be less potent (see p. 289 & 376).

Reverberations can also be of some benefit to the singing bird, depending on the acoustic design of its song. They affect the spectral degradation of song notes in a way that depends on the shape of the note (**Fig. 6.1**). For a typical frequency-modulated note, a signal receiver will hear both the most direct sound wave, and also at the same time echoes of the preceding part of the note. The echoes lead to the presence of delayed lower frequencies for upward slopes, and delayed higher frequencies for downward slopes. In contrast, for sustained whistle-type notes of constant frequency, the delayed sound waves are all of the same frequency, and do not lead to spectral degradation. On the contrary, due to accumulation of scattered sound waves of the same frequency, the tail following the last direct sound wave will be longer and louder for notes with a narrow-frequency range. As a consequence, under the right conditions, birds that use such simple song notes without frequency changes produce a longer and louder signal with the same amount of energy (Slabbekoorn et al. 2002).

The green hylia, a small bird of the rainforest of West and Central Africa, uses song notes that remain at one constant frequency (**CD1 #53**). The long-range song of males consists of two loud notes tuned to a very narrow-frequency range at about 3.8 kHz (**Fig. 6.3**). Playbacks show that territorial males respond more strongly to signals of longer duration (Slabbekoorn et al. 2002). One of the changes reverberations impose on the original signal is elongation, and consequently reverberations may be beneficial,

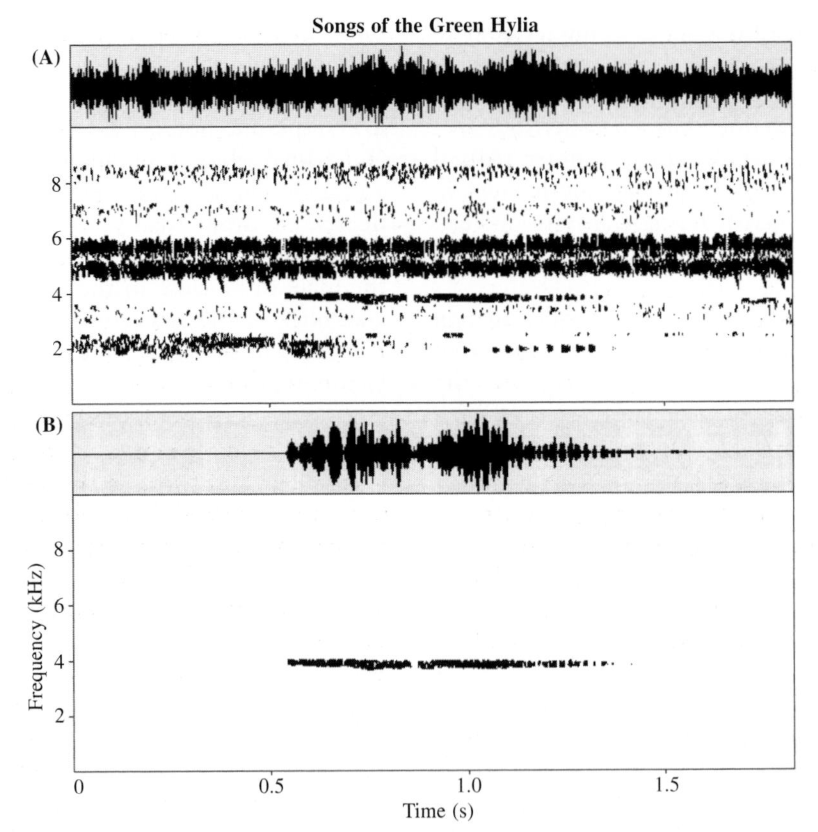

Figure 6.3 Song of the green hylia. Amplitude envelopes and sonograms of **(A)** the two-note song embedded in the natural background of ambient noise in the tropical rainforest of Cameroon, and **(B)** same song with the ambient noise removed by filtering above and below it (**CD1 #53**).

assisting the male hylia in transmitting its song to potential rivals. So, song degradation is not always detrimental to the singing bird.

Two other factors that could improve the active space of a song note restricted to a narrow-frequency range are related to transmission and perception. First, concentrating energy within a narrow-frequency band, rather than spreading the same energy through a broader range of frequencies, may lead to an increase in transmission distance (Morton 1975; Marten et al. 1977; Wiley & Richards 1982). Furthermore, longer sound elements are easier to detect than shorter ones due to a perceptual mechanism called temporal summation (Klump & Maier 1990), again increasing the active space of the signal.

Narrow-frequency notes are not restricted to the territorial songs of bird species living in dense tropical forests (Slabbekoorn et al. 2002). The alarm calls of small passerines to aerial predators (Marler 1959; see Chapter 5) are of a similar design; also alerting components in songs of many North American bird species have a distinct narrow-frequency whistle-type character (Richards 1981a). Transmission advantages of this acoustic design probably help in drawing attention to the more complicated parts of the song to follow (**Fig. 6.2**).

Distance Assessment Based on Acoustic Cues

Another remarkable benefit of signal degradation is the ability of birds to retrieve detailed information about the caller's distance from the changes that accumulate in a song during propagation. This is termed 'ranging' (Morton et al. 1998; Naguib & Wiley 2001). The most obvious distance cue is the amplitude. We know how loud the talking and laughing of people would be if they were standing right next to us, and by the degree of attenuation we are able to roughly judge how far away they are. Mostly we make such calculations unconsciously, and we are taken by surprise if the loudness of a sound does not match our expectations. It is an eerie experience when in a dome-shaped hall, reflections make the sound of somebody talking in the distance as loud as if he were standing right next to you. However, ranging does not rely on amplitude alone. Birds use other cues that change predictably with transmission distance, particularly the accumulation of reverberations and frequency-dependent attenuation.

One of the most detailed studies on ranging has been on North American Carolina wrens (Richards 1981b; Naguib 1995, 1996), which live in semideciduous forests with a dense understorey of shrubs. Each male vigorously defends a territory in the spring to safeguard his female, nest site, and food resources against potential rivals. If another male wren announces its presence by singing, ranging allows the territory holder to determine whether he has to deal promptly with an intruding rival or with a male outside his territory boundaries. Correctly assessing the distance to the sound source prevents unnecessary spending of time and energy on aggressive approaches. Experimental presentation of songs recorded previously, either clean nearby recordings or long distance recordings of degraded songs, reveals that Carolina wrens perceive the level of song degradation and respond accordingly (Naguib 1995). They do so even though both degraded and undegraded songs are played from the same location and at the same volume. They approach the loudspeaker quickly and directly after hearing undegraded song, but they often overfly the loudspeaker after hearing degraded song as if they expect the simulated singer to be further away. After the initial flight beyond the loudspeaker, males often sing loudly and move around nearby, as if looking for the intruder. With even more degradation, response levels are usually low, either because the territorial male judges the singer to be too far away to be a threat, or because he fails to recognize the song.

Ranging is a more complicated task than it appears. A Carolina wren judging the distance of a singer needs to take several things into account (Morton et al. 1998). First, it needs to possess some sort of representation of undegraded wren song. How loud is the song of another wren when perched right next to you? What are the general spectral characteristics of conspecific song, and what do clear unreverberated wren songs sound like? If a bird hearing a degraded song has experience with exactly the same song of the same individual at close range, it is likely that the ranging task is easier and the distance judgment more accurate. Therefore, familiar neighbors are likely to be monitored more precisely than complete strangers. Still, in both cases the bird has to compare a complex set of acoustic characteristics of the degraded song to those of an undegraded memorized song.

Another requirement for ranging is an ability to take account of how degradation develops with transmission distance in the local habitat. Different habitats can lead to different degrees of degradation over the same distance, and the density of obstacles and reflective surfaces varies with the season. In the semideciduous forest habitat of the Carolina wren, for example, males defend their territories throughout the year, both when there are no leaves on the trees and shrubs and when they are in full foliage. The trees come into leaf quite quickly in the spring and sounds suddenly become more and more degraded when transmitted over the same distance. Playbacks show that the wrens adjust their ranging criteria and respond differently to the same amount of degradation before and after foliage development (Naguib 1996).

THE HABITAT AS A SPACE FOR ACOUSTIC PERFORMANCE

Habitat Heterogeneity and Acoustically Distinct Layers

The habitats in which birds live and communicate are very complex auditoriums. The variety of physical processes associated with sound transmission, attenuation, and degradation account for only a small part of this complexity. Few habitats have just one type of leaf and one type of twig, homogeneously distributed between sender and receiver. There is often a variety of shrub and tree species present that will affect sound transmission differently. There may be a dense shrub layer at one location and more open patches in another, both irregularly interspersed with thick tree trunks. The ground may be bare or carpeted with vegetation, flat or hilly. There may be rocky outcrops, alpine meadows, densely foliated fern canyons, or hedge-type vegetation along a stream, all contributing to changing patterns of sound absorption and reflection depending on the positions of sender and receiver in their heterogeneous environment.

There is often also more stereotypic variation in the form of acoustically distinct layers. One may find a relatively closed canopy in a temperate forest together with a densely-foliated shrub layer with only bare tree trunks in the open space between. Savanna vegetation may be dominated by ground-covering grasses with only occasional, sparsely distributed acacia trees and an occasional baobab reaching into the sky. A similar structure can be found in marshy areas, with a mosaic of willow tree patches sticking out of otherwise contiguous reedbeds. Of all habitats, the tropical rainforest may be the most complex, with many distinct strata between the dark forest floor and the often 50-meters-high canopy, with emergent tree giants of over 80 meters. In addition, all habitats have open air above available for acoustic communication by birds able to fly up above the meadows, reedbeds, and forest canopies.

Habitat layers will differ in their sound transmission characteristics depending on the vegetation type and density. In addition to the intrinsic differences related to the vegetation characteristics within a habitat layer, horizontal sound transmission may also be affected by adjacent layers. For example, song beamed through a relatively open forest layer may be subject to relatively low attenuation, but reverberation rates may be high due to the distinct reflective surfaces of adjoining layers. Besides vegetation layers (Morton 1975), reflections may also originate from ground or water surfaces (Wiley & Richards 1982; Cosens & Falls 1984). Distinct reflective layers may cause irregular patterns of sound 'tail' formation in songs. They may even produce discrete echoes, if the usually chaotically distributed delay times of reflected sound waves become more synchronized. This is similar to what birds may experience in more extreme form in urban environments, with brick buildings and asphalt reflecting their songs strongly. A study in the African rainforest revealed that sound tails are louder and longer for artificial tones transmitted over a narrow forest trail than when transmitted over a nearby transect straight through the understorey vegetation (Slabbekoorn unpublished). Tail duration and relative amplitude are determined by the interplay between attenuation and accumulation of reverberations. Lower attenuation rates due to absence of vegetation, with similar reverberation rates, may cause the increase in tail length and amplitude over the forest trail transect, as through a relatively open forest layer. Birds may benefit from singing through such 'acoustic tunnels' in the forest, experiencing less attenuation than expected ('negative excess attenuation;' Embleton 1963; Aylor 1971; Martens 1980).

Communication Through Layers: Choice of Song Post

Many birds optimize their active space by picking out a perching height for singing within a habitat-layer best suited for sound transmission. In urban habitats, collared doves coo from chimneys and lampposts, while black redstarts can be heard singing on the rooftops of the highest buildings.

The red-winged blackbirds of North America advertise their song and display visually with their red shoulder epaulets at various places in their territory, but always at emergent perches above the reedbeds and not within them (Brenowitz 1982a, b). Singing corn buntings in Dutch meadows preferentially perch above grass level on barbed wire. This trend is typical for grassland species in general, although they do not all necessarily choose the highest spot available (Harrison 1977). Several forest bird species are also known to climb up to an elevated song post, which can lead to a considerable increase in the transmission distance of their song.

The winter wren occupies the lowest layer in the forest, and is often hidden in dense shrub foliage. By using a song post at a height of 5 meters instead of 0.2 meters above the ground, a wren is able to double its active space (Mathevon et al. 1996). The amplitude at 40 meters for a song sung at ground level is the same as the amplitude at 80 meters for a song sung at an elevated perch 5 meters high. At the same time, the elevated song post also improves the bird's ability to hear the songs of competing neighbors, maybe explaining why climbing to a higher perch is often observed in response to playback of degraded song (Mathevon & Aubin 1996; Holland et al. 1998). In line with this, a blackbird in the shrub layer, hearing a conspecific in the distance, has to move almost 90 meters horizontally, approaching the singer, to improve audibility in the same degree as when moving up in the forest to a perch at about 9 meters (Dabelsteen et al. 1993). The greater attenuation in the lower layer of the temperate forests where these two birds live is due mainly to greater leaf density and attenuation through interference by sound waves reflected by the ground (Marten & Marler 1977; Embleton 1996).

Most birds have preferred perch heights both for singing, and for other activities (Ficken & Ficken 1962; Lemon et al. 1981; Nemeth & Winkler 2001). A species will experience different transmission characteristics for their territorial songs depending on the forest layer they use most of the time. Furthermore, if singing and foraging or nesting typically occur at different heights (Harrison 1977; Hunter 1980), horizontal and diagonal transmission will be important, potentially crossing different vegetation layers. Such a case has been studied in temperate forest with male and female European blackbirds, which have somewhat different habits and use consistently different forest layers (Dabelsteen & Pedersen 1988a, b). Males spent a lot of time in relatively open areas, at an emergent song post or flying around, patrolling territorial borders. Females have a more secluded lifestyle, often under cover in a lower shrub layer. Blackbird song consists of relatively low-pitched 'motifs,' that transmit far, and higher-pitched 'twitter' elements, which have a much shorter acoustic reach (Dabelsteen 1984; Dabelsteen et al. 1993). Both convey information important to territorial males and to receptive females (Dabelsteen & Pedersen 1988b; 1992). However, as song is usually emitted from an elevated song post, other males, being at more or less the same height, will hear the song after transmission through a relatively open habitat layer, while females in dense understorey will hear a more heavily degraded song. As a consequence, females have to deal with relatively weaker high-frequency components of male song, due to the frequency-dependent filtering of the vegetation. This may have led to greater sensitivity to high frequencies in females. An experimental increase in frequency that led to significant reductions in the male response during playback in the field (Dabelsteen & Pedersen 1985) still triggered strong female responses in a laboratory test (Dabelsteen & Pedersen 1988a). Although test conditions varied between the sexes, these results are at least consistent with the hypothesis, with females appearing less discriminating and thus maybe more sensitive than males in their response to increased song frequencies.

It should be noted that changes in transmission characteristics with the forest layer or with horizontal or diagonal propagation pathways will also affect detection of singing birds by human observers. Variations in the likelihood of song detection due to habitat heterogeneity may be important factors when sounds are used in monitoring bird populations (Desante 1986; Haselmayer & Quinn 2000; see Chapter 12).

Singing in the Sky

Choosing a particular song post is not the only option in exploiting a favorable habitat layer for optimal song propagation. Many species not only call during flight (see Chapter 5), but also sing during a ritualized and stereotypic flight display, using a favorable habitat layer without a fixed song post (**Box 22, p. 189**). The list of potential examples is long and varies widely with respect to taxonomy and habitat. Sedge warblers often leave their hidden perch to perform a short song flight above a dense layer of reeds (Buchanan & Catchpole 1997, 2000). The wood warbler is a common forest species in western Europe, typically found in old beech woods with a relatively open understorey. The males of this species often produce their accelerating series of song notes in a song flight parachuting down from a more densely foliated layer into the 'open space' underneath (Slabbekoorn unpublished). The skylark is probably the most famous aerial singer and certainly one of the most impressive. The three stages of its vocal performance in the air, climbing, level flying, and descending flight, may last more than half an hour (Hedenström 1995). Broad-billed sandpipers breed on bogs in northern Europe, where they perform their vocal display flights. They alternate between rapid wing fluttering and longer glides while uttering a rhythmically repeated call, before they deliver their song during a descending flight (Svensson 1987). Blue-breasted kingfishers produce repeated series of loud song notes of descending frequency (**Fig. 6.9**) while circling in undulating flight just above the canopy of the tropical rainforest in Central Africa (Slabbekoorn unpublished).

The use of a different habitat layer for singing in flight than for foraging and hiding does not necessarily mean that song flight has evolved just to optimize sound transmission. It may also add a conspicuous visual component to the song that is obviously most apparent in an open habitat layer. Furthermore, singing flyways may function to demarcate territory boundaries, as a more dynamic strategy compared to songbirds that travel silently between song perches. Male pintail snipes perform undulating flights, alternating gliding, fluttering, and diving, thereby outlining their territories, which may be up to 200 meters across (Byrkjedal 1990). In these flights they produce a variety of vocal sounds in addition to the mechanically-produced 'fizzing' of their tail feathers. Similarly, a close relative in western Europe, the woodcock performs a display flight at dusk and dawn with slow wing beats just above the treetops. The territory boundaries of neighboring males can be tracked by ear, following the audible trail of alternating muffled grunts and high-pitched notes.

Many species perform both flight song and perched songs. Interestingly, the song features sung in these different settings may differ considerably, for example with respect to the ratio of distinct song components present (Schmidt et al. 1999), or through the use of a phonologically distinct set of song notes (Nowicki et al. 1991). This may be related to the difference in environmental selection pressures for the two methods of song display, but it also raises the question of whether flight and perched songs serve the same functions. Research on several species with conspicuous song flights has already indicated that their function may vary with species and context (**Box 22, p. 189**).

Weather Conditions

Climatic factors add another layer of complexity to acoustic communication outdoors. A rain shower can completely change the acoustic world from one minute to another (Lengagne & Slater 2002) and wind and temperature gradients continuously modify signaling conditions (Wiley & Richards 1982). The consequences can be dramatic, particularly when a bird uses the more exposed layers in its habitat through the choice of an elevated song post or by singing in flight. Whether a singing bird is heard upwind or downwind affects the apparent loudness of the

BOX 22

THE FUNCTIONS OF PERCHED SONG AND FLIGHT SONG

Some birds sing on a perch, some in flight, and some do both. Why does this variation occur? Field observations and experiments in a natural setting can give us some clues, but the answers may vary with the species and, in any case, the interpretation of such data is not always easy, as we see with flight song. The bluethroat, a close relative of the nightingale, has both perched song and flight song. A detailed study of their function in Finnish Lapland showed that both are restricted to a short period in the spring, before and during pair formation (Merila & Sorjonen 1994). No paired males were heard singing after clutch completion, and an unpaired male continued to sing throughout the season, both perched and in flight. Also, males that lost their clutch resumed their singing activity immediately. The authors concluded at that time that perched song functions primarily in mate attraction. But there was also more flight song activity with an increasing number of singing males, suggesting a function in territory defense. Then in a follow-up study, simulation of territorial intrusions with a stuffed bird and song playback triggered an increase in both types of song (Sorjonen & Merila 2000), showing that in fact both perched song and flight song are important for territory defense, at least in the period before pair formation, when the experiments were done. So, having a territory may help males to gain access to females, but the seasonal pattern of song and song–flight activity is not necessarily evidence for a direct role in mate attraction. Correlations do not necessarily mean causation. The only remaining argument for song being attractive to females in this species is the fact that males with high song–flight activity acquired a mate more quickly and the anecdote of a female approaching the speaker when her mate was accidentally absent (Merila & Sorjonen 1994).

The common yellowthroat, a North America warbler, also has perched and flight songs (see below; **CD2 #38**), but their functions clearly differ. In contrast to perched song, song–flight activity in yellowthroats only increases after the period of establishing territories in the early spring (Kowalski 1983) and there is no difference in song–flight activity before and after pairing (Ritchison 1991). The most likely function of flight song here seems to be a signal for distracting potential predators. In a study of seven males in Kentucky, the song–flight rate was independent of the location and activity of other yellowthroats, but increased considerably when the investigator entered a bird's territory (Ritchison 1991). Males usually only initiated a song flight some distance from the female or the nest, and landed even further off, as if to draw a predator away. The flight song is acoustically similar to the perched song, but with additional introductory and terminal notes that are short and of a wide frequency range, an acoustic design that may facilitate locating the sound source.

Skylarks illustrate another deviant function of flight song; they sing when a bird of prey is pursuing them (Hasson 1991; Cresswell 1994). Singing may signal that a bird is in good health and difficult to catch, much as stotting in gazelles has been interpreted. Although skylarks do not usually sing in winter, almost a hundred observations were made in Scotland of birds singing during attack by merlins. Sometimes, if a merlin approached a flock, several flock members began singing; individuals varied in whether and how well they sang when chased. Merlins chased for a longer time those birds that sang poorly or more sporadically, compared with those that sang a full song; some sang for up to 80 seconds with the predator in hot pursuit. Those that did not sing were caught more often than birds that uttered some phrases, while birds that sang full, continuous song escaped at the highest rate (Cresswell 1994).

Hans Slabbekoorn

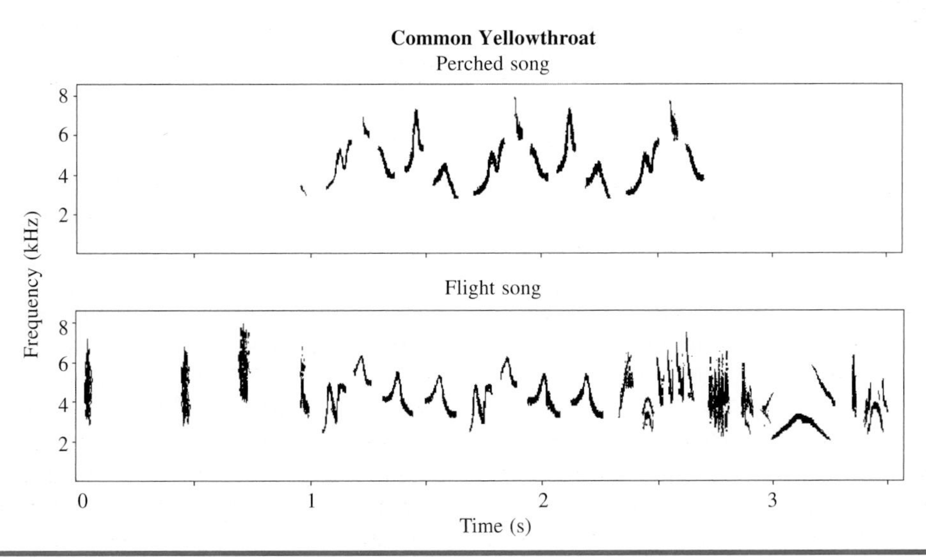

song. In general, wind leads to increased degradation rates through the increase in air turbidity, adding refraction and reflection effects to the reverberations caused by the vegetation.

The most thorough investigation on the impact of wind on acoustic communication concerns a study on the African elephant (Larom et al. 1997a, b). A combination of wind and temperature measurements throughout the diurnal cycle and computer modeling of omni–directional low-frequency sound propagation revealed dramatic fluctuations of the active space of the extremely low-frequency, infrasound calls of elephants; they have an audible area of only 50 km^2 during most of the day, with a high directionality depending on the wind. However, low wind velocities, minor air turbidity, and the temperature inversions mentioned below create an optimal time window in the early evening, allowing communication with an omni-directional calling area frequently exceeding 200 km^2. Such fluctuations in directionality and active space must also affect long-distance communication in many birds, particularly those living in open habitats (Henwood & Fabrick 1979).

Air temperature affects the speed of sound: a higher temperature leads to a higher sound velocity. This causes an interesting phenomenon where large bodies of air with a different temperature meet (Wiley & Richards 1982). Sound arriving at the boundary of such a temperature gradient will either be refracted back, if sound travels from cold to hot air, or will be largely absorbed, if sound traveling through hot air meets cold air. Such thermal gradients can have dramatic consequences for song audibility. For example, at the end of the night, there is often a layer of cold air close to the ground forming a so-called 'temperature inversion.' Birds that sing in the dawn chorus benefit because sound is reflected back from the higher and warmer air layers and beams their song loud and clear through the tunnel of cold air. For the same reason sound generally travels well above cold water or ice. The opposite is true when the sun heats up the soil during the day, and the highest temperatures are close to the ground: sounds will now be deflected into the sky instead of back to the ground. The resulting high attenuation rate is probably one of the reasons that many birds are less active vocally after midday (Staicer et al. 1996; Brown & Handford 2003).

NOISE ANNOYS: COMPETITION FOR ACOUSTIC SIGNAL SPACE

Natural Habitats are Noisy

Birdsongs are usually heard against a background of other sounds. In natural environments birds are continuously exposed to a stream of noise from which they have to extract the songs and calls and other sounds relevant to them. Ambient or background noise in natural habitats arises from environmental factors such as wind or rain and insects and other biotic sources (Morton 1975; Ryan & Brenowitz 1985; Waser & Brown 1986). The background noise apparent at the time of singing affects the signal-to-noise ratio for a listener, and may interfere with the efficiency of the signal. Interfering noise makes signals more difficult to detect, and information encoded within the signals can be harder to extract (Brenowitz 1982a; Wiley & Richards 1982).

The level of signal interference depends on whether the frequencies of the signal and the background noise overlap (Dooling 1982a; Klump 1996; Langemann et al. 1998; Hulse 2002). Rustling leaves and twigs produce sounds over a wide range of frequencies; air passing over the ground and around buildings or rocks causes low-frequency noise. Signals of one species become noise for another, filling the air in natural habitats, each adding frequencies according to their body size. Insects are mainly restricted to relatively high frequencies, while vocalizations of frogs, toads, birds, and mammals extend into lower frequencies. Besides its spectral characteristics, noise can be subdivided roughly with respect to duration into continuous and discrete. Continuous noise bands can be a serious constraint on acoustic signaling by masking a particular frequency range, often for long periods of time (Slabbekoorn & Smith 2002a). Discrete sound events, including songs and calls of other animals can cause masking problems, with effects not only on detection, but particularly on signal discrimination. A bird species' own song has to be recognized among the other sounds it hears; this is a special perceptual task, a part of what has become known as auditory scene analysis (Bregman 1990; Hulse 2002).

Urban Rumble and Tropical Jumble

Each habitat has its own typical pattern of ambient noise (**CD1 #54**). The exposure to wind and the composition of local animal communities is strongly influenced by habitat-specific climatology and vegetation (Waser & Brown 1986; Slabbekoorn & Smith 2002a). Low-frequency noise is generated in almost all environments and will often be relatively loud because it transmits with little attenuation. Long-lasting echoes of wind-generated sound and noise from rocky streams determine the acoustic background that prevails at alpine meadows bounded by mountain peaks. Water flow can generate significant noise levels up to 4 kHz in riparian scrub, a favorite habitat of many birds (**Fig. 6.4**). Urban environments have many low-frequency sound sources in the form of cars, trucks, and all sorts of machinery in addition to wind (Slabbekoorn & Peet 2003). Reflections between buildings and concrete floors give rise to the continuous rumble of cities, a typical sound environment of many urban birds.

Even in more natural habitats birds may be

Figure 6.4 Different habitats have different sound transmission characteristics. **(A)** Leaflayers in dense forest; **(B)** bare tree trunks in aspen forest; **(C)** marshland; **(D)** urban habitat.

affected by the low-frequency noise of human activities, in particular from aircrafts (Brown 1990) and the ever-increasing presence of highways. In the United States, there are more than 6 million kilometers of public roads with over 200 million vehicles. Ten percent of these roads are within national forests, while road densities in European countries are even higher (Forman & Alexander 1998). In the Netherlands, bird-breeding densities are significantly lower close to highways and traffic noise is one potentially disturbing factor (Reijnen et al. 1995, 1997). A study on bird communities across different riparian habitats in the Rocky Mountains showed the impact of human-induced noise (Stone 2000). Increasing noise levels were correlated with lower species richness and lower overall bird density in two of the four habitat types investigated. In addition, extreme anthropogenic noise pollution may affect the song characteristics of the birds that remain; Il'ichev et al. (1995), for example, report on deformed song produced by chaffinches that live in a Russian forest near a very noisy road. In the Dutch city of Leiden, great tits in noisy territories sing with a higher minimal frequency than birds in relatively quiet territories (Slabbekoorn & Peet 2003). Presumably, great tits adjust their songs to local noise conditions through selective copying of song elements from neighbors when settling in a territory. Low-frequency notes may be less often copied with heavy perceptual interference from low-frequency noise. Flexibility of species in song development and frequency range requirements varies, and, consequently, species may differ in whether they can successfully breed close to anthropogenic noise sources.

Particularly in Mediterranean or tropical areas, forests are often home to a variety of insects that produce high-frequency sounds of considerable amplitude. Each forest type has its own insect community and therefore different noise characteristics (Brenowitz 1982a, b; Ryan & Brenowitz 1985; Waser & Brown 1986). Humid rainforest and dryer gallery forest in Cameroon show significant and consistent differences in their highly complex noise spectra (Slabbekoorn & Smith 2002a). Ambient noise in the rainforest is characterized by many varying but distinctive frequency bands, generally getting louder with higher frequencies. Gallery forest noise has fewer frequency bands and similar amplitude levels throughout the frequency range. Cicadas alter this pattern during some periods of the day by producing loud noise bands at frequencies above 6 kHz (**Fig. 6.5**). These spectral profiles vary at a site with the seasons, but the ambient noise in two rainforest sites studied in Cameroon nevertheless remained distinct across seasons (Slabbekoorn unpublished data). In addition to noise differences between major habitat types, even subtle changes in forest composition can result in consistent differences in ambient noise, as between littoral rainforest along the coast of Cameroon and similar forest more inland (**Fig. 6.5**). In general, the incredibly high and geographically variable biodiversity of the sound-producing community in these tropical forests leads to a spectacular range of tropical noise spectra, although they can be surprisingly stereotypic within a habitat type.

Acoustic Interference Among Birds

Songs of other birds singing nearby are potential sources of dramatic interference. The degree of masking will depend on the similarity of the competing sounds. Different species living in the same habitat often make use of overlapping frequency ranges. Their songs may have different species-specific features, but individual song components may sound very similar when picked up out of a cacophony of singing birds. These birds that sing together can be seen as competing for 'signal space' (Nelson & Marler 1990). A depiction of signal overlap among species in a multi-dimensional space illustrates how masking and discrimination problems depend on community composition and vary among species (**Box 23**, p. 194). In addition to acoustic similarity, other factors such as perch height and diurnal fluctuations in singing activity, affect the strength of competition for signal space. It might seem that 24 hours in a day would be enough to give

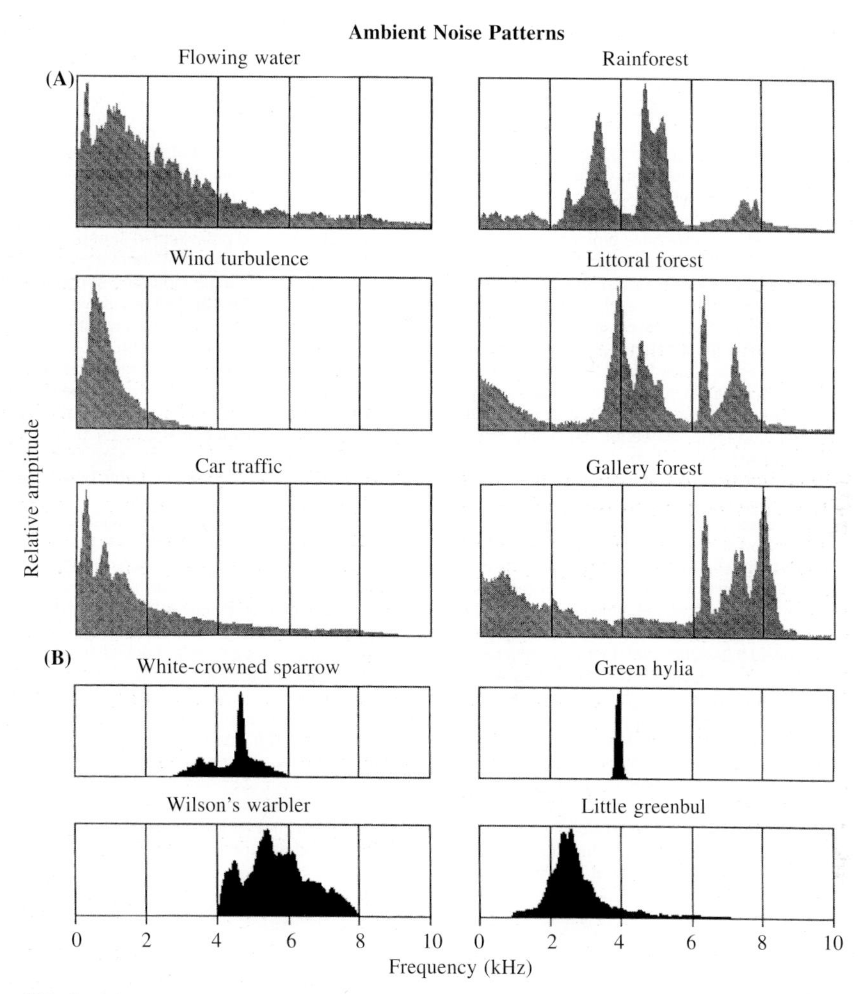

Figure 6.5 **(A)** Ambient noise spectra in six different habitats: riparian scrub in the Sierra Nevada, California with flowing water; a montane meadow in the Rocky Mountains, Colorado with wind turbulence; car traffic noise on 19th Avenue, San Francisco, California; tropical rainforest with several cicada and bird species in Cameroon; littoral forest with cicadas and birds, Cameroon; and gallery forest with loud cicadas, also from Cameroon (**CD1 #54**). **(B)** Power spectra of the songs of four bird species: white-crowned sparrow, as in **Fig. 6.2**; Wilson's warbler, audible on the recording from the riparian habitat; green hylia, as in **Fig. 6.3;** and little greenbul as in **Fig. 6.10**.

broadcasting time to all bird species. But, although peak song activities of different species are often somewhat separated, many species prefer a similar time window at dawn, leading to high chances of overlap (Staicer et al. 1996; Mace 1987). In such cases, birds singing at the same time of day may still be able to reduce acoustic interference by active overlap avoidance.

Singers within audible distance of each other and therefore with a potentially overlapping audience, may adjust their song timing to minimize interference. For example, in California chaparral, wrentits seem to avoid simultaneous singing with loud neighboring Bewick's wrens. Males of the two species sing asynchronously, with wrentits singing predominantly after

BOX 23

A PLACE FOR BIRDSONG IN SIGNAL SPACE: THE JUST MEANINGFUL DIFFERENCE

To communicate efficiently with their songs, birds must be both audible and recognizable – not always easy when a dozen or more species are singing at the same time. We can think of birds living together as competing with each other for vacant places in the 'signal space' that characterizes a given location (Nelson & Marler 1990). It is a practical undertaking to record samples of all bird species that sing in a particular ecological community and to represent their physical characteristics in multidimensional space. Nelson (1988, 1989b) tackled this problem in an old-field community of scrub and grassland in upper New York State. He recorded song samples from thirteen species, took twelve different measurements, and assembled them all in a giant matrix. His first step was a statistical analysis to determine which of the twelve features did the best job in sorting songs into species. He focused especially on two common birds, chipping sparrows and field sparrows (**CD2 #39–40**). The results were surprising. Only three measurements were needed to classify almost all of the chipping sparrow songs correctly. For field sparrows, seven features were needed. Why the difference? A 'signal space' approach provides a possible explanation.

To characterize the signal space for birdsong in this habitat, Nelson plotted the first two discriminant functions derived from seven acoustic features, for all thirteen of the species analyzed. Most species occupied a distinct place in the signal space (**A**). The location of chipping sparrow and field sparrow song is especially interesting. The former is on the edge of the signal space, hence the small number of features needed to classify its songs. In this community, chipping sparrows have few close 'acoustic neighbors' in the signal space that might create recognition problems. Field sparrow song is located more centrally, resembling several other species and thus requiring more information for accurate classification. Of all measurements taken, maximum song frequency was the single most accurate feature in a statistical classification of the songs of all species.

What relationship does this have to song recognition by the birds themselves? Do they care more about song frequency than other characteristics? How do they rate other high-ranking classificatory features such as number of phrases, note duration, and inter-note interval? To address these questions, Nelson went back into the field armed with recordings of normal songs and a large set of synthetic field sparrow songs, in which the four best discriminators were systematically varied by small increments (**B–F; CD2 #41**). Playbacks to territorial males showed that, although sensitive to change in several features, field sparrows were more intolerant of change in song frequency than any other feature, as predicted. By titrating male responsiveness to small changes, Nelson showed that within each feature males tolerated a change of two or three standard deviations around the normal mean value, before a sudden decline. This constitutes what we have called a 'just meaningful difference' or jmd, comparable to the 'just-noticeable difference' or jnd of the psychophysicist, which defines a minimum for discrimination (Nelson & Marler 1990). A jnd applies in all contexts, whereas a jmd is expected to vary with the context. The same signal can have different meanings with different jmds, depending on the functional situation and receiver identity. The appropriate signal space will then take different forms, and jmd values will therefore change. Birdsongs are vehicles both for species recognition, as discussed here, and for individual recognition. The frequency jmd for individual recognition in male field sparrows is estimated at around 400 Hz, compared with 900 Hz for species recognition. Jmds for dialect recognition might fall somewhere in between. Discrimination of song renditions that relate to motivational state may have even smaller jmds. At each level the nature of the relevant signal space will depend on the function being served. When the primary concern is quick and accurate species recognition, birds can afford to overlook subtleties that become crucial at another level.

Peter Marler & Douglas A. Nelson

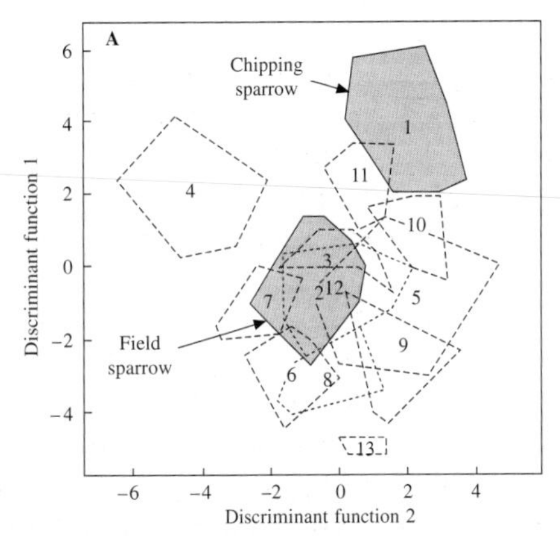

A plot of song measurements in a space defined by the first two discriminant functions, based on seven acoustic song features. Species in addition to chipping and field sparrows are: 3-prairie warbler; 4-northern cardinal; 5-rufous-sided towhee; 6-tufted titmouse; 7-white-throated sparrow; 8-indigo bunting; 9-yellow warbler; 10-dark-eyed junco; 11-swamp sparrow; 12-song sparrow; 13-black-capped chickadee

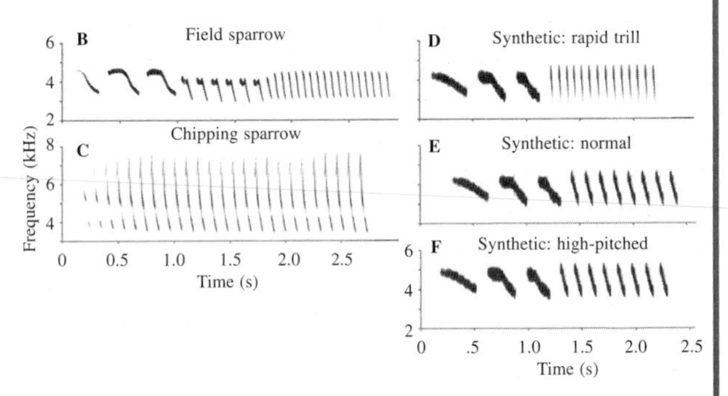

Two normal field and chipping sparrow songs (**B** & **C**) and three synthetic field sparrow songs, one replicating normal song (**E**), the other two shifted from the norm by one jmd, one with a more rapid trill than usual (**D**), the other higher pitched than usual (**F**). Both are weaker playback stimuli than normal to territorial male field sparrows. From Nelson 1988, 1989b.

Bewick's wrens slow down, keeping quiet with outbursts of Bewick's wren song (Cody & Brown 1969). Similarly, the temporal overlap in the songs of a set of five male pairs of the red-eyed vireo and the least flycatcher was lower than expected by chance, the flycatcher most actively avoiding interference with the longer song of the vireo (Ficken et al. 1974). This phenomenon of temporal avoidance also occurs among males of the same species. Male tufted titmice that counter-sing with competing males, or with playbacks of conspecific song, adjust their timing to avoid song overlap (Schroeder & Wiley 1983). In another study, a set of five ovenbirds also avoided songs played back to them by singing their own songs preferentially in the silent periods shortly after each song they heard (Ficken et al. 1985). Song overlapping among males of the same species sometimes plays a more complicated role, not just in hindering detection and recognition through masking, but as an aggressive signal between rivals, for example (Todt & Naguib 2000; see Chapters 2 & 3).

Despite the active avoidance of temporal overlap, jamming still occurs, by song or other sounds. Not all species adjust actively to environmental noise. The winter wren, with its wide-frequency song, has many different species competing for the same signal space, but does not seem to adjust the timing of singing to that of others (Brémond 1978). Green hylia in coastal Cameroon, presumably adapted to life and communication in the noise of the tropical rainforest, seem to utter their songs without any reference to the loud and highly predictable sound bursts of pulsating cicadas. By looking at their tails moving up and down with the coo rhythm, collared doves can be seen vocalizing above busy streets where traffic noise completely masks their song to our ears. Overlapping sounds must reduce the audibility of a song, resulting in lower rates of detection or recognition or both. This will result in lower signal efficiency, as is found in winter wrens that respond less or later to playback of conspecific song when mixed with a set of other songs that overlap in frequency and time (Brémond 1978). However, operant conditioning experiments in the laboratory show that birds are also very clever in singling out a target song among a mixture of songs (Hulse 2002; see Chapter 7). European starlings are still capable of detecting not only their own songs in a medley of other species (Hulse et al. 1997), but can also focus on one individual in a concert of starling songs (Wisniewski & Hulse 1997). A study on zebra finches indicated that such tasks of detection under noisy conditions may be easier if the target is a conspecific song instead of an alien one (Benney & Braaten 2000).

Calling in a Crowd

Some of the most challenging conditions for communication by sound are found in bird colonies. Many birds breed with nests side by side in neighboring trees, reedbeds, or burrows. Rooks, common grackles, European bee-eaters, and bank swallows are obvious examples. The most famous colony breeders are found on the coast: many seabirds, like gulls, terns, murres and gannets have large colonies with hundreds or thousands of breeding pairs. The sounds of courtship calls, territorial fights over nest sites, and alarm calls, can all add up to an almost continuous noise, sometimes bursting out with extreme amplitude levels exceeding 100 dB (Blokpoel & Neuman 1997). Nevertheless, many aspects of social life in a colony depend on acoustic communication, especially the ability to recognize individuals acoustically, something that may be crucial for relocating mates or relatives among many similar-looking individuals (Mathevon 1997; Charrier et al. 2001).

Locating someone in a crowd is easier with some knowledge of their approximate location. For example, a partner is usually expected to be on or near the nest, which can be located with the help of general landscape cues. However, high nest densities may still lead to scores of birds that look alike, and how can a bird be sure that a stranger has not taken over the nest from its partner? In all the colonial seabirds studied, it turns out that individuals have clearly distinct calls and that this vocal fingerprint is used for identification. When mates return from a foraging flight at sea there is usually an interactive vocal display. An impressive series of studies revealed that penguins in their arctic colonies recognize individuals by voice despite very noisy conditions (Aubin & Jouventin 2002; **CD1 #55**).

The king penguin breeds in colonies of up to 40,000 pairs. Standing on wet beaches or snowy surfaces they brood a single egg on their feet, while moving around a bit waiting for the other parent to return from sea. Smell or visual cues are not sufficient to recognize and find each other. Some experiments showed that pairs had no idea where to go when their hearing ability was temporarily sabotaged or when their beaks were temporarily closed with adhesive tape (Jouventin 1982; Aubin & Jouventin 2002). Similarly, after the chick hatches from the egg, it will wait together with all other chicks for the parent to return with fish. A chick can expect an aggressive response when approaching the wrong adult: only its own parents will provide food. Accurate recognition of the voice of their parents guide chicks through a safe and well-fed childhood. When a bird returns from the sea and makes its way into the colony, reciprocal calling and recognition serve both to bring mates together and to reunite parent and offspring.

However, two factors make the colony less than ideal as a place to communicate detailed information acoustically. First, penguins standing in a crowd form a barrier of penguin flesh and feathers that has dramatic impact on sound transmission. The penguins partly avoid this problem by raising their beaks to a vertical position and calling with their necks stretched to the fullest extent (Jouventin 1982). The second obstacle is the high amplitude level of noise from the colony itself, matching each bird's calls in spectral and temporal characteristics. Average sound amplitude levels in a colony can be 74 dB, while calling individuals may exceed 95 dB. Playback experiments in the colony revealed that king penguins are still able to detect, recognize, and localize the natural display call at more than 10 meters (Aubin & Jouventin 1998).

The individually distinct frequency modulation pattern in the king penguin call plays a crucial role in vocal identification under these noisy conditions. The basic structure of syllables is always the same with many harmonics that go first up then down in frequency with the inflection point always in the first half of the syllable (**Fig. 6.6**). Within these species limits, each individual penguin has its own typical variant of the frequency modulation shape. Individual variation in a sound fragment as short as 200 ms is enough to allow recognition (Jouventin et al 1999). In addition, the calls are two-voiced, consisting of two simultaneous series of harmonics with slightly different frequencies (see Chapter 9). This results in a typical pattern of amplitude modulations or 'beats' (Aubin et al. 2000), which not only improve sound propagation of the calls in the colony environment, but also contribute to the potential for individual recognition. One-voice calls, artificially produced from king penguin recordings, were not as efficient as the natural two-voice calls in identification tests (Lengagne et al. 2001).

Coping Strategies for High Noise Levels

Several behavioral strategies and physiological adaptations can help birds to cope with high noise levels. In general, successful communication under noisy conditions depends on the sender or the receiver improving the signal-to-noise ratio in one way or the other. First of all, songs are generally relatively complex with multiple features that contribute to the response-eliciting capacity of a signal, with some features being more important than others (Gaioni & Evans 1986; Slabbekoorn & ten Cate 1998a). With multiple cues potentially conveying a message, there is more chance that the signal-to-noise ratio for one of them is sufficient to reach a receiver. Similarly, repeating the same signal over and over again raises the chance that one repetition coincides with a quiet gap in the noise and consequently the probability of getting the

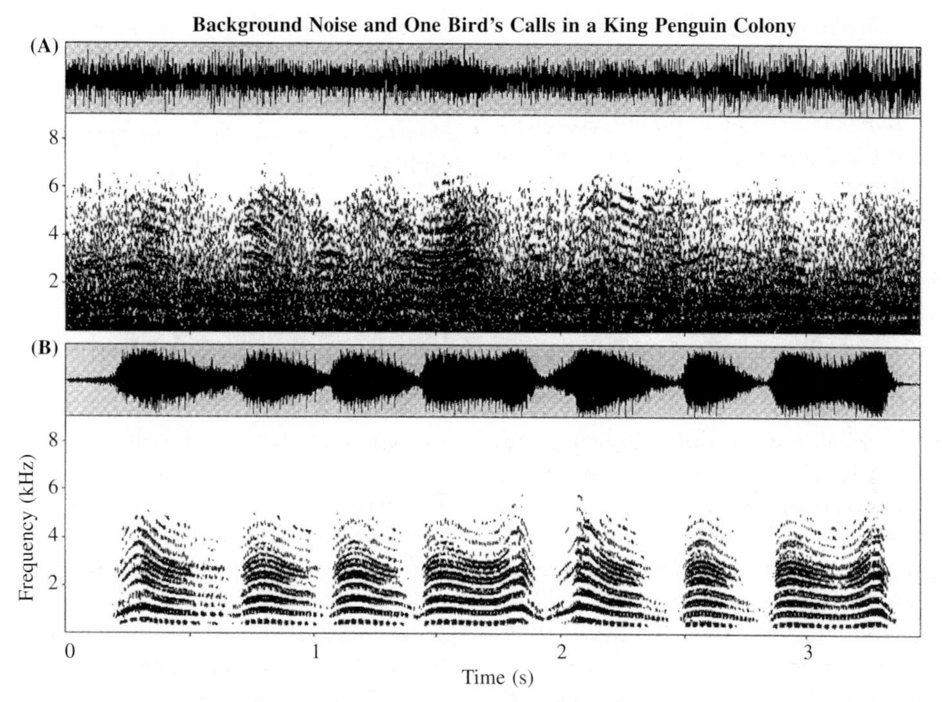

Figure 6.6 **(A)** The amplitude envelope and a sonogram of background noise in a king penguin colony. **(B)** Calls of an individual king penguin. The species-typical frequency modulation patterns provide multiple cues for individual recognition under noisy conditions (**CD1 #55**).

message across. In line with this hypothesis, king penguins increase redundancy in response to high noise levels, producing calls of longer duration with more syllables under windy and noisy conditions (Lengagne et al. 1999).

Birds may also adjust their song amplitude to the ambient noise, like people speaking louder to overcome competing sound at a pop concert. The more the noise the louder a bird will sing. Context-dependent adjustment of song amplitude, called the Lombard effect, appears to be widespread among birds. The Lombard effect has been demonstrated both in birds that learn their vocalizations, such as budgerigars (Manabe et al. 1998) and in those that do not, such as Japanese quail (Potash 1972). Furthermore, zebra finches can adjust the amplitude of both learned and unlearned parts of their vocal repertoire in response to varying noise levels (Cynx et al. 1998). The effect is also found in territorial species such as the nightingale (Brumm & Todt 2002). Interestingly, nightingales that were singing in an aviary increased the amplitude of the low-amplitude notes of their songs more than the high-amplitude notes in response to noisy conditions. This altered the amplitude modulation pattern of their songs and may have reduced the potential of encoding information. A study on blue-throated hummingbirds revealed the Lombard effect for the first time in the field (Pytte et al. 2003). Male hummingbirds adjusted the amplitude level of their 'chip' vocalizations in response to both natural and experimental noise changes. They produced louder song when noise levels were raised by playback of noise from a stream.

On the receiver side, there are a number of physiological mechanisms that mitigate the impact of noise (Klump 1996; see Chapter 7). The auditory system can be viewed as a bandpass filter that allows birds to tune into the frequency range used by their species. Noise beyond this range can be filtered out and will have less impact on signal detection. In addition, birds are often surprisingly good in extracting a signal of interest even when frequencies of signal and noise overlap, as shown for the penguins in the field. There is an analogy with the ability of humans to concentrate on one particular speaker in a noisy room: the 'cocktail party effect.' We can switch from listening to one speaker to eavesdropping on another, often with rather low signal-to-noise ratios, when several conversations are going on at the same time. When we switch focus to a neighboring speaker we are less able to comprehend what our previous conversation partner is saying, although the signal-to-noise ratio remains unchanged.

Several perceptual phenomena contribute to the cocktail party effect. Birds benefit from 'spatial release' from masking noise (Larsen et al. 1994); this refers to the improved detectability of song when the singer is separately located from the noise source. This ability depends on directional hearing, which, in birds that generally lack true pinnae, is largely based on amplitude differences in the sound heard by both ears in timing and phase, and especially the sound shadow of the head (Lewald 1990; Klump & Larsen 1992). The location of the noise source with respect to sender and receiver is crucial to its masking impact. For example, two counter-singing birds that sing at equal amplitudes and have the same auditory sensitivity may experience dramatically different noise conditions when one is perched closer to the noise source than the other. Switching to a different listening perch (Mathevon & Aubin 1996; Holland et al. 1998) can improve signal-to-noise ratios, in the same way that a song becomes louder relative to the ambient noise when approaching a singer. The signal-to-noise ratio improves both when the song gets louder, and also when the noise is attenuated by some obstacle that causes a 'sound shadow' such as a dense shrub layer, a hill, or walls and buildings in urban areas.

ENVIRONMENTAL SELECTION AND SIGNAL EVOLUTION

Habitat-dependent Songs in Species Comparisons

Birds communicate vocally in almost every habitat type; tropical kingfishers send their mellow song notes through humid African rainforests, the greater roadrunner produces dove-like coos in desolate places like the desert of Death Valley, while rock ptarmigans growl and croak over arctic snow and ice. Signaling conditions vary with the type and density of vegetation, wind and temperature gradients, and the spectral characteristics of competing noise. In bird species that have been associated over evolutionary time with one particular habitat, song characteristics may have been shaped by selection related to the local signaling conditions. For this reason, the acoustic design of birdsong across taxa may show habitat-dependent similarities, even though the pattern will be blurred by other factors that have also influenced signal evolution, such as body size and morphology, ancestral song characteristics, and female song preferences (see Chapters 2, 10, & 11).

Comparisons among tropical bird species reveal that 'closed' forest songs typically consist of relatively low-pitched and often long-drawn tonal notes (**Fig. 6.7; CD1 #56**), while the birds of more open country generally produce higher-pitched songs with repetitive trills of short and stereotypic notes (Chappuis 1971; Morton 1975; **Fig. 6.8; CD1 #57**). This pattern can be discerned even if you take body size and phylogenetic relationships into account (Ryan & Brenowitz 1985). Similar differences in acoustic design are also found in temperate zones between songs of birds from forest and 'open' habitats (Sorjonen 1986; Wiley 1991; Badyaev & Leaf 1997). The explanation for these consistent song differences lies in contrasts in habitat-dependent sound transmission characteristics due to vegetation

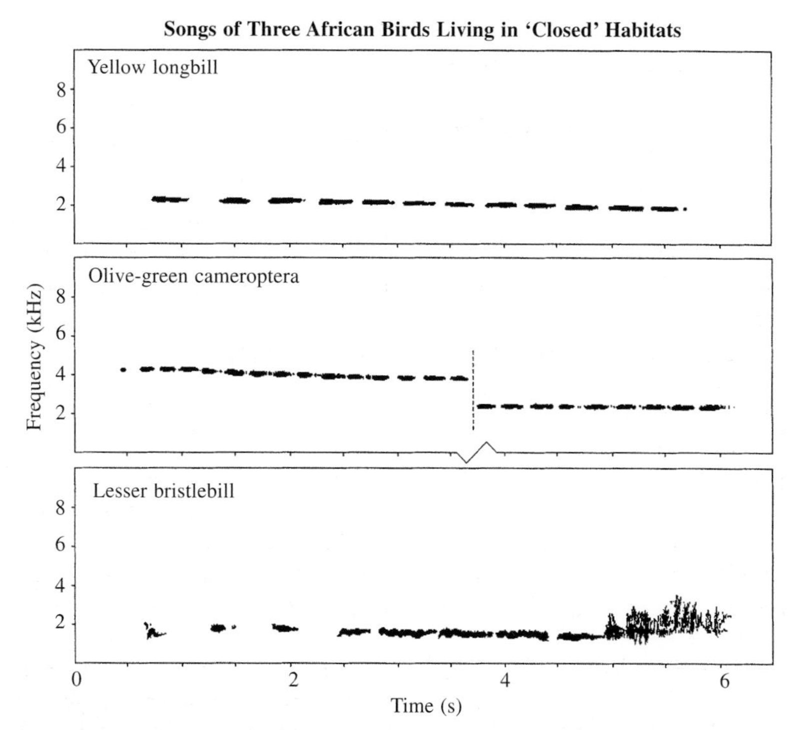

Figure 6.7 Three typical songs from small passerines living in 'closed' habitats in dense rainforest in Cameroon: a yellow longbill, an olive-green cameroptera, and a lesser bristlebill. All three have a relatively low frequency and long notes with only minor frequency modulations. The middle part of the cameroptera song is not depicted (**CD1 #56**).

density and wind exposure. Dense vegetation leads to more extreme frequency-dependent attenuation favoring the use of low-frequency sounds for communication. In addition, the use of long-drawn tonal notes at one frequency leads to additional benefits in closed forests due to the accumulation of reverberations (Slabbekoorn et al. 2002). In open habitats, the lack of shelter from wind hinders communication with long-drawn notes, while stereotypic songs with a strong repetitive temporal pattern still succeed in conveying their message (Brown & Handford 1996, 2000; **Box 24**, p. 201).

The patterns of habitat-dependent song characteristics revealed by broad-scale species comparisons are also found in smaller groups of close relatives. Species that we know to have split up relatively recently, based on phylogenetic reconstruction using molecular data (see Chapter 10), sometimes live in acoustically distinct habitats. There are several cases of species that have diverged acoustically in a way that would be expected, given the transmission properties of their respective habitats (*Acrocephalus* warblers – Jilka & Leisler 1974; Heuwinkel 1982; *Sylvia* warblers – Bergmann 1978; Darwin's finches – Bowman 1979, 1983). Thus, habitat-dependent selection pressures may lead both to convergence of song characteristics in species living in the same habitat, and to divergence of songs of closely related birds living in different habitats (**Fig. 6.9; CD1 #58**).

Song divergence driven by differences in environmental selection pressures may even occur within the same habitat when species occupy different forest layers. In American parulid warblers, song frequency is positively correlated with singing height: high-pitched songs are sung at high song posts where interference by vegetation is expected to be relatively low (Ficken

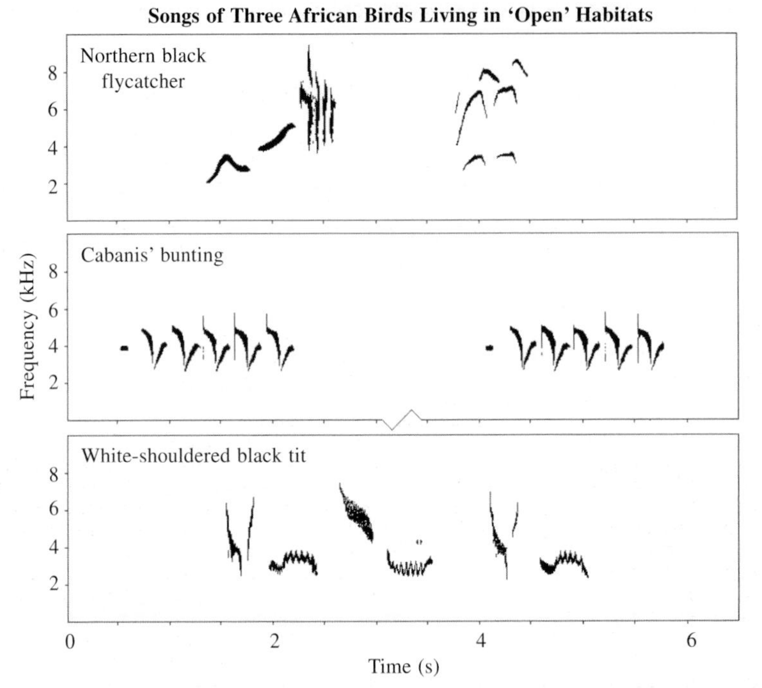

Figure 6.8 Three typical songs from small passerines living in 'open' habitats in savanna woodland in Cameroon: a northern black flycatcher, a Cabanis' bunting, and a white-shouldered black tit. Note the wide frequency range and complex temporal structure. The interval between songs has been shortened for the Cabanis' bunting (**CD1 #57**).

& Ficken 1962; Lemon et al. 1981). In another example, five closely related species of antbird tend to use song perches at different heights in South American tropical rainforests, each with different layer-specific transmission and noise characteristics (Nemeth & Winkler 2001). Their songs are relatively simple and show a negative correlation between perch height and song element length. Relatively long notes with a low- and narrow-frequency range are used near the forest floor (as in the examples from Africa in **Figs. 6.7 & 6.9**), while very short elements produced in a fast trill are typical of the canopy-living species. The songs of three of the five species degraded less during transmission experiments at the species-specific song perch height than at the perch height of the other species. Again, the acoustic environment appears to have influenced the direction of evolutionary signal changes, although other benefits or constraints may have prevented a perfect match between the signal and environmental selection pressures in all five species.

Song Divergence Within a Species

Song differences between closely related species may have emerged before or after populations of the ancestral species became reproductively isolated and thus separate species. Song may play a role in the process of speciation when acoustic differences among populations affect dispersal and interbreeding. Subgroups of one species that live in different habitats may diverge in morphology or reproductive behavior, and interbreeding between them may lead to a reduction in reproductive output or survival of offspring. If song characteristics diverge in parallel with traits that affect fitness, because both are associated with being in a distinctive habitat, songs may be used to identify individuals that are well adapted to the local environment

BOX 24

SONG DIALECTS, HABITATS, AND GENETICS: THE RUFOUS-COLLARED SPARROW

The chingolo of South America, properly called the rufous-collared sparrow, provides us with one of the clearest examples of a widely distributed, habitat-related, song dialect system **(CD2 #42–47)**. Over 30 years ago, Nottebohm (1969) first described the extensive geographic variation in the song of this species in Argentina. He suggested that the dialects perhaps serve to enhance the genetic integrity of local populations, but direct investigation of this possibility (Handford & Nottebohm 1976), provided no support for what came to be called the Genetic Adaptation Hypothesis (GAH; Rothstein & Fleischer 1987). However, the distribution of song dialects (defined by the interval between the repeated syllables of the terminal trill) proved to be closely associated with the many distinct habitat types occupied by this ecologically catholic songbird. The trill interval tends to vary from short (~50 ms or less; rapid trills) in open grasslands to long (100–200 ms or more; slow whistles) in woodlands and forests (see fig.), a contrast much like that first described by Morton (1975) when comparing songs of other birds occupying different habitats.

The relationships between song dialects and habitat were explored further across the diverse habitat arrays of north-western Argentina (Handford 1981, 1988; Handford & Lougheed 1991). My students and I surveyed the ecological ordering of dialect variation over a huge geographical area (1,200 × 350 km), across a dramatic sweep of structurally – distinct habitats, including puna scrub, grassland, desert scrub, thorn scrub, woodlands and forests. Throughout, rapid trills were associated with open grasslands and scrub, and slower trills with closed woodland and forest habitats (see below). The spatial patterns also showed remarkable temporal stability, over at least 30 years; stability on the order of centuries is implied by the persistence of certain habitat dialects long after local agriculture has eliminated the native vegetation (Handford 1988). This massive demonstration of habitat-based song variation strongly supports what became known as the Acoustic Adaptation Hypothesis (Rothstein & Fleischer 1987) with song features apparently adapted to the sound transmission characteristics of the environment. The work also provided a firm base for a final evaluation of the GAH on a similar geographical scale (Lougheed & Handford 1992). The substantial genetic variation shown by the chingolo proved to be organized largely by distance, with song dialects imposing no further structure: it seems that for this species the GAH has no explanatory value.

Recent work on this bird confirms that the clear ecological segregation of vocal dialects extends from 22 °S at the Bolivian border to 42 °S in northern Patagonia. Across this vast space, the greatest song diversity is concentrated in the ecologically diverse north-west; in the ecologically more uniform central and southern regions, great song uniformity is encountered; habitat islands, such as montane grasslands, are represented by song dialect islands. On the other hand, a tropical population in Ecuador shows no such pattern: instead, individuals have song repertoires (from 1 to 7 trill types; mean = ~4) and local populations can show nearly as much trill variation as is known from all Argentina. The explanation for this contrast is still being sought.

Paul Handford

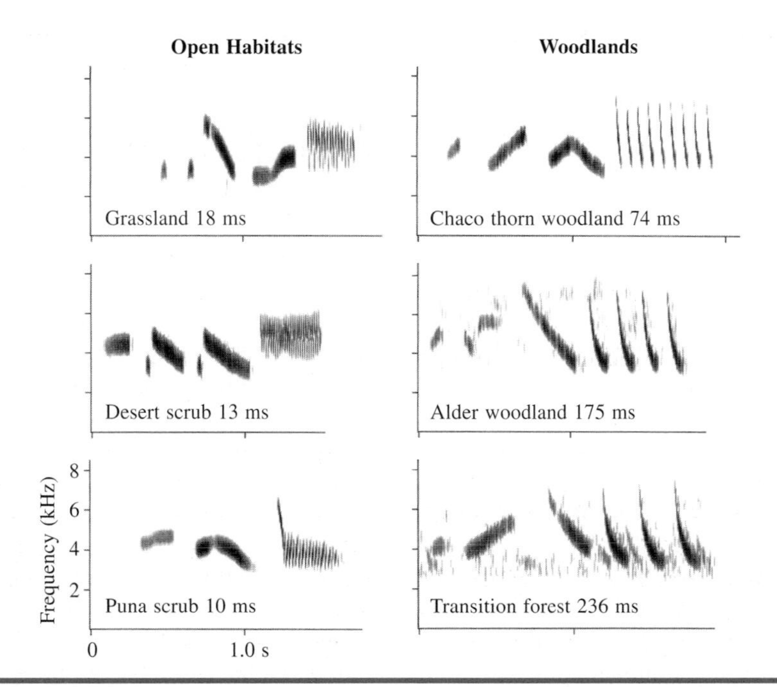

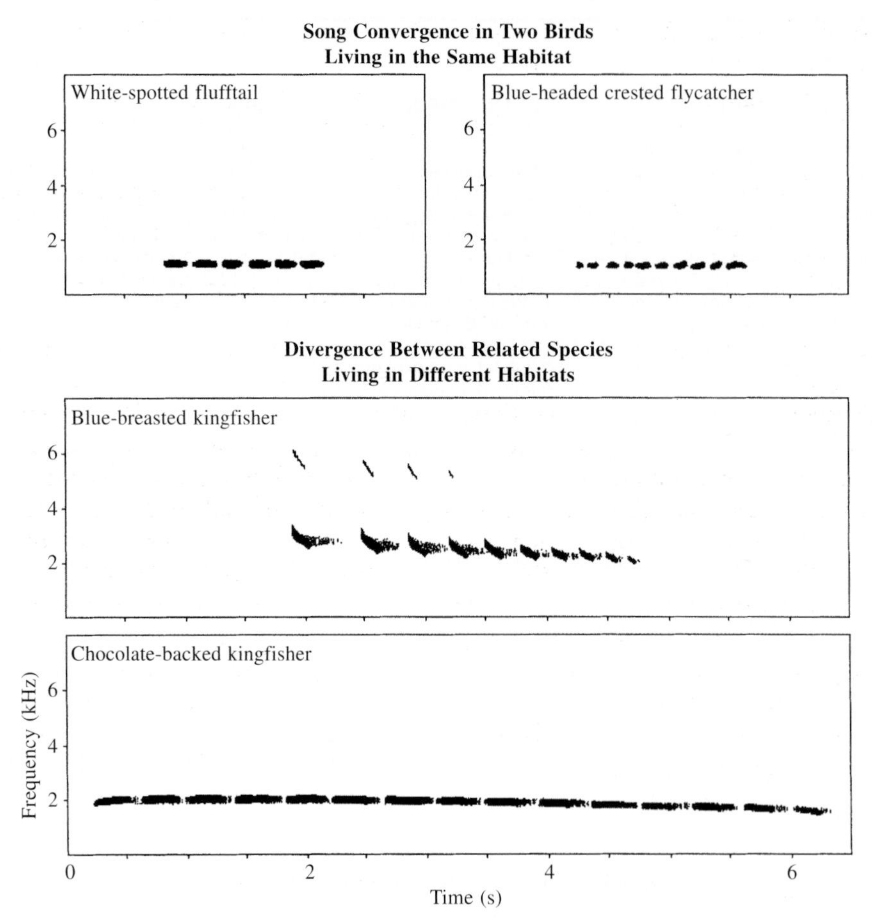

Figure 6.9 An example of convergence in song characteristics of two unrelated species living in the same habitat: a rail, the white-spotted flufftail and the blue-headed crested flycatcher. Both are small birds and live near the forest floor in the dense rainforest in Cameroon. At the bottom is an example of divergence between two closely related species: the blue-breasted kingfisher and the chocolate-backed kingfisher. Both occur in the rainforest, but the first often sings in flight above the canopy, while the second is typically perched within the forest (**CD1 #58**).

(Slabbekoorn & Smith 2002a, b). As a consequence, song divergence within a species may favor within-group matings between birds from the same habitat, leading to assortative mating. In this way, song divergence can promote reproductive isolation between subgroups, in particular when it is associated with different habitats.

The little greenbul is a common African Passerine that lives in both dense tropical rainforest and in gallery forest that forms ribbon-type patterns through a savanna landscape. Birds living in gallery forest are larger, have longer wings, and have a different bill shape compared to those that live in contiguous rainforest (Smith et al. 1997). Despite a tendency for their breeding seasons to be non-overlapping, genetic data revealed current gene flow across habitats, indicating interbreeding between individuals from the rainforest and gallery forest. Song also differs between habitats: several features such as maximum and minimum frequency and song note rate show habitat-dependent variation, in populations that are geographically close

(Slabbekoorn & Smith 2002a). Another song characteristic that varies among populations is the sequence in which distinctive song types are sung (**Fig. 6.10; CD1 #59**). Song type sequencing is independent of the spectral and temporal characteristics of the songs and can be regarded as environmentally neutral. Thus, whether a bird sings the same four song types in the order I-II-III-IV or IV-III-II-I has no impact on song transmission through the environment, and will be affected similarly by interfering ambient noise. As would be expected, song sequencing shows no habitat-dependence; it is similar in populations that are in different habitats but geographically close (Slabbekoorn & Smith 2002a).

Various studies have revealed intraspecific song divergence correlated with acoustic properties of the habitat (Slabbekoorn & Smith 2002b). In cases where habitat-dependent selection seems to drive divergence in both song and morphology, such as in the little greenbul, speciation may be imminent. However, the process of speciation takes place over an evolutionary time scale, and we cannot predict whether cases of intraspecific song divergence will eventually lead to the formation of new species or not. A special opportunity occurs when divergences in song and behavioral response are not hidden in evolutionary time, but are spread out in geographic space, as in 'ring species.' In the greenish warbler, two populations that overlap in distribution but do not interbreed are connected by a ring-shaped chain of interbreeding populations (Irwin 2000; Irwin et al. 2001a; **Box 25**, p. 204).

There may be many cases in which song differences prevent hybridization in the wild among birds that are capable of producing viable and fertile offspring (Grant & Grant 1996a, 1997a). However, song learning, and the possibility of culturally transmitted song characteristics, adds another layer of complexity to these issues (Ellers & Slabbekoorn 2003). Song copying may affect patterns of song divergence at the population level, with the potential to accelerate habitat-dependent song divergence. Birds may preferentially learn what is heard best under local acoustic conditions (Morton et al. 1986), thereby shaping habitat-dependent songs (Hansen 1979). However, the

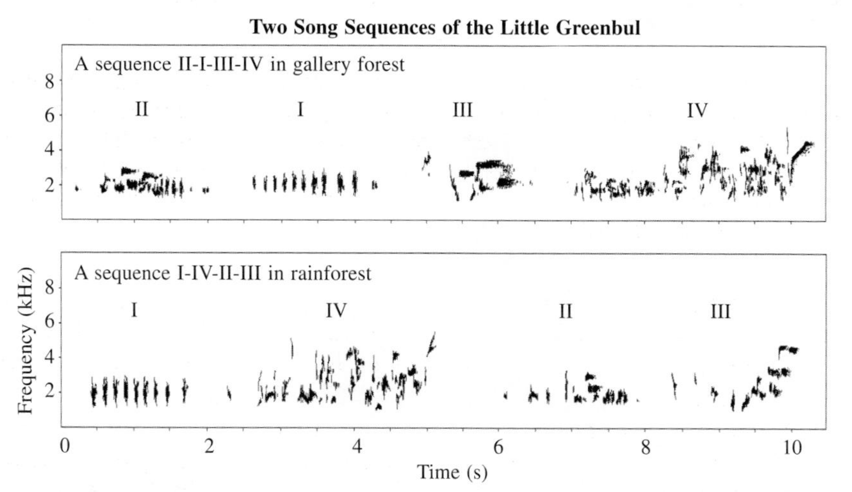

Figure 6.10 Two singing sequences of the little greenbul, typical for certain populations and reflecting what can be called 'sequence dialects' (**CD1 #59**). Little greenbuls sing continuously alternating songs of four types. Illustrated are two strings of four songs each. Above is a sequence of song types II-I-III-IV, recorded in a gallery forest, also typical for several rainforest populations, and below is a song type sequence I-IV-II-III typical for two other rainforest sites. Songs have a significantly lower minimum frequency in rainforest populations (as measured for song types III and IV; Slabbekoorn & Smith 2002a).

BOX 25

SPECIATION IN A RING: THE ROLE OF SONG

The evolutionary divergence of male birdsong could cause two groups to diverge and become reproductively isolated and hence separate species. However, because speciation in birds can take a great deal of time, there is little direct evidence for a central role of song divergence in speciation. An unusual opportunity to use spatial variation to infer how speciation occurred in time is presented by a ring species, in which two reproductively isolated forms are connected by a chain of intermediate populations (Irwin et al. 2001b). The greenish warbler species complex stretches across most of Eurasia. Two reproductively isolated forms coexist in central Siberia, but they are connected by a chain of intergrading populations to the south, encircling a large region of deserts, including the Tibetan Plateau. Ticehurst (1938) studied morphological variation in the complex and concluded that an ancestral species in the south had expanded northward, diverging along two pathways on either side of the Tibetan Plateau, eventually giving rise to the reproductively isolated forms in Siberia. Molecular variation in the complex supports this hypothesis. The two Siberian forms are ecologically similar but sing distinctly different songs, and playback experiments show that males generally respond aggressively only to songs of their own form (Irwin et al. 2001a). Presumably females would only be attracted to songs of their own form. Songs change gradually through the chain of populations (**CD2 #48–55**). As they diverged in structure during the two northern expansions, they apparently became longer and more complex. All songs are constructed out of a limited set of distinct 'song units' (Irwin 2000). The apparently ancestral populations in the south sing simple songs consisting of one song unit repeated four to six times. In west Siberia, songs are longer and more complex, made up of long song units with a wide frequency range. In east Siberia, songs are also longer and more complex, but they consist of many short song units with a narrower frequency range. The parallel northward increases in song complexity are probably the result of ecological changes experienced by the expanding populations. More dense vegetation and lower population density in the north likely increase the importance of song in mate attraction, and greater food abundance and longer day lengths in the north may have decreased the cost of singing. Thus, parallel ecological changes appear to have caused parallel selection for greater song complexity. This still resulted in song divergence, because there is more than one way to evolve greater complexity.

Darren E. Irwin & Jessica H. Irwin

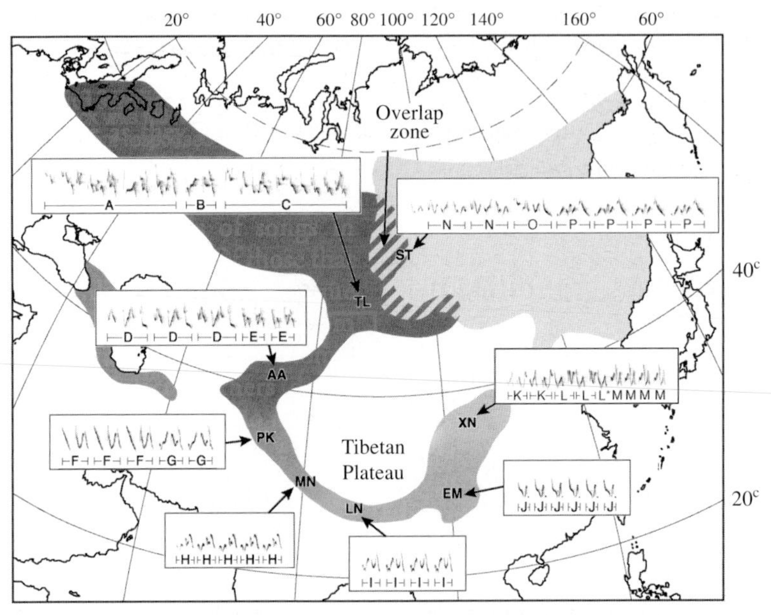

The breeding distribution of greenish warblers with eight representative song spectrograms. Shades of gray correspond to the amount of differentiation between different populations. Letters indicate distinct song unit types. Songs in the south are simple. Moving north toward west Siberia, the length and frequency range of songs and song units increases. Moving northward toward east Siberia, the number of units and unit types increases, but unit length and frequency range does not. Gradual evolution of song during two northward expansions has resulted in distinct differences in the overlap zone in central Siberia.

flexibility associated with learning could at the same time counteract assortative mating by enabling birds that learn the local song after dispersal to move across habitats and avoid being recognized as non-adapted immigrants (Slabbekoorn & Smith 2002b). There is obviously still a lot to explore here before we have a complete understanding of the impact of singing behavior on population dynamics and speciation.

CONCLUSIONS

It is surely clear from this review of the ecology of birdsong that singing in the wild is not a simple process. We need more, detailed knowledge about song production, sound transmission, noise interference, and signal perception for a complete understanding of the current potential for communication in a given environment. The relationship between signal design and environmental selection pressures is crucial for gaining insights into the evolutionary origin and current functioning of acoustic signals. It is possible to acquire such knowledge only through a combination of field and laboratory studies, striving to integrate data combining behavior, ecology, and environmental physics, within an evolutionary framework. The ways in which song varies between and within species may also tell us something about the origin of species and help to explain current biodiversity across habitats and its geographical distribution. Human influences are changing the world, removing habitats and creating new ones. The enormous variety of natural noise spectra in the tropical rainforest is disappearing at an alarming rate, as the urban rumble of human settlements spreads throughout the world. The loss of natural habitats may leave us forever with unanswered questions about how signals have evolved in association with these environments. As we study the ecology of birdsong in the tropical forests that remain today, we become painfully aware of the beauty and complexity of ecosystems that are being destroyed. On the other hand, new environments lead to new research opportunities. Urban areas are dramatically different from any other natural habitat and are likely to drive acoustic signals into new directions. Therefore, new studies may yield insights into how ancient natural environments have led to the current birdsong diversity and how we are changing evolutionary pathways. The anthropogenic impact on the acoustic environment of today will surely leave its imprint on birdsongs of the future.

Chapter 7

Audition: can birds hear everything they sing?

ROBERT DOOLING

INTRODUCTION

The complexity of birdsong has fascinated us throughout the ages. Scientists have long suspected that birds are capable of producing, perceiving, and learning very complex features of their songs that are beyond the capabilities of human hearing. Students of birdsong have pointed out the parallels between birdsong learning and another complex acoustic communication system that develops through learning, our own speech (Marler 1970b; Doupe & Kuhl 1999). We know in the case of speech that the complexity in production is sometimes accompanied by unusual capabilities in perception, attention, memory, and learning. Laboratory studies of human speech perception have been crucial in identifying some of these special processes, and they set the stage for much of the work on birds.

It is still the case that much more is known, from both field and laboratory studies, about the production of songs and calls of birds than is known about their hearing and auditory perception, and what little we know comes from different points of view that are not always easy to integrate. One approach tries to define the limits of sensitivity for detecting and discriminating sounds. Simple sounds are used because they are easy to describe and control, and researchers often seek to understand the mechanisms of hearing and the proximate mechanisms of detection and discrimination by comparing birds with, say, mammals. The relevance of this approach to understanding a bird's behavior in its natural habitat is not always apparent.

Another perspective focuses on the role of hearing and auditory perception in vocal learning and acoustic communication in birds. Here, complex sounds such as natural vocalizations, or synthetic models of them, are used as test stimuli in an effort to better understand the relationship between hearing and vocal production especially in acoustic communication under natural conditions. Laboratory behavioral studies of hearing and auditory perception in birds are beginning to identify some rather special capabilities manifest in the perception of song by birds in their natural world. These studies tend to focus on a few species that are especially tractable laboratory subjects such as budgerigars, canaries, starlings, and zebra finches but the case is strong that the principles uncovered are quite general.

How do we Know What Birds can Hear?

It is a relatively easy matter to ask humans if they hear a sound, whether they can discriminate two sounds, or even whether a set of markedly different sounds belongs in one category, such as speech, while another collection of very different sounds belongs in a different category, such as music. But, how can we determine what birds hear? Surely, birds must be able to hear the sounds they themselves produce, which gives us a rough starting point. The anatomy of the bird ear also gives some clues but, like the analysis of vocalizations, this provides only an indirect approach to understanding hearing. A more direct physiological approach measures the electrical responses, or evoked potentials, that are generated at various levels of the nervous system in response to sound. But hearing is, after all, a behavioral response to sound involving the whole, awake organism (Stebbins 1970; Klump et al. 1995). Thus an even better method is to use behavioral techniques, training birds to respond to the presentation of a sound or to a difference between two sounds. Over the past several decades, such behavioral procedures have given us a fairly good idea of what birds can hear and what their acoustic world is like.

The Bird Ear

The basic principles of hearing are similar in birds and mammals. Acoustic pressure is transmitted to the fluids of the inner ear through the outer ear, the auditory canal, and the middle ear, via the ear drum and the middle ear bones, which match impedances between air and water and protect delicate inner ear structures. Vibrations transmitted to the inner ear fluids cause movement of the hair cells on the sensory epithelium, which in turn elicit neural discharges in the auditory nerve. Beyond these basic principles, the ear of birds is very different from the more familiar mammalian ear and this has led to much speculation about the hearing of birds. Unlike most mammals, most birds do not have external pinnae, and this almost certainly affects their ability to localize sounds. Birds have a single bone in the middle ear, the columella, rather than the three-bone middle ear characteristic of mammals, which may limit the ability of birds to hear very high frequency sounds (Saunders et al. 2000). In both birds and mammals, the sensory surface of the inner ear, the auditory epithelium, is made up of several types of specialized sensory cells called hair cells. The auditory epithelium of mammalian ears shows a pattern of one inner and three outer hair cells and a constant orientation of hair cell bundles across the entire length of the membrane. Bird ears, by contrast, are very complex showing many hair cells across the width of the auditory epithelium (**Fig. 7.1A**) and rather remarkable variation across the auditory epithelium in the pattern of orientations of hair cell bundles, the shape of the bundle, and the number and height of stereovilli on each hair cell (Gleich & Manley 2000; **Fig. 7.1B**).

As shown schematically in **Figure 7.1C**, the sensory surface in birds is not coiled and is much shorter than that in mammals, about 2 mm in the canary and the zebra finch versus 30 mm in humans, which may also place limits on the ability of birds to hear at both low and high frequencies, compared with mammals. This extraordinary complexity of the bird ear is not completely understood and raises many issues about how well birds can detect, discriminate, and learn complex sounds such as vocalizations and even whether some of the basic hearing mechanisms are the same in birds and mammals. Recent evidence shows that in spite of its diminutive size, the bird ear is a highly specialized organ capable of supporting very fine auditory discrimination and perception which, in some cases, exceeds the acuity of many mammals, including humans.

AUDITORY DETECTION

The Avian Audiogram and Hearing Range

The audiogram is the most basic measure of hearing. It is a graph of the minimum audible sound pressure at various frequencies throughout an organism's range of hearing. Over the past several decades, behavioral or physiological audiograms have been measured in over 50 species of birds (Dooling et al. 2000). On average, birds hear best between about 1 and 5 kHz with absolute sensitivity often approaching 0–10 dB

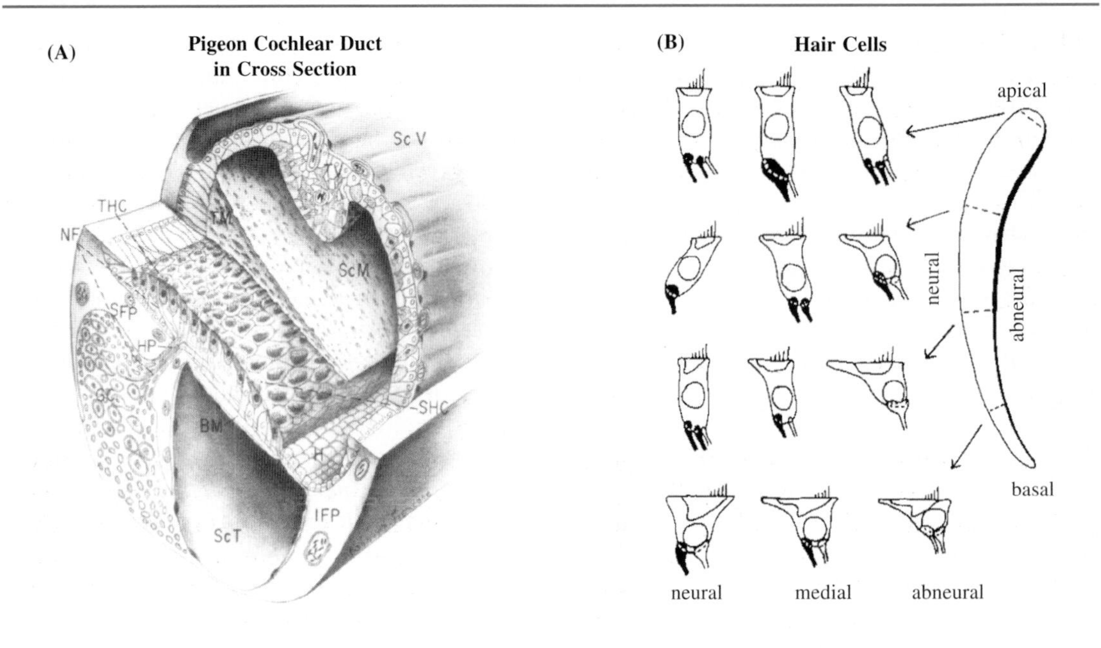

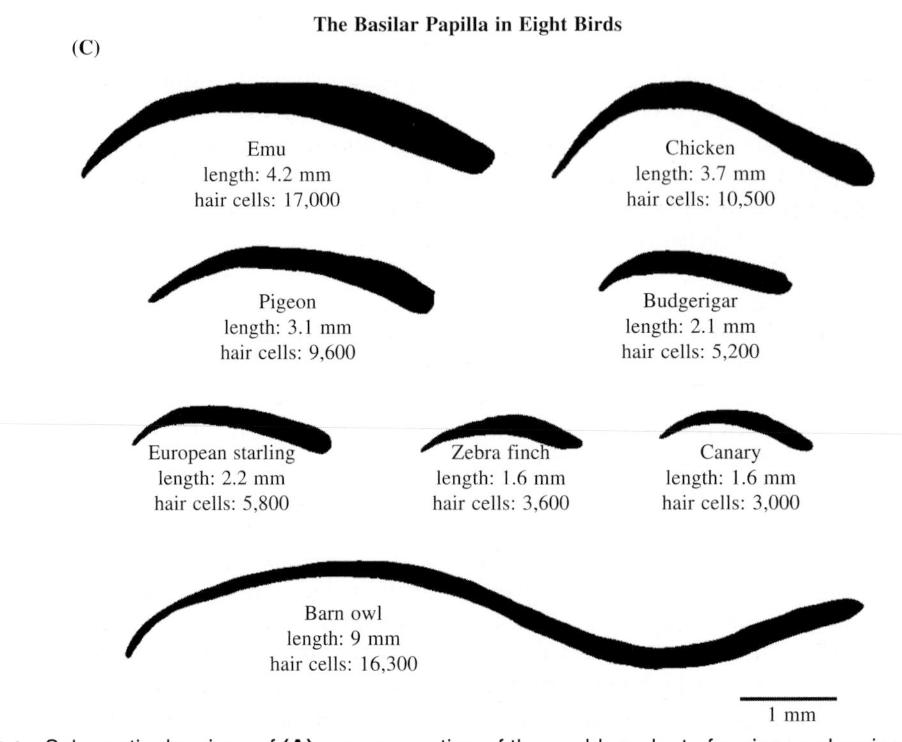

Figure 7.1 Schematic drawings of **(A)** a cross-section of the cochlear duct of a pigeon showing the major features of the avian inner ear (Takasaka & Smith 1971) and **(B)** hair cells and their innervation on the basilar papilla showing morphological hair cell gradients across the length and width of the avian papilla (Fig. 3.5 from Gleich & Manley 2000, © Springer Verlag). **(C)** Sketches of the basilar papilla of eight bird species along with hair cell counts (Fig. 3.2 from Gleich & Manley 2000, © Springer Verlag).

SPL at the most sensitive frequency, usually in the region of 2–3 kHz (**Fig. 7.2**). Though there are exceptions, we can discern several general trends. Nocturnal predators such as owls generally hear better than either songbirds or non-songbirds over their entire range of hearing, meaning that they can detect softer sounds. Songbirds tend to hear better at high frequencies compared with non-songbirds, which hear better at low frequencies.

On average, the limits of 'auditory space' available to a bird for vocal communication extends from about 0.5 kHz to 6.0 kHz. This is arbitrarily defined as the frequency range or bandwidth 30 dB above the most sensitive region of the audiogram, so that sensitivity declines as the frequency is lowered below or raised above 3 kHz. So, at around 500 Hz and around 6.0 kHz, sensitivity is approximately 30 dB worse than it is at the average bird's 'best' frequency. If we measure the main energy of bird vocalizations intended for distance communication such as songs and loud calls, by looking at the long term average power spectrum, we find that it falls well within this frequency region (see Chapter 6). While there tends to be a correlation between hearing sensitivity at high-frequencies and the highest frequencies contained in the species'

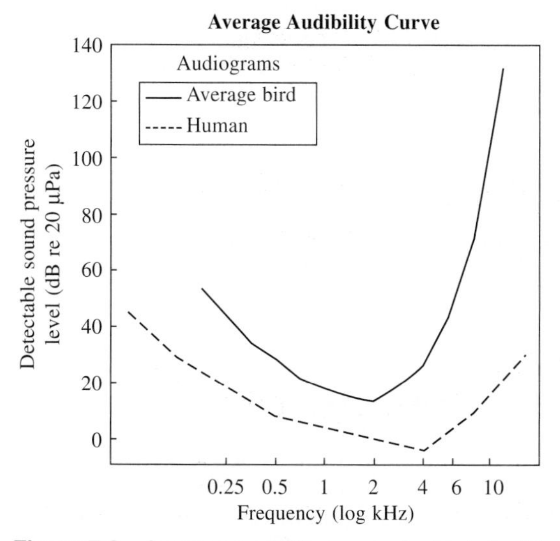

Figure 7.2 Average audibility curves for birds (solid line) and humans (dashed line).

vocalizations (Dooling 1980, 1982b; Dooling et al. 2000), hearing has yet to be studied in birds known to have unusually high-pitched vocalizations such as grasshopper sparrows and hummingbirds. There is good evidence that birds sometimes exploit the edges of their auditory sensitivity as with high-pitched aerial predator alarm calls of many small birds (see Chapter 5), which are clearly designed to be transmitted over short distances and to be hard to locate (Klump et al. 1986; McGregor & Dabelsteen 1996).

For all the work on song learning in songbirds, surprisingly little is known about how hearing develops, especially in altricial birds. Studies of the development of hearing in birds have focused almost exclusively on the precocial domestic chick and on developmental neurobiological mechanisms in hearing rather than on the relation between hearing and vocalization. It is important to know how hearing develops since the beginning of the sensitive period for song learning appears to begin quite early in some songbirds (see Chapter 3).

A recent approach uses the auditory brainstem response (ABR) recorded from electrodes under the skin to measure hearing in young birds (Brittan-Powell & Dooling in press; Brittan-Powell et al. 2002). The ABR is a scalp-recorded potential that results from synchronous neural discharge. It is manifested as a series of waves occurring within the first 10 ms following stimulation and represents the progressive propagation of auditory neural activity through the ascending auditory pathway. Since the recordings are not invasive and the nestling is sedated for only a short time, it is possible to track the development of hearing in individual nestling budgerigars and canaries from a few days after hatching until adulthood. Results show that hearing thresholds in both of these birds reach adult levels by 20–25 days of age.

Whether this is typical of other altricial species remains to be seen. But the similar time course of hearing development is interesting since, in the vivarium, canaries leave the nest much earlier, at about 18 days, than do budgerigars, at about 30–35 days. Poor hearing in the first week or two of life may be one factor that sets a lower limit on the sensitive periods for song learning. After that period, there is good evidence both from vocal learning studies as well as more direct assays of perception, that some species of sparrow show a perceptual preference for species-specific songs, a preference that may guide the earliest stages of vocal learning (Dooling & Searcy 1980; Nelson et al. 1997; see Chapter 3).

Auditory Feedback and Vocal Behavior

We know from many studies of birds deafened early in life or in adulthood that auditory feedback plays a critical role in song development, learning, and maintenance (Konishi 1965b; Konishi & Nottebohm 1969; Marler & Waser 1977; Marler & Sherman 1983; Nordeen & Nordeen 1992; Okanoya & Yamaguchi 1997; Woolley & Rubel 1997; Lombardino & Nottebohm 2000; **Box 26,** p. 211). Several recent studies asked more refined questions about auditory feedback. Budgerigars were trained by operant conditioning to produce contact calls at a fixed loudness (Manabe et al. 1998). When background noise was introduced, there was an increase in vocal intensity. This effect, first studied in humans, is known as the Lombard effect (Lane & Tranel 1971). With earlier results from quail (Potash 1972), and zebra finches (Cynx et al. 1998), these experiments show that birds can control their vocal intensity over at least a limited range of 5–10 dB.

In another study, budgerigars were trained by operant conditioning to produce a contact call and the precision with which this call was produced was controlled by requiring the birds to match a digitally-stored template of the call. Then the hair cells of the inner ear were temporarily damaged using a high dose of antibiotics. The birds lost the ability to match the template precisely after they experienced severe hearing loss but quickly regained precision as newly regenerated hair cells restored hearing (Dooling et al. 1997). Interestingly, while the loss of precision in vocal production is initially the consequence of hair cell damage, precision is acquired long before the papilla is fully repopulated with new, functional hair cells (Dooling et al. 1997). This suggests that only a small amount of hearing, or more accurately, recovered hearing, is necessary to guide nearly normal vocal precision in vocal production, as was found in young male canaries with hearing loss after being reared in white noise (Marler et al. 1973).

The relationship between vocalization and hearing in birds raises significant questions. For one, the intensity with which birds vocalize, and the cacophony accompanying large flocks and social groups, makes one wonder about hearing damage from such vocalizations. Exposure to intense sound causes damage to the delicate hair cells of the inner ear, though controlled studies show that birds are relatively resistant to hair cell damage from noise exposure compared to mammals (Ryals & Rubel 1985a, b; Ryals et al. 1999). The exact mechanism underlying this resistance is unknown but the eustachian tube may play a role (Larsen et al. 1997; Ryals et al. 1999). Interestingly, birds show an acoustic reflex to self-produced vocalizations. As in mammals, this reflex dramatically reduces the efficiency with which sound power is transmitted to the inner ear thereby protecting the hair cells during vocal production (Borg et al. 1982; Counter & Borg 1982). In another contrast to mammals, birds can regenerate the sensory hair cells of the inner ear following damage (Ryals & Rubel 1988). While there are species differences, hair cell regeneration generally appears to result in almost complete recovery of the ability to detect soft sounds and to discriminate small intensity, frequency, and time differences (Hashino et al. 1988; Hashino & Sokabe 1989; Marean et al. 1993; Niemiec et al. 1994; Saunders & Salvi 1995; Dooling et al. 1997; Ryals et al. 1999).

Masking: When Noise Interferes with Hearing

The audiogram represents the best an organism can accomplish as far as sound detection is concerned. Noise can interfere with hearing and usually does. This is almost always the case in natural environments where the level of noise

BOX 26

AUDITORY FEEDBACK AND BENGALESE FINCH SONG

In many species of songbirds, song learning only takes place during a short period early in life. In these so-called closed-ended learners, it was long believed that auditory feedback is no longer necessary once a bird completes its song development. This dogma was based on studies on a limited number of species; when white-crowned sparrows and chaffinches were deafened in adulthood, they still retained their normal songs (Konishi 1965a; Nottebohm 1968). However, recent studies demonstrate that, in fact, songbirds continue to rely on auditory feedback to maintain their song structure, but the degree of auditory dependence in adulthood varies among species. Especially striking results were obtained with Bengalese finches. When adult males were deafened, most began to jumble the syllable sequence (see below) and to 'stutter' in as little as five days, by repeating the same syllable without the normal transitions to other syllables (**CD2 #56**). Quantitative analyses of the probability of these syllable transitions in pre- and post-deafening songs confirmed that real-time auditory feedback is necessary for adult Bengalese finches to maintain the normal syntactical structure of their songs (Honda & Okanoya 1999; Okanoya & Yamaguchi 1997; Woolley & Rubel 1997). The auditory dependence of song production is likely to be linked to the fluid nature of the species-specific structure of Bengalese finch song. Unlike zebra finches that produce highly stereotyped songs with all utterances virtually identical, Bengalese finch songs are made up of syllables that are strung together under a complex set of rules; syllables or trills of syllables can be sequenced in more than one way, each transition being associated with a certain probability (Honda & Okanoya 1999; Okanoya 2000). Thus, real-time auditory feedback may be especially critical for the generation of this fluid style of syllable syntax.

The complexity of the song syntax of Bengalese finches seems to have emerged after they were domesticated from their wild ancestral species, the white-backed munia, which has a more stereotyped syntax. Interestingly, two independent strains of Bengalese finches, one European, the other Japanese, appear to have acquired their syntactical complexity somewhat independently, with the song of the Japanese strain being more complex. Comparative studies of these two strains and the ancestral white-backed munia provide a valuable opportunity to understand how higher order syntactical organization is related to auditory feedback mechanisms. Also, like zebra finches, their contact calls are sexually dimorphic (Yoneda & Okanoya 1991), inviting study of their underlying physiology.

Ayako Yamaguchi & Kazuo Okanoya

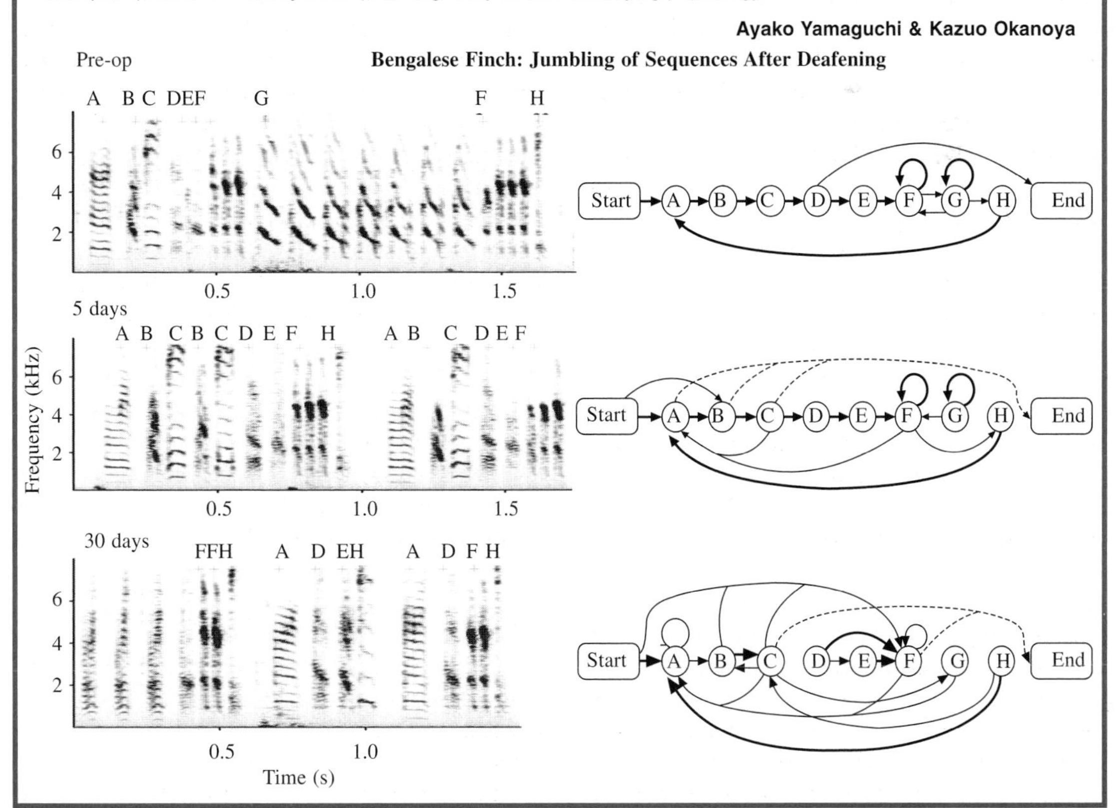

and its temporal and spectral characteristics exert a major influence on the active space available for acoustic communication between birds. This is difficult to study in the field (Brenowitz 1982a; Klump 1996) but laboratory studies, where both signals and noise can be controlled, can provide guidelines about the effect of noise on hearing in the real world.

Laboratory studies with pure tones and white noises show just how intense a pure tone must be relative to background noise in order to be heard (Dooling et al. 2000). Though there can be large species differences, there are also some clear consistencies. Generally speaking, in the frequency region of best hearing, around 3 kHz, for most birds, tone levels must be on average about 25 dB above the spectrum level of noise to be detected. Studies such as this have established a number of basic principles regarding the masking of signals by noise (Klump 1996; Dooling et al. 2000). First, it is energy in the frequency region of the signal that is most effective in masking. Noise at other frequencies has much less effect. Second, the signal-to-noise ratio needed for detection stays relatively constant over a wide range of noise levels. Third, the spatial relationship between the signal and the noise is important in determining the degree of masking. If the signal and the noise come from different directions, much less masking occurs. And fourth, it is one thing to *detect* a signal such as a vocalization and quite another to *discriminate* between one vocalization and another, or to *recognize* a particular vocalization. These general principles help us to understand how well birds can communicate in noisy environments in nature.

There is a long standing interest in environmental effects on the design of bird vocalizations (Morton 1970, 1975; Marten & Marler 1977; Marten et al. 1977; Waser & Waser 1977; Wiley & Richards 1978). The factors influencing the maximum distance over which a biologically meaningful sound may be heard include the location and intensity with which the signaler vocalizes, the sound-attenuating and sound-modifying characteristics of the environment, and the location and sensitivity of the receiver (see Chapter 6). Some aspects of this problem can be addressed in the laboratory.

Recently, Lohr and his colleagues (2003) measured the ability of budgerigars and zebra finches to detect and discriminate among own species and other species contact calls when presented in different noise types. One purpose of this study was to determine whether there are special, species-specific sensitivities to particular design features of calls that would enhance detection. Results showed that budgerigar calls tend to be more readily detectable than zebra finch calls for both species simply because the total acoustic energy is spread over a broad frequency range and concentrated in a narrower frequency range in budgerigar calls than in zebra finch calls. When signal-to-noise ratios at threshold were calculated, combining the peak sound pressure level of calls and the noise spectrum level in the narrow-frequency 1/3 octave band surrounding the peak frequency of the call, threshold differences among call types disappeared. In other words, the frequency band with the greatest signal-to-noise ratio is what governs detection no matter who the listener is or which vocalization is under consideration. Moreover, the task of *discriminating* between two contact calls, rather than simply detecting a call in noise, required slightly better signal-to-noise ratios.

As a test of whether noise in the spectral region of the call is what primarily governs masking, tests were repeated with the same overall level of noise but with a different spectral shape. The shape chosen is characteristic both of traffic noise and of natural noise levels in temperate forest and grasslands (Warring 1972; Awbrey et al. 1995; Klump 1996). The birds were able to detect and discriminate calls better in traffic noise than in white noise or in flat noise – confirming that it is the amount of energy in the frequency region of the signal, and the signal-to-noise ratio, that is most important in masking the signal (Lohr et al. 2003). This conclusion is reinforced by a three-way comparison of detection thresholds for a pure tone, a frequency-modulated pure tone, and an amplitude-modulated pure tone, all presented at the same average intensity. Amplitude-modulated tones are easier for both species to detect because the peak signal to noise ratio is greater relative to the average intensity within a given frequency band.

Coupled with knowledge of how loudly birds call (Brackenbury 1979), these data can be used to derive a family of functions that describe the maximum distance for vocal communication distance between two birds in different noise types. **Figure 7.3** shows the straightforward situation for budgerigars and zebra finches detecting own-species versus the other species' contact calls in white noise and in traffic-shaped noise. White noise has equal energy at all frequencies while traffic-shaped noise has more energy at low than at higher frequencies (see Chapter 6, **Fig. 6.5**). In this example, we assume a source level at the bird of 95 dB SPL, as the intensity at which the bird vocalizes, and excess attenuation of 5 dB/100 m. Keeping noise constant, detection was always easier than discrimination, as would be expected, and both tasks were easier in traffic noise than in flat noise. Though the difference was small, when translated into distance following the inverse square law governing sound attenuation, with a 6 dB drop in sound pressure level with each doubling of distance, birds can *detect* another calling bird at a much greater distance than they can *discriminate* between different vocalizations.

Several other factors bear on how well birds can communicate acoustically in noisy conditions (see Chapter 6). The so-called 'cocktail party' effect that helps listeners to focus attention on specific sounds to the exclusion of others under adverse listening conditions has several aspects. One already mentioned that received little attention until lately is called by experts 'spatial release from masking.' More masking occurs when the signal and the noise come from the same location in space rather than from two different

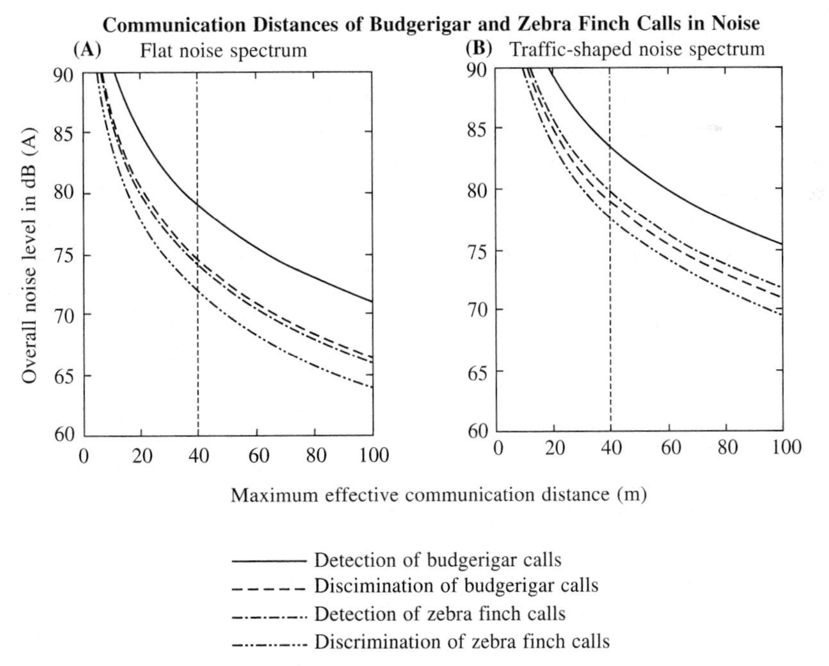

Figure 7.3 Estimates of the maximum communication distances of budgerigar and zebra finch calls, with backgrounds of **(A)** broadband flat noise, and **(B)** traffic-spectrum noise thresholds, based on detection and discrimination thresholds; they assume excess attenuation of 5 dB/100 m and a source intensity of 95 dB SPL. Modified from Lohr et al. 2003. The vertical dashed line represents a distance of 40 m. Discrimination threshold distances, in dotted lines, approximate the maximum limit for communication distance with a given call type, thus defining the 'active space' of that call under those conditions (see Chapters 5 & 6).

locations. Because birds have small heads and closely spaced ears, there was originally some doubt about whether they enjoyed a spatial release from masking. A recent experiment measured masking of pure tones in budgerigars when the noise and the tone came from different directions (Dent et al. 1997b; **Fig. 7.4**). When the tone and the masker were separated by 90 degrees, birds improved their threshold in noise by as much as 10–15 dB, comparable to the spatial masking release experienced by humans (Hirsh 1971; Durlach & Colburn 1978). Again, considering that sound pressure level decreases about 6 dB with every doubling of distance, this amount of spatial release from masking could have a huge effect on how much better birds can hear one another if they position themselves away from a natural noise source.

A final consideration is whether birds can control their vocal output level. The loudness of songs and calls of a number of songbirds and non-songbirds has been measured, and maximum peak sound pressure levels of 90–100 dB a meter away from a bird are not uncommon (Brackenbury 1979). Moreover, it is likely that the control of vocal intensity is both under voluntary control and reflexively responsive to environmental noise conditions, in at least some birds (Cynx 1999). Budgerigars were trained to produce contact calls for food, and to match with precision a computer-stored template (Manabe et al. 1998). Then the birds were reinforced for increasing their vocal intensity; they increased call loudness over a 5–10 dB range. In terms of transmission distance under natural conditions, a 10 dB change in vocal output could have a significant effect on communication in noisy environments – recall that for humans a 10 dB increase in sound pressure level is approximately a doubling of loudness.

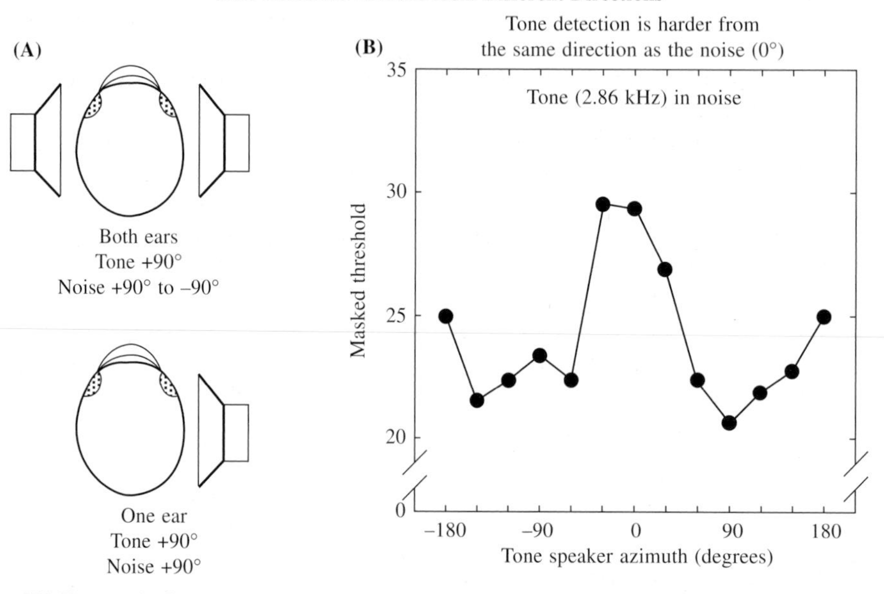

Figure 7.4 **(A)** The setup for an 'unmasking' experiment. It is easier (5–10 dB advantage) to detect a tone when a noise is presented to both ears rather than one. **(B)** Average masked thresholds for tone detection by three subjects with the noise source directly ahead (0°) and the tone moved around the bird in 30° increments. Thresholds less than 30 dB represent a release from masking. Detecting a tone is easier when the tone and noise come from different directions. From Dent et al. 1997b.

AUDITORY DISCRIMINATION

Hearing is more than just the detection of sound. Even a cursory consideration of acoustic communication in humans and animals reminds us that being able to discriminate one sound from another is critically important. Bird vocalizations are exquisite examples of precise, complex, multidimensional acoustic signals, and birds learn these sounds and exploit many of these dimensions in acoustic communication. Information on this ability is available from two sources; discrimination studies that use simple sounds to probe the limits of hearing, and those using more complex sounds such as vocalizations, with an eye toward understanding the role of hearing and auditory perception in vocal communication.

Frequency, Intensity, and Time Discrimination Using Simple Sounds

Pitch is a particularly important characteristic of many communication signals, including human speech and bird vocalizations. Field studies have shown repeatedly that frequency is an important cue for song recognition in many species (Falls 1963; Bremond 1968, 1976; Fletcher & Smith 1978; Nelson & Marler 1990; Slabbekoorn & ten Cate 1998a, b). Psychophysical studies of frequency discrimination in birds using pure tones show that they are quite sensitive to changes in the frequency of acoustic signals. In their frequency region of best hearing, birds on average can discriminate a little less than a 1% change in the frequency of a pure tone (Dooling et al. 2000). For example, frequency variation in a 3 kHz tone must be greater than about 20 Hz to be discriminated by a bird.

While this is not better than the limits of human frequency resolution (Wier et al. 1977), it is still remarkably good. Under controlled conditions birds can hear smaller frequency variations than those typically found in bird vocalizations; this includes changes that are smaller than the limits of our techniques for measuring frequency changes in bird sounds, and certainly smaller than the amount of frequency change that is biologically relevant. This prompted Nelson and Marler (1990) to make the useful distinction between a traditional 'jnd,' or just noticeable difference, and a 'jmd,' a just meaningful difference, in attempting to understand song-based species and individual recognition in the field. As Klump (1996) points out, the significant meaningful frequency difference from field studies is usually around 2–3 standard deviations from the local population mean, while the jnd for frequency in laboratory psychophysical tests is usually less than 1% (Dooling et al. 2000). However, it should be born in mind that these two sets of values may not be all that different with respect to the value of d', which is an index of discrimination from signal detection theory. This is because the background that forms the basis for the decision process in birds in the field is much broader than that in the laboratory, where the background and the target are single fixed-frequency tones. Moving to and fro between laboratory and field studies is important in showing that a wider application of signal detection theory in the field can strengthen comparisons with laboratory data (Klump 1996).

Birds are not nearly as adept at discriminating changes in intensity. They are much worse than humans and on a par with other mammals (Dooling & Saunders 1975; Hienz et al. 1980; Okanoya & Dooling 1985; Fay 1988; Dooling et al. 2000). Where humans can normally discriminate a 1 dB change in the intensity of a pure tone, birds typically require a difference of 2–4 dB, depending on the task. This is perhaps not too surprising; fine intensity gradations are probably an unreliable way to code biologically relevant information in the real world, given problems of absorption and reflection as sounds travel through the natural environment; temporal characteristics are another matter.

The stereotyped, punctate nature of many birdsongs has long fueled speculation that birds may be particularly sensitive to the temporal aspects of song. Data from related domains including song learning (Greenewalt 1968), cochlear anatomy (Pumphrey 1961; Schwartzkopff 1968, 1973), and single-unit recordings (Konishi 1969) have all been used to argue that birds are capable of an unusual degree of temporal resolution. The question is, in what aspect of temporal resolving power are birds superior? Judged by many measures of temporal processing, such as the improvement in detection as sounds are made longer, providing for maximum temporal integration, the abilities to discriminate changes in duration, and to detect a discontinuity in an ongoing sound, with tasks like gap detection, and study of modulation transfer functions, birds are no better than other vertebrates including humans (Viemeister & Plack 1993; Dooling et al. 2000). What is common to all these measures is that they involve changes in the overall level of the sound as a function of time; in other words, the amplitude envelope. But there is another way to impose temporal changes on sounds that do not involve the envelope. As described below, it is in discrimination at this level of temporal fine structure that birds may excel (Dooling et al. 2002).

Discrimination Tests with Complex Sounds

Bird calls and songs, and other complex sounds are, by definition, multidimensional. A variety of multidimensional sounds have now been used to probe the sensitivity of the bird auditory system as a basis for comparison with the more extensive information we have on mammalian hearing, including that of humans. Questions of interest for the avian auditory system include the shortest time interval that can be resolved, the longest duration over which the ear can integrate energy, and the size and shapes of the auditory filter bandwidths (Dooling et al. 2000). In general, birds are good at spectral resolution and do well in discriminating changes in tonal quality and timbre (Amagai et al. 1999; Cynx et al. 1990; Lohr & Dooling 1998), and changes in the rate at which sounds vary in frequency and amplitude (Dooling et al. 2000), in detecting signals in masking noise that varies in time (Klump & Langemann 1995; Hamann et al. 1999), and in pattern perception in general (Hulse et al. 1997).

Take the case of pattern perception. Do birds show some of the more complicated perceptual strategies that define the phenomenon of auditory scene analysis, as described by Bregman and his colleagues (Bregman & Campbell 1971; Bregman 1990)? In complex, real-life situations, humans routinely segregate sounds that occur together into separate auditory objects. These auditory objects can be identified by a unique set of features, whether it be spatial location spectral quality, pattern, or pitch; the 'cocktail party effect' already discussed is a result of a combina-tion of these.

Hulse and his colleagues (1997) tested starlings for such abilities using both natural and synthetic stimuli. They could readily be trained to identify a birdsong of one species presented at the same time as another species' song (Hulse et al. 1997). They could learn to discriminate among many samples of the songs of two individual starlings and they maintained that discrimination when songs of a third starling were added; they did well even when songs from additional starlings were added as further background distracters (Wisniewski & Hulse 1997). Testing the birds on tonal patterns, starlings maintained discrimination between fixed and changing temporal patterns in the face of large frequency differences in some of the components (MacDougall-Shackleton et al. 1998). Results like these suggest that auditory scene analysis is not unique to humans but may play a role in the auditory perception of birds, and other nonhuman vertebrates that benefit from the ability to parse a noisy world into auditory objects.

Another surprising similarity between birds and humans concerns the perception of the sounds of human speech. Since some birds are excellent mimics of speech, it stands to reason that they can hear the acoustic distinctions that are involved. A series of studies have shown that budgerigars place vowels /i/, /a/, /e/, and /u/ in phonetically appropriate categories in spite of variation in who is talking, and their gender

(Dooling & Brown 1990; Dooling 1992). In working with consonants, speech researchers have synthesized continuous series of speech sounds that span the acoustic boundary between two different consonants. Sounds drawn from one end of the continuum are perceived as one consonant and sounds at the other end are perceived as another; humans, under most conditions, show a rather sharp boundary between the two categories of speech sounds. Given some of these synthetic phoneme continua, budgerigars exhibit perceptual phonemic boundaries near the human boundaries for /ba/-/pa/, /da/-/ta/, /ga/-ka/, /ra/-/la/, and /ba/-/wa/ (Dooling et al. 1989, 1995; Dent et al. 1997a), suggesting they hear the categories of speech as humans do.

The perception of speech sound categories is not unique to budgerigars. Zebra finches, starlings, Japanese quail, red-winged blackbirds, and pigeons also discriminate and categorize speech sounds in ways similar to humans (Hienz et al. 1981; Kleunder et al. 1987; Dooling 1992; Dooling et al. 1995). These data from various birds suggest that, if there is anything 'special' about speech, it has as much to do with the acoustics of the spoken word as it does with the natural complexities of the human auditory perceptual system.

But, just as there are some remarkable similarities between birds and humans in the perception of complex sound patterns, there are also unexpected differences, as with music. Hulse and his colleagues have shown that starlings are quite adept at the kind of serial pattern perception that appears to underlie the recognition of melodies in humans (Krumhansl 1990). In categorizing rising and falling pitch patterns, starlings respond to both relative and absolute pitch features of the patterns but, in contrast to humans, absolute frequency cues generally predominate (Hulse & Cynx 1985, 1986; Page et al. 1989; MacDougall-Shackleton & Hulse 1996; Ball & Hulse 1998). When budgerigars were trained to discriminate between simple melodies consisting of three kinds of pitch patterns: rising, falling, and constant frequency, they proved to be quite sensitive to the frequency range of the tone sequences, but relatively insensitive to whether successive tones rose in frequency, stayed the same, or fell – the exact opposite of what humans report when listening to these patterns (Dooling et al. 1987b). Evidently some changes in frequency are more salient to birds than others even if they are all easily discriminable (**Box 27**, p. 218).

PERCEPTION OF VOCAL SIGNALS

A large body of evidence from song learning and from field studies of singing behavior shows that birds are sensitive, both to such details as note quality, and to larger scale features such as ordering of notes and syllables, tempo, and the number of syllable repetitions, all features of their species' songs (Kroodsma & Miller 1982, 1996). Moreover, as with human infants and speech, so with birds some of these sensitivities are present quite early in the bird's life, in some serving as a guide to song recognition in vocal learning (Marler 1997). This is one of several remarkable parallels between speech and language learning in humans and song development in birds (Marler 1970b; Doupe & Kuhl 1999). The effects of hearing loss on vocal production, in both young and adults, are comparable in songbirds, parrots, and humans (Dooling et al. 1987a; Oller & Eilers 1988; Nordeen & Nordeen 1992; Heaton & Brauth 1999; Heaton et al. 1999; Lombardino & Nottebohm 2000). In each case, severe hearing loss early in life drastically affects the development of a normal vocal repertoire while hearing loss later in life is less disruptive, although there are still effects.

In another direction, most of what we know about the sophisticated perceptual abilities of humans subserving speech development and communication comes from the laboratory where speech sounds, or just as often their synthetic analogs, are used as stimuli, under rigorous experimental control. This approach is favored because it often yields a clearer picture of what is special about speech and speech perception that cannot be obtained in any other way. And

BOX 27

ARE SONGBIRDS PYTHAGOREANS? ABSOLUTE AND RELATIVE PITCH PERCEPTION

Musicians tend to speak of absolute or perfect pitch, and relative pitch. Those with absolute pitch rely on octaves; they hear notes at octave intervals as similar (Ward & Burns 1982). People with relative pitch either classify sounds as relatively lower or higher, or they recognize ratios between notes. This latter finding can be traced back to the sixth century BCE when Pythagoras discovered that pitch intervals vary with frequency ratios. Whole number ratios determine the intervals on a musical scale and in harmonic structures. More recent research has shown that the Pythagorean rule applies even across different human cultures (Krumhansl 1990), although this account skirts the twentieth century blossoming of non-Pythagorean musical scales and harmonics (Pierce 1983). Given that humans are generally Pythagoreans, is the same true of songbirds? In the first attempt to examine pitch perception in songbirds, starlings were trained on an operant task to discriminate between two very simple melodies (**CD2 #57–60**). One consisted of four tones rising in frequency, the other, of four descending tones. There were four examples of each, and the rising and falling melodies all overlapped within the same octave. After learning this task, the starlings were presented with the identical melodies, except they had been transposed up or down an octave. The starlings failed to show any discrimination between the rising and falling melodies, and in fact required as many trials to learn the new task as with the original (Hulse & Cynx 1985). Clearly pitch perception in starlings cannot be predicted from what we know of human pitch perception.

The finding that starlings are predisposed towards absolute pitch has been replicated in other species (Hulse & Cynx 1985; Weary & Weisman 1991) and verified in the field. If a territorial field sparrow's song is slightly transposed up or down in frequency, neighboring birds cease to recognize it as familiar (Nelson 1988). Similarly, the perception of abrupt frequency changes as meaningful vanished in collared doves when tested in field playbacks that were beyond their natural frequency range but well within their audible range (Slabbekoorn & ten Cate 1998b). However, avian absolute pitch, which also can be found in non-songbirds (Cynx 1995), is not like human absolute pitch perception. Starlings do not appear to use the similarity of notes separated by an octave (Cynx 1993). They seem to actually memorize the true frequency. This tendency to learn by rote may help explain the thousands of trials required to train them to distinguish between melodies.

As mentioned, there are at least two kinds of relative pitch. Starlings and white-throated sparrows will respond appropriately to tones or songs that have been transposed, although their ability to deal with relative pitch changes appears to be limited by the absolute pitch range in which it was previously learned (Hulse et al. 1984; Hurley et al. 1992). Discrimination generalization near these limits is close to the sensory limits for frequency discrimination in starlings (Cynx et al. 1986). This is somewhat odd. If they have carefully memorized individual frequencies, why would they extend this discrimination within a certain training range, yet precipitously lose it at the limits of the frequency range? As to recognizing the ratios between notes, starlings, like humans, still respond to the 'missing fundamental' when the fundamental frequency of a harmonically structured sound is removed (Cynx & Shapiro 1985). This suggests that a bird's auditory system can fill in what is missing, based on Pythagorean ratios. There have also been reports that black-capped chickadees and white-throated sparrows can recognize song structure based on the frequency ratios between song notes (Weary & Weisman 1991; Hurley et al. 1992). The variety in use of absolute and relative pitch cues may reflect species differences, or it may result from variation in the motivation to discriminate rather than actual capability.

However, we can conclude that many songbirds are Pythagoreans and are able to hear frequency ratios, but also that they perceive absolute pitch in a different way than we do. Could it be that some songbirds are predisposed to hear sound frequencies the way we see colors? For example, any number of frequencies in the visible light spectrum stand in whole number ratios to each other, but we perceive them as qualitatively different, not as mathematically related. Birds may be predisposed to perceive sounds in the same qualitative way. While we avoid being confused by the overlapping frequencies of human voices by our ability to hear out harmonics and the relative pitch of formants, birds may have evolved a different system in which each bird species 'owns' certain frequencies, based on a form of absolute pitch perception.

Jeffrey Cynx

tests with simple sounds such as pure tones and noises are somewhat unreliable in predicting how humans will perceive the sounds of speech or, more importantly, how well they understand them. Might the same be true of birds? In fact, when the songs and calls of birds, or artificial sounds like them, are used in laboratory tests, evidence of special or exceptional sensitivities does sometimes emerge. The following examples serve to illustrate the point.

Species-specific Perceptual Specializations

In the search for special processes underlying avian acoustic communication, and possible parallels with human speech and language learning, song has received the most attention. Because of their length songs are difficult to use in psychophysical tests, but calls are much shorter, and some are learned (see Chapter 5), as they are in budgerigars (Farabaugh et al. 1994; Farabaugh & Dooling 1996; Hile et al. 2000; Hile & Striedter 2000). Calls were used in a psychoacoustic study of discrimination (Dooling et al. 1987b; Dooling 1992). The contact calls of budgerigars, canaries, and zebra finches are all fairly short with peak energy around 2–5 kHz. In other respects, the calls of the three species are quite different (**Fig. 7.5A; CD1 #60**).

Budgerigar contact calls contain rapid frequency modulation but virtually no harmonic structure. Canary calls have much less frequency modulation, a simple harmonic structure, if any, and a relatively high fundamental frequency. Contact calls of zebra finches, on the other hand, are rich in harmonic structure with a relatively low fundamental frequency and little if any frequency modulation.

Birds were trained to detect a change inserted into a background of repetitive sounds. One call was used for the background and another as the 'target.' Detection of the target was rewarded with food, using many different pairs of calls as both background and target. Confronted with an easy discrimination, we know that birds respond more quickly than when discrimination is difficult (Dooling et al. 1992). So, response time or latency provides an index of similarity between two complex sounds. In other words, if two calls are perceptually similar, response latencies will be longer than with two dissimilar calls.

When the response latencies from these within-species and cross-species tests were analyzed by the techniques of multidimensional scaling (MDS) and cluster analysis, several interesting patterns emerged. MDS produces a map of multidimensional space in which calls that are perceptually similar are grouped close together in one category, and those that are different are placed elsewhere. All three species showed evidence of somewhat separate perceptual categories for budgerigar, canary, and zebra finch calls (**Fig. 7.5B**). In terms of response latencies, this means that the birds responded more quickly when a call from one species was the background and the target was a call from another species. It is important to note that the birds had not been trained previously with particular combinations of these calls. They responded as described on the very first test. Thus, in a very real sense, these are *natural* perceptual categories, not artificial categories induced by training.

All three species also responded more quickly when discriminating among own species calls than among calls of the other two species (**Fig. 7.5C**). In other words, budgerigars discriminated among different budgerigar calls faster than among different zebra finch or canary calls. Canaries discriminated among canary calls faster than they did among budgerigar or zebra finch calls, and zebra finches discriminated faster among calls of their own species than they did among own species calls or canary calls. In all three species, own species calls were discriminated most efficiently. This result, obtained in a completely balanced experimental design where every subject was tested in exactly the same way, suggests that each species is specialized for the perceptual processing of its own calls.

How do such specializations arise? One possibility is that the basic hearing capabilities of each species are different and that species-specific contact calls have evolved to match these

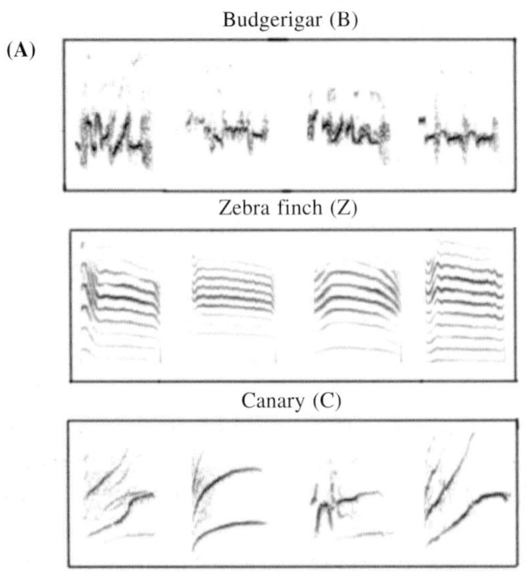

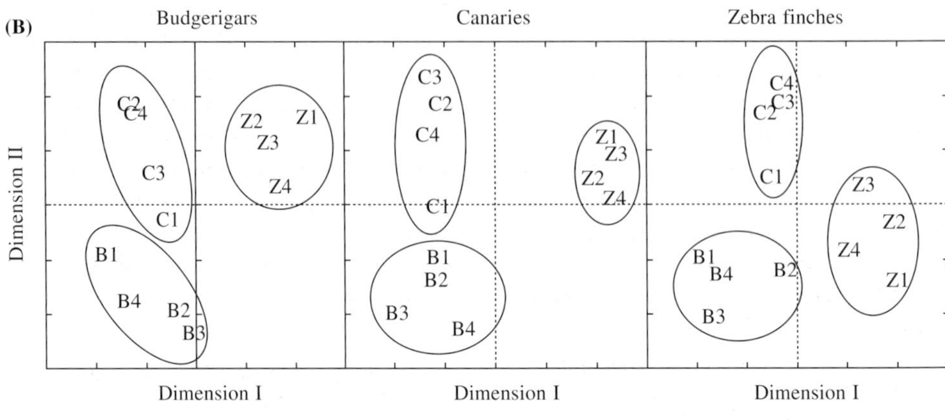

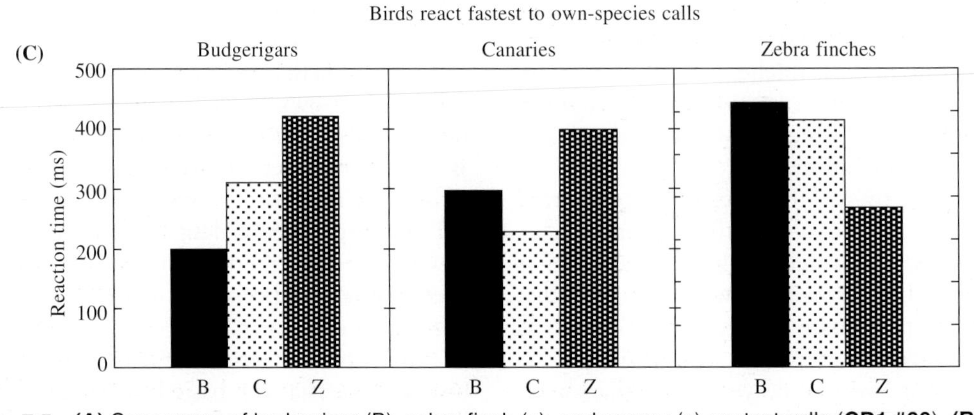

Figure 7.5 **(A)** Sonograms of budgerigar (B), zebra finch (z), and canary (c) contact calls (**CD1 #60**). **(B)** Maps of the first two dimensions from a multidimensional scaling analysis. Each map represents reaction time data for one species discriminating among a set of 12 contact calls of three species (B, C & Z; modified from Dooling et al. 1992). **(C)** Average reaction times for each species making within-category pair-wise discriminations. Each species reacted faster when discriminating between own-species calls, showing that this was easier for them.

basic innate auditory capabilities. The evidence for this interpretation is not compelling, given that the basic hearing capabilities of these three species are rather similar. Another possibility is that species-specific perceptual sensitivities in these birds begin as a set of broad, basic auditory capabilities that are similar across species, then are quickly and profoundly influenced by early, selective exposure to own-species sounds, perhaps as nestlings or fledglings. This would then lead to perceptual preferences for the learning, production, and use of conspecific sounds in adulthood.

As a variation on this theme, new perceptual preferences could be acquired and new perceptual sensitivities developed during social interactions throughout adulthood. In budgerigars, this seems to be the case. Three budgerigars that lived together for several months and came to share similar contact calls indistinguishable by sonographic analysis, were tested on their ability to discriminate among their own calls and those of their cagemates. The speed with which budgerigar contact calls can be learned (see Chapter 5, **Fig. 5.27**) made it possible to test the plasticity of the perceptual processes of these adult birds. The results showed that birds that were cagemates could discriminate among the calls of all three, but other budgerigars, unfamiliar with the cagemates or their calls, could not (Brown et al. 1988). Evidently, through the experience of living together, cagemates learned to extract the essential call differences among cagemate calls required for individual recognition while ignoring variation in irrelevant acoustic characteristics – something inexperienced birds could not do.

The Role of Attention in Call Perception

Perception, memory, and attentional processes must all be involved in how birds acquire sensitivity to species and individual call differences It appears that budgerigars focus their attention in special ways when perceiving complex vocalizations or sounds modeled on them. Dent and I (1998) created 7-tone patterns designed to capture some features of budgerigar contact calls (**CD1 #61–62**). Each 20 ms tone, with an inter-tone interval of 5 ms, was of a different frequency within the range of 2–4 kHz where most of the energy of budgerigar contact calls lies. The tonal complex was presented repeatedly at a rate of 2/s and birds were asked whether they could detect a frequency change in one of the components. Eventually both birds and humans showed thresholds that were as good as they could do on conventional tests using longer tones presented in isolation (**Fig. 7.6**). If, however, subjects were confronted with a high uncertainty task, the location of the frequency change being varied from trial to trial, human performance deteriorated dramatically – presumably because the task now required them to distribute their attention across all seven tones in the pattern when listening for a change. Remarkably, budgerigars tested on these same tonal complexes did just as well as when they knew where to listen for a change and even as well as when long tones were presented in isolation. One interpretation of these results is that budgerigars process these synthetic 'contact calls,' as a whole perceptual unit, much as humans hear entire words. Otherwise said, perhaps the birds were listening 'synthetically' rather than 'analytically' (Dent et al. 2000).

Perceiving Temporal Fine Structure – A Special Capability of the Avian Auditory System?

The zebra finch vocal repertoire contains predominantly harmonic sounds. Zebra finches and budgerigars are able to discriminate slight mistuning of one of the harmonics in a simulated female zebra finch contact call (Lohr & Dooling 1998). Mistuned harmonics (**CD1 #63**) produce a variety of sensations in human listeners and there is a recent literature, including that focused on a perception of roughness, changes in the subjective pitch of the mistuned components, and other specific cues to inharmonicity (Moore et al. 1985a, b, 1986; Hartmann et al. 1990). But mistuning a single harmonic also causes

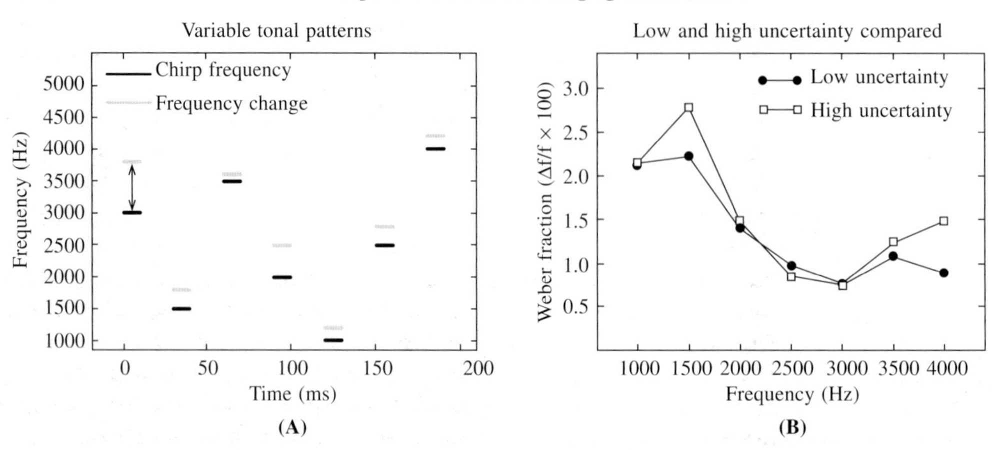

Figure 7.6 **(A)** Schematic diagram of a variable frequency tonal pattern **(CD1 #61–62)**. During a low-uncertainty test session, frequency changes on each occurred at only one position, as shown by the arrow. During a high-uncertainty test session, the change can occur at any position from trial to trial, by any amount. **(B)** Average frequency-difference-limens (FDLs) for the variable frequency tonal patterns are plotted as Weber fractions for two birds. The budgerigars were run on three different variable frequency tonal patterns under both the low uncertainty (closed circles) and high uncertainty conditions (open squares). From Dent et al 2000.

changes in temporal fine structure. **Figure 7.7** shows a schematic sonogram of a female zebra finch contact call, the corresponding spectrum, and the thresholds (in Hertz) that zebra finches, budgerigars, and humans showed for discriminating the mistuning of several harmonics. Zebra finches, whose repertoire is characterized by complex harmonic sounds, are exceedingly sensitive to changes in these harmonics – more so than budgerigars, and much more sensitive than humans. Zebra finches are ten times more sensitive to harmonic mistuning than they are to a frequency change in a pure tone presented in isolation So there is a strong indication that the birds are actually responding to changes in temporal fine structure rather than frequency.

To explore this idea, birds and humans were tested on specially synthesized sets of sounds called Schroeder complexes (**CD1 #64**). These are harmonic complexes in which the phases of the individual components have been adjusted so as to produce waveforms with a flat envelope and identical long-term spectra. The term 'envelope' refers to changes in the level of a sound over time and 'spectra' refer to the distribution of energy over frequency. These waveforms are special in that they have similar envelopes and spectra and differ only in temporal fine structure. If birds can discriminate among these, they can only do so on the basis of temporal fine structure.

In **Figure 7.8A** there are two Schroeder harmonic complexes with a fundamental frequency of 200 Hz – the positive Schroeder complex being just the reverse of the negative Schroeder complex, as if played backwards. The overall spectrum and the envelope are the same for these two waveforms, but they differ in the temporal fine structure within each period, which is 5 ms for a harmonic complex with a 200 Hz fundamental. Birds and humans were tested on the ability to discriminate between positive and negative Schroeder complexes as a function of fundamental frequency. Humans lose the ability to discriminate between a positive and a negative Schroeder complex when the fundamental frequency is higher than about 300 Hz (**Fig. 7.8B**). Birds, especially zebra finches, can still discriminate these waveforms even when fundamental frequencies reach 1000, Hz which corresponds to a time interval of only 1 ms

Red-winged blackbird Elliott

Common yellowthroat Elliott

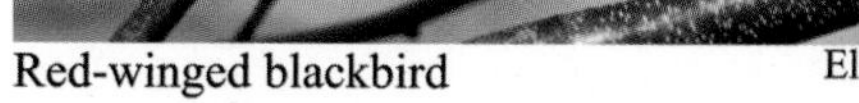

Green hylia Slabbekoorn

Rufous-collared sparrow Paz-Soldán

Plate VIII

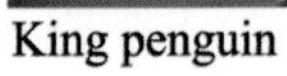

King penguin — Ostfeld

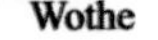

Great tit — Wothe

Budgerigar — Dent

Zebra finch — Berkhoudt

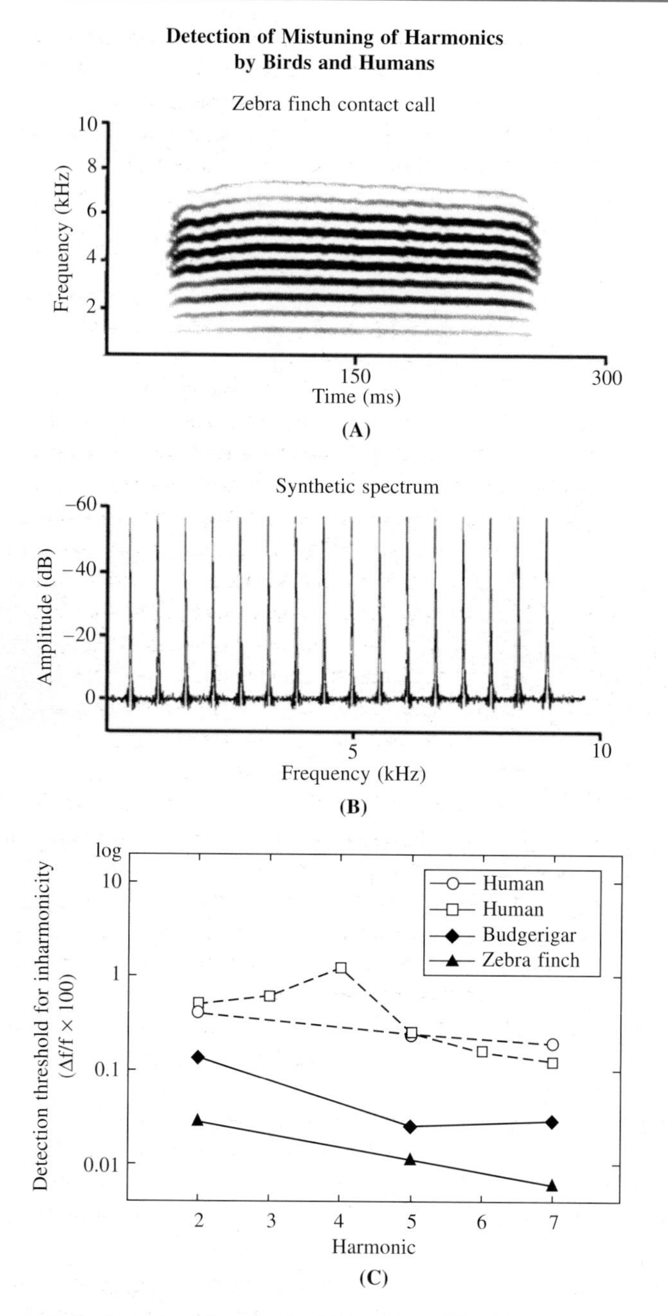

Figure 7.7 **(A)** Sonogram of a female zebra finch contact call. **(B)** Power spectrum of a synthesized call consisting of a complex harmonic tone. **(C)** Thresholds for detecting inharmonicity in humans and two species of birds. Bird data are for a fundamental of 570 Hz (Lohr & Dooling 1998). Human data are for 570 Hz (Lohr & Dooling 1998) and for 400 Hz (Moore et al. 1985a). The threshold for detecting mistuning is plotted as a percentage of harmonic frequency (log scale); when sensitivity is high, the threshold is low. Modified from Lohr & Dooling 1998.

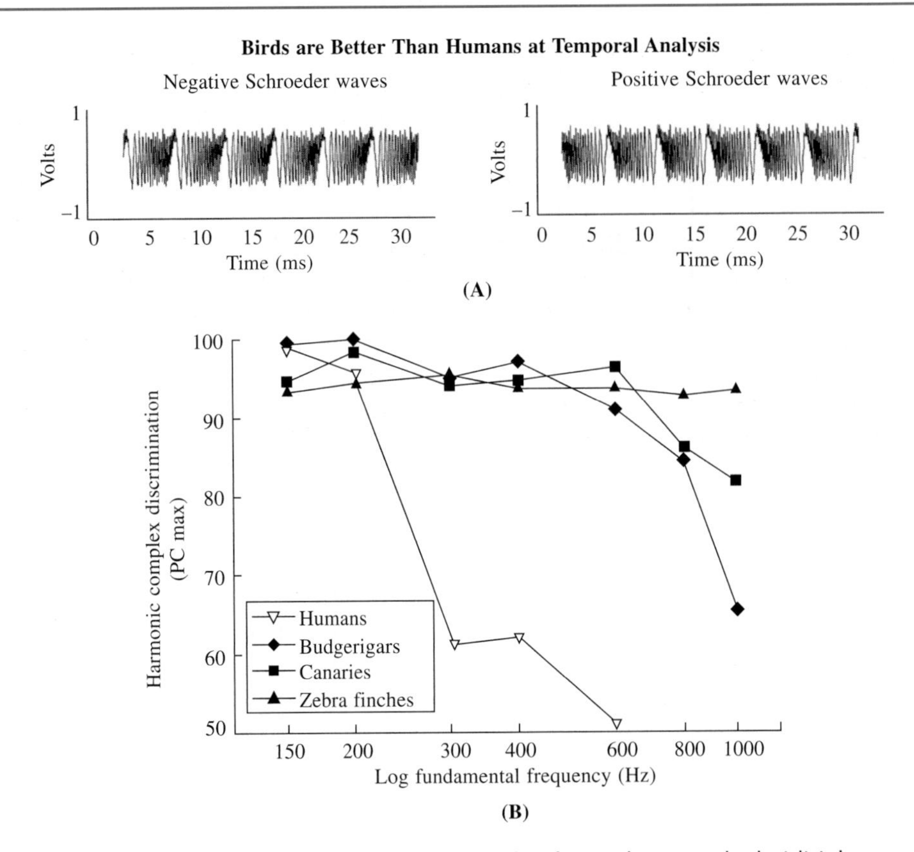

Figure 7.8 Birds are good at detailed temporal analysis of complex sounds, but it takes sophisticated experimentation to demonstrate it. **(A)** Schematics of several periods of a Positive and a Negative Schroeder complex (**CD1 #64**) at a fundamental frequency of 200 Hz. The fundamental frequencies used ranged from 150 Hz to 1000 Hz, producing fundamental periods ranging from 6.6 ms to 1 ms. Phases of each tone component were adjusted to eliminate envelope cues that might be used for discrimination, selected according to an algorithm by Schroeder (1970). This algorithm produced a waveform with instantaneous frequencies that rise or fall across each period. **(B)** All the birds were able to discriminate between harmonic complexes with much higher fundamental frequencies (800 to 1000 Hz) than humans; this requires temporal analysis over fundamental periods as short as 1 ms in duration. Humans were unable to discriminate between complexes with fundamental frequencies higher than about 250 Hz. Modified from Dooling et al. 2002. Enhanced time processing of complex sounds by birds, relative to humans, may be a general characteristic of the avian auditory system.

(Dooling et al. 2001, 2002). Since the only difference between these stimuli is in temporal fine structure, it must be the basis of discrimination. Thus, as ethologists have long suspected, birds do indeed have an enhanced capacity for resolving the temporal fine structure of complex sounds – a capacity that clearly extends their auditory capabilities, enabling them to discriminate very subtle differences between vocalizations (Lohr et al. 2000).

CONCLUSIONS

Studies of hearing in birds reveal that they hear well over a range of frequencies that embraces most of those used in their songs. Within a narrower range of frequencies, where they hear best, the ability to discriminate between two sounds approaches the level of acuity often reported for humans. However, there is also a major difference. Birds excel in discriminating between two complex sounds which differ only in temporal fine structure. Several decades of research has shown that the role of hearing in acoustic communication in birds is extraordinarily sophisticated and complex. Take the case of vocal learning. Perceptual studies in birds show an early role in focusing a bird's attention on the correct stimuli for learning and guiding subsequent vocal productions. Later in life, a host of rather special capabilities become evident when one probes the properties of the avian auditory system with species-specific sounds or with complex synthetic sounds modeled on natural vocalizations. These capabilities reinforce our impression of the general sophistication of the avian auditory system. They also point to particular perceptual specializations, subserving the development, learning, and maintenance in adulthood of the complex acoustic communication systems that are an integral part of the social life of so many birds.

Chapter 8

Brains and birdsong

ERICH D. JARVIS

INTRODUCTION

The special brain structures for singing, and learning how to sing, were discovered in 1976. Since then, with still growing momentum, there has been a wealth of fascinating discoveries on the structure, function, and evolution of the brains of birds, especially those that engage in vocal learning. Less than half of all birds possess the ability to learn and reproduce new sounds. These vocal learners, parrots, hummingbirds, and songbirds belong to only 3 of the 23 major bird groups. They have the necessary forebrain anatomy for producing learned vocalizations. All other birds use only basal brain structures for vocal production and their vocalizations are innate, and genetically inherited from their parents.

Of the birds that are vocal learners, most is known about songbirds, especially the canary and the zebra finch. These are subjects of choice because they breed easily in captivity and display opposite extremes of vocal learning behavior. In zebra finches, only the males sing, have vocal learning, and possess the appropriate forebrain structures for the purpose; they are closed-ended vocal learners, developing one song motif as juveniles and singing it for life. In canaries, both males and females sing, and both have vocal learning brain structures; they are open-ended learners and, like humans, continue to learn new songs as adults. An understanding of these contrasts and their underlying principles holds promise of new and fundamental insights into how brains make learning possible. The discovery of similar sets of brain structures in parrots and hummingbirds throws new light on how brain structures for vocal learning have evolved. The shared features imply that there are strong epigenetic constraints placed on the brain and behavior during evolution, and suggest that perhaps humans evolved somewhat similar brain structures for speech and singing. Although the cognitive side of song learning hardly bears comparison with human speech, 'bird-brained' need no longer be a pejorative. If a parrot imitates its own and other species' sounds so readily and a child can hardly be prevented from learning to speak, how is it that a chimpanzee does neither? The answer perhaps lies in the fact that no brain structures have been found in chimpanzees equivalent to those present in songbirds and humans.

Other important findings about brains were first made in the songbird vocal learning system. These include the startling discovery of large sexual dimorphisms in the brain, the role of hormones in regulating brain structure and learning, and the ability of the adult songbird brain to generate new neurons, paving the way for current brainstem cell research. This chapter will review some of these pioneering studies, and how the brain controls singing behavior, which is the theme of this book.

GENERAL BRAIN ORGANIZATION

Understanding the science of birdsong ultimately requires a basic understanding of the vertebrate brain. Birds and mammals, including humans, are vertebrate animals. Regardless of brain and body size, all vertebrates share a general brain organization that consists of five basic regions. These are the spinal cord within the vertebral backbone, the hindbrain in front of it, then the midbrain, the thalamus, and the cerebrum, with the cerebellum (meaning 'small brain') on top of the hindbrain (**Fig. 8.1A**). Among those vertebrates that are amniotes, having embryos that develop within amniotic fluid-filled sacs, that is reptiles, birds, and mammals, all have a similar structural organization in these five brain regions, except for one, the cerebrum (Veenman et al. 1995; Reiner et al. 1998). In reptiles and birds the cerebrum is organized into large cell clusters; in mammals only the bottom part is organized into clusters, called the basal ganglia, whereas the upper part is arranged in layers that together form the cortex. Because mammals have long been regarded as more intelligent than birds or reptiles, it was assumed that one of their major anatomical differences, the mammalian cortex, was responsible for their more intelligent behavior (Herrick 1956). Many viewed evolution as proceeding linearly with its ultimate goal the creation of 'man,' 'his' language, civilization, large brain size, and folded cortex. Accordingly, almost all subdivisions of the cerebrum of birds and reptiles were given names that designate the mammalian basal ganglia, which were thought to be primitive (Edinger et al. 1903; Edinger 1908; Herrick 1956; **Fig. 8.1B**, old nomenclature). We now know that the organization of the bird cerebrum is much more similar to the mammalian cerebrum than previously thought (Reiner et al. 1998 in press A, B; Puelles et al. 1999). In 2002, at a Nomenclature Forum held at Duke University, the subdivisions of the entire bird cerebrum were renamed to reflect more accurately the many homologies that exist between avian and mammalian brains (Jarvis et al. 2002; Reiner et al. 2004a; **Fig. 8.1B**, new nomenclature).

The cerebrum of the bird and the mammal consists of three major cell zones, the pallium, the striatum, and the pallidum. The striatium and pallidum portions combined are the basal ganglia, and in most animals they sit at the base of the cerebrum (Marin et al. 1998). The pallium zone sits above the striatum at the top of the cerebrum, and includes the cortex in mammals and four major brain subdivisions in birds (Reiner et al. 1998; Puelles et al. 1999; Swanson 2000b). Brain size and folding are more related to animal size than to behavioral complexity (Van Essen 1997). The smaller the animal, the less folding there is of the cerebrum. Hummingbirds with one of the smallest bird brains and no folding have more advanced types of behavior, such as vocal learning, than many mammals with large and folded brains, such as horses.

Neurons, the communicating cells of the brain, typically have a cell body with axons that make long distance connections to other neurons (**Fig. 8.1C**). Neurons also have dendrites that receive these axonal connections. The meeting places of the axons and dendrites are the synapses. In the basic connectivity plan of both birds and mammals, sensory information from the outside world, and from the inner body, destined for the cerebrum, first passes through sensory receptor neurons that have one axon projecting out into the body or the exterior, and another into the spinal cord for the lower body or into the hindbrain for the upper body. Here they synapse onto their appropriate cell groups, which in turn project up to the midbrain. The midbrain neurons send projections into the thalamus. The thalamic neurons process and project the information to the cerebrum, where they synapse with a network of cells involving the pallium, striatum, and pallidum. Neurons that control movement form a network that traverses in the opposite direction, from the cerebrum, to the midbrain, and spinal cord. Neurons of the spinal cord, lower motor neurons, send their axons to muscles. The cerebrum controls complex learning and voluntary behaviors, and 'conscious'

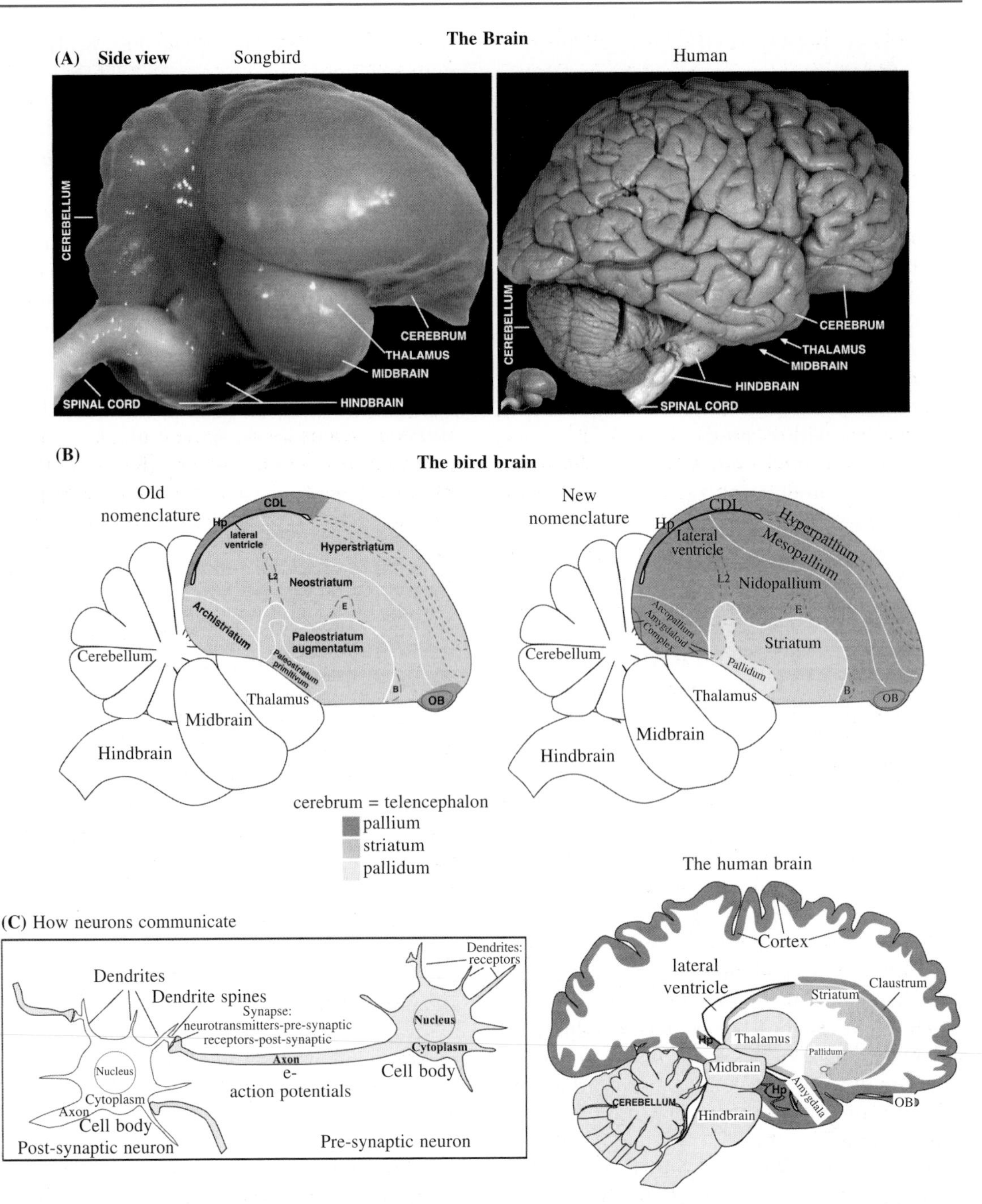

Figure 8.1 **(A)** Songbird and human brains compared. **(B)** Old and new nomenclature for the bird brain. Dashed gray lines separate regions that differ by cell density and size, such as the six layers of the cortex. The darkest gray area of the human cerebrum are all axon pathways, typically called white matter, connecting the cortex with various cerebral and non-cerebral areas. Abbreviations: E, entopallial nucleus; B, basorostral nucleus; L2, field L2 nucleus; OB, olfactory bulb; H_P, hippocampus; CDL, Corticoid dorsal lateral **(C)** A diagram of neurons and their connections.

processing, whereas the midbrain is involved in innate behaviors and 'non-conscious' processing. Glial cells typically provide physical support and nourishment to neurons.

THE SEARCH FOR VOCAL LEARNING BRAIN AREAS

Vocalizations are learned in some species and innate in others, and this behavioral contrast is expected to reflect a brain difference. As humans, however, we often have difficulty in understanding its true basis because, in everyday language, we do not distinguish auditory learning from vocal learning. Vocal learning is the ability to imitate the sounds that you hear or improvise. It is a rare trait, found so far in only six animal groups: three birds (parrots, hummingbirds, and songbirds) and three mammals (humans, bats, and cetaceans, the whales and dolphins [Thorpe 1961; Marler 1970b; Wiley 1971; Caldwell & Caldwell 1972; Nottebohm 1972; Kroodsma & Baylis 1982; Dooling et al. 1987a; Guinee & Payne 1988; Baptista & Schuchmann 1990; Rubsamen & Schafer 1990; Esser 1994; Gaunt et al. 1994; McCowan & Reiss 1995]). Auditory learning is the ability to form memories of the sounds that you hear, and to associate sounds with objects and living things in the environment. Auditory learning is much more widespread. It is present in nearly all land and many aquatic vertebrates. An example helps in understanding this distinction. A dog can learn the meaning of the sounds sit (in English), sientese (in Spanish), or osuwali (in Japanese). Dogs are not born with this knowledge of human words. They acquire it through auditory learning. However, a dog cannot imitate the sounds sit, sientese, or osuwali. Humans, parrots, and some songbirds can. This is vocal learning, and though it depends upon auditory learning (Konishi 1965a), it is distinct from it. Aside from mimics, most vocal learners only imitate sounds of their own species.

The history of the search for vocal learning brain areas in songbirds has its roots in the search for language brain areas in humans, beginning nearly 200 years ago. Vocal learning is the behavioral substrate for spoken language. In the first recognized breakthrough, in 1861, a French physician, Paul Broca, published a paper about an autopsy of one of his patients who had a stroke (Broca 1861). The patient could only speak one word, 'tic,' and had extensive brain damage, with a central location for the lesion in the now famous Broca's Area of the cortex. From this and several subsequent patients, Broca concluded two things: (i) that language within the brain was localized, not spread out and (ii) its location was on the left side of the frontal lobe. Forty-four years later in 1905, Oswald Kalischer, a German scientist, attempted to determine if parrots have a Broca's-like Area (Kalischer 1905). He performed left and bilateral hemisphere lesions, and stimulation experiments, on Amazon parrots. He found some vocal effects, but this was preliminary work and was never followed up. Studies on vocal non-learning animals, including mice and monkeys, revealed no cerebral vocal regions (Kuypers 1958b; Jürgens 1995, 1998). Only midbrain vocal regions that control innate vocalizations were found. It was not until 115 years after Broca's discovery that Fernando Nottebohm, and two of his colleagues, Tegner Stokes and Christiana Leonard, in 1976 published a pioneering report, announcing the discovery of cerebral vocal structures for learned vocalizations in a non-human species, the canary, a songbird (Nottebohm et al. 1976).

Nottebohm began his neuroanatomical investigations in the peripheral nervous system outside the spinal cord. He was studying the organization and function of the songbird syrinx and its connection with the axon nerve bundle that controls it, the tracheosyringeal nerve (Nottebohm 1971a, b), so-called because its axons innervate both the trachea and syrinx. The songbird syrinx is the main organ that produces learned vocalizations (see Chapter 9). Nottebohm made the interesting finding that after surgical disconnection of the left tracheosyringeal nerve from the syrinx, canaries and other songbird species had more difficulty in producing their learned songs than after disconnecting the right nerve (Nottebohm 1971b; Nottebohm & Nottebohm 1976). He then reasoned that this tracheosyringeal nerve dominance might originate in the central nervous system, perhaps in the cerebrum, as is the case for human language. He and his colleagues made what turned out to be a lucky guess and created surgical lesions in brain regions next to the auditory pathway. They figured that a vocal system might be next to an auditory system (Nottebohm et al. 1976). The region they focused on included what is now called the auditory pallium. In their very first try, they found that the canary ceased to sing normally, particularly with lesions to one side of the brain.

This was the first demonstration of lateralization in the cerebrum of a non-human species; like humans, canaries were left-side dominant. Repeating their lesions in more birds, they narrowed down the location that led to loss of learned song to one region. The structure in this region was given the name HVC, for hyperstriatum ventrale pars caudale, later renamed as the High Vocal Center. Lesions to the left canary HVC resulted in greater song deficits than to the right HVC. They then labeled degenerating axons from the HVC lesions and determined that HVC projected to another structure in the pallium involved in production of learned vocalizations, that is now called Robust nucleus of the Arcopallium, RA, because of its robust appearance in stained tissue sections. HVC also projected to a structure in the striatum that they called Area X, because at the time lesions there did not lead to noticeable song deficits (Nottebohm et al. 1976). In stained brain sections, these vocal nuclei can be seen under low magnification, and sometimes with the naked eye. It is remarkable that they were not identified earlier in anatomical investigations of the avian brain.

Soon thereafter, there was an explosion of studies on the cerebral vocal structures of the songbird brain. Mark Konishi and his students performed the first electrophysiological experiments in the brain of awake songbirds (Katz & Gurney 1981; McCasland & Konishi 1981), in search of the brain regions that integrate learned auditory information with learned vocalizations. Nottebohm and Konishi also introduced into neurobiology the zebra finch, a bird in which vocal learning had been described in detail by Immelmann (1969) and his students. It was believed that study of the differences between zebra finches and canaries would reveal the underlying brain mechanisms for continued adult brain plasticity. Zebra finches pass through the juvenile phase of vocal learning relatively fast, within 90 days after hatching, instead of a full year as in some other species, and can breed several times a year. Since then, zebra finches and canaries have become the mainstay of songbird neurobiology research. In 1981, John Paton, Kirk Manogue, and Nottebohm used electrophysiological and neuronal connectivity approaches to describe several similar brain structures in a parrot (Paton et al. 1981). In 2000, Claudio Mello and I used behavioral and molecular approaches to reveal the entire set of cerebral vocal structures in hummingbirds (Jarvis et al. 2000). Because we had used the same approach with all three vocal learning groups, we were able for the first time to compare cerebral vocal structures in hummingbirds, parrots, and songbirds. For studies in humans, the invention of the new imaging techniques of Positron Emission Tomography (PET) and Magnetic Resonance Imaging (MRI) in the 1980s and 1990s, and their use to study language-activated brain areas in humans, also propelled the brain and language field forward (Poeppel 1996; Binder 1997). However, it is still not possible for scientists to access human brains as readily as we can do with birds.

Hundreds of papers later, a new research field has emerged, focused on the neurobiology of songbird vocal communication, with over 98 laboratories worldwide as of 2003, resulting in many detailed analyses of the brain network involved in the learned vocal communication of songbirds. Thus, over 140 years after Broca's discoveries in humans and 26 years after Nottebohm's discoveries in canaries, we now know more about brain pathways for vocal learning in birds than we do in humans.

THE SONGBIRD VOCAL COMMUNICATION BRAIN NETWORK

Networks of neuronal connectivity can be determined by a variety of methods. The two best are based on neuronal tract-tracing and electrical activity. For tract-tracing, a colored and/or fluorescent dye is surgically injected into the brain region of interest; over several days the dye is transported through axons and/or dendrites to connecting regions; the brain is then dissected from the animal, sectioned and examined underneath a microscope for presence of the tracer in the connecting regions. Alternatively, electrical activity in two or more neurons can be recorded simultaneously and the circuitry is deciphered mathematically based on the relative timing of their firing. Tract-tracing is good at defining the global circuitry, whereas electrical activity is better at defining the microcircuitry. Both methods have been used in defining the brain pathways responsible for songbird vocal communication.

The key network consists of three different, interconnected pathways, one posterior, one anterior, and an auditory pathway. Of several versions, the one I favor most is presented in **Figure 8.2** (**Box 28**, p. 233). The posterior and anterior vocal pathways are often called collectively the 'vocal control nuclei,' 'song control

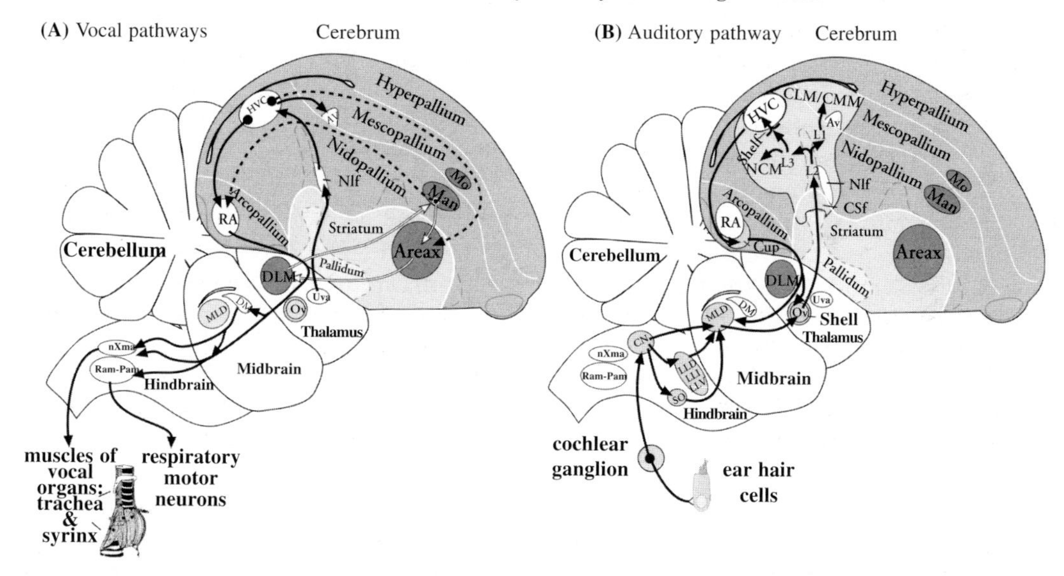

Figure 8.2 **(A)** The anterior and posterior vocal pathways in the songbird brain. **(B)** There is one main auditory pathway. Abbreviations not in the main text: CN, cochlear nucleus; LLD, lateral lemniscus, dorsal nucleus; LLI, lateral lemniscus, intermediate nucleus; LLV, lateral lemniscus, ventral nucleus; SO, superior olivary nucleus; PAm, para-ambigualis nucleus; RAm, retroambigualis nucleus. Black arrows connect the nuclei of the posterior vocal pathway (light background). White arrows link nuclei of the anterior vocal pathway (dark background). Dashed arrows connect nuclei between the two pathways. Within the cerebrum: dark gray is pallium, medium gray is striatum, and light gray is pallidum. Connectivity was extrapolated from the following studies: Nottebohm et al. 1976, 1982; Okuhata & Saito 1987; Bottjer et al. 1989, 2000; Wild 1994, 1997; Johnson et al. 1995; Nixdorf-Bergweiler et al. 1995; Vates & Nottebohm 1995; Livingston & Mooney 1997; Vates et al. 1997; Wild et al. 1997a, 2000; Iyengar et al. 1999; Luo & Perkel 1999a, b; Perkel & Farries 2000. Syrinx drawing from Suthers 1997.

nuclei,' or the 'song control system,' nuclei here referring to collections of neurons.

The posterior vocal pathway consists of four cerebral nuclei all located within the back part of the pallium (**Fig. 8.2A**), HVC and RA, and two less commonly studied nuclei, the Interfacial Nucleus, NIf, and Avalanche, Av. They are integrated in a circuit with the final output from RA to the midbrain vocal nucleus DM and to a set of hindbrain areas that includes respiratory nuclei RAm and PAm, and the tracheosyringeal nucleus, the so-called 12th nucleus, nXIIts. From there, axons of the respiratory nuclei make synapses onto other motor neurons of the spinal cord, which in turn control muscles of the chest wall and air sacs involved in respiration. Axons of the tracheosyringeal nucleus synapse onto the muscles of the trachea and syrinx. Thus, the posterior vocal pathway controls motor neurons, producing sounds and modulating breathing while doing so. In contrast to birds, mammals use the larynx to produce vocalizations (see Chapter 9). So far, no connections have been found in songbirds from the posterior vocal pathway onto motor neurons that control the tongue and beak movements although we expect to find connections there because song is modulated by beak movements (see Chapter 10). In general, cerebral projections connecting directly to hindbrain and spinal cord motor neurons are a telltale sign of involvement in learned movements.

The anterior vocal pathway consists of three cerebral nuclei located towards the front of the cerebrum, Area X, the Magnocellular nucleus of the Anterior Nidopallium, MAN, and the oval nucleus of the Mesopallium, Mo, which is the most recently discovered (Jarvis et al. 1998), and one nucleus in the thalamus (**Fig. 8.2A**). Little is known about the function of Mo. Connections between the others form a loop from MAN to Area X to the dorsal part of the thalamus and back to MAN, which we can abbreviate as MAN → Area X → Dorsal Thalamus → MAN. This pathway is divided into two parallel parts, lateral (l) and medial (m) as lMAN → lArea X → DLM → lMAN and mMAN → mArea X → ?DIP → mMAN. The posterior pathway sends input into the anterior pathway's medial and lateral parts by way of a projection from HVC → all of Area X. The anterior pathway in turn sends output to the posterior pathway by way of a projection from lateral and medial MAN to different parts of the posterior pathway; lMAN → RA and mMAN → HVC (**Fig. 8.2A**; **Box 28**, p. 233). Thus, the anterior vocal pathway has little direct interaction with vocal motor neurons of the brainstem (hindbrain and midbrain), but through its interactions with the posterior pathway is poised for other functions, including the learning of vocalizations.

The microcircuitry is important in understanding how these pathways operate. The HVC has three major neuron types: (i) the RA-projecting neurons that send axons from HVC → RA, (ii) the X-projecting neurons that send axons from HVC → Area X, and (iii) the interneurons that make connections within HVC, between its RA- and X-projecting neurons. Thus, the X-projecting neurons are the input cell type to the anterior pathway. The output cell type of the anterior pathway in MAN has two axons going in different directions; for lMAN, one axon projects to lArea X, staying within the anterior pathway, and the other to RA, in the posterior pathway. Presumably this cell type in mMAN has one axon projecting to mArea X and the other to HVC. In this manner, the lateral part of the anterior pathway influences RA in the posterior pathway and the medial part of the anterior pathway influences HVC in the posterior pathway.

The auditory pathway, processing sound as a first step in song learning, begins at the hair cells in the cochlea of the ear (**Fig. 8.2B**; see Chapter 7), activated by sound (Hudspeth 1997). They send axons to cochlear, CN, and leminiscal, LL, sensory nuclei in the hindbrain. These in turn project to the midbrain auditory nucleus MLd. MLd projects to Ovoidalis, Ov, in the thalamus. Ov projects to the primary auditory cells of the pallium, L2, then to secondary areas, L1 and L3, and then onto tertiary auditory areas,

BOX 28

THE SONG SYSTEM: A MINIHISTORY AND SOME CONNECTIVITY DETAILS

As is often the case in science, discoveries are made in bits and pieces. The order of discovery influences thinking about how things work. In the song system, the pathways were named differently at first. The posterior circuit was considered a 'direct pathway,' by which HVC sends information to RA; the anterior circuit was called an 'indirect pathway,' from HVC to RA (Okuhata & Saito 1987), terminology that was well established by the early 1990s. The anterior pathway was also sometimes called the recursive loop (Williams 1989) or accessory loop (Doupe & Konishi 1991). The direct pathway was said to begin in HVC with neurons projecting to RA. The indirect pathway was also said to begin in HVC with neurons projecting to Area X; Area X to DLM, DLM to lMAN, and finally lMAN back to RA. In this view, HVC belonged to both pathways. In the mid-1990s, a connection from lMAN to lArea X was found, effectively forming a loop (Nixdorf-Bergweiler et al. 1995; Vates & Nottebohm 1995), later shown to be a closed loop (Luo et al. 2001). The pathway then began to look more like the so-called cortical–basal ganglia–thalamic–cortical loops of the mammalian brain (Bottjer & Johnson 1997; Perkel & Farries 2000). In the late 1990s, the indirect pathway was renamed the anterior forebrain pathway (AFP), because of its location (Doupe 1993), and this has become common usage. In addition, connections between the medial part of the pathway were discovered and it was realized that they parallel the more commonly studied lateral part (Foster et al. 1997; Jarvis et al. 1998). To complement the anterior circuit's name, the direct pathway was renamed the posterior pathway (Jarvis et al. 1998). Finally, comparisons with other vocal learners and mammals (Jarvis unpublished) led to the view presented in this chapter, with the loop of the anterior pathway considered as the basic pattern, also present in the brains of other vocal learners and in nonvocal pathways of mammals. I argue that differences between species arise primarily in the connections between the anterior and posterior nuclei.

Some connectivity details

There are many complex connections within the vocal pathways of songbirds. RA projects to UVa in the thalamus, DM in the midbrain, the tracheosyringeal motor nucleus (nXIIts) in the brainstem, and to at least four premotor nuclei in the brainstem that control breathing, abbreviated PBvl, IOS, RVL, rVRG, and RAm (Wild 1997). UVa projects up to HVC, providing a direct route for feedback from RA to HVC (Striedter & Vu 1997). DM, like RA, projects to the tracheosyringeal motor nucleus and the same four breathing-related brainstem nuclei, providing a means for DM to modulate activity from RA onto these nuclei. It is this DM connectivity that coordinates the syringeal and respiratory muscles during vocalizing in non learning species. In songbirds, it appears that RA has taken over the innate vocalizing system by synapsing onto DM and all of its downstream targets (Vicario 1994). The respiratory nucleus RAm makes synapses onto motor neurons of the spinal cord which, in turn, control muscles of the chest wall and air sacs involved in breathing. In order to coordinate the two sides of the brain, UVa and DM send axons to their connecting counterparts on the other side of the brain.

Within the cerebrum, the HVC and lMAN neurons contact the dendrites of the same cells in RA (Mooney & Konishi 1991), but not always in the same location. This is how lMAN may modulate HVC activity into RA. In HVC, the interneurons contact both its X-projecting and RA-projecting neurons, coordinating their activity (Mooney 2000). In the anterior vocal pathway, the connections form closed loops (Luo et al. 2001). This means that there will be contact from a given neuron in lMAN to lArea X, and from there, to DLM, which will send its axon back to the same lMAN neuron within that loop. Adjacent will be another neuronal closed loop parallel to it. An unusual feature of the anterior vocal pathway, and perhaps of the avian brain generally, is that in Area X, in the striatum, there is a smaller number of pallidal-like neurons to which the striatal neurons are thought to make contact (Perkel & Farries 2000). If this is correct, HVC and lMAN project onto the striatal neurons in Area X which, in turn, converge onto the pallidal neurons in Area X and, from there, to the DLM of the thalamus. In the mammalian striatum, the pallidal cells appear so far to be entirely separate from the striatum.

Erich Jarvis

Caudal Medial Nidopallium, NCM, Caudal Medial Mesopallium, CMM, the HVC shelf of the nidopallium and the RA cup of arcopallium, and a Caudal part of the lateral Striatum, Cst. The auditory pathway also has a descending feedback projection with connections similar to the vocal pathway, from the HVC shelf to the RA cup to the shells of thalamic and midbrain auditory regions. One key feature of the auditory pathway is that, once projections from the auditory thalamus, Ov, reach the cerebrum's primary auditory receiving cells, L2 in the pallium, the L2 cells send projections that spread out to many other pallial and one subpallial auditory areas for further processing. Sensory input ascending into cerebral regions is a telltale sign of sensory learning.

It is an open question, where auditory information enters the vocal pathways. It was long thought that the main auditory input was directly from L2 into HVC (Kelley & Nottebohm 1979). However, this view was based on non-specific tract-tracing leakage into the adjacent NIf vocal nucleus that projects to HVC. Instead, L2 projects to an auditory region called the HVC shelf, which in turn sends only a few axons into HVC (Fortune & Margoliash 1995; Vates et al. 1996). Given that the HVC shelf projects to the RA cup, one might wonder if this is how auditory information enters, but the cup also sends very few if any axons into RA (Mello et al. 1998). Two other less well-studied locations may be potential sources of significant auditory input. Preliminary results show that auditory electrical activity into HVC requires input from NIf, and not from the adjacent field L2 region (Boco & Margoliash 2001). Input to NIf comes from Uva of the thalamus (**Box 28**, p. 233). Another potential source of auditory input is a region called para-HVC, a thin layer of cells medial to HVC, on the surface of NCM, connected with NCM, and projecting into Area X of the anterior vocal pathway (Foster & Bottjer 1998). Taken together, the major source of auditory input into the posterior vocal pathway may come directly from brainstem auditory areas into Uva into vocal NIf; direct auditory information into the anterior pathway may come from the NCM shell into Area X. The final answers on the linkages between the auditory and vocal networks await further investigation.

FUNCTIONS OF VOCAL PATHWAYS AND NUCLEI

Use of Lesions, Electrophysiological Recordings, and Gene Activation

In total, the songbird vocal communication network consists of seven vocal cerebral nuclei, one thalamic nucleus, one midbrain nucleus, one hindbrain nucleus, and a comparable number of auditory nuclei (**Fig. 8.2**). The layout suggests that different parts serve distinct behavioral and physiological functions. To decipher these functions, three basic approaches have been used: (i) the creation of lesions as performed in songbirds by Nottebohm and colleagues (1976), (ii) recording electrical activity as first performed in songbirds by Leppelsack (1978), Konishi and students (Katz & Gurney 1981; McCasland & Konishi 1981), and (iii) examining the molecular biology of the circuitry as first performed by Clayton, Mello, myself, and colleagues (Mello et al. 1992; Jarvis et al. 1995; Clayton 1997). To understand how the vocal pathways function, it is useful to review the underlying logic of these three approaches, all widely used in what is collectively called neuroethology, a term invented by Jerram Brown (Brown & Hunsperger 1963), studying the neural basis of mammalian vocalizations.

1. With lesions, results are interpreted in terms of loss of function. If a brain region is destroyed and a particular aspect of behavior is affected, then the interpretation is that the brain region is responsible for that aspect of the behavior.

2. With electrical activity, current changes across neuron membranes are measured. Action potentials are large currents that travel down axons to communicate with connecting neurons. There are also sub-threshold potentials; small local current changes on the membranes that

communicate within and between neurons. To measure activity, metal or glass electrodes are inserted into the brain regions or cells of interest. The electrical potentials picked up are sent by wire to detection devices. Easiest to measure is multiunit activity, registering the summed activity from several neurons near the electrode tip. This can be performed in awake animals. In contrast, single unit activity, the activity of a single neuron, is very difficult to measure, whether inside or outside the cell. It is feasible with tissue slices or anaesthetized animals, but difficult in awake animals; with each movement, the electrode attached to the head shifts slightly and loses contact with the cell. One solution has been to use movable electrodes that can be adjusted with a small micro motor attached to the bird's head (Fee & Leonardo 2001). Sometimes it is possible to extract single unit activity out of multiunit recordings. Multiunit measurements are sufficient to establish general relationships between activity and behavior, whereas single unit measurements are better for exploring mechanisms of communication within a neural network.

3. Molecular studies in songbirds have had their biggest impact by measuring the synthesis of gene products, messenger RNA (mRNA), and proteins. The order of events for gene product synthesis is that gene in the DNA is used to synthesize mRNA by a process called transcription and the mRNA is then used to translate the genetic code and synthesize the protein by a process called translation. *In situ* hybridization is used to detect mRNA synthesis and immunocytochemistry for detecting protein synthesis. Most useful for songbird studies have been activity-dependent genes; their products are synthesized as a result of neurons firing action potentials. The activity-dependent gene most studied is ZENK. The molecular function of the ZENK protein is to bind to the DNA regulatory regions of select genes and either enhance or repress their mRNA transcription. Because of this, ZENK is called a transcription factor. It is also called an immediate early gene, or IEG for short, because in the brain ZENK is normally synthesized at very low levels and then is briefly up regulated in cells after a short period of increased brain electrical activity (Worley et al. 1991). Once the activity ceases or returns to baseline levels, new synthesis stops. Because the half-lives of the mRNA and protein are short, about 15 min and 30 min respectively, the accumulation and presence of the gene product is short-lived (Herdegen & Leah 1998). In this manner, the IEGs can be used in ways somewhat similar to functional magnetic resonance imaging (fMRI), in the sense that fMRI activity implies recent electrical activity. However, like fMRI, the relationship of ZENK expression with neuron electrical activity is not one-to-one. Many areas of the thalamus, the pallidum, and primary sensory neurons of the pallium do not synthesize ZENK, regardless of the level of activity. Other IEGs are differently distributed in the brain, and other factors besides electrical activity can affect IEG synthesis (**Box 29**, p. 236).

The Posterior Vocal Pathway

When either HVC or RA, the two main cerebral structures of the posterior pathway, is lesioned bilaterally, on both sides of the brain, songbirds are unable to produce learned vocalizations (Nottebohm et al. 1976). Lesioned canaries still attempt to sing as judged by their posture and throat movements, but they are silent or produce only faint sounds (**Fig. 8.3A; CD1 #65**). Zebra finch become unable to produce learned calls, or lose modifications that they have learned, reverting to the innate version (Simpson & Vicario 1990). Innate vocalizations are retained. When Uva or NIf are lesioned, most learned syllables are retained, but syntax, the ordering of learned vocalizations, is affected and becomes more variable (Williams & Vicario 1993; Hosino & Okanoya 2000). With large lesions incorporating the midbrain vocal nucleus DM or the tracheosyringeal nucleus, birds can no longer produce either learned or innate vocalizations, becoming mute (Brown 1965; Nottebohm et al. 1976; Seller 1981). Smaller lesions of DM can reduce the motivation to vocalize. Taken together, these experiments show

BOX 29

ACTIVITY-REGULATED GENE EXPRESSION IN THE BRAIN: A HYPOTHESIS

The association between electrophysiological activity and the synthesis of so-called immediate early genes (IEGs) is well established, but the exact relationship is not known. IEGs may be directly responsive, not to electrical potentials in neurons, but to the neurotransmitters that are released at synapses. The neurotransmitters, such as glutamate and dopamine, then bind to their respective receptors. If enough receptors are occupied, two independent events are hypothesized: (i) depolarization of the post-synaptic neuron and subsequent firing of it's own action potential to generate behavior or process sensory information; and (ii) activation of second messenger pathways that leads to synthesis of ZENK, c-fos, BDNF, and other IEGs also in the post-synaptic neuron. Second messengers are molecules such as calcium ions that enter the cell or cyclic AMP molecules that are formed within the cell, in both cases after the transmitter binds to its particular cell receptor. The second messengers then attach to and activate protein transcription factors ready to go into the cell. These then bind to the promotor regions of select IEGs, such as ZENK, to turn on their mRNA and subsequent protein synthesis. Soon after performing their various functions, the IEGs are rapidly degraded by enzymes. Methods that reveal the presence of IEG activation provide valuable insights into which brain areas and circuits were engaged immediately beforehand.

Post-synaptic neurons that do not express the necessary receptor or second messenger systems to activate synthesis of a particular IEG, such as ZENK, will still fire action potentials in response to pre-synaptic input on its receptors, but will not turn on ZENK expression. One can also experimentally dissociate post-synaptic activity of neurons, preventing them from firing, and still get neurotransmitter-induced ZENK synthesis (Keefe & Gerfen 1999). However, in most parts of the brain, in contrast with other tissues, electrophysiological activity and IEG synthesis are co-induced by synaptic neurotransmitter release. I hypothesize that the level of IEG expression is controlled by the firing rate of the presynaptic neurons. A literature analysis suggests that the higher the rate of action potentials, the greater the amount of IEG synthesized, at least for ZENK (Chew et al. 1995; Mello et al. 1995; Stripling et al. 1997; Jarvis et al. 1998; Hessler & Doupe 1999b). Brain areas that do not express ZENK, such as the pallidum, primary sensory neurons of the pallium, and parts of the dorsal thalamus (Mello & Clayton 1994; Jarvis et al. 1998), presumably lack the appropriate receptors, or have receptors that inhibit its expression. Other IEGs are synthesized in response to neuronal activity in different subsets of brain regions. For example, the IEG c-fos, also a transcription factor, is synthesized in the same areas as ZENK, but it has a higher threshold for induction by activity, and it is synthesized at relatively higher levels in pallial song nuclei by singing (Kimpo & Doupe 1997; Wada & Jarvis unpublished). The mRNA of BDNF, a brain growth factor, is synthesized after singing *only* in pallial regions of the cerebrum, including the song nuclei, and not in the striatal nucleus Area X (Li & Jarvis 2001). Presumably the striatum, including Area X, does not have the necessary receptor combination to induce BDNF by neurotransmitter release. The multiplicity of receptor types for different neurotransmitters means that stimuli may trigger many different electrical and molecular responses throughout the brain.

Erich D. Jarvis

Proposed Model of Activity-regulated Gene Expression

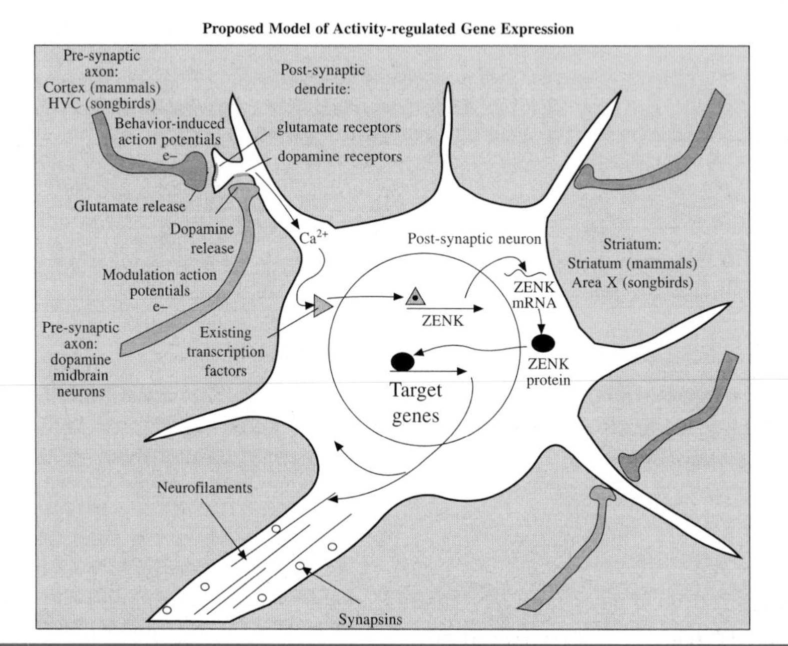

that to utter learned vocalizations, most cerebral nuclei of the posterior pathway are required; midbrain and hindbrain nuclei are needed for production and contextual modulation of all vocalizations, innate and learned.

Electrophysiological recordings show that immediately before singing, neurons in NIf fire first, followed by HVC, and then RA (McCasland 1987). This type of premotor activity occurs milliseconds before sound output (**Fig. 8.3B**). During singing these nuclei continue to fire action potentials and stop milliseconds before the sound output ceases. This firing pattern indicates that the posterior pathway neurons are the direct brain generator of the learned vocalizations. Molecular studies show that the act of singing induces a large increase in synthesis of ZENK and other IEGs in the posterior vocal pathway nuclei (Jarvis & Nottebohm 1997; Kimpo & Doupe 1997). Low levels of ZENK mRNA appear within the first 5–10 min of singing, peak after 30 min, and stay at a steady state as long as the bird continues to sing at a regular rate. The amount synthesized is related to the number of songs a bird utters (Jarvis & Nottebohm 1997). This means that either the mRNA is stabilized, or it is degraded and re-synthesized at a steady rate as the bird continues to vocalize. The interactions between behavior, electrical activity, and gene expression, are such that electrical activity in the brain network leads to muscular activity, in turn producing vocal behavior and at the same time inducing gene expression in that pathway (**Box 29**, p. 236). The gene products synthesized then regulate the expression of other genes.

Two open questions are: how does the electrical activity of the posterior vocal pathway actually generate singing behavior and what are the cellular

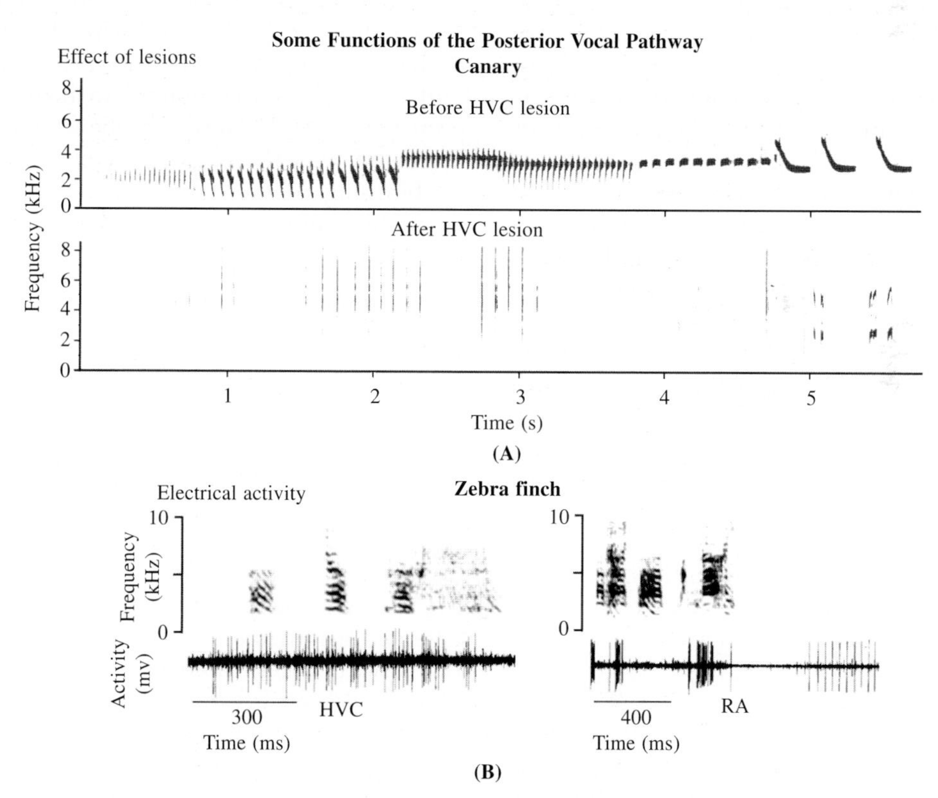

Figure 8.3 **(A)** Sonograms of canary song before and 9 months after bilateral lesions to HVC (**CD1 #65**). Modified from Nottebohm et al. 1976. **(B)** Electrical activity during singing in HVC and RA of two different male zebra finches. Modifed from Yu & Margoliash 1996.

and behavioral consequences of regulated gene expression. Insight into the first question comes from single unit recordings of individual HVC and RA neurons in zebra finches while singing, or in other states with neurons firing as they do during singing. In HVC, each RA-projecting neuron produces a burst of action potentials once per song motif, the repeated unit of zebra finch song, at a precise time within the motif. These HVC neurons burst in a sequence (Hahnloser et al. 2002). In contrast, each HVC interneuron, as well as neurons in RA, fires multiple times during a song motif, in synchrony with each other and with the single burst of the RA-projecting neurons (Yu & Margoliash 1996; Hahnloser et al. 2002). The X-projecting neurons of HVC fire at rates intermediate between the RA-projecting and interneurons. These findings indicate that each HVC interneuron and each neuron in RA connected to the tracheosyringeal nuclei receives convergent input from multiple HVC RA-projecting neurons. They also suggest that HVC has many subpopulations of RA-projecting neurons, each active at different times in a song motif and each activating a different ensemble of RA neurons (Hahnloser et al. 2002). The different RA ensembles then produce patterns of activity in the muscles of the syrinx and respiratory apparatus, controlling the timing of muscle contraction and relaxation appropriate for the sounds produced. That is, RA has to fire in a manner that coordinates timing of both syringeal muscles for vocalizing and abdominal muscles for breathing (Wild 1997). This appears to be achieved by two different but interconnected sets of neurons in RA, one projecting to the tracheosyringeal motor neurons and the other to respiratory premotor neurons (Mooney et al. 2002). During breathing without singing, the respiratory neurons regulate breathing independently of RA. During singing, RA as well as DM take over (Vicario 1994; Wild et al. 1997a; Mooney et al. 2002). The electrophysiological findings further suggest that HVC interneurons make strong inhibitory contacts on the X-projecting neurons, modulating their firing during singing and thus HVC's signals to the anterior vocal pathway. However, more investigation is required to decipher the exact nature of the relationship of electrical signals from Uva to NIf to HVC to RA to the tracheosyringeal and respiratory nuclei, and onto the syringeal and respiratory muscle, as well as signals to the anterior vocal pathway.

Insight into the cellular and behavioral consequences of regulated gene expression comes from cell culture and gene blocking experiments in other species. In cultured mouse cells, the ZENK protein binds to the regulatory regions of genes involved in modulating the structure of neurons and transport of molecules inside neurons (**Box 29**, p. 236). It has also been hypothesized that IEGs like ZENK act as molecular switches that convert short-term memories into long-term ones (Goelet et al. 1986). However, songbirds that are singing well-learned songs still produce ZENK (Jarvis & Nottebohm 1997). I hypothesize that in the posterior vocal pathway, perhaps every time the bird sings, ZENK is induced to help replace proteins that get used up during the act of singing, and maintaining the song motor memories; when the bird sings 30 min later, the pathway is ready to produce song again.

The Anterior Vocal Pathway

In contrast to the posterior vocal pathway, many aspects of the anterior vocal pathway have been enigmatic since its discovery as, for example, the naming of 'Area X' (Nottebohm et al. 1976). In 1999, the neurobiology graduate students of Duke University captured the essence of this enigma with a play they wrote for the departmental retreat, called the 'Area X-Files.' Interestingly, the brain region in which Area X is located, the striatum, puzzled scientists studying humans and other mammals years before it was officially accepted at the Duke University 2002 nomenclature forum that this area in birds is homologous to the mammalian striatum (Wilson 1914; DeLong & Georgopoulos 1981; Parent & Hazrati 1995; Brown & Marsden 1998). Nevertheless, the anterior vocal pathway of songbirds has special functional properties that help us gain a general understanding of how the cerebrum works.

When nuclei of the anterior vocal pathway are bilaterally lesioned, birds of all ages still produce some form of learned vocalizations (**CD1 #66**). They can sing, but ongoing vocal learning is disrupted and they can no longer imitate new sounds. For example, when juvenile zebra finches are lesioned in lArea X during the sensitive period for vocal learning, the bird's song remains plastic and thus nothing new can be imitated or crystallized. When lesioned in lMAN, the bird's song rapidly becomes stereotyped and again nothing new can be imitated (Bottjer et al. 1984; Sohrabji et al. 1990; Scharff & Nottebohm 1991; Nordeen & Nordeen 1993). Note that most so-called Area X lesions, are actually in its lateral portion lArea X. In zebra finch adults that have mastered their song and are no longer imitating anything new, lesions to lMAN and lArea X have no apparent effect (**CD1 #66**). However, in species that still learn as adults, either by imitation or improvisation (canaries) or re-develop old songs after a non-singing seasonal lapse (white-crowned sparrows), lesions in lMAN do have effects (Nottebohm et al. 1990; Suter et al. 1990; Benton et al. 1998). In the plastic song of adult canaries, which occurs in the fall, lesions to lMAN reduce vocal plasticity, whereas during stereotyped singing, in the spring, lesions have no effect. In white-crowned sparrows before the yearly re-initiation of singing in the fall, lesions to lMAN result in stereotyped song first becoming plastic and then crystallizing to a different song; but lesions during the highly stereotyped singing season in the spring have no effect. And in zebra finches, adult song can be rendered plastic again either by partial cutting of the tracheosyringeal nerve or by deafening the animal. These manipulations lead to gradual modification and deterioration of the song, but prior lesions to lMAN prevent this experimentally induced plasticity (Williams & Mehta 1999; Brainard & Doupe 2000).

Thus, regardless of age, the anterior vocal pathway is not needed for song production, but is necessary for naturally occurring or experimentally induced song learning or modification. If the pallial structure lMAN is removed when song is plastic, song becomes stereotyped; but if song is already stereotyped, then there is no change. If the striatal structure lArea X is removed when song is plastic, it remains so; but if song is already stereotyped, then there is also no change. This suggests that during vocal learning lMAN of the pallium adds variability whereas lArea X of the striatum adds stereotypy to the vocalizations. A balance between the two enables learning to occur. For these reasons, it appeared at first that during non-imitative stages of life, such as the adult phase in zebra finches, the anterior vocal pathway is no longer used.

Then a challenge emerged. In both juveniles and adults the act of vocalizing induces ZENK gene expression in nuclei of the anterior vocal pathway of zebra finches, in all of MAN and of Area X (Jarvis & Nottebohm 1997). In fact in adults as well as young, ZENK synthesis in Area X was the highest of all vocal nuclei. Then electrophysiological studies of adults revealed action potentials milliseconds before and throughout singing in both lMAN and lArea X (Hessler & Doupe 1999a). In lAreaX, tonically active neurons were recorded and, besides increased activity, some decreased activity during singing. Neither a change in electrophysiological activity nor ZENK expression occurred when the birds simply heard another bird's song; small increases of activity occurred in half of the animals when the birds heard playbacks of their own song. In contrast, large increases of activity and ZENK expression occurred when deafened birds were actively singing, at levels not detectably different from intact birds singing. These findings suggest that electrophysiological activity and gene expression in the anterior vocal pathway are motor-driven, as in the posterior vocal pathway. This still begs the question of why lesions of the anterior vocal pathway, at least the lateral half, do not affect singing in adults with well-learned song, when the pathway is highly active during singing.

Social Context

One answer came when the vocalizing-driven

BOX 30

DIRECTED AND UNDIRECTED SONG: ZEBRA FINCHES IN CAPTIVITY AND THE FIELD

Zebra finches sing frequently in the field and in captivity. Throughout the year, at almost any time of the day when birds are stationary in trees and shrubs, one can hear the cheerful, mechanical sounds that constitute song phrases of the zebra finch. During pre-copulatory courtship, directed song is emitted by a sexually aroused male when he sings directly at the female a few centimetres away as he dances towards her. The visual and vocal components combine to form a powerful sexual signal that is often ignored, or avoided, by most females; but if conditions are right, she responds with a tail vibration display that is an invitation for the male to mount and copulate. In wild flocks, males will confront and sing to any new female that lands near them, but copulation is rarely invited except by their own mated female at a private location in the few days before egg-laying. When choosing sexual partners laboratory females prefer males with a high rate of singing and with complex song phrases (Collins et al. 1994). Undirected song, originally called solitary song, is more frequently heard. Males, often perched alone on the tops of bushes, will stare straight ahead, and sing many phrases that usually appear to be completely ignored by other zebra finches. Undirected singing is also common in resting flocks when males seem to find enough solitude for a few phrases. In captivity, visual isolation from conspecifics frequently increases bouts of undirected singing. During undirected song, males remain stationary and never make any courtship movements. The two versions of song are equally loud, and directed song is a more intense performance: faster, more notes and longer bouts, though the differences are subtle (see below; Sossinka & Böhner 1981). While directed song is clearly a sexual signal, the function of undirected song is far from clear. When a female partner was experimentally removed, a wild male immediately increased his undirected song rate and reduced it when she returned (Dunn & Zann 1996a). This suggests males are advertising their quality and unmated status via undirected song and the presence of their mate inhibits such performance. Close proximity of male companions also inhibits undirected song but the functional significance of this behavior is unknown. During breeding, undirected song is most commonly performed during nest building when the female has just entered the partly built enclosed nest and he is just outside on the way to collect more nesting material. If he does not sing there is a good chance she will leave the nest shortly afterwards and this could result in extra-pair mating during her fertile period or allow other females to dump eggs in her nest. Thus undirected song in this context appears to be a form of mate and nest guarding (Dunn & Zann 1996b).

Richard A. Zann

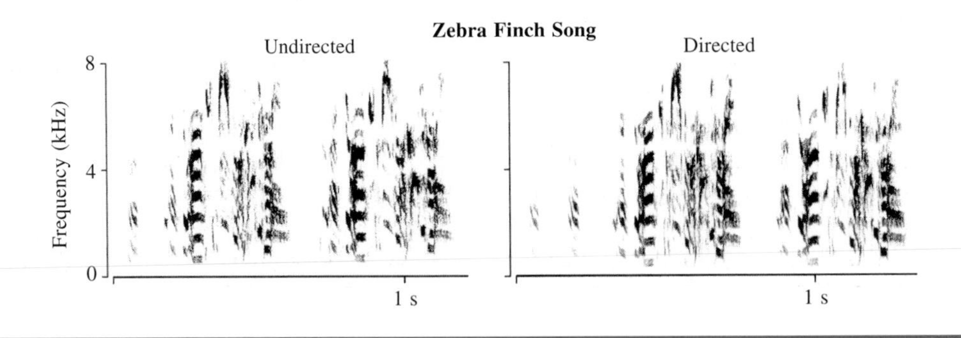

synthesis of ZENK in lMAN and lArea X, as well as the lower two-thirds of RA, was found to be dependent upon social context (Jarvis et al. 1998). In zebra finches, undirected singing, given while not facing another bird, increases ZENK synthesis in lMAN, lArea X, and RA. Directed singing, facing another bird while singing, usually a female, induces much less ZENK synthesis in these nuclei. In contrast, ZENK expression in the medial part of the anterior pathway, mMAN and mArea X, and in both RA-projecting and X-projecting neurons of HVC is similar during directed and undirected singing. Electrophysiological recordings are consistent with these results, showing that in lMAN, lArea X, and RA, electrical activity is different during undirected and directed singing, whereas in HVC it does not differ (Hessler & Doupe 1999b; Dave & Margoliash

BOX 31

ANTERIOR FOREBRAIN PATHWAY LESIONS DISRUPT SONG IN ADULT BENGALESE FINCHES

Lesioning Area X or LMAN in the brains of juvenile zebra finches had profound effects on song learning and performance, but no effect was detected in adult zebra finches (see p. 239). Nevertheless, the nuclei of the anterior forebrain pathway do not regress in adulthood. Furthermore, a part of this system is apparently active while singing, as shown by electrophysiological recordings and gene expression studies. Thus, the function of the anterior forebrain pathway in adulthood remains an enigma. We used adult Bengalese finches to re-examine the real-time involvement of the anterior forebrain pathway in song production. We selected this species because Bengalese finches are critically dependent upon real-time auditory feedback when producing the adult song (Okanoya & Yamaguchi 1997; Wooley & Rubel 1997). We reasoned that the feedback control might be mediated by the anterior forebrain pathway. When a partial lesion of Area X was made in adult Bengalese finches, a marked deficit was observed; the number of song note repetitions increased dramatically after the lesion (**CD2 #61**). Curiously, the effect occurred only in the portion of the song where the number of repetitions was naturally variable; the part of the song where the number of repetitions was fixed was not affected at all. The effect of surgery lasted up to two weeks, after which the original song was recovered, identical with that in the preoperative recordings (Kobayasi et al. 2001). We suspect that the symptom observed here might be somewhat similar to Huntington's disease in humans in that this behavior becomes difficult to stop once it has started. It seems that the anterior forebrain pathway, including the basal ganglia, may be involved in the real-time control of song production, especially regarding the temporal precision of song duration.

Kohta I. Kobayasi & Kazuo Okanoya

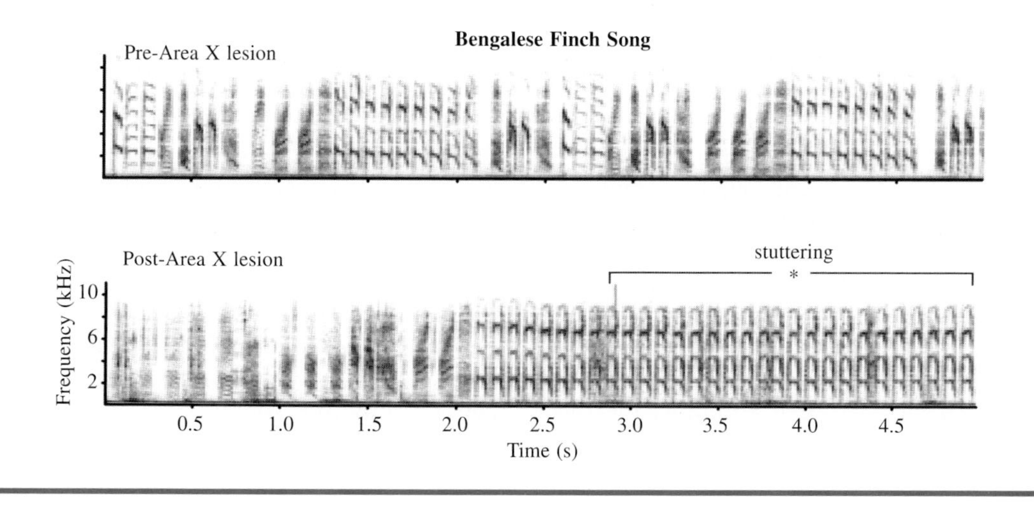

2000; Jarvis et al. 2002). During undirected singing, both lMAN and lArea X, the two regions best studied electrophysiologically, show robust firing throughout song bouts in a relatively noisy manner (Hessler & Doupe 1999b). During directed singing, much less firing occurs and in lMAN the noise drops out leaving a pattern more matched to individual song syllables. Because lMAN projects to RA and mMAN projects to HVC, and because there is a social context difference in RA and not in HVC, I propose that the lateral part of the anterior vocal pathway regulates RA and the medial part regulates HVC.

The comparison of activity during directed and undirected singing has given rise to various alternative hypotheses about the functional role of the anterior vocal pathway in adults. One hypothesis is that, besides learning, the anterior vocal pathway is also used to produce small moment-to-moment differences in singing output to modulate the meaning of vocalizations for listening birds (Jarvis et al. 1998). In zebra finches, there are small differences between directed and undirected singing (**Box 30**, p. 240), including the presence of more introductory notes, slightly faster motifs, and longer song bouts in directed song (Sossinka & Bohner 1980). Another hypothesis is that the anterior vocal pathway is somehow connected with the attention that the bird gives to its surroundings during directed versus undirected singing (Hessler & Doupe 1999b). Still another suggests that undirected song is simply practice, and that use of the anterior vocal pathway and the production of ZENK maintains the pathway's health and sustains motor memories (Jarvis & Nottebohm 1997; Jarvis et al. 1998). During directed singing, stimulating the female, as the object of desire, is presumably more important than song practice. A proposed role for auditory feedback will be discussed later. However, if there is a maintenance function, lesions to the pathway should result in slow deterioration of song. This has not been found in adult zebra finches. An effect has been seen in well-learned song of a close relative, the Bengalese finch (Kobayasi et al. 2001). Lesions to a large portion of Bengalese finch Area X resulted in temporary effects on song syntax. The birds stuttered when producing syllables that were often repeated in normal song (**Box 31**, p. 241), suggesting a role for the anterior vocal pathway in the generation of some adult syntax, but the role with regard to social context is not clear. There were no differential effects on directed versus undirected singing.

Interestingly, lesions to the medial part of the anterior vocal pathway, the part that is always highly active during both directed and undirected singing, results in an immediate effect on syntax in adult zebra finches (Foster & Bottjer 2001). Lesions to adult mMAN caused increased syntax variability, but had very little effect on song syllable structure. Lesions to mMAN in young birds did not appear to affect their early plastic song, but the birds could not crystallize a stereotyped syntax, though they still had relatively stereotyped syllable structure. It should be noted that these nuclei are very tightly packed together, and a number of the lesions encompassed mMo above and/or mArea X below mMAN. Generally, the medial part of the anterior pathway appears to be involved in syntax learning and production, with mMAN possibly influencing HVC to help form stereotyped syntax. Taken together with the effects of lesions in the posterior pathway suggests that removal of any of the vocal nuclei inputs to HVC, such as NIf, Uva, and mMAN, results in the inability of HVC to produce normal stereotyped syntax (Foster & Bottjer 2001).

Despite the progress, the basic function of the anterior vocal pathway remains rather elusive. It surely plays a role in vocal learning, and it is very active during singing at all stages of life. But beyond that, the situation is less clear, as is also true of the anterior forebrain-basal ganglia pathway of mammals (Wilson 1914; DeLong & Georgopoulos 1981; Parent & Hazrati 1995; Brown & Marsden 1998). Deciphering the basic functions of the anterior vocal pathway may yield new insights into cerebral functioning in general. But to understand vocal learning, we must also consider how the anterior vocal pathway and the song system in general gain access to auditory information.

THE BRAIN AND AUDITION

Functions of the Auditory Pathway

For some years after discovery of the songbird vocal control system, the search in the brain for the locus of auditory song processing focused on the vocal nuclei. There was a conviction that the place to look was where auditory and motor information meet. In the first electrophysiological studies of HVC, action potentials were detected in response to sound, especially to playbacks of the bird's own song (Katz & Gurney 1981; McCasland & Konishi 1981), thus identifying the vocal nuclei as possible auditory processing stations. It came as a surprise when molecular biologists played songs to zebra finches and canaries and found induced ZENK synthesis, not in the vocal nuclei, but in another set of cerebral brain areas, including NCM (Mello et al. 1992; **Fig. 8.4A**). This finding set NCM on the map and eventually six other cerebral auditory responsive areas as a network (**Fig. 8.3**) important in the processing of song as stimuli (L3, L1, HVC shelf, CMM, RA cup, and Cst; Mello & Clayton 1994). The electrophysiology of some of these auditory areas had been studied earlier (Leppelsack 1978), mostly in the context of non-song sounds (Müller & Leppelsack 1985; Müller & Scheich 1985), but their potential relevance to song processing and learning was not appreciated at the time.

We now know that when male or female songbirds hear songs of their own species, high rates of action potential firing and strong induction of ZENK synthesis occur in these seven auditory cerebral areas. When they hear songs of other species, less ZENK is synthesized in these areas (**Fig. 8.4A**). When they hear short duration pure tones, ZENK is not induced, and electrical activity is actually inhibited below baseline (Mello & Clayton 1994; Chew et al. 1995; Jarvis & Nottebohm 1997; Stripling et al. 1997; Jarvis et al. 2002). Thus, electrophysiological and molecular responses of the songbird auditory cerebrum are highly sensitive and species-specific. One area in the pallium, L2, and one in the thalamus, Ov, are known to be auditory as judged by their sound-induced electrical activity and connectivity, but ZENK is not expressed in them. In the midbrain auditory region, MLd, ZENK synthesis does occur in response to hearing song.

The species-specific molecular response in

Gene Induction in the Auditory Pathway (NCM)

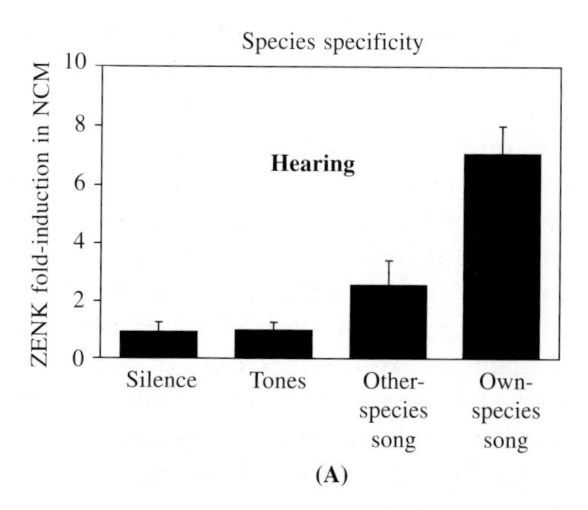

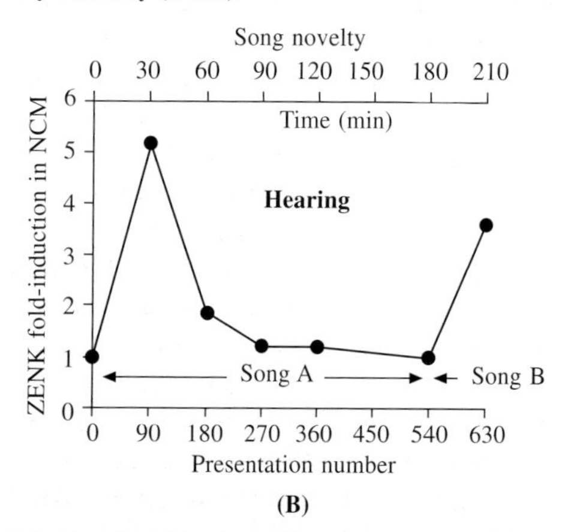

Figure 8.4 Hearing-induced gene expression in the zebra finch brain. **(A)** Accumulated ZENK mRNA in NCM in the auditory pathway, relative to silence, when zebra finches hear various stimuli for 30 min; the other species was canary. From Mello et al. 1992. **(B)** Accumulated ZENK mRNA in NCM when birds heard first a familiar song (song A), then a novel one (song B). Modified from Mello et al. 1995.

NCM is learned while the birds are growing up. When young zebra finches have been raised in social isolation, playback of their own-species songs does not induce ZENK synthesis in NCM (Jin & Clayton 1997). When zebra finches are raised by canaries, as adults, the most potent stimulus at activating ZENK synthesis in the auditory cerebrum is not zebra finch song but canary song (Ribeiro 2000). Thus, what a bird hears as a juvenile can influence its brain molecular responses as an adult.

Electrical activity and ZENK synthesis in these cerebral areas is also influenced by song familiarity. When males or females hear playbacks of novel zebra finch songs, ZENK synthesis is strong in the cerebral auditory regions, but is reduced when they become familiar with songs after repetitions, as if they have a molecular memory of recently heard songs (Mello et al. 1995; **Fig. 8.4B**). Novel songs also induce electrical activity, with an initial burst of high firing rates in NCM neurons that also decreased as the songs were repeated. When hearing a familiar song however, the firing rate never decreased to silent baseline levels. If the now familiar song is presented several hours or a day or two later, the firing rates start where it left off (Chew et al. 1995; Stripling et al. 1997. This is a form of neuronal memory called long-term habituation. It appears that auditory memories of song are being stored in NCM and other connected cerebral auditory areas.

Lesion studies also show that long-term song memories can involve NCM. Ikebuchi & Okanoya (2000) trained Bengalese finches to discriminate between songs and measured heart rates when the birds listened to novel and familiar songs (Ikebuchi et al. 2003). Female heart rates increased when they heard novel songs. As they became familiar with the song, their heart rates decreased to a steady level. When NCM was lesioned, the females' heart rates no longer increased on hearing novel songs, suggesting that NCM is required for discrimination of novel from familiar songs. They could learn a behavioral discrimination task in which birds had to discriminate between two songs heard, but could not remember later what they had learned. It appears that NCM is not required for the short-term auditory memory of song, but is needed for long-term memory of them. These memory deficits of NCM lesions are somewhat reminiscent of the symptoms of Alzheimer's disease in humans.

ZENK and other genes may be involved in the formation of long-term memories. When general inhibitors of mRNA or protein synthesis are injected into NCM at the time of novel song playback, they do not affect the short-term habituation of neuronal activity that occurs with repeated stimulation. However, when the songs are played later, after 3 hours, the electrical activity rate is as high as with a novel song (Chew et al. 1995). Inhibiting mRNA and protein synthesis specifically at 6 hours after hearing novel songs also prevents further long-term maintenance of the habituated memory of songs. Since ZENK and other IEGs are expressed quickly, within the first hour that birds hear the novel songs, and because blocking synthesis of gene products at this time affects long-term memory, it appears that these genes may act as molecular switches that convert short-term memories into long-term memories (Goelet et al. 1986). The effect at later times suggests that there are several waves of gene expression involved in the process.

The electrophysiological mechanisms by which the auditory pathway processes and learns species-specific sounds are only just beginning to be understood. As information is conveyed from the midbrain to the auditory regions of the cerebrum, each station shows more complexity in its response to sounds, including responsiveness to species-specific songs (Chew et al. 1995, 1996). In the cerebrum, field L2 responds first, followed by L1, L3, CMM, and NCM, in the order in which they are connected. L2 responses are evoked by many types of sounds, in a linear fashion, and in a tonotopic manner, with low-frequency neurons at the top of L2 and high frequency at the bottom (Capsius & Leppelsack 1999; Sen et al. 2000). At each subsequent station, the neurons then respond less linearly, registering specific features such as frequency modulation, syllable combinations, and down-sweeps and up-sweeps. These more sophisticated responses are also evident in the ZENK gene expression response, with different species-specific syllable types being processed in particular parts of NCM (Ribeiro et al. 1998). The perception of song probably involves all of these anatomical regions, with those furthest removed from L2, in the Nidopallium (L1, L3, NCM), Mesopallium (CMM) and Caudal Striatum (Cst), possibly serving as the prime locations of the song 'percept'.

Although we now know that NCM and other areas are auditory centers for processing songs, this does not explain the lack of induced ZENK expression and other IEGs in the vocal control nuclei after hearing songs. This lack of increased expression is elucidated by the later discovery that there was less hearing-induced electrical activity in vocal nuclei when the animals were awake, than when they were anesthetized or sleeping (Dave et al. 1998; Schmidt & Konishi 1998; Hessler & Doupe 1999a; Nick & Konishi 2001). Most earlier electrophysiological work was conducted on anaesthetized subjects, whereas the gene expression studies were done with awake animals. This new paradox, awake versus sleeping-induced hearing activity, still left a major question unanswered: where do the functional interactions between the auditory and vocal pathways take place?

Auditory Feedback and the Template

As is intuitively obvious, vocal learning, and vocal imitation in particular, requires that a bird hear the tutor's song that he will imitate. In addition, Konishi (1965a) found that a songbird also needs to hear himself practice that song in order to imitate it accurately. If a songbird is allowed to listen and form an auditory memory of the tutor's song, but then is deafened before he physically practices that song, the bird will not learn to accurately produce what he had earlier heard. From these behavioral experiments arose the idea of the song template (Marler & Tamura 1964; Konishi 1965a; **Fig. 8.5**). This hypothesis proposed that during song learning, a young bird first forms an auditory memory, a template somewhere in the brain, based on the tutor's songs that he hears. Sometime thereafter, days to almost a year depending upon the species, the bird begins to practice singing and by listening to his own song, he will try to match his produced vocal output with the auditory template in his brain. Since the formulation of this hypothesis, the search for the auditory template and the site for auditory feedback in the songbird brain has been a major area of research. The search has lead to findings that are both interesting and elusive.

As already mentioned, the exact source of auditory input into the vocal pathways is unresolved, even though some auditory cell populations, L2, HVC shelf, and RA cup, are directly adjacent to vocal nuclei cell populations, NIf, HVC, and RA. However, the search for sites of auditory–vocal integration, or sensorimotor integration more generally, was not limited to connectivity studies. It was thought that HVC, showing electrical activity both while vocalizing and when the bird hears his own song played to him, should be a good candidate site for such integration and the song template (McCasland & Konishi 1981; McCasland 1987).

But soon thereafter auditory responses were found throughout all known cerebral vocal nuclei (HVC, RA, lArea X, lMAN) as well as in the tracheosyringeal motor neurons (Williams & Nottebohm 1985; Williams 1989; Doupe & Konishi 1991). Depending upon the anesthetic used, auditory responses were induced not only by playbacks of the bird's own song, but also by other birds' songs and non-biological sounds, such as tones (Katz & Gurney 1981; Williams & Nottebohm 1985), leading Williams and Nottebohm (1985) to propose the Motor Theory of Song Perception, as a counterpart to the Motor Theory of Speech Perception for humans (Liberman et al. 1967; Liberman & Mattingly 1985). They hypothesized that the same brain areas used to produce sounds (vocal–motor areas) were also used to perceive them (hearing–sensory

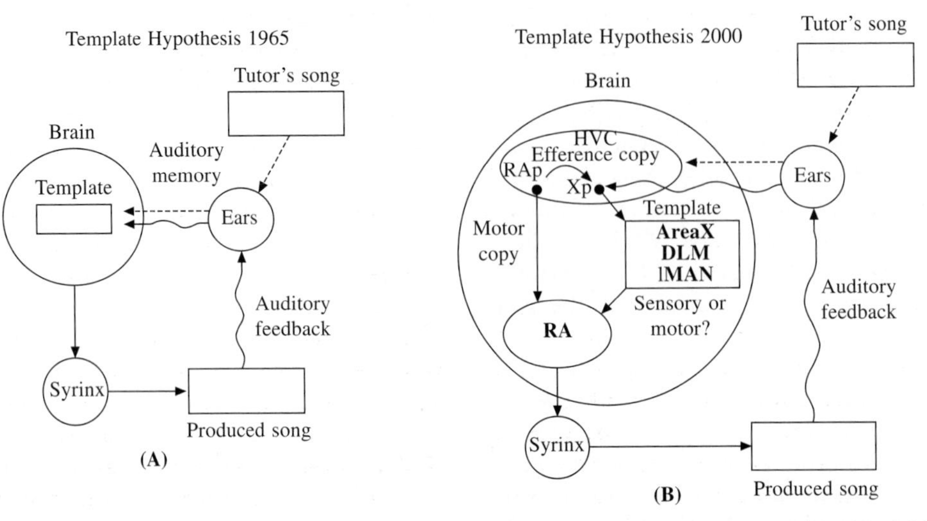

Figure 8.5 Template models of song learning. **(A)** Template hypothesis as proposed by Konishi 1965a. Tutor's song is heard and stored as an auditory template. The bird tries to produce an imitated song, hears himself sing and compares that with the template. **(B)** A template hypothesis proposed by Troyer & Doupe 2000. First, the anterior pathway receives an efference copy of the song from the X-projecting neurons of HVC, coming in turn from RA-projecting neurons of HVC. Second, auditory feedback is sent to the HVC's X-projecting neurons to the anterior pathway and compared there with the efference copy.

areas). This theory was part of a larger debate begun even before Broca's time, between the 'diffusionists' and the 'locationalists.' Diffusionists believed that brain functions were spread out and that the same area is used for more than one function. Locationalists believed that specific regions of the brain served specific functions, as propounded by Gall & Spurzheim (1810–1819; Benson & Ardila 1996). For songbirds, the firing of neurons throughout the vocal control system in response to any sound brought the debate closer to the diffusionist's model. This finding also indicated that the vocal pathways could process more than just the birds' own song, throwing some doubt on whether vocal pathways store only a song template for auditory feedback comparison or simply process sounds for auditory perception in general.

The next candidate was the anterior vocal pathway (Doupe & Konishi 1991), which at the time did not seem to do anything in adult animals after learning was complete. It was proposed that when a juvenile zebra finch forms an auditory memory of a tutor's song that the auditory template is stored in the anterior vocal pathway. The pathway would then act as a coincidence detector between the song produced and the song heard. Specifically, it was proposed that during juvenile practice, when vocal motor activity originating in HVC generates song, the bird would hear itself sing, and this information would be transferred from the auditory pathway to HVC. HVC in turn would send this auditory feedback signal into the anterior vocal pathway via its X-projecting neurons. The anterior pathway would then compare the produced feedback song with the expected song as represented by the template. With a mismatch the anterior pathway would then correct that discrepancy by changing the activity of RA neurons via lMAN (Williams 1989; Doupe & Solis 1997; **Fig. 8.5**).

A major feature of this hypothesis is the presumption of responsiveness of the anterior vocal pathway to auditory stimuli, particularly of the bird's own song. Three kinds of results began to change this view. First ZENK synthesis in the anterior vocal pathway, including MAN and Area X, was found to be activated not by hearing playbacks of the bird's own song, but by singing (Jarvis & Nottebohm 1997). Moreover, when birds were deafened, singing still drove increases in ZENK synthesis in the anterior and posterior vocal pathways to the same levels found when they could hear. Second, like the posterior vocal pathway, auditory activity in the anterior vocal pathway was found to be shut off or highly diminished when zebra finches were awake (Margoliash 1997; Hessler & Doupe 1999a). In addition, like the posterior vocal pathway, the singing electrical activity in lMAN and lArea X was premotor, firing before song output, and occurring whether the birds could hear or were deaf (Hessler & Doupe 1999b). Thus, in the anterior vocal pathway there is motor electrophysiological activity and gene expression with no requirement for auditory feedback. Third, were the discoveries of closure of the anterior forebrain loop via the lMAN to lArea X connection (Nixdorf-Bergweiler et al. 1995; Vates & Nottebohm 1995; Luo et al. 2001), and medial and lateral parallel connections of the anterior pathway, one with an output to RA and the other to HVC (Foster et al. 1997; Jarvis et al. 1998). This suggested the more global view, that the anterior vocal pathway serves, not as an indirect route of transferring information from HVC to RA, but as a closed-loop processing station receiving information from HVC and sending that processed information or instructions to HVC and RA. Debate continues about whether there is low auditory activity or none at all in the anterior and posterior vocal pathway nuclei. However, there may be species differences; preliminary data reveal robust hearing-induced activity in HVC of awake song sparrows (Nealen & Schmidt 2002).

Striving to reconcile these contradictions, Troyer & Doupe (2000) proposed a theoretical model, showing how the anterior vocal pathway may still function in auditory feedback with a stored song template. They argue that whereas a motor song signal is transported from HVC to RA, via HVC's RA-projecting neurons, at the same time an 'internal sensory efference copy' of song is transported in parallel to the anterior vocal pathway from HVC to Area X, via HVC's X-projecting neurons (**Fig. 8.5B**). The term 'efference' refers to activity directed away from a central location in the brain, in this case HVC. The two signals, one motor, the other an internal sensory efference copy, would arrive in RA and be compared there, but auditory feedback from the bird's own song would be compared in HVC's X-projecting neurons. Problems confronting this model are the requirement that the signals from HVC to Area X should be sensory, when there is no known sensory input for the efference copy in this model, and the fact that in anesthetized birds all HVC neuron types, X-projecting, RA-projecting, and interneurons, are equally responsive to auditory activity (Mooney 2000). In addition, in awake birds, all HVC neuron types show motor-driven ZENK expression and premotor electrical activity during singing (Jarvis & Nottebohm 1997; Hahnloser et al. 2002). However, it is possible that the X-projecting neurons send a 'motor efference copy' of song (Margoliash 1997; Mooney et al. 2002).

If the auditory responses in vocal nuclei occur mainly when the birds are anesthetized, what is their role in real life? Interesting answers were also found in sleeping birds (Dave & Margoliash 2000). When sleeping zebra finches hear playbacks of their own song, the hearing-induced electrical responses in HVC and RA suddenly appear, as if the birds were anesthetized. When the birds wake up, the hearing-induced responses cease. In HVC, these auditory responses were most prominent when the birds heard song playbacks during deep, slow-wave sleep (Nick & Konishi 2001). When sleeping zebra finches are not hearing playbacks of song, HVC and RA sometimes display spontaneous complex patterns of electrical activity that match the singing activity which the bird produced earlier that day (Dave et al. 1998; Hahnloser et al. 2002), as though the birds' vocal motor pathway is replaying the singing activity produced during the day, but without the bird actually singing or hearing song. Replay during sleep might stabilize motor memories of a bird's songs so that it can sing them again later (Dave et al. 1998), or the bird may simply dream about singing (**Box 32**, p. 249).

The best evidence for a role of the anterior vocal pathway in some form of auditory feedback is that lesioning of lMAN prevents the song deterioration resulting from tracheosyringeal nerve cutting and deafening (Williams & Mehta 1999; Brainard & Doupe 2000). Given the limited auditory responses of awake animals, this role might only be permissive, with actual template instructions coming from elsewhere, possibly from the auditory pathway. The species-specific molecular responses of ZENK and electrophysiological activity in NCM and related areas have led to a focus on them as a potential location of the template. When zebra finches raised alone with a tutor became adults, induction of ZENK synthesis was highest when the birds heard song of their tutors (Bolhuis et al. 2000). Moreover, the closer a tutor was matched, the higher the levels of ZENK synthesis in NCM. This response is counterintuitive, given that familiar songs of non-tutor birds induce less ZENK expression (Mello et al. 1995; **Fig. 8.4B**). It is possible that hearing tutor songs in novel social contexts may induce high levels, but this has yet to be tested. Nevertheless, responsiveness of NCM and related areas is clearly selective, especially to sounds that resemble the birds' own song, so that they could be involved in auditory feedback.

Another finding points in this direction. When songbirds hear themselves sing, ZENK is induced in the bird's NCM and other auditory areas in a much more restricted pattern than when they hear other birds sing, or their own song from a tape recorder (Jarvis et al. 1998 unpublished). Apparently awake birds identify playbacks of their own song as a novel song from some other bird, sounding similar to their own. Thus, tests of the auditory feedback and template hypotheses using tape recorded playbacks of the birds' own song may be deficient because this method does not mimic the state of the animal while actually singing himself.

In summary, the search for mechanisms underlying auditory feedback and the hypothesized template continues, and has the potential to yield exciting findings. It is not yet known if the interactions between auditory–sensory information and vocal–motor information occur in the vocal pathway, the anterior vocal pathway loop, the auditory pathway or somewhere not yet studied. Wherever they take place, the sensorimotor interactions might involve sub-threshold electrical responses, or some non-electrical mechanism. It looks as though the answers will be found by studying animals while they hear themselves vocalize, rather than while hearing song playbacks, especially perhaps during sensitive periods for vocal learning when birds are actively using auditory feedback to generate imitated songs.

SENSITIVE PERIODS

'Critical' or 'sensitive' learning periods are stages in life when there is a limited window of heightened ability to learn new information. They occur in all animals, and are prominent in human development (Rauschecker & Marler 1987). Songbirds have become a premier model for understanding the neuroethology of sensitive periods, and similarities with those for language learning in humans (Doupe & Kuhl 1999) make songbirds especially interesting. Vocal learning in both songbirds and humans has four phases. In songbirds, the first is called the auditory acquisition phase in which they form auditory memories of the songs that they hear; the second is a babbling-like subsong phase in which they engage in vocal motor practice, but do not produce imitations; in the third, plastic song phase, they practice imitating sounds with increasing precision; fourth is a crystallization phase in which they go through a period that is puberty-like, the voice becomes stabilized and well learnt, and the birds become adults and ready to breed. The timing of these phases varies in different species and they sometimes overlap (see Chapter 3).

BOX 32

DO BIRDS SING IN THEIR SLEEP?

Sleep, as defined by specific postures, elevated sensory thresholds, and brain states, appears to be ubiquitous in birds and mammals, yet its functions are poorly understood. It has been studied extensively in mammals but only recently has the phenomenon of sleep in songbirds begun to be addressed. All birds show REM and non-REM periods of sleep (Rattenborg et al. 2002), but the pattern and frequency may be different than in mammals. Small birds have short periods of REM sleep (circa 4 s.; Szymczak et al. 1993). This may reflect the trend relating the period of sleep cycles to animal size but, in general, birds may have less REM sleep than do mammals (Siegel 1995). Recent observations have given the study of birdsong an unexpected focus in sleep research. We have known for some twenty years that some neurons in the song system of the brain respond best to playback of song, especially the bird's own song; most detailed studies were conducted on anesthetized white-crowned sparrows and zebra finches (Margoliash 1983; Margoliash & Fortune 1992; Theunissen & Doupe 1998). The situation in unanesthetized birds is more complicated. In awake zebra finches, neurons in parts of the song system, such as RA, respond more weakly than expected to song playback, or hardly at all. The same RA neurons respond vigorously to song playback in sleeping birds (Dave et al. 1998). Whereas sensory thresholds in general are elevated during sleep, there are clearly exceptions to this rule. The implications of this phenomenon for basic and perhaps even clinical research have yet to be fully appreciated. In birds, one hypothesis is that suppression of auditory activity in awake animals in the motor pathways, including RA, is the expression of an already established, well-controlled adult song, and that auditorily-driven activity in RA during singing represents an error signal. The expression of auditory activity during sleep may reflect a singing-like state the song system enters, enabling it to access the hypothesized error signals that guide adaptive changes to re-stabilize song during sleep.

The initial RA recordings during sleep were from clusters of neurons, but eventually techniques were developed to record from single cells in RA while birds sang, and to re-record them later during undisturbed sleep. The result was a remarkable finding (see below). In virtually every cell encountered, there was bursting activity during undisturbed sleep that matched the same bursting patterns observed for that cell when the bird was singing while awake (Dave & Margoliash 2000). Analysis of electrical brain activity during sleep is challenging because there is no immediate, reliable time referent linked to behavior, calling for some sophisticated statistical analysis. Nevertheless, obvious matches were found between the sleeping and waking states (see below), as for example where an RA neuron emitted a long train of 30 or more spikes organized into multiple bursts, each of which matched the sequencing of the activity patterns of that same neuron during singing. Are birds singing in their sleep? The occurrence of coordinated bursting in song system neurons suggests that song is being replayed in the sleeping brain. A similar phenomenon has been reported in the hippocampal spatial memory system of sleeping rats after they have been exploring a maze (Wilson & McNaughton 1994). Replay concepts are at the heart of some theories postulating that memory consolidation takes place during sleep, but a role for sleep in the acquisition or maintenance of birdsong is still speculative. Do birds dream of singing? Owners of pet budgerigars and other birds will not need much convincing, but the truth is that we still have no objective access to the mental imagery that birds may experience during sleep. In some cases RA neurons give trains of bursts during sleep representing whole syllable sequences or motifs, and apparently the population of RA neurons bursts synchronously during sleep. If we could learn how to release brainstem inhibition of the bird's vocal system, we might find that sleeping birds would actually burst into song.

Daniel Margoliash

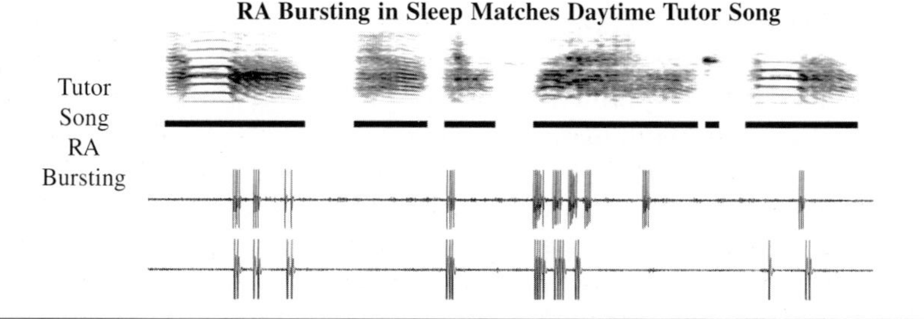

Critical periods in vocal learning behavior are controlled by hormones and the brain. Songbirds hatch from the egg presumably with cerebral vocal nuclei already present, but they cannot yet produce learned vocalizations. Day 10 after hatching is the earliest that the vocal nuclei have been sought and found (Nixdorf-Bergweiler 2001). At this age, they produce the innate begging calls that induce feeding by the parents (see Chapter 5). As the hatchling bird grows so does its brain. The cerebral vocal nuclei grow at different rates (Nixdorf-Bergweiler 2001): in zebra finches, lMAN size peaks before the subsong phase and then decreases; Area X, HVC, and RA sizes peak right after the subsong phase and are maintained to adulthood.

Not all connectivity is in place after hatching. The connectivity of the anterior vocal pathway is present, but that for the posterior vocal pathway is not yet completed (Mooney & Rao 1994). During the auditory acquisition phase, which overlaps with the subsong phase in the zebra finch and several sparrows, neurons from HVC to RA wait outside of the RA nucleus. As the bird begins to produce subsong, these HVC neurons grow into RA and find connections there. At this time, the lMAN axons in RA are spread throughout RA. Later as the bird begins to produce plastic song, for the first time practising imitations in earnest, the axonal spread is pruned back until the appropriate connections are made (Herrmann & Arnold 1991). During this pruning process, electrical activity in the anterior vocal pathway, from lMAN to RA, also undergoes changes. Before subsong, the speed of electrical signaling between the two regions is fast. During and after subsong, the speed decreases (Livingston & Mooney 1997; Mooney 1999). The growth of connections from HVC into RA is thought to prepare the posterior vocal pathway for producing sounds. The pruning of connections from lMAN to RA and the decrease in signaling speed are thought to aid in providing RA with the necessary information from the anterior vocal pathway on how to sing the imitated song.

Changes in gene expression also occur during sensitive periods. In young animals, before the subsong phase, the amount of ZENK synthesis in the auditory regions, in vocal nuclei, and in other brain areas is high, without the need for hearing song or for singing (Jin & Clayton 1997; Jarvis et al. 1998; Whitney et al. 2000; Stripling et al. 2001). During the subsong phase, the basal levels start to decrease, and ZENK can then be induced by hearing song, in the auditory pathway, and by singing subsong, in the vocal pathway (Jarvis & Nottebohm 1997; Jin & Clayton 1997). By adulthood, the basal levels are dramatically reduced and the induction of ZENK by singing and hearing is further amplified. The higher levels of basal ZENK in juveniles are thought to reflect their higher brain plasticity compared with adults. During these successive developmental stages, a host of other genes undergo changes in expression in the vocal nuclei. These include genes for synaptic transmission, affecting the glutamate receptors that are responsible for changing the speed of communication between neurons, and genes for generating connectivity and hormone receptors (Gahr & Kosar 1996; Soha et al. 1996; Singh et al. 2000). It is not clear how these changes in gene regulation influence sensitive periods, but some insight has been gained by examining species and sex differences.

SPECIES AND SEXUAL DIFFERENCES

There is just one vocal learning primate, humans. The vocal learning suborder of oscine songbirds includes over 4000 species (Nottebohm 1972; Sibley & Ahlquist 1990), with a wide spectrum of variation in vocal learning behavior, from closed-ended vocal learners, such as zebra finches, to highly plastic opened-ended vocal learners, such as canaries (Nottebohm et al. 1990; Catchpole & Slater 1995; see Chapter 3). As juveniles, most species go through similar phases of song learning leading up to crystallized adult song. However some, such as the zebra finch crystallize one song as an adult, and others, such as canaries, go through seasonal phases of plastic song, and learn new song themes as adults; yet others can learn new songs any time of the year. Like humans, however, most open-ended vocal learners still learn song most easily as juveniles. Other variations include differences in the ability to imitate other species' sounds, the ability to improvise, and to increase repertoire size, and syntactical complexity (see Chapter 4). We can assume that there is something significant about the mechanisms operating in songbird brains that make this great variation in vocal learning behavior possible.

Species Differences

Could there be different sets of vocal nuclei from one species to another? Nearly a hundred songbird species have now been examined, and they are remarkably consistent. All have the four large cerebral vocal nuclei, HVC, RA, MAN, and Area X (Brenowitz 1991; DeVoogd et al. 1993; Brenowitz 1997). When carefully examined, many also have the smaller cerebral vocal nuclei NIf, Av, and Mo (personal observations). Thus, the behavioral variability is not explained by the presence or absence of certain vocal nuclei.

A second possibility is differences in anatomical connectivity. This is more challenging; it takes years of study to determine connectivity in one species. Nevertheless, connectivity between zebra finches and canaries has been compared in detail, and although comparisons are not complete, no major differences have been found (Vates & Nottebohm 1995; Vates et al. 1997).

The third source considered was variation in the size of vocal nuclei (Nottebohm et al. 1981; DeVoogd et al. 1993; MacDougall-Shackleton et al. 1998). Size differences between species and between individuals of the same species do occur, but the data are contradictory and controversial. Some reports show that both across and within species, those with large vocal repertoires, measured as the number of songs a bird has, or the number of different syllables or phrases it employs, have relatively larger HVCs (Nottebohm et al. 1981; DeVoogd et al. 1993; Airey & DeVoogd 2000; Airey et al. 2000; **Fig. 8.6**). In some species with seasonal learning periods, HVC size increases during learning periods (Nottebohm et al. 1986). However, others have failed to replicate the finding that repertoire

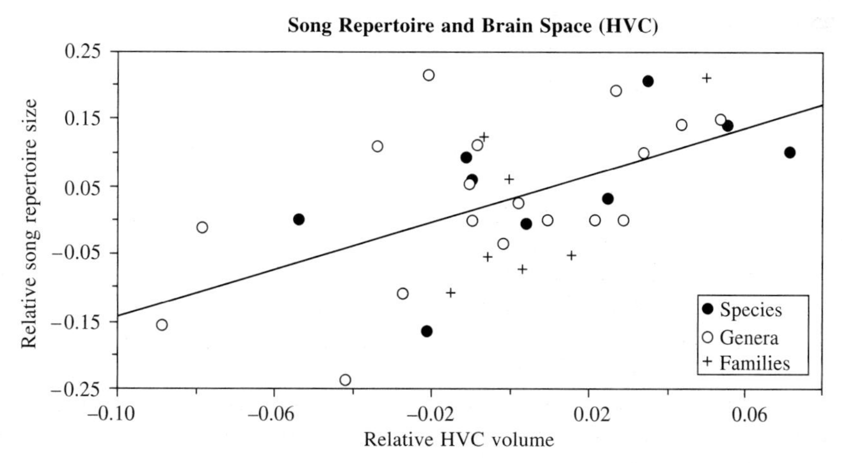

Figure 8.6 Song repertoires and brain space. X-axis values were calculated as relative HVC/cerebrum volume of one species minus relative HVC/cerebrum volume of another species (●), average HVC of volume of one genus minus that of another (O), or HVC volume of one family minus HVC of another (+). The same was done for y-axis values, calculating song repertoire size differences between species, genera, or families; males of ~40 species were used for these calculations. There is a positive correlation at all three levels of analysis. Modified from DeVoogd et al. 1993.

size is related to vocal nuclei size, at least within a species (MacDougall-Shackleton et al. 1998; Ward et al. 1998). The interpretation of these contradictory results may be complicated by the finding that the amount of singing a bird performs influences the size of the vocal nuclei and the number of cells in them (Li et al. 2000; Sartor & Ball 2001; Alvarez-Borda & Nottebohm 2002). Thus, the changes seen in seasonally singing birds may occur because they sing more during learning periods. Differences between species or individuals of a species may reflect the amount of song a particular species or individual is prone to produce. As with muscles of the body, the more the bird sings, the greater the potential increase in size of the cerebral vocal nuclei, as discussed later. Yet there is some supportive evidence that the relative size of the vocal nuclei is important; zebra finches have a larger lArea X relative to lMAN than do canaries (unpublished observations), a difference that is consistent with the hypothesis that lArea X endows a bird's song with stereotypy and lMAN endows it with plasticity.

A fourth compelling source of species differences is genetic, especially the differential expression of genes for synaptic transmission. Synaptic transmission involves neurotransmitters released at synapses, and glutamate is the most abundant neurotransmitter in the brain. The 'pre-synaptic neuron' releases the glutamate from its axons and the 'post-synaptic neuron' receives the glutamate on receptors in its dendrites. Each type of neuron has one or more of the 24 different kinds of glutamate receptors. The receptors transmit information about the electrical signals between pre- and post-synaptic neurons, to influence in different ways the expression of genes such as ZENK, in post-synaptic neurons (**Box 29**, p. 236). Compared to the rest of the brain, the patterns of expression of glutamate receptors are unusual in vocal control nuclei, apparently either higher or lower than the surrounding brain subdivision to which they belong (Wada et al. 2004). Furthermore, each species examined had its own unique pattern of expression in the vocal nuclei (Wada & Jarvis unpublished). Specializations of mGluR2, NR2A, and NR2B glutamate receptor distribution were strongly correlated with the type of syntactical complexity displayed in the song of each species. Lower mGluR2 and NR2A, and higher NR2B were related to higher syntax variability. Syntactical complexity and vocal learning style are often related, with open-ended vocal learners having more variable syntax than closed-ended species (see Chapter 4). It may be that one source for some of the variability in vocal learning is to be found in differences in the distribution of ancient gene families.

Sex Differences

There was once a widespread belief that, except for breasts, genitalia, and overall body size, men and women were basically the same, particularly their brains. It came as a shock to many that some male songbirds have large cerebral vocal nuclei and females have smaller ones, or even none at all (Nottebohm & Arnold 1976; Arnold & Mathews 1988). Although small sex differences had been detected earlier in other brain areas, this first demonstration of significant gender differences, seen even with the naked eye, in the brain of an evolutionarily advanced vertebrate, challenged the notion that there are no significant differences between males and females (Raisman & Field 1971; Ball & Macdougall-Shackleton 2001). Soon thereafter, scientists began to find significant sexual differences in select brain areas of other birds, and of mammals, particularly in parts of the hypothalamus that control reproductive functions (Ball & Macdougall-Shackleton 2001), and cerebral areas that control language (Harasty et al. 1997; Cooke et al. 1998).

In those songbird species in which the females have few or no detectable vocal nuclei, such as zebra finches, the females do not engage in vocal learning. Female zebra finches are born with cerebral vocal nuclei, but as they mature their vocal nuclei atrophy. As adults their cerebral vocal nuclei are hardly noticeable (Nixdorf-Bergweiler 2001). One of them, lMAN, shrinks less dramatically than others and in adult females of some species, destroying it hampers the perceptual processing of the male's song (Hamilton et al. 1997; Burt et al. 2000). In canaries, females have vocal nuclei, but they are smaller than those in males, and females do not learn as much song or sing as much as males (Nottebohm & Arnold 1976). In other species, such as bay wrens, vocal nuclei are comparable in size in males and females, and they counter-sing with each other (Arnold et al. 1986; Ball & Macdougall-Shackleton 2001; Balthazart & Adkins-Regan 2002).

These gender differences may be related to environmental constraints. Most songbird studies have focused on birds living in temperate climates, especially in North America and Europe, where there are more species in which only males sing (Morton 1996a), leading some to assume that the natural order of vocal behavior in birds is that only males learn vocalizations and sing. However, many more species live in the tropics and often both males and females produce learned song (Kroodsma et al. 1996). Morton (1996a) proposed that the changing seasons and harsher winters could have led to selection for more division of labor, with males singing and females specializing more in selecting mates and tending to the young. This temperate–tropical contrast is apparent even when comparing different populations of closely related species: At one extreme, the Carolina wren lives in a temperate zone climate, and females have no detectable cerebral vocal nuclei and do not sing; at the other extreme are its tropical cousins, the bay wren and buff-breasted wren, whose females have vocal nuclei and song repertoires equal in size to those of males (Arnold et al. 1986; Brenowitz & Arnold 1986; Morton 1996a; Nealen & Perkel 2000). Thus, the environment may in a sense have had an influence on the evolution of gender differences in brain and behavior. Not surprisingly, such differences are in part hormonally controlled.

The Role of Hormones

Testosterone and estrogen are found in all vertebrates; testosterone is often assumed to regulate male sexual behavior and estrogen assumed to regulate female sexual behavior. However, this still popular notion has become outdated, partly because of surprises about the interrelationships of hormones, brains, and sex uncovered in songbirds. The first surprise was when Gurney & Konishi (1980) found that if they injected a young female zebra finch with extra doses of estrogen she would grow male-like vocal nuclei. Testosterone did not have this effect, but did cause estrogen-injected females to sing as adults later in life (Pohl-Apel & Sossinka 1984). These females did not display male plumage, but were sufficiently masculinized that they attempted to breed with other females. Thus, hormonal treatment of a developing songbird can convert a vocal non-learner into a learner. However, it is *not the case* that injecting hormones induces cerebral vocal nuclei and vocal learning in non-vocal learning species, such as quail; nor does injecting estrogen into vocal learning males increase the size of their vocal nuclei; similarly, blocking estrogen and testosterone in young vocal learners or removing their gonads does not prevent formation of cerebral vocal nuclei (Adkins-Regan & Ascenzi 1990; Balthazart et al. 1995; Wade 2001), but does lead to a decrease in size of the vocal nuclei and reduction in the amount of singing, though the birds still learn to imitate song and sing (Wade 2001; Alvarez-Borda & Nottebohm 2002). In addition, female canaries, which sing less than males, sing more when given testosterone, and their vocal nuclei increase in size during the process (Nottebohm 1980; DeVoogd & Nottebohm 1981). Debates ensued about the roles and sources of testosterone versus estrogen in the development of cerebral vocal nuclei and song learning (Wade 2001; Balthazart & Adkins-Regan 2002). The search for solutions is ongoing and not without contradictions (**Box 33**, p. 254), but the following is a synthesis of the facts as they now stand.

Testosterone is actually a pro-hormone, a hormone precursor. It is converted by three different enzymes, aromatase, 5α-reductase, and 5β-reductase, into estrogen, 5α-dihyrotestosterone, and 5β-dihydrotestosterone

BOX 33

PATHWAYS FOR HORMONAL INFLUENCE ON BIRDSONG

Birdsong is seasonal, and one of the earliest observations made by field biologists and natural historians is that singing behavior is positively correlated with various measures of reproductive physiology. Experiments established that testosterone from the gonads enhances seasonal changes in song production. Both estrogenic and androgenic metabolites of testosterone appear to be involved in these effects of testosterone on song. They exert their effects on behavior by binding to intracellular receptors. Pioneering autoradiography studies in the early 1970s revealed that receptors for testosterone are found in several of the forebrain song control nuclei including HVC, RA and IMAN (Arnold et al. 1976). This observation was somewhat unexpected in that androgen and estrogen receptors in the brain otherwise seemed to be restricted to the limbic regions and selected areas in the diencephalon and mesencephalon. This finding, along with the observation that the syrinx itself has androgen receptors (Lieberburg & Nottebohm 1979), suggested that testosterone might activate song by binding directly to receptors in the vocal control system as well as in the syrinx. Anatomical studies employing either immunohistochemical methods for the localization of androgen receptor (AR) and estrogen receptor (ER) proteins as well as *in situ* hybridization studies of the messenger RNA for AR and ER confirmed and extended the initial autoradiographic studies (Balthazart et al. 1992; Bernard et al. 1999). However, at least in temperate zone birds, song is part of a still larger suite of male reproductive behaviors, extending beyond the domain of the song system in the strict sense. In both birds and mammals the preoptic region of the brain is known to coordinate the effects of testosterone on male sexual behaviors. In European starlings, lesions to the preoptic region block song behavior as well as other reproductive behaviors (Riters & Ball 1999). Catecholamine cell groups that project to the forebrain song control nuclei are also known to express both androgen and estrogen receptors (Appeltants et al. 1999; Maney et al. 2001). It is therefore possible that the effects of testosterone on song are at least in part mediated by its action in the preoptic region and/or in brainstem catecholamine cell groups that, in turn, project to forebrain song control regions. Although it seems likely that direct action of testosterone in song control nuclei regulates the *quality* of song produced, testosterone effects on the *motivation* to sing may involve other parts of the brain as well.

Gregory F. Ball

(Ball & Balthazart 2002; Balthazart & Adkins-Regan 2002). The synthesized estrogen crosses cell membranes and binds to estrogen receptors inside the cells. The bound receptor acts as a transcription factor causing, amongst other actions, changes in gene expression. Likewise, the androgen 5α-dihyrotestosterone crosses cell membranes and binds to androgen receptors inside the cell. The bound receptor then causes a different set of changes in gene expression, amongst other actions. The synthesized 5β-dihydrotestosterone is the inactive form of the testosterone pro-hormone, and is treated as waste.

There are two sources of these hormones: the gonads and adrenals, and the brain itself (Holloway & Clayton 2001; Schlinger et al. 2001). Testosterone is produced and released into the blood stream by the testes in males and by the adrenal glands in males and females; estrogen is produced by the ovaries. Testosterone crosses the blood–brain barrier to act in the brain, as indicated, where it is converted into estrogen, in both males and females. Brain-synthesized estrogen is released back into the blood stream at levels as high as those released from the gonads (Schlinger & Arnold 1992). Within the brain, distribution of hormone receptors and the hormone-synthesizing enzymes differs with the brain region, sex, species, season, and stage of development, somewhat like the glutamate receptors. The androgen-synthesizing enzyme, 5α-reductase, is present throughout much of the brain of songbirds, but the estrogen-synthesizing enzyme, aromatase, is selectively expressed in NCM and related auditory areas. This is true of both male and female songbirds, whether or not females have vocal nuclei (Metzdorf et al. 1999; Soma et al. 1999; Saldanha et al. 2000). In the cerebral vocal pathways, only HVC has estrogen receptors. In adult canaries, the receptor levels are high during each breeding season, whereas in zebra finches, they are high only once in life, during early juvenile development, in the HVC of both males and females (Gahr & Konishi 1988; Gahr & Metzdorf 1997; Jacobs et al. 1999). In adults of most species, Area X lacks androgen receptors, whereas all pallial vocal nuclei contain androgen receptors but at different levels in different species (Jacobs et al. 1999; Tramontin & Brenowitz 2000; Ball et al. 2002). These hormone and brain differences have a genetic basis.

In mammals, the XX chromosome pair makes a female and the XY pair a male. In birds, the opposite occurs; a ZZ pair makes a male and a ZW pair makes a female. Thus, in mammals, the female is the 'genetic default;' a Y chromosome is needed to make a male body from a female. In birds, the male is the genetic default, and a W chromosome is needed to make a female body from a male (Balthazart & Adkins-Regan 2002). Estrogen synthesized in the egg helps mold the male testes into female ovaries, a change probably facilitated by the W chromosome. In birds as in mammals, circulating testosterone converted to estrogen in the brain masculinizes the brain. Strong evidence for sex linkage came from the brain of a zebra finch that was gynandromorphic, half male, half female (**Box 34**, p. 256).

Bringing this knowledge together, the following tentative picture emerges (**Fig. 8.7A**). During early development, an as yet unknown genetic program that includes genes on the Z chromosome starts the formation of the cerebral vocal nuclei in both sexes. Sometime around hatching, circulating testosterone that enters the brain is converted to estrogen via aromatase; the estrogen binds to its receptor in HVC, facilitating growth of HVC and connecting nuclei. The circulating testosterone is also converted in the brain to the androgen 5α-dihyrotestosterone using 5α-reductase, which then binds to its receptors in pallial vocal nuclei to modulate not tissue growth, but crystallization of connections and later enhancement of singing output.

For those species in which the females do not engage in vocal learning, it is probable that genes on their W chromosome actively prevent or reverse the growth of cerebral vocal nuclei. This genetic mechanism presumably evolved after the evolution of songbird vocal learning in tropical zones, for species living thereafter in temperate zones. However, when these females are given extra doses of estrogen early in life, as in the zebra finch, inhibition of vocal nuclei growth is overcome. In addition, if one of the two Z chromosomes of males is not fully inactivated, they have the potential like mammals to synthesize extra doses of gene products responsible for the formation of vocal nuclei.

During the juvenile-to-adulthood transition, an androgen surge helps crystallize the song to adult form and readies the bird for breeding. There is a parallel with human puberty; large hormonal increases occur, an adult-like voice crystallizes, and the young teenager, male or female, is biologically ready to have children. The underlying mechanisms in the two sexes are not the same, and in birds the sex differences are not yet fully understood. In species where only males imitate song, female vocal nuclei have already shrunk in size by this time; so these females go through a puberty-like phase, but without engaging in vocal learning. In species that learn new songs seasonally, a similar process reoccurs; song becomes plastic, learning occurs, and then testosterone products recrystallize the new songs and ready the bird's singing for the upcoming breeding season.

Androgens are thought to crystallize the voice by binding to pallial vocal nuclei receptors, activating genes that stabilize synapses between vocal nuclei and helping the survival of new neurons arriving in HVC (Rasika et al. 1994; White et al. 1999). A castrated male gives plastic song but fails to crystallize until given a seasonally appropriate dose of testosterone (Marler et al. 1988). Artificially high doses of testosterone in juveniles or in the plastic phases of adult singing also induce song crystallization (Korsia & Bottjer

BOX 34

WHAT MAKES A BIRD BRAIN FEMALE OR MALE?

Even casual observations of birds reveal a profound sex difference in the ability to sing. Song is a quintessential example of male sexual advertisement and, in many species, females sing little or not at all. The male's greater ability is directly caused by a marked sex difference in the structure of the neural song circuit. In zebra finches, for example, several brain regions controlling song are about five times larger in males than in females. This large sex difference in brain structure, the first to be discovered in any vertebrate (Nottebohm & Arnold 1976), invites an obvious question; what causes brain development to differ in the two sexes?

The first answer was suggested by studies in mammals, in which hormones secreted by the testes act on the fetus to make the brain male. The important hormones are testosterone and its metabolite, estradiol. When female zebra finches were treated with estradiol on the day of hatching, they developed a much more male-like neural song circuit and a fairly good male song (Gurney & Konishi 1980). Androgens also seem to play an important role because the masculinizing actions of estradiol are blocked by an anti-androgen, flutamide (Grisham et al. 2002). Thus, although estrogen induces a masculine pattern of development, it appears to require an androgen-dependent step. These studies suggest that, in birds also, male sex hormones make a male brain. Some doubts about this idea arose, however, because blocking the action of these hormones does not prevent a male from singing and developing a male-like song circuit (Arnold 1997). The hormonal theory was further challenged when genetically female zebra finches, induced by a drug injection to develop large testes rather than ovaries, were found to have a feminine neural song circuit (Wade et al. 1999), contradicting the earlier findings.

An alternative idea is that the genetic sex of brain cells plays a role in determining whether the brain has male or female characteristics. The strongest support for this view comes from the analysis of an unusual bilateral gynandromorphic zebra finch that arose by mutation (Agate et al. 2003). This bird was genetically male on the right half of its body, and genetically female on the left half (see fig.). Male plumage covered only the right half of its body, and the right gonad was a testis and the left gonad an ovary. Amazingly, the neural song circuit was more masculine on the right half of the brain than on the left, indicating that the differences in genetic sex on the two sides of the brain influenced their sexual characteristics. Although the origin of sex differences is not fully understood, both hormonal and intrinsic genetic factors appear to play critical roles. One intriguing, as yet unsupported, idea is that genetically male brain cells may themselves produce more sex hormones than female brain cells, and that sex hormones originating in the brain are important for sexual differentiation as well as for the control of adult behavior (Schlinger et al. 2001).

A Gynandromorphic Zebra Finch

Arthur P. Arnold

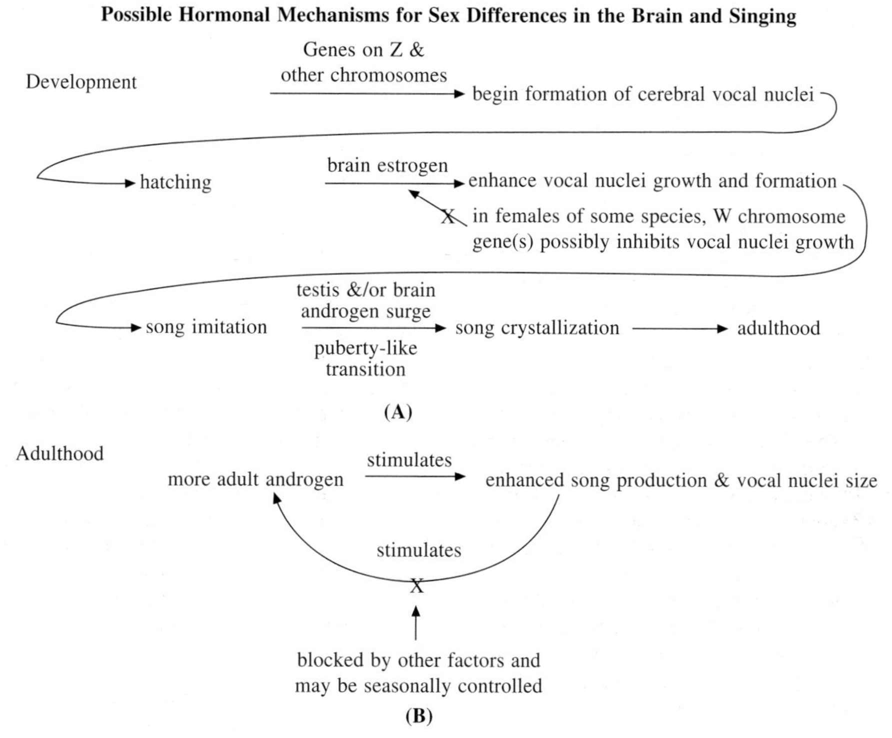

Figure 8.7 Diagrams of possible interactions between **(A)** sex hormones, brain development, and **(B)** the formation and activation of the vocal learning system in youth and adulthood.

1991; Nottebohm 1993). Androgens stimulate singing, and a positive feedback mechanism is thought to be involved, with androgen binding to vocal nuclei receptors enhancing singing, which in turn enhances testosterone release by the gonads, which then encourages more singing (**Fig. 8.7B**). This positive feedback loop can be broken in the non-breeding season of seasonally singing species, when androgen receptor levels in the vocal nuclei are low, and the birds just sing for short 10-min bouts (Ball et al. 2002). As a possible mechanism by which androgen enhances singing, binding to androgen receptors in pallial vocal nuclei neurons may reduce the threshold for electrical activity, as occurs in mammals (McEwen 1994; Ramirez et al. 1996), making the neurons associated with singing more ready to fire. The general picture is that hormones have three types of facilitating effects on the songbird vocal communication system: they help form the system, they stabilize synapses and neurons involved in the learning and crystallization of song, and they induce song production.

However, hormone receptors are not present in the cerebral vocal nuclei of all vocal learners. The HVC-like vocal nuclei of hummingbirds have high levels of androgen receptors, but no estrogen receptors (Gahr 2000). In the one species of parrots examined, budgerigars, there are no androgen or estrogen receptors in their vocal nuclei (Gahr 2000). Only songbirds have high aromatase in their auditory regions (Metzdorf et al. 1999; Soma et al. 1999; Gahr 2000; Saldanha et al. 2000). Thus, it looks as though hormones may have different roles in the vocal systems of different vocal learners.

ADULT NEUROGENESIS DISCOVERED

One of the major outgrowths of these hormone

studies in songbirds was the discovery that the adult brain is able to make new neurons.

Nearly 100 years ago, Ramon y Cajal, an early Nobel Laureate in neuroscience, stated that, unlike skin, which is constantly renewed, in the adult brain "the nerve paths are something fixed, ended, immutable. Everything may die, nothing new may be rejuvenated" (Ramon y Cajal 1913). This view became dogma, epitomized in the statement "do not destroy your brain cells because you will not make any more." Over the years, various attempts were made to challenge this doctrine by new methods, trying to induce new neuronal growth in adult brains, but without success. The view prevailed that warm-blooded vertebrates do not generate new neurons as adults, perhaps because it would be too costly to hold onto long-term memories (Rakic 1985). In the 1960s, Altman (1962) found new neuron proliferation in the adult rat brain. However, his work met with strong criticism, and was difficult to repeat, discouraging further investigation.

Song Plasticity and Rejuvenation of the Brain

Goldman & Nottebohm (1983) were startled when, twenty years later, as they studied the relationship between seasonal re-growth in the size of HVC and hormones in adult canary brains, they found newly labeled neurons (**Fig. 8.8A**). Nottebohm had found earlier that injection of testosterone in female canaries resulted in a 90 percent increase in the size of HVC. To identify dividing cells, the animals were injected with radioactively labeled thymidine, one of the nucleotides of DNA. When cells replicate, the radioactive nucleotide is taken up into the DNA of the new cell and can be detected. Goldman and Nottebohm went to great lengths to prove that the new cells were actually neurons. Perhaps not surprisingly, their results were met with skepticism, and at first were difficult to get published. Later Paton, Alvarez-Buylla, and Kirn decisively proved that the newly generated cells in the songbird brain were indeed new neurons that were functional in the song system (Paton

New Neurons in the Adult Brain

Discovered in songbird HVC in 1983

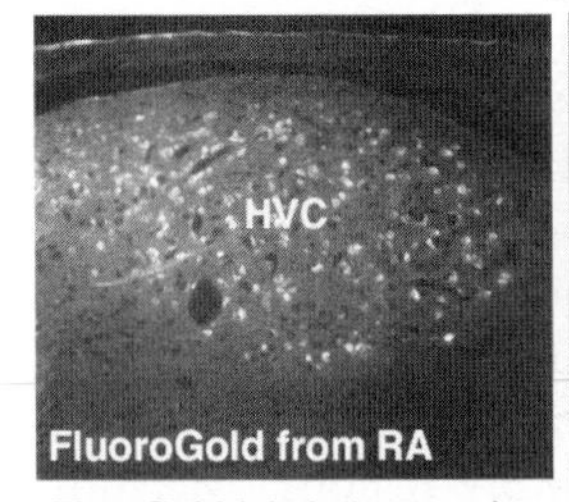

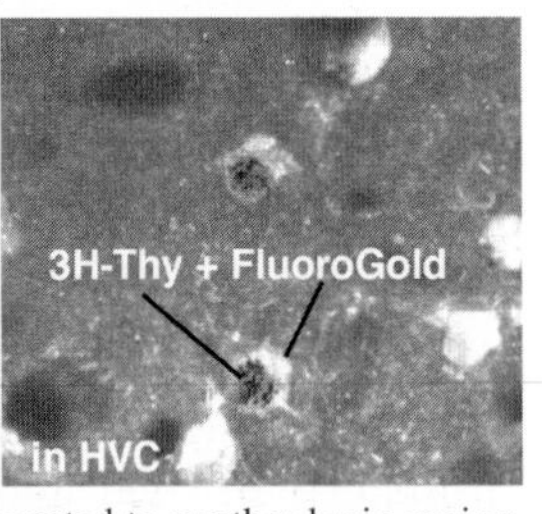

FluroGold labels 'neurons' connected to another brain region. Tritiated thymidine, ^{3}H-Thy, labels newly divided cells.

(A)

Discovered in human hippocampus in 1998

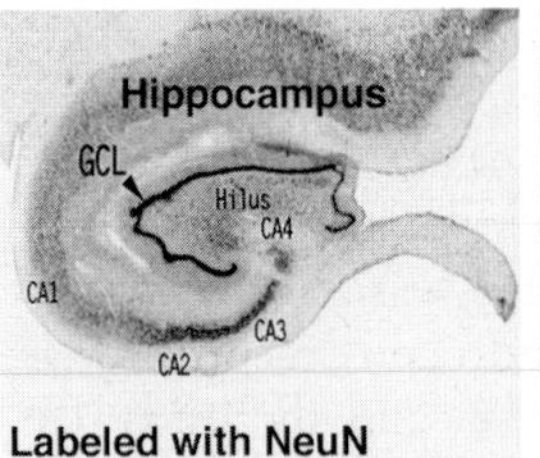

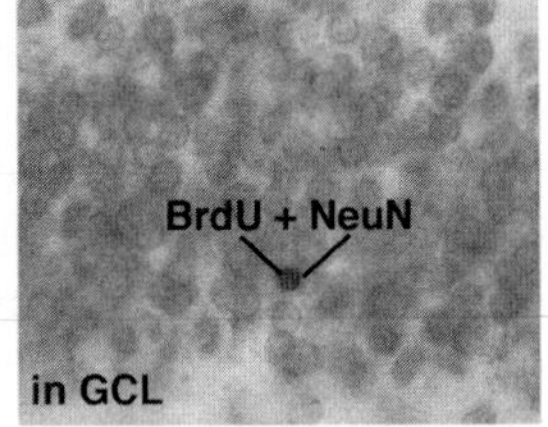

NeuN, a neuron-specific gene, identifies cells as neurons. Bromodeoxyuridine, BrdU, also labels newly divided cells.

(B)

Figure 8.8 **(A)** Evidence of newly formed neurons in the adult songbird brain from Alvarez-Buylla & Kirn 1997. Left: canaries were injected with ^{3}H-thymidine; after allowing new neurons to reach HVC, RA was injected with FluoroGold, which was taken up by the axons of HVC's RA-projecting neurons. Right: at high power, Flo-labeled cells containing ^{3}H-Thy are seen. **(B)** Evidence of newly formed neurons in the adult human brain (Eriksson et al. 1998). Left: frontal sections of a human hippocampus labeled with a marker that identifies the neuron-specific gene NeuN (dark staining areas). Right: high power of the granule cellular layer (GCL) double-labeled with NeuN and bromodeoxyuridine (BrdU) showing a neuron born in the brain of this patient when he/she was an adult. Pictures kindly provided by John Kirn **(A)** and Fred Gage **(B)**.

& Nottebohm 1984; Alvarez-Buylla et al. 1988 1990, 1992). Remarkably, neuron death and replacement were found to occur in HVC, at a high rate on a daily basis, with nearly all of the neurons of a given cell type replaced each year (Kirn et al. 1991; Nottebohm 2002). In retrospect, it is hard to imagine how neurogenesis in the adult brain was missed.

But the skepticism persisted. The vocal nuclei are highly specialized brain tissues, and it was argued that perhaps new adult neurons are found only in these unusual structures. Alvarez-Buylla and collaborators then showed that new neurons were present throughout the songbird cerebrum, and in the cerebrums of vocal non-learning birds as well (Nottebohm & Alvarez-Buylla 1993; Ling et al. 1997). Then, using the same techniques, they found new neurons in the cerebrum of a mammal, the mouse (Lois & Alvarez-Buylla 1993; Lois et al. 1996; Doetsch et al. 1997), finally confirming Altman's (1962) findings. The new neurons they found were mostly transported to the olfactory bulb.

Skeptics retreated to another line of defense, arguing that the olfactory bulb is a primitive structure and of little importance in holding onto complex memories. However, around this period, studies of mice using the same approaches as those used on songbirds, demonstrated the incorporation of new neurons in the mammalian hippocampus, a region involved in learning and memory (Cameron et al. 1993; Gould & McEwen 1993). The numbers of new neurons entering the hippocampus each day were staggering, in the thousands, and were significantly reduced by stress and enhanced by learning (Gould et al. 1999a; Gould & Tanapat 1999). Most animals kept in cages are stressed and this may be one reason why adult neurogenesis was overlooked in the past. Gould then also found new neurons entering the adult primate cortex, an area certainly responsible for complex learning (Gould et al. 1999b). Finally, Gage, Eriksson, and colleagues in 1998, studying the autopsied brains of patients who had been injected days or years earlier with a cell division marker (bromodeoxyuridine) used to diagnose the severity of their cancer, found evidence of new neurons in the human hippocampus that survived in the patients until their death (Eriksson et al. 1998; **Fig. 8.8B**). So, the 100-year-old doctrine of no new neurons in adulthood gradually crumbled (Specter 2001; Nottebohm 2002), though still not without challenges (Kornack & Rakic 2001; Rakic 2002a, b).

The discovery of new neurons in the vocal nuclei of songbirds led Nottebohm to propose a role for them in the formation of new song memories (Nottebohm 1984). Several findings appeared to support this view. In canaries, old neurons die and are replaced by new neurons at a higher rate during times of the year when the birds learn new song syllables, after the molt in the autumn and around the breeding season in April to May (Kirn et al. 1994; **Fig. 8.9A**). However, other studies seemed to challenge this notion. Zebra finches continue to show neuronal death and new neuron addition in HVC well after song learning is complete (Ward et al. 2001). Other species that undergo seasonal changes in breeding and singing behavior accompanied by changes in death and incorporation of new vocal nuclei neurons do not add new songs to their repertoires (Tramontin & Brenowitz 1999, 2000).

If there is a link to learning, it may be indirect, as suggested by Li and colleagues (2000), who found that just the act of singing without any evidence of learning new songs, drives the increased survival of new neurons in HVC within a few days (**Fig. 8.9B**), with the number of newly arrived neurons proportional to the amount of singing. This motor-driven survival of new neurons adds further to testosterone-induced enhancement of survival (Alvarez-Borda & Nottebohm 2002). Similarly, in the rat brain, wheel running for extended periods of time increases proliferation and incorporation of new neurons in the hippocampus (Van Pragg et al. 1999). It seems that one of the most salient factors influencing the addition of new neurons is exercise, the act of performing a behavior rather than a need to incorporate new memories. A relationship to exercise does not necessarily exclude learning; however, the new neurons

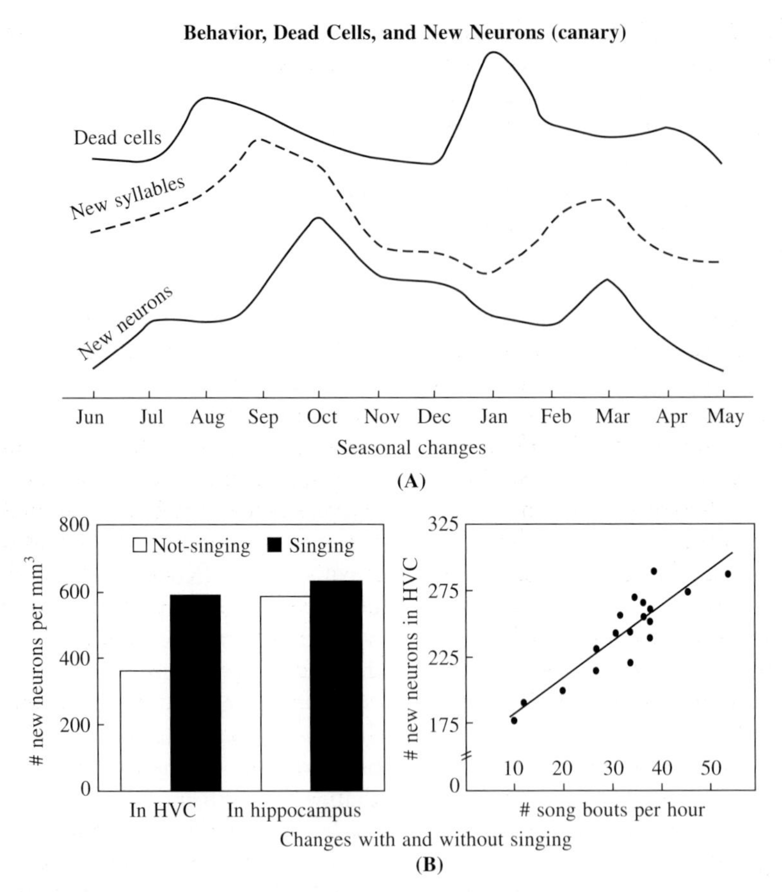

Figure 8.9 **(A)** Changes in cell death, song learning, and new neuronal survival across the year in brains of males of a seasonal breeding species, the canary. Peaks in cell death are followed by peaks in new song syllables and new neurons. Modified from Alvarez-Buylla & Kirn 1997, Kirn et al. 1994, and Nottebohm et al. 1986. **(B)** Singing-driven enhancement of new neuron survival in HVC. Left: canaries were either allowed to sing for 7 days, or prevented from singing by interrupting them. Singing increased the survival of neurons that entered HVC but had no affect on new neurons that entered the hippocampus, a non-vocal area. Right: the enhancement of new neuron survival is proportional to the rate of singing over a 30-day period. This enhancement was independent of the gonads; these animals had their testes removed. Modified from Li et al. 2000 and Alvarez-Borda & Nottebohm 2002.

presumably inherit or acquire the memories embodied in the old neurons.

Not all neuronal types are replaced. Neurons with short-distance axonal connections, such as local neurons, are more likely to be replaced than those that traverse long distances, such as those from the cortex or pallium to the spinal cord. Neurons in the adult thalamus and midbrain are not replaced (Nottebohm & Alvarez-Buylla 1993). Interestingly, the latter are areas involved mainly in innate behaviors. In the songbird vocal system, only neurons of HVC and Area X are continually replaced in adulthood, and not those of lMAN and RA (Kirn et al. 1999). The arcopallium in which RA resides generally does not undergo neuron replacement and, like RA, is the part of the avian brain that sends long descending axons to the brainstem and spinal cord (Zeier & Karten 1971; Kirn et al. 1999). Within HVC, its RA-projecting neurons are replaced throughout adult life, but its X-projecting neurons are not (Alvarez-Buylla et al. 1988; Kirn et al. 1999). Using a laser-lesion procedure, Scharff et al. (2000) selectively labeled the RA-projecting and X-projecting neurons of HVC with a dye, which kills the cells when zapped with the laser at a particular wavelength of light. When they killed off the X-projecting neurons in adult zebra finches, song was temporarily affected for a day or two and then returned to normal. This is similar to what happens when lArea X is lesioned. The birds showed no induced regeneration of the X-projecting neuron type in HVC. HVC just became smaller, due to the neuron loss. When they killed off the RA-projecting neurons, however, the birds could not sing, as happens when the entire HVC is lesioned. After several months, the birds slowly recovered the same or a very similar song to that produced before the lesion, and newly incorporated RA-projecting neurons were found throughout HVC. Thus, when damaged, the RA-projecting neuron population of HVC can be restored, and remembered behavior can be recovered. This shows that the song template is not stored in the RA-projecting neurons.

The discovery of neurogenesis in adult bird brains helped to revitalize the field of adult mammalian brainstem cell research, and the search for new ways to repair brain injuries (Gage 2000). Although scientists have a long way to go, there are exciting prospects. Adult brainstem cells of both birds and mammals have been found within the walls of the cerebral ventricles, in areas called the ventricular zones, which line the cavity where the cerebral spinal fluid flows (Doetsch et al. 1999). The stem cells divide and give rise to daughter cells that then migrate to their destination in the brain and become neurons. One of the problems in taking advantage of this mechanism to induce repair of damaged brain tissue, is that there will be a need to control the incorporation of cells that normally get replaced, to induce and control those not normally replaced, where they migrate to, and their eventual connectivity. Vocal learning birds can serve as a useful animal model for the development of these kinds of brain repair techniques.

EVOLUTION OF VOCAL LEARNING: SONGBIRDS, HUMMINGBIRDS, AND PARROTS

Can vocal learning birds serve as useful animal models for language? To answer this question we need to gain a better understanding of the evolution of vocal learning. All songbirds examined, including birds that are not typical singers, such as corvids – crows, ravens, jays and magpies – have cerebral vocal nuclei (DeVoogd et al. 1993; Brenowitz & Kroodsma 1996). In corvids they are presumably used for learning and producing their complex calls and low volume songs (see Chapter 5). Thus, vocal learning was probably present early in songbird evolution. How did this come about? Nottebohm (1972) proposed that songbirds, hummingbirds, and parrots evolved their vocal learning abilities independently. His reasoning was that most of their close relatives, including the sub-oscine songbirds (but see Chapter 3), are all vocal non-learners, and presumably represent the ancestral condition for birds. After the discovery that some parrot vocal nuclei have similarities with those of songbirds (Paton et al. 1981), Brenowitz (1991) proposed that just as the behavior they control is similar, the cerebral vocal nuclei of songbirds and parrots evolved independently but with shared properties. In his detailed study of budgerigar vocal nuclei connectivity, Striedter (1994) gave parrot vocal nuclei different names than in songbirds, assuming that if they evolved independently, then they should not have the same names. My colleagues and I did the same for hummingbirds, and used molecular mapping of vocal areas to provide new perspectives (Jarvis et al. 2000).

The late Luis Baptista first proved that hummingbirds are vocal learners (Baptista & Schuchmann 1990). He was a lover of hummingbirds, and his work was in part the inspiration for the direction of my own research on vocal learning. Unlike songbirds and budgerigars, hummingbirds are difficult to work with. They are among the smallest and fastest flying birds. They do not breed readily in captivity, and they are very territorial and often kill each other if they are kept in close quarters. Because of this, we took our molecular brain mapping experiments into the field. In the Atlantic Forest region of Brazil, near Santa Teresa, we found a site with one of the highest populations of hummingbirds in the world. Unlike songbirds and parrots, hummingbirds live only in the Americas. They consist of two main lineages, the Trochilinae and the Phaethornithinae. The two species we studied were the somber hummingbird, an ancient Trochilinid, and the rufous-breasted hermit, an ancient Phaethornithinid (**Box 35**, p. 263). After we observed a bird, either singing, or listening without singing for 30 min in the morning, we attracted him into a cage with a sugar water bottle, captured him, and examined ZENK expression in his brain. This approach, similar to the one we had used earlier on songbirds and parrots (Jarvis & Mello 2000), helped us to identify vocal nuclei that were not revealed by other methods.

Comparisons of the ZENK expression patterns in the three taxa of vocal learners revealed some remarkable similarities. All had seven cerebral vocal structures that showed gene activation during singing. We found three of the seven in nearly identical brain locations, in anterior parts of the cerebrum, though they have different shapes. In songbirds, these structures are part of the anterior vocal pathway. The remaining four of the seven are all in different brain locations in each of the vocal learning groups, but are still within the same brain subdivisions relative to each other. Two of these, the HVC-like and the RA-like structures have similar shapes. In songbirds, these structures are part of the posterior vocal pathway. All three vocal learning groups had similar auditory areas, showing gene activation after hearing species-specific vocalizations, all in the same relative brain locations. The similarity of these auditory areas is not surprising, since vocal non-learning birds were already known to have such auditory regions (Wild et al. 1993). Interestingly, the location of the four posterior nuclei relative to the auditory areas differs, in accordance with the relative age of each order. Parrots are the oldest, and their posterior vocal nuclei are far away from the auditory areas; hummingbirds are the next oldest, and their posterior vocal nuclei are closer and adjacent to auditory regions; songbirds are the most recent, and their posterior vocal nuclei are shifted further back, embedded within the auditory areas.

Lesion studies confirmed that the cerebral vocal structures of parrots, at least those studied, are required for producing and learning vocalizations (Heaton & Brauth 2000a, b). Interestingly, lesions in the parrot nucleus parallel to HVC, NLc, revealed that it is required for the birds to speak English words (Lavenex 2000). This was the first time anyone had established the role of a structure in a non-human brain for the production of imitated human speech. Electrophysiological studies in anaesthetized budgerigars show that their vocal nuclei also have auditory responses, and this may turn out to be a basic feature of vocal learners. Neither lesioning nor electrophysiological studies have yet been conducted in the vocal nuclei of hummingbirds, though there have been preliminary hummingbird connectivity studies (Gahr 2000).

The comparisons show that all three vocal learner groups have similar posterior vocal pathways, connecting a region of the nidopallium to the arcopallium, then to the tracheosyringeal motor neurons, and then to the syrinx and the abdominal respiratory muscles. Connectivity of the anterior vocal nuclei has only been studied in the songbird and the parrot. Both have similar anterior vocal pathway connectivity, forming a pallial–striatal–thalamic–pallial loop (Striedter 1994; Durand et al. 1997). Where the connectivity differs greatly, is in the connections between their posterior and anterior vocal pathways. In the parrot brain, the anterior vocal pathway does not receive its posterior pathway input from the HVC-like structure, as occurs in

BOX 35

COMPLEX SONGS WITH A SMALL BRAIN: HUMMINGBIRDS

Hummingbirds have many fascinating characteristics, including one of the highest brain-to-body size ratios in existence, and a large species-specific diversity of vocalizations, some of which display local dialects (Wiley 1971; Snow 1973; Stiles 1982; Rehkaemper et al. 1991; Gaunt et al. 1994; Ventura & Takase 1994; Vielliard 1994; Rusch et al. 1996; Schuchmann 1999; Ficken et al. 2000). Given Luis Baptista's demonstration of vocal learning in hummingbirds (Baptista & Schuchmann 1990), we wonder how such a small brain, the size of the tip of a man's pinky could do what many animals with bigger brains cannot. With the discovery of the hummingbird cerebral vocal system, of a kind not found in non-vocal learning animals (Jarvis et al. 2000), the answer appears to lie in the patterning of neuron connectivity rather than absolute brain size. We have been studying two species, the sombre hummingbird **(A)** and the rufous-breasted hermit (**B**). Sombre hummingbird song has a relatively stereotyped and rhythmic syntax, like a zebra finch, but the syllables are complex and two-voiced, with rapid frequency modulations (**CD2 #62**). Rufous-breasted hermit song has a complex syntax, consisting of rolling sequences of increasing and decreasing pitches and modulations, generating a great variety of different syllables (**CD2 #63**). Like songbirds, these species appear to produce their songs in affective contexts, either in an undirected manner or directed to the opposite sex during courtship. A wealth of information may be gained by studying the brains of the smallest birds in existence.

Erich D. Jarvis & Adriana R. Ferreira

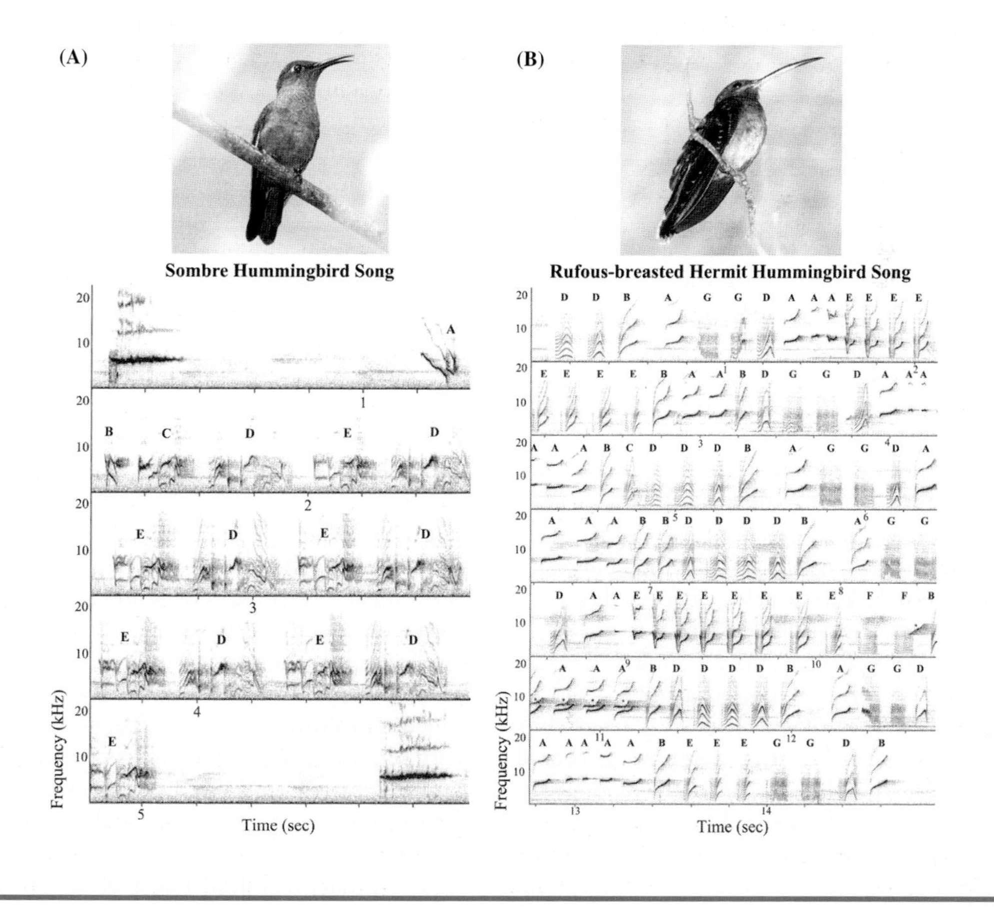

songbirds. Instead, part of the posterior vocal pathway's RA-like nucleus projects into the two anterior vocal nuclei, Mo-, and MAN-like nuclei. In the parrot, the output of the anterior pathway does not have well-separated medial and lateral divisions. Instead, the same region of the MAN-like nucleus projects to both its RA-like and its HVC-like vocal nuclei. Since connectivity of the songbird Mo nucleus has not been yet determined, a complete comparison with parrots is not possible (but see below).

In parrots and songbirds auditory pathways from the ear to the auditory pallium (NCM and other areas) are very similar to those in vocal non-learning birds. Thus, auditory pathway connections appear to be an ancient inheritance. However, there are differences in the mode of interaction of the auditory and vocal pathways. Besides the major difference in location of the posterior vocal nuclei relative to the auditory regions, the parrot vocal pathways receive auditory input from two different auditory pathways: one is like that in songbirds, from Ov of the thalamus to field L2 to L1 and L3; the other is from a more frontal pathway, less well characterized, that involves a projection from the lateral lemniscus intermediate (LLI) nucleus of the hindbrain to nucleus basorostralis. In parrots, both pathways, L1 and L3, and nucleus basorostralis then project to the NIf-like vocal nucleus, which then projects into the parrot nuclei of the posterior (HVC-like) and anterior (Mo) pathways.

By combining the connectivity and gene expression findings for all three vocal learning groups, we get an indication of some basic requirements for avian vocal learning. There are seven cerebral vocal nuclei organized into a posterior vocal pathway that projects to the lower vocal motor neurons and an anterior vocal pathway that forms a loop and is involved in other aspects of vocal learning. Another requirement is that every major cerebral subdivision except the pallidum and hyperpallium has at least one vocal nucleus. This in turn suggests a basic requirement for avian cerebrally controlled behaviors where, as in mammals, each cerebral function involves some part of a six-layered cortex plus the basal ganglia (Swanson 2000a). Variation seems to be permissible, in the shapes of the vocal nuclei, the absolute brain location of the posterior vocal pathway relative to the auditory pathway, the connectivity between the two vocal pathways, and in the connectivity between the vocal and auditory pathways. There are also variations in the relative sizes of vocal nuclei. The parrot Mo-like and MAN-like vocal nuclei are relatively much larger than their songbird and hummingbird counterparts; like canaries, with a relatively large MAN to Area X ratio, parrots are thought to display more vocal plasticity than most songbirds or hummingbirds.

Three Evolutionary Hypotheses

Now that we have all of this information in hand, we can begin to answer questions about the evolution of vocal learning brain areas in birds. Modern birds are said to have evolved from a common ancestor sometime around the cretaceous–tertiary boundary at the time of the extinction of dinosaurs (Feduccia 1995; **Fig. 8.10**). How did seven similar brain structures with somewhat similar connectivity evolve in three distantly related vocal learning bird groups in the past 65 million years? To put the question in perspective, the phylogenetic distance between parrots and songbirds (Sibley & Ahlquist 1990) is as great as that between humans and dolphins (Novacek 1992; **Fig. 8.11**). To explain the shared similarities, three hypotheses have been proposed (Jarvis et al. 2000).

Hypothesis 1: Three out of the 23 avian orders evolved vocal learning independently (**Fig 8.10**, solid circles). Each time, they evolved seven similar brain structures to serve the purposes of learned vocal communication. This would suggest that the evolution of brain structures for vocal learning, and for complex behavior in general, is under strong epigenetic constraints. If true, then a similar scenario can be made for the vocal learning mammals: humans, cetaceans, and bats (**Fig. 8.11**, solid circles). The evolution of wings

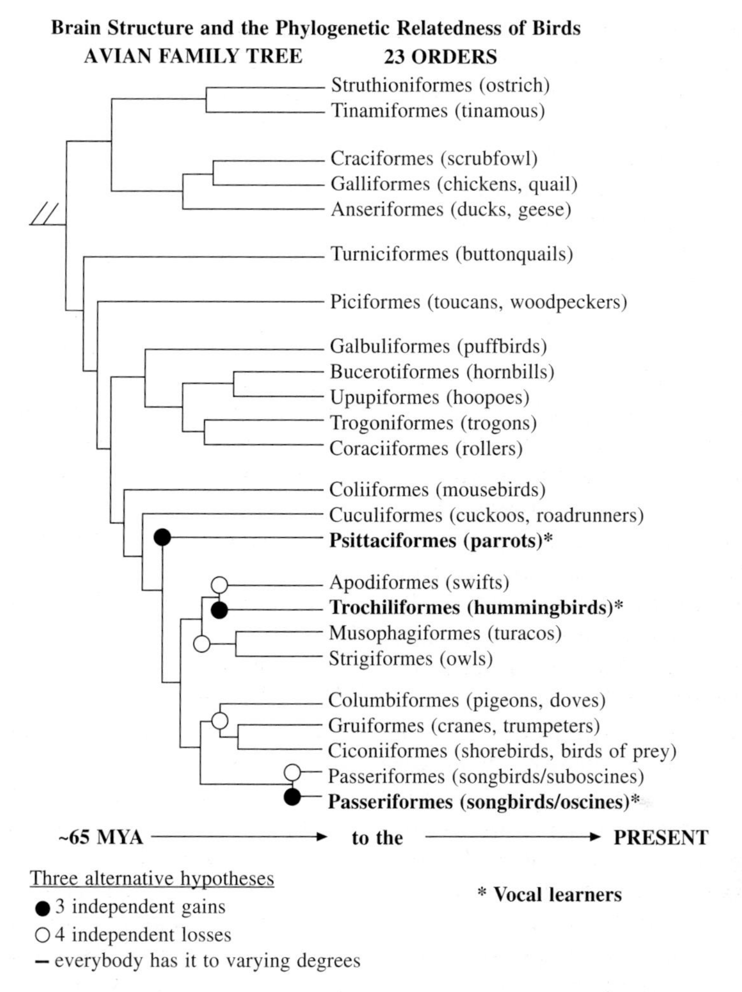

Figure 8.10 A phylogenetic tree of living birds according to DNA relationships. The Passeriform order is divided into its two suborders, suboscine and oscine songbirds. Lines at the root of the tree indicate extinct orders. Open and closed circles show the minimum ancestral nodes where vocal learning could have evolved or been independently lost. Vocal learners are in bold. Modified from Sibley & Alhquist (1990).

provides an analogy. Wings evolved independently at least four times, in birds, bats, pterosaurs (ancient flying dinosaurs), and insects. In each case, they evolved at the sides of the body, usually one on each side, and not one on the head, the other on the tail, or elsewhere. One hypothesis is that wings evolved in similar ways because of a strong constraint, the center of gravity of the body, dictating the most energetically efficient manner for flight. According to this logic, one can predict that if, in another half a million years or so, pigeons were to evolve vocal learning, then they too would have seven similar brain regions in well-defined locations.

Hypothesis 2: An alternative hypothesis is that there was a common avian ancestor with vocal learning, possessing the seven cerebral vocal nuclei. These traits were only retained in the three current vocal learning orders and lost at least four times independently in the interrelated vocal non-learning orders (**Fig. 8.10**, open circles). Repeated losses would suggest that maintenance of vocal learning and the underlying cerebral vocal structures may be under another type of epigenetic constraint; there would presumably be considerable costs to retaining vocal learning or evolving in adaptive circumstances that did not require vocal learning. If the losses occurred independently, several times, then again a similar scenario can theoretically be advanced for the vocal learning mammals: humans, cetaceans, and bats (**Fig. 8.11**, open circles). However, chimpanzees and other primates would have had to lose the trait recently, multiple independent times.

Hypothesis 3: A third hypothesis is that avian vocal non-learners have some rudimentary system of cerebral vocal nuclei that scientists have missed previously, and these systems were independently amplified in the vocal learners. If true, this would present a challenge to the hypothesis that cerebral vocal nuclei are unique to vocal learners. It implies that all birds have at least the primordia for the necessary brain structures, with the potential for vocal learning to varying degrees. A similar scenario could be argued for mammals and perhaps all advanced vertebrates; they all have at least the primordia for vocal cerebral brain structures and for vocal learning, including chimpanzees, lions, tigers, and bears, to varying degrees, and that these were amplified in the vocal learners.

Whichever hypothesis is correct, singly or in combination, the answer will be fascinating. They all suggest that the evolution of brain pathways for complex behaviors is constrained by factors as yet unknown. All lead to the same prediction, that vocal learning mammals, including humans, may have evolved vocal brain pathways with features held in common with vocal learning birds.

PARALLELS BETWEEN BIRDS AND HUMANS?

It has been appreciated for some time that the vocal learning behavior of songbirds shares developmental characteristics with human language learning (Thorpe 1961; Marler 1970b). It may seem bizarre to even suggest that bird brains can teach us something new about how human brains work. But in an era when geneticists have shown us how profitable it can be to move from humans to fruit flies to bacteria and back, I am encouraged to offer some suggestions about how human/avian comparisons might proceed.

The precise relationships between songbird and human vocal brain regions remain unexplored. The omission exists in part because of the erroneous historical belief that the avian brain is one large basal ganglion, that humans are much more special than they really are, and because of the obvious limitations on the experimental study of humans. Now that we have a new picture of the relationships between avian and mammalian brains (Jarvis et al. under review; Reiner et al. in press B) and given our expanding knowledge of the brain pathways for learned vocal communication amongst distantly related birds, the time may be ripe for a reappraisal of the neurobiology of human language as viewed from a comparative perspective.

For comparisons to be productive, some translations need to be made between the terminologies used in bird and human vocal communication research (Doupe & Kuhl 1999). We can separate human vocal communication into three categories: comprehension, speech, and language. Comprehension is the perception and understanding of language; speech is the production of language sounds, including phonemes and words; and language is the syntactical sequencing of these sounds into meaningful phrases and sentences. The term 'language' has also been applied to non-vocal communication, such as reading and signing. Most neurobiologists interested in language have strived to project these behavioral terms onto different brain structures, comprehension and language to the cortex, in Wernicke's and Broca's areas, and speech to lower brainstem motor areas, the periaqueductal gray and 12th nucleus (Hollien

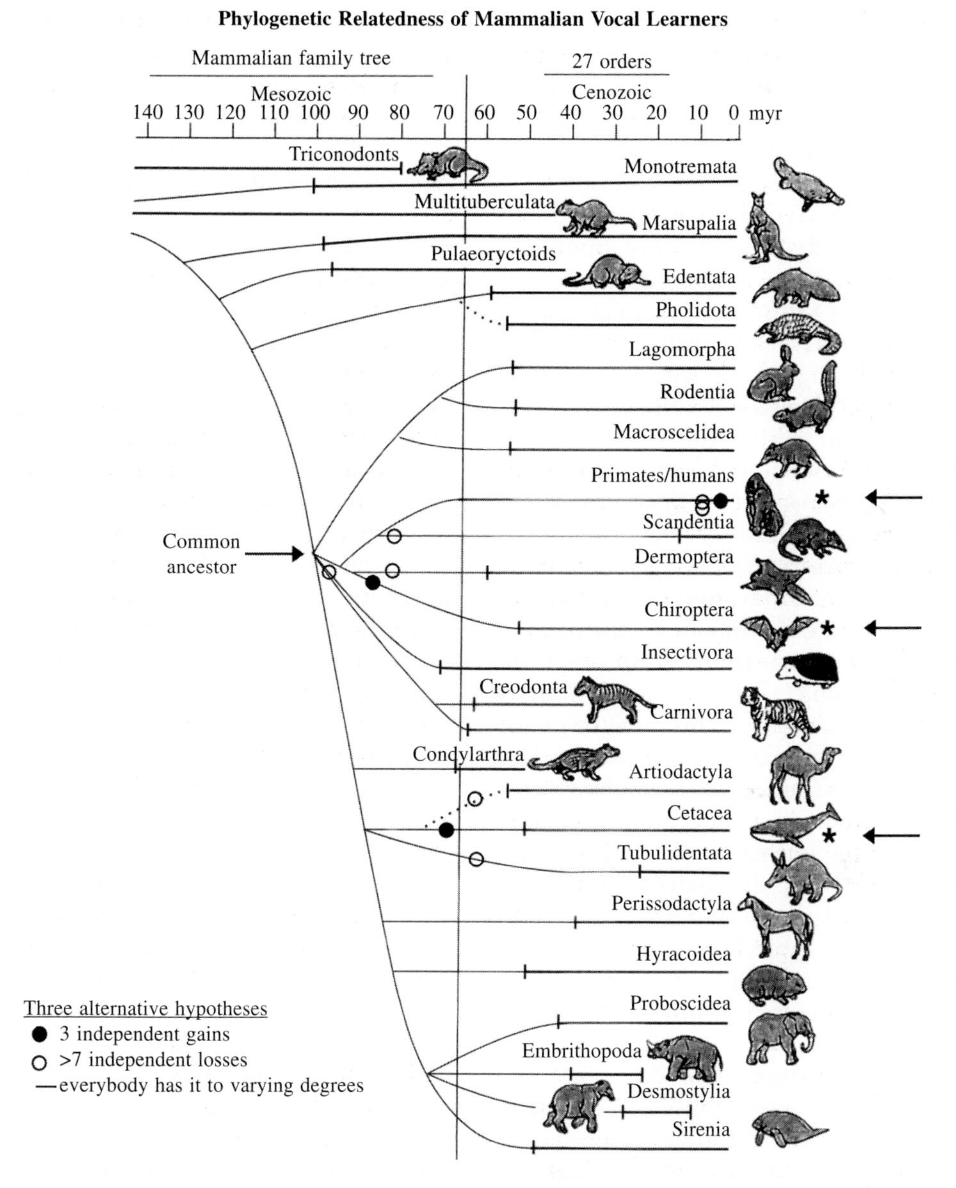

Figure 8.11 The mammalian family tree and the evolution of vocal learning. The phylogenetic tree of living mammals is based on relationships compiled by Novacek (1992). Except for primates, the Latin name of each order is given with examples of some common species. Vocal learners are asterisked. Open and closed circles show the minimum ancestral nodes where vocal learning could have either evolved or been independently lost. Independent losses would, at a minimum, have required a common vocal learning ancestor node located by the arrow. Within the primates, there would have to be at least 6 independent losses (tree shrews, prosimians, new and old world monkeys, apes, and chimps), and one loss (tree shrews), followed by a regain of vocal learning (humans). This assumes that all other primates are vocal non-learners. Compare with **Fig. 8.10**.

1975; Geschwind 1979; Fitch 2000a; Kent 2000), though linguistic definitions can vary widely depending upon the subfield or the investigator. In songbird research, there is no equivalent distinction between language-like and speech-like properties. One word, 'song,' represents both production of learned sounds and their sequencing. In this manner, the linguistic definition of 'language' for humans is at least somewhat similar to the avian neurobiologists' definition of 'song;' the linguistic definitions of 'speech' and 'singing' for humans are similar to the avian neurobiologists' definitions of 'call and song production.' Using the rules of avian neurobiologists, speech and singing combined would be called 'language production.' Rather than projecting predefined behavioral terms onto different brain structures, however, students of birdsong have tended to project definitions in the opposite direction, from brain structures to the behaviors they appear to control.

The terminology for behavioral deficits resulting from brain lesions also varies from one scientist to another. For humans, we tend to speak of dysarthria (slurred speech), verbal aphasia (poor syntax production), and auditory aphasia (poor language comprehension; Benson & Ardila 1996). For songbirds, the equivalent terms are not well formalized, but they include song degradation, disrupted syllable structure, disrupted sequence or syntax production, and impaired song discrimination, the latter being equivalent to a comprehension deficit. With the benefit of translations like this, the new comparative brain nomenclature, and a detailed literature review (Jarvis 2001), I propose the following speculative thesis on the neurobiology of human language.

We can think of humans, like vocal learning birds, as having three basic cerebral pathways for vocal behavior, one posterior, one anterior, plus an auditory pathway; together they are responsible for human vocal learning, speech, and singing. The posterior vocal pathway begins with facial motor cortex in the cerebrum and is the only one in which some connections have been determined. Using an old tract-tracing method, staining stroke-induced degenerating axons from autopsied human brains, Kuypers (1958a) showed that the human face motor cortex projects directly to the lower motor neurons that control vocalizations, the nucleus ambiguous. As the tracheosyringeal nucleus in birds projects to the syrinx, so the nucleus ambiguous in mammals sends axons that control muscles of the larynx. However, a connection from the cortex to the nucleus ambiguous is lacking in our closest relative, the chimpanzee, and in other non-human primates (Kuypers 1958b; Jürgens 1995) as well as pigeons and other vocal non-learning birds (Wild et al. 1997a). Thus, these properties of the human face motor cortex are analogous to those of the pallial nuclei HVC- and RA-like combined, in vocal learning birds (**Fig. 8.12**).

I propose that the human anterior vocal pathway consists of the classical Broca's area of the cerebral cortex, plus an entire band of cortex on either side of Broca's area. Human patients suffering damage in these areas can produce the sounds of language, but they have difficulty learning new speech sounds and in producing correct syntax (Benson & Ardila 1996). In humans, these two deficits are identified as poor repetition and verbal aphasia; in songbirds the equivalent terms used are poor imitation or disrupted song learning and syntax deterioration. Thus, the properties of this strip of cortex, for which I have suggested the name 'the language strip' (Jarvis 2001), most resemble those of a combination of the avian pallial MAN-like and Mo-like vocal nuclei. Damage to two other major brain regions in humans also leads to language deficits in imitation and syntax, without eliminating production entirely, though sometimes the effects are worse than those after damage in the language cortex. These are the anterior portion of the human striatum and a region of the anterior human dorsal thalamus (Damasio et al. 1982; Naeser et al. 1982; Graff-Radford et al. 1985; Alexander et al. 1987; **Fig. 8.12**). The worst deficits may perhaps occur because, as in the brains of non-human mammals, connections from the cortex converge in the striatum, and then into the thalamus, concentrating function into smaller and smaller areas (Parent & Hazrati 1995). These properties of the human anterior striatum and thalamus are analogous to those of the striatal Area X-like and thalamic DLM-like vocal nuclei of vocal learning birds. My interpretation is that the human anterior pathway, consisting of a language cortex strip, anterior striatum, and anterior dorsal thalamus, is connected in a cortical–basal ganglia–thalamic–cortical loop, much as in non-vocal areas of non-human primates (Parent & Hazrati 1995) and in the anterior vocal pathways of vocal learning birds.

I suggest that the human auditory pathway for language comprehension is a sensory perceptual learning pathway with projections from the hair cells in the ear that reach the midbrain (the inferior colliculis), the thalamus (the medial geniculate), the primary auditory cortex, and the secondary auditory cortex (Wernicke's Area). All mammals, birds, and reptiles examined have a similar set of connections (Carr & Code 2000). When tested, the cerebral part of this pathway in various species has proved,

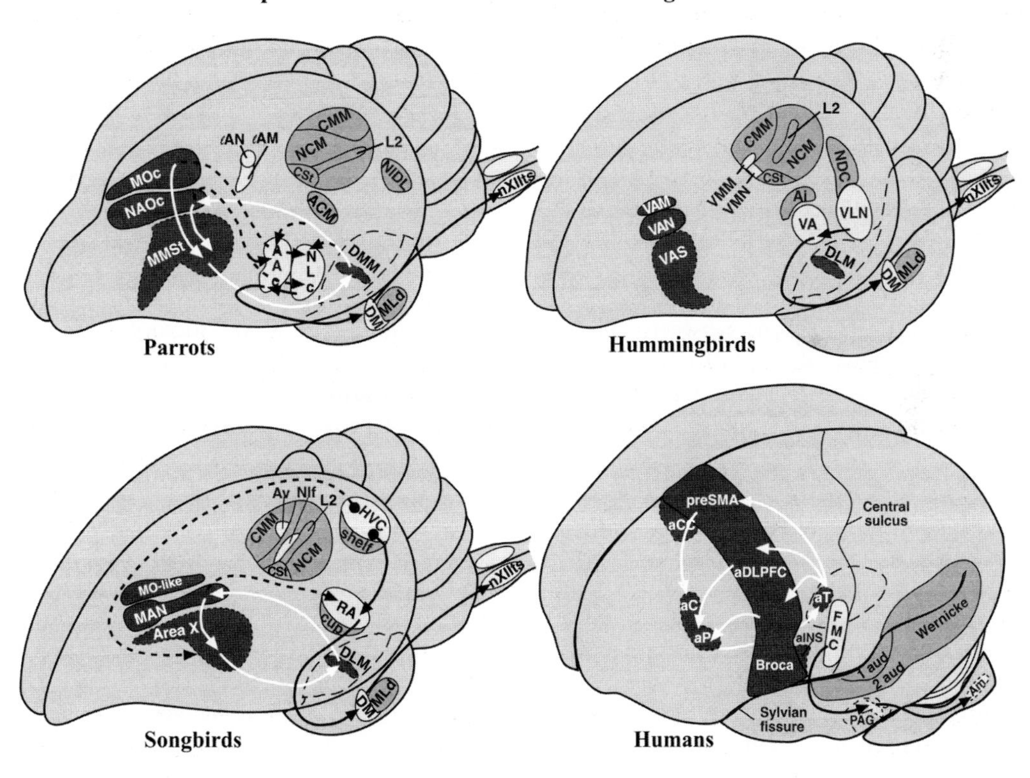

Figure 8.12 A speculative comparison of left hemisphere brain organization and connectivity of vocal and auditory brain areas of parrots, hummingbirds, humans and songbirds. **White arrows** indicate proposed anterior vocal pathways and black arrows indicate proposed posterior vocal pathways. Connectivity was extrapolated from the following sources: songbird (cited in **Fig. 8.2** legend); parrot (Paton et al. 1981; Brauth et al. 1987, 1994, 2001; Brauth & McHale 1988; Striedter 1994; Durand et al. 1997; Farabaugh & Wild 1997; Wild et al. 1997b); hummingbird (Gahr 2000). Connectivity in hummingbirds is not known, and humans is conjectured, or based upon findings in non-human mammals and vocal learning birds (Kuypers 1958a).

as in songbirds, to be tuned to the perception of species-specific sounds (Wang et al. 1995).

There is an additional parallel in the occurrence of cerebral dominance. As mentioned earlier, in humans and in many songbirds, the left side of the brain is dominant over the right in vocal communication. According to my proposed model, the left human posterior vocal pathway controls the production of speech phonemes and words and the left human anterior vocal pathway controls the sequencing of phonemes into words and words into sentences. The human vocal pathways on the right side control singing in a similar manner. When the left language areas are damaged, humans have trouble producing or sequencing speech (aphasia), but when the right language areas are damaged they have trouble producing and sequencing singing (amusia; Wertheim 1969; Benson & Ardila 1996). Like vocal learning birds, there is a sex difference in the language areas of humans, but in the reverse direction. Women, although left dominant, are more prone than men to utilize both sides of their brains for speech and language; they have more connections between the two hemispheres, a higher density of language area synapses, and larger language brain regions than men (Shaywitz et al. 1995; Witelson et al. 1995; Harasty et al. 1997). Such differences are thought to explain why, from a young age, women have better language skills than men (Bradshaw 1989; Mann et al. 1990; Halpern 1992). These brain differences between the sexes, although less extreme than in many birds, are also thought to be under the influence of testosterone and estrogen (Kimura 1996).

From this strictly anatomical point of view, and leaving aside all of the cognitive requirements for language, I offer the speculation that the 'main' difference between the vocal communication systems of humans and vocal learning birds lies, not in the presence or absence of particular pathways, or in their pallial organization, but in the relative sizes of structures that make up their vocal pathways. In the anterior vocal pathway, humans appear to have a much larger pallial representation (MAN- and Mo-like regions) relative to the striatum (Area X-like region), compared with vocal learning birds. Such a huge size difference is consistent with the prodigious variability and complexity of human vocal behavior that appears to be greater than all of the vocal learning birds combined. The pallial organizational differences between mammals and birds, a layered cortex in humans and a globular pallium in birds (**Fig. 8.1**), must mean that there is more than one type of brain organization that can produce vocal learning behavior.

One wonders then how humans, members of a relatively young, approximately 120,000-year-old species, evolved brain pathways for learned vocal communication that appear to be somewhat similar to those used by the very distantly related vocal learning birds? As far as their auditory pathways go, birds and humans probably inherited them from a common reptilian-like ancestor. The general ability to engage in auditory learning can explain why dogs, chimpanzees, and other animals are able to acquire an understanding of human speech sounds. Dogs presumably have a Wernicke's-like area for learning to process complex sounds.

With regard to the vocal pathways, humans obviously must have evolved their cerebral vocal pathways independently of a common ancestor with birds and reptiles, given that vocal learning is so rare. This conclusion calls for some caution, given the occasional reports of other animals, such as seals, chimpanzees, Japanese macaques, and prairie grouse said to have imitated either human speech or calls of what were thought to be a different dialect of their species (Hayes & Hayes 1951; Sparling 1979; Ralls et al. 1985; Masataka & Fujita 1989; Perry & Terhune 1999). But we need more experimental study before these anecdotes can be properly assessed; most have been difficult, if not impossible, to replicate (Owren et al. 1993; Marler 1999).

With independent evolution as the most plausible hypothesis, then the similarity of the vocal pathways of humans and vocal learning birds would suggest that they evolved under some

Northern cardinal Read

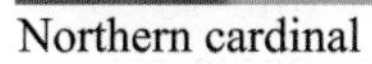

Brown thrasher Cassady

Northern mockingbird Elliott

Swiftlet Suthers

Plate X

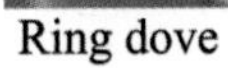
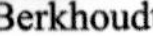
Ring dove Berkhoudt

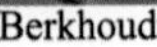
Collared dove Berkhoudt

Red-eyed dove Janse

Feral pigeons Slabbekoorn

type of environmental, epigenetic constraint. One advantage of learned vocalizations is the ability to adapt sounds for more efficient communication in different environments (see Chapter 6). Humans are thought to have evolved in the East African Rift Valley where within limited areas you can find tropical forest, savannah, and desert. Perhaps vocal learning evolved in both birds and humans to cope with diverse environments, favoring diverse vocal behavior, exploiting the potential of brain pathways, based upon a design that was already part of the basic vertebrate plan. To make the necessary changes in the brain, a motor system would be needed for producing learned movements, with posterior and anterior pathway components already present, to be directed to the lower motor neurons that control the muscles of the vocal organs; once connected to a sensory pathway for hearing, this could become a recipe for vocal learning.

CONCLUSIONS

Scientists have a come a long way in their studies of brains and birdsong. As we acquire new information, the ideas and concepts presented here will be subject to change. Some of the exciting advances we can anticipate will come from deciphering the role that the anterior vocal pathway plays in learned vocal communication. Another profitable area will be the role of replayed electrical activity in the brain associated with singing during sleep. We still have much to learn about the mechanisms of auditory–vocal integration and the search for a template. The discovery of new neurons in the adult brain has revolutionary implications for medical science. The molecular biology of vocal learning is helpful in understanding genetic mechanisms of behavior, and in resolving the great mystery of how vocal learning evolved. Some areas as yet unexplored include the study of vocal brain areas in the other mammalian vocal learners, cetaceans, and bats. The extensive knowledge we now have about vocal learning in birds may provide a useful guide in how best to approach the study of these mammalian vocal learners, though cetaceans will always be a challenge. New techniques may emerge for exploring brain connectivity and behaviorally-driven gene expression in human brains in an ethically responsible manner, though it is not yet clear how best to proceed. Students of both human and non-human vocal behavior need to share mindsets and develop a shared vocabulary, if the translation of discoveries from one field to the other is to be facilitated. This is especially important if research on vocal learning birds is to be of any help in understanding the underpinnings of language in the human brain. Last, but not least, we must not lose sight of the continuing need for studies on the peripheral structures, the beak, the syrinx and its muscles, which, like our lips, tongue, and throat, play a key role in the physical production of learned sounds.

Chapter 9

How birds sing and why it matters

RODERICK A. SUTHERS

INTRODUCTION

Songbirds have both an esthetic and a scientific impact on our lives. Birdsong adds beauty and vitality to our environment. The possibility of its absence due to increasing levels of environmental toxins, so eloquently described by Rachel Carson (1962) in her groundbreaking book, *Silent Spring*, was a potent factor in mobilizing public opinion to become better stewards of our environment. Scientifically, the highly developed vocal communication of this group, sharing as it does a number of parallels with speech, has made songbirds especially suited for the study of complex, learned vocal communication at every level, from its molecular biology to its evolution.

This chapter is about how birdsong is produced. How do birds generate such a variety of sounds and what vocal gymnastics are required to produce them? The vocal system is the interface between the bird's brain and his song. Knowing how it functions, and coming to appreciate its limitations as well as its capabilities, is important in understanding vocal communication. Birdsong is the product of the carefully coordinated activity of many different muscles. Some of these are associated with the vocal organ but many belong to other muscle groups such as the respiratory system and the vocal tract. Together they convert nerve impulses from the brain into song.

Less is known about the vocal mechanisms of non-songbirds, and they will be mentioned only briefly to indicate the variety of avian vocal systems, and to provide a perspective from which to focus on the oscine songbirds. Readers interested in additional information on songbirds should consult reviews of this subject (Nowicki & Marler 1988; Suthers 1997, 1999a, b; Gaunt & Nowicki 1998; Doupe & Kuhl 1999; Suthers et al. 1999; Goller & Larsen 2002).

AN ORGAN FOR SINGING

The avian vocal organ, the syrinx, is located deep in the chest in an air sac that is connected to other air sacs and the lungs. It is present in all birds except vultures, but its exact location and structure varies considerably (King 1989). In some species, including doves, pigeons and parrots, the syrinx consists of modified rings of cartilage near the lower end of the trachea (**Fig. 9.1**).

In other species, such as penguins, many owls, nightjars, oilbirds and some cuckoos, the syrinx exists as two semisyrinxes located in the primary bronchi (**Fig. 9.1**). In many birds, however, including but by no means limited to the oscine songbirds, the syrinx is at the junction where the two primary bronchi join to form the trachea. This tracheobronchial syrinx includes modified cartilages from both the upper end of each bronchus and the lower end of the trachea (**Fig. 9.1**).

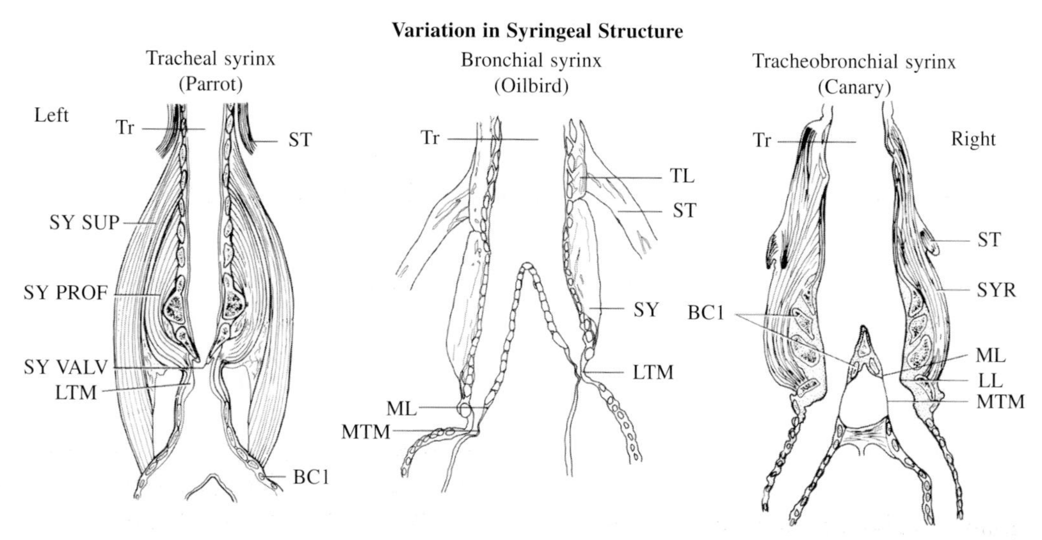

Figure 9.1 Examples of variation in syringeal anatomy. The tracheal parrot syrinx has two syringeal muscles and a pair of lateral tympaniform membranes. The bronchial syrinx of the oilbird has one pair of syringeal muscles and a pair of medial and lateral tympaniform membranes in each bronchus. Songbirds have several pairs of syringeal muscles in their tracheobronchial syrinx. Tr, trachea; ST, sternotrachealis muscle; SY SUP, superficial syringeal muscle; SY PROF, deep syringeal muscle; SY VALV, pneumatic valve; LTM, lateral tympaniform membrane; MTM, medial tympaniform membrane; BC1, first bronchial cartilage; SY, syringeal muscle; ML, medial labium; LL, lateral labium; SYR, muscles of syrinx. (Oilbird modified after Suthers & Hector 1985; parrot and canary modified after King 1989).

The Respiratory Rhythm

The timing of vocalization and the tempo of song begins with the respiratory cycle (Vicario 1991). Expiratory muscles compress the bellows-like air sacs in the thorax and abdomen, increasing respiratory pressure so air flows out through the syrinx and trachea (Hartley 1990). During song, a small inspiration called a 'mini-breath' (Calder 1970) is usually taken after each syllable, except at very high syllable repetition rates (Hartley & Suthers 1989). Inspiratory muscles expand the thorax and abdomen, reducing pressure in the air sacs and reversing the direction of airflow through the syrinx (Wild et al. 1998). Vocalization occurs during expiratory airflow, aside from rare exceptions when it occurs during inspiration, as in the 'wah' sounds of doves (Gaunt et al. 1982), and certain syllables in the songs of some zebra finches (Goller & Daley 2001). Most birds sing with their beaks open, but doves and pigeons coo with their beaks and nares closed. Sound is produced during airflow in an expiratory direction through the syrinx and into the esophagus. After each coo there is a short expiration followed by a brief inspiration (Gaunt et al. 1982).

The Source of Sound

It is now generally agreed that vocalizations are generated by airflow-induced oscillation of elements in the wall of the syrinx that convert some of the air's kinetic energy into acoustic energy. These oscillations are presumably sustained by interaction between Bernoulli forces and the inertia of air in the vocal tract, in much the same way as are oscillations of human vocal folds (Titze 1994; Gardner et al. 2001). An alternative mechanism for sound production based on the principle of an aerodynamic whistle was put forth (Gaunt et al. 1982) to explain pure tone vocalizations in birds such as doves. However, this aerodynamic whistle hypothesis is not supported by the observation of tympaniform membrane vibration in pigeons (Goller & Larsen 1997a), or by experiments with collared doves vocalizing in light gas mixture (Ballintijn & ten Cate 1998). In both collared doves and ringdoves, there is evidence that the sound generated in their syrinx contains prominent harmonics, which must be filtered out by the vocal tract to produce the tonal properties of the coo (Beckers et al. 2003b).

The anatomical structures in the syrinx that produce the sound differ with the syringeal anatomy. Goller and Larsen used a fiberoptic endoscope to make direct observations of syringeal motion during vocalizations elicited by stimulating vocal control regions in the brain of anesthetized birds (**Box 37**, p. 277). In the tracheal syrinx of pigeons they showed that the lateral tympaniform membranes (LTMs), on each side of the trachea bulge toward each other into the tracheal lumen during vocalization where they form a slit-like aperture and vibrate during respiratory airflow (Goller & Larsen, 1997a, b; Larsen & Goller 1999). The LTMs (**Fig. 9.1**) also can be seen to fold into the tracheal lumen and vibrate during vocalization in the tracheal syrinx of the cockatiel (Larsen & Goller 1999, 2002).

In songbirds, the relatively well-developed medial tympaniform membrane (MTM) at the cranial end of each bronchus was long assumed to be the source of sound (Miskimen 1951; Greenewalt 1968; Fletcher 1988; Gaunt 1988; **Figs. 9.1 & 9.2**). However, endoscopic observations of the songbird syrinx during vocalization reveal that it is the connective tissue forming the internal and external labia at the anterior end of each bronchus that is adducted into the syringeal lumen and vibrates during vocalization (Larsen & Goller 1999). Surgical removal of both MTMs has only a minor effect on the vocalization, demonstrating that whatever their function, they are not essential for song production.

Syringeal Muscles Have Separate Functions

The muscular apparatus capable of acting on the syrinx to control vocalization receives input from the brain by the hypoglossal nerve. In some species, such as swiftlets, there are no intrinsic syringeal muscles, meaning muscles having both their origin and insertion on the syrinx (**Box 36**, p. 275). In these birds, syringeal function during vocalization is controlled by muscles that have at least one of their attachments outside the syrinx. Other birds, such as parrots, have one or more pairs of specialized syringeal muscles.

Doves have two pairs of extrinsic syringeal muscles (see **Fig. 10.5**, p. 310). One, the sternotrachealis, reduces the tension across the syrinx and allows the lateral tympaniform membranes to fold into the tracheal lumen where airflow causes them to vibrate. Contraction of the other muscle, the tracheolateralis, abducts the tympaniform membranes out of the lumen and terminates phonation (Goller & Larsen 1997a). The role of these muscles, if any, in frequency modulation of coos is uncertain. Gaunt (1988) reported that during a coo both muscles are simultaneously active in ring doves, and Beckers et al. (2003c) demonstrated that gradual frequency modulation in the coo, but not abrupt frequency jumps, closely parallels pressure changes in the interclavicular air sac, suggesting expiratory muscles may be important regulators of sound frequency. Amplitude modulation of coos is associated with interruptions of tracheal airflow presumably due to the closing and opening of a valve, perhaps formed by the adducted tympaniform membranes or the glottis (Beckers et al. 2003c).

Songbirds have the most complex syringeal musculature of all. They have 5 pairs of muscles that attach to the syrinx and a sixth pair that acts on it indirectly via its insertion on the trachea (King 1989). In general, more muscles provide more degrees of freedom in configuring the syrinx to produce different sounds (Gaunt 1983). The complexity of song is roughly correlated with the number of syringeal muscles, but this relationship is not a simple one. Parrots, for example, succeed in producing complex vocalizations with only two pairs of syringeal muscles (**Fig. 9.1A**) and the Australian lyrebird,

BOX 36

AVIAN SONAR: CLICKING IN THE DARK

Two groups of birds, some *Aerodramus* Asian swiftlets (Medway 1967; Griffin & Suthers 1970; Griffin & Thompson 1982; Coles et al. 1987) and the neotropical oilbird (Griffin 1953; Konishi & Knudsen 1979) make clicking sounds with their syrinx that enable them to navigate in the dark by echolocation. Although swiftlets and oilbirds have quite different lifestyles, their sonar allows them to nest in the relative safety of dark caves. Swiftlets are diurnal insectivores and spend the day on the wing catching insects which they locate visually. Oilbirds are nocturnal frugivores with excellent night-time vision. They spend the daytime resting on ledges in their nesting cave and use vision, supplemented by echolocation on dark nights, to feed on the fruits of tropical trees.

Clicks make good sonar signals because they contain energy over a wide range of frequencies and have a short duration, making it easier to hear faint echoes reflected by surrounding objects (**CD2 #64**). Furthermore, clicks can be produced by an anatomically simple syrinx. Although details of click production differ between oilbirds (Suthers & Hector 1985) and swiftlets (Suthers & Hector 1982), in both it involves rapid, synchronous opening and closing of both sides of the syrinx. Each cycle typically produces a pair of clicks (**A**). The first click occurs when the syringeal valve, located at the labia and tympaniform membranes, briefly vibrates as it narrows the syringeal lumen just before closing it completely. After a short, silent interval while the valves are closed, a second click occurs as the valve begins to open, again allowing airflow to vibrate the membranes until they are withdrawn from the syringeal air stream.

Swiftlets have no intrinsic syringeal muscles so they operate their syringeal valves by using extrinsic muscles in sequence to first close the valve by pulling the tracheobronchial syrinx towards the bronchi (accomplished by the sternotrachealis muscles) and then opening the valve by stretching bronchi (through contraction of the tracheolateralis muscles; Suthers & Hector 1982). The mechanism is similar in oilbirds with two important differences. Oilbirds have a bronchial syrinx (Müller 1841; Garrod 1873; see p. 273; **Fig. 9.1**), in which each syringeal valve (i.e. semisyrinx) is in the primary bronchus and the valve is opened by a syringeal muscle that lies along the surface of each bronchus and attaches to a bronchial cartilage bordering the lateral tympanic membrane (Suthers & Hector 1985). One of the beauties of this simple mechanism for producing sonar clicks is that two sonar pulses are generated during each cycle of muscle contraction, thus doubling the number of echoes produced.

Why do some birds have a bronchial syrinx? Oilbirds may provide a clue. A puzzling feature of their syrinx is that the two semisyrinxes are placed at different positions along the bronchus, with the distance from the trachea varying on the right and left sides, and between individuals. By inserting a plug in either the left or right bronchus to seal it off from the trachea, we showed that each bronchus had a quarter wave resonance, like that of a stopped tube, allowing frequencies with a wavelength 4 times the length of the bronchial tube to pass while filtering out other frequencies (Suthers 1994; **B**). The distribution of sound energy as a function of frequency thus depends on the length of the section of rigid bronchus between the trachea and semisyrinx. Because this length differs between the left and right bronchus and varies between individuals, the vocalizations of each oilbird contain a pair of frequency bands of high intensity sound, or formants that are unique to that individual. Detectable in some echolocating clicks (Suthers & Hector 1988), these formants are prominent in many of the bird's long duration broadband 'social' vocalizations, and might provide a way for individuals to recognize each other in the dark (**CD2 #65**). Interestingly, most bronchial syrinxes occur in nocturnal species (Suthers 1994).

Roderick A. Suthers

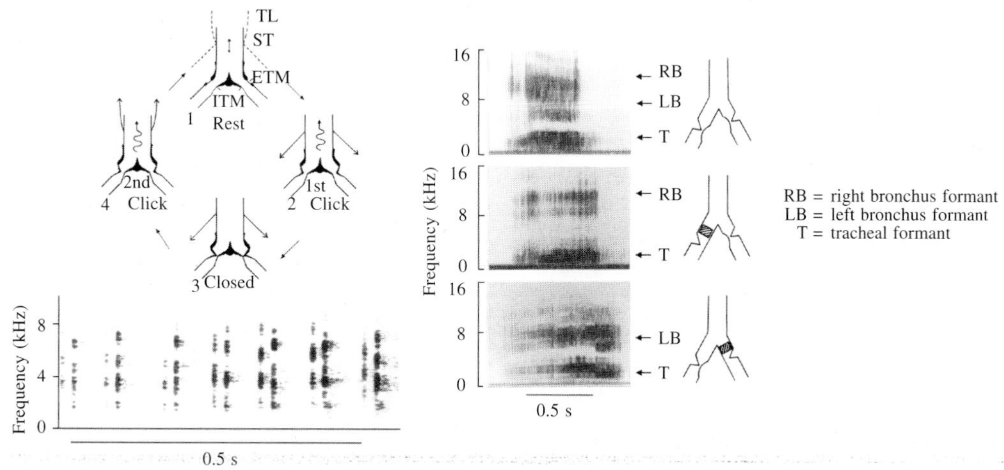

known for its vocal versatility and now generally considered to be a songbird (Higgins et al. 2001), has but 3 pairs. The remainder of this chapter focuses on the oscine songbirds.

The function of syringeal muscles in songbirds has been studied by recording their electrical activity in the form of electromyograms (EMGs), while measuring the rate of airflow through each side of the syrinx, and monitoring respiratory pressure as the bird moves about freely in its cage and sings spontaneously in front of a microphone (Suthers 1990, 1997; Suthers et al. 1994, 1996, 1999; Goller & Suthers 1996a, b; **Box 37**, p. 277). Additional insights into syringeal function have been obtained by using an endoscope to make direct observations on the motion of syringeal structures in response to electrical stimulation of individual muscles in anesthetized birds (Larsen & Goller 2002). Although different muscles may have more than one role in sound production and interact in complex ways that are still poorly understood, two of the most basic aspects of song, its timing and its frequency, are primarily controlled by the dorsal and ventral syringeal muscles, respectively, acting in conjunction with the muscles that mechanically ventilate the respiratory system during breathing.

The dorsal muscles, the syringealis dorsalis and the tracheobronchialis dorsalis, play a key role in operating a pneumatic valve at the upper end of each bronchus that controls the timing of phonation on each side of the syrinx (Goller & Suthers 1995b, 1996b). Each valve is formed by the medial and lateral labia that are also the structures generating sound (**Fig. 9.2**). When these dorsal muscles are relaxed during quiet respiration, and during mini-breaths between syllables, the labia are abducted out of the air stream. In this position they do not oscillate and the syringeal lumen is open, presenting minimal resistance to airflow. Contraction of the dorsal muscles initiates sound production by adducting the labia into the lumen where they form a slit and oscillate, as air flows across them (Goller & Suthers 1995b; Goller & Larsen 1997b; Larsen & Goller 1999). Stronger contraction of these muscles, perhaps assisted by other muscles (Larsen & Goller 2002), terminates or prevents phonation on the same side of the syrinx by moving the labia still closer together and closing the slit, stopping airflow and arresting labial oscillation.

The large ventral syringeal muscles, the syringealis ventralis, appear to be involved in regulating sound frequency (**Fig. 9.2**). The amplitude of their EMG is correlated in a positive way with the fundamental frequency of the vocalization (Goller & Suthers 1995b, 1996a). They presumably control fundamental frequency by varying the tension or elasticity of the oscillating labia. Upward-sweeping frequency modulation (FM) is accompanied by a corresponding increase in ventral muscle activity, which is also high during high-pitched notes with a high fundamental frequency but little FM.

There is great variety in the frequency structure, or 'syllable morphology' of birdsongs. Producing such a variety of sounds would seem to require very complicated patterns of syringeal muscle activity to generate a different vocal gesture for each syllable type, although some theoretical models of song production suggest the vocal gestures required may be much simpler than previously assumed (Gardner et al. 2001; Laje et al. 2002; Suthers & Margoliash 2002).

Sensory Feedback: Fine-tuning the Song Motor Program

To achieve or maintain the optimum respiratory pressure and the appropriate configuration of the syrinx during song, the bird needs to monitor these variables as it sings. Using sensory feedback to achieve this, it could then adjust the ongoing motor program being sent from the brain to the muscles to correct any deviations from the intended vocal gesture. Such deviations might occur as a result of changes in posture or ongoing physical activity. We know that auditory feedback is essential during song learning (Konishi 1965a), and continues to play a role in the long-term maintenance of adult crystallized song (Nordeen

BOX 37

RECORDING FROM THE SYRINX OF A SINGING BIRD

Much can be learned about song production by monitoring respiratory dynamics and the activity of syringeal and respiratory muscles during song. The bird is anesthetized and a tiny thermistor bead is placed in the lumen of each primary bronchus, just below the syrinx (Suthers 1990; Suthers et al. 1994). The thermistors are connected to a circuit that maintains them at a constant temperature of about 60°C. Air flowing through the bronchus during either inspiration or expiration will conduct heat away from the thermistor and the additional current required to maintain its temperature provides a measure of the rate of airflow through each side of the syrinx.

The respiratory pressure driving airflow through the syrinx is measured with a miniature piezoresistive pressure transducer mounted on the bird's back and attached to a small tube inserted into one of the air sacs. The bird's respiratory system is different from that of mammals. Birds have several air sacs that act like bellows pumping air through the lungs. Air sacs are connected to the lungs and to each other by complex passages and openings. A midline interclavicular air sac connects the two sides of the respiratory system and contains the syrinx, so respiratory pressure should be the same in both bronchi. By measuring the respiratory pressure, inspirations and expirations can be distinguished. The movement and position of the labia on each side of the syrinx can be monitored by using bronchial airflow and respiratory pressure to calculate changes in syringeal resistance. A positive air sac pressure with no airflow indicates the labial valve on that side of the syrinx is closed, whereas a high flow rate with low pressure indicates the labia are drawn back out of the air stream.

The function of various syringeal and respiratory muscles can be deduced by inserting a pair of very fine hair-like stainless steel wires into them and recording the relationship of their electromyograms (EMG) while monitoring syringeal airflow, respiratory pressure and the acoustic properties of the song at a microphone in front of the bird (Goller & Suthers 1996b). All of these physiological signals are carried by fine wires to a velco tab on the bird's back where they connect to other wires that travel up through the top of the cage to recording instruments. Birds instrumented in this way are free to move around in their cages and sing spontaneously. At the end of an experiment, which may last a week or more, the bird is again anesthetized, the apparatus is removed and he is returned to the aviary.

Observing syringeal motion directly, Goller and Larsen did some elegant experiments using an angiofiberscope to view the mechanical action of the syrinx in response to electrical stimulation of the song control system in the brain of anesthetized birds (Goller & Larsen 1997b; Larsen & Goller 2002). A 1.4 mm diameter optical fiber was inserted, either down the trachea to view labial movements, or into the air sac containing the syrinx to see the muscles on its external surface. In some experiments they electrically stimulated individual muscles, and in others (Larsen & Goller 1999) they used a laser optical system to detect the vibration of sound-producing structures. Although the vocalizations produced by this method are not normal song, they provide valuable information about how sound is produced.

Roderick A. Suthers

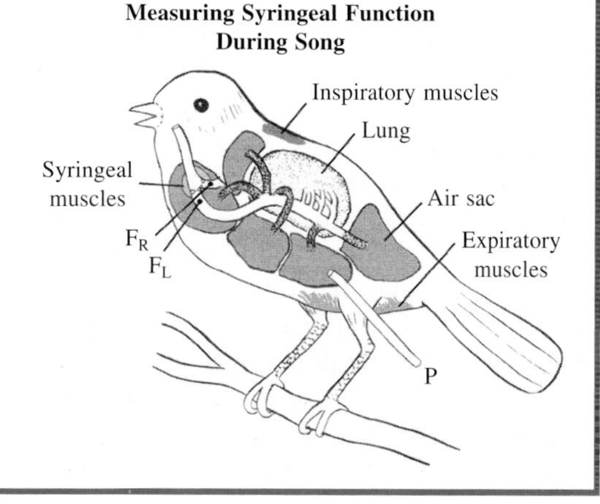

Location of the transducers in the respiratory system used to study song production. F_R, thermistor measuring airflow through the right side of the syrinx; F_L, thermistor measuring airflow through the left side of the syrinx; P, cannula in the air sac to measure respiratory pressure. EMG electrodes may be placed in syringeal, thoracic inspiratory and/or abdominal expiratory muscles. Inspiratory muscles enlarge the chest, reducing air sac pressure and drawing air in. Expiratory muscles compress the chest and force air out through the syrinx.

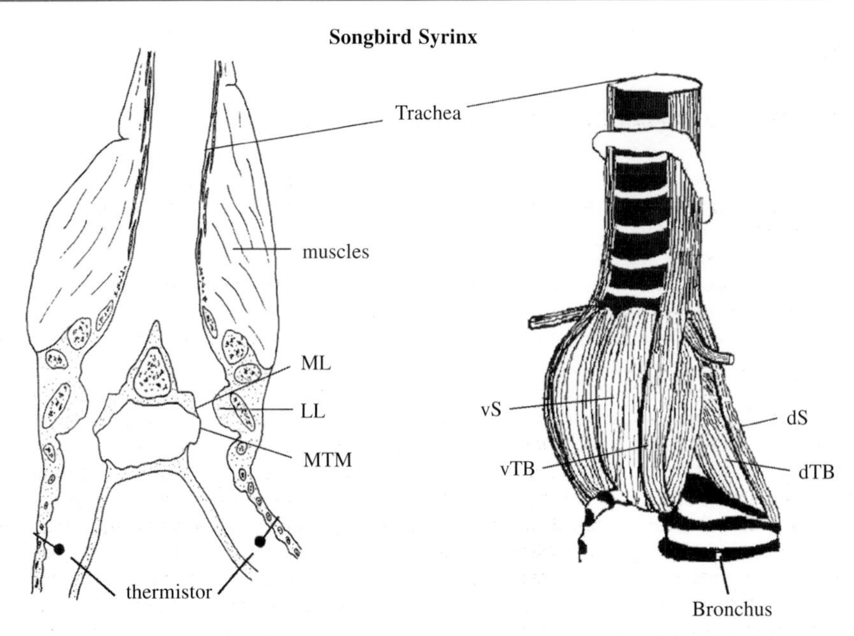

Figure 9.2 A songbird syrinx showing its bipartite structure with 2 sound generators and multiple pairs of muscles. On the left is a frontal section through a brown thrasher syrinx showing the location of microbead thermistors used to sense airflow through each side. The medial and lateral labia form pneumatic valves at the cranial end of each bronchus and oscillate to produce sound. On the right is a ventrolateral view of the same syrinx depicting the syringeal muscles. vS, ventral syringeal muscle; vTB, ventral tracheobronchialis muscle; dS, dorsal syringeal muscle; dTB dorsal tracheobronchialis muscle. For other abbreviations see **Figure 9.1**. (Modified after Goller & Suthers 1996a).

& Nordeen 1992; Okanoya & Yamaguchi 1997; Leonardo & Konishi 1999; **Box 38**, p. 279).

Non-auditory feedback from mechanoreceptors or proprioceptors that may sense the tension can provide a rapid response to correct a parameter that is too high or too low. Air pressure or the relative position of structures in the respiratory and vocal system are also important, even in adult crystallized song. For example, if the respiratory pressure is momentarily increased experimentally during a syllable by injecting a small, randomly timed puff of air into the air sac of a singing cardinal, it elicits a compensatory partial relaxation of the abdominal expiratory muscles that returns the pressure to near its original level (Suthers et al. 2002). Since this muscle response does not require extensive processing at higher levels of the brain, it can occur rapidly – often within the same syllable during which the pressure perturbation occurred. Similar recordings from syringeal muscles indicate that they also adjust their activity to compensate for unexpected perturbations in subsyringeal pressure during phonation (Suthers & Wild 2000). The sensory receptors that provide the feedback underlying these responses have not been identified, but likely include mechanoreceptors or proprioceptors in the muscles, air sacs or walls of the respiratory and vocal systems.

Lateral Independence of Motor Control

Each side of the syrinx is capable of acting independently, since it is innervated separately by the tracheosyringeal branch of the hypoglossal nerve on the same side, which in turn receives most of its input from the same side of the brain (Nottebohm & Nottebohm 1976; Nottebohm 1977; Vicario & Nottebohm 1988; Wild et al.

BOX 38

SONG IS DISRUPTED BY DELAYED FEEDBACK AND THEN RECOVERS

Songbirds use auditory feedback to both learn and maintain their songs, and the dynamics of this process may be investigated by manipulating the auditory feedback heard by singing birds. Deafening adult zebra finches results in a complete loss of auditory feedback and, after roughly four weeks, a marked degradation in song structure (Nordeen & Nordeen 1992). However, it is difficult, though not impossible (Woolley & Rubel 2002), to restore auditory feedback to deafened birds. In contrast, by using a computer-controlled system to continuously monitor a bird's vocalizations, specific types of artificial auditory feedback may be generated in real-time, played back to the bird as he sings, and started and stopped at any moment (Leonardo & Konishi 1999). If the artificial feedback signal is a 100 ms delay of the bird's song (ABCD) as he sings it, then when the bird sings syllable B he will hear B+A, with the normal feedback for B superimposed with the delayed feedback for A. Two to four weeks of exposure to delayed auditory feedback caused a dramatic loss of the spectral and temporal stereotypy seen in crystallized song, consisting of stuttering, deletion, and distortion of song syllables and the creation of new ones (**CD2 #66**). Stuttering sometimes increased the length of a song bout of 2–6 s to more than 60 s. Restoration of normal feedback enabled the recovery of the original songs of each bird. All of the changes in song structure gradually became more infrequent, and were eventually replaced by the temporal and spectral organization characteristic of the original song. A complete recovery took 2–4 months. Thus adult zebra finches appear to retain a great deal of potential song plasticity, even though they do not normally modify their songs after crystallization. Furthermore, despite destabilization of the behavior, a memory of the original song persists, and can be used to correct the vocal output upon restoration of normal auditory feedback. This indicates that the song is maintained by an active control process which requires both a memorized song model and the auditory feedback generated during singing. Elucidating the neural mechanisms underlying this feedback control process remains an area of vigorous research.

Anthony Leonardo

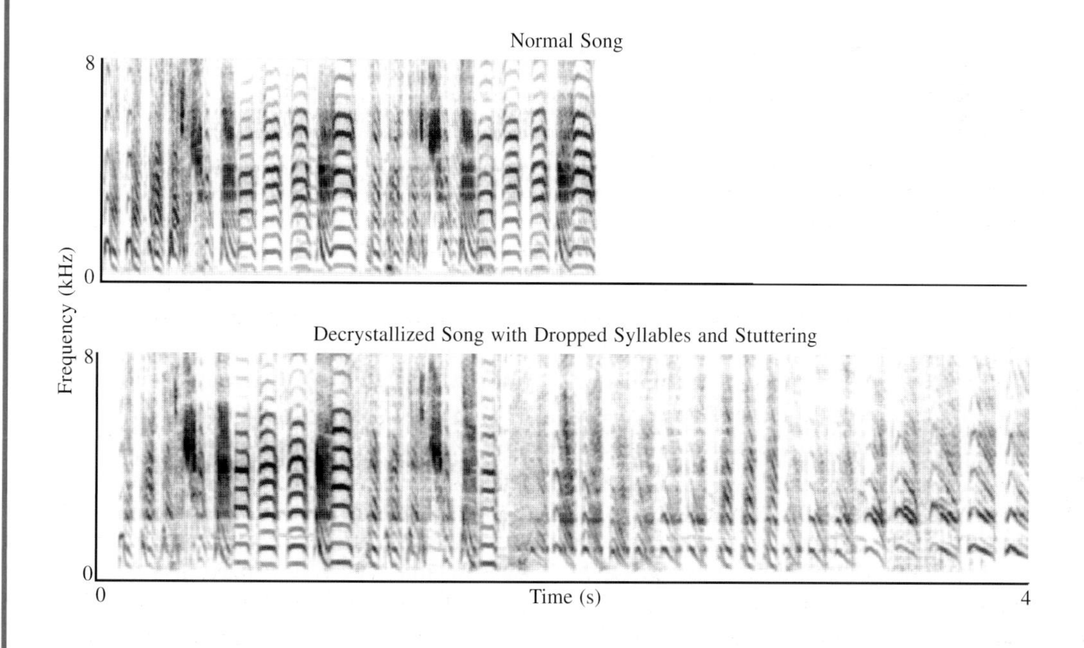

2000). The implications of this independence of right and left sides of the syrinx for vocal communication are far reaching, for it gives songbirds the possibility of two, virtually independent sound sources. Although both right and left sides are subjected to similar respiratory pressures generated by the respiratory muscles and both share the same supra-syringeal vocal tract, the dorsal syringeal muscles can switch sound production from one side of the syrinx to the other very rapidly, silencing one side by closing it while holding the labia on the opposite side in a phonatory position. Likewise, the ventral syringeal muscles can independently vary the fundamental frequency generated on each side. Songbirds have taken full advantage of this lateral independence to greatly increase the variety and complexity of their songs (Greenewalt 1968; Stein 1968; Suthers 1999a).

The Stereotypy of Adult Song

Most adult songbirds sing a stereotyped repertoire in which reiterations of the same syllable type vary little in their acoustic properties. This acoustic stereotypy, together with the fact that there is, at least sometimes, no immediate change in the vocalization if auditory feedback is removed by deafening, is part of the evidence suggesting that the pattern of syringeal and respiratory muscle activity required to produce a particular syllable is stored in the brain as a 'motor program' (Konishi 1985; Vu et al. 1994). According to this hypothesis, the syllables of crystallized song might be produced by activating the appropriate network of neurons in the brain; these are presumably wired together in such a way that they automatically generate the basic pattern of coordinated muscle contractions, the vocal gesture, needed to produce that syllable. Auditory or somatosensory feedback may not be required, at least in the short term, to produce the basic vocal gesture, but it could adjust or modify it to correct for errors that might otherwise occur if, for example, conditions vary in the peripheral vocal or respiratory system. Although central pattern-generating circuits for different syllable types have not yet been identified in songbirds, they have an important role in controlling many rhythmic behaviors in other animals (Getting 1989; Pearson 1993).

Whatever the neural mechanism underlying syllable production, syllables rather than whole songs, seem to represent functional units of song production. Birds normally stop singing after finishing a syllable, as opposed to in the middle of a syllable. If one interrupts the song of a zebra finch by flashing a strobe light, these stops are almost always between syllables (Cynx 1990; Franz & Goller 2002). The stereotypy of syllables is reflected in the stereotypy of the pattern of expiratory pressure and airflow that accompanies each syllable (Allan & Suthers 1994; Suthers et al. 1996; Franz & Goller 2002). These respiratory parameters depend on both the syringeal muscle activity influencing the aperture of the labial valve in each side of the syrinx and on the activity of the expiratory muscles that compress the air sacs and generate the positive subsyringeal pressure that drives airflow. Each syllable type has a relatively invariant pattern of pressure and airflow produced by its vocal gesture. When zebra finches copy a syllable they typically also copy the pattern of respiratory pressure that the tutor used to produce that syllable (Franz & Goller 2002; **Box 42**, p. 322).

CARDINAL SONG: A CASE STUDY

Even relatively simple songs like that of the northern cardinal require skillful coordination of syringeal and respiratory muscles. An adult male cardinal assembles its songs from a repertoire of between 8 and 21 different syllables (Halkin & Linville 1999). A song, a few seconds long, typically consists of from one to several of these syllables, each repeated several times to form a series of phrases (**CD1 #67**). There are distinctive patterns of airflow and pressure associated with each syllable type. These patterns reflect the fact that each is produced by a stereotyped motor pattern coordinating the respiratory and syringeal muscles to produce the particular vocal gesture for that syllable (**Fig. 9.3**). Many syllables include extended upward or downward frequency sweeps that sometimes exceed 2 octaves. Cardinals are unusual in that females also sing, though their song is less stereotyped and harmonics are more prominent compared to the songs of males (Lemon & Scott 1966; Dittus & Lemon 1969; Yamaguchi 1998).

Except during a rapid trill, each syllable is followed by a mini-breath that replenishes respiratory air. Since both inspiration and expiration are active processes, each syllable–mini-breath cycle is accompanied in sequence by contraction first of the abdominal expiratory muscles (Hartley 1990), followed immediately by the thoracic inspiratory muscles (Wild et al. 1998). During trills with high syllable repetition rates there is not enough time for a mini-breath between syllables. Instead, expiratory muscles continue to contract and maintain a positive pressure below the syrinx during the entire trill. One side of the syrinx is kept closed and the other side produces the trill by repetitively opening and closing so that each puff of air generates a syllable (**Fig. 9.3**).

To our ears, the long FM sweeps of a cardinal sound smooth and continuous. It was therefore surprising when closer examination revealed that each side of the syrinx contributes to a different part of the same syllable. Fundamental frequencies below about 3.5 kHz are consistently sung on the left side, and those above this frequency are sung on the right side (Suthers & Goller 1996, 1997; Suthers 1997, 1999a; Suthers et al. 1999; **Fig. 9.3**). The two sides of the cardinal syrinx are thus specialized to operate over different frequency ranges, facilitating the production of extended FM sweeps that cross the boundary between these registers, and thus contain sequential contributions from each side. A syllable that sweeps from 2 to 7 kHz, for example, begins in the left syrinx and is switched to the right syrinx midway in the upward sweep. This sequence of lateralized production is reversed in syllables that sweep from high to low frequencies. Most impressively, the song generally contains no hint of this switch from one set of oscillating labia to the other. To the human ear, and when viewed as time–frequency sonograms, there is usually no obvious interruption or discontinuity in the frequency sweep at the moment of change from one side to the other.

This is an extraordinary feat of virtuosity. Consider the sequence of actions that a cardinal must execute with precision to produce even an acoustically simple note lasting less than half a second. A single broadband FM sweep from high to low begins with: (i) closure of the left syrinx; (ii) expiratory muscle contraction; (3) opening the right syrinx and configuring it to produce the first portion of the sweep; (iv) closing the right syrinx while (v) opening the left syrinx and configuring it to continue the sweep; (vi) closing the left syrinx to terminate phonation; (vii) relaxing expiratory muscles, and (viii) contracting inspiratory muscles; (ix) opening both sides of the syrinx for an inspiratory mini-breath to replace the air used to produce the syllable. Depending on the repetition rate of the syllable, this entire sequence of events may take place in scores of milliseconds and be repeated with precision up to 16 times per second to produce a phrase in the song!

LEARNING TO SING

Juvenile songbirds must meet many developmental challenges on their way to adulthood. Not the least of these is learning to sing their species' song. Males must become expert vocalists before they are a year old if they are to have a chance of reproducing during their second year. How does a young cardinal, of either sex, go about learning to execute this 'checklist' of actions that are necessary to produce even a simple wideband frequency sweep (see Chapter 3)?

Vocal Learning

Cardinals are age-limited, closed-ended learners, and their adult song repertoire is learned during their first year (Lemon & Scott 1966; Dittus & Lemon 1969; Halkin & Linville 1999). Shortly

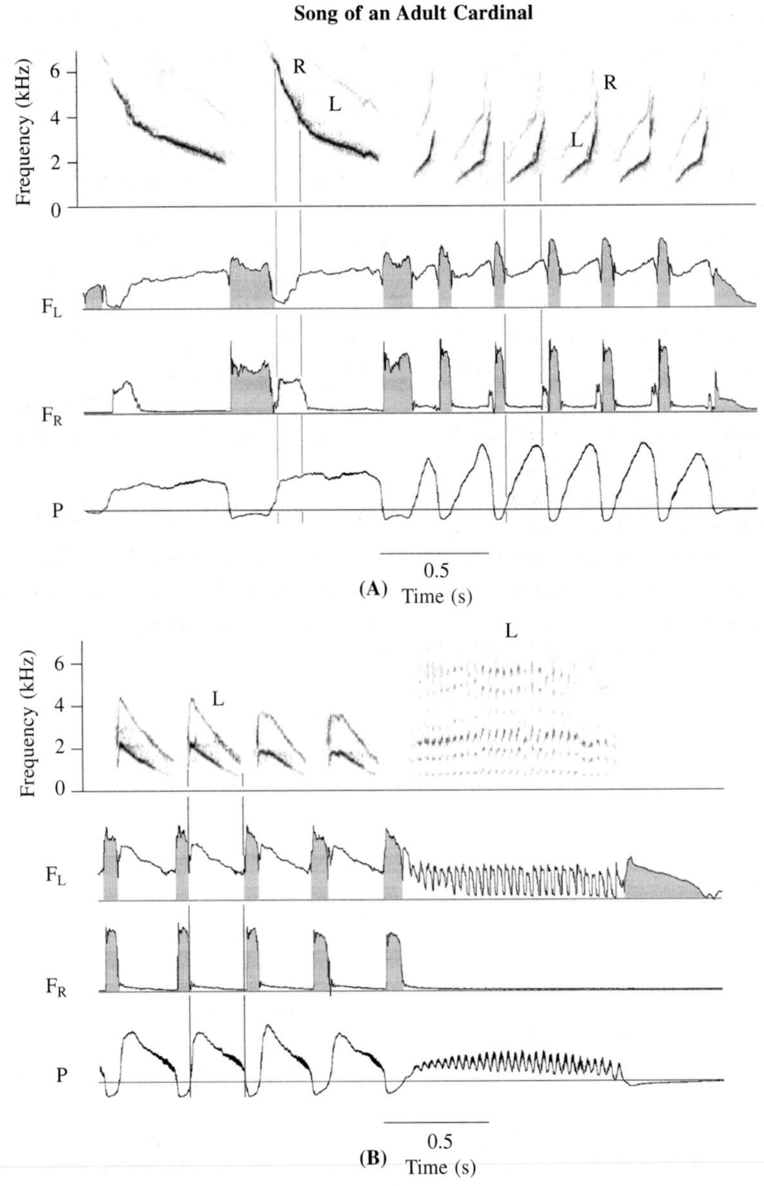

Figure 9.3 Phrases from songs of an adult northern cardinal. (**A**) Two long, downward sweeping FM syllables followed by faster upward FM syllables. In each syllable, frequencies below about 3.5 or 4.0 kHz are sung on the left side (L) as indicated by airflow through the left syrinx; higher frequencies are produced on the right side (R). Absence of flow through either side during a positive respiratory pressure indicates that the labial valve is closed, preventing both airflow and sound production on that side. Note the mini-breaths between each syllable and the stereotypical patterns of airflow and pressure that are characteristic of each syllable type. (**B**) Another syllable with a low frequency fundamental produced entirely on the left side (L). This is followed by a trill using pulsatile expiration instead of mini-breaths. Expiratory pressure remains positive during the entire phrase. F_L and F_R show the rate of airflow through the left and right syrinx, respectively, P is the pressure in the cranial thoracic air sac. Horizontal lines indicate zero airflow and ambient air pressure. Shaded portions of the airflow are inspiratory mini-breaths.

after fledging, juveniles begin to produce sequences of low intensity, highly variable notes having little resemblance to adult song. Gradually during the autumn, and then again early in the following spring, these vocal ramblings become louder and some notes begin to recur in a recognizably similar form. This early stage of vocal development, 'subsong,' (**Fig. 9.4**) is slowly transformed into the next stage, 'plastic song,' as a recognizable, though initially poorly controlled, syllable morphology develops (**CD1 # 067**). Eventually, when the bird is about 10 months old plastic song becomes indistinguishable from adult song. Instead of being plastic or malleable, the song becomes fixed or 'crystallized.' Subsong and plastic song are thought to involve trial and error motor learning during which the juvenile relies on auditory feedback of his own vocalizations (Konishi 1965a), and presumably somatosensory feedback (Suthers et al. 2002); he learns the muscle actions underlying the motor patterns that are required to produce adult song based on an auditory 'template' in the brain, which may be either innate or shaped by experience early in life (Marler 1997). Experiments on zebra finches, in which this process of sensorimotor integration was temporarily disrupted at different stages of song development by using botulinum toxin to reversibly block transmission of nerve impulses to the syringeal muscles, suggest that in addition to the sensitive period for song memorization, there is another one for sensorimotor integration during the latter part of song development (Pytte & Suthers 2000).

Milestones in Vocal Motor Control

Little is known about how the respiratory or vocal motor skills of adult song develop. At what stage in their development are young birds able to coordinate respiratory and syringeal muscles,

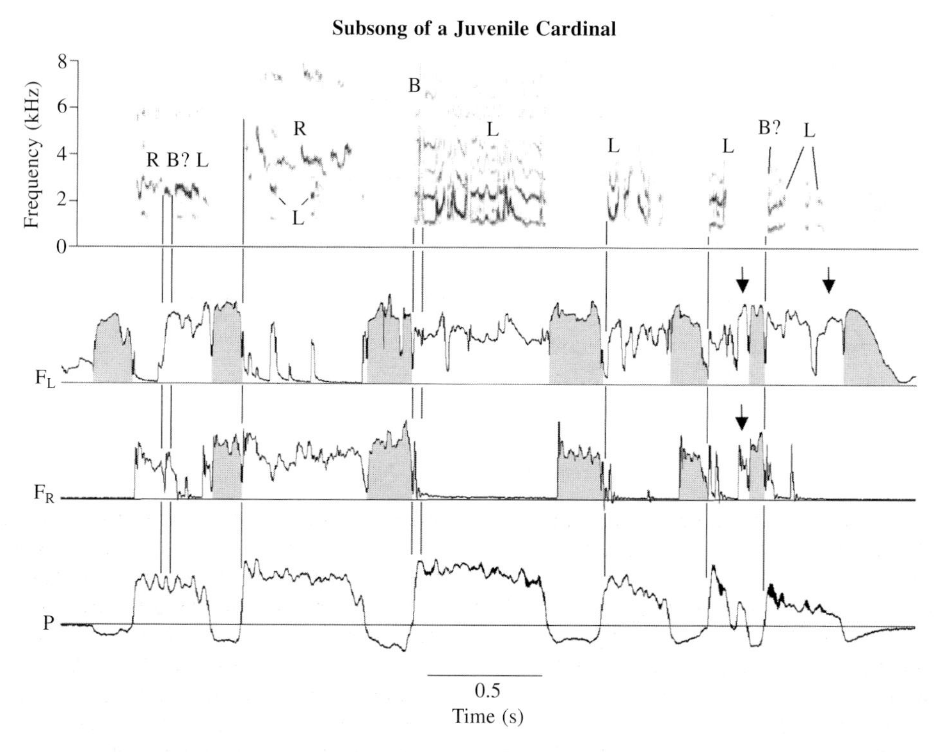

Figure 9.4 Subsong from a 92-day-old juvenile cardinal. Arrows indicate periods of expiration without sound production. Abbreviations as in **Figure 9.3**. In those marked B, both sides contribute.

or achieve independent control of their two syringeal sound sources? Are there bottlenecks in the process underlying motor development and coordination that limit song acquisition? Does the progression through successive stages of juvenile song correspond to the mastery of specific hurdles in vocal production? To better understand the motor development of song, we studied its production in juvenile male cardinals at various stages of vocal learning (Suthers & Goller 1998a, b).

Even young nestlings may produce different sounds simultaneously on each side of the syrinx. Two-voice elements in the begging calls of young chaffinches indicate that separate sounds are being generated on each side of the syrinx in nestlings only a few days old (Nottebohm 1971b; Wilkinson 1980). It is not clear, however, if these two-voice vocalizations are simply a passive byproduct of poorly controlled airflow through both sides of the syrinx, or if they are under active control of the syringeal muscles.

In cardinals, the ability to control which side of the syrinx produces sound is present in fledglings singing subsong when they are less than 6 weeks old. Juveniles at this age are already capable of unilateral sound production by closing one or the other side of their syrinx, as do adults. Sometimes both sides of the syrinx produce sound at the same time, but often a note is generated only on the left or the right side, the silent side being closed to airflow (**Fig. 9.4**). Unlike adult song, each note is quite variable and the frequency structure of successive left and right notes is not coordinated or linked (Suthers & Goller 1998a).

When adult cardinals sing, the timing of expiratory airflow is carefully coordinated with vocalization. Each syllable typically begins as soon as the syringeal valve opens and ends when the valve closes (**Fig. 9.3**). As a result, nearly all of the exhaled air contributes to phonation and very little is 'wasted' between syllables, even during the fast pulsatile trills. The respiratory–syringeal coordination necessary for this efficient use of air during song is not present at the beginning of juvenile song, and emerges only gradually during the early stages of song development. During early subsong and even into plastic song, expiration sometimes continues into or through the silent intervals between notes (**Fig. 9.4**, arrows).

As subsong progresses, there is a tendency to alternate sides and to produce a pair of notes, one from each side, during a single expiration. This foreshadows the adult motor pattern of using the two sides in sequence to make a single continuous frequency sweep. But in early juvenile song the pair of notes within a single expiration are not joined, nor is there any clear, consistent coordinated sequence in their frequency patterns. The linking of a high frequency sound from the right with a lower frequency contribution from the left side of the syrinx to form a coordinated, seamless frequency sweep occurs in the latter part of song development.

PUSHING THE ACOUSTIC ENVELOPE

In adult song, the respiratory and syringeal muscles act together like experienced partners in a dance choreographed by the brain, but the dance steps differ between species. The possession of two parallel sound sources, each with its own pneumatic valve and neural control, gives oscine songbirds an additional degree of freedom absent in species that have a tracheal syrinx or that lack independent control of the two sides of the bronchotracheal or bronchial syrinx. Studies of how different species use the two sides of their syrinx during song suggest that the distinctive styles of singing among various songbird taxa are achieved by using their duplex vocal organ in different ways. These may optimize particular acoustic effects at the cost of sacrificing virtuosity in other aspects of sound production (Suthers 1997, 1999a, b; Suthers & Goller 1997).

Longer Song Bouts Versus Faster Tempos

Increasing Song Duration

Respiratory air is an essential, but potentially limited, resource for vocalization. Sustained song or even the duration of individual notes might be limited by the volume of air that can be exhaled to produce them. This possible limitation is particularly acute in small birds that sing big songs. Canaries are small birds that can sing in a seemingly continuous fashion for up to about a minute. During a single song they may produce scores or hundreds of syllables, yet their resting respiratory tidal volume is only a fraction of a milliliter! Calder (1970) showed that the depth of the thorax increases slightly between most notes in a canary song. From this he hypothesized that the bird takes a mini-breath between syllables, to replenish its air supply. Measurements of respiratory pressure and tracheal airflow during song showed that each mini-breath replaces about the same volume of air that is exhaled to produce the syllable (Hartley & Suthers 1989). As a result, there is probably little or no net change in respiratory volume during mini-breath song.

The mini-breath respiratory pattern is not unique to songbirds. It appears to be used by many species, including non-songbirds as diverse as parrots and oilbirds (Suthers 2001). In our own species, shortened inspirations, analogous to mini-breaths, are also present during conversational speech (Hixon 1973). In all these cases brief inspirations permit a continuity of vocal communication that is free from periodic interruptions by long inspirations.

Increasing Syllable Repetition Rate

Although mini-breaths allow a bird to sing longer songs, they limit the song's tempo to rates of syllable repetition having silent intervals long enough to accommodate an inspiration. Some birds get around this constraint by switching to a different respiratory technique involving pulsatile expiration, described above for cardinals, which does not require the reversal of airflow between syllables. Pulsatile expiration can almost double the maximum syllable repetition rate that is possible with mini-breaths. This increase in syllable repetition rate is achieved at the cost of depleting the volume of respiratory air available for sound production. This limits the length of phrases based on pulsatile expiration, but it is unclear how often this limit is actually reached, though this may sometimes occur (see Chapter 11).

The syllable repetition rate that requires a bird to switch from mini-breaths to pulsatile expiration is probably determined by the physical properties of respiratory structures – particularly the mass and elasticity of the thorax and abdomen – that must oscillate at the syllable repetition rate to drive ventilation. Small birds can sustain a mini-breath respiratory pattern at higher syllable repetition rates than can large birds. The 'cut-off' repetition rate for mini-breaths for an 18 g canary is about 30 syllables per second compared to about 16 syllables per second for a 35–40 g cardinal. Pulsatile expiration is not subject to this limit since it requires little movement of the body wall, and the labia, whose valve-like action gates airflow, have little mass.

Some zebra finches attempt to get the best of both worlds by vocalizing during some mini-breaths between 'normal' expiratory syllables (Goller & Daley 2001). These inspiratory syllables have a distinctive high fundamental frequency and are copied by young zebra finches exposed to them in tutor songs. It is not known if inspiratory syllables have special perceptual significance in communication or why more birds don't use them. It is possible that they may be energetically expensive to produce since the labia must remain in the air stream during inspiration, resulting in a higher resistance to airflow and presumably requiring more effort from the inspiratory muscles.

Increasing Spectral Diversity

Bandwidth: Left and Right Vocal Registers

Songbirds use the two sides of their syrinx in different ways to extend the bandwidth of their song or to produce syllables that vary widely in their tonal quality. The range of frequencies a song encompasses, its bandwidth, is increased by having each side of the syrinx specialized to cover a different frequency band. In all species studied, the fundamental bandwidth of the right side is shifted upward compared to that of the left (Suthers 1999a), but there is often substantial overlap in the frequencies produced by each side. Cardinals are unique, among the small sample of species studied, for their limited overlap between frequencies produced on each side and their habit of combining sound from the two sides to produce a single FM note. The anatomical or physiological basis for these separate left–right vocal registers is not known. In many songbirds there are anatomical asymmetries between the two primary bronchi that might include subtle differences in the labia or muscles on each side, as yet unidentified. Whatever its mechanism, the increased overall bandwidth available for vocalization increases the possibilities for spectral diversity in songs.

The influence on song of lateralized specialization for particular frequency ranges is evident in different strains of canaries. One of these is the Waterschlager canary, which was inbred for its distinctive low-pitched song. Another strain is the outbred domestic canary. Nottebohm and Nottebohm (1976) conducted a series of experiments on Waterschlagers in which either the left or right side of the syrinx had been denervated by cutting the tracheosyringeal nerve on one side. They demonstrated that this strain produces about 90% of its syllable repertoire in its left syrinx. As a result of the strong lateral dominance of the left side, most of this bird's song is below 3 kHz, except for an occasional right side syllable that has a median frequency range between 2.5 and 4.2 kHz (Nottebohm & Nottebohm 1976). This left dominance is in contrast to song of the outbred domestic strain in which each side of the syrinx contributes about an equal proportion of the repertoire, and some syllables include notes from both sides (Suthers et al. 2001). In these domestic birds, syllables frequently have fundamental frequencies as high as 5 or 6 kHz, and a single bilaterally generated syllable may have a bandwidth of 6 or 7 kHz compared to 3 or 4 kHz in the case of unilaterally produced syllables.

Subsequent studies show that Waterschlager canaries have an inherited auditory defect that decreases their sensitivity to sounds above 2 kHz by as much as 40 dB (Okanoya & Dooling 1985; Gleich et al. 1994a, b; Gleich & Klump 1994). Their strong left syringeal dominance may well be the result of being partially deaf to most of the sounds from the right syrinx.

In the family *Fringillidae*, which includes sparrows, cardinals and Darwin's finches, there is an inverse relationship between syllable bandwidth and repetition rate – rapid trills have narrower bandwidths (Podos 1996; see Chapter 11). It is suggested that syllable bandwidth might be constrained at high repetition rates by the time required to change the shape and tuning of the mouth and throat to track the rapidly changing fundamental frequency (Nowicki et al. 1992; Podos 1997, 2001). In many cases beak opening is correlated with sound frequency and appears to be a factor in tuning resonances of the vocal tract by altering its effective length and suppressing harmonics. Podos argues that beak size and syllable repetition rate have evolved together in Darwin's finches. He suggests that the relatively low syllable repetition rate of finches with large beaks is due to the difficulty of moving larger mandibles rapidly to quickly change the tuning of the vocal tract filter (Podos 2001). The role of vocal tract resonance and beak movements in birdsong is covered in Chapter 11 (Nowicki 1987; Westneat et al. 1993; Moriyama & Okanoya 1996; Suthers & Goller 1997; Suthers 1999a; Suthers et al. 1999; Hoese et al. 2000; Williams 2001).

Inharmonic Sounds

The songbird syrinx is well suited to achieve dissonant sounds by singing two overlapping notes that are different in pitch but not harmonically related to each other. The widespread occurrence of these 'two-voice' notes and their likely origin from separate sides of the syrinx was pointed out by Greenewalt (1968) and Stein (1968). Two-voice syllables require airflow through both sides of the syrinx, while different motor programs to the muscles on each side generate unrelated sounds that can be independently modulated in frequency and amplitude. Two voice syllables are especially prominent in songs of some mimic thrushes such as grey catbirds and brown thrashers (Suthers 1992; Suthers et al. 1994, 1996; Goller & Suthers 1995a, b, 1996a, b, 1999). The separate, simultaneous 'voices' that form the distinctive inharmonic elements in these songs (**Fig. 9.5; CD1 # 68**) tend to differ less from each other in their fundamental frequency than when each side sings alone, suggesting that simultaneous two-voice phonation comes with constraints on the degree of frequency separation between the two sides that is possible when they operate simultaneously. This limitation might be due to biomechanical interactions between the two halves of the syrinx or to a need to keep both fundamentals within the frequency pass band of the vocal tract, with which both sides connect.

Whereas two-voice syllables depend on the capacity of the two sides of the syrinx to function independently during sound production, inharmonic vocalizations can also result from the nonlinear interaction between the sounds generated on each side. The 'dee' syllable of the black-capped chickadee's song consists of a series of overlapping frequency components that differ from those produced by either side of the syrinx alone. It is not the sum of the sounds produced on each side after unilateral syringeal denervation. It appears instead to be composed of the sum and difference heterodyne frequencies generated by nonlinear interactions between the sound produced by each side of the syrinx (Nowicki & Capranica 1986a, b). The coupling between left and right sound generators might be mediated by air pressure fluctations associated with sound or transmitted mechanically through the syringeal

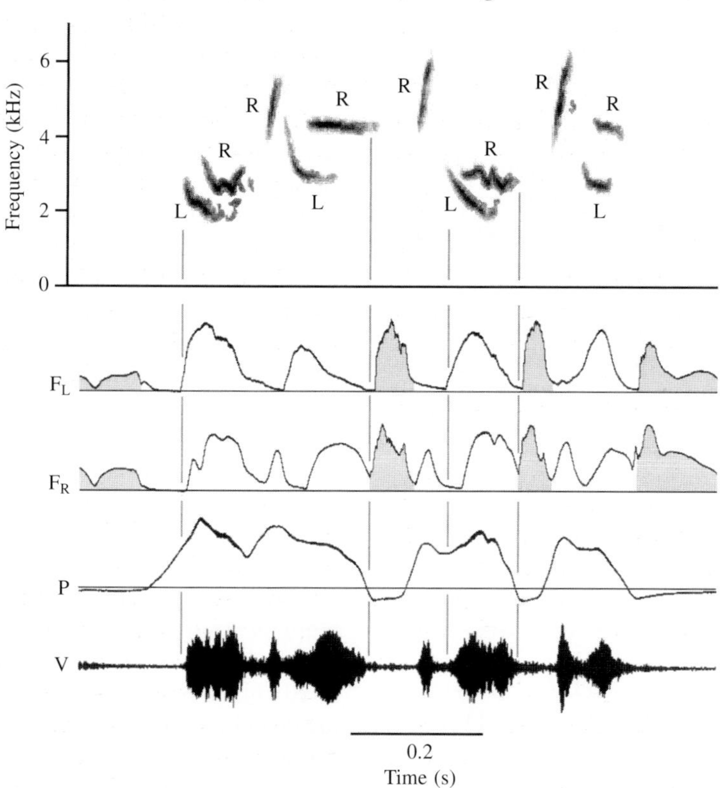

Figure 9.5 Brown thrasher song illustrating inharmonic, two-voice syllables (**CD1 #68**). V is the time waveform of vocalization. Abbreviations as in **Figure 9.3**. Modified from Suthers et al. 1994.

tissue. Similar nonlinear bilateral coupling has not yet been noted in other species, perhaps because bilateral interaction is less prominent in the larger syrinx of most other birds studied.

Other kinds of nonlinear phenomena may also account for the abrupt transitions from periodic to chaotic sounds, as well as period doubling and other acoustic effects that are present in some songs, such as those of the zebra finch (Fee et al. 1998). The realization that non-linear vocalizations may have a role in social communication has led to an increased interest in their production (Fitch et al. 2002).

Spectral Contrast

The duplicate sound sources in the oscine syrinx facilitate the introduction into song of abrupt changes in fundamental frequency. The song of the brown-headed cowbird, for example, begins with 2 or 3 expirations each containing a 'cluster' of notes, rapidly produced using pulsatile expiration (Allan & Suthers 1994; **Fig. 9.6**; **CD1 # 69**). Successive notes are produced on opposite sides of the syrinx. In each case, the first note is sung on the left side and is immediately followed, or may partially overlap, the next note, which is sung on the right side at a higher frequency. The frequency of these alternating notes increases in a staggered step-wise sequence until the end of the expiration. Abrupt frequency steps between successive notes that follow each other with little or no silent interval between them are possible if separate generators are involved. While one side of the syrinx is producing a note, the muscles of the other closed, silent side can adjust the tension on its labia so that it starts the next note on pitch without a slurred FM between the end of one note and the beginning of the next, something considered again below.

PERFORMANCE CONSTRAINTS: INSIGHTS FROM A VOCAL MIMIC

Have these species-specific patterns of song production evolved because they are the best way to generate that species' style of song, perhaps the only way, or are they simply independent modes of singing, each with the potential to generate a wide range of song types? To what extent have the evolutionary forces shaping the acoustic properties of a species' song influenced the evolution of production mechanisms, and vice versa? Here I approach from a different viewpoint, some of the issues considered by ten Cate (see Chapter 10), and by Podos and Nowicki (see Chapter 11).

Vocal mimics, such as mockingbirds, provide an opportunity to investigate these questions. When a mimic copies another species' song does it also use the same mechanism employed by that species to produce the song, or does it invent a different way to produce the song? The northern mockingbird is renowned for its ability to mimic other species (Baylis 1982), but the vocal mechanism it uses to accomplish this is unknown. When a mockingbird imitates a cardinal's song, for example, does it do so by joining sounds produced sequentially on each side of the syrinx to generate extended FM sweeps as the cardinal does, or does it produce an unbroken sweep on one or both sides together?

Zollinger and I (2004) tutored hand-reared juvenile mockingbirds with recorded songs of other species. We found that when a mockingbird sings a cardinal song it usually mimics the vocal mechanism of the cardinal, switching between sides of the syrinx in mid-frequency sweep, but the switch is seldom as seamless or the frequency sweep as smooth as when the cardinal sings it. When the mockingbird's motor pattern differed from that of the cardinal, the mockingbird's vocalization also differed from that of the model. In a similar experiment, juvenile mockingbirds were tutored with recorded cowbird songs. The cowbird's song covers an exceptionally wide frequency range that often extends from a few hundred Hz for the lowest introductory notes to 10 kHz or more in the final whistle. Although mockingbirds were unable to reproduce the highest or lowest frequencies of this tutor, they mimicked both the sound and the cowbird's vocal production mechanism for other notes in the note clusters, singing alternate notes on opposite

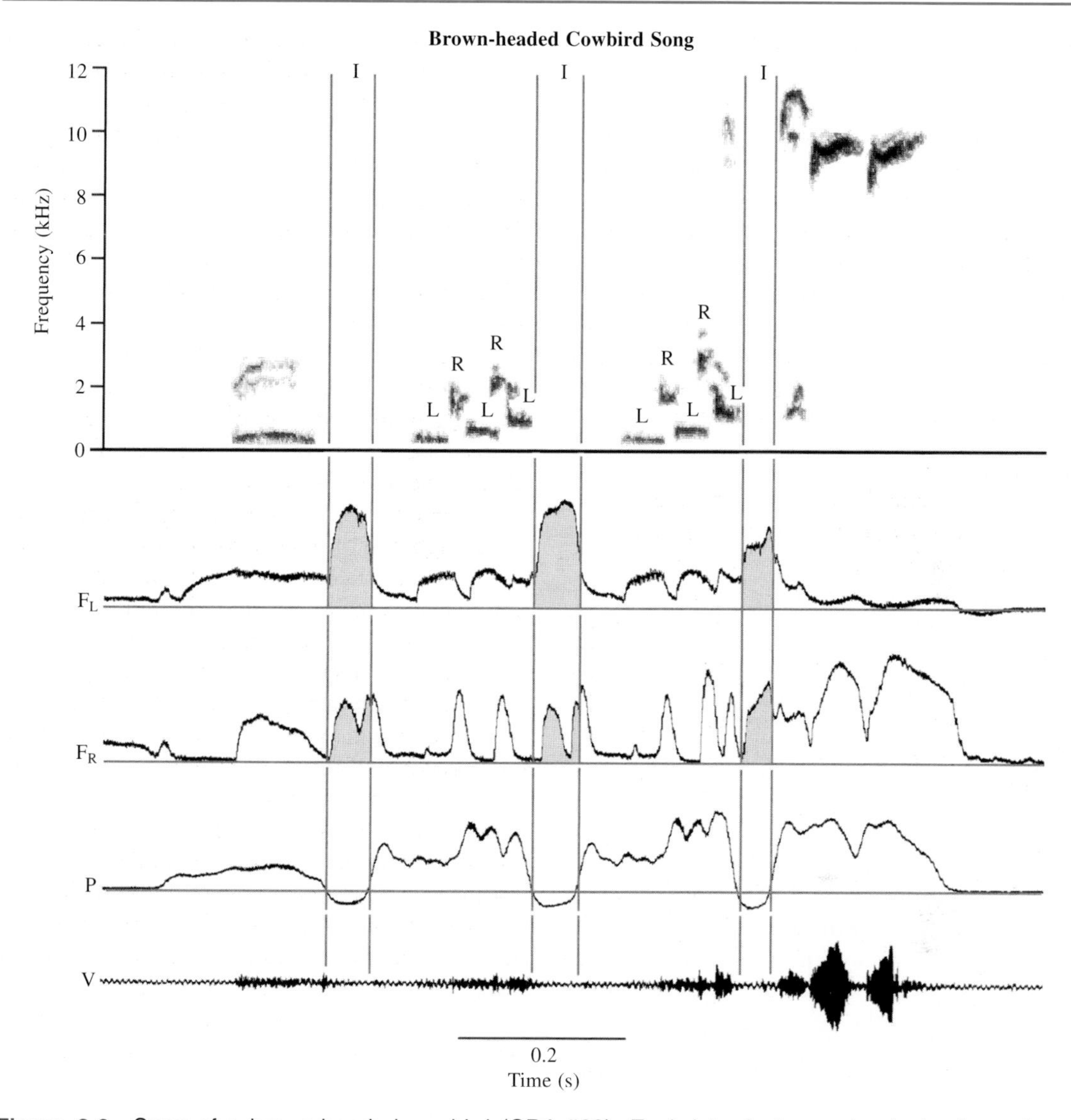

Figure 9.6 Song of a brown-headed cowbird (**CD1 #69**). Each introductory note cluster is produced during a single expiratory phase by pulsatile expiration that alternates between sides, permitting abrupt frequency changes between successive notes. The last expiration produces the final whistle and is always on the right side. Abbreviations as in **Figure 9.3**.

sides, beginning with the lowest frequency note on the left.

To test the possibility that the taped tutor songs of cardinals or cowbirds contained some subtle cues such as brief pauses or discontinuities, other than frequency, that might influence when the mockingbird switches from one side of the syrinx to the other, some juvenile mockingbirds were tutored, not with natural songs, but with computer synthesized cardinal-like FM sweeps or cowbird-like note clusters. When as an adult, the mockingbird mimicked the synthesized cardinal-like FM sweeps, it sang the high frequency portion on its right side and the low frequency portion on its left, switching sides at 3 or 4 kHz, during both upward and downward sweeping sounds (**Fig. 9.7; CD1 #70**). When copying a sequence of frequency-stepped tones similar to the cowbird's note cluster, the mockingbird likewise switched back and forth between sides of his syrinx producing each tone on the same side as would a cowbird.

Perhaps the mockingbirds determine which side of their syrinx to use simply on the basis of the note's frequency. In mockingbirds, like other songbirds, the frequency range of the right syrinx is shifted upward compared to that of the left. Although the absolute frequency of a note is a factor in determining which side of the syrinx will sing it, the context and other acoustic properties of the song are also important and can override frequency bias. This is shown by an experiment in which mockingbirds were tutored with two pairs of synthesized tones similar to those often present in cowbird's note clusters. In each pair, the second tone was at 2 kHz. In one pair this sound was preceded by a lower frequency tone whereas in the other pair, it was preceded by a higher frequency tone, (**Fig. 9.8**). In both cases, the frequency difference between tones was 300 Hz and the second tone started immediately after the first. When the mockingbird mimicked these tone pairs, the side it used to produce the 2 kHz note differed, depending on whether the immediately preceding tone was higher or lower. If the first tone was higher, it was sung on the right and the 2 kHz tone was sung on the left. If, however, the first tone was lower it was sung on the left and the 2 kHz tone was produced on the right (**Fig. 9.8**).

It is clear from this experiment that a 2 kHz note can be produced on either side of the syrinx. Why then did the mockingbird change sides between the tones in each pair? Why not sing the second tone on the same side that was used to produce the first? The answer is apparent from the occasional times a mockingbird failed to change sides between tones. When both tones were sung on the same side of the syrinx, the abrupt step-wise frequency change between tones was lost, and the end of the first note became slurred into the frequency of the second note so that the two were connected with an FM sweep. The ability to achieve abrupt step-wise frequency changes without a significant silent period between notes may depend on exploitation of the two voices of the bipartite syrinx. Perhaps it is only by switching sound production from side to side during the rapidly produced notes in their note clusters that cowbirds are able to achieve the clean frequency steps, or spectral contrast, between notes. This use of the bipartite syrinx circumvents the limited ability of a single sound source to abruptly change the frequency at which it oscillates. To achieve this acoustic effect without slurring, the mimic must exploit the vocal flexibility made possible by two sound sources, as the cowbird does in its song.

Our findings on how mockingbirds copy other species support the view that the songs of different species have been selected to maximize the performance of certain species-specific acoustic features, such as the abrupt frequency steps between cowbird notes, or the smooth extended FM sweeps of cardinals. Most songbirds have become vocal specialists who push the performance limits of singing in certain directions, often at the expense of some other kinds of acoustic skills. To be a successful vocal mimic, the mockingbird must remain to some extent a vocal generalist, a jack-of-all-trades but master of none, who does a fairly good job of mimicking other species, but usually with less expertise than the rightful owner.

VOCAL GYMNASTICS: ARE FEMALES IMPRESSED?

Two important functions of song are, a role in competition between males for breeding territories, and in attracting a mate. Exactly what aspects of song are most important in fulfilling these needs is poorly understood, and may differ between species (Catchpole & Slater 1995; Gil & Gahr 2002; see Chapter 2). Most studies of the relationship between singing and mating success have focused on variables such as repertoire size, or the frequency or duration of singing. These are all amenable to measurement,

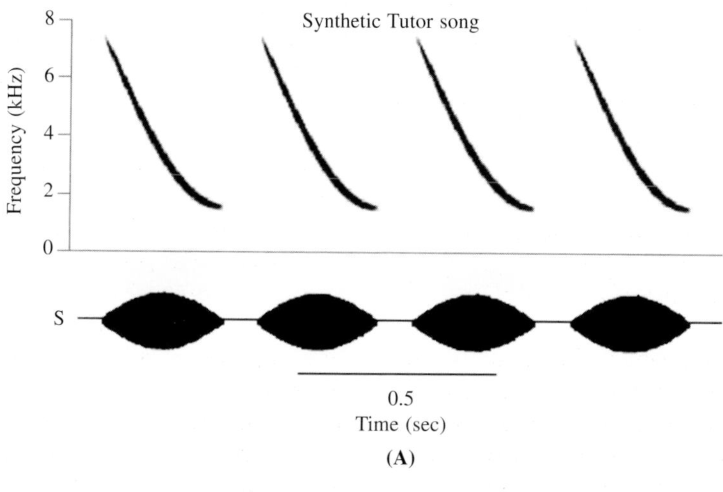

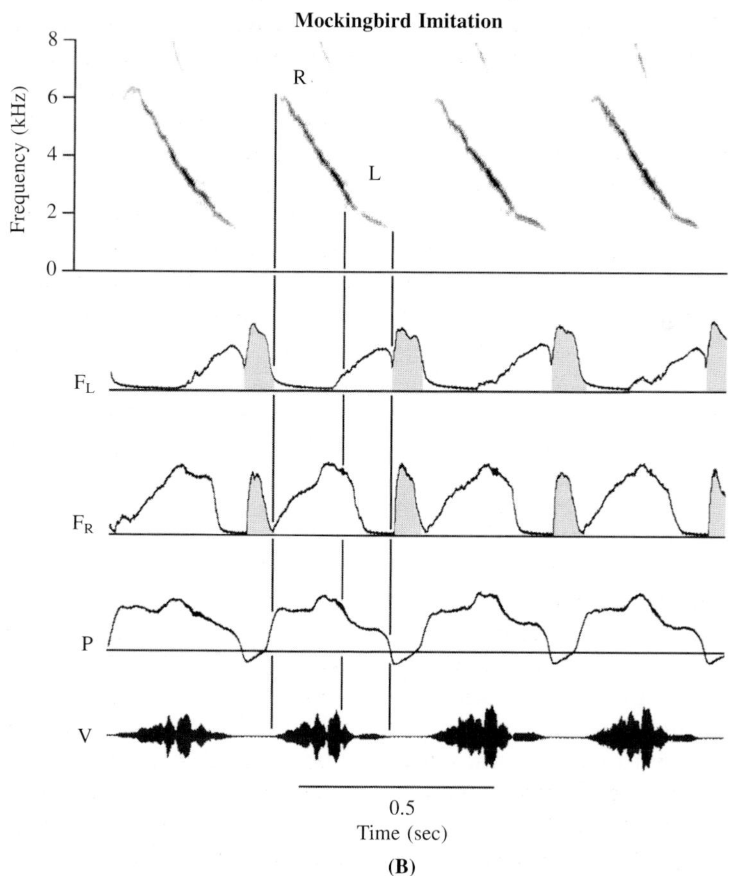

Figure 9.7 (**A**) A tutor 'song' consisting of a series of computer synthesized downward FM sweeps similar to syllables sung by cardinals. (**B**) A mockingbird's reproduction of these sweeps (**CD1 #70**). The timing of airflow through each side of the syrinx shows that the mockingbird begins each sweep on the right side and changes to the left side for the portion of the sweep that is below about 3 kHz. Abbreviations as in **Figure 9.3**.

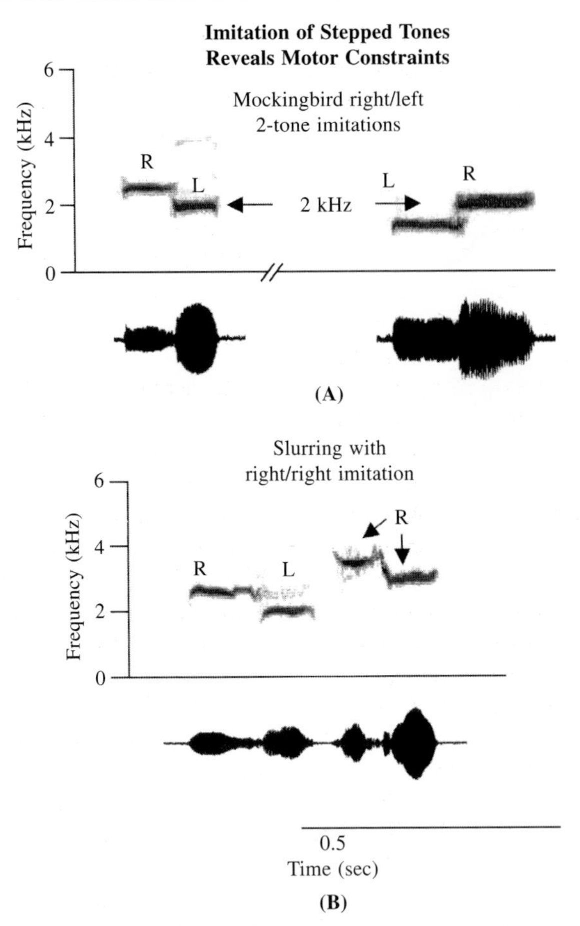

Figure 9.8 (**A**) A mockingbird's imitations of pairs of computer-synthesized tones showing that the context as well as the frequency of a note influences the side on which it is produced. (**A**) When the 2 kHz tone is preceded by a tone at a slightly higher frequency, the first tone is sung on the right and the 2 kHz tone is sung on the left side. If the preceding tone is at a lower frequency, it is sung on the left, and the 2 kHz tone is now sung on the right side. (**B**) Switching sides of production enhances the spectral contrast between successive notes. When singing the second pair of synthesized tones, the mockingbird produced both on the same side. As a result, the two tones are joined by an FM sweep. R and L indicate side of production. Tutor tones and syringeal airflow and pressure are not shown. (**A** modified after Zollinger and Suthers 2004).

but are not necessarily correlated with the level of motor skill, and are unlikely to reflect the difficulty that is required to execute the vocal performance. A male's quality might be reflected not only in his stamina for sustained singing, but also in his ability to flawlessly and repeatedly sing syllables or phrases that require special motor skills (Suthers & Goller 1997; Nowicki et al. 1998a; ten Cate et al. 2002). Among the species in which the motor mechanisms of song production have been studied, an aspect of singing that seems likely to require special vocal motor skills is coordination between the two halves of the syrinx. Examples are the acoustically seamless switch between right and left sides during-wide band frequency sweeps by cardinals or the concurrent production of independently modulated sounds forming two-voice syllables in brown thrashers. It is not yet known whether females pay attention to these particular aspects of cardinal or thrasher song. However, in brown-headed cowbirds and domestic canaries, there is some evidence suggesting that parts of their songs requiring a high degree of motor coordination may have a special perceptual significance to prospective mates.

Male brown-headed cowbirds have a vocal repertoire that includes several different song types. In the eastern subspecies, each song begins with 2 or 3 note clusters followed by a whistle (**Fig. 9.6**). West, King, and their colleagues showed that some songs are more effective than others in eliciting copulation solicitation displays (CSD) from female cowbirds (King & West 1977, 1983; West et al. 1979, 1981; see Chapter 14). A male's breeding success is correlated with the potency of his songs. Experimental manipulation of songs indicates that the note clusters are more effective than the final whistle in eliciting CSDs. A high-frequency note, designated the 'interphrase unit' at the end of the last note cluster appears to be particularly important. Songs of dominant males were more effective than those of other males. The inclusion of stereotyped note clusters in juvenile plastic song triples the effectiveness of juvenile song in evoking a female copulatory response (West & King 1988a). Further experiments with computer-edited a songs are needed to identify the specific temporal or acoustic aspects of note clusters that are most important in stimulating a female. But it already seems clear that a male cowbird's mating success is influenced by his bilateral motor skills in singing stereotyped patterns of notes that are rapidly produced, with alternating steps in frequency.

The domestic canary provides another example of a possible relationship between vocal motor dexterity and female choice. Male domestic canaries have a song repertoire of about 20–30 different syllables. Individual songs contain a subset of these syllables, repeated to form a sequence of phrases. Each syllable type is always sung in the same way and at the same repetition rate, but there are large differences in the spectral and temporal properties of a phrase, depending on its syllable type. Some syllables may be almost a pure tone repeated a few times a second, others include upward or downward frequency sweeps of varying complexity sung at various repetition rates which may exceed 30 per second.

Researchers at the University of Paris (Vallet & Kreutzer 1995; Vallet et al. 1998) found that certain kinds of 'type A' syllables are more effective than others in attracting the interest of a prospective mate (**Box 4**, p. 58). They played different phrases, each consisting of a single syllable type, to sexually receptive female canaries and recorded their CSD responses. The female invites the male to mate with her by arching her back, raising her tail, and fluttering her partially spread wings. Vallet and his colleagues found that wide-band syllables that were more complex than simple frequency modulated sweeps, composed of two or more notes, and sung at a repetition rate of 16 per second or higher evoked significantly more female displays than other syllables. Similar type 'A' syllables recorded from wild canaries also elicit CSDs from domestic canaries when played back to them in the laboratory (Leitner et al. 2001), indicating they are not an artifact of domestication.

Studies of how domestic canaries produce their songs suggest that 'A' syllables may be relatively difficult to sing (Suthers et al. 2001). Domestic canaries sing some syllables in their left syrinx, others in the right, and still other multi-note syllables contain contributions from both sides of the syrinx. The wide-band 'A' syllables of three birds studied, included contributions from both sides of the syrinx (**Fig. 9.9**; **CD1 #71**). The syllables that are most effective in eliciting CSDs from a female canary thus require the male to produce stereotyped notes from each side of his syrinx at note repetition rates at least twice the syllable rate. Furthermore, behavioral playback experiments (Vallet et al. 1998) suggest that the effectiveness of 'A' syllables in stimulating a female is greater if the silent interval between the notes within each syllable is short (between 16 and 23 ms, as opposed to 23–30 ms). This is consistent with the hypothesis that bilateral coordination involving very rapid left–right switching between

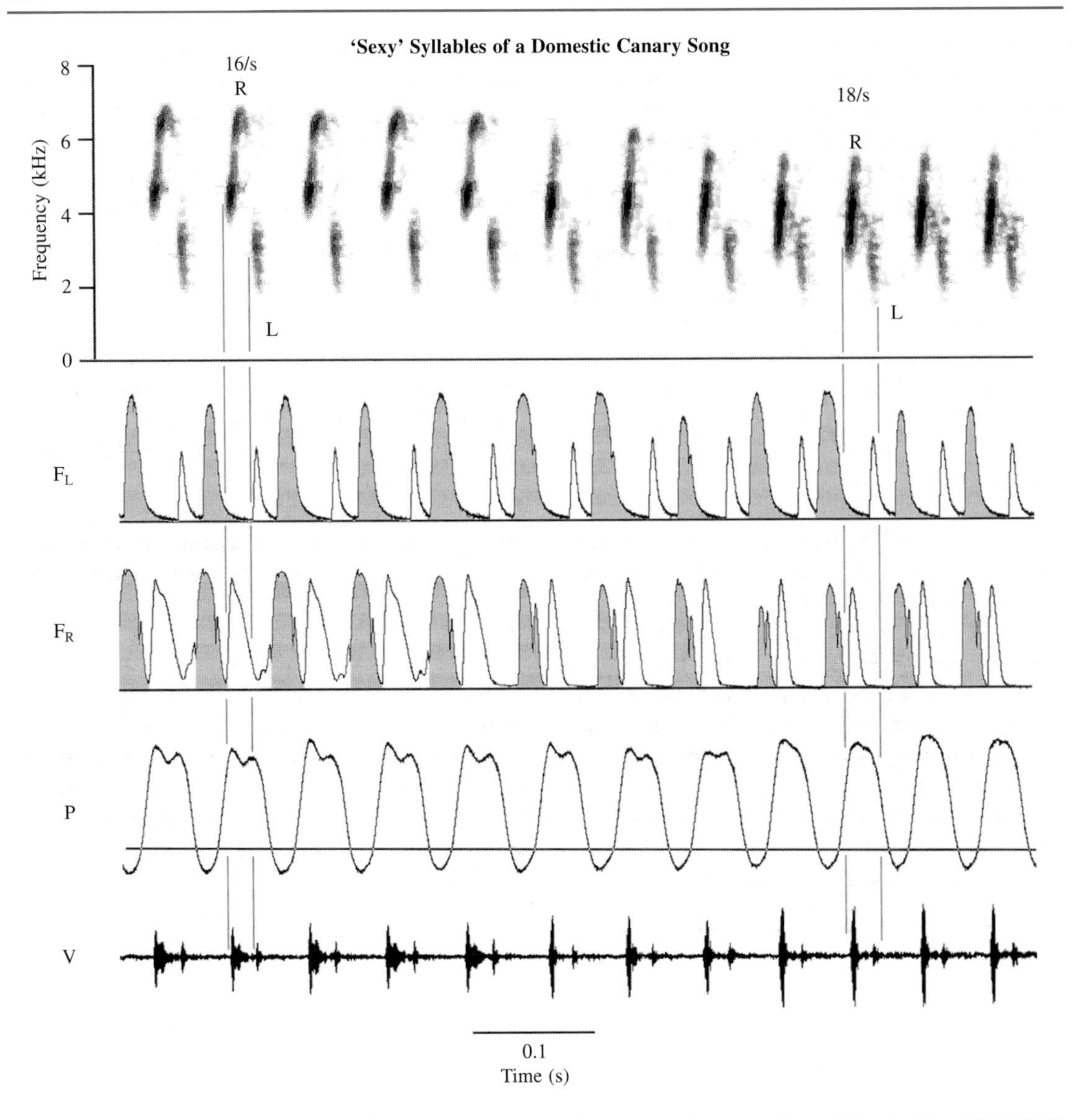

Figure 9.9 A segment of a domestic canary song containing two phrases of type 'A' syllables (**CD1 #74**). Each syllable is composed of two notes. The first note sweeps upward on the right side for about 20 ms and is followed by an approximately 10 ms downward sweep at a lower frequency on the left side. To be maximally effective in eliciting copulation displays from a receptive female, syllables must be complex and sung at repetition rates greater than 16 per second. Abbreviations as in **Figure 9.3**.

sides of the syrinx may have special perceptual importance in the vocal communication of birds.

The kind of information females get from 'A' syllables remains to be determined. One possibility is that they provide an indication of the male's health or fitness. Of all the syllables in a canary's repertoire, 'A' phrases may demand the greatest precision in bilateral motor coordination within the syrinx, and also between the syrinx and the respiratory muscles and other vocal tract or cranial muscle groups involved in singing. It may be that males who are not in top physical condition have difficulty producing these phrases. If so, the 'A' phrase may be an honest indicator of male fitness (see Chapter 2).

Another possibility is that the presence of 'A' phrases in song may indicate a male's reproductive state – that he is ready to breed. Field studies of wild canaries show that the syllable repetition rate of their repertoire, but not repertoire size, changes seasonally. 'A' syllables sung during the breeding season are replaced during the winter with syllables at lower repetition rates (Leitner et al. 2001). Seasonal changes in steroid levels can affect syringeal muscles and vocal behavior (DeVoogd 1991; DeVoogd et al. 1991; Beani et al. 1995; Hartley et al. 1997; Tramontin et al. 2000). The presence of 'A' phrases in a male's song might indicate his readiness to mate. Domestic canaries sing longer strings of 'A' phrases in the presence of other male or female canaries than when they are singing alone, even though the number and duration of songs is similar in both situations (Kreutzer et al. 1999).

These two possible functions of 'A' syllables in intersexual communication are not mutually exclusive. Other factors may also play a role in mate selection, and it is not clear if there is a correlation between the presence of 'A' syllables and reproductive success. Whatever the case, a better understanding of how birds sing promises to be a valuable tool in deciphering the communicative functions and evolution of song.

CONCLUSIONS

Twenty years ago, in a review on avian sound production, Brackenbury (1982) lamented that, "many ideas about the functioning of the passerine syrinx are based on guesswork." In the past decade new techniques for recording and observing syringeal function during song have yielded significant progress toward removing some of this guesswork and advancing our understanding how birds sing, though the number of species sampled is still very small. There is every reason to believe the next decade will be equally productive. As we learn more about how the avian vocal system works, and its acoustic possibilities and limitations, we will gain new insights into the neural organization, behavioral significance, and evolution of a behavioral system that is unique in the animal kingdom.

Chapter 10

Birdsong and evolution

CAREL TEN CATE

INTRODUCTION

The song of a European nightingale and the cooing of an Eurasian collared dove sound completely different. Nightingale song is highly variable, with long strings of differently structured songs incorporating notes of many different types and spanning a wide range of frequencies. The collared dove 'song' is a monotonous repetition of a stereotyped three-note coo with limited variation in frequency. Yet, both are produced by territorial males and serve similar functions: deterring rivals and attracting mates. Why, then, are they so different? And how does the enormous variety in the vocal signals of birds arise? What are the processes and mechanisms that constrain or enhance this variation? Can we reconstruct the pattern of vocal evolution and understand why and how vocalizations evolve in such different ways? These are the questions addressed in this chapter. In doing so, the focus is on 'song' rather than other vocalizations. Songs are broadly defined as the more or less complex vocalizations that play a prominent role in attracting mates or deterring competitors, thus including both nightingale songs and dove coos. I start with an overview of the factors that affect the evolution of vocalizations. Next I address the question of whether and how we can reconstruct vocal evolution. Having thus introduced the main conceptual issues, I go into more detail in a case study on the evolution of the coo-vocalizations in doves of the genus *Streptopelia* (turtle doves). Finally I discuss briefly whether the findings in doves can be generalized to other groups, in particular to songbirds where learning is important in generating vocal variety.

EVOLUTION AND SONG: A TWO-WAY INTERACTION

So, why are the sounds of nightingales and doves, or birds in general, so different? One might say that this is because the songs serve in species recognition; they have to be distinct to enable conspecifics to recognize each other. As we shall see, this definitely is a factor that reinforces differences between species. However, the vocal differences between species seem far beyond what would be necessary to provide species specificity. Another reason may be that although the functions of songs in various species are more or less similar, they are not identical. The relative importance of deterring rivals or attracting mates may differ between species and the features of dominant importance for these functions may vary. Also, if, for instance, male–male competition is the prime function of song within two species, they may differ in the precise qualities that song needs to convey. These different qualities may be expressed in different vocal parameters. Species also live in different environments and, as shown in Chapter 6, the sound characteristics providing optimal transmission differ between habitats, creating another source of species differences. All of these factors and more must mean that, when studied in greater detail, the selection on vocal signals will turn out to be in some degree different for each species, and different again if they are subjected to selective breeding in captivity (**Box 39**, p. 298).

But suppose we know all the different selection pressures, would we then be able to predict all aspects of signal structure? The answer is 'no'. Selection may mold a signal through evolutionary time, but signals do not arise *de novo* for each species. Instead, they are based upon those of their ancestors. Species carry with them a phylogenetic history. This is reflected in similarities between closely related species in morphological characteristics and color patterns: bird watchers who have never seen a collared dove before, will immediately recognize it as a species of dove due to morphological similarities shared with other doves. The same applies to behavioral and vocal features. Just as new morphological characteristics are inevitably contingent upon those of an ancestral species, so are vocal characteristics. For this reason, species from different taxonomic groups can be expected to differ in the evolutionary outcome when coping with similar selection pressures. For instance, bird species from different taxa all use lower frequency vocalizations in 'low-forest' habitats compared to congeners in grassland or edge habitats (Morton 1975; Marten & Marler 1977). However, the details of the precise relationship differ from one taxonomic group to another (Ryan & Brenowitz 1985).

To understand the evolution of the enormous variety in vocal signals used by birds, we need take into account not only the selection pressures, but also the historical component. The mechanisms that birds use to produce signals or to detect and perceive them are a product of evolution, resulting from selection on signal structure and function. On the other hand, the same mechanisms provide constraints as well as possibilities for the evolution of novel signals or signal features (see Chapter 11). The signals a species uses today are the outcome of such interactions. The aim of this chapter is to demonstrate how these can be disentangled.

Reconstructing the Past?

The Evolutionary Origin of Song

Efforts to reconstruct the evolution of a particular morphology can benefit greatly from the availability of fossils, making it possible to trace ancestral features and subsequent changes. However, behavior and sound patterns do not fossilize. Although their origin is thus hidden in the past, some inferences about that past can be obtained from comparative research among present-day birds. All existing avian orders possess a syrinx–the sound-producing organ in birds (see Chapter 9). In those few bird species where it seems to be virtually lacking, such as the New World vultures, this is considered to be a derived trait, with the syrinx present in ancestors, but lost later in evolution. Also, the simple syrinx present in the otherwise primitive ratites is interpreted as a regressive variant (King 1989).

Omnipresence of the syrinx is strong evidence that it is an ancestral trait. Sound production is the only function of the syrinx, suggesting that the common ancestors of modern birds were already using vocal signals. Although a thorough phylogenetic analysis of syringeal structure is lacking, several authors have suggested that the 'archetypal' syrinx was of the tracheobronchial type (King 1989), having membranes and two pairs of so-called 'extrinsic' muscles (see Chapter 9). This type of syrinx is present in many bird species, in particular in non-songbirds such as gulls and doves, and allows production of a wide variety of sounds. Hence, it is likely that vocal signals had already evolved in the ancestor of all modern birds, or may be even earlier in 'protobirds' or their ancestors, the feathered saurian Maniraptorans (Prum 2002).

Darwin (1871) suggested that the evolutionary starting point for vocal signals may have been the incidental sounds produced during breathing, as it probably was for other higher tetrapodes. Companions may have responded to hearing these sounds in such a way that the individual producing them gained some advantage. For instance, breathing sounds may have been

BOX 39

THE ART OF SELECTION: EVOLUTION IN THE AVIARY

Humans have 'played evolution' for ages by selectively breeding domesticated animals. Inspired by the success of aviculturists in molding domesticated rock doves, Charles Darwin used artificial selection to support his theories on natural selection (Darwin 1860, 1875). He noted that traits vary dramatically among breeds, often exceeding the differences between true species. He argued that this variation must have emerged during domestication because it is greater than that in surviving wild populations of rock doves. Also, there are no closely related dove species that match the typical morphology, plumage, flight displays, and vocal behavior found among domesticated pigeons. The strange coos of the so-called trumpeter and laugher breeds show that selection can alter species-typical vocalizations over time. Whereas feral pigeons produce relatively simple and monotonous series of coos (**CD2 #67**), English trumpeters have a more complex cooing behavior including changes in pace and amplitude (**CD2 #68**), while Thailand laughers elaborate their coos with unusually exaggerated, harmonically rich 'wok-notes' (Baptista & Abs 1983). All are still considered one species.

Birds have long been caged and selected for vocal traits to please the human ear. In many societies there are traditional competitions in which pet birds are judged for things like singing rate and voice quality. In central Asia, black partridges are trained to call at high rates; an untrained partridge may call 35–40 times in an hour, while a trained one in competition can exceed 245 times, perhaps partly attributable to the reported use of steroids (Birmani 1999). Cooing contests with zebra doves (**A**) are popular in south-east Asia; hundreds of birds in beautiful cages are hoisted up on six-meter poles (**B**) and evaluated both on cooing quantity and quality (Stephens 2000). Some breeders restrain the mobility of the air sacs of their bird by applying a little external string, which supposedly makes the coo pattern more beautiful. In several Dutch villages, roosters compete in yearly crowing contests with the champions crowing up to 200 times in half an hour (Tuttel 2000). There is also a long history in Europe of singing competitions with passerines, especially for chaffinches and canaries. The earliest reported chaffinch song contest is from Belgium in 1593 (DeClerq et al. 1995), and this is still a popular Flemish sport. Birds used to be blinded by sticking the eyelids together with a red hot metal wire, in the belief that birds placed in the dark or blinded would sing best (DeClerq et al. 1995). In canary song contests, birds are judged for both the vigor and, more importantly, the quality of singing. Selective breeding has led to four breeds with distinct singing styles (Güttinger 1985; Wolnik 1994; Mundinger 1999). The German 'roller' is selected for low-amplitude and low-frequency 'tours' or 'rolls,' which are trills of repeated syllables, supposed to be sung with the beak closed. The Belgian 'waterslager' adds 'bubbling water-notes,' while the 'American singer,' derived from crosses with waterslagers, is selected for a pleasant and variable song without clear acoustic criteria. The newest breed is the Spanish 'timbrado' (**CD2 #69**), originating from backcrosses between Spanish domesticated canaries and their wild ancestors from the Canary Islands (**CD2 #70**). The preferred characteristics are loud, high pitched songs, with metallic notes, some reminiscent of bells and castanets. Interestingly, the recommended way to improve timbrado song is to separate juvenile males from adult tutors in the sensitive period (Espada 2002). In this way, the breeders try to avoid learned convergence on a single 'perfect song,' aiming instead to select for 'purely genetic' song virtuosity. Although variation in reproductive success is the driving force that shapes vocal behavior in both domesticated and wild songsters, the consequences can clearly be very different when human preferences prevail over those of the birds themselves.

Hans Slabbekoorn

(A)

(B)

particularly prominent during physical exertion, as during a fight. If this sound somehow intimidated an opponent, it would give an individual an advantage over others and anything making the sounds more conspicuous or intimidating would be beneficial. In this way, a vocal signal evolved as a sound affecting the behavior of another individual and contributing to the fitness of the sender. At first, such signals would have been simple ones, with a straightforward and direct connection to morphology and physiology of the sender. Acoustic variation among individuals may have reflected differences in size or physiological state, perhaps linked in turn to the signaler's motivational state. For instance, an animal in distress may tense its muscles, resulting in higher pitched sounds. Such processes of 'expressive size symbolism' and 'motivation–structural rules' (Morton 1982; Owings & Morton 1998) may have preceded the emergence of the more elaborate vocal signals of today's vertebrates. They may explain why many animals use harsh, low frequency vocalizations when aggressive and more tonal, higher pitched ones when fearful (see Chapter 5). Such vocalizations are usually rather simple in structure, but they may have contributed to the origin of the songs that are the focus of this chapter. In line with this, comparative evidence suggests that some male vocal signals used in female choice may have been derived from signals evolved in the context of male–male competition (Berglund et al. 1996; Borgia & Coleman 2000). To go beyond these 'educated guesses' about the early evolution of vocal signals, it is necessary to go into much more detail. When the aim is to try to understand the processes underlying the great variety of birdsongs, comparative analysis of the vocal signals of closely related species provides an entry to this problem.

Using Vocalizations to Reconstruct Phylogeny

Past researchers have often been skeptical about the possibilities of reconstructing how behavioral or vocal signals took shape, or about using such signals to examine the relationships between species. Behavior in general was considered to be so much more plastic than morphology that specific traits might appear or disappear quickly. If so, the distribution of behavioral traits among related species might not bear any relation to their phylogenetic relationship, and reliable reconstructions of signal evolution would be impossible. Nevertheless, some early studies showed a relation between taxonomy and behavioral variation. For instance, Konrad Lorenz (1941) demonstrated that a classification of ducks and geese on the basis of presence or absence of various behavior patterns showed a remarkable congruence with classifications based on morphological features. He also pointed out that analyzing the form of particular courtship displays and comparing them across species is often helpful in postulating hypotheses about the behaviors from which they were derived. However, these early studies relied more on intuitive argumentation than on quantitative analysis. Also, they were often *phenetic* classifications, based upon grouping species according to the number of traits they shared or their overall similarity. However, for an evolutionary approach the aim should be to arrive at a *cladistic* or *phylogenetic* classification, in which organisms are classified according to common ancestry.

In the Past 15 years or so, the subject of reconstructing the evolution of behavior patterns and vocalizations, and using these traits to infer the phylogenetic relationships between species has gained new momentum (Martins 1996). Statistical techniques have been developed to reconstruct and test the reliability of phylogenetic patterns. Using these techniques, several studies have shown that phylogenies reconstructed by using morphological or behavioral data provide similar results (Gittleman et al. 1996; Wimberger & de Queiroz 1996), demonstrating the validity of behavioral variation as a tool for reconstructing phylogenies.

This outcome is not that surprising. Any type of behavior is dependent upon the presence of a specific morphology to make a particular movement, including not only the nervous system, but also bones, muscles, membranes, and other tissues. Also, behavior is closely linked to a neural substrate, a specific pattern of innervation, and a particular physiological state to produce the behavior. Closely related species are more likely to share similarities in these features than more distantly related ones, which may be reflected in behavioral similarities. The same argument applies to vocal signals and several studies support this view. Payne (1986), for instance, examined the relationship among *Dendroica* wood warblers, based on their songs. This led to a hypothesis of speciation events that showed great similarity to a reconstruction based on plumage patterns, current distributions, and historical climate changes. Others have used vocal characteristics to shed new light on the phylogenetic relationship between species. For example, Chappuis & Erard (1993) re-examined the relatedness among several species and subspecies of African forest bulbuls of the genus *Bleda*, inspired by unexpected variation in vocal characteristics. They arrived at a new hypothesis concerning the phylogeny of these birds. Vocal variation also made Whitney et al. (2000) aware that the *Herpsilochmus pileatus* complex of antwrens consists of three rather than two species. Another example concerns an analysis of the vocal characteristics of purple sandpipers, rock sandpipers and dunlin, demonstrating the untenability of the previously held view that rock and purple sandpipers are variants of one species (Miller 1996).

Vocal analysis can thus give rise to new and robust classifications. In the case of the warblers and the bulbuls, this finding is all the more interesting since these are songbirds that learn their songs. Although learning provides the potential for substantial evolutionary plasticity, in which phylogenetic information may be lost quite easily, this apparently need not be the case (Slabbekoorn & Smith 2002b). One reason for this is that faithful copying may limit the amount of change in song features from one generation to the next, preserving species identity. Nevertheless, cultural mutation rates are considered higher than genetic ones (Lynch 1996) and it may be that songs are less suitable for assessing phylogenetic relationships above the genus level. Another factor is that even in songbirds vocal variability among species may be hampered by constraints on the plasticity (see p. 318; see Chapter 11).

Vocal signals thus can be used as a tool to establish the relationship between species. At the same time, reconstructing such a phylogeny based on song features also indicates gains, losses, and changes that have occurred in the vocal features themselves during evolution. It thus provides a way to arrive at a phylogenetic hypothesis of song evolution. Although informative, such an approach can also be misleading. For instance, closely related species can show strong divergence in vocal features, suggesting that they are less related than they really are. Also, non-related species may converge in their vocal features, for instance, when experiencing similar selection pressures. This will suggest a closer relatedness than is the case. Both phenomena may distort inferences about vocal evolution. Finally, comparing vocal traits is usually based on the assumption that sounds that differ greatly in their trace on a sonogram will also differ greatly in underlying mechanisms. This may be true, but it need not be so (Beckers & ten Cate in press; Fee et al. 1998; see p. 312). Such evolutionary patterns may go unnoticed if the reconstruction of vocal evolution is based on a phylogeny derived from the same vocalizations. Other methods are required to uncover the patterns and processes underlying vocal differentiation.

Using Phylogenetic Trees to Reconstruct Vocal Evolution

One powerful way to reconstruct vocal evolution is to plot vocal variation on a phylogenetic tree based on non-vocal characteristics (Price & Lanyon 2002a, b; **Fig. 10.1**). This method avoids the pitfalls of confusing similarities by convergence with those due to common descent, and of assumptions about the ease of evolutionary transitions from one type of vocalization into another one. To construct a tree based on traits that have no relationship to vocalization may

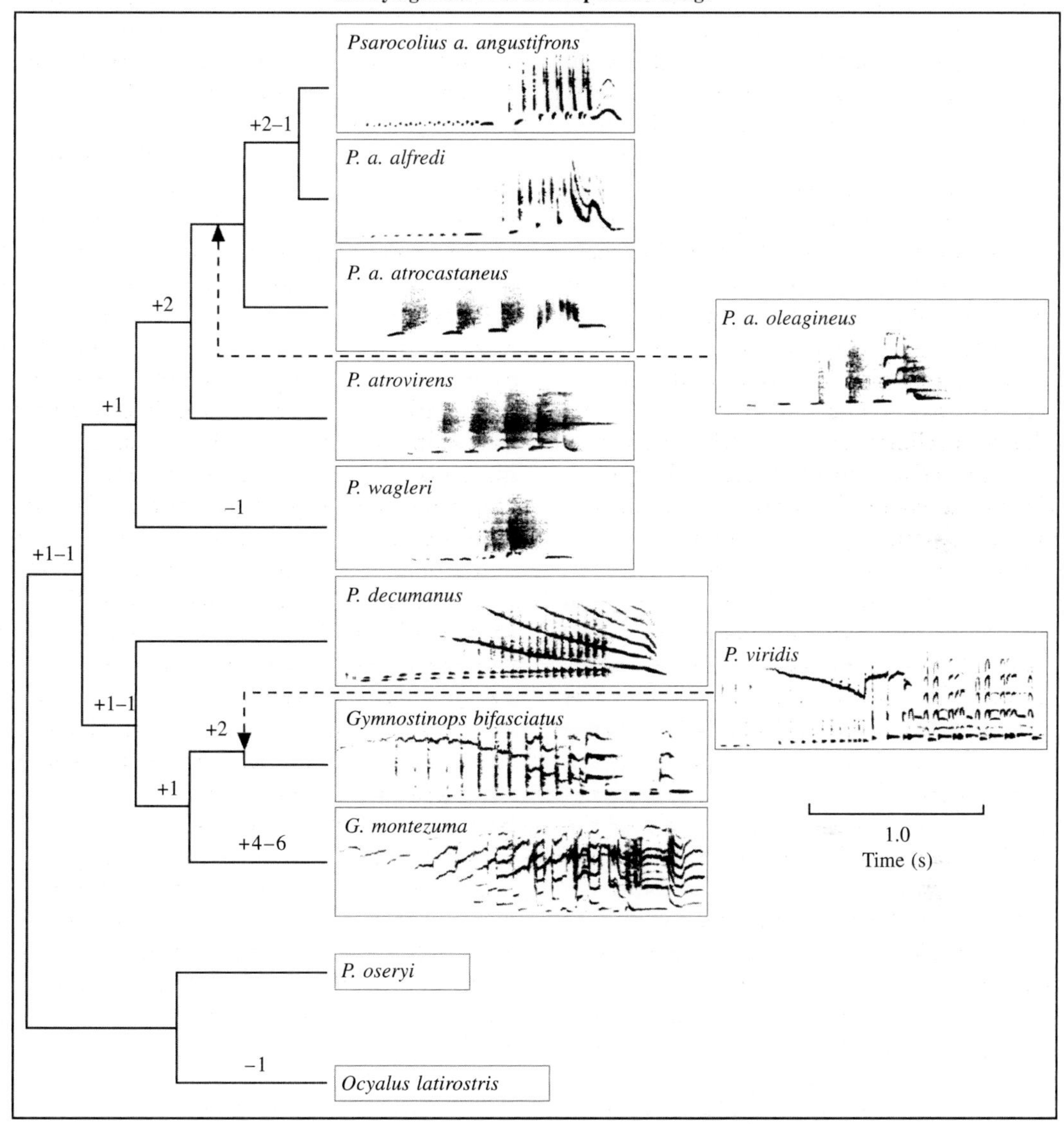

Figure 10.1 Song differentiation in *Psarocolius*, *Gymnostinops*, and *Ocyalus* oropendolas. The phylogenetic 'tree' demonstrating relatedness is derived from an analysis of mitochondrial DNA (Price & Lanyon 2002a). The columns show sonograms of the song of several species and subspecies from the genera *Psarocolius*, *Gymnostinops*, and *Ocyalus*. The numbers show how many vocal features differ between a particular species or group and its presumed ancestral species (+*n* indicates the number of gains in features; –*n* the number of losses; Price & Lanyon 2002b). These numbers are calculated by relating the song differentiation to the molecular tree. No molecular data were available for *Psarocolius angustifrons oleagineus* and *P. viridis*, but their estimated position in the tree, shown with hatched lines, is based on the most important vocal features (Price & Lanyon 2002b).

not be as simple as it sounds. For instance, the taxonomy of several avian groups, like the Anatidae (Delacour & Mayr 1945) and passeriformes (Ames 1971), is based partially upon syrinx structure. Syringeal morphology was also used as a prime character for a revision of the suboscine family of manakins (Prum 1992). Although vocal structure has no straightforward, one-to-one relationship with syringeal structure, these taxonomies cannot be considered as fully independent of the vocal features of the species concerned. In such cases caution is needed when using the phylogeny to reconstruct vocal evolution.

Nowadays, comparative researchers usually prefer DNA-based trees. Under the assumption that the variation in the DNA markers used for reconstructing the phylogeny is independent of vocal variation, one can reconstruct which branches of the tree gave rise to which type of evolutionary transitions in vocal structure. This may indicate which traits are evolutionarily labile, and may be lost or gained relatively easily, and which are more stable, and resistant to change (Price & Lanyon 2002b). Such knowledge can then guide further research on underlying mechanisms or adaptive significance. Also, comparing the distribution of vocal characteristics over a phylogenetic tree with the distribution of other factors, such as habitat or mating system, can reveal their importance in shaping vocal differences. These comparisons also allow an assessment of the relative impact of phylogenetic and other factors on the variation in vocal characteristics.

McCracken & Sheldon (1997), for instance, examined the effects of phylogenetic relatedness and habitat structure on different aspects of the vocalizations of 14 heron species, using the glossy ibis as an outgroup. They used a phylogenetic tree based upon DNA hybridization, and examined the distribution of vocal variation on this tree. The results suggest that the ancestral state for the clade as a whole was a series of repeated syllables with a harmonic structure and of low frequency. Bitterns, day-herons, and night-herons share some derived features. These groups do not emit syllables in series and have higher fundamental frequencies. A common ancestral trait for bitterns is that they use tonal notes and possess a simple syntax for combining different notes. Day- and night-herons are united by vocalizations of one to three syllables lacking tonal notes. This analysis shows that ancestral traits can be identified and also shows some of the transitions that have taken place. Especially interesting is the discovery that among herons, features such as peak frequency and frequency range show no relation to the phylogeny, but are correlated with ecology. Species in dense forest vegetation have low peak energy and a narrower frequency range compared to open savanna and grassland species, matching with the sound characteristics that give the best transmission in the respective habitats.

In another comparative study in which an independent phylogeny was used, VanBuskirk (1997) demonstrated that the overall song structure of North American wood warblers is correlated with habitat, while the structure of individual notes within songs shows a stronger association with phylogenetic history. These two studies suggest that similar principles apply to both non-songbirds (herons) and songbirds (wood warblers).

RECONSTRUCTING VOCAL EVOLUTION IN STREPTOPELIA DOVES

We see that vocal signals in birds are not only subject to various selection pressures, but also reflect phylogenetic history. These factors operate in concert, but may vary in importance at different levels of vocal structure. To disentangle the various processes involved, an integrative study is required in which these various processes are studied in a coherent way. Thus far, there have been few such studies. This is the background of our investigations of vocal variation in turtle doves of the genus *Streptopelia*. We chose non-songbirds for study to avoid the additional complications that song learning introduces into studies on vocal evolution. We are examining both the selective forces operating on vocal signals, and the mechanisms underlying sound production and perception. A full understanding of the potential constraints or biases for evolutionary change provided by these mechanisms is essential to prevent research on the interaction of vocal evolution, and the mechanisms of production and perception from becoming narrative 'just-so' studies. Our grasp of these issues is still incomplete, but the following represents the state of the art with respect to some vocal features that are relevant in communication.

Phylogeny and Vocal Variation in Doves

There are 17 species of turtle doves and their phylogeny, based on both nuclear and mitochondrial DNA, is well resolved; one has a domesticated form (Johnson et al. 2001). One species, the pink pigeon, has been only recently recognized as a *Streptopelia* turtle dove (Johnson et al. 2001) and officially still bears the old genus name. The English names generally refer to plumage pattern and color, or to the species' distribution range. In addition, they are named 'dove,' 'turtle dove,' or 'pigeon,' none of which reflect phylogenetic relationships. According to an anonymous cook in Hong Kong, the distinction between a dove and a pigeon in general is a matter of taste (de Kort 2002). It would make sense to replace 'dove' and 'pigeon' by the label 'turtle dove' for all species of this genus. Here is the current nomenclature.

S. decaocto	Eurasian collared dove
S. roseogrisea	African collared dove; its domesticated formis known as the 'ring dove,' *S. risoria*
S. decipiens	African mourning dove
S. semitorquata	red-eyed dove
S. capicola	ring-necked dove
S. vinacea	vinaceous dove
S. reichenowi	white-winged collared dove
S. hypopyrhra	Adamawa turtle dove
S. lugens	dusky turtle dove
S. turtur	European turtle dove
S. orientalis	Oriental turtle dove
S. bitorquata	island turtle dove
S. tranquebarica	red turtle dove
S. (Nesoenas) mayeri	pink pigeon
S. picturata	Madagascar turtle dove
S. chinensis	spotted dove
S. senegalensis	laughing dove

Turtle doves, like many other dove genera (Goodwin 1983), use coos in association with three different contexts, each with its own distinctive postures. Most common is the 'perch-coo'. This is an advertisement signal, used mainly by territorial males on conspicuous singing posts within their territory. It functions both in territorial defense and in mate attraction. If other individuals enter the territory and land somewhere, males will approach them and show 'bow-coos,' in which the cooing is accompanied by a bowing display directed towards the intruder. The third type is the 'nest-coo,' produced by both members of a pair preparing for nesting or when breeding. **Figure 10.2** provides the sonograms of the three coo-types for those species for which phylogenetic and acoustic data are available (**CD1 #72–86**).

In some species, the three coo-types have the same basic structure, although more detailed measurements may reveal quantitative differences between them as, for example, in the Eurasian collared dove (Ballintijn & ten Cate 1999a, b; ten Cate 1992). For other species, two coo-types are the same and differ from the third type as in the ring-necked dove, with a deviant bow-coo, and the red-eyed dove, with an aberrant perch-coo. Finally, there are some species for which all coo-types differ in structure as with the Adamawa turtle dove. A comparison with other dove and pigeon genera indicates that the most likely ancestral state is that turtle doves had differentiated coo-types (de Kort & ten Cate 2004), and that some species thus lost the contrasts between coo-types during evolution.

Despite the variation between and within species, all vocalizations share a number of features

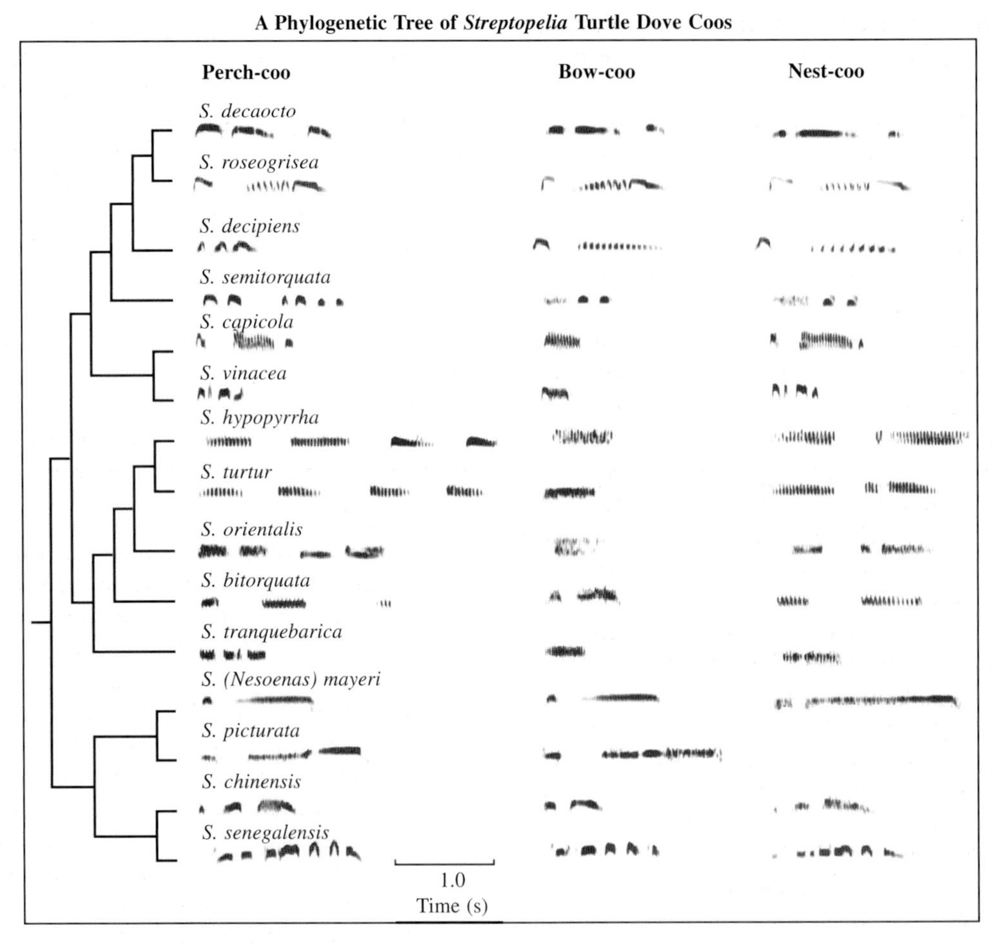

Figure 10.2 Inter- and intraspecific variation in the coos of turtle doves (**CD1 #72–86**). The phylogenetic tree demonstrating relatedness is based on mitochondrial as well as nuclear DNA analyses (Johnson et al. 2001). For each species, three coo-types are shown except for no *S. picturata* nest-coo. Frequencies range from 210–1100 Hz (de Kort & ten Cate 2004). For two species in the genus (*S. reichenowi* and *S. lugens*) data are lacking.

and some patterns seem characteristic of certain subgroups. For all turtle doves, and also related genera (Mahler & Tubaro 2001), the frequency range of coos is at the low end of the spectrum for birds. For turtle doves this range is between 210 and 1100 Hz (de Kort & ten Cate 2004; Slabbekoorn et al. 1999). Even if the overall sound structure differs between the coo-types within a species, the frequency range for that species remains similar. Frequency changes can occur during coo elements but, if so, are usually limited in magnitude, commonly consisting of a rise at the beginning of an element and a fall towards the end. The most common sound patterns are prolonged tonal sounds or amplitude modulated trills but in some species, such as the red turtle dove and the oriental turtle dove, noise-like patterns occur. Trill modulation rates can vary between species; they are low in the second element of the African collared dove, high in the red turtle dove, but the variation is limited between the coo-types within a species. Harmonics are absent or weak. Both the variation between coo-types and between species concerns mainly parameters such as element numbers, duration, and whether there are tonal or trilled elements.

Commonalties are present among close relatives. For instance, in the European turtle dove, the Adamawa turtle dove, and the oriental turtle dove, all three coo-types have a similar temporal structure. All have four elements in their perch-coos and, in all, the first two elements are very similar, but slightly different from the latter two. For two of the three species, the sound structure is a trill, for the oriental turtle dove the structure is a noisy. These patterns suggest that the temporal structure of the coos may be similar to that of the ancestral species, while sound characteristics may have changed. Two species in this cluster – the island turtle dove and the red turtle dove – also have trilled or more noise – like features in their elements, but generally lack tonal components, suggesting that trilled elements might be the ancestral features for this group as a whole. The pink pigeon and the Madagascar turtle dove also have a very similar vocal structure, both in temporal and other parameters. In other sister species pairs, the Eurasian collared dove and the African collared dove; the ring-necked dove and the vinaceous dove; the spotted dove and the laughing dove, species differ in the temporal patterning of coos, while for the perch-coos one of the two species has only tonal and the other a combination of tonal and trilled elements.

Variation in coo structure within the genus *Streptopelia* thus provides an ideal opportunity to examine the factors affecting both the selection processes that may give rise to the variation as well as the mechanisms that determine the types of change that occur.

Selection on Signal Structure?

Coos serve in communication with conspecifics. The perch-coo is a long distance signal and, since it is not accompanied by visual displays, it may be especially subject to selection on vocal characteristics. Three main factors might be of importance. First, there may be environmental selection for optimal transmission of long distance signals (see Chapter 6). Second, within a species inter- and intrasexual selection may occur, leading to evolutionary changes in acoustic characteristics shaped by the responses of other conspecific males and females (see Chapter 2). Third, other species are always present and, especially when these are closely related species with similar signals, there will be selection for species specificity.

Although environmental selection may have some effect on signal structure in turtle doves (de Kort & ten Cate 2004), it is unlikely to be important enough to give rise to species differences in signals, as most species occur in similar habitats. Therefore, sexual selection and interspecific competition may be more important.

Sexual Selection

An extensive study of the Eurasian collared dove demonstrated that the perch-coo has a prominent role in communication between males (Slabbekoorn & ten Cate 1998b, 1999; ten Cate et al. 2002). Some of the variation among males, in particular, the presence or absence of upward 'jumps' in frequency, usually in the first part of coo elements, are linked to male weight: heavier males produce more jumps. Playback experiments demonstrated that territorial males respond more strongly to coos with jumps (**Fig. 10.3**). These are apparently perceived as more threatening. Thus perch–coo structure is subjected to sexual selection through male-male competition in at least one species. Similar jumps occur in other dove species, such as the African collared dove (Beckers et al. 2003a), although their communicative significance has not yet been tested.

Interspecific Competition

Interaction between species is another factor affecting signal evolution. Two or more turtle dove species often occur together in a similar habitat. In some areas in Africa up to five different species are all within hearing range of each other. If they had similar coos, this might well give rise to fitness-reducing interactions. Males might respond to the territorial signals of related species that are no competitors for the respondent, or females might be attracted to males of other

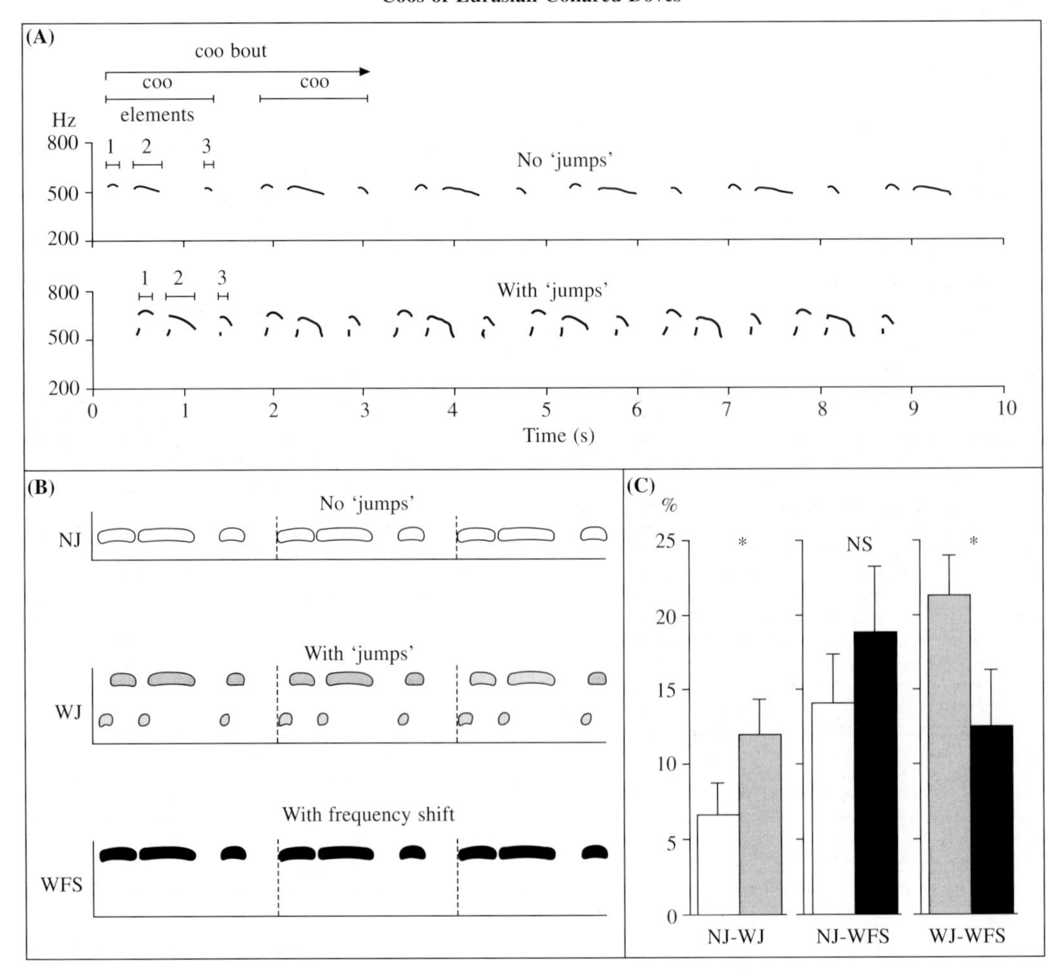

Figure 10.3 Coos of Eurasian collared doves with frequency jumps. At the top (**A**) are two series of perch-coos from different birds, one with frequency jumps, the other without. Below (**B**) are schematic sonograms of three playback stimuli based on a coo without a jump (NJ), edited to either include a jump (WJ) or to increase frequency to the level of the jump (WFS). At the bottom right (**C**) are the results of playbacks, showing a stronger territorial response to the coo with jumps. It is the presence of a jump and not the related increase in frequency that induces the response (Slabbekoorn & ten Cate 1997; ten Cate et al. 2002).

species, leading to detrimental hybridization (**Box 40**, p. 307). That sympatry may be responsible for interspecific divergence in perch-coos was revealed by playback experiments with two species, the African collared dove and the vinaceous dove. Both of these species responded more strongly to the perch-coos of allopatric congeners than sympatric ones in an experiment that controlled for phylogenetic distance of these congeners (de Kort & ten Cate 2001; **Fig. 10.4**). This outcome suggests the presence of selection towards divergence on either the coos of sympatric congeners or their perception.

Interspecific competition for unambiguous signals also seems to affect the coo-type differentiation within species. An analysis

BOX 40

WHAT DO HYBRID BIRDS SOUND LIKE?

In many birds, a preference for own-species song results in a behavioral mating barrier between species. Nevertheless, crosses between species do occur, generally under three conditions: (1) birds that breed far outside the usual range may have no choice but to mate with another species (Poot et al. 1999; Veen et al. 2001); (2) interbreeding may occur in a geographic zone where two species overlap (Emlen et al. 1975; de Kort et al. 2002); or (3) when one species expands into another's breeding range, and behavioral barriers to hybridization are absent or weak (Gill 1997; Panov et al. 2003). Hybrids between the golden-winged warbler and the northward-expanding blue-winged warbler in North America are so numerous that hybrids and backcrosses have their own common names.

The genetic make-up of birds contributes to certain species-specific markers in the vocal repertoire and whether they learn to sing or not (Payne 1980; Baptista 1996). Hence, songs of hybrids provide a window on the inheritance patterns controlling song structure and development. For example, the 'wet-my-lips' or 'fürchte-Gott' ('fear God') of the European quail is not learnt and is clearly distinct from the call of the Japanese quail that has a long, noisy, vibrato third note (**CD2 #71–72**). The two hybridize readily where they meet in a natural hybrid zone in eastern Siberia, and in western Europe where captive-bred Japanese quail have been released repeatedly. Hybrid calls tend to be more variable among individuals compared to either parental species. Frequency characteristics are often intermediate, whereas temporal features usually resemble the father's species (Moreau & Wayre 1968; Guyomarc'h & Guyomarc'h 1996; Collins & Goldsmith 1999; **CD2 #73**). Similarly in doves, coos of hybrids are intermediate or like those of one of the parental species (Lade & Thorpe 1964; Davies 1970; de Kort et al. 2002). Interestingly, the outcome of aviary crosses between doves depends on their relatedness; hybrids of less closely related species have more distorted coos, not necessarily intermediate between those of the parents (Baptista 1996).

Learning adds another layer of complexity. Among the variations recorded in natural songbird hybrids, there are males that sing either the song of the mother's or the father's species, an intermediate song, or a repertoire with a mixture of both. A hybrid between a male reed warbler and a female great reed warbler sang a mixture of the songs of both, with a bias to the latter (Beier et al. 1997). A hybrid between a male dark-eyed junco and a female white-throated sparrow sang a junco-type trill followed by a varying number of white-throated sparrow-like notes (**CD2 #74–76**), while its individual call repertoire appeared larger than that of either parent species and contained calls specific for both (Jung et al. 1994). When exposed only to song of other species, many songbirds will copy heterospecific song accurately, although copied elements are usually restricted to species–species spectral and temporal characteristics (Baptista 1996), and this is likely to be true for hybrid songbirds as well. Whether hybrids learn more from one or the other parental species may depend largely on the relative density of singing males (Gelter 1987). Hybrids raised in captivity from a cross between a male European goldfinch and a female greenfinch incorporated song features of both, and even of a third species, the chaffinch, which had been singing near the aviary (Hinde 1956b).

Whether hybrids are successful in reproducing in the wild will depend on whether they are fertile, their health and vigor and plumage characteristics, and also on their vocal attractiveness to either of the parental species or to other hybrids. Birds of the parental species may learn to reject hybrids through experience (Irwin & Price 1999), but the evolution of hybrid avoidance is only expected when there are negative fitness consequences to hybridization, which is not always the case (Grant & Grant 1992, 1996a; Arnold & Hodges 1995; Deregnaucourt & Guyomarc'h 2003). Doves and galliforms that hybridize will be ideal for further experimentation, and crosses between species that depend on learning are an exciting challenge for the future, especially with birds raised under controlled auditory and social conditions.

Hans Slabbekoorn & Sarah Collins

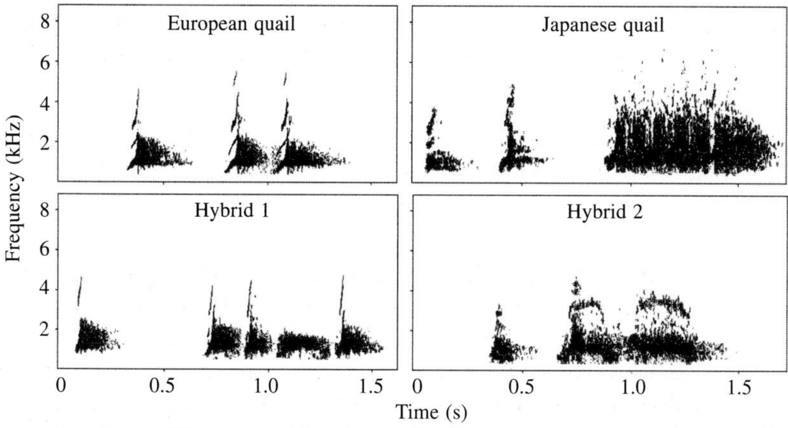

Sonograms of European and Japanese quail calls and of two hybrids between a male European and a female Japanese quail.

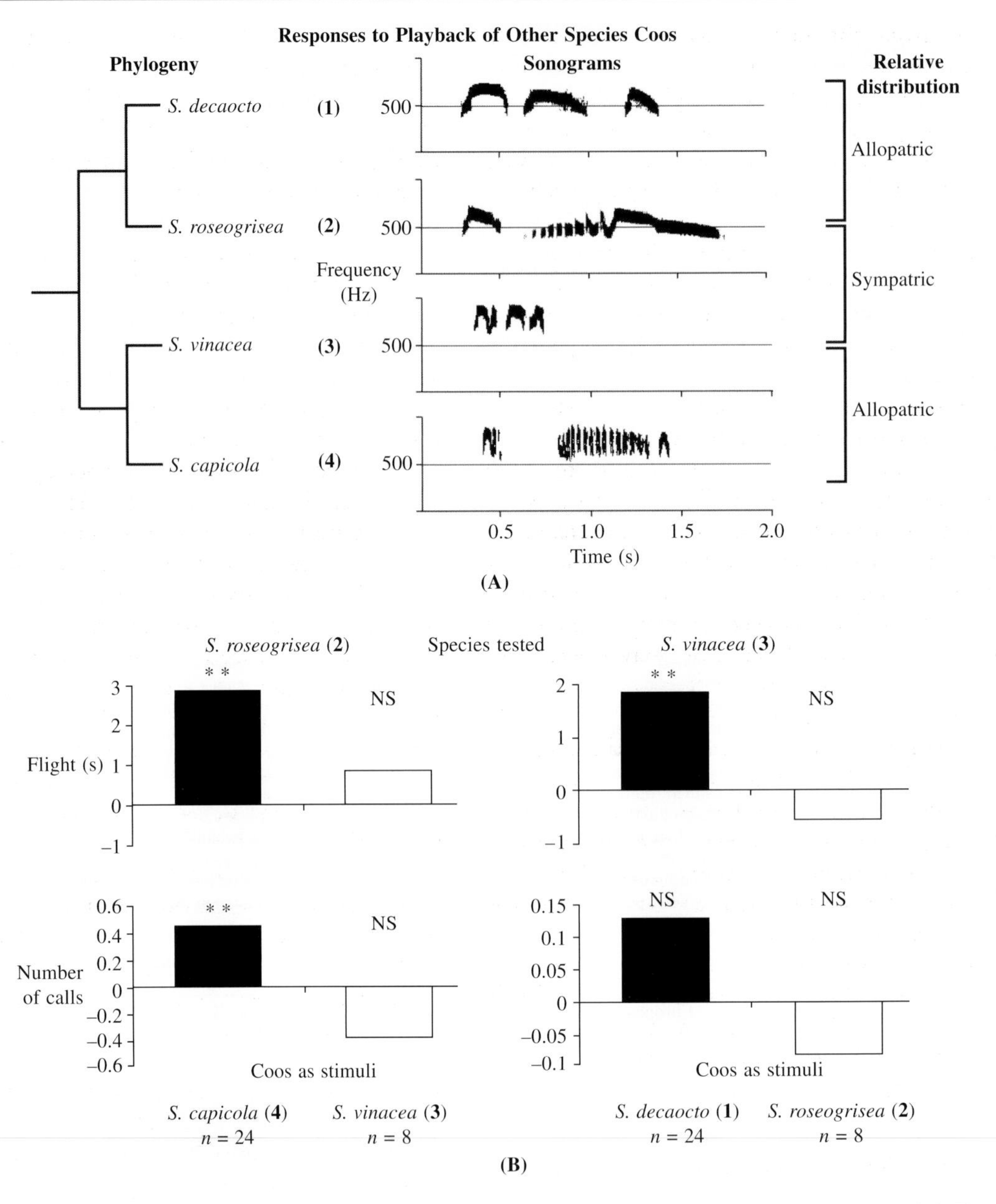

Figure 10.4 Cross-species responsiveness to perch-coos of sympatric and allopatric doves. (**A**) Two pairs of sister species (*S. decaocto* and *S. roseogrisea; S. vinacea* and *S. capicola*) were used. Their relationships are shown in the phylogenetic tree. The focal species (*S. roseogrisea* and *S. vinacea*) were tested with the perch-coos of the other pair member, one allopatric, the other sympatric. (**B**) The stronger responses to the coos of allopatric congeners compared to those of sympatric congeners show that these are perceived as more similar to the conspecific coo (de Kort & ten Cate 2001); either the coos or the perceptual tuning has diverged more in sympatric species than in allopatric species.

comparing the rate of evolutionary change between perch-coos and bow-coos showed that perch-coos evolved more rapidly (de Kort & ten Cate 2004). Perch-coos are the prime signal for long distance communication about species identity and doves generally are sympatric with one or more related species. Therefore we interpret this finding to mean that the need for species specificity in the vocal signals of doves seems to promote differentiation between species and, at the same time, signal uniformity within a species.

We can conclude that various selection pressures operate on dove coos. With little habitat variation, environmental factors are expected to be rather similar between species. With respect to selection for species specificity, the main factor is that the sound becomes distinct from others with no prior directionality. The effects of sexual selection may be somewhat in between. Sexual selection may operate on those vocal features that reflect some aspect of quality of the sender, driven by male–male competition or female choice. If the qualities that males advertise and the vocal parameters in which they are reflected are similar between related species, their vocal features may converge. But what determines which aspects of a signal become subject to sexual selection? Why, for instance, should frequency jumps be a signal in male–male competition? And what are the vocal dimensions that are distinctive to the birds themselves? Such questions can only be addressed by studying the mechanisms of vocal production and perception in doves.

Vocal Variation: Limits and Opportunities

Vocal differences between related species may arise from changes at various levels, from the structure and size of the vocal organ, to changes in the motor program for vocal production (see Chapters 9 & 11). It is hard to imagine that mutations in these mechanisms would be such that vocalizations are equally likely to change in all directions. Some changes might be difficult to realize, even under strong selection, because of severe structural or organizational limitations that may only be altered through a larger scale restructuring of the vocal system. Such limitations then become 'constraints' for evolutionary change in signal characteristics. Sometimes such a constraint can be broken by a major change, an 'innovation,' which overcomes the limitation. On the other hand, there may also be biases that promote certain vocal changes over others, for instance, changes that require only marginal shifts in the sound-producing mechanisms to generate strong effects on vocal characteristics. Such changes might provide special opportunities for evolutionary change.

The mechanisms underlying perception may also constrain signal evolution. To fulfill their communicative function, vocalizations need to be perceived and recognized. Thus, if a difference in vocalization between two populations of one species arises, the only way for this difference to play a role in subsequent speciation is for it to be detected and to affect receiver responses. Perceptual systems can be biased and, as a result, some vocal changes or differences are more readily detected than others. Therefore, the constraints and biases present in signal receivers will affect which changes in vocal features are most likely to acquire communicative significance. To understand these constraints and biases and how they may affect vocal evolution, we need an understanding of the mechanisms of vocal production and perception. Again, we can draw on our dove studies.

Determinants of the Frequency Range of Dove Coos

Dove coos are low pitched, with a more or less tonal quality. They lack the complex frequency modulations present in songbirds. The ability of songbirds to produce more elaborate vocalizations may be linked to their more complex syrinx with various membranes and additional muscles (King 1989; Larsen & Goller 2002; see Chapter 9). Dove syringes, like those of several other groups of birds, have fewer muscles (King 1989). Whatever it takes to develop a more complex syrinx over the course of evolution, it just has not happened in doves. Doves thus cannot produce complex frequency fluctuations. Nevertheless, they do show intra- and interspecific variation in the frequency level and range of their vocalizations. So, which factors affect the frequency of dove coos?

Studies of the way doves produce their vocalizations (Gaunt et al. 1982; Goller & Larsen 1997b; Ballintijn & ten Cate 1998; Beckers et al. 2003b, c) show that the prime source of dove coos are the lateral membranes (**Fig. 10.5**). Coos lack harmonic overtones, but when the sound is measured inside the bird, near its source, the spectrum shows many harmonics (**Fig. 10.5**). The tonal quality of cooing seems to result from a resonator tuned to low frequencies which filters out the high pitched harmonics and amplifies the fundamental frequency (Beckers et al. 2003b).

The low end of the frequency range of a species may be constrained by body size. Lower frequencies may require a larger resonator, and therefore a larger body, to be produced (Fitch 2000b; see Chapters 5 & 11). For *Streptopelia*, this relationship is unclear (Slabbekoorn et al. 1999), but New World pigeon species do show a relation between body size of a species and use of low frequencies (Tubaro & Mahler 1998). Also, larger breeds of domestic pigeons tend to have lower frequency coos (Baptista 1996).

Within individuals, there are both gradual and abrupt changes in frequency. These two types of changes seem to originate in different ways. Gradual changes in frequency, such as

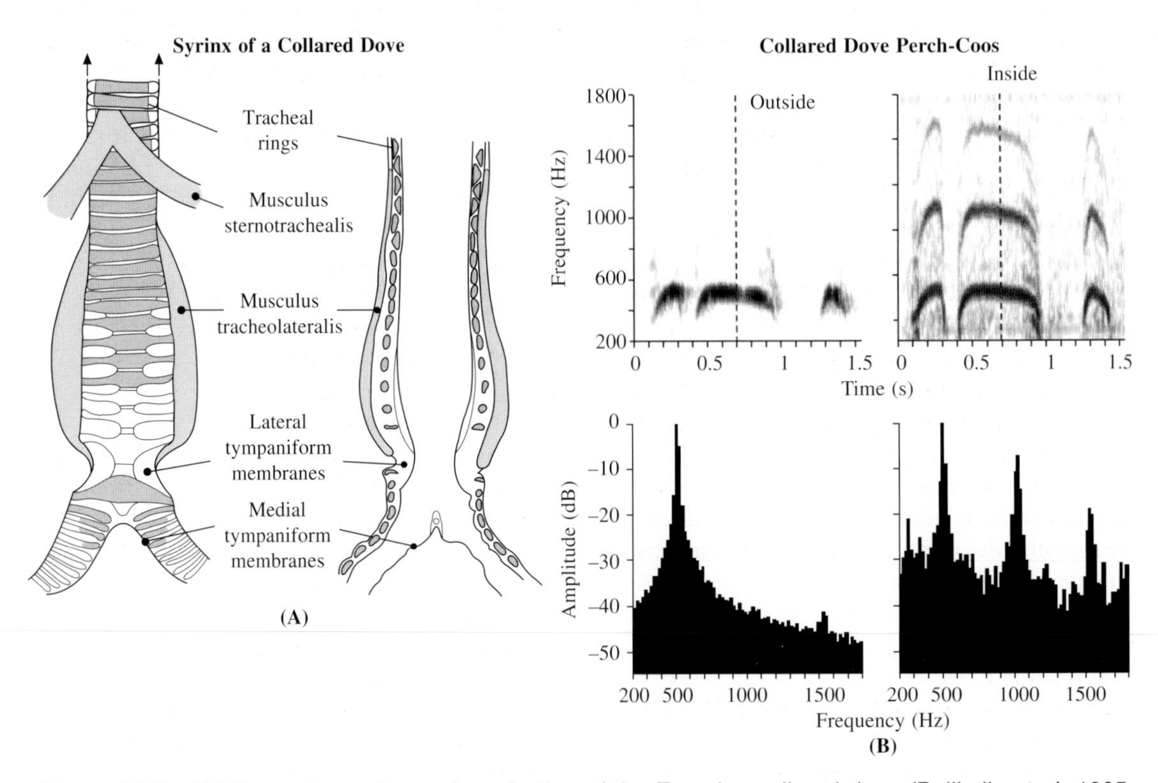

Figure 10.5 (**A**) The syrinx and sound production of the Eurasian collared dove (Ballintijn et al. 1995; Ballintijn & ten Cate 1997). The lateral tympaniform membranes are considered to be the primary sound source. (**B**) Sonograms (top) and amplitude spectra (bottom) measured at the dotted line in the sonograms. Perch-coos were recorded simultaneously 'outside' the bird and 'inside,' in the trachea, close to the sound source. The 'inside' recording shows harmonics that are filtered out before the sound leaves the body (Beckers et al. 2003b). In the middle element of the 'inside' signal a nonlinear transition from a pure tone harmonic structure to one containing side-bands can be seen towards the end.

the increase or decrease in frequency that many species show at the start and end of coo elements, are closely linked to correlated pressure changes in the interclavicular air sac. This suggests that the pressure exerted by the air sac on the vocalizing membranes controls the gradual changes in frequency (Beckers et al. 2003c). If so, physiological limits on the range of pressures that can be achieved in this way may constrain the range in frequencies that can be produced.

Of particular interest are the jumps in frequency, because of their clear, communicative significance (Slabbekoorn & ten Cate 1997, 1998b; ten Cate et al. 2002). Study of the ring dove, the domesticated version of the African collared dove (Beckers et al. 2003b) and the collared dove (unpublished data) showed no evidence that the occurrence of jumps is due to sudden changes in flow and pressure patterns, although jumps occur only with higher flow rates and pressure levels. Abs (1980) also noted sudden jumps in the frequency level of the sounds produced by increases in the speed with which air was blown through the syrinx of a dead domestic pigeon. The most likely explanation for such jumps is that they are caused by nonlinear dynamics of the vocal tract; gradual increases or decreases in pressure or flow, or muscle actions, may at some threshold value destabilize the vibratory frequency of the vocal membranes and lead to a new, stable equilibrium at a different frequency. The result of such a phenomenon is an extension of the range of frequencies in dove vocalizations beyond what would otherwise be possible.

Nonlinear phenomena in general are important for creating vocal variation in some birds (Banta Lavanex 1999; Fee et al. 1998) and various mammals (Fitch et al. 2002; Wilden et al. 1998). Nonlinear transitions may underlie various changes within vocal structures, for instance, the shift from a tonal to a chaotic noise-like structure, or the period doubling of a harmonic sound (Fitch et al. 2002; Fletcher 1992). Some of these phenomena occur in doves. **Figure 10.5** shows a transition from a pure-tonal sound to one with side bands, a phenomenon known as 'bi-phonation;' **Figure 10.6** shows a transition from a bi-phonated sound to a chaotic one. Such phenomena may underlie not only structural differences between or within different vocalizations of a species, but possibly also those between species (Beckers & ten Cate in press). For instance, the noisy coos of the red turtle dove and the oriental turtle dove sound very different from the tonal coos of other turtle doves. However, as shown in **Figure 10.6**, just a slightly different tuning of some parameters of the vocal system or its activation in a tonal ancestral species might suffice to cause a shift to a different stable equilibrium, giving rise to the evolution of noisy sounding species.

Thus, the properties of vocalizing mechanisms can impose physical limits on the frequency range and fluctuations in frequency, but nonlinear processes can also enable variation in vocal characteristics. Evolution may exploit the latter to create vocal differences within and between species without substantial changes in apparatus or motor pattern.

Determinants of the Temporal Structure of Dove Coos

Analysis of turtle dove perch-coos shows that temporal rather than frequency characteristics, such as the duration of coos, elements, pauses, and numbers of elements are most important in distinguishing between species (Slabbekoorn et al. 1999), as is true of other genera of doves and pigeons (Mahler & Tubaro 2001). Cross-fostering and hybrid studies show that such species differences have a clear genetic basis (Lade & Thorpe 1964; Baptista 1996).

Temporal variation may be a function of motor control mechanisms such as control of the airflow through the vocal tract, syringeal muscle activity, or of the interaction of both (Podos 1996). Although we know little about the functioning of the brain areas controlling dove vocalizations, changes in the neural circuitry may be subject to fewer constraints than changes in the vocal apparatus. Parameters such as repetition rate, number and duration of elements and pauses, and the rhythm of the sound pattern may all be achieved by the gating of flow or 'on–off' cycles

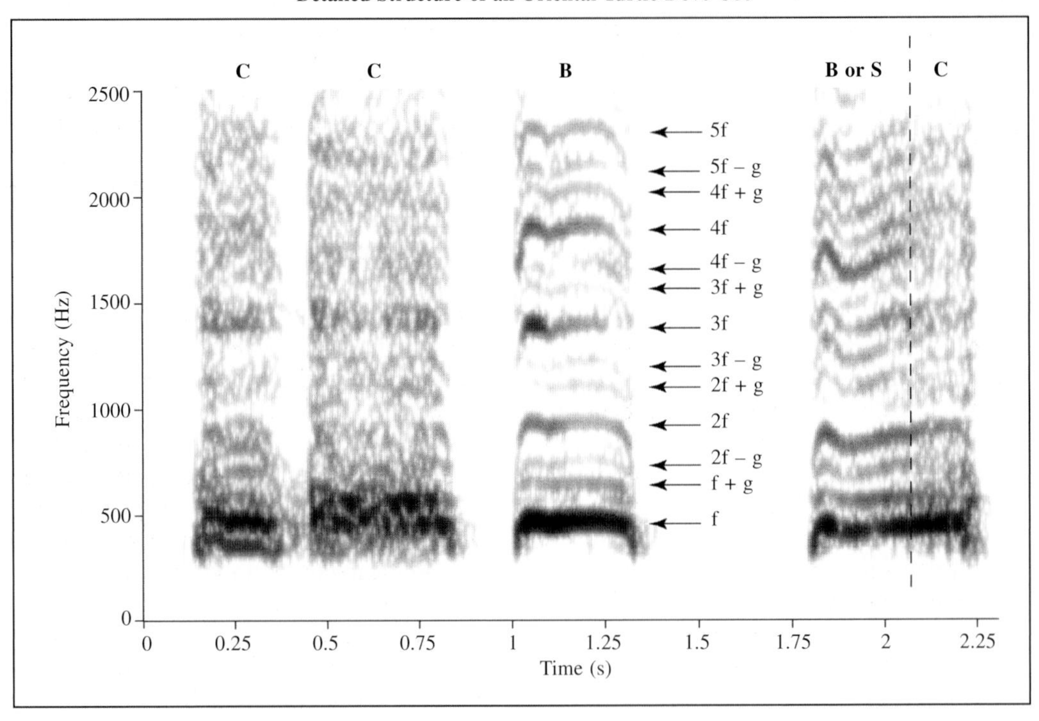

Figure 10.6 Nonlinearity and biphonation in the coo of the oriental turtle dove. The first two elements of the four-element coo sound 'noisy,' but the structure is 'chaotic' (C), not random. Element three shows biphonation (B) a harmonic signal (with frequency *f*) is modulated by a lower frequency component (*g*), which generates so-called 'side-bands' around each harmonic of *f*. The fourth element starts as either a biphonated sound (B) or as one with so-called 'sub-harmonics' (S – weak frequency bands regularly spaced between the dark dominant frequency bands). Part way this element assumes a chaotic structure. From Beckers & ten Cate in press.

in syringeal muscle action (Gaunt 1988). The precise nature of the motor pattern generator may also be a factor. In the Eurasian collared dove, the three elements of the perch-coo seem to be subject to separate control (ten Cate & Ballintijn 1996). If so, uncoupling the linkage between these units may enable such changes as the addition, doubling, deletion, or reordering of coo elements. Eurasian collared doves, for instance, sometimes produce coos with four, rather than three elements. This seems to originate from a 'stuttering' repetition of the first element (Ballintijn & ten Cate 1999b). Such variation within a species might provide the basis for evolutionary change. This possibility is consistent with the finding that vocal patterns in 'trumpeter' and 'laugher' breeds of domestic pigeons, artificially selected for their vocalizations, show these kinds of changes in coo structure. They owe their peculiar long drawn-out coos mainly to the repetition of elements or element combinations that are present, but not repeated, in their wild ancestor (Baptista & Abs 1983). In addition, a biophysical model (Gardner et al. 2001) suggests that variations in a few parameters can induce seemingly complex variation in starts, stops, and pauses in bird vocalizations. Hence, like some of the phenomena occurring in the control of the frequency domain, some of those in the time domain also seem to be due to the inherent properties of the mechanism involved.

The temporal structure of dove coos may thus be evolutionarily plastic, allowing the emergence of individual and, ultimately, of species differences. However, this does not mean that there are no constraints. As outlined by Podos & Nowicki (see Chapter 11), parameters such as duration, duty cycle, and repetition rate of vocal units may be constrained by limitations in respiratory volume or the need to coordinate respiration and vocalization. Whereas songbirds are able to use mini-breaths to sustain prolonged vocalizations, it is not clear whether any non-songbird is able to do this. Thus ring necked doves breathe only during the pause between two coos and not during the pauses within a coo (Gaunt et al. 1982). The same seems to apply to the Eurasian collared dove (ten Cate & Ballintijn 1996). If the respiratory cycle is indeed tightly linked to the coo cycle, this sets a limit to the duration or duty cycle of the coo. Whether any dove species can use mini-breaths during a coo cycle, hence breaking that constraint, deserves further study.

Vocal Perception in Doves

A change in vocal structure can only become established as a new trait, either within or between species, if it is perceptible and meaningful to others, requiring us to examine the receiving end of signal exchanges. With respect to frequency, the hearing range of birds is, broadly speaking, comparable to that of humans (Dooling 1982a; see Chapter 7), although birds are better at assessing temporal details in vocal structure (Dooling et al. 2002). However, species differ somewhat in the frequencies they hear best. Psychophysical studies show that for the domestic pigeon, a species from the sister clade to the turtle doves, optimal hearing is centered around lower frequencies than it is for songbirds, as is true for nonoscines generally (Dooling 1982a). The frequency range of dove coos is well within their area of optimal hearing, as one would expect. Like many other species, pigeons can detect differences in the frequency of two sounds down to 1–3% (Sinnott et al. 1980). We can relate this finding to the readily perceived frequency jumps in the coos of the collared dove, which are on average changes of about 20% (Slabbekoorn & ten Cate 1998b). If modulations were initially a byproduct of the way the vocal system in doves operates, made in particular by heavier doves and hence providing a message about the sender (ten Cate et al. 2002), their easy detection has surely been a factor in their evolution as a signal.

Between bird species, vocalizations differ both in frequency and temporal parameters. This does not mean that these differences are of equal importance for communicating species identity or other messages. Psychophysical studies have demonstrated that animals in general are better able to detect differences in frequency between two sounds compared to equivalent differences in sound duration. For budgerigars, for instance, the minimum difference they can detect between two frequencies is less than 1% (Dooling & Saunders 1975), while for duration it is 10–20% (Dooling & Haskell 1978). As a consequence, the finding that dove coos differ in temporal parameters does not mean that the birds use these differences to discriminate between species. On the other hand, if characteristics of the vocal system make it easier for differences in temporal features to evolve, the perceptual system may evolve greater sensitivity to such parameters, even when the mechanisms underlying these differences in perceptual sensitivity show limited scope for evolutionary change. For collared doves, field studies show that at least some aspects of temporal parameters are important in triggering a response (Slabbekoorn & ten Cate 1999).

To test the relative importance to those sound parameters that differ between dove species, some laboratory experiments have taken into account the natural vocal structure and interspecific variation in dove coos. One examined the parameters used by ring-necked doves to distinguish own-species from other-species vocalizations (Beckers et al. 2003a). Using a so-called 'Go–NoGo' design, doves could peck a key which next exposed them to either a sound ('Go') or a silent period ('NoGo'). A second peck, after the 'Go' sound, led to a food reward. Initially the sound was a conspecific perch-coo. When the doves were trained, some coos were replaced with coos of other turtle dove species, testing the prediction that the ring-necked doves would peck more for coos they perceive as more similar to conspecific ones. The outcome was that the doves treated the coos of some species, like the spotted dove, as perceptually similar to their own, while others, like those of the European turtle dove, were treated as being very different (Beckers et al. 2003a). A further analysis showed that virtually all variation in perceptual similarity was due to three factors: duration, minimum frequency, and the 'randomness' of a sound (the latter as expressed in a factor called 'Wiener entropy;' Tchernichovski et al. 2000). The analysis also allowed an assessment of the sensitivity of doves to the various parameters, indicating that they were equally sensitive to differences in minimum frequency and coo duration. As such data on other species are lacking we cannot make comparisons, but the results suggest that doves are more sensitive to variation in temporal parameters of vocalizations than suggested by psychophysical studies in other bird species on the discrimination thresholds for differences in duration. Evolution may thus have sharpened the perception of temporal parameters in doves.

The results also show that the discrimination between conspecific and heterospecific coos is based on a combination of various parameters (Beckers & ten Cate 2001). It appears that signal evolution has been guided by, on the one hand, the options that the sound-producing mechanism of senders provides for variation in sound structure and, on the other hand, by the sensitivities to specific parameters inherent to the perceptual system of receivers. The evolution of vocal signals is truly the product of sender–receiver co-evolution!

Vocal Evolution in Turtle Doves: Conclusions

Our studies have identified sexual selection and the interaction with related species as the main selection pressures operating on the characteristics of dove vocalizations. Sexual selection operates on the frequency jumps, favoring their presence in a signal involved in territorial defense. Species interactions may operate on both frequency and temporal characteristics, but with no clear directionality.

As the production of frequency jumps in coos is correlated with higher body weight, and therefore presumably with fighting ability (ten Cate et al. 2002), jumps may be an indication of quality. It is tempting to suppose that there is a direct linkage between the feature that apparently has been selected as a quality signal, and the finding that frequency jumps may be a common consequence of higher pressure or flow levels in the vocal tract. The possibility that stronger birds more readily achieve higher flow levels, and the ready perceptibility of jumps may explain their evolution as quality signals. We can speculate that, at an early stage of dove vocal evolution, jumps arose as an artifact of vocalizing with greater strength and that, because this artifact was perceptually salient, it next evolved into a true signal. Thus, inherent properties of the mechanisms of vocal production and perception may have provided directionality to signal evolution. Similarly, the temporal features that distinguish between the various species, like element number, may have required only marginal changes in the motor program of cooing and thus are readily available for the selection leading to species differences.

However, some caution is necessary in interpreting these connections as causally related to the direction of evolutionary change. Thus far, the full 'vocal space' potentially available to doves for vocal evolution has yet to be explored. Further experimental and comparative studies are needed to identify the crucial parameters determining this vocal space. In combination with biomechanical or physical modeling, this will allow exploration of what may occur if the system is altered beyond the limits existing in present species and will shed light upon the vocal transitions that the system permits.

BOX 41

SONG LEARNING AS THE MOTOR FOR SPECIATION IN VIDUID FINCHES

Indigobirds, whydahs, and other Vidua finches are African brood parasites which lay their eggs in the nests of other species. Most are specialists, parasitizing one particular estrildid finch. Some show specific adaptations to these host species with the mouth markings and plumage of their young resembling those of the host species' young, and sing songs that incorporate sounds of their host (Nicolai 1964; Payne 1973b); females are attracted to that particular song type (Payne 1973b; Payne et al. 2000). The song of the host species also stimulates ovarian development and attracts parasitic females to host species' nests. Nicolai (1964) suggested that parasite species had evolved jointly with their hosts, but DNA studies show that the most likely branching patterns of hosts and parasites do not match. The evidence suggests that Vidua finches speciated more recently than their hosts, supporting a 'colonization' model for evolution, with parasitic lineages switching from one host species to another, leading to subsequent adaptation to the new host (Payne et al. 1998, 2000). How does a parasitic species colonize a new host and become reproductively isolated from its ancestors? In viduid finches, as in other songbirds, learning is important in developing songs and song preferences. Payne et al. (1998, 2000) demonstrated that when village indigobirds, which normally parasitize firefinches, were cross-fostered to Bengalese finches (a domesticated species from Asia) they developed a distinctly different Bengalese finch-like song, instead of the usual firefinch-like song (see below; **CD2 # 77–78**). Cross-fostered females, which normally prefer village indigobirds singing firefinch-like songs, now preferred mates singing Bengalese finch-like songs. When these females were placed in aviaries containing several potential host species, including firefinches and Bengalese finches, they preferentially laid their eggs in the nests of Bengalese finches, the species that raised them! Thus, young raised in a novel host nest differ substantially from their biological parents in their behavior, taking an important step towards colonizing the new host; they also achieve reproductive separation from ancestors raised by the traditional host. Once successful in the new situation, traditional Darwinian mechanisms of genetic mutation and selection might subsequently lead to morphological adaptations, such as the nestling mimicry of mouth markings, eventually resulting in a new parasitic species. So, the rapid speciation of Vidua finches is the outcome of an evolutionary process in which song learning plays a critical role (Payne 1973b; Payne et al. 2000; Sorensen & Payne 2001), as it may do in other evolutionary processes (ten Cate 2000).

Robert B. Payne & Carel ten Cate

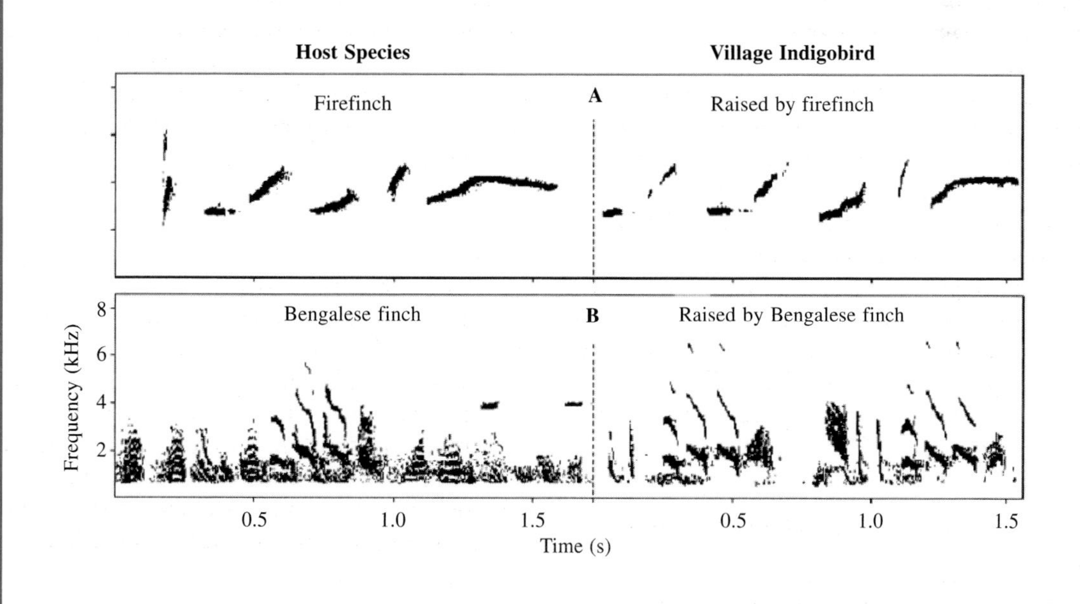

HOW ABOUT OTHER BIRDS?

The mechanisms underlying dove vocalization influence the direction of vocal evolution by constraining or biasing certain types of change over others. Similar principles will operate in other groups of birds, each of which will have their 'vocal space' determined by somewhat different parameters. This even applies to songbirds (Slabbekoorn & Smith 2002b), considered to be the most plastic of all with respect to the mechanisms underlying vocalization (**Box 41**, p. 315). Songbird vocalizations have been taken to be nearly free from constraints on the mechanism of vocal production. For instance, Prum (1992) contrasts the striking diversification in the structure and musculature of the syrinx of suboscine manakins with the comparatively uniform, albeit complex, syrinx present in true songbirds. He suggests a relationship between the differences in syrinx variability within these two groups, and the fact that oscines learn their songs, while suboscines generally do not (Prum 1992; see Chapter 4). Selection operating on interindividual variation in non-learned songs will, according to Prum, mean a strong selection on the apparatus making these songs, while syringeal morphology will be less subject to selection when vocal variation is the result of learning. While syringeal variation in manakins might originate from selection on vocal variation in that group, the dove studies indicate that there are other ways to get vocal variation among non-learning species than by changes in syrinx morphology. Also, although the incredible vocal variation in some songbirds, such as the nightingale, indicates that vocal learning may increase vocal variation, the appropriate vocal apparatus and motor control mechanisms to produce this variation are still required. Therefore, with selection for vocal variation induced by learning, it is not inevitable that there would also be a reduction in the impact of selection on syrinx morphology. One may even argue that the presence of vocal learning will select for a syrinx or vocal tract that is capable of producing more and more diverse sounds. At the same time, even in songbirds able to produce many different song types, the evolution of vocal characteristics can be constrained by the vocal apparatus (see Chapter 11). The morphology of the syrinx and the vocal tract, for instance, may limit the frequency, and shape, or number of notes that a particular bird can produce, as when males of the two sub-species of zebra finches cross-fostered to the other sub-species alter note phonology, but not the frequency range of their songs (Clayton 1990a).

Apart from the vocal apparatus, central processes involved in song learning may also be subject to constraints, with a bias favoring some vocal changes over others. Birds may, for instance, prefer to copy the notes or songs of conspecifics over those of other species, as in song and swamp sparrows (Marler & Peters 1988b). Also, songbirds exposed to songs of other species may copy some elements, but retain much of their species-specific song structure in singing them (Baptista 1996).

At another level, the characteristics of the learning process itself can be under selection. Here also, specific types of modifications may be more likely to occur than others. For instance, songbird species with continuous songs, like the brown thrasher and other Mimidae usually also have a large repertoire of syllable types (Kroodsma & Parker 1977). In contrast, species with discrete songs, like the chaffinch, often possess small repertoires (Slater 1981). Irwin (1988) suggests these changes are due to heterochrony, with a shift in the relative timing of different phases of the song learning process. In other words, the evolution of this aspect of the diversification of song types in songbirds may originate from a change in the developmental process.

So, even though learning mechanisms may provide songbirds with more options for vocal change than non-learning species have available, this does not mean that they are free of constraints, or that song learning is completely free of biases. The many relationships between vocal variation and phylogeny also imply that vocal learning does not introduce the prospect for limitless vocal change. The same principles operate as in non-songbirds, albeit with more degrees of freedom. Our fascination with the mechanisms underlying vocal learning and the resulting vocal variation in songbirds should not blind us to the existence of limits, the study of which is important if we are to gain a full understanding of songbird vocal variety.

CONCLUSIONS

Bird species differ greatly in their songs. A full understanding of the evolution of this variety requires us to combine the study of selection and phylogeny with that of the mechanisms involved in their production and perception. This requirement is not unique to the study of vocal signals, but can be extended to evolution of signals in general (Endler 1992; Endler & Basolo 1998). The vocal signals of birds, however, offer especially attractive model systems for such studies. This is so, not just because many aspects of bird vocalizations – their communicative significance, their development, their production and perception are amenable to study in great detail – but also because of their intrinsic attractiveness to us as human listeners.

Chapter 11

Performance limits on birdsong

JEFFREY PODOS AND STEPHEN NOWICKI

INTRODUCTION

The musical virtuosity of birds, celebrated throughout this book, is made possible by a vocal mechanism that we are just beginning to understand. The question of how birds produce sound has a rich history (Hérissant 1753) and traditionally has been explored indirectly by drawing inferences from anatomical description (Ames 1971), analyses of sound patterns (Greenewalt 1968), and analogies to human music and speech (Nowicki & Marler 1988).

More recent work on avian vocal production mechanisms has shifted emphasis to the study of live, singing birds, an approach that has produced some of the key discoveries of the past quarter century. Here is a sample of such discoveries (see Chapters 8 & 9): (i) the brain nuclei that mediate vocal production and learning are widely distributed, hierarchically arranged, and developmentally and seasonally plastic; Yu & Margoliash 1996; Doupe 1998); (ii) the left and right sides of the syrinx contribute asymmetrically to sound production (Nottebohm 1971a, b; Suthers 1990); (iii) the vocal source at the syrinx may comprise not one but multiple pairs of membranes and other tissues (Goller & Larsen 1997b; Larsen & Goller 1999); (iv) the vocal tract, including the trachea, larynx, and beak, contributes to the pure-tonal structure of songbird song, by damping harmonic overtones (Nowicki 1987; Westneat et al. 1993); and (v) syringeal control and acoustic output are related in a nonlinear manner, thus accounting for some otherwise puzzling features of birdsong complexity (Fee et al. 1998; Gardner et al. 2001). Most of these findings could not have been anticipated by inference; the discovery component of the field of avian vocal mechanics thus remains alive and well.

Our principal goal here is to explore how recent advances in our understanding of vocal production mechanisms might inform our understanding of the evolution of bird vocalizations. Analysis of the relationship between mechanism and evolution in biology has a venerable history for morphological systems (Raup 1966; Gould 1980; Nijhout 1991), but this approach has been applied to animal behavior only sporadically (Garland & Losos 1994; Prum 1998). In animal behavior, proximate mechanisms such as those concerned with development, control, and mechanics (Tinbergen 1951) can shape evolution by imposing constraints or biases on the direction or magnitude of evolutionary change (Wake & Larson 1987; Perrin & Travis 1992; Wenzel 1993).

Some of the broader questions that motivate our interest in the interface of mechanisms and evolution for bird vocalizations include: What anatomical systems contribute to vocal behavior in birds, and how do they operate? To what extent can diversity in the anatomy and function of the avian vocal apparatus explain the extraordinary diversity of sounds that birds can make? And, to recall a debate in which Luis Baptista was a key participant (Baptista & Trail

Plate XI

Swamp sparrow Elliott

Medium ground finch Podos

Vegetarian finch Podos

Cactus finch Podos

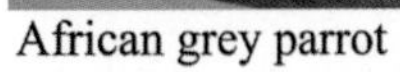

African grey parrot — Pegg

Brown-headed cowbird — King

Aratinga parrot — Spotswood

European starling — Elliott

1992), what are the consequences of vocal diversity for bird speciation? Research into these questions is still in its early phases. Nevertheless, as we show here, some interesting conclusions have started to emerge.

We divide the chapter into two sections. The first focuses on the morphology and function of the avian vocal 'instrument.' The central message is that vocal mechanisms are limited in their potential, and thus the vocal apparatus circumscribes the range of sounds birds can and cannot produce. The second section examines three specific examples: the evolution of species diversity in the Passeriformes (the largest taxonomic grouping of birds); the evolution of vocal novelty in one songbird, the swamp sparrow; and the relationship between morphological evolution and vocal diversity in Darwin's finches. With these examples, we illustrate some of the ways by which mechanisms of vocal production influence patterns of vocal diversity in birds. We also hope to gain insight into the processes underlying the evolution of vocal behavior, and their implications for the evolutionary diversification of bird groups.

MECHANISMS OF VOCAL PRODUCTION

The Avian Vocal Instrument

The anatomical apparatus birds use to sing is like most other systems in biology: the more we look, the more complex it appears to be. If we had to choose a key point that has emerged from recent work, it would be that the avian vocal apparatus consists of multiple, interacting components that are extraordinarily closely and precisely coordinated during sound production.

The primary sound-generating organ of birds is the syrinx, an organ that is unique to the class Aves (Greenewalt 1968; Gaunt & Nowicki 1998). The syrinx generates sound in a manner that is roughly analogous to the way the human larynx works during speech production: airflow from the respiratory tract causes tissues to vibrate, and these vibrating tissues generate sound (see Chapter 9). In songbirds, a pair of thin membranes, the 'medial tympaniform membranes,' have long been thought to act as dual sound sources (Greenewalt 1968), although it now appears that a variety of syringeal tissues contribute to sound production (Goller & Larsen 1997b; Fee et al. 1998; Larsen & Goller 1999). Early models suggested that the syrinx alone is responsible for all aspects of sound production (Greenewalt 1968). It is increasingly apparent, however, that sound production depends not just on the syrinx, but also upon a suite of complementary motor and neural systems. Patterns of breathing, for example, are finely coordinated with syringeal activity, and thus appear to be essential for controlling the overall temporal pattern of a vocalization (see Chapter 9; **Box 42**, p. 322).

The portion of the vocal tract anterior to the syrinx, including the trachea, larynx, and beak, also plays a central role in sound production. The vocal tract acts as an acoustic resonance chamber, modifying sounds produced by the syrinx. The mechanism by which vocal tract resonances modify syringeal output is not fully resolved, although it seems likely that in birds, as in humans, the vocal tract selectively attenuates some frequencies while allowing others to pass undiminished (Nowicki 1987; Nowicki & Marler 1988; Fitch & Hauser 2003). In human speech, this selective filtering accounts for the complex frequency characteristics of different vowel sounds (Fant 1960); in the production of birdsong, the vocal tract filter enables birds to produce highly pure-tonal sounds, in which acoustic energy is concentrated at single frequencies (Nowicki & Marler 1988).

For the vocal tract to serve as an acoustic filter, its resonances must be dynamically modified during song so as to track changes in the frequency output of the syringeal acoustic source. We know that changes in the position of the tongue, jaw, and lips all contribute to changing vocal tract resonances during speech (Fant 1960). In the production of birdsong, there also may be a number of factors that influence vocal tract resonance, but the role of the beak in particular is central (Westneat et al. 1993; Podos et al. 1995; Hoese et al. 2000). As a songbird opens or closes its beak, it effectively shortens or lengthens its vocal tract, with the acoustic result being a shift of vocal tract resonances to higher or lower frequencies, respectively (Nowicki & Marler 1988; Westneat et al. 1993). Simply put, birds need to open their beaks more widely when singing high-pitched sounds than with low-pitched ones. Further, they need to open and close their beaks during song in register with changes in the output of the syrinx, which are often very rapid. Thus, movements of the vocal tract must be coordinated with syringeal and respiratory activity for a bird to sing normally.

Nowicki and Marler (1988) likened the motor challenges birds face in the production of a complex vocalization to those faced by a human musician playing a rapid, complex score on a musical instrument. To illustrate this point, consider what a songbird needs to do in order to produce a trilled song (Nowicki et al. 1992; Podos 1997). A 'trill' is a repeated series of identical acoustic units (**Fig. 11.1; CD1 #87**). These themselves may be single motor gestures ('notes,' represented as a single continuous trace on a sonogram), or perhaps groups of notes that repeat as a unit ('syllables'). Each note has a characteristic frequency contour and amplitude profile generated by a vibrating membrane in the syrinx (or perhaps two interacting membranes), the behavior of which is determined by the tension set by the syringeal muscles. Airflow necessary to excite membrane vibration is produced by activity of respiratory muscles (**Box 42**, p. 322). This flow may correspond to one continuous breath cycle, although in at least some cases 'mini-breaths' will be taken with each repeated element in the trill (see Chapter 9). To produce pure-tonal elements, jaw musculature must open and close the beak in register with changes in the frequency contour of individual notes. It is likely that movements of the trachea and larynx also are involved in modulating vocal tract resonances during song. All of these movements must be repeated precisely and in rapid succession, often for dozens of cycles, and in a synchronized manner. Finally, as is true for an instrumentalist, the operation of these motor systems is probably affected by many other factors having little to do with song production directly, such as whether a bird is flying or not, its posture when perched, or movements associated with other behaviors performed simultaneously such as visual wing displays. The bottom line is that the production of birdsong must now be viewed as a fundamentally multifaceted, nearly beak-to-foot effort.

Vocal Performance

Given the complexity of avian vocal production mechanisms, including the need to coordinate many interacting systems, it follows that the act of singing can be physically or physiologically challenging to birds (Nowicki et al. 1992; Lambrechts 1996; Podos 1997; Suthers & Goller 1997; ten Cate et al. 2002). Physical limitations on the production and control of vocalizations are particularly interesting as a source of insights into patterns of vocal evolution. By definition, vocal evolution entails changes in the acoustic structure of vocalizations that accumulate across generations within a given lineage (Slater 1986). Most of the evolutionary changes that have been observed in long-term field studies, such as in the structure of individual elements ('notes'), the presence or absence of those elements, or the order in which they occur in sequence (Jenkins 1978; Grant & Grant 1996a; Payne 1996), seem unlikely to tax the vocal apparatus. But there are certain modifications, for example increases in the rate at which units are repeated, or changes in the acoustic frequency range of elements, that may challenge the performance limits of the production system. There may be limits on the speed or force the vocal system musculature is able to exert, or constraints on neural control capabilities (Nowicki et al. 1992; Podos 1996, 1997; Gil & Gahr 2002; ten Cate et al. 2002). Limits on motor proficiency could have a number of evolutionary consequences, including stasis

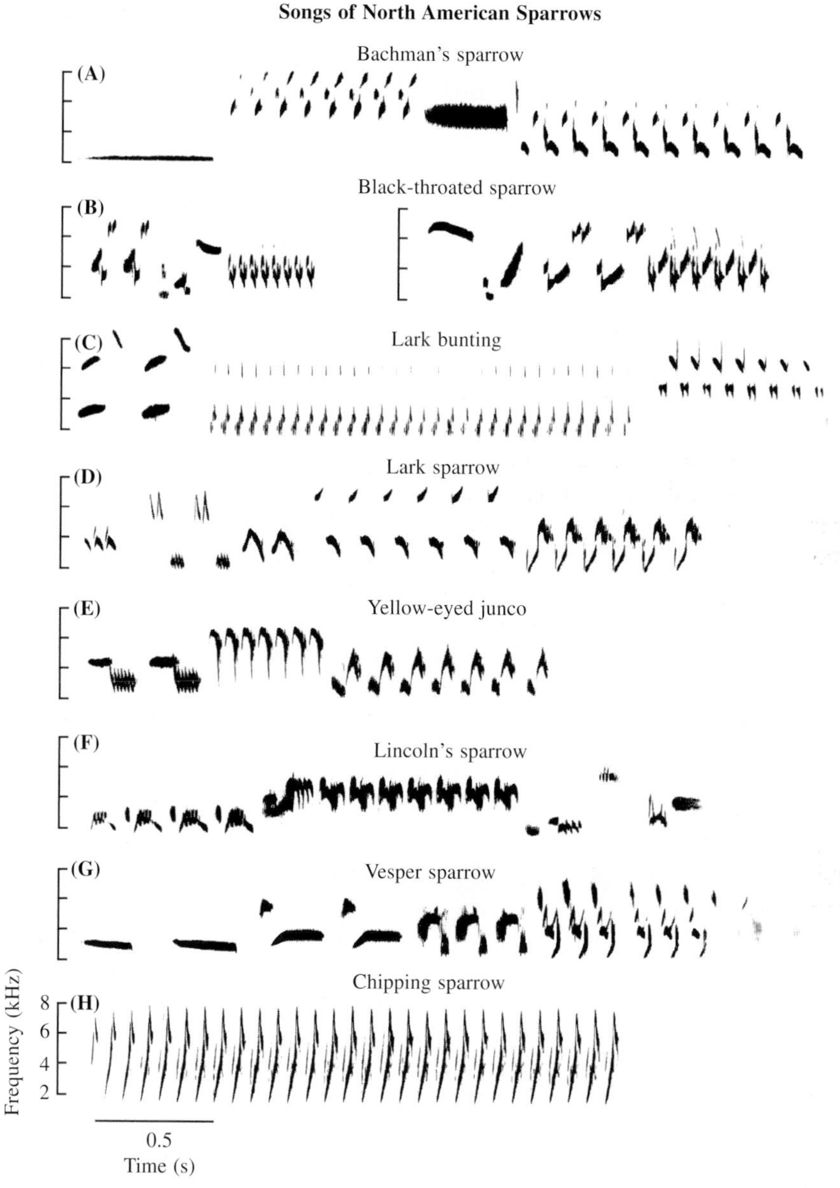

Figure 11.1 Songs of eight North American Emberizine sparrows, illustrating the pronounced song differences that can occur even between close relatives (**CD1 #87**). (**A**) Bachman's sparrow; (**B**) black-throated sparrow (two examples shown); (**C**) lark bunting; (**D**) lark sparrow; (**E**) yellow-eyed junco; (**F**) Lincoln's sparrow; (**G**) vesper sparrow; (**H**) chipping sparrow. Each of these songs contain trilled sequences. Reprinted from Podos 1997.

in the evolution of vocal characters despite selection pressures for change, or the emergence of tradeoffs among vocal characters (Wells & Taigen 1986).

The fact that songbirds open and close their beaks repeatedly while singing, maintaining close coordination with changes in the fundamental frequency of the signal from the syrinx, is one

BOX 42

BREATHING PATTERNS DEFINE UNITS IN ZEBRA FINCH SONG

Many birdsongs consist of a variable number of distinct sound segments, either syllables or notes, separated by silent intervals. In a zebra finch song motif, 4–8 different syllables are strung together in a stereotyped sequence. To understand the motor control of song production, we need to know how these segments are represented in the motor program. Indirect evidence suggests that syllables may constitute modular units. Young zebra finches copied song by incorporating entire syllables, but not syllable fragments, into their emerging song (Williams & Staples 1992). Also, when singing zebra finches were interrupted by a light flash from a stroboscope, they stopped at the end of syllables and not in the middle (Cynx 1990).

In a more direct examination of this question, the respiratory motor patterns of song were studied in a large number of male zebra finches (Franz & Goller 2002). Zebra finch song is generated by a stereotyped series of distinct expiratory pressure pulses, with each breath corresponding to a different song syllable (**A**). Short, typically silent inspirations (mini-breaths) occur in between these expiratory pulses. The temporal sequence of song, with its cadence of phonation and silent periods, is therefore determined by alternately breathing in and out (Wild et al. 1998). Males that sang similar song syllables tended to do so using remarkably similar respiratory patterns (**B**). In three quarters of the imitated syllables, the similarity included detailed modulation of the expiratory pressure pulse and the occurrence of mini-breaths following the syllable. In the few exceptions where pressure patterns differed but song was similar, a mini-breath was omitted or inserted. These observations indicate that respiratory motor patterns used to generate a particular sound may be dictated at least in part by the requirements for producing that sound, as well as by physiological needs, such as air supply and gas exchange. It appears that silent inspirations provide a temporal definition of song syllables and, thus, of corresponding components of the song motor program, underlying the organization of the expiratory air sac pressure pulses. It follows that the fine-grain patterning of breathing must play a defining role in the temporal organization of the song motor program.

Franz Goller

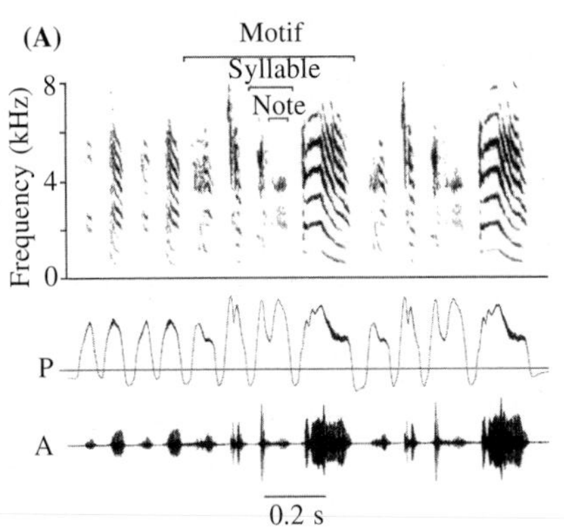

Air sac pressure pattern of song (P) in a zebra finch with inhalations below the line and exhalations above. (A) is the amplitude envelope.

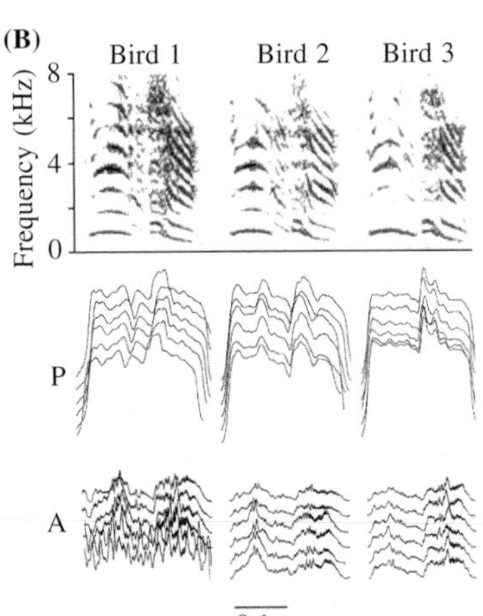

A syllable imitated by 3 individuals illustrates the similarity between birds in air sac patterns. P and A show six examples of each.

source of a specific constraint on patterns of vocal evolution. There is likely to be a tradeoff between how fast syllables can be repeated ('trill rate') and how broad a range of frequencies each repeated unit can encompass ('frequency bandwidth;' Westneat et al. 1993; Podos et al. 1995; Hoese et al. 2000). Given the role of beak movements in song production, we would predict that, while slow trills can be associated with both narrow or wide frequency bandwidths, rapid trills will be limited to narrow bandwidths. This is because of the difficulty of opening and closing the beak both widely and rapidly (**Fig. 11.2A**). A rough analogy can be made here to handclapping (Podos 1997). When clapping slowly, it is possible to produce both soft claps (where the hands are not necessarily widely separated), or loud claps (where the hands must necessarily be pulled far apart before making each clap). When clapping at faster rates, however, there is not enough time to separate the hands widely and bring them back together again quickly. Thus, beyond a certain rate, one cannot clap both quickly and loudly.

Evidence for this kind of tradeoff comes from analyses of the songs of 34 species in the songbird family Emberizidae, looking especially at trill rate and frequency bandwidth (Podos 1997). The results of such an analysis are plotted out as triangular patterns, not only within species but also within genera and across the entire sample (**Fig. 11.2B**). This result clearly supports a role for a mechanical tradeoff in trill production. Songs fill the lower left section of the plot, but do not occur in the region of fast trill rates and wide bandwidths. Presumably, a species' songs could only evolve into this constrained region of the 'acoustic space' portrayed in this graph if there was corresponding evolution of beak morphology, jaw musculature, or other aspects of the vocal tract that would allow birds to move their beaks more rapidly than is presently possible (Hoese et al. 2000).

Other aspects of birdsong might similarly be explained by performance tradeoffs limiting vocal evolution. In canaries, for example, there is a shift in the pattern of breathing associated with the rate of syllable repetition in trilled vocalizations (Hartley & Suthers 1989). At slower rates, canaries breathe cyclically while singing, making small inspiratory 'mini-breaths' between each song element; above a certain rate, birds simplify this pattern to a continuous expiration (see Chapter 9). This shift is consistent with the idea that birds are able to use mini-breaths only up to particular rates, beyond which they must change their breathing pattern to compensate for performance limits. Similarly, birds may shift from a cyclical pattern of beak gape changes to a single, continuous gape change over the course of particularly fast trills (Podos 1997).

Two patterns observed in the singing behavior of great tits may be caused by similar tradeoffs. First, the length of repeated phrases in these birds increases with increasing song durations (Lambrechts & Dhondt 1987). Second, the relative stereotypy of notes ('song frequency plasticity') decreases with increasing trill rates (Lambrechts 1997). Although the physical basis of these apparent tradeoffs is not understood, both seem likely to reflect performance boundaries imposed by limitations of vocal production mechanisms (Lambrechts 1996). The song length/phrase length relationship in particular may derive from a performance tradeoff similar to that described by Wells & Taigen (1986) for gray treefrogs, in which calls are produced at slower rates as call durations increase. Whereas the vocal performance limits described in birds seem to arise from biomechanical sources as opposed to the energetic causes thought to be responsible in gray treefrogs, a similar tradeoff between acoustic features emerges.

Diversity in Mechanisms Underlying Vocal Production

Thus far, we have focused on patterns in the structure of vocalizations observed within species or between close relatives. Is it also possible to explain larger-scale vocal patterns in terms of differences in morphology or other mechanisms involved in production? At some level, the answer to this question must surely be yes. Chickens

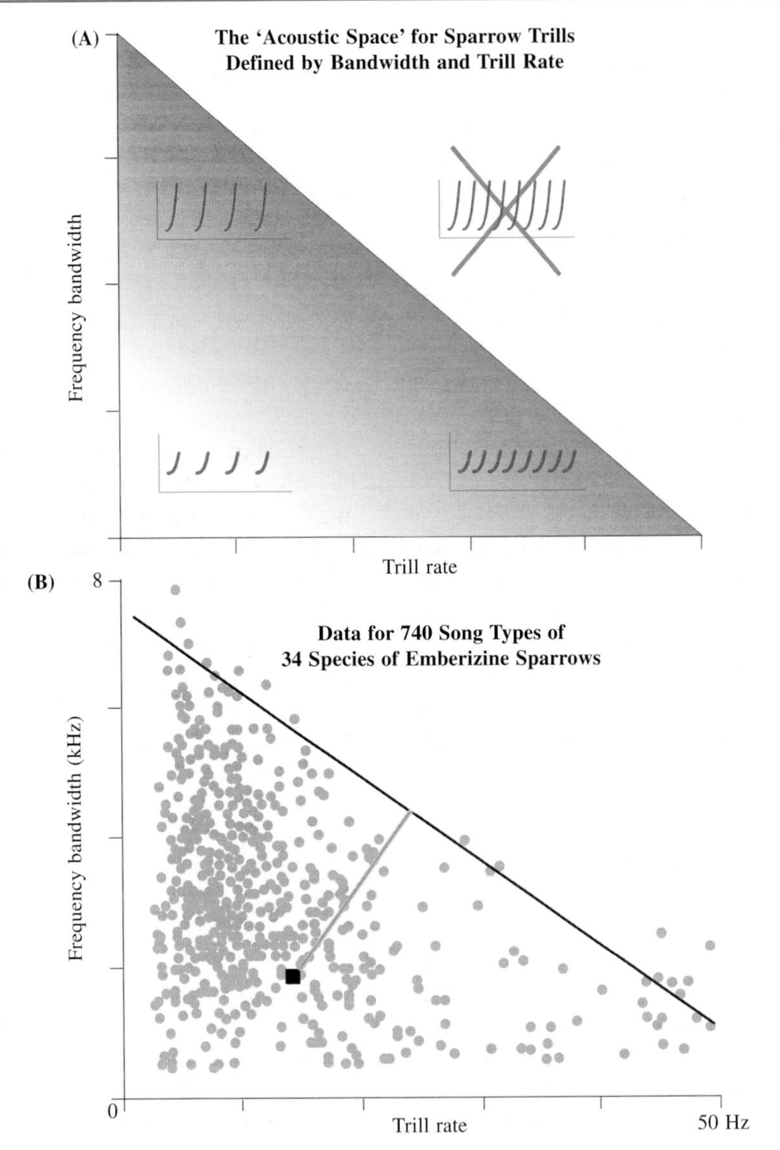

Figure 11.2 (**A**) Trill rates and frequency bandwidths can be used to depict the 'acoustic space' that a vocal trill occupies. Songs with low trill rates and a limited frequency bandwidth (lower left sonogram) require minimal vocal tract movements and thus should be easily produced. However, as songs increase in either frequency bandwidth (upper left sonogram) or in trill rate (lower right sonogram), the demands required of vocal performance are elevated. Increases in bandwidth require wider gapes, and increases in trill rates require more rapid beak opening and closing. The physical challenge of producing songs with faster trill rates and/or broader frequency bandwidths should increase steadily, indicated here by the light-dark gradient, until a performance limit is reached. Songs beyond this boundary, on the upper right cannot be produced because they would require vocal tract movements exceeding the bird's performance capabilities. (**B**) Plot of trills for 740 song types from 34 species from the family Emberizidae. The triangular distributions evident for individual species and genera support the vocal tract constraint hypothesis (Podos 1997). Vocal performance for any given trill as, for example, the filled square, can be measured as the minimal distance (light line) to an upper bound regression (dark line; Podos 2001).

(Galliformes) and parrots (Psittaciformes), for example, produce distinctly different kinds of sounds, and these differences must stem in part from the fact that syringeal anatomy and vocal production mechanisms are radically different in these two groups (Nottebohm & Nottebohm 1976; Gaunt & Gaunt 1977). Syringeal structure is so diverse across different orders of birds (Beddard 1898; King 1989; Prum 1992) that it seems self-evident that anatomical variation must contribute to vocal variation at this level. But with our limited knowledge of production mechanisms across different groups of birds with diverse syringeal anatomy, it is difficult to find detailed examples to support this point.

It seems equally clear that differences in vocal anatomy have little to do with some of the other dimensions of vocal diversity we observe in nature. Consider the New World *Empidonax* flycatchers. The different species in this group are so similar morphologically that trained ornithologists have a difficult time distinguishing them, even when held in the hand (Pyle 1997), yet their songs are quite distinct. In some cases diverse vocal characters have proven to be a key diagnostic character in formal systematic analyses where morphological characters do little to separate species (Whitney et al. 2000). The oscine suborder of the Passeriformes, the 'true songbirds,' stands out as exhibiting the most notable vocal diversity within a taxon of birds (Catchpole & Slater 1995; see Chapter 4), while at the same time being noted for an exceptional *lack* of variation in syringeal structure across all 4,000 or so species in the group (Cutler 1970; Ames 1971). On the other hand, bird watchers often comment that different species in a genus or even a family somehow sound alike, in spite of more obvious differences in their songs. To what extent can we attribute differences in how birds sound to more subtle differences in morphology and, of greater interest to us, to what extent does morphology constrain or enable the evolution of diversity in bird sounds, even within a group such as the songbirds, which is so uniform from a syringeal point of view?

A straightforward example of how morphology influences vocal behavior concerns body size. Birds with a relatively large body have a correspondingly large syrinx (Cutler 1970) and, as a general rule, produce lower-pitched vocalizations (Bowman 1979; Wallschläger 1980; Ryan & Brenowitz 1985; Ballintijn & ten Cate 1997). The reasons for this relationship seem clear. The membranes in larger syrinxes will themselves be correspondingly large and thus tend to vibrate across a lower range of frequencies than the correspondingly smaller membranes of smaller syrinxes. A similar example is the predicted relationship between body size, respiratory tidal volume, and the maximum duration of a single call. Longer calls, at least those for which the lungs' tidal volume is continuously depleted (Hartley 1990), should be restricted to larger individuals with larger lungs (Fitch & Hauser 2003). The fact that the anatomy of the syrinx is so highly conserved within oscines, however, suggests it may be hard to find additional functional correlation between syringeal structure and vocal diversity in this group, beyond general relationships such as these.

Another possible origin of vocal diversity, not necessarily related to body size, is variation in the size and shape of the vocal tract. This variation can be impressive in birds, especially within groups for which there is extensive variation in beak morphology because of feeding specializations (Bowman 1961). Perhaps the most extreme example of diversity in vocal tract morphology comes from those birds, including some cranes, guans, curassows, swans, and geese, that possess an unusually elongated trachea, extending up to 20 times their expected length based on body size (Roberts 1880; Fitch 1999; **Box 43**, p. 326). This extra length is physically accommodated by coiling of the trachea in the sternum or thorax. Many hypotheses have been advanced over the years to explain the function of this unusual trait, categorized by Fitch (1999) as either physiological or acoustic in nature. Physiological hypotheses have focused on gas exchange in respiration. Schmidt-Nielsen (1972), for instance, suggested that the trumpeter swan, a migratory species with low rates of breathing,

BOX 43

ACOUSTIC EXAGGERATION OF SIZE BY TRACHEAL ELONGATION

At least 60 bird species possess an elongated trachea which loops or coils within the sternum or thorax, termed 'tracheal elongation' (TE). This peculiar trait, found in eleven families and six orders, has been known for centuries, and has evolved repeatedly (Frith 1994; Fitch 1999). The function has been the subject of much speculation. Various physiological functions have been suggested, including CO_2 or temperature regulation, or water retention, but all of these hypotheses are of limited applicability, and have not been supported when tested empirically. Thus, most recent commentators have agreed that TE serves some acoustic function, often suggesting that TE lowers the pitch of calls, including the fundamental frequency. However, this notion contradicts current models of bird vocal production, since the fundamental frequency of bird vocalizations is not controlled by tracheal length (Gaunt et al. 1987; Nowicki 1987; Nowicki & Marler 1988).

I propose that TE does not modify the fundamental frequency, but serves to lower the frequency of formants and, thus exaggerating the bird's apparent size (Fitch 1999; **CD2 #79**). Formants are overtones emphasized in the spectrum. Usually, tracheal length is correlated with body size (Hinds & Calder 1971), and formant frequencies, as the acoustic correlates of tracheal length, thus convey information about body size. Birds easily learn to perceive formants in human speech (Dooling & Brown 1990), and whooping cranes spontaneously respond to formants in their species-specific calls (Fitch & Kelley 2000). Thus, in ordinary birds, formants provide an acoustic indicator of body size. This baseline sets the stage for exaggeration: by elongating its trachea a small bird can duplicate the formants of a larger bird, and thus sound bigger than it actually is. Although low formants might sometimes also be useful for enhancing sound propagation, in ground-calling species, lowering formant frequencies via TE actually decreases propagation, due to ground interference (Wiley & Richards 1978).

The 'size exaggeration' hypothesis is consistent with available ecological and behavioral data. First, the theory of evolutionarily stable deception (Wiley 1994) suggests that exaggerators must be prepared for perceivers who 'call their bluff.' Thus, species who exaggerate their size should also be larger, on average, than those which don't. This prediction is met in TE species, which are significantly larger than related species which lack TE. Thus birds with TE have apparently been subjected to selection for large body size, as required by the size exaggeration hypothesis (Fitch 1999). Second, to be effective, vocal exaggeration should occur in species where other cues to size are typically absent; most TE species live or nest in dense, low-visibility environments, and/or call primarily at night. Third, the observed patterns of sexual dimorphism support the size exaggeration hypothesis: in species where both sexes show TE, both are involved in territory and nest defense, while in species with sexually dimorphic TE, only the sex involved with territory defense or lek display exhibits TE. The 'exception that proves the rule' is the Indian painted snipe. In this polyandrous species, large females defend territories and compete among themselves for the smaller males who tend the nest (Johnsgard 1981). Only female snipes have TE.

W. Tecumseh Fitch

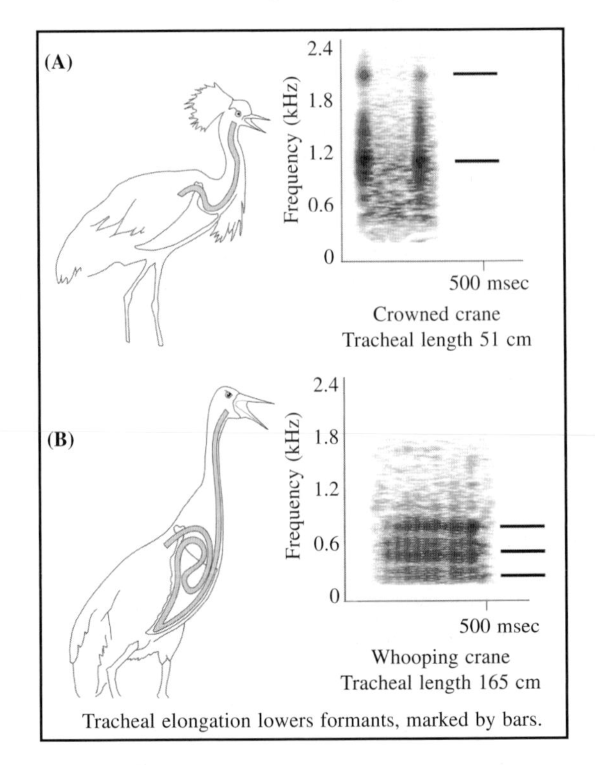

Tracheal elongation lowers formants, marked by bars.

requires an elongated windpipe to provide the volume of respiratory 'dead-space' needed for mixing carbon dioxide with outside air. Acoustic hypotheses have linked tracheal elongation to aspects of sound production. Numerous authors have suggested, for instance, that an extended windpipe helps birds to vocalize at a lower pitch. Fitch (1999) used a comparative approach to argue that these hypotheses, while perhaps functionally plausible, are not sufficient as a general explanation. For example, in an evaluation of Schmidt-Nielsen's (1972) idea that tracheal elongation in the trumpeter swan has evolved to facilitate carbon dioxide mixing, Fitch (1999, pp. 36–37) notes that "similarly sized migratory birds lack this adaptation, while some non-migratory species have extensive coiling . . ." Furthermore, not all species with tracheal elongation have unusually low fundamental frequencies.

Instead, Fitch argues that tracheal elongation enables birds to produce calls with a specific acoustic modification, in which the resonant frequencies or 'formants' are more closely spaced than would be otherwise possible. It appears that species having an elongated windpipe are indeed more likely to produce vocalizations with closely spaced formants (Fitch 1999). In this way, they can produce sounds with the tonal quality of a larger bird, thus exaggerating their advertised body size (Fitch 1999; Fitch & Hauser 2003). A playback study with whooping cranes demonstrated that birds are indeed able to perceive this variation in formant spacing (Fitch & Kelley 2000).

The Role of Learning

In three groups of birds – parrots (Psittaciformes), hummingbirds (Trochiliformes), and songbirds (oscine Passeriformes) – normal vocal development depends on the imitation of adult models to which young individuals are exposed early in life (Marler 1976; Kroodsma & Baylis 1982; see Chapter 3). Many authors have drawn a causal connection between vocal learning and the evolution of vocal diversity (Marler & Tamura 1964; see Chapter 10). The logic here is straightforward: learning, as an open-ended process, would seem to be able to generate novel vocal patterns with relative ease. Novel vocal patterns could then be retained and built upon in diverse ways in different bird groups. A creative role for song learning in the evolution of vocal diversity is supported by a number of observations. In the laboratory, hand-reared birds develop vocal innovations, by inventing novel notes or combining segments from different models (Ince et al. 1980; Marler & Peters 1988b). Long-term field studies have documented these kinds of changes in vocal structure between generations (Jenkins 1978; Grant & Grant 1996a; Payne 1996). Other lines of evidence suggest that, once introduced, changes could be maintained across generations. Captive birds learn non-natal dialects and, on some occasions, songs of other species, for example, when reared with them as social tutors (Baptista & Petrinovitch 1984), or tutored with 'hybrid' training songs containing both own-species' and other-species' song elements (Marler & Peters 1977; Soha & Marler 2000). Field studies have also documented occasional heterospecific song learning (Grant & Grant 1996a), as well as a diverse array of natural mimics (Baptista & Catchpole 1989; Chu 2001).

These open-ended properties of song learning might suggest that the functional morphology of the vocal apparatus, obviously very versatile, plays little or no role in shaping the evolution of vocal diversity. If a white-crowned sparrow can faithfully reproduce the vocalizations of another species such as a strawberry finch (example from Baptista & Petrinovitch 1984), then how could anatomical differences in the vocal mechanisms of these birds possibly explain all the differences in their normal vocal behavior (Marler 1976; Konishi 1985)? The ability of one bird to learn the songs of another would seem to prove that sound production mechanisms of the two species are functionally equivalent.

Song learning is not entirely open-ended, however. Notably, many birds show an innate propensity to learn only the songs of their own species. If a young swamp sparrow male is tutored with songs recorded from an adult song sparrow, for example, he learns little from this exposure, almost as though he had heard nothing at all (Marler & Peters 1987, 1989). Marler (1976) thus suggested that the range of models a young bird can learn is determined by perceptual filters or other neural mechanisms that innately specify a set of parameters characterizing acceptable sounds to copy (see Chapter 3). Considerable experimental evidence supports this view (Marler & Peters 1989; Marler 1990a, b, 1997). The implication, then, is that species differ in the songs they sing because of divergent neural mechanisms, yet to be fully described, which underlie selectivity in song model preferences.

It may not be necessary to invoke perceptual mechanisms or neural predispositions, however, to account for all the ways that songs differ between even closely related species. Consider again the songs of swamp and song sparrows. When young of both species are raised in captivity without the benefit of any adult song model, or after being deafened at a young age so that they cannot hear even themselves sing, they develop highly abnormal songs that bear little resemblance to normal adult songs of either species (Marler & Sherman 1983, 1985). This result is hardly surprising, given the importance of imitation and auditory feedback during song development (see Chapter 3). More surprising is that, even in the complete absence of song models or auditory feedback, the two birds sound different from each other as adults and, further, that some of the differences in their songs mirror those that distinguish normal songs of the two species. When song sparrows are isolated or deafened, for example, they produce songs with a complex syntax of several parts, whereas similarly deprived swamp sparrows typically produce continuous trills, providing strong evidence that these song features are somehow encoded innately (Marler & Sherman 1985). At the same time, some of the differences Marler and Sherman observed might be anatomically based. For example, the songs of isolate and deafened song sparrows are generally lower in frequency and have lower note repetition rates (number of notes/total song duration) than those of swamp sparrows. Both of these differences could equally well reflect the fact that song sparrows are about 20 percent larger than swamp sparrows, with correspondingly larger syrinxes, vocal tracts, and jaw muscles.

Morphology may guide vocal development and learning in even more specific ways. Pure-tonal, 'whistle-like' sounds are a ubiquitous characteristic of many birdsongs (Nowicki & Marler 1988), including the songs of swamp and song sparrows. When Marler and Sherman (1985) deprived young birds of all song models, these isolate birds produced not only pure-tone notes, but also many unusual notes with harmonic overtones, suggesting that tonal quality might be learned. Peters and Nowicki (1996) tested this idea by training one group of song sparrows with a selection of tutor songs that were normal in every respect except that they included added harmonic overtones; these were produced by recording birds in a helium-enriched atmosphere (Nowicki 1987). A second group was trained with those same songs produced in a pure-tonal fashion, recorded in normal atmosphere. As expected, birds trained with pure-tonal songs copied those songs and reproduced them in a pure-tonal fashion. The birds trained exclusively with harmonic songs also imitated the songs they heard but, interestingly, they subsequently reproduced them in a pure-tonal fashion, without harmonics. If tonal quality is learned, we would have expected just the opposite result – harmonic-tutored birds should have produced their copies with harmonics, or at least they should have produced more harmonic sounds than usual.

To explain the elimination of harmonics, Peters and Nowicki (1996) suggested that the propensity to produce pure-tonal sounds might be an intrinsic physical property of the sound-generating apparatus, including the syrinx and vocal tract, but that this propensity is only realized when the vocal tract attempts to reproduce conspecific sounds. When an isolate bird produces invented material, the vocal tract may or may not be in a posture typical for the production of conspecific sounds and so the tonal quality of sounds produced becomes more variable. By contrast, configurations of the vocal tract or syrinx associated with production of species-typical sounds may be most stable if the sound is produced in a pure-tonal fashion. Thus, if a bird attempts to copy conspecific song material, its song will tend to be pure-tonal, regardless of the tonal quality of the models being copied. Vocal tract constraints of this sort also have been proposed to account for the limited set of phonemes found across all human languages (Stevens 1972; Ohala 1983).

VOCAL MECHANICS AND EVOLUTION: CONSTRAINT AND OPPORTUNITY

Evidence that the diversity of bird vocalizations is shaped by the morphology and mechanics of the sound-producing apparatus may help to explain why the songs or calls of birds have not evolved in certain acoustic dimensions. An understanding of vocal mechanics might also help to explain additional patterns of vocal evolution within and between specific bird groups. Indeed, the vocal virtuosity of birds tends to make us focus not only on how songs and calls are constrained, but also on why they are so diverse, how this diversity has arisen, and how it influences the evolutionary diversification of birds. To explore such questions we focus on three examples: the evolution of species diversity in passerine birds, the origin of vocal novelties in swamp sparrows, and the interaction of vocal mechanics, song diversity, and speciation in Darwin's finches.

Syringeal Complexity and the Evolution of Passerine Diversity

Raikow (1986) initiated a debate concerning the possible relationship between evolutionary mechanisms underlying vocal diversity and taxonomic diversity in birds. The Passeriformes comprise just one of about two dozen orders of birds, yet this group includes over half of all avian species. This disproportionate number led Raikow to suggest that some anatomical or behavioral features unique to passerines may have served as 'key adaptations' that facilitated the extensive radiation of this group. A survey of anatomical features unique to passerines, such as spermatozoon morphology, hallux size (the big toe), and thigh muscle position failed to reveal a likely candidate for such a key adaptation (Raikow 1986; Fitzpatrick 1988; Vermeij 1988). Candidate behavioral characters specific to passerines include generalized capacities for learning (Raikow 1986), high metabolic rates, short generation times (Fitzpatrick 1988), the ability to move into new environments (Baptista & Trail 1992), plasticity in nesting adaptations (Olson 2001) and, of greatest interest to us here, an unusually high degree of vocal diversity. In contrast to anatomical features, it is not difficult to envision how behavioral features may have enhanced passerine species diversification (Fitzpatrick 1988; Vermeij 1988). A direct role for vocal diversity is particularly plausible; because vocal communication is used in mate and species recognition, diversity in vocal signal structure can lead directly to reproductive isolation and speciation (Baker & Cunningham 1985; Martens 1996; Slabbekoorn & Smith 2002b).

In an informal test of this possibility, Raikow (1986) made two distinctions among bird groups. First, he noted that morphologically 'specialized' syrinxes have evolved within three particularly species-rich passeriform groups, namely the oscine songbirds, the Furnarii (ovenbirds), and New World Tyrannid flycatchers. By contrast, groups with morphologically primitive syrinxes appear to include fewer species. Implicit here is the idea that morphologically more complex syrinxes are capable of producing a correspondingly wider diversity of sounds. The idea that signal complexity and species diversity might be correlated recalls Ryan's (1986) demonstration that anuran diversity correlates with the morphological complexity of the frogs' auditory perceptual systems. In general, greater diversity in a signaling system should provide greater opportunities for pre-mating reproductive isolation. In Raikow's second distinction, whether birds learn their songs or not, he suggested that oscine Passeriformes show greater vocal diversity, including more geographic variation within species, than suboscine groups that do not generally seem to learn their songs, though bellbirds appear to be an exception (see Chapter 4). Greater diversity would presumably occur because of enhanced opportunities for rapid vocal diversification afforded by learning, which should in turn enhance the occurrence of reproductive isolation among diverging lineages.

Baptista & Trail (1992) responded to Raikow by arguing that available data actually do not support a role for vocalizations as a causal agent in passerine diversification. Their primary line of reasoning was that correlations between syringeal complexity and vocal complexity are not all that robust. Many groups of closely related species, including estrilid and emberizine finches, hummingbirds, and doves, have highly conserved syringeal morphology yet show high within-group variation in 'vocal virtuosity' (Baptista & Trail 1992, p. 243). The duration of the courtship coo in doves may vary from approximately a quarter of a second in some species, such as domestic pigeons, to over 20 seconds in other dove species, in spite of the apparent similarity of the vocal machinery of these species in structure and function (Baptista & Abs 1983).

Baptista & Trail (1992) also pointed out that some groups which do not learn their songs, such as the neotropical flycatchers, are as numerically diverse as some groups that do learn their songs, casting doubt on Raikow's second distinction. Baptista and Trail also made a third point, that in mate choice in some species, vocal cues may be superseded by visual signals. Female zebra finches, for instance, can be convinced to mate with other subspecies when males are disguised with the appropriate plumage (Clayton 1990b).

Yet to be considered in this debate is the role of vocal performance limits. Baptista and Trail's (1992) refutation of Raikow's hypothesis was based in part on the observation that the similarly structured syrinxes of different species can produce a wide range of sounds. At first glance, this observation appears to undermine any attempt to assign a causal link between vocal morphology and sound production. If vocal morphology drives vocal evolution, how can we explain vocal mechanical systems as flexible as those of doves, for example? This is a variant of the argument, outlined earlier, that the ability of some birds to learn other species' songs implies a limited role for mechanical constraints in vocal evolution.

Wide variations in the vocal output of a given vocal mechanism are not unexpected, however, because the maximal performance abilities in any given group will not always be attained. Some species or individuals will only rarely vocalize at or near their full mechanical potential. To take the dove example, maximal durations of calls are likely determined by maximal tidal volumes specific to the respiratory systems of these birds, and this limit may only be approached in a subset of species. Similarly, the family-wide tradeoff in emberizid finches between trill rate and frequency bandwidth (Podos 1997) only becomes apparent through analysis of a large sample of trills; this is so because most songs fall below performance boundaries. Recognition that there are vocal performance limits, even if they are only rarely expressed, solidifies our confidence in the link between vocal morphology and sound production. When Slabbekoorn and Smith (2002b) distinguish between realized and potential acoustic variation, they articulate a similar point, noting that evolutionary divergence among populations in vocal potential can facilitate evolutionary changes in song characteristics. With respect to the syrinx of passerines, and particularly of oscines, we suggest that its anatomical complexity has enabled a particularly wide range of potential vocal outcomes, as compared to what can be accomplished with less complex syrinxes, such as those of parrots and hummingbirds (Baptista & Trail 1992). As such, we believe that Raikow's hypothesis merits further consideration. Characterization of performance limits among different bird groups will be a valuable goal for future studies.

Swamp Sparrow Song: Motor Constraints and the Evolution of Vocal Novelty

Performance limits imposed on the sound-producing apparatus help to explain why bird vocalizations have not evolved to fill all of the acoustic space they could conceivably occupy. We have already used as an illustration the limited, triangular distribution of emberizine sparrow songs in a graph of trill rates and bandwidths (**Fig. 11.2**). Understanding the physical basis of performance limits may also help to explain how diversity arises in the evolution of song structure. Scrutiny of vocal production mechanisms in the swamp sparrow provides insight into how song might evolve in new directions, especially with regard to large-scale features of song organization.

The group that includes white-crowned and song sparrows (**Fig. 11.3**) illustrates how closely related species often differ radically both in the structure of the elements that compose the song ('phonology'), and in the pattern in which these elements are presented ('syntax'). How has this diversity arisen? One commonly accepted answer to this question is that song variation arises in the form of copy error, as song is transmitted across generations through learning; species differences in song are thought to evolve gradually as the result of drift or selection acting on this variation (Nottebohm 1972; Slater 1989; Kroodsma 1996; Lynch 1996). Indeed, long-term field studies have documented evolutionary changes in such song features as the structure of acoustic elements (Jenkins 1978; Grant & Grant 1996a) and the addition, deletion, or recombination of elements (Ince et al. 1980; Slater 1989; Payne 1996). Ring species can be especially illuminating in this regard (**Box 25**, p. 204).

It is not clear, however, whether, when large inter-species structural differences are observed in a group of birds, these are entirely attributable to the processes of drift or selection acting on copy error. First, the changes in song structure observed across generations, both in the field and in the laboratory, tend to occur as relatively minor phonological differences, and not as the larger-scale syntactical differences that can be so apparent across species. Given enough time, small variations could conceivably accumulate to yield large differences between species. But there are often enormous differences between the songs of even closely related species as, for example, with swamp and song sparrows (**Fig. 11.3**), and it is not clear how small phonological changes alone could transform one into the other. This problem is analogous to the difficulty of explaining large-scale macroevolutionary patterns on the basis of small-scale microevolutionary processes.

Furthermore, the effects of both cultural selection on song learning, and biological selection on song function are strongly stabilizing. As discussed earlier, young songbirds are predisposed to learn songs that fall within a circumscribed range of variation, so songs that are farther removed in their acoustic characteristics from a species-typical norm are less likely to be passed on to future generations. Similarly, functional responses to song diminish in both male–male and male–female interactions as a song's acoustic structure diverges from the species norm (King et al. 1980; Baker et al. 1987; Nelson 1988; Searcy 1990; Searcy & Yasukawa 1996). Selection for optimal transmission in acoustic habitats can also render some song variants more effective signals than others (Morton 1975; Wiley 1991; see Chapter 6). Thus, we are left with the puzzling question of how major variations in song structure could evolve even within a closely related group of birds, in the face of stabilizing selection and in the apparent absence of a mechanism for generating large-scale mutations.

Our work on performance limits and song production in one species, the swamp sparrow, suggests a possible answer to this question. Swamp sparrow song typically consists of a single string of syllables, each composed of a cluster of 2–5 individual notes (**Fig. 11.4; CD1 #88**). Although this species is widely distributed across the North American continent (Mowbrey 1997), all swamp sparrow songs appear to share a relatively narrow range of acoustic characteristics. On the

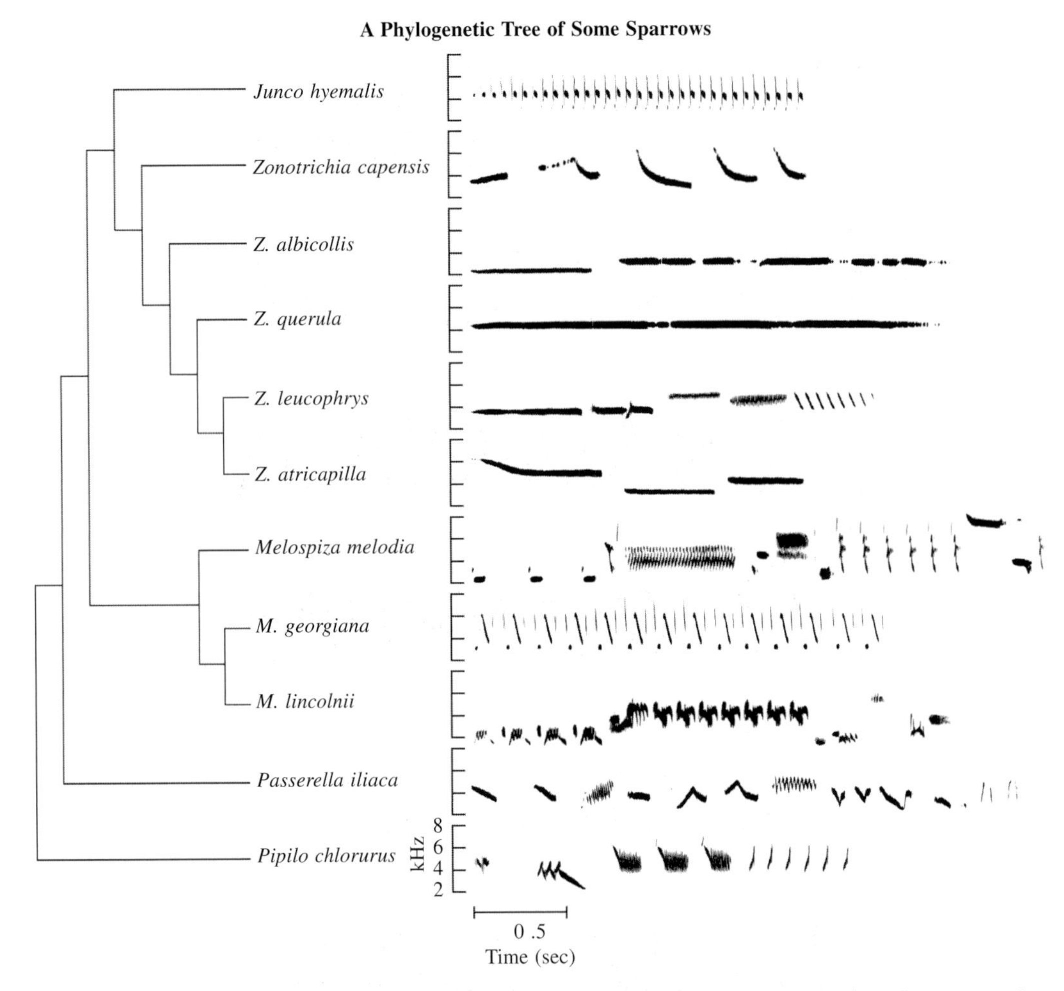

Figure 11.3 Song diversity in three genera of sparrows and two other close relatives. From top to bottom they are: yellow-eyed junco, rufous-collared sparrow, white-throated sparrow, Harris' sparrow, white-crowned sparrow, golden-crowned sparrow, song sparrow, swamp sparrow, Lincoln's sparrow, fox sparrow, and green-tailed towhee. As the examples show, songs vary along many levels of organization including the fine structure of notes, or phonology, the arrangement of notes into clusters, and the higher level organization of song, the syntax. The phylogenetic hypothesis of Zink & Blackwell (1996) on the left provides a framework for assessing evolutionary transformations in song structure. Phylogenetic continuities in some features can be detected, such as the structure of whistles in songs of the white-crowned sparrow and its cogeners, and the frequency-timing relationships within trills (Podos 1997). Still, song appears to be strongly labile in evolution, as illustrated by large-scale differences in song features among closely related species, as with song syntax in swamp and song sparrows. Performance limits may provide means by which novel forms of vocal syntax arise. From Podos et al. 1999.

phonological level, all notes produced by individuals across the entire species range can be classified into six categories of perceptually distinct note types (Marler & Pickert 1984; Clark et al. 1987; Nelson & Marler 1989). On the syntactical level, all swamp sparrow songs are produced as a continuous trill of syllables repeated at a relatively constant rate. In the laboratory,

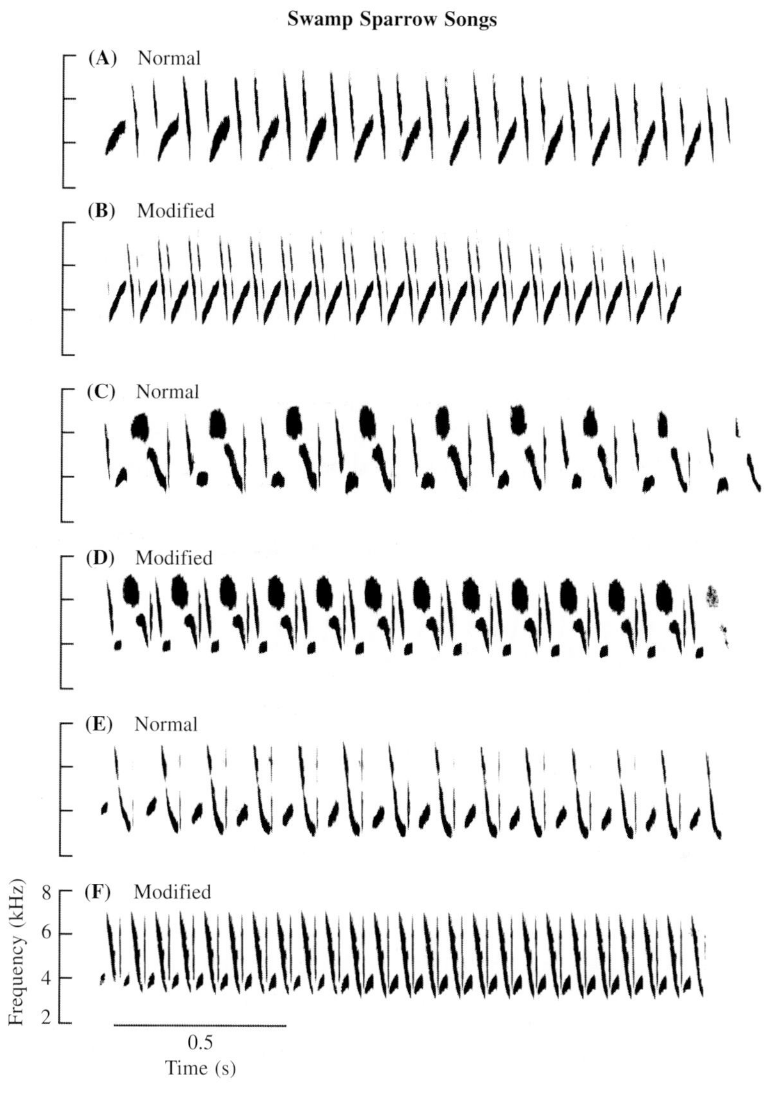

Figure 11.4 Swamp sparrow songs, from a population in Dutchess County, New York. They normally consist of single trills, shown in **A**, **C**, and **E**. **B**, **D**, and **F** are modified versions of songs used to train young birds, with the intervals between notes manipulated in various ways (**CD1 #88**). From Podos 1996.

swamp sparrows show a strong preference for imitating conspecific songs and are especially resistant to learning phonology that deviates from the set of six species-typical note types, even when given no other choice of models (Marler & Peters 1977, 1989). When young birds learn from swamp sparrow phonology presented in a more complex syntactical structure, they typically reproduce this phonology as a simple, continuous trill (Marler & Peters 1989). Even birds raised in total isolation with no exposure at all to song models produce songs as continuous trills (Marler & Sherman 1985). Finally, as expected from our argument about vocal tract limitations on song production, there is a tradeoff between trill rate and bandwidth in swamp sparrow songs, with faster trill rates being constrained to relatively narrow bandwidths (Podos 1997; Ballentine et al. 2004).

Dealing with Tutor Songs Near Performance Limits

To test whether performance limits do indeed set an upper limit on the evolution of trill rates in this species, hand-raised young male swamp sparrows in captivity were tutored with songs that had been digitally manipulated to produce faster trill rates than those normally produced by wild birds (Podos 1996; **Fig. 11.4; CD1 #88**). Their singing was compared to a control group trained with a comparable set of songs having normal trill rates. Birds were able to memorize these fast-trill models, but they were unable to reproduce them in fully accurate detail. Furthermore, the kinds of inaccuracies the young birds introduced were consistent with the hypothesis that the vocal apparatus placed constraints on vocal production. Some models were reproduced with diminished rates, others with notes omitted, and yet others with a 'broken' syntax, in which birds produced songs as short bursts of syllables, reproduced accurately and at the fast rate of the model, but separated by silent gaps, suggesting that the motor system could not sustain the high syllable rate for the entire duration of the song (**Fig. 11.5; CD1 #89**). With these 'errors' birds produced simpler and, in effect,

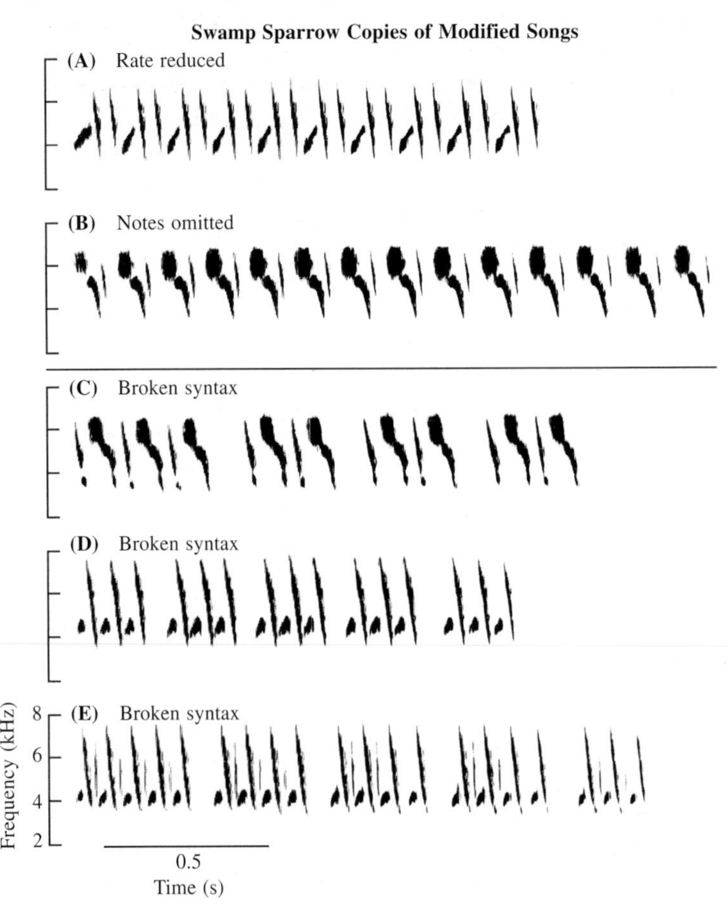

Figure 11.5 Copies by tutored swamp sparrows of manipulated training songs (see **Fig. 11.4B, D, & F**). Song **A** was copied, but at a reduced trill rate, and song **B** was reproduced, but with two notes per syllable omitted. Songs **C**, **D,** and **E** were reproduced with a 'broken' syntax (**CD1 #89**). From Podos 1996.

motorically less challenging versions of the tutor models they attempted to copy. Interestingly, the type of inaccuracies introduced into copied songs depended upon the degree to which models had been manipulated; only the most severely modified models were reproduced with broken syntax, the most extreme form of inaccuracy (Podos 1996).

The broken syntax songs of these birds is of particular interest because they represent a pronounced deviation from the restricted syntactical organization normally observed in swamp sparrow songs. The mechanism responsible for the broken syntax innovation can be compared to that which underlies gait changes in a running animal (Podos 1996): a continuous increase in an output parameter, such as running speed or trill rate, leads to a discontinuity in the function of the underlying motor system. A gait change in locomotion occurs when that parameter passes some threshold or constraint as with the production of song syntax. Such production shifts allow animals to maintain motor competency and efficiency across different output rates (Alexander 1984). In the case of birdsong, discontinuities arising from performance limits could serve as the source of large-scale innovations, thus providing a potential mechanism for the generation of macroevolutionary change in song structure.

Conditions for Evolutionary Song Innovation

If the occurrence of broken syntax in swamp sparrow song is to serve as an example of how large-scale innovations in signal structure might evolve, three conditions must be met. First, the innovation must be transmittable to subsequent generations. For birdsong, this means that young birds must accept songs with the innovation as suitable models for learning, and must be able to reproduce those models with the innovation intact. Second, songs with the innovation must function as communication signals at least as well as normal songs that lack the innovative trait. Because song is a dual function signal (Catchpole & Slater 1995), it is important to measure the effect of an innovation such as broken syntax on both male–male and male–female communication. Third, there must be selection on males to produce songs close to the performance limit to generate the novel syntax in the first place. Having observed broken syntax in the laboratory when birds are pushed to their performance limit by tutoring with artificially accelerated songs, would we also expect males in natural field conditions to learn and produce songs at or near their performance limits?

Captive studies demonstrated that the first of these conditions could be met (Podos et al. 1999). A group of hand-reared swamp sparrows was exposed to tutor songs with both normal and broken syntax. The birds copied broken syntax models in the same proportion as they heard these songs during training, demonstrating that young swamp sparrows do not have an initial bias against broken syntax songs as models to be learned. More importantly, these birds crystallized some of their songs with a broken syntax closely matching the syntax of the models they copied. Thus, broken syntax song structure may be both memorized and reproduced faithfully. This result was surprising, given that young swamp sparrows are normally highly conservative as song learners (Marler & Peters 1977, 1989). In particular, prior work had demonstrated that swamp sparrows rarely if ever deviate from the production of continuous trills, irrespective of early experience, leading Marler (1984) to suggest that an innate central motor program must be responsible for this species-typical song feature. An alternative speculation (Podos et al. 1999) is that the readiness of young swamp sparrows to learn and reproduce broken syntax may be explained by the coupling of sensory and motor mechanisms in song learning and production (Nottebohm et al. 1990; Marler & Doupe 2000; see Chapter 8). Given this coupling, perceptual mechanisms might be inherently sensitive to broken syntax as a song feature, even if it rarely or never occurs in normal songs, because it arises as a natural 'solution' when the motor system is pushed beyond normal performance capabilities.

Evidence is mixed as to whether broken syntax

songs function as well as species-typical songs lacking the innovation. Territorial playbacks in the field (Nowicki et al. 2001) showed that male swamp sparrows respond similarly to broken syntax songs and a set of identical, but continuously trilled songs. Playback experiments provide only an indirect measure of function, however, because song functions in male–male interactions not so much to trigger aggressive responses in territory owners, as is typically measured in playbacks, but to repel intruders (Searcy & Nowicki 2000). Nonetheless, the fact that swamp sparrow males responded to broken syntax as strongly as they did to normal syntax suggests that, should it arise in a population, this innovation would function equally well in male–male communication.

Female swamp sparrow responses to broken syntax were also tested using the copulation solicitation display assay (Searcy 1992a; Nowicki et al. 2001). This assay is a better indicator of function than field playback, because the measured response – the performance of a precopulatory posture and associated displays by the female – is directly related to a male's ability to stimulate females to court and copulate. In these tests, females responded strongly to broken syntax songs, clearly recognizing them as signals to which they would respond sexually. However, females also showed a significant preference for normal songs when directly compared to broken syntax songs. Taken at face value, this result suggests that broken syntax in swamp sparrow song is not a suitable paradigm for the evolution of song innovations, because the condition that song with the novel trait functions as well or better than natural songs, is not met. Two important caveats must be considered, however. First, a well-established tenet of population genetics is that less advantageous traits can become established in populations in which selection is weak and effective population size is small (Wright 1931). In a small breeding population, in which females have fewer opportunities to express their preferences, males singing broken syntax songs may well be able to attract mates, at levels sufficient to maintain the trait. Second, female preferences themselves are likely to be influenced by learning. The females that were tested presumably had never been exposed to broken syntax songs during their lives. However, females raised in a nest within earshot of a male singing broken syntax songs might not discriminate against the innovation, and perhaps might even prefer it. In this case the innovation would be more likely to be maintained and spread in the population. At present we can only speculate on this possibility.

The final condition that needs to be met if our swamp sparrow example is to serve as a model for the evolution of syntactical innovation in song is that males would need to experience selective pressures to produce songs at or near performance limits, to initiate the production of broken syntax in the first place. This possibility is only now being explored. In general, females are expected to prefer mating signals that provide them with accurate information about male quality (Andersson 1994; see Chapter 2). Females might pay particular attention to signal features that are costly or challenging to produce, because only the highest quality males will be able to produce those features. In many species of songbird, females base their preferences for males on song characteristics, most notably the amount that a male sings or the size of his song repertoire (Searcy & Yasukawa 1996). The cost of performing more singing is clear, both in terms of energy and the time taken away from other activities, so the amount a male sings may indeed be a reliable indicator of his condition (Greig-Smith 1982a). The cost of producing a larger repertoire is less obvious, although evidence is accumulating that there may be relevant costs associated with developing a brain adequate to this task (Nowicki et al. 1998, 2000, 2002b).

Another mechanism by which females could evaluate potential mates is by assessing their ability to produce motorically challenging songs (Vallet & Kreutzer 1995; see Chapter 9; **Box 4**, p. 58). For swamp sparrows and other species that produce trilled songs, where there is a predicted performance boundary in the acoustic space defined by trill rate and frequency bandwidth, females may prefer songs that lie closer to this boundary over songs that are more distant from it, if 'higher performance' songs are indeed indicative of higher quality males. In this way, sexual selection by female choice could push males up against their performance limits. Recent work has demonstrated that female swamp sparrows do indeed prefer songs that more closely approach the tradeoff limit for trill rate – bandwidth performance in their population, as compared with the same song types that approach the performance limit less closely (Ballentine et al. 2004).

SONGS OF DARWIN'S FINCHES

Ecology, Mechanics, and Vocal Diversity

If physical limitations on vocal production mechanisms do indeed influence the evolution of birdsong, what happens when some components of the vocal apparatus are under strong selection for some behavior other than song (Nowicki et al. 1992)? Bird beaks provide an especially interesting case, given the essential role they play in song production, and their frequent specialization for feeding, as illustrated by Darwin's finches of the Galápagos Islands. These birds represent a premier example of adaptive radiation, in which a single ancestor has diversified into multiple descendent species inhabiting a variety of ecological niches (Schluter 2000). Adaptive radiation in Darwin's finches has centered around the diversification of feeding niches, and includes birds that eat seeds, insects, flowers, fruits, pollen, nectar, leaves, cactus pods, and even blood (Lack 1947; Grant 1999). This diversification is matched by an equally impressive array of specializations in beak form and function. The ground finches, for instance, have evolved heavy beaks analogous to a linesman's pliers, suitable for crushing hard seeds, whereas the warbler finches have light, slender beaks analogous to needlenose pliers, more efficient for probing and manipulating insect larvae (Bowman 1963; **Fig. 11.6**, left).

The evolutionary link between food type and beak evolution in Darwin's finches is firmly established. Peter and Rosemary Grant and colleagues (Gibbs & Grant 1987; Grant 1999; Grant & Grant 2002b) have demonstrated that even short-term fluctuations in food availability, driven largely by seasonal variations in weather patterns, lead to evolutionary changes in beak morphology via natural selection. During a severe drought, for instance, beak size in the medium ground finch increased over a single generation, presumably because of the advantage held by larger-beaked birds in husking large and hard seeds (Boag & Grant 1981). Beak evolution may also be driven by competition among species for common resources. Evidence for competition comes from patterns of morphological variation that suggest a history of character displacement between sympatric species (Lack 1947; Schluter et al. 1985). Directional selection on descendent species, together with broad-scale morphological responses to shifts in the ecological environment, appear sufficient to explain species differences in beak form and function in this group of birds (Grant 1999; Grant & Grant 2002b).

The diversity of beaks among Darwin's finches reflects not only the diversity in feeding adaptations but also, in light of what we now know about beaks and song production, differentiation in vocal production capacities. To the extent that beak characteristics shape the acoustic properties of song, then species diversity in beak morphology should be matched by corresponding diversity in vocal behavior. Early discussions of song diversity in Darwin's finches gave no support to this view, because it was believed that the songs of different species were mostly indistinguishable (Orr 1945; Lack 1947; Grant 1999). With the advent of sonographic analysis, however, differences among species in song structure became apparent (**Fig. 11.6**, right; **CD1 #90**), in spite of the statistical overlap among species for song features (Ratcliffe 1981; Bowman 1983). The birds themselves seem highly skilled, although not unerringly so, in using song for species recognition (Ratcliffe & Grant 1985; Grant & Grant 1996b).

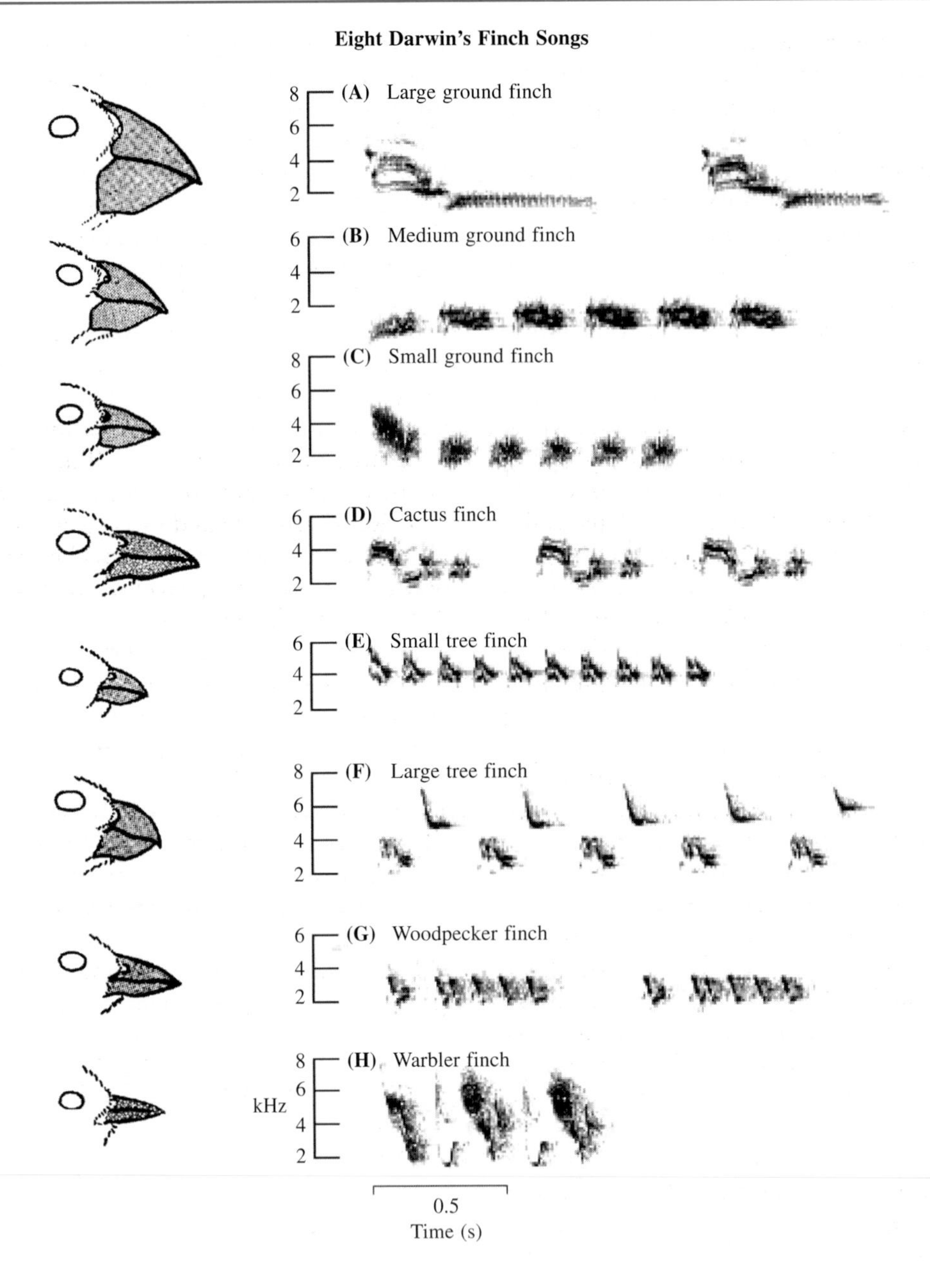

Figure 11.6 Beak morphology and representative sonograms of songs from eight Darwin's finch species on Santa Cruz Island (**CD1 #90**). Species shown are, from top to bottom: (**A**) large ground finch, (**B**) medium ground finch, (**C**) small ground finch, (**D**) cactus finch, (**E**) small tree finch, (**F**) large tree finch, (**G**) woodpecker finch, and (**H**) warbler finch. Species differences are apparent in both morphology and song structure. The beak sketches are reprinted from Bowman (1961). Vocal performance (see **Fig. 11.2B**) correlated with beak morphology across these species, as predicted by the vocal tract constraint hypothesis. From Podos 2001.

How might the kind of variation in beak morphology expressed by Darwin's finches translate into vocal variation? As a starting point, it seems likely that species with an ability to apply large forces with their beaks, for example, those with beaks adapted to crush hard seeds, would face particularly strong constraints on the speed at which they could change the configuration of their vocal tract during sound production (Podos 2001). There is an intrinsic tradeoff in vertebrate motor systems between force and speed (Vogel 1988), which suggests that a beak and associated musculature adapted for strength would necessarily be compromised in their speed of movement. Furthermore, the intrinsic speed of muscle declines with increasing muscle size, because greater numbers of contractile units require more time to activate (Vogel 1988). Species thus 'encumbered' by large beaks with comparatively massive jaw musculature are expected to face comparatively severe restrictions on the types of songs they could sing. By comparison, birds with smaller beaks should suffer fewer motor constraints on beak dynamics and thus should be free to produce songs at higher performance levels.

To test this prediction, the songs and morphology of individually marked birds from eight Darwin's finch species were examined on Santa Cruz, one of the central Galápagos Islands (Podos 2001). Measurements taken included three dimensions of beak size and three dimensions of body size. Banded birds were then recorded and their songs analyzed, specifically to measure the two variables we have linked previously to vocal performance capacities – trill rate, and frequency bandwidth (Podos 1997; **Fig. 11.2A**). The vocal performance of a song was defined by its deviation from a performance boundary determined as the upper bound regression of the relationship between trill rate and bandwidth (**Fig. 11.2B**).

Consistent with the vocal tract constraint hypothesis, vocal performance of Darwin's finch songs correlated with measures of beak morphology, with lower performance songs being associated with larger beaks (Podos 2001). This correlation was significant not only across different species, for which variation in beak morphology is expected to be pronounced, but also among individuals in a single population of the medium ground finch (Podos 2001; **Fig. 11.7**). This finding does not of course mean that variation in vocal performance caused by differences in beak morphology and musculature is the sole driver of vocal diversification in these birds. Other causes of vocal evolution not directly related to performance, such as copying inaccuracies during song learning, might better explain variation in other vocal features such as note structure or song syntax (Ratcliffe 1981; Grant & Grant 2002b), and may also explain variation in trill rate and frequency bandwidth in some instances. For example, the two forms of the warbler finch, recently recognized as distinct species based on genetic analyses (Petren et al. 1999), differ in vocal performance and in one beak measure (beak length); here, vocal performance diverges in a direction opposite to that predicted by the vocal constraint hypothesis (Grant & Grant 2002a). Nevertheless, some proportion of the variation in the songs of Darwin's finches, particularly across species and also within some species, appears to be accounted for by differences in beak morphology (Podos 2001).

Relationships between beak morphology and song in other groups of birds appear to be more

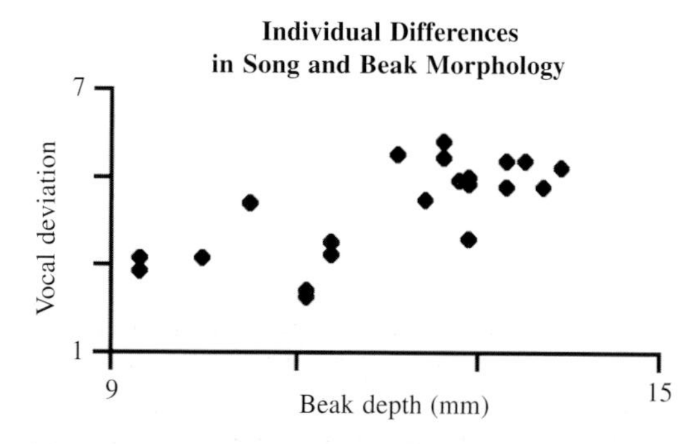

Figure 11.7 Individual differences in vocal performance were found to correlate with beak morphology within the medium ground finch, in accordance with the beak constraint hypothesis. From Podos 2001.

varied. In reed buntings, several song parameters, including the number and diversity of syllable types, correlate positively with a beak measure (beak depth) that is highly variable across this species' geographic range (Matessi et al. 2000). Further, the degree to which reed bunting populations are distinct in song structure corresponds to levels of morphological divergence, but not to geographical divergence. These patterns imply a functional role for the beak in song divergence (Matessi et al. 2000), although a mechanical link between syllable diversity and beak morphology has not been established. Among the Neotropical woodcreepers (Dendrocolaptidae), beak length correlates negatively with peak acoustic frequencies (Palacios & Tubaro 2000). This finding is consistent with expectations drawn from our understanding of vocal mechanics; birds with larger beaks should be able to filter harmonic overtones from source sounds with comparatively low fundamental frequencies.

A contrasting example comes from the work of Slabbekoorn & Smith (2000), who analyzed the songs of large- and small-billed forms, or 'morphs' of the black-bellied seed-cracker in Africa. Large-billed morphs are specialized to eat comparatively hard seeds, and thus are expected to face comparatively severe constraints on vocal performance. However, no differences were detected between morphs of this finch in a wide range of acoustic frequency and timing parameters (Slabbekoorn & Smith 2000). One possible explanation for this lack of difference is that the song of this species, described by Chapin (1954) as a short, pleasant warble, might not be particularly challenging to produce; thus differences in potential vocal proficiency among morphs may exist but are not expressed. For example, only a subset of songs contained trills, in which birds need to quickly repeat the same set of sounds. Another possible explanation for a lack of difference among morphs is that the pressure of sexual selection is weak and thus that birds do not necessarily approach potential performance limits in their realized vocal output (Slabbekoorn & Smith 2002b).

Recognition of the dual role of the beak in both feeding and singing provides new insights into mechanisms of speciation. Darwin's finches are a classic system for studying how ancestral species split into multiple descendent species, and consensus has been reached on a number of points (Swarth 1934; Lack 1947; Bowman 1961; Grant 1999). First, speciation in Darwin's finches appears to have been primarily if not exclusively allopatric (Grant 1999), which is not surprising given the many opportunities for geographic isolation on the Galápagos archipelago. Second, pre-mating isolation mechanisms such as song are thought to play a particularly strong role in speciation, as is illustrated by the occasional production of successful hybrids among various Darwin's finch species (Grant & Grant 1997b, 1998). This argues against the existence of genetic or physiological post-mating isolation mechanisms. Third, trait differences among populations and species are thought to be driven by rapid and precise adaptation to divergent local ecologies (Grant 1999). Adaptation is manifest largely as evolutionary changes in beak morphology, and is driven by natural selection for feeding ecology, (Gibbs & Grant 1987) and by selection against interspecific competition (Schluter et al. 1985). On the basis of these consensus views, Grant (1999) and collaborators have argued that ecological adaptation in Darwin's finches is the primary driving force in their diversification, that evolutionary changes in mating signals occur as incidental correlates of ecological adaptation, and that the diversification of mating signals, including song, plays an important role in speciation, by providing cues about species identity (Grant & Grant 2002b).

Beak Morphology and Reproductive Isolation

The finding that beak morphology and acoustic features of song are correlated suggests that, as populations adapt to diverging feeding ecologies, birds will experience direct and rapid changes in song structure (Podos 2001). This mechanism of vocal evolution is distinct from others previously proposed, such as cultural evolution through copying inaccuracies (Grant & Grant 2002b), in that it is directly, if only incidentally, linked to the primary locus of adaptation in these birds – beak morphology. Changes in song structure driven by selection for trophic specialization might in turn enhance reproductive isolation among incipient species, and may thus lead to speciation (Podos 2001; Slabbekoorn & Smith 2000). This is because the effectiveness of reproductive isolating mechanisms such as song generally depends on the extent to which signals are distinct among incipient species, with more distinct signals increasing the probability of 'correct' matings.

Probabilities of pre-mating isolation are also shaped by the extent to which birds rely on specific signal features in species recognition. If birds are not attentive to changes in song, such as those that might occur in Darwin's finches in response to beak adaptations, then reproductive barriers among interacting populations should not persist and incipient species will not diverge. This might happen if, for example, birds rely on acoustic cues which are not influenced by performance limits and beak morphology. If, on the other hand, Darwin's finches discriminate among mating signals based on acoustic features that are linked to beak morphology as, for example, if they no longer respond to songs with slight alterations in trill rate, then reproductive isolation should be accelerated by morphological adaptation and corresponding vocal evolution. Through extensive morphological adaptation and subsequently pronounced changes in song structure, the high diversity of ecological niches in the Galápagos Islands may have promoted abundant opportunities for reproductive isolation and speciation in these birds (Podos 2001).

Several key questions remain. Young male Darwin's finches learn to sing through imitation of adult males, normally their fathers (Grant & Grant 1996a). On rare occasions, however, males come to be reared by 'fathers' of another species, and learn the wrong song as a consequence (Grant & Grant 1997b, 1998). This phenomenon raises a potential critique of the vocal constraints model of finch song evolution. If birds have the potential to learn the songs of other species, then how can vocal mechanics limit vocal production to species-specific ranges? One possible answer to this question may come from the analysis of patterns of morphological variation and interspecies vocal imitation. Copying across species often occurs between pairs of species that share similar body and beak sizes (Bowman 1983; Grant 1999). For example, copying is known to occur between large species pairs, such as the large ground finch and the large cactus ground finch; medium-sized species pairs, such as the medium ground finch and the cactus finch; and small species pairs, such as the small tree finch and the warbler finch. Contingency analysis could be applied to determine whether copying tendencies are significantly biased towards similarly sized birds with similar beaks. If so, such a bias might be explained by the comparatively similar vocal capacities based on similar morphology. We further predict that songs learned from species that differ notably in size will reflect this mismatch, for example, by being copied with substandard levels of accuracy and precision.

Little is known about the role of vocal signals in mate selection and reproductive isolation in Darwin's finches. Songs vary among species (Ratcliffe 1981; Bowman 1983), and appear to be the primary cue for long-distance assessment of species identity (Ratcliffe & Grant 1985; Grant & Grant 1998). However, finches also rely on visual cues, notably beak size and shape, to assess species identity, particularly in close interactions (Ratcliffe & Grant 1983a, b). As pointed out by Baptista & Trail (1992), the role of song in reproductive isolation becomes harder to pin down in bird groups that also rely on morphological cues. Nevertheless, responses of males to playbacks indicate that Darwin's finches can use song in species and population recognition, in the absence of distinguishing visual cues (Ratcliffe & Grant 1985; Grant & Grant 2002a). The specific vocal traits that enable species recognition have not yet been identified.

It is likely that birds will focus on acoustic features that differ the most among species, or are invariant within species, because such traits often provide the most reliable indicators of species identity (Nelson 1989a; Nelson & Marler 1990).

Ideally, a complete understanding of reproductive isolation also requires data about female preferences, given the fact that females are choosier than males in their selection of mates (Ratcliffe & Otter 1996; Searcy & Yasukawa 1996). Available data in Darwin's finches, however, are from playbacks to male birds (Ratcliffe & Grant 1985; Grant & Grant 2002a). Most relevant to our discussion here is the need to learn whether variation in song features that correlate with vocal performance, such as trill rate, influences female preferences in mate selection. Demonstrating such an influence would provide further support for a link between the vocal mechanics of singing, morphological adaptation, and the emergence of new species.

CONCLUSIONS

Our understanding of vocal mechanics in birds has advanced rapidly in recent decades, largely through the success of physiological studies of live birds. Information about vocal mechanics helps to specify the kinds of sounds birds can and cannot produce. In particular, evidence that birds must coordinate multiple motor systems to vocalize suggests a broad role for performance limits in shaping the evolution of bird vocalizations. Such performance limits can be observed in the form of tradeoffs between different acoustic features of vocalizations, such as that between the rate and the frequency bandwidth of repeated elements in trilled songs. Other performance-related factors that might shape patterns of vocal evolution include variation in the size and shape of the vocal apparatus, and limits on the plasticity of vocal learning.

We have argued that an exploration of vocal mechanics, and the performance limits they impose, can provide novel insights into the evolution of birds and their songs. In the debate about the causes of adaptive radiation in perching birds, there has been repeated reference to the wide variation in acoustic patterns often found within closely related bird groups. Such variations are not surprising, however, given what we know about vocal performance. For this reason, we recommend further examination of the nature and complexity of vocal morphology as a possible causal agent in species diversification. In sparrows, we have shown that performance limits on trill production can lead to a novel form of vocal syntax which, in turn, suggests a mechanism for the evolution of vocal novelty. Studies of song learning and function provide some support for this hypothesis. In Darwin's finches, the evolution of diverse beak shapes and sizes appears to have influenced vocal evolution, such that birds with larger beaks, of the type used for crushing hard seeds, face the greatest vocal performance constraints. The existence of beak-related performance constraints in these birds implies in turn that there are associations between ecological diversification, morphological adaptation, and behaviorally mediated processes that have a bearing on species recognition and reproductive isolation.

Chapter 12

Birdsong and conservation

SANDRA L.L. GAUNT AND D. ARCHIBALD MCCALLUM

INTRODUCTION

Conservation involves the planned management of natural resources in such a manner as to protect those resources from loss or injury. Biological conservation works to implement these measures with the additional intent of retaining natural balance, diversity, and evolutionary change in an environment (Lincoln et al. 1998). This is a hugely complex undertaking that demands a diversity of disciplines working in collaboration and often with synergistic, even catalytic, consequences. Systematic and evolutionary biology are centrally important in understanding the diversity and relationships of organisms that exist. Ecology shows how those organisms subsist, interact, and coexist. Ethology focuses on individuals and how their behavior results in survival, reproduction and, ultimately, death. The list is as ever-expanding as ripples on a glassy pond, yet each level is as essential to conservation as the next.

Earlier chapters have documented how birdsong studies contribute to our understanding of bird behavior, especially of the cognitive processes of learning and communication in birds. This chapter documents the application of our current understanding of bird communication behavior to conservation, restoration, and management of wild nature. Just as Luis Baptista not only studied nature, but also worked to protect and restore it, so the entire field of bioacoustics completes the circle: science contributing its understanding to protecting and restoring the very resource that inspires it.

How can the songs birds perform be instruments in conservation efforts? Fundamentally, because these songs are indeed acts of communication, they can be received by both unintended as well as intended receivers. Two-way intentional communication by definition involves the transfer of information from one brain to another. This requires that the receiving brain be configured in a manner similar or complementary to that of the sender, a configuration that is not likely to evolve *de novo*, according to current understanding of natural selection. For example, song is widely believed to communicate information, used by birds to recognize, attract, or repel their own kind (see Chapter 2) and about the singer's ability to hold a territory against would-be rivals and for would-be mates. The information content of these messages is thought to have evolved incrementally (Harper 1991), and can lead to exaggerated signals (see Chapter 2). Exaggerated signals become difficult to hide from unintended recipients. Natural selection readily favors those who intercept the signal and exploit it for their own uses. This occurs without benefiting the signaler or the intended receiver. The signaler can no longer live without the signal, but using it has unintended consequences.

Enter the conservationist who assumes the role of both eavesdropper and deceiver but, in this case, the intention is to benefit the signaler and its intended recipients (**Box 44**, p. 344). Conservationists *listen in* because birdsongs, or more precisely the whole range of communicative sounds birds make (see Chapter 5), allow us to identify the species, and sometimes the sex, age, and even individual identity of the maker. All of this encoding of information has evolved because it is of benefit to the sender to transmit the information to certain rivals, associates, potential mates, or offspring. Identification of species-specific, sometimes sex-specific, vocalizations is possible for a skilled naturalist without seeing the bird, providing a useful 'sound vision.' A skilled naturalist can use the information encoded in the vocalizations of birds to increase the number of individuals detected in a given area by a factor of 10 (Buckland et al. 2001). In the near future, the science of bioacoustics will allow others, who lack the 'good ear' of the trained or truly gifted, to bring the same kind of high-quality information home.

Conservationists can also *deceive* when they communicate to a bird that one or more other birds of a selected species, sex, or fitness state is present or wishes to interact. Ethologists have used 'playbacks,' broadcasting of the same or another bird's recorded vocalizations back, in numerous ways, from tests of territorial behavior in the field (see Chapter 2), to tutoring in learning experiments (see Chapter 3). As we shall see, playback is also used by conservation biologists to monitor populations, incite breeding activity, and attract individuals to settle in appropriate but under- or unused habitats.

The possibilities for both passive and active use of bird sounds are extensive, but they are under-exploited for conservation work, making the first decade of the new millenium an exciting and pivotal time in the application of bioacoustics to conservation. But, as with conservation itself,

BOX 44

SURVEYING ENDANGERED RAILS BY VOICE

Hearing is better than seeing when it comes to surveying and monitoring secretive or visually inconspicuous birds such as rails. Broadcasting recorded vocalizations to elicit vocal responses can be especially useful because natural calling rates vary with bird density, and passive listening surveys can make dense populations seem too large and less dense ones seem too small. Playbacks can help moderate this density effect. Rail populations are widely believed to be declining (Eddleman et al. 1988; Conway et al. 1994), with the black rail and the western subspecies of the clapper rail attracting special concern. However, reliable data on population trends are meager. By eliciting vocalizations from hidden birds, call-broadcast methods garner more data than passive listening surveys (Gibbs & Melvin 1993; Hinojosa-Huerta et al. 2002) – and more data provide better estimates of population sizes or trends. It is estimated that playbacks increase detections as much as 185% for black rail, 260% for sora, 540% for Virginia rail, 650% for clapper rail, and 925% for king rail (Conway & Gibbs 2001).

However, there are drawbacks to the call-broadcast approach. It can draw birds in (Legare et al. 1999), making density estimates based on birds' locations unreliable. Species differ in their responsiveness to recordings, complicating efforts to compare counts of responsive species such as the Virginia rail with less responsive ones such as the black rail. Broadcasting multiple species' calls at each visit can distort data, since some species (e.g. black rail) may turn silent once the calls of others (e.g. Virginia rail) are played. Furthermore, choosing which vocalization to play can be tricky, since for most rails the functions of different calls are not well understood. The geographic source of the recording may also matter, if birds respond differently to subspecies or regional variants. And, of course, individuals can habituate to tapes, limiting their repeated use. Nevertheless, even given all these limitations, most biologists feel playbacks are justified by the increase in data, especially when used with standardized methods and monitoring protocols (Ribic et al. 1999). Conway & Gibbs (2001) advocate beginning each survey point with a passive listening period without playback, followed by an equal-length playback period. In this way, naturally occurring and playback-elicited calling frequencies can be compared and used separately or together for different purposes.

Jay Withgott

complexities abound. In this chapter, we will explore potential contributions of bird voices to conservation, and evaluate the instruments, both available and needed to capture, analyze, and manage sound in conservation efforts.

LEVELS AND ISSUES IN CONSERVATION

Biological conservation focuses on two levels of organization: species and ecosystem. Species receive the most attention because they are assemblages of like kinds, which necessarily interbreed, evolve, and eventually go extinct. Prevention of extinction has received the bulk of publicity and funding, both actual and incipient, because extinction is rampant (Myers 1987), mostly human-caused, and final. Ecosystems are communities of local populations of different species interacting with each other and with the physical environment. Species are not only components of ecosystems, they rely on the functioning of ecosystems for their survival. In the past decade, conservation biologists have focused increasingly on 'ecosystem management' as a way to preserve not only the functioning whole, but also the component parts.

Inventories of biological diversity or censuses and life history studies of one or more species within a specific habitat precede and assist conservation measures to identify healthy habitats, providing evidence for changing conditions, and suggesting measures to manage unhealthy alterations. Inventories based on conventional sighted methods and capture/marking techniques involve many hours of intense labor, may miss or overlook species, and may affect survival or reproductive success of birds (Croll et al. 1996). Recorded sound inventories can cover more area in less time and eliminate adverse capture, handling, and tagging of animals while providing a permanent record not available from surveys conducted purely by ear.

Identifying optimal areas for management or protection as reserves requires extensive, often time-consuming investigation of large areas. An approach to assessing areas for protection with maximum efficiency is to identify species 'hotspots,' areas with rarities or rich in species (Myers 1988; Prendergast et al. 1993; Tardif & Des Granges 1998). In a large area of disturbed ecology, rarity hotspots, especially when coupled with other taxa, can serve to pinpoint areas of maximum priority for management.

A potential managed area should contain populations of rare or endemic species that are large and coherent enough to persist over time. The question of minimum viable population size then becomes an issue. Viability has traditionally been assessed with disruptive and labor-intensive mark–recapture studies, but sound recording is a promising alternative. Sound recording surveys are especially effective both for quickly locating rare species and for determining population size. Simple, ambient recording using manned or unmanned microphones can be augmented by the use of 'playbacks.' 'Record–rerecord' approaches may also make it possible to acquire data more economically and with less disruption to the target species.

Not only should conservation plans protect present species diversity, but they also need to consider future diversity, and the evolutionary processes that spawn new species (Smith et al. 1997). From this perspective habitat, not simply particular habitat types, but also gradients, such as rainforest and bordering areas of other forest types, need protection. Recent findings that birdsong divergence may promote speciation over ecological gradients (Slabbekoorn & Smith 2002a, b) add credence to the inclusion of evolutionary processes in conservation plans.

APPLICATIONS OF BIRD COMMUNICATION TO CONSERVATION

Defining Evolutionary Limits

Until recently, species limits were defined by an interruption of gene flow, assuming that members of different species do not interbreed, but species were actually described primarily on the basis of visual characters. The revolution in DNA sequencing and its application to the reconstruction of evolutionary trees on the basis of genetic dissimilarity has led to the emergence of a new definition of species as independently evolving lineages (Zink & McKitrick 1995). Thus, although Baltimore and Bullock's orioles do indeed interbreed in certain places, the two populations are clearly evolving independently, and their former status as separate species has been reinstated (American Ornithologists' Union 1998).

Some populations such as the oak titmouse of California and the juniper titmouse of the Great Basin, formerly thought to be too similar in appearance to rate separate species status, have been shown by genetic studies to have been separated genetically for several million years. In these titmice, as in many other recently split species pairs, the two populations have long been known to have distinctively different vocalizations (Cicero 2000). Indeed, voice has been the partner of genetics in the description of many new species within the past decade (**Fig. 12.1**). It is now generally clear that two geographically isolated populations that differ primarily in vocalizations are good candidates for species status, as long as these differences are not merely learned variations of the same song or call types, and should be the subject of genetic studies.

For example, Luis Baptista used voice in arguing for the recognition of the Socorro dove as a full species endemic to the Socorro Island in the Revillagigedo Archipelago, unique from the mainland mourning dove (**Fig. 12.1**), but extirpated from the island for over 20 years (Baptista et al. 1983). From captive populations, pure Socorro doves are being identified using DNA fingerprinting techniques by Juan E. Martínez-Gómez, of the University of Missouri. Reintroductions of descendents from this strain are planned once the island's native habitat is restored, an effort being undertaken by an organization Baptista founded, the Island Endemic Institute, in cooperation with Grupo de Ecologie y Conservacion de Islas A.C.

Eavesdropping Applications

Listening to Census

Bird voices are so conspicuous that we frequently locate them by that modality alone and, as discussed above, they can usually be identified in that way. This is especially true for crepuscular, nocturnal, and otherwise secretive birds, living in dense vegetation such as in tropical rainforests, mountain cloud forest, and marshes everywhere. Moreover, many birds have peaks of twilight singing at dawn and dusk in mixed species choruses (Staicer et al. 1996), conditions under which sound can be used to better advantage than vision to distinguish species.

It follows then that census schemes often rely on singing birds for estimates of species diversity and population size. To address the fact that the ranges of many species of birds are not adequately known (Remsen 2001), Birdlife International's field ornithologists (Collar 1999) have been crisscrossing the tropics and other poorly explored areas, looking and *listening* for members of over 1,000 globally threatened species (IUCN 2000). A more focused listening census followed a credible report in 1999 of a sighting of the ivory-billed woodpecker in the Pearl River basin of Louisiana, a species long thought to be extinct. An international team was assembled in early 2002 to corroborate the sighting. The existence of sound recordings from the 1930s, archived at the Macaulay Library of Natural Sounds (MLNS), made it possible to familiarize the searchers with the distinctive 'toy horn-like' calls of this species, and twelve passive recording devices were also deployed (see p. 353, **Fig. 12.3D**).

Even for species whose ranges are well known, primarily in Europe and North America, population trends require monitoring. In the late 1980s, conservationists in eastern North America began to suspect that migratory songbirds were in decline. The Breeding Bird Survey (BBS), a joint program of United States

Figure 12.1 A gallery of some recently described new species whose voices were instruments in their recognition (**CD1 #91**). (**A**) The subtropical pygmy-owl from the eastern Andes and, below, the sympatric Andean pygmy-owl (Robbins & Howell 1995). (**B**) The Emei leaf warbler from Mount Emei Shan, Sichuan Province, China and, below, a close relative, Blyth's leaf warbler (Alstrom & Olsson 1995). (**C**) Lulu's tody-tyrant from northern Peru and, below, a close relative, rufous-crowned tody-tyrant (Johnson & Jones 2001). (**D**) The Socorro dove formerly from Socorro Island, Revillagigeledo Islands off western Mexico and, below, a mourning dove (Baptista et al. 1983).

and Canadian conservation agencies, provided the hard evidence for this decline, leading to enhanced conservation efforts on behalf of dozens of affected species, a crisis clearly calling for ecosystem management. The BBS is conducted by hundreds of volunteers each year, following a very rigorous protocol and, in woodland habitats, up to 90 per cent of birds detected are identified by voice (Buckland et al. 2001).

For the unlucky species that become rare enough to require active, species-specific management, more precise population estimates are needed to assess the effectiveness of management decisions. Species whose populations are difficult to survey by visual counts are often estimated by indirect means, such as owl pellets, mammalian scats, and markings. These can be unpredictable, may underestimate the actual population, and are not sex specific. On the other hand, sound recordings under some situations can better census the population size as well as proportion of male and females, because many species exhibit individual signature calls or sex-specific calls (**Fig. 12.3E**). Various techniques have been developed to do this with some European bird species (Terry et al. 2001), for studies of the nocturnal endangered African forest elephants whose numbers had been underestimated by the method of dung mound counts (Payne 2001), for devising the management plans for the threatened spotted owl (Thomas et al. 1990), and for other threatened owl subspecies (K. Otter personal communication).

Monitoring Birds to Evaluate Habitat Health

Though most bird monitoring is done to evaluate bird populations, habitat health can also be tracked by monitoring birds. We are all too aware of the rapid rate of habitat destruction throughout the world owing to human activities such as agriculture, logging, building, drilling, and mining. In some areas, such as tropical and boreal forests, the rate of destruction exceeds the rate at which we can document and study poorly or undocumented species. For example, the huge diversity of hummingbird species in the neotropics includes entire groups whose vocalizations and breeding biology remain little known. Thus, every carefully conducted sound recording effort can provide invaluable documentation of an area's soundscape as well as the range of species. This 'sound vision' opens insights to general habitat structure, such as an increase in wind due to loss of vegetation, to specific animal types present, and their numbers for comparison across time. Perhaps even more than conventional specimens, field notes, and publications, these sound recordings provide permanent records of the condition of an area at a specific point in time, providing an invaluable baseline for future comparisons.

A combination of written census work and recent sound recording surveys was employed by avian biologist and pioneering sound recordist, Joe Marshall, to track the changes and loss of breeding avian species on Redwood Mountain in the Sequoia Forest of the southern Sierra Nevada, California, surveyed by him previously in the early 1930s (Marshall 1988). His field notes and bird specimens were deposited in research museums. Fifty years later, aided with sound recording equipment, he was able to monitor the changes that had taken place. Some changes documented by him in species composition were clearly the result of local environmental changes, mostly attributable to logging and 'conservation' efforts to control for fire. But populations of the olive-sided flycatcher and Swainson's thrush had decreased in areas where the forest structure and climate had not changed. The declines could not be explained by the summer habitat, and Marshall, based on his recordings on wintering grounds in South America and reports from other investigators throughout the forested highlands of the neotropics, suggested that the loss of forest, not in the northern breeding habitat, but in the southern wintering habitat was the cause (Marshall 1988).

Few areas have the benefit of long-term baseline data in the form of field notes or sound recordings deposited in archives and, where baseline data are lacking, we no longer have the luxury of time and funding to obtain those data. Thus, alternative means must be developed to evaluate the impact of disturbance on a region's ecology to make informed conservation management decisions. Such evaluation procedures need to be cost-effective and yield reliable results that are obtainable in a timely fashion.

Indicator Taxa

One method is the use of carefully chosen species and groups that can serve as biodiversity indicators. These surrogates predict and identify changes in species as related to vegetation diversity, structure, and complexity. Indicator taxa need to be abundant and ecologically diverse throughout the area, must have a functional role in the ecosystem, and should respond to environmental change in quantitative and predictable ways (Noss 1990; Mooney et al. 1995). For example, habitat or diet specializations are behaviors that have been correlated with rarity and predilection to local extinction within perturbed environments (Leck 1979; Laurance 1991). Butterflies in Ontario, Canada (Kerr et al. 2000), bats in neotropical rainforest (Medellín et al. 2000), and birds in urban settings (Blair 1999) have been used as indicator species to assess habitat health. For this technique to be effective, sampling techniques need to be reliable, standardized, and cost-effective. Sampling birds that might be used as indicator species can be done by sound recording, as a reliable way to sample species richness, composition, and abundance. Sound recordings are equal or superior to visual sightings in the neotropics, can be standardized, and avoid the complications and biases of capture techniques (Haselmayer & Quinn 2000). There are of course also biases intrinsic to auditory sampling; census timing must be coincident with the time of song production and, for many species, only males will be detected.

Untested avian candidates for this method are neotropical hummingbirds. There are 328 recognized species of hummingbirds, making this the second largest avian family, and they are particularly diverse and abundant in the neotropics. They feed on nectar, insects, and pollen, and some have specific habitats (Schuchmann 1999). As a consequence, they are also important in ecological processes such as pollination, insect regulation, and seed dispersal. During the breeding season, which is prolonged in the neotropics, males can be detected by song, and some species may be detected with techniques for recording high-frequency ultrasound (M.S. Ficken personal communication; S.L.L. Gaunt, personal observation; E.D. Jarvis personal communication).

While some species are known to increase in abundance in disturbed habitats, others will be found to decrease or be extirpated. Once indicator taxa are identified, gathering data for species and habitat can be a relatively quick way to assess habitat degradation and therefore sites for immediate conservation concern (Landers et al. 1988). However, making useful choices of surrogates is not without difficulty. Often in the past, choices have been ad hoc and burdened with the implicit assumption that habitat for the indicator will be suitable and sufficient for the species for which it represents (Niemi et al. 1997; Andelman & Fagan 2000). More than one potential animal or plant taxon should be identified (Mac Nally et al. 2002), and statistically valid, objective methods must be employed in the taxon selection process (Mac Nally & Fleishman 2002).

Deceiving Applications

Playback to Census

Playbacks of bird sounds are an essential tool for censusing rare and secretive birds found in low densities (Johnson et al. 1981; Marion et al. 1981; Johnson & Dinsmore 1986). A bird may respond to playback of its own species or to that of a predator such as an owl by moving into the open, as when the bird watcher 'pishes.' The former is more specific, attracting particular species that may reveal themselves by voice or movement into the open (Johnson et al. 1981). Territorial species, and often both sexes, respond readily to conspecific vocalizations at all times of the year.

Of course, if sound recording is done concurrently, birds can be recorded and their own song played back to them. In Peru, Theodore Parker used this technique to aid positive identification of flocks of parrots in flight. He recorded the parrots' flight calls, immediately played these back, which often caused the flock to veer toward the playback allowing for identification (Budney & Grotke 1997). Luis Baptista always used playback when teaching in his Sierra Nevada summer bird course. In the neotropics he found playback to be a very effective way to positively identify wrens, which are highly vocal but extremely elusive in dense underbrush. However, this technique was almost useless with certain hummingbird species such as green violet-ears that tend to flee from playback (Baptista & Gaunt 1997b).

Species Recovery with Playback

Identifying optimal management areas where species have not yet been lost is not always possible. Often, disturbed areas have already lost most or all of the bird species they once supported. Recovery then becomes an issue. Islands are especially prone to natural species loss, and such loss is hastened by human impact through deforestation, introduction of species, and disease. The Socorro dove is a prime example of the complete loss of an island endemic due to both deforestation and mammal introduction. Recovery of species lost may be possible by first returning the habitat to its former condition, followed by the reintroduction of birds from captive breeding programs, as with the California condor, the Hawaiian goose or nene and, potentially, the Socorro dove. Alternatively, when available, recruits from extant populations can be attracted to suitable habitat.

Bioacoustics may play a somewhat unexpected role in the captive breeding approach. Breeding has been augmented in captive doves (Cheng 1992) by introducing tape-recorded conspecific sounds, which have been shown to stimulate gonadal development and thus enhance breeding (Baptista & Gaunt 1997b). Acoustic signals can thus be as useful as visual stimuli, such as decoys, conspecifics or reflections of these in mirrors (Ellis et al. 1998), for stimulating breeding, and they are a lot easier to produce.

The broadcasting of sounds was recently used in the wild to successfully induce reproduction in a reintroduced population of Caribbean flamingos (O'Connell-Rodwell et al. 2002). Flamingos require a minimum population size to stimulate breeding (Stevens 1991), and the reintroduced population on Guana Island in the British Virgin Islands was well below the minimum at only six individuals. Earlier investigators in zoos had used mirrors to trick captive flamingos into thinking that their colony was larger (Stevens & Pickett 1994). O'Connell-Rodwell and colleagues used a combination of broadcast colony sounds and decoys on nests to simulate a larger colony and improve reproductive effort.

Possibly one of the most successful reintroduction programs using bioacoustic tools has been the recovery of seabird colonies previously abandoned due to climate change or human activity. Early in this remarkable effort, abandoned colonies that were still suitable for breeding or that could be restored for breeding had been located, but getting the birds back was not easy. Ethologists had documented that many species, including seabirds, cue on the presence of conspecifics in selecting breeding habitat (Smith & Peacock 1990; Reed & Dobson 1993). Kress and colleagues recognized the potential that these studies presented. They played sound recordings of breeding colonies coupled with decoys of seabirds on the colony and attracted migrating birds to the abandoned colonies, including terns (Kress 1995), puffins (Kress & Nettleship 1988), storm-petrels (Podolsky & Kress 1989), and the endangered dark-rumped or Galapagos petrel (Podolsky & Kress 1992; **Fig. 12.2**). This technique has also been successful in the restoration of common murre colonies off coastal California (Parker et al. 2000). In an interesting twist, this technique was used in the successful *relocation* of a large Caspian tern colony with over 9,000 pairs, whose feeding activity had hampered the survival of endangered juvenile

Figure 12.2 The dark-rumped petrel (**A**) was recolonized to Santa Cruz Island (**B**) in the Galapagos by playing colony sounds (**C**) from speakers (**D**), and by placing decoy birds on the site (Kress & Nettleship 1988; Podolsky & Kress 1989; Podolsky 1990). It has been renamed the Galapagos petrel (**CD1 #92**).

salmon in the Columbia River estuary (Roby et al. 2002).

Controlling Bird Populations with Playback
As pointed out in Chapter 6, noise annoys. When bird populations become very large, their noise or presence can create all kinds of problems for property owners, and safety problems for some installations such as military and commercial air space. Every year the Borror Laboratory of Bioacoustics (BLB) receives dozens of requests for alarm calls of one species or another, most notably common grackles and European starlings, and even for such non-vocal birds as turkey vultures to disperse them by playback. Two recent resources from the US Department of Agriculture and Canadian Civil Aviation (Hygonstrom et al. 1994), the latter available on the web, summarize techniques for the control of wildlife.

Unfortunately, acoustic harassment with alarm calls or predator calls to disperse birds from roosting, nesting, or foraging sites is not very effective. Birds readily habituate so that the sounds soon become ineffective (Transportation Canada 2002). Sometimes the combination of sound with some other stimulus to increase the relevance of the sound, such as bird decoys, or loud cannons can improve the results. But even with very noxious combinations, birds habituate. Dispersal of European starlings by harassment with harmonic components of their alarm calls, which may contain ultrasound, has been reported (Aubin & Bremond 1992). Interestingly, recent reports from the Netherlands, not yet corroborated by published results, suggest that the use of general ultrasound dispersed large roosts of European starlings. Details are limited and the full procedure may involve a mixture of stimuli.

Nevertheless, for those who find the birds' presence simply annoying, we usually advise that they sit it out – the season will change and the birds will move on. For more serious threats, such as nesting colonies on military airstrips or loss of commercial crops, we try to be as helpful as we can, recognizing that safety officers cannot just sit by and do nothing. The recent success of relocating tern colonies from salmon runs discussed above, suggests that, for colonial seabirds at least, relocation with sound playback rather than acoustic harassment is a viable option.

INSTRUMENTS AND METHODS

Listening

The various monitoring schemes we have described require extensive knowledge of bird sounds. Luckily, avian biological diversity is among the best documented of any group of animals. Though there are continuing debates and controversies in avian systematics (Welty & Baptista 1990), the identification and naming of living birds is greatly facilitated by abundant field guides, courses, workshops and, above all, experience gained from simply watching birds. Today many publishers are producing field guides to birds not only for temperate regions but also for many well-traveled tropical regions of the world, *The Birds of North America* species accounts are now complete (Poole & Gill 2002), and the *Handbook of Birds of the World* is up to the sixth volume (del Hoyo et al. 2001) with an introduction by Luis Baptista.

The unique characteristics of each bird's vocalizations are important for identification (Becker 1982). They can be learned and used by experienced observers to identify species, often on voice alone. Some field guides come with companion birdsong tapes or compact disks (CDs, Cornell Laboratory of Ornithology 1992; Colver et al. 1999), some publications such as *The Birds of North America* species accounts and *Stimmen der Vögel Europas* by Bergmann & Helb (1982) are available with sonograms, and stand-alone CDs are available by the dozens. Major sound archives will soon have their sound libraries available on-line for auditioning.

Playback

To use animal sounds for playback, the conservation biologist need not necessarily train to become a sound recordist or bioacoustician if the sounds can be obtained from a public archive of existing sounds or from other sources. Many archives today have websites with inventories of the animals in their holdings (**Fig. 12.3**). The biologist will need to know the behavior of the birds of interest to recognize what vocalization types, produced by what age and sex class, under

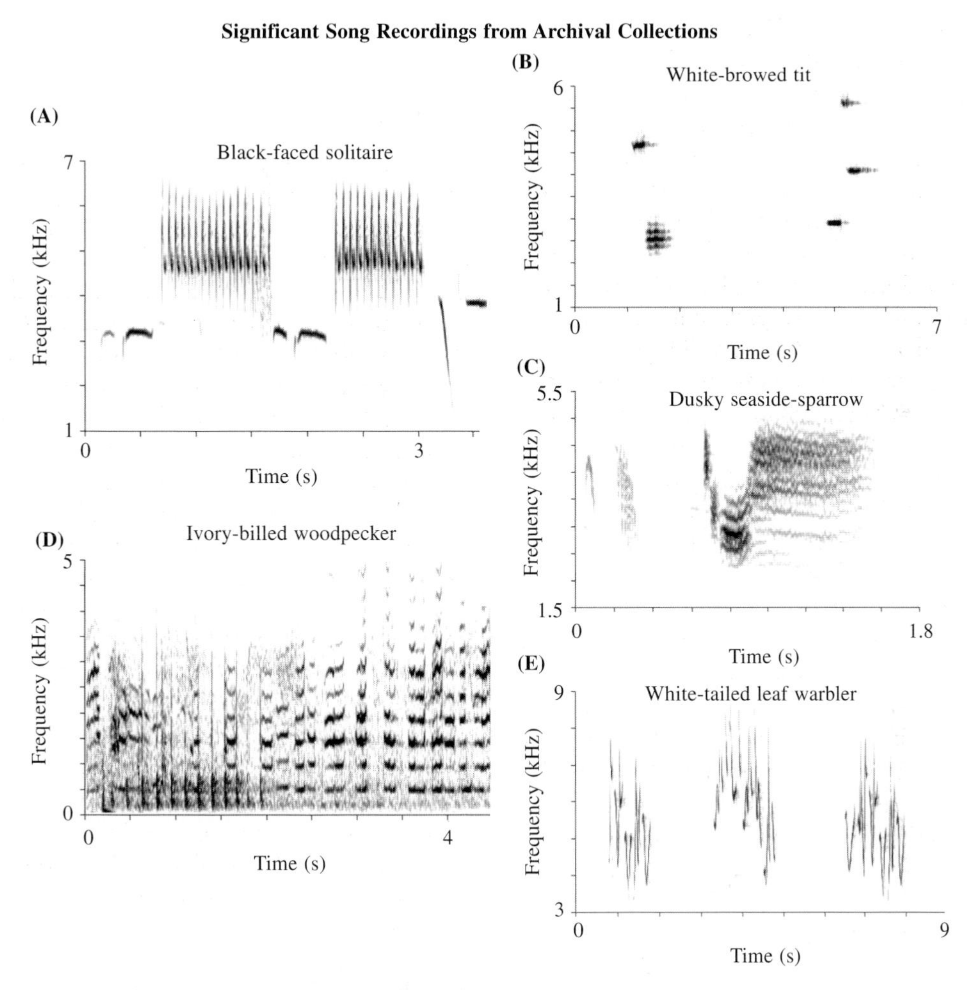

Figure 12.3 Five examples of public sound archives with contact information below and a sound file that is unique or important for that collection (**CD1 #93**) (**A**) A black-faced solitaire song recorded in Costa Rica (Luis Baptista master field tape #LB9201, BLB 21615. California Academy of Sciences, Department of Ornithology and Mammalogy, 50 Golden Gate Park 94118. http://www.calacademy.org). (**B**) A song of a white-browed tit endemic to western China (BLB 18073. Borror Laboratory of Bioacoustics, Museum of Biological Diversity, The Ohio State University, Columbus OH 43212 http://blb.biosci.ohio-state.edu). (**C**) A song of the extinct dusky seaside-sparrow (FLSMNH 724 specimen 2. Bioacoustics Laboratory and Archive, Florida State Museum of Natural History, University of Florida, Gainesville, FL 32601 http://www.flmnh.ufl.edu/birds). (**D**) The voice of an ivory-billed woodpecker, now extinct on the Gulf Coast (MLNS 6784. Macaulay Library of Natural Sounds, Cornell Laboratory of Ornithology, 159 Sapsucker Woods Road, Ithaca NY 14850, http://birds.cornell.edu). (**E**) A white-tailed leaf warbler song, from China (BLNSA Kina94I#82. The British Library, National Sound Archive, 96 Euston Road, London NW1 2DB UK, http:www.bl.uk/nsa).

what conditions, should be used in a project. Most often this will be male breeding season song. The Borror Lab of Bioacoustics inventory is linked to the BLB database such that once an animal is identified as present in the collection, the dataset for each recording for that species can be accessed to identify those meeting the needed criteria.

Generally, repeated playback should be avoided, as animals habituate and there can be deleterious side effects (Baptista & Gaunt 1997b). A suggested protocol is to broadcast one or two stimuli and wait for a response. Some species, such as titmice and wrens, respond almost immediately; others, such as rails, may require additional playback. If capturing for marking, sexing, aging, or collecting is a part of the conservation effort, then playback can be used to attract birds into mist nets or other traps. If the research is addressing a behavioral question, choice and presentation of playback signals are critical for appropriate statistical analysis (McGregor 1992). In all cases, a field-reliable portable means to broadcast the playback must be available.

Recording

Active

Although conservationists can learn birdsongs and perform playbacks with existing material, most of the applications we have described involve active recording, or are facilitated by it. Recording bird sounds has two main uses for conservation, one immediate and one protracted. Recordings can be used immediately to improve the accuracy of censuses and other monitoring efforts, both by playing them back to oneself for a second listen, and by playing them back to the target birds to confirm identification (see p. 352). The protracted benefit comes from depositing the recordings in a recognized archive, whence they can be used by others for playbacks, and by scholars to refine our understanding of geographic variation and species-specific features of bird sounds. This cumulative scholarship in turn makes it easier to correctly identify sounds heard and recorded in the future.

Since sound recording can be so instrumental in conservation work, it is cost-effective to have recording and playback equipment combined in one unit. The simplest, least expensive, and most portable machines are still analog cassette recorders with built-in speakers. Auxiliary speakers may be needed to obtain sufficient sound amplitude. A detailed discussion of both analog and digital recorders, microphones and accessories, cables, recording media, field cases, and more is provided by Budney & Grotke (1997). Specific brands and models are described and compared on various websites (http://biosci.ohio-state.edu for the BLB site and links therein to other sites). Additionally, portable computers with appropriate software can be used to digitize directly to hard disk in the field, and to interactively play the recording back on the spot: SINGIT! (Bradbury & Vehrencamp 1994), SIGNAL™ and SIGWIN™ (Beeman 2002; Burnett 2001; Burnett et al. 2001), SYRINX (Mennill & Ratcliffe 2001), to name a few. This is a boon to experiments in ethology, and a useful tool in conservation, both for capturing and for surveying. A review of sound analysis and playback equipment can be found in McGregor & Ranft (1994), and can be updated by consulting review articles as they appear in the journal *Bioacoustics* and to linked websites on the BLB website, especially that by S.L. Hopp.

Animal sound recording is both an art and a science. Some of the greatest avian recordists such as D.J. Borror (founder of the BLB), W.W.H. Gunn (Federation of Canadian Ornithologists), P.P. Kellogg (co-founder of the Laboratory of Ornithology), and L. Koch (British Broadcasting Company), were both artists and scientists. A sense of how to approach a target bird, experience in optimizing the target's signal against a background of interfering noise, and an understanding of how microphones and recorders function, are but a few of the required elements. Sound recording techniques under natural conditions, and especially in the neotropics, are outlined in Budney & Grotke

(1997). But getting beautiful or usable recordings from a target bird is only the beginning of sound recording for scientific purposes — the proverbial tip of the iceberg.

Accurate documentation of the event recorded is as essential as the recording itself. Fortunately, with longer playing time on the equipment and media of today, documentation can be narrated directly onto the master field tape. What to narrate and how to coordinate the narration with the bird's performance should be carefully planned. The basic data to accompany recordings can be found in Kroodsma et al. (1996) and on the BLB website.

Remote

Most commonly, sound recordings of specific target or focal birds are directly recorded sequentially on a master field tape. Another approach is to survey remotely using a stationary recording set-up, with a microphone and recorder, or a microphone and a laptop computer, or an array of unmanned microphones connected by cable or radio link to conventional multichannel tape recorders or directly to computer hard disk. Free-running recordings in particular, designed to augment or substitute for point-count censuses, are collected as samples of the entire soundscape, and a fair representation is necessary of ambient noise levels as well as the relative amplitudes of all individuals and species. Whether the entire community is recorded or only selected species, as with the Cornell Bioacoustics Research Program's (BRP) study of two endangered species at Fort Hood, Texas (http://birds.cornell.edu), the key factor is detection and discrimination of different individuals. Such recordings are usually edited with an automated or semi-automated procedure, and correct identification of the events on the recording require reference sound samples of high quality. Conversely, remote recordings may have embedded within them excellent, archivable samples, which may prove to be invaluable, particularly if they are rare vocalizations unlikely to be obtained by a mobile, opportunistic recordist.

Microphone arrays have been used for decades to track, identify, and count many groups difficult to see or attract by playback, including anurans (Heyer et al. 1993) and, using hydrophones, fish and marine mammals (Fish & Mowbray 1970; Norris & Evans 1995). Recent work using microphone arrays to follow the behavior of song sparrows suggests that this approach may also prove useful in bird census work (Bower 2000). Arrays allow localization in space of individual singers through calculation of differences in arrival times of the same sounds at microphones with known coordinates, making it possible to count individuals with high confidence, and also to describe their behavior in great detail.

Currently, microphone arrays are constrained by the necessity for a cable connection between microphones and recorder. Portable microphone–transmitters that radio their input back to a central recording location have been used successfully for elephants in Africa (Payne 2001). Once the technical challenge of transmitting the high frequency sounds of birds is solved, portable microphone–transmitters will make it possible to set up an array one afternoon, record the dawn chorus the next morning, and then move on to another site. This will be invaluable in surveying large areas, extensively or intensively, by surveying sites that are difficult to access by foot, or are highly sensitive to repeated human visitation.

Nocturnal bird migrations have already been tracked by microphone arrays in grasslands (Evans & Mellinger 1999), and from rooftops in the Delaware Valley (MLNS Project Birdcast). Though more individuals of certain rare species, such as the Bicknell's thrush, than one could see in a lifetime can be detected by this method, the amount of data that must be processed to do so is staggering (Evans 1994). The computer programming department at the Cornell Laboratory of Ornithology is working to keep up with this data sorting challenge by automating the recognition of signals, in this case the flight calls of sparrows and warblers, on the basis of duration and frequency. Dealing with massive amounts of raw recorded data is the continuing challenge for all recording efforts.

Survey Strategies

Census results can vary between surveys over time because of different techniques as well as actual changes in distribution of species. In the tropics, surveys are plagued by this dual problem because of the large number of species, the rarity of many, and the large numbers of variables that govern the presence or absence of birds. As a census technique in temperate habitats, counting territorial singing males is standard, but is less effective in the tropics because more than two-thirds of the bird species are not territorial (Karr 1981). Other survey techniques include the observer count of animals from a fixed point (point counts), within a defined area (quadrants), or as encountered while traveling through an area (transects). Playback is sometimes used during point counts to increase detectability (Lynch 1995), and Parker (1991) promoted the use of sound recordings as an alternative to traditional specimen collection. Recording sounds as an alternative to observer report in a point count survey had not been tested until recently, but offers some advantages.

How effective can recorded "sound" vision be, compared to observer counts based on both sighting and hearing, when you consider the decreased number of sensory cues available in evaluating recordings at a later time? That effectiveness was evaluated in Peru by comparing the species lists from observer point counts, and from simultaneously obtained sound recording at the same points. The conservation mission was to assess the effects of ecotourism on animal communities surrounding tourist lodges (Haselmayer & Quinn 2000). The tests paired 136 recording samples to the same number of observational samples. Recordings were auditioned once, and sections could be listened to a second time to generate the species list. In addition, an outside expert was used to verify the accuracy of the tape editor, something that could not be done for the point count observations. The expert agreed with 96% of the identifications, failed to detect 4% and added 9% to the species list; the results showed that sound recordings were more effective in sampling stations with high species richness (see Haselmayer & Quinn 2000, Fig. 1).

Sound recording provides' the advantages of a permanent record that can be re-examined by both editors and distant experts, not to mention the added specimens available to archives. It also reduces the need for skilled observers in the field. With the addition of automated microphone arrays, vast areas can be easily monitored. But again, data are generated quickly, and methods must be established to facilitate the organization and analysis.

Acoustic Analysis

Sonograms and Beyond

Recordings of a bird census or survey can be used just for re-listening, essentially to roll back time, improving both accuracy and detection rate in a bird census or survey. But the uses of sound recordings do not end there. Simple analyses can greatly augment both accuracy and detection rate.

For the past half-century, bioacousticians have used visual representations of sounds, or sound spectrograms, for analysis. A sonogram is a picture of sound in time that can be read somewhat like a musical score. Technically it is a transformation of the time-varying amplitude of the sound wave to a graph of frequency versus time (**Box 1**, p. 5; **Box 2**, p. 6; **Box 45**, p. 358). The mix of frequencies that one encounters simultaneously while listening is spread out on the vertical axis allowing for easy inspection. Once you learn to interpret these graphs, they can be read easily and are useful in species identification. Bird vocalizations that sound similar to most listeners can be distinguished by differences in their fine structure, which are often quite distinct on the sonogram, even though imperceptible to the human ear. In addition, simultaneous sounds can be easier to distinguish than by ear. If they are at different frequencies, they will be completely separated on the sonogram. Even if they overlap in frequency, as when two individuals of the same species sing, the pattern recognition virtuosity of the human visual system makes it possible to distinguish them. In the past, the process of sonogram inspection required bulky, single function, expensive units that were available in only a few laboratories and archives, but it has been made much more accessible by computer programs that produce sonograms in real time, as one listens to the sounds.

Beyond usage after the fact, sonograms can be used before a census effort to train census-takers. Although the most ambitious experiment so far in making sonograms a part of the identification process in a field guide to birds (Robbins et al. 1966) was not a success, the modern real-time spectrogram programs overcome early limitations. Most of us have better visual memory than auditory memory; most birders are more competent at identifying birds visually than by sound. Once the squiggles on the graph can be associated with a particular familiar, but hard-to-remember, sound, visual memory is enlisted in the task of distinguishing and identifying songs. For example, the simple trill of the dark-eyed junco has a quality that is hard to characterize in words, but is easy to remember once it is seen that the repeated 'note' is actually two separate notes at different frequencies that are too close in time for complete resolution by the human ear.

Advanced acoustic analysis methods such as spectrogram cross-correlation (**Box 45**, p. 358) or discriminant function analysis require a pre-existing library of sounds with which to compare the input sample. This may be workable where we have libraries of sounds for most species. But for many localities, the neotropics especially, those libraries are incomplete. There is, however, growing interest in and success with species and individual recognition using so-called neural network programs that can be trained as data become available. Research using such neural network identification systems have promise with groups of animals that are very difficult to census by sight, for example, bats (Stocker 1998; Burnett & Masters 1999), whales (Deecke et al. 1999), deer (Reby et al. 1997), and marsh birds (May 1998).

These techniques work better for some bird groups than others. The sounds of most birds are not strongly influenced by learning (see Chapter 4). There is variation, but it is quantitative rather than qualitative. For example, even though individual great bitterns may be identified by their signature calls (McGregor & Byle 1992), all bitterns are easily recognized as such. The development of algorithms to distinguish bitterns from other herons, and other marsh birds should be a straightforward task. On the other hand, parrots, hummingbirds, and oscines are capable of learning their songs and calls, and require auditory exposure to models to develop species-specific song (see Chapter 3). Songbirds in particular are censused largely by their advertising songs, but the complexity and diversity of their repertoires make developing automated species-recognition algorithms very challenging. Comparison of an unknown song with a library of identified songs becomes an especially daunting computational task for species that improvise their songs, such as the sedge wren; or have huge repertoires of song elements, such as the brown thrasher; or very small local dialects, and just a few males per dialect, as in the indigo bunting (Payne et al. 1981; Boughey & Nicholas 1990; Kroodsma et al. 1999b).

It is more elegant and, in the long run more accurate, to take another tack, based on the *species-specific* characteristics of each repertoire. After all, even mimics seem to know their own species (Becker 1982). Probably millions of birders can identify the northern mockingbird, the gray catbird, and the brown thrasher, all common backyard birds, despite the huge size of their repertoires. These three species, for example, can be distinguished by the phrasing of their songs, the brown thrasher typically delivering each short motif twice before moving on to another, while the catbird does not repeat, and the mockingbird often gives more than two renditions per motif. Moreover, species that have bewilderingly complex songs, sometimes difficult to identify, have distinctive species-specific calls. Simple notes or differences in phrasing are much easier to identify with computerized methods than are the thousands of song elements in a thrasher's song (Kroodsma & Parker 1977), or the subtly varying construction of the song of the song sparrow (Searcy et al. 1995). The best bets for the comparison-to-library approach with songbirds are first, to focus on a limited number of target species, as in the BRP's Fort Hood and ivory-billed woodpecker projects, where the targets are only one or two endangered species; and second, to make more use of calls, as in the Laboratory of Ornithology's Birdcast Project, in which the calls of a few dozen species are selected in real time by a computer algorithm, and then identified to species after the fact with more sophisticated tests.

Manual processing of the huge quantity of acoustic data from even a one-hour recording session can take an investigator out of the field for several days. In the near future, it may be possible for conservationists to obtain off-the-shelf mobile monitoring systems that can acquire

BOX 45

CROSS-CORRELATION ANALYSIS

Digital cross-correlation analysis is a way of comparing pairs of sonograms objectively; two digital sonograms are moved past each other in time and compared. The cross-correlation coefficient values generated are plotted versus the time offset (**A**). The peak correlation value from this curve is a measure of the similarity between the two sonograms; the higher the value the greater the similarity. In a hummingbird, the green violet-ear, males display with song from traditional sites (or leks) to attract females (**CD2 #80**). We suspected that all males on a given lek performed the same song type. To test this we used sonogram cross-correlation to make all pairwise comparisons of notes from members of a lek and between nearby leks. Cluster analysis resulted in a tree linking all birds with similar songs, having similar peak correlation coefficients, on the same branches (**B**); songs are composed of phrases of two notes (A–B) or three notes (C–G) that are repeated for 10's of minutes at a time, and only one phrase from each bird is illustrated. Birds that were nearest neighbors in a lek always clustered together and separate from birds in other leks. In **B**, the lower branch, Lek I, had two birds, the upper branch, Lek II, had four birds, and a third branch of one bird for whom no neighbor was observed (Gaunt et al. 1994).

Sandra L. L. Gaunt & D. Archibald McCallum

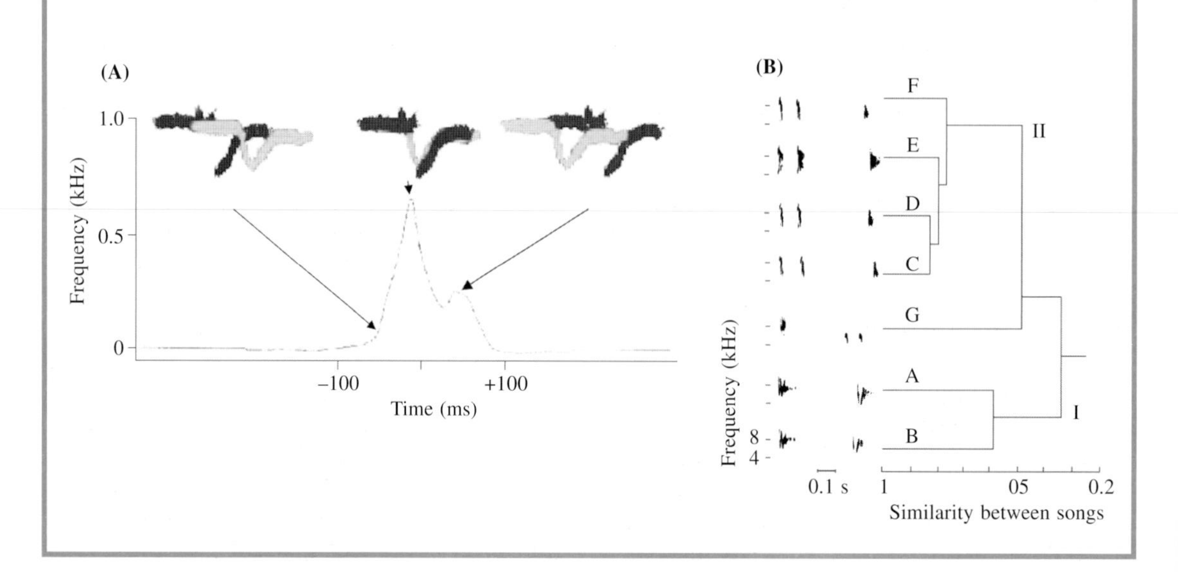

sounds continuously in a variety of remote locations, and radio the data back to a central processing center. Such data will be so voluminous as to be practically useless unless procedures are available to make the input stream manageable. A series of technological solutions to this problem are available or under development.

Individual Recognition by Voice

Our discussion thus far has focused on identification of species. Identification of individuals by voice can be important to conservation (McGregor & Peake 1998; Terry et al. 2001), permitting conservationists to compare the accuracy of census techniques and determine causes of differences between them. There are more thorough and subtle ways to determine the overall health of the population than just counts, as discussed above, in assessing minimum viable population size. For example, some members of a population reproduce more successfully than others (Newton 1989, 1995), and identifying the underlying sources of variation can aid in directing efforts to conserving these individuals and enhancing environmental factors favoring their success.

Traditionally, individuals, sex, and age classes have been determined with mark-recapture or mark-resighting techniques, which require at least an initial capture of each animal for marking. Individual vocal signatures have been documented for many passerine and non-passerine bird species (Lambrechts & Dhondt 1995), and vocal identification of sex and age classes has also been demonstrated for several species (Baptista & Gaunt 1997b; **Fig. 12.4**). Therefore, the use of a record–rerecord approach, with individual recognition by vocal signatures, has been proposed as preferable to indirect methods, such as collecting nests, pellets, and droppings; and intrusive techniques, such as banding and tagging with markers or transmitters. Individual recognition by voice is potentially powerful as a conservation tool as the basic research of ethologists has demonstrated, but collaborations between behaviorists and conservationists to evaluate the utility of those methods have been few, and actual implementation in any sustained conservation program has not been realized (McGregor et al. 2000).

The major challenge with this approach appears to be a technical one. When individual signature calls are sufficiently distinct, samples can readily be assigned to a known individual either by visual inspection and sorting of sonograms, or by more unbiased and automated computer approaches (Terry et al. 2001). For example, discriminant function analysis has been used successfully to match a sample of unknown calls with a library of identified sounds (Peake 1997; Peake et al. 1998; Peake & McGregor 1999). The record–rerecord method must, however, be able to recognize as such those sounds not made by any of the individuals represented in the library. Although a challenge, recent computer simulations have shown that, under some circumstances at least, a novel application of discriminant function analysis can correctly assign a set of individual sound samples to already known and previously unrecorded individuals (McCallum unpublished). This technique is being tested with a population of Bonin petrels, which are not amenable to more conventional techniques because they are active outside their deep nesting burrows only on dark nights, and because capture could unduly disturb them (M. Steinkamp personal communication).

Future of Automated Acoustic Species Recognition Systems

We can anticipate many new tools for acquiring sounds as the power of digital sound recording expands, the miniaturization of components increases, and costs for new technologies decline. The automated acoustic species-recognition systems, mention of which has punctuated this chapter, are awaited eagerly, although their availability, applicability, and generality are still a long way off. At this time, event detectors can be programmed to scan a digital sound file such as , for example, the downloaded one-hour sample from a digital tape recorder, and identify acoustic events meeting specific criteria of frequency, amplitude, and phrasing. These criteria can be species-specific, perhaps limited to the high

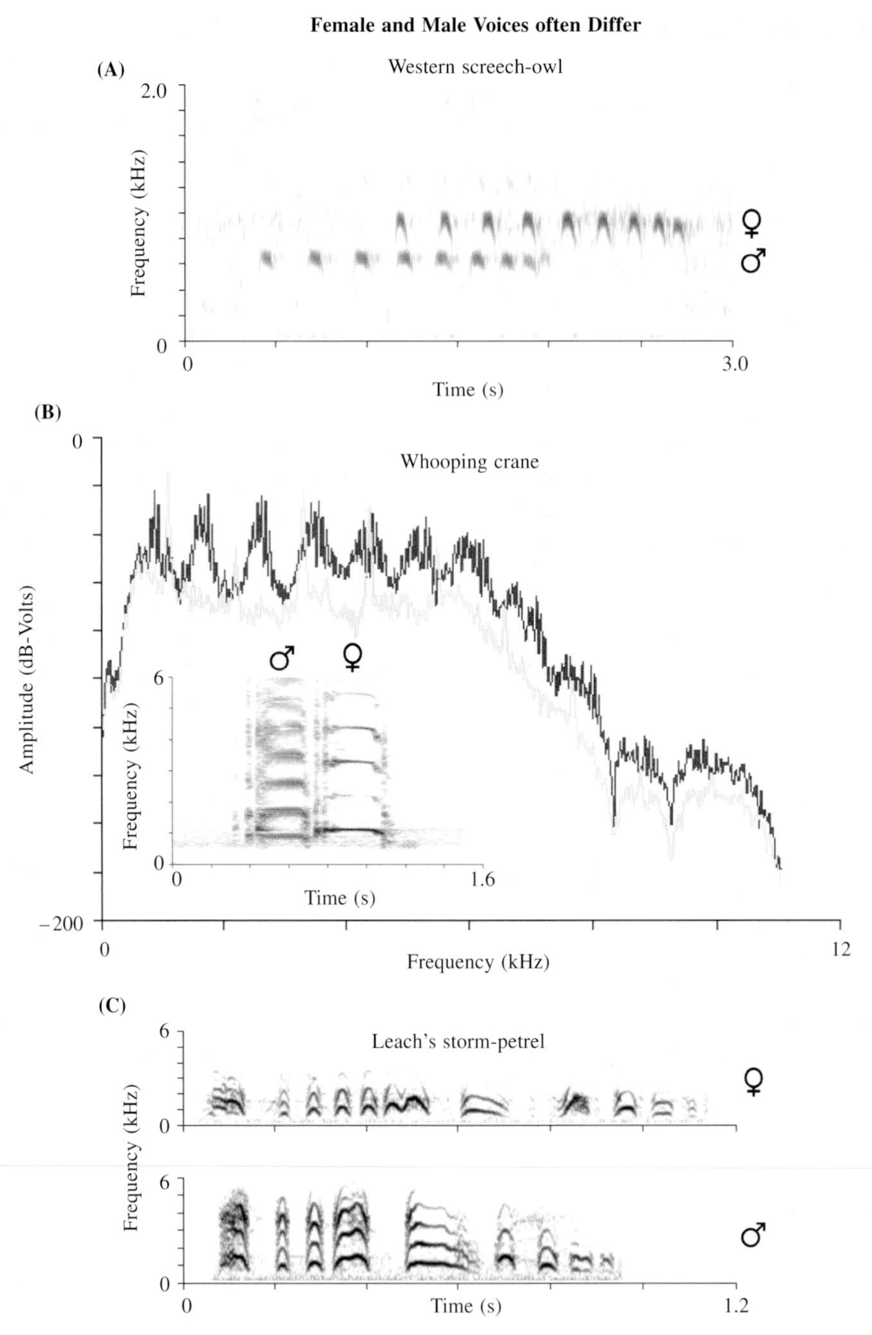

Figure 12.4 Sex by voice (**CD1 #94**): (**A**) the western screech-owl bounce call; the male is at a lower frequency, joined by a female at a higher frequency (Herting & Belthoff 2001). (**B**) The whooping crane guard call in the sonogram; a male call is followed by one from a female (Carlson & Trost 1992). The differences in the fundamental frequencies are multiplied in the harmonics, producing sounds that are easily distinguished by us, and presumably by cranes. The contrast is better seen in the power spectra above the sonogram. (**C**) Leach's storm-petrel; calls from the burrow of a female (top) and a male (bottom; Taoka et al. 1989).

frequencies characteristic of the song of the golden-crowned kinglet, or to phrasing, in which each motif is delivered in couplets, as in the brown thrasher. They can also be more generalized, as say events of 1.0–2.5 seconds in the frequency range of 2–6 kHz. The start/stop times registered by the event detector can be used to estimate the total singing time by a given species or to identify events of interest for direct inspection and further analysis.

The next evolutionary stage is the real-time event *detector*, which registers and saves to disk events meeting specified criteria as they occur. This approach saves disk space, because only desired events are recorded. The down side is that there is no way to check the accuracy or precision of the event detector's decisions. Spot checks must be made to make sure that the real-time event detector is performing as desired. Such detectors are still in the developmental stage and have been used only in a few pilot projects, all of which were targeted at a small number of species.

The holy grail of automated sampling is a real-time event *analyzer* that could detect and analyze all sounds passing through it. Many people unfamiliar with the details of acoustic analysis, and the numerous complexities of bird sounds, assume that progress in human voice-recognition systems portends the imminent availability of real-time event analyzers. Actually, there is little connection between the two. Despite the complexity of human language, it is a single language, and each of the approximately 9,000 bird species has its own, albeit simple, 'language.' Perhaps two-thirds of these, including most non-passerines except parrots and hummingbirds, plus most suboscine passerines, are so simple that they could, in principle, be programmed into a real-time event analyzer working on a supercomputer. The other third are much more varied and complex. Nonetheless, once years of basic research have been accomplished on the salient cues that identify a vocalization as to species, a real-time event analyzer might be able to identify any sample played to it in a laboratory. Even then, there will remain major technical hurdles, among them so-called overprint – the simultaneous vocalizing of more than one individual at similar frequencies. The human mind is adept at separating simultaneous vocalizations, but it will take major advances in artificial intelligence for computers to meet the human standard, which itself is not perfect.

Data Management and Archiving

We are a long way from having automated systems for recordists to manage their amassed data. The real work is not in the collecting but in dealing with the material collected. For the serial recording census techniques, it is imperative that written edits be made of the content of the tape. Information on the nature and timing of each event in minutes:seconds, as opposed to tape recorder indexes – which vary between machines, should be documented for easy retrieval. Manual edits can be simplified by narrating keywords directly onto the master field tape, like 'start,' 'pause,' and 'stop' during the recording sessions, and then using event detection to find these keywords and their associated narration while editing.

Today there are many commercially available database programs that can assist in the management of specimens and data on your accumulating collection of field tapes. The Macauley Library of Natural Sounds makes available a database management system developed in the program FoxPro^{tm} specifically for sound recordists. The Borror Laboratory of Bioacoustics is developing methods that will allow recordists to enter data directly on-line to its database for sounds destined for that archive.

No matter what the nature of the conservation project using sound recording techniques, those recordings obtained from field tapes are specimens and should be treated as any other biological specimen, including deposition for safe storage and general availability in a public museum or archive (Hardy 1984). Your recording efforts, once they have served your needs, may prove invaluable to others in unexpected ways, but only if they are documented and organized in such a way that they can be discovered and made available, which is exactly the mission of the sound archive. For example, recordings with adequate documentation could be used to extend the reach of the habitat–occupancy mapping pioneered with museum skins (see The Species Analyst, The University of Kansas Museum of Natural History, http://nhm.ku.edu), which would be especially useful because the bulk of the skins were collected many decades ago. Unfortunately, many recording efforts end up unusable because of lack of documentation, organization, and availability.

CONCLUSIONS

Although birdsong has played a major role in avian census work since the beginning of efforts to monitor bird populations, the introduction of formal bioacoustic analysis and electronic technology to the problem of monitoring is a recent and evolving phenomenon. As this review shows, the need for more and better data exists, the potential is great, the problems to be overcome are significant but not insuperable, and the work has begun in several parts of the world. We have the following recommendations to speed the coming of the revolution in bioacoustic monitoring. First, bioacousticians need to continue to educate the conservation community about the tools of the bioacoustic trade. Second, sound archives should continue to strive for more samples per species and more comprehensive collections, particularly with regard to non-song vocalizations, and all sounds of species from beyond the north temperate base of ornithological expertise. This requires support from government funding sources as a full player in the conservation biology-initiated resurgence in support of natural history museums. There is also a need to expand the collection of representative samples of natural soundscapes from around the world, and to increase efforts to make available database management and sound analysis techniques. Third, academic bioacousticians should enrich their research programs with projects that have conservation applications. Finally, we recommend that the recreational birding community should begin collecting and archiving sound recordings, thereby contributing even more significantly to the conservation enterprise.

Chapter 13

Grey parrots: learning and using speech

IRENE M. PEPPERBERG

INTRODUCTION

One might wonder how a chapter on parrots fits into a book about the intricacies and beauty of bird song. The raucous squawks of wild parrots and their reputation as mindless mimics in captivity would seem to exclude them from consideration on several grounds. In fact, parrots are intriguing candidates for many kinds of research. Unlike the subjects of most scientific experiments, they have been known, at least until recently, more for their role as companion animals than for their natural behavior patterns. They grace (or deface, depending on one's point of view) millions of homes worldwide yet, unlike other domesticated species, they are usually either wild-caught or only a single generation removed from the wild. The amazing facility of these creatures to adapt to captivity, to accept humans as their adopted flock, and to use their vocal and cognitive abilities to integrate themselves into what might be regarded as totally alien environments suggest that these birds have some hitherto unappreciated abilities.

Little is known about the natural behavior of the grey parrot of Africa. We do know that wild greys mimic other species (Cruikshank et al. 1993; **Box 46**, p. 364), perhaps more commonly than has been thought. Old World parrots such as the grey may or may not share some of the vocal characteristics of the more frequently studied species of the New World (Wright 1996; Bradbury et al. 2001; Wright & Dorin 2001; Bradbury 2003; May unpublished), but my studies suggest that the cognitive abilities of grey parrots, at least in captivity, have been underestimated, their mimetic abilities misunderstood, and that they share many behavior patterns in their vocal learning and production with both songbirds and even humans. In the following pages, I briefly describe some of my findings on the cognitive and communicative abilities of grey parrots.

COGNITIVE AND COMMUNICATIVE ABILITIES

Similarities with Other Birds and Humans

Whereas the phylogenetic closeness of great apes to humans and the large brains of cetaceans lead us to expect that their communicative and cognitive capacities may share characteristics with humans, we rarely expect analogous abilities in birds. Nevertheless, birds such as parrots, with brain architectures, evolutionary histories, and vocal tracts so different from those of humans (see Chapters 8 & 9) can learn to comprehend and produce meaningful human speech, and some of their information processing abilities are, I will argue, comparable to those of humans and other mammals. What is the extent of this behavior? Studies in the wild are difficult for a number of reasons; what can we learn about parrot commu-nication and cognition in the laboratory?

Although phylogenetically remote from

BOX 46

VOCAL MIMICRY BY WILD AFRICAN GREY PARROTS

The ability of captive African grey parrots to imitate heterospecific sounds is well known, from both popular anecdotes and scientific studies (Todt 1975; Rauch 1978; Pepperberg 1981; see Chapter 13). It has been assumed that this vocal mimicry does not occur in the wild (Nottebohm 1972; Kroodsma & Baylis 1982; Brosset & Erard 1986; Forshow 1989). This paradox was demolished by an opportunistic recording made in Zaire at Botshima (1°15' S, 22°E) in the Salonga National Park. A wild pair of African greys vocalizing together in an isolated tree included sounds of several other species in their vocal repertoire (**CD2 #81–86**). We distinguished 203 motifs of 88 different types in the four minutes of recording, analyzed with the help of Alick Cruikshank. Among them we recognized 46 cases of vocal mimicry, 23% of the entire sequence. We categorized them into sets associated with ten different model species; one is the call of a bat, the other nine belong to various species of birds (Cruikshank et al. 1993). In addition, analyses of four recordings made previously in Gabon and Ivory Coast indicate that vocal mimicry is widespread in wild populations of African grey parrots. From them we can add two more bird species to the models imitated in the Zairian recording.

Three classes of mimetic motifs can be recognized, according to whether parrots reproduce from the model song a single note (**A**), a part of it (**B**), or an altered version (**C**). This discovery is interesting from several points of view. The diversity of imitated models, and the widespread nature of the phenomenon confirms that parrots do indeed display this skill in the wild as they do in captivity. It also raises the question of why this discovery is so belated, in spite of the familiarity of grey parrots, and their wide distribution range in Africa. It is surprising that vocal mimicry has been overlooked by previous observers. The observation has some socio-sexual significance: the prolonged vocal sequences of the two birds may indicate that they are singing or duetting as paired birds; we have frequently made similar observations, notably in Gabon, suggesting that this type of vocal exchange between pairs of parrots is not rare. Finally, the fact that two imitated species are allopatric birds (**A-1 &-2**) signifies that African grey parrots either travel over long distances to wherever they hear these species, or learn from other parrots in communal roosting sites. Above all, these facts attest to the paucity of our knowledge about the vocal behavior of wild parrots, and the need for new observations using modern techniques.

Jean-Pierre Gautier & Claude Chappuis

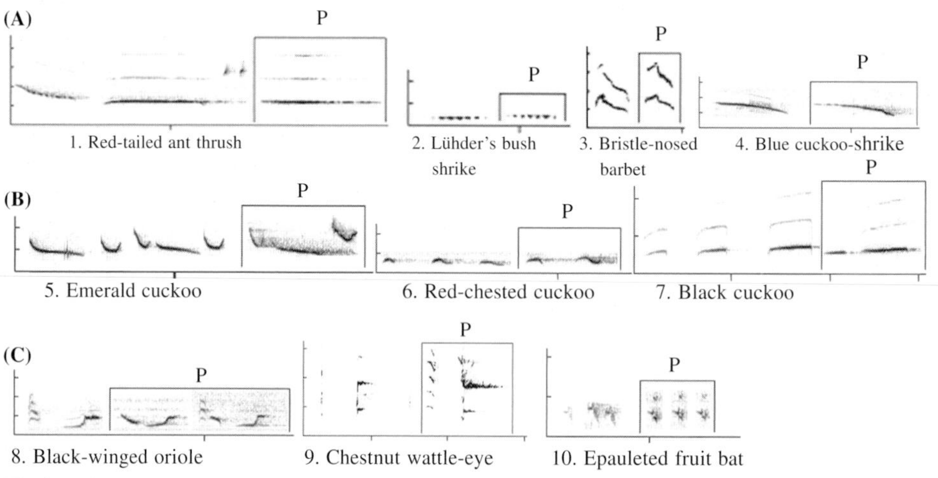

The boxed sonograms are imitations by a pair of African grey parrots (P) of ten other species in Zaire, also illustrated, taken from Chappuis 2001, including one bat. The frequency axis is in 2 kHz increments. The time marker is one second, varying between sonograms.

humans, grey parrots have several cognitive abilities that resemble those of humans. I have evidence that greys learn simple vocal syntactic patterns and can use English words in a referential fashion; their abilities to deal with certain tasks, whether these involve label acquisition, numerical competence, conjunction, or recursion, are in some ways comparable to those of young children (Pepperberg 1981, 1994a, 1996, 1999; Pepperberg & Wilcox 2000; Pepperberg & Shive 2001). The oldest subject, Alex, uses words to label more than 50 objects, 7 colors, 5 shapes, quantity to 6, 3 categories such as material, color, and shape, and uses "no," "come here," "wanna go X," and "want Y," X and Y being labels for an appropriate location or object (**Fig. 13.1; CD1 #95**). Altogether he uses labels for about 100 items (Pepperberg 1990a). He responds appropriately to queries from caretakers about the category of an object, its relative size, the absolute quantity of a collection, and the presence or absence of similarities and differences between

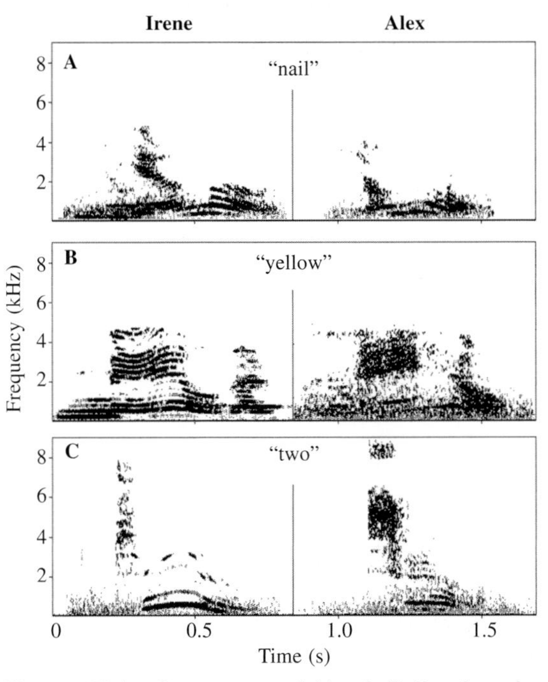

Figure 13.1 Sonograms of Alex imitating Irene's speech (**CD1 #95**).

objects; he comprehends the labels he produces and understands the difference between 'labeling' and 'requesting' (Pepperberg 1999). He and other greys understand object permanence, the fact that a hidden item does not cease to exist (Pepperberg & Kozak 1986; Pepperberg & Funk 1990), and are able to track complex movements of hidden items (Pepperberg et al. 1997). Grey parrots can also use mirrors to find hidden objects (Pepperberg et al. 1995). Cumulatively, the evidence suggests that Alex and other greys exhibit capacities often presumed to be limited to humans and other primates.

Parrots and humans, like songbirds, share aspects of sound production. Grey parrots engage in considerable sound play, including phonetic 'babbling' and recombination (Pepperberg et al. 1991), and use such abilities to construct new speech patterns from existing ones (Pepperberg 1990c); they appear to represent their labels acoustically as humans do, and develop phonetic categories (Patterson & Pepperberg 1994, 1998). Data from such sources as X-ray video (Warren et al. 1996) suggest that Alex uses 'anticipatory coarticulation,' setting up the vocal tract for the next speech sound as the previous one is completed; thus he appears to separate specific phonemes from the speech flow *and* to produce them in such a way as to facilitate production of upcoming phonemes; so, for example, his /k/ in "key" differs from that in "cork" (Patterson & Pepperberg 1998). In humans, these abilities are taken as evidence for *top-down processing* (Ladefoged 1982). Alex also can recombine labels in novel ways to respond to novel situations and transfer them to new contexts (Pepperberg & Brezinsky 1991).

In the laboratory we know that parrots can learn from each other (Pepperberg et al. 2000), but little is known as yet of the vocal behavior of grey parrots in the wild. Preliminary evidence (May unpublished) suggests that greys' behavior patterns match those of some other parrots. Other parrots appear to have dialects (**Box** 47, p. 366) and to use vocalizations to maintain flock cohesion (Wright 1996; Baker 2000; Bradbury et al. 2001; Bradbury 2003; Vehrencamp et al.

BOX 47

CALL DIALECTS OF PARROTS

The ability of parrots to mimic sounds from their environment is widely known, mostly from observations of captive birds. In wild parrots, such an evolved ability must serve an important function in communication, and the most pertinent environmental sounds must come from conspecific companions. In many parrot species there are large fluctuations in group size through the course of a day. They are notoriously noisy, and it seems likely that their vocalizations serve to signal group identity, and to maintain social cohesion. Within-group imitation of vocal signals often leads to local dialects, with sharing of signal features among individuals within a population, and distinct differences between populations. Parrot repertoires usually include a vocal signal designated as a 'contact' call, used by a mated pair to stay together, and also uttered by all the birds when in proximity to the night roost of the flock. As a vocalization often heard and loudly delivered, the contact call is an inviting target for field investigations.

One of the first major efforts to look for geographic variation in vocal signals in parrots was carried out on the yellow-naped Amazon in Costa Rica (Wright 1996). Over two large geographic areas, these birds are distributed in a number of different flocks. Each flock (20–200 birds) returns to its traditional roosting location in the evening, and from there the birds range out each morning to feed. Wright found that these two regional populations, each encompassing multiple roosts, had distinct contact calls. In addition, within a region, the roosts sampled differed more subtly in the acoustic features of their contact calls, exhibiting roost-specific dialects (**CD2 #87**).

Contact calls of ringneck parrots in western Australia also differ regionally (Baker 2000), with two large dialects in one study area (**CD2 #88**). Within a region, different roosts were vocally distinct, as in the yellow-naped Amazon. In a sense, both species have dialects within dialects, as occurs, of course, in human speech. As far as is known, both of these parrot species mate monogamously for life, and the male and female of a pair stay in close proximity through the course of each day, ranging over many kilometers as they forage. In the ringneck parrot, the pair often forms temporary associations with other pairs and unmated birds and, during such forays, there are many vocal interactions. The contact call is uttered whenever the pair flies from place to place. Typically, as one bird flies out from a foraging site, or as a group breaks up, the contact call is uttered and the mate quickly follows also giving the contact call. There is some evidence that these calls have features unique to each pair, allowing mates to recognize each other even in a crowd of other ringneck parrots, as occurs at the roost in the evening. Such vocal signatures also seem to convey roost affiliation for the group as a whole. Thus, within a single call type, significant social information may be communicated about pair identity, roost affiliation, and membership in a regional dialect.

Myron C. Baker

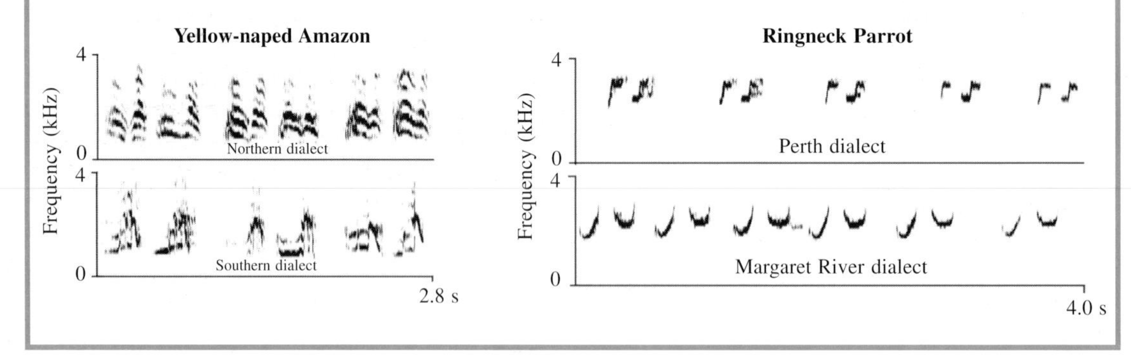

2003), establish pair-bonds and recognize specific individuals (Gnam 1988); they have vocal sentinel behavior (Levinson 1980; Yamashita 1987), and are known to engage in complex pair-bond duets (Nottebohm 1972). Whether greys, like some Amazon parrots, have dialects, and alter calls when changing dialect areas (Wright & Dorin 2001) is as yet unknown. Long-lived, they reside in large groups (May & Lynn unpublished), the social organization of which has yet to be established. The apparent complexity of parrots' communicative abilities led me to examine not only their competence but also the learning processes involved in achieving such competence.

Vocal Practice and Monologue Speech

Like children (Kuczaj 1983; Nelson, K. 1989) and songbirds (Baptista 1983), Alex engages in active vocal practice. In addition to sound play, involving repetitions and variations on a set of tones, this practice includes something termed 'monologue speech' – active practice of a communication code in situations without overt social stimulation (Kuczaj 1983; Pepperberg et al. 1991). There are two kinds of monologue speech, *private speech* produced in solitude and *social-context speech* produced in the presence of potential receivers but without obvious communicative purpose. Alone in his cage, Alex engages in the former (Pepperberg et al. 1991), which explains some of his acquisition data. Although most of his words appear initially as rudimentary patterns that improve over the course of training sessions – first as a vocal contour, then with vowels, finally with consonants (Patterson & Pepperberg 1994, 1998) – sometimes completely formed new words materialize after minimal training and without apparent overt practice in his trainers' presence. By eaves-dropping, we found that he was practicing these labels privately. In contrast, social-context speech appears when, with trainers nearby, he often recombines labels or label parts in their corresponding orders, keeping the beginnings and endings of labels appropriately. If we provide acceptable corresponding objects for these novel vocalizations, thereby *referentially mapping* his speech, these innovations become part of his repertoire (Pepperberg 1990c).

Such behavior patterns are presumably integral to development and, because they occur across species, may be taken to support or contribute to an evolutionary theory of language play (Kuczaj 1998). Like most play, monologue speech facilitates learning by allowing experimentation with adult systems free of the consequences of failure. It allows full freedom to choose topics and contexts, to attempt novel words, and compare familiar and novel forms to extract regularities and discover differences (Kuczaj 1983). The lack of negative feedback, with no need to use correct forms in monologues, may also encourage practice and accelerate learning.

In private, Alex also sometimes reconstructed and reinvented scenarios absent from formal training, similar to behavior observed in children (Weir 1962). Monologues included utterances from daily routines, such as "you go gym," "want some water," as well as strings involving often-heard patterns, such as "you be good, gonna go eat lunch, I'll be back tomorrow." Sometimes, question–answer dialogues also emerged, such as "snap, snap, snap, snap," "how many?," "Four." Interestingly, Kyaaro, another grey in training who was less accomplished than Alex, used entire dialogues in solitude, reproducing two different trainers' voices and synthesizer sounds, as well as using his own voice; a typical example would be as follows: "listen, Kyo" (in the voice of trainer 1). "Click, click, click, click" (replicating the synthesizer). "How many?" (in the voice of trainer 2). "Four" (in his own voice). "Good boy!" (trainer 1 or 2).

Referential Mapping

In addition to monologue speech, Alex also produced new vocalizations by recombining parts of existing utterances. His spontaneous novel phonemic combinations often occurred outside of testing and training procedures (Pepperberg 1990c). Juvenile greys behaved similarly (Neal 1996), with combinations appearing in contexts and forms reminiscent of children's play. These utterances were rarely if ever used by trainers, but sometimes resembled both existing labels and separate human vocalizations, such as "grain" from "grey." Interestingly, after analyzing over 22,000 imitations of words by Alex, we never observed 'backwards' combination, such as "percup" rather than "cupper" (Pepperberg et al. 1991). We infer that Alex combined pieces of labels by abstracting rules for beginnings and endings of utterances. His behavior thus suggests, although it does not prove, that he parses human sound streams in human-like ways, acoustically representing labels as do humans, using similar phonetic categories (Patterson & Pepperberg 1994, 1998): this is additional behavior that some regard as consistent with top–down processing (Ladefoged 1982).

When we followed up on spontaneous utterances by referentially mapping them, providing relevant objects to which they could refer, Alex rapidly began to use these labels routinely to identify or request appropriate items. So, for "grain," trainers gave Alex seed, something not normally available, and talked to one another about "grain" and identified it; later sprouted legumes were substituted. Alex received a ring of paper clips for saying "chain" spontaneously, and was given the appropriate fruit for "grape" and a nutmeg grater for trimming his beak after he uttered "grate." "Cup" (from "up") was rewarded with metal cups and plastic mugs, "copper" (first produced as "cupper") with pennies, and "block" with cubical wooden beads. "Chalk" (from "talk") was mapped to blackboard chalk of various colors and lengths; "truck" to toy cars and trucks. After we ignored "cane," "shane," and "cheenut;" he abandoned these words (Pepperberg 1990c).

Thus spontaneous utterances that initially lacked language value appeared to acquire such value if caretakers interpreted them as meaningful and intentional and, our subjects, just like children, reacted positively to this interpretation (Pepperberg 1990c). The birds behaved as though repeated interactions serve to 'conventionalize' both the sound patterns and the sound-meaning connections in the direction of standard communication. Somewhat like children engaged in testing the meaning of their newly acquired labels (Brown 1958, 1973), grey parrots use humans to provide referential information for relatively novel labels (Pepperberg 2001). For example, our youngest subject, Arthur, took a label used in a very specific context – "wool" for a woolen pompon – and pulled at a trainer's sweater while uttering it. The bird seemed to be testing the situation, and our responses, in this case signs of high affect and excitement, seemed to stimulate him further, revealing the potential power of an utterance and encouraging him in early categorization attempts. Even if birds err in initial categorizations, they are reinforced when a correct, new label for the item is provided: we state that an almond isn't "cork," but a "cork nut" (Pepperberg 1999). My view is that parrots, like children, have a repertoire of desires and purposes, driving them to form and test ideas in dealing with the world; these ideas may be the first stages of representation and categorization in cognitive processing (Pepperberg 2001).

TRAINING PARROTS TO LEARN AND USE REFERENTIAL SPEECH

The popular view of parrots is that they effortlessly acquire new sounds of any origin, but such is not the case. Pet parrots do indeed sometimes mimic microwave beeps and their owner's morning greeting, but several previous researchers (Mowrer 1950; Grosslight & Zaynor 1967; Pepperberg 1999) were unable to teach mimetic birds to speak using standard conditioning techniques. The reason for this failure may be that parrots engage in more than one type of vocal learning, each requiring different types of input. I view the simple mimicry of a sound used only to attract the attention of another individual as a different process from the acquisition of vocalizations that have communicative content (Pepperberg 2002).

Like young children (Hollich et al. 2000), grey parrots acquire communication skills most effectively when the method for teaching them is referential, functional, and socially rich (Pepperberg 1997; Pepperberg et al. 1998, 1999, 2000). The term reference relates to the meaning of an utterance – the relationship between labels and the objects to which they refer – and it is exemplified by the presentation to the bird, during training, of the specific objects that it names; this object thus serves as the primary reward for vocalizing appropriately. A second key factor is functionality, which involves the context in which an utterance is used and effects of its use; the opportunity to use labels during training as requests for objects gives the bird reason to learn the unique, unfamiliar sounds of human speech. The third factor is social interaction between subject and trainer during training. This behavior helps to focus the bird's attention on certain components of the environment, emphasizes common attributes of objects or actions and thus possible underlying rules for their use, and allows trainers to adjust procedures to a learner's level. Interaction with trainers engages subjects directly, provides a contextual explanation for actions, and demonstrates the consequences of those actions. After establishing the effectiveness of my primary training method, the model/rival technique, a decade-long effort was made to understand why it was so much more successful than other procedures.

The Model/Rival Technique

In the 1970s, after reviewing what was then known about vocal learning in parrots, language acquisition in children, and psychological constructs of social learning in humans (Pepperberg 1999), I devised a method of instruction, termed the *model/rival* (M/R) training system. The approach efficiently teaches a parrot both how to speak and how to use its speech appropriately. It is based on methods developed by Bandura (1971) and Todt (1975), and it uses three-way *social* interactions among two humans and a parrot, as a way to demonstrate the vocal behavior that is to be learned (Pepperberg 1981). The parrot observes two humans handling and speaking about one or more objects and how these individuals interact with each other. As the parrot watches and listens, one trainer presents objects and queries the other trainer about them, with such expressions as "what's here?," "what color?," giving praise and transferring the named object to the human partner as a reward for correct answers. Incorrect responses are punished by scolding and by temporarily removing items from sight. Thus the second human serves both as a model for the parrot's responses and its rival for the trainer's attention, and also illustrates the consequences of errors. The model must try again or talk more clearly if the response was deliberately made incorrectly or garbled; that is, the model is subject to the process of corrective feedback and the bird observes it. The parrot is also included in the interactions: it is queried and rewarded for successive approximations to correct responses, and training is adjusted to its performance level. If a bird is inattentive or its accuracy regresses, trainers threaten to leave.

Unlike M/R procedures others have used (Pepperberg 1999; Pepperberg & Sherman 2000), we *interchange* roles of trainer and model, and include the parrot in interactions. This procedure emphasizes that a questioner is sometimes a respondent, and demonstrates that the procedure can effect environmental change. Role reversal also counteracts an earlier methodological problem: birds whose trainers always maintained their respective roles responded only to the particular human questioner (Todt 1975). With our technique, birds will respond to, interact with, and learn from any human.

M/R training uses only *intrinsic reinforcers*: to ensure the closest possible linkage between labels or concepts to be learned and their appropriate referent, the reward for uttering "X" is access to X, the object to which the label or concept refers. Earlier unsuccessful attempts to teach birds to communicate with humans used *extrinsic* rewards (Pepperberg 1999): a single food was used that was neither related to, nor varied with, the label or concept being taught. This procedure delayed label and concept acquisition by confounding the label of the targeted exemplar or concept with that of the food reward. We never used extrinsic rewards. Use of the label to request the item from the start also demonstrates to the bird that uttering labels is functionally useful.

To examine Alex's cognitive capacities, we sometimes had to teach him to reclassify familiar objects or introduce items in which he had little interest. In such instances, Alex might fail to focus on the designated objects. We therefore trained him on "I want X," encouraging him to separate labeling and requesting (Pepperberg 1988); his reward then became the right to request something more desirable than the one he identified, a technique that provides flexibility but maintains referentiality. To receive X after identifying Y in this protocol, Alex must state "I want X" and trainers will not comply until the task concerning Y is completed. Thus the words he uses as labels are true identifiers. Adding "want" provides additional advantages: first, trainers can distinguish incorrect labeling from appeals for other items, particularly during testing, when a bird unable to use "want" might not be erring but asking for treats and its scores might thus not reflect true competence. Second, birds may demonstrate low-level intentionality: Alex rarely accepted substitutes when requesting X, but rather continued his original demands (Pepperberg 1987b, 1999).

M/R training successfully enabled label and concept acquisition, but did not specify which elements of training were *necessary and sufficient*. The earlier unsuccessful attempts to teach birds to speak referentially were clearly lacking in at least some of these elements, but did not attempt to deconstruct the learning environment, or test the significance of different input elements individually and in various combinations. To try to learn why the M/R technique was so successful, we began to test how changes might affect learning not only with respect to sound reproduction, but also with respect to comprehension and appropriate use, focusing on actions that require cognitive processing.

Eliminating Aspects of M/R Training

To provide training input that varied with respect to reference, functionality, and social interaction, we contrasted M/R tutoring with sessions using different forms of human interaction, videos, and audiotapes. Note that most of these tests required additional subjects: had we significantly changed Alex's training protocols, he might have ceased to learn simply because we had made a change, not because of the quality of the change. Our new subjects had no history of prior training. We performed one M/R experiment (variant 1) with Alex and several others with Kyaaro, and two other naïve greys, Alo and Griffin.

'M/R-variant 1' eliminated as much reference and functionality as possible while still retaining the bare bones of a two-trainer system (Pepperberg 1994b). Two humans enacted the same roles as in the basic M/R procedure, but interacted with each other without referring to specific objects or collections. Humans uttered a vocal sequence based on Korean number labels, without reference either to specific items or to Alex's previously acquired English numbers (Pepperberg 1994a, 1997); he saw only a string of pictured numerals, and no sound was attached to a particular numeral. The Korean-based words, required for a simultaneous study on numerical competence and serial learning (Silverstone 1989; Pepperberg & Silverstone unpublished), provided a maximal contrast with English. Training lacked functional meaning and all but minimal referentiality, because the pictured numbers were not specifically or individually referenced, and were used simply as a focus of joint attention. As in the basic M/R procedure, roles were reversed and Alex was included in the interactions. Although Alex is usually rewarded either with an item he has labeled *or* the opportunity to request a favored item ("I want X," Pepperberg 1987a), here we used only the latter reward and vocal praise; errors elicited scolding and time-outs. This is the only nonstandard training procedure in which new words were acquired, but, importantly, they were acquired without any reference.

In experiments with the juveniles Alo and Kyaaro (Pepperberg 1994b), each bird received training in a set of two conditions in social isolation: one set of labels was presented via audiotapes of Alex's sessions, so that input was non-referential, not contextually applicable, and noninteractive; another set of labels was presented via videotapes that demonstrated how Alex and trainers receive referential rewards for labeling, such that input was referential, non-interactive, and minimally functional and contextually applicable. In both cases, nothing the juveniles uttered could result in a reward. Here, despite many precautions to ensure that other aspects of training remained constant, no new words were acquired. In contrast, both birds learned referential labels in concomitant M/R sessions, showing that the competence of the birds was not an issue.

Several experiments then explored video training in more detail. One investigated how interaction with a co-viewer might affect learning from videos (Pepperberg et al. 1998). A co-viewer provided social approbation for viewing, and pointed to the screen while making comments like "look what Alex has!" but did not repeat targeted words, ask questions, or relate content to other training. The plan was that birds' attempts to utter a label would garner only vocal praise, but they didn't make any attempts. Social interaction was limited; referentiality and functionality were the same as in earlier videotape sessions. In a second video experiment the co-viewer was more interactive, uttering targeted labels and asking questions (Pepperberg et al. 1999). In yet a third, a socially isolated parrot watched videos while a student in another room monitored its utterances through headphones and could deliver intrinsic rewards remotely (Pepperberg et al. 1998). In a fourth experiment, we used live video from simultaneous sessions with Alex to minimize habituation. None of these procedures resulted in the acquisition of new words though, again, the parrots simultaneously learned labels presented in traditional M/R sessions.

We also examined the role of joint focusing of attention between bird and trainer; Baldwin (1991) provides parallels with children's learning. A single trainer faced away from the bird, which was within reach of, for example, a key, and talked about the object, emphasizing its label, "look, a shiny *key*!," "do you want *key*?," etc. The trainer used sentences that framed the label, allowing for repeated use of the label while minimizing possible habituation (Pepperberg 1981), but had no visual or physical contact with the parrot or the object. If a bird attempted to utter the targeted label it would receive only vocal praise, so that some functionality and considerable social interaction were eliminated (Pepperberg & McLaughlin 1996). Again, no new word learning occurred.

In another experiment, with Griffin, we just eliminated interactive aspects of modeling. A single student labeled objects and queried the bird while jointly attending to objects, interacting fully with the bird and the objects (Pepperberg et al. 2000). Griffin did not utter any of the targeted labels during or immediately after 50 such sessions, but produced them after two or three subsequent M/R sessions. Because birds that were switched to M/R training after 50 video sessions needed only about 20 sessions before beginning to use targeted labels appropriately (Pepperberg 1999), which is approximately the same number of sessions needed when labels are introduced initially in M/R training, I suspect that Griffin had in fact acquired labels during the solo-trainer sessions, but did not understand how to use them until he had observed their usage modeled. But, importantly, during the experiment itself no new words were uttered.

In sum, the parrots did not acquire referential speech by any technique other than the original M/R training. For grey parrots, at least, learning to speak meaningfully seems to occur most readily when training involves two humans who demonstrate the referentiality and functionality of a label to be learned, socially interact with each other and the bird, exchange roles of questioner and respondent, portray the effects of labeling errors, provide corrective feedback, and adjust the level of training as the subject begins to learn. When training lacks some or all of these elements, birds typically fail to learn much at all.

If learning does occur under conditions of impoverished input, it seems to be limited to simple mimetic speech, as exemplified by the results of Alex's M/R-variant 1 training, in which we removed reference and, with it, some functionality. He learned to produce vocalizations – the string of vocal labels that were modeled – but the results of the experiment differed from those of M/R studies in two important ways. First, acquisition took nine months, which was unusually long (Pepperberg 1981; Silverstone 1989; Pepperberg 1994b). Second, and most strikingly, he was not immediately capable of using these sounds properly as labels; he could not learn to use them in a referential manner, with respect to either serial labeling or quantity. Even after we modeled one-on-one correspondence between eight objects and the string of labels, he was unable to partition the string into its individual elements to refer to smaller quantities, as when presented with four items and asked to "say number."

Apparently, Alex had thus learned to produce, but not to comprehend, the use of these sounds (Pepperberg 1994b), a bit like a child who, having learned the 'alphabet song' thinks 'LMNOP' is a single letter. Given his previous success on both production and comprehension of words after M/R training (Pepperberg 1990a, b, 1992, 1994a), we assume that Alex's failure was a consequence of the training protocol (Pepperberg 1994b). These experiments suggest that goal-directed imitation may be different from 'mere mimicry.' The M/R results suggest that training that engenders simple mimicry is usually incapable of engendering full comprehension, with the implication that different cognitive processes are involved in the two types of learning.

Mutual Exclusivity

Changes in input patterning may have other, subtler, effects. There is some evidence from Griffin of what in children is called mutual exclusivity (ME; Pepperberg & Wilcox 2000). ME refers to the assumption, briefly held by most children during early word acquisition, that each object has one, and *only* one label. Thus children will argue that their pet cannot be an 'animal' because it is a 'dog' (Merriman 1991; Liittschwager & Markman 1994). Along with the 'whole object assumption,' which argues that labels generally refer to entire objects, and not to some feature of them (Macnamara 1982), ME supposedly guides children in the earliest stages of acquiring labels for objects. Later, ME may help children interpret novel words as feature labels, overcoming the whole object assumption. For example, ME may help a child understand that the novel label 'taupe,' when used to refer to a ball, can be interpreted as its color (Markman 1990). Nevertheless, very young children sometimes find second labels for items more difficult to acquire than the first, because the second label, such as its color, is viewed as an alternative. Input, however, affects ME; children (Gottfried & Tonks 1996) and parrots like Alex, who receive inclusivity data (X is a kind of Y; e.g. color labels taught as additional, not alternative, labels, "here's a key; it's a green key"), generally accept multiple labels for items and form hierarchical relations. Thus, shown a wooden block, Alex can answer a number of different questions about it, such as "what color?," "what shape?," "what matter?," *and* "what toy?" (Pepperberg 1990a). But parrots given color or shape as an alternative label (e.g. "here's key", later "it's green"), like children, exhibit ME (Pepperberg & Wilcox 2000). Thus Griffin, given the latter input, exhibited ME and answered "what color?" with previously learned object labels when shown variously hued woolen pompons, wooden sticks, and other items in over 50 training sessions; for him, the pompon couldn't be labeled 'blue' because it was already labeled 'wool.' Similarly, while learning an object label – *cup* – he answered "what toy?" with colors of the various cups used in training and he had difficulty acquiring the label 'cup' (Pepperberg & Wilcox 2000).

CONCLUSIONS

These studies of the acquisition of referential vocal communication by parrots may ultimately provoke more questions than answers. How do creatures so phylogenetically distant from humans as parrots learn to pick up on human behavior and learn to speak meaningfully, both in training and in less structured social situations? Are such skills part of a rich convergent evolution in a wide variety of social creatures, adaptable to the particular social circumstances in which they find themselves? Could it be that other social creatures than ourselves are capable of more than we expect? Only an open mind on the part of researchers and intensive investigative study will answer these questions. Parrots present us with a unique opportunity to re-examine some of our beliefs, prejudices, and expectations about avian abilities.

Chapter 14

Singing, socializing, and the music effect

MEREDITH J. WEST, ANDREW P. KING, AND MICHAEL H. GOLDSTEIN

INTRODUCTION

Close Encounters of the Third Kind

Scene: the Wyoming desert, with military and scientific experts standing near a gigantic keyboard while gazing at a hovering spaceship (Spielberg 1977). The conversation is as follows:

> *Give her six quavers, then pause. She sent us four quavers, a group of five quavers, a group of four semi-quavers...*
> *What are we saying to each other?*
> *It seems they're trying to teach us a basic tonal vocabulary.*
> *It's the first day of school, fellas.*
> *Take everything from the lady.*
> *Follow her pattern note for note.*
> *Start with the tone:*
> *Up a full tone.*
> *Down a major third.*
> *Now drop an octave.*
> *Up a perfect fifth.*

In this chapter, we consider the ways in which vocal and other forms of music may function to foster communication between otherwise disconnected individuals. We begin with a metaphorical example. The quotation above is from Steven Spielberg's movie and book, *Close Encounters of the Third Kind* (Spielberg 1977). In the story, scientists and government agents attempt to communicate with the alien crew of an approaching spaceship to effect an actual meeting after several close sightings. A 'third' kind of encounter is one involving actual contact, as opposed to seeing an alien ship or finding physical evidence of its presence on Earth. Their link is a synthesizer playing a simple five-note melody to the spaceship. When the space crew first matches melodies with the crew on Earth, a ripple of recognition overtakes the people on the ground, telling an outside observer that embryonic communication is stirring. When the space ship then negotiates a riff on part of the melody, both in sound and in light, a sense of communion settles over the crowd. The improvisation reveals the higher order properties of the melody, unmasking a common code and conveying the frisson felt at this fictional intergalactic moment.

The word 'communication' has at its roots the idea of community and sharing. In this chapter, we contend that part of the fascination with nature's sounds and music is the opportunity for shared moments. We do not try to solve the riddle of why such sharing is important or feels good. Instead, we focus on some components of the process by which that sensation of sharing is achieved. In all cases, we focus on individuals getting to know one another. First, we discuss the social dynamics of singing and listening in brown-headed cowbirds. In cowbirds, we see how a brief song, such as a five-note melody, can serve as the foundation for messages as simple but influential as "You and I are alike", or "You're the one." Second, we turn to human infants and their caregivers as they go through the process of getting acquainted, where individuals with very different skill levels use musical sounds not to signal recognition, but involvement. Musical behavior by parents and children is a means of sharing knowledge and action, a way to say, "I am paying attention to you." Finally, we look at humans and European starlings in an effort to dissect further how music can function to forge bonds between alien taxa, in this case, humans and birds. Here, the important concept is that of creating a shared context where the sounds or music serve as the link for the text, "I understand and I remember."

SONG, GESTURE, AND COWBIRD RELATIONSHIPS

The Role of Song in a Brood Parasite

Interesting problems of recognition and communication occur on Earth without the need to consider the extraterrestrial dimension. When traveling to another country, especially if the language is different, it can seem like the first day of school all over again. Nonverbal communication becomes an even more important medium during everyday speech – from pointing to pantomime to body posture. These codes, however, are not immune from cultural influence: rules exist regarding proximity and the meaning of certain gestures (Eibl-Eibesfeldt 1972). Successful use requires social learning, including observation, imitation, and trial-and-error (Galef 1988). Nonverbal, or at least non-vocal, codes are also not unique to humans – we have been studying a songbird that uses a combination of melodies and non-vocal gestures (West & King 1988b; King et al. in press). The combination bears striking resemblances to the ways in which humans compound these communicative forms.

Brown-headed cowbirds are brood parasites. The females lay their eggs in the nests of many other species and the young are never fed or cared for by their natural parents (Friedmann 1929). There are many questions that stem from the cowbird's unusual start in life and its deviation from parenting its own young. Many questions involve close encounters (Mayr 1974). How do cowbirds meet? How do males and females get together? Here we focus on the dilemma it poses for communication among cowbirds, especially with respect to courtship and mating: how do cowbirds assess members of their own species? And, given that males provide no resources other than sperm, how does a female choose a healthy and vigorous male? She cannot compare territories or nest sites, the quality of food, or signs of paternal behavior.

As it turns out, the one-second song of the male cowbird is part of the answer to both of these questions: finding a mate and discriminating among potential competitors (King & West 1977). Another part of the answer is how the song is combined with non-vocal behavior on the part of the singer and the recipient. In essence, what cowbirds orchestrate are performances, context-specific presentations of a species-typical melody, along with species-typical visual signals. To do so requires more than the male's ability to sing or the female's ability to discriminate among songs; it requires careful staging and sensitivity to each other's social reactions.

The cowbird's song is the beginning of the story: it is a one-second vocalization with unique features. According to Greenewalt (1968), "this undistinguished bird, of unprepossessing appearance and habits is the undisputed winner in the decathlon of avian vocalization" (p. 119; **Fig. 14.1; CD1 #96**). Among the cowbirds' achievements in Greenewalt's extensive survey are: (i) the widest frequency margin between the two song phrases, covering nearly four octaves; (ii) the highest maximum frequency of about 10.7 kHz; (iii) a frequency spread of two octaves between the two voices, exceeded only by the American bittern; (iv) one of the shortest notes, less than 2 ms; (v) the most rapid glissando covering 5–8 kHz in 4 ms; and (vi) a modulating frequency in the 'high' voice exceeding 700 Hz. Allan & Suthers (1994) and Suthers (see Chapter 9) give more information on how singing comes about. Greenewalt (1968) confessed he had no idea as to the functional need for such

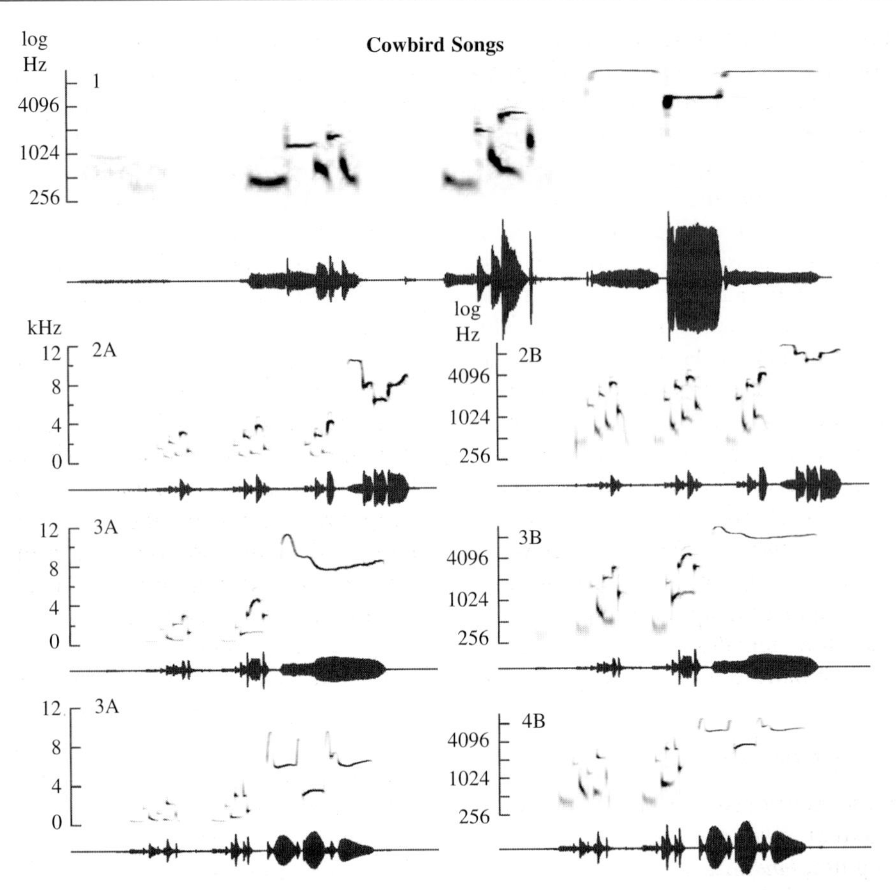

Figure 14.1 Sonograms and amplitude traces of brown-headed cowbird song. Song (1) is calibrated on a log scale to show the fine detail in the introductory notes, the part of the song to which the female is most sensitive: songs 2, 3, and 4 show linear (**A**) and log (**B**) renditions of three additional male cowbird songs (**CD1 #96**).

accomplishments, deferring to 'madam cowbird' as a source of possible answers.

As it turns out, female cowbirds can give some answers to Greenewalt's question regarding function, but the song is not a lock-and-key mechanism to insure reproduction between males and females. In 1977, we thought we had discovered such a failsafe mechanism, one that seemed to solve many of the cowbird's problems of mate identification, as well as show the song's critical connection during courtship (King & West 1977). In our work, we studied captive female cowbirds, in breeding condition, as judged by egg laying, but deprived of male companions. In typical courtship, a female would hear hundreds of songs in the first hours after dawn. But, in our setting, the females lived without males in sound attenuation chambers (picture a 1 m^3 FM radio booth) and heard only six songs per day. When we played back cowbird song to a female, as opposed to the song of another species, the sound literally went in her ear and down her spine. She arched her body and separated the feathers on her back, assuming the stereotypical stance used when mating (see Chapter 2). If one were looking for a graphic example of a melody creating chills or thrills, one need look no further. The response was unambiguous – females either responded with copulatory postures or paid so little attention to the playback song that their behavior did not reveal that a song had been played at all. Females did not respond to the songs of other species and did not respond equally often to the songs of all male cowbirds: some songs elicited far more response. We defined such differences in the eliciting properties of males' songs as differences in song potency (West et al. 1979).

We have used the female bioassay to answer many questions about the nature of mate identification and its acoustic basis in cowbirds and met with considerable success (King & West 1990). But it would be wrong to think that knowing all one can about the song by playback will decipher male–female communication. In Spielberg parlance, the playback procedure achieves an encounter of only the first or second kind – our cowbirds were not actually encountering one another. Despite the power of the female bioassay to force answers about song perception, it requires highly abnormal conditions for its elicitation: females live with no males and hear only a handful of songs/day.

Non-vocal Responses Shape Vocal Development

Our data also now make it clear that it is not just song content, w*hat* males sing, that matters. Rather, song use, *how* males use their song, also plays a necessary role in competition or courtship. Indeed, the pragmatics of song use has come to dominate our attention in recent years because we discovered that male cowbirds do not necessarily know how to use their song to be effective communicators, even if they possess powerful song material. Returning to the spaceship analogy, how the two crews used the five-note melody, the details of timing and emphasis, were also important. What if the spaceship had played its melody and the Earth crew had waited too long to reply? But what is too long? Too short? How do we learn the pragmatics of communicating with melodies?

An important approach we have used to understand the importance of performance has been to videotape groups of males and females to look at interactions involving song, measuring behavior moment by moment. This view has proven invaluable for three reasons: it has revealed a tutorial role for female cowbirds; it has shown that courtship and competition have their roots in social encounters occurring many months prior to the breeding season; and it has allowed us to have a close encounter, that is, to experience what one second's worth of sound can accomplish.

Before explaining the full experience, some words about its parts. We have found that non-vocal responses from females to male song facilitate vocal development itself (Smith et al. 2000; King et al. in press). We allowed young male cowbirds to interact with one of two groups of adult females, a local one or a geographically distant group. We found that the non-singing females displayed two rapid responses to song, wing stroking and gaping. Wing stroking is a rapid and silent response to song in which the female flicks her wing away from her body: when gaping, a female arches her head and quickly opens and closes her beak. Both responses occur more when young males are with females from their local group, and both occur more often when the male has shown attentiveness to the females by orienting his song in their direction. Females do other things as well, most noticeably, showing no change in behavior when a male sings, thus ignoring his overtures. The high rate of ignoring may explain why behaviors such as wing strokes and gapes seem so prominent to males when they occur.

To understand the function of such female stimulation, we correlated female responsiveness with measures of vocal ontogeny. We found that juvenile males, exposed to more visual cues from local adult females, developed mature song faster and thus showed an earlier decline in practice than did males receiving less social feedback from the distant group of females. In addition, males housed with local females produced songs of higher potency. Even small differences in attentiveness within the local group had large effects: only very close encounters modulated song development and function. Thus, song ontogeny, and possibly the formation of local dialects, relies on cross-modal communication, a non-imitative and cognitive means of song acquisition.

The greater use of wing strokes and gapes was also associated with the males and females who interacted in longer social bouts, suggesting that perhaps these responses also act to sustain attention between pairs, allowing communication to occur. It may be that the female's behavior reinforces such non-vocal aspects of performance even more than the song structure itself. Males who receive wing strokes repeat the song pattern that worked, but perhaps the reinforcement is for the act of repetition itself. A singular feature of many kinds of music is of course astute repetition.

The apparent look of disinterestedness on the female's face during many interactions may also be important. Males become very excited when the female departs from this demeanor and does something even so small as turning her head. We have found that even brief wing strokes or gapes can lead a male cowbird to levitate off his perch, hop excitedly toward the female, and sing whatever song elicited the female's movements. Thus, the contrapuntal use of acoustic and visual signals between males and females may serve to orchestrate the sustained kinds of interaction necessary for each sex to profit from the encounter.

Indeed, in all contexts, playback tests, aviary observations, and videotaped interactions, it is clear that there are rules to the ceremony of singing and listening. For the male's part, his performance is not only a rigorous outpouring of sounds but of other behaviors as females do not simply sit still and allow the male to entertain them. The pursuit of singing involves incredible effort to achieve a goal, often with no obvious success. Males run, fly, and walk after females, attempting to get close before they sing. They also use other behaviors to 'sneak' in a song, such as the head down display, a greeting display performed by both sexes. In the display, the male's and female's head can come into actual physical contact, at which point the male may try to switch rapidly into singing mode.

The sensitivity of the non-singing females to male song is also seen when females approach males, or fail to do so. Whether in aviaries or small enclosures, females are far, far less likely to get close to males than are males likely to approach females. Indeed, in general, the most 'positive' responses shown by a female, outside of the context of mating, are the absence of any 'negative' response. She does *not* fly away, sidle down the perch, walk off, or peck him. In nature, positive responses to mating songs are also rare. Males may sing thousands of songs between copulations even to a female with whom he has been paired for weeks.

Taken as a whole, the movements involved in singing interactions physically and psychologically resemble movements used by humans in non-vocal communication, where nods of the head can have several different meanings, and where seating orientation at 90° or 180° degrees has different meaning within and across cultures. Male cowbirds also seem to remember what they were singing when they received positive responses and retain the songs eliciting female responses, for the future, just as humans show learning if head-nods are used strategically as reinforcement (Argyle 1972).

The overall quality of these interactions cannot, however, be truly appreciated in real time – events occur too swiftly for the human ear or eye to take in the amount of acoustic/harmonic detail (Dooling et al. 2002). The presence of rhythm is, however, clearly apparent when audiotapes are slowed down, allowing us to share in some of the temporal fine structure. The one-second male vocalization, when slowed to quarter speed, sounds to humans like a melody produced with multiple instruments with a long exchange of frequencies, and ending with a series of undulating whistles, changing quickly in pitch and intensity. The female's social reaction also appears tied to the waves of sound: when responding with a mating posture, the sound first claims her head, then her back, and then her tail. Other movements seem part of the performance as well. For example, females often wipe their beak against a perch. The action is hardly noticeable in real time, and the sound of the motion only comes alive when interactions are played frame-by-frame. Are these beak movements incidental to the sound or part of a response, like tapping one's foot, snapping a finger, or humming? The ability of music to seize one's whole body would argue for considering the possibility that these movements are not independent of the sound events.

We highly recommend this route of watching or listening to singing interactions, at half or quarter speed, as a means to 'experience' song more in the way that the birds may see and hear it, given their superior temporal resolution of light and sound (Hailman 1977; Dooling et al. 2002 **CD1 #6, #69 & #71; CD2 #9**). Such procedures have also been useful for researchers of human nonverbal behavior (Eibl-Eibesfeldt 1972). The value is seeing the experience as a dynamic flow of events across participants and across sensory modalities. Nelson & Marler (1990) have talked of the concept of 'just-meaningful differences' in the 'signal space' for song perception (**Box 23**, p. 194); this may also be true for perception of interactions. Our work suggests that scientists must beware of their own reductionistic tendencies to make sure they do not take apart the very thing they want to study, in this case, the higher order construct of a perceptual grouping of sights and sounds that comprises a unified performance.

In sum, the effects of melodious sounds in cowbirds are at once simple and complex. Simple, in that the female responds with an unambiguous response to sound during a crucial context, but complex in that the synchrony seen at that time is built out of thousands of cross-modal interactions throughout the year.

SONG, MUSIC, AND HUMAN DEVELOPMENT

The Mozart Effect

In cowbirds, social recognition develops by way of small, seemingly mundane social reactions to the ubiquitous singing attempts of males. These social mechanisms of development have often been overlooked because they are not special or specialized. The everyday behavior of males and females creates contingencies – infrequent but reliable social feedback – in which gradual learning can take place. As we will see, musical elements in the speech of human adults to infants creates opportunities for social learning, but the effects of exposure to musical speech are more subtle than might be supposed.

We stress the possible subtlety because currently popular musical methodologies with humans suggest there are simple ways to use music to inoculate children against low IQ, poor spatial reasoning, and other cognitive problems. These techniques fall under the mesmerizing name of the 'Mozart effect' (the htpp www.mozarteffect.com website's logo is "Explore, Learn, Shop"). The websites promise brain enrichment for infants, youngsters, and adults. But the 'Mozart effect' is most effective in telling us how vulnerable humans are to using any aids to facilitate communication when dealing with a creature so unlike themselves in terms of size, experience, sensory capabilities, and emotional tendencies. Who wouldn't want a pre-packaged solution with so august a name? No wonder, the 'effect' in which exposure to classical music purports to enhance performance on intelligence tests has attained the status of an urban myth, a myth sustained by toy companies, technology gurus, and writers of parenting manuals.

But mere exposure to music is not a catalyst of infant cognitive development. A look at the actual experiments reveals a very different picture of the so-called effect. In the original study, college students (not infants) participated in three spatial reasoning tasks from an IQ test (Rauscher et al. 1993). Before they participated in the reasoning task they heard either 10 minutes of (i) a Mozart sonata (Sonata for Two Pianos in D Major), (ii) a relaxation tape, or (iii) silence. Students who had heard Mozart performed better on the reasoning task than students who had heard the relaxation tape or nothing. Their performance corresponded to about an 8–9 point increase in spatial IQ. The effect disappeared within about 10–15 minutes. The same authors (Rauscher et al. 1995) replicated the effect with a measure of spatial reasoning: subjects viewed diagrams of a piece of paper which is folded and cut; they were asked to select a diagram which represented the appearance of the paper after it is unfolded. In this replication, subjects showed better spatial reasoning ability after listening to the same Mozart sonata than if they had listened to 10 minutes of Philip Glass's *Music With Changing Parts* or after 10 minutes of silence.

The findings garnered widespread attention, including everything from the development of a 'Baby Mozart' video to former Georgia Governor Zell Miller requesting funding for classical music CDs for each infant born there. Despite the attention, attempts to replicate the original findings have failed (Steele et al. 1999). The observed short-term changes in spatial reasoning appear to be the result of changes in arousal or mood (Steele 2000; Thompson et al. 2001). College students who listened to an Albinoni adagio, music of similar complexity to Mozart but slower and sad, did not show increases in spatial reasoning ability (Thompson et al. 2001). In the original study by Rauscher and colleagues, listening to Mozart probably increased arousal and positive mood while hearing the relaxation tape, or silence decreased arousal. Boredom and negative mood decreased performance on the tasks. To test this hypothesis, Nantais & Schellenberg (1999) assessed spatial reasoning in college students after listening to 10 minutes of music, by either Mozart or Schubert, and after 10 minutes of silence. Music improved spatial reasoning performance relative to silence. However when, in a second experiment, the silence was replaced with a narrated Stephen King story, students performed equivalently on the spatial reasoning task in both conditions. In addition, students were asked to rate whether they had preferred listening to the music or the story. Spatial reasoning scores were higher for the preferred condition.

Other research by Rauscher and her colleagues has investigated the influence of music lessons on spatial reasoning. In this study, preschool children were given 6 months of piano lessons or computer lessons (Rauscher et al. 1997). Relative to controls, children who received piano instruction showed more improvement in solving jigsaw puzzles, an age-appropriate spatial reasoning task. This research, in which children played an active role in their development, has been largely ignored by the media.

To summarize, we wish to argue that, in humans, music is about active modulation of behavior, not the regulation of intelligence. The so-called Mozart effect is the epitome of a passive effect and is also a good example of a non-social approach to cognitive influence. As noted earlier with cowbirds and playbacks, the effect relies on Spielberg's encounters of the first or second kind; only indirect exposure with no active involvement.

Infant-directed Speech

The distinction is important because human caregivers and infants are very different creatures in the same way that male and female cowbirds are different in terms of sensory and motivational capacities, and thus the task of negotiating encounters is neither an automatic nor stress-free experience. Infants' sensory and motor systems are immature; they do not see, hear, react to, or remember their world as adults do. As an indication of the differences between adults and infants, cultures around the world have evolved a different vocal code to use in communication. Adults and children use a special form of speech, called infant-directed or ID speech, when talking to infants. As compared to adult-directed (AD) speech, ID speech is characterized by slower rate, higher mean fundamental frequency (F_0), exaggerated prosody (greater F_0 variability), simplified syntax, and shorter utterances (Fernald & Kuhl 1987; Fernald 1992; **CD1 #98**). This register is used in most cultures when addressing infants, though it is also used with pets and institutionalized elderly adults (Culbertson & Caporael 1983), organisms that also have different sensory and cognitive demands.

The pitch contours of ID speech give it a melodious quality, and from the first days of life infants prefer to listen to ID over AD speech

(Cooper & Aslin 1990). Many experiments have assessed the relative influence of ID and AD speech by measuring how long infants will look at a visual stimulus, usually a black-and-white checkerboard, when exposed to recordings of an adult speaking in each style. Selective visual orientation produced by ID speech has been shown to persist over at least the first nine months (1-month-olds, Cooper & Aslin 1990; 4-month-olds, Fernald 1985; 4–5 and 7–9 month-olds, Werker & MacLeod 1989).

Across languages and cultures, the pitch contours of ID speech convey similar messages to infants (Fernald 1989, 1992; Papousek 1992). Prohibition such as "don't do that," is signaled by abrupt rising or falling pitch. Soothing an infant is accomplished through the use of lower, slowly falling pitch contours. Lullabies contain similar acoustic characteristics (Trehub & Trainor 1998). In play with babies, we use phrases containing high, exaggerated pitch modulation to get infants' attention and to label objects. The consistency with which we adjust our pitch to match a message allows prelinguistic infants to learn that all the noise coming from adults' mouths has functional consequences – that certain kinds of sounds predict particular emotions and actions on the part of adults.

Infant-directed speech serves to organize an infant's attention. The acoustic characteristics of ID speech create arousal, which is evidenced by increased looking time to a visual stimulus when listening to speech (Kaplan et al. 1995b). ID speech also creates a special form of arousal called 'sensitization.' Sensitization is a cause of 'dishabituation' in which a visual stimulus that has become boring suddenly, becomes interesting again after exposure to ID speech, (Kaplan et al. 1995a). ID speech has been shown to help infants learn about the relationships between things. For example, infants demonstrated better learning of sound–checkerboard and sound–face contingencies when arousing tones or ID speech predicted the visual stimuli, versus AD speech (Kaplan & Fox 1991; Kaplan et al. 1996).

These findings led to an investigation of the ID speech of depressed mothers and the possible consequences for their infants. Mothers diagnosed as depressed produced speech that was flatter and less variable in pitch – they had less F_0 variability than did non-depressed mothers (Kaplan et al. 2001). Infants listening to depressed ID speech did not learn associations between the speech and a face as well as infants listening to non-depressed ID speech (Kaplan et al. 1999, 2002). ID speech may also make the sound structure of language, known as phonology, more salient to young learners. When we speak to infants, we exaggerate our vowels, making them more distinct from each other (Kuhl et al. 1997).

In social interactions, the melodious vocalizations of adults function to regulate the interactions of infants and their caregivers. When engaged in face-to-face interactions, mothers and infants take turns vocalizing (Anderson et al. 1977; Papousek et al. 1985). The synchrony of vocal turn-taking is regulated by both mother and infant, with each partner's behavior predictive of the other's (Jaffe et al. 2001). In two recent studies we have investigated the role that siblings play in infant development by measuring infants' reactions to sibling-produced ID speech. We have found that the ID speech produced by siblings is more interesting to infants than that produced by mothers. Children with siblings also react differently than singletons (Goldstein et al. in press b). Seven-month-old infants will also crawl to be closer to their older siblings when the siblings are vocalizing than when they are not (Goldstein et al. in press a). ID speech thus enables caregivers, older siblings, and infants to share attentional space. Creating a cognitive and social space that is occurring at the right sensory speed for the infant may bring information into the realm of 'just-meaningful differences' (Nelson & Marler 1990) as we noted earlier for birds.

In summary, musical speech to infants has many important functions, as it predicts the behavior of caregivers, organizes infant attention, facilitates associative learning, and regulates social interactions. ID speech and singing create a framework in which adults and babies, despite being very different organisms, can learn about each other and establish the earliest building blocks of communication. As any parent can tell you, not all close encounters between parent and child are mutually satisfying. What infants want can be fantastically puzzling and if music helps in only a small percentage of interactions, it serves a conspicuously important function.

BIRDSONG AND MUSIC: CROSS-SPECIES INSPIRATIONS

A Starling in a Human Household

In this section, we explore further the idea of music and sounds as ways of creating relationships but, in this case, the participants are European starlings and humans. It may seem odd or futile to some to ask about starlings' effects on another species or vice versa, but in the context of this book, exploring the function of musical sounds in nature, it is an adventure not to be missed.

Cowbirds and human infants appear to establish a shared pattern of recognition through social interactions – so do starlings (Chaiken et al. 1997; Hausberger 1997). But the focus in cowbirds on getting a song just right does not seem to be the basis of vocal effects in starlings. A mature male cowbird produces less than 10 seconds of sound he can call his own, mostly in the form of song types; stable acoustic patterns. A starling may have 60–80 song types called motifs embedded in long strings of sounds lasting many minutes (Chaiken et al. 1993, **Box 10** p. 83). An entire cowbird song would represent one starling motif.

We did not set out to study European starlings – we were seduced by this species when we attempted to use it as a heterospecific companion for cowbirds in studies on song learning. Starlings could sing, imitate, and interact, but presumably provide no species -typical experiences to young cowbirds. We obtained our first starling about the time we raised our first human infant, and thus we were coping with all the learning necessary to communicate with our son.

Starlings are talented mimics increasing their repertoire by using the sounds of other birds, other animals, man-made machines, and humans themselves (West et al. 1983; West & King 1990). The Latin scholar Pliny, for example, reported that he taught European starlings to repeat Latin and Greek phrases. When we began our work with starlings, we did not know of Pliny's attempts. Until life with our first starling, we were totally unaware of the species' ability to mimic human sounds. As will be seen, these sounds include human words, but we did not and do not consider starlings' mimicry of human speech sounds as fundamentally different from any other sounds starlings choose to imitate. At no time were we trying to understand if starlings could be taught an anomalous communication system as has been done in a fascinating way with parrots (Pepperberg 1988; see Chapter 13). We were using the knowledge that starlings will include human sounds as a way to bring some control to their dazzling array of mimicking possibilities. What we wanted to know is the same as what we wanted to know in cowbirds and humans: how do starlings use sounds in social interactions? How do they choose them? And what effect do they have? But, in this case, the social partner was a human (West et al. 1983; Engle 2000).

In our view, singing or vocalizing is to starlings what foraging is to cowbirds or fussing is to human infants, a way to find new sources of stimulation. One reason for such replenishment in starlings is to improve the condition of their repertoire. In nature, large repertoires confer an advantage on male starlings' fitness; they produce more offspring and grand offspring (Eens 1997, see Chapter 2, p. 59).

Our first starling did not learn to mimic sounds under our guidance, although he was under our care. Until he produced something that sounded like "good morning" we were completely unaware of how much he had observed us as we were busy observing infant cowbirds and humans. But it was easy to forget the bird's scientific role as he sat with us while we had our morning coffee, took a shower with someone in the household, mimicked human infant cries, and attempted to join a lab meeting (**Fig. 14.2; CD1 #97**). His growing repertoire also showed a decidedly human influence, as did those of other birds we studied. Endearing terms and domesticity were especially strong themes, a reflection perhaps of the family 'zeitgeist'.

The starlings' mimicry suggested to us that the motivating conditions for vocal learning in this species and others are not simple: sounds heard most frequently by the birds, such as "no" or "here's some lettuce," were not present in their mimicry and sounds only rarely heard seemed over-represented, such as "defense, defense" chanted after overhearing a televised basketball game. Our impression was that the starlings threw out sounds and watched what happened. They learned quickly that mimicking "hi there" led to human approach more often than sounds that were unintelligible from humans' point of view. The shaping was bi-directional: if humans did not pay attention to the sounds, the starlings' experimentation and improvisation continued. The key to beginning to understand the rules for acquiring human mannerisms has several parts: access, acquired salience, and seeing the consequence of one's actions.

An intriguing part of life with starlings was being able to trace the course of sound from its original condition to its 'starlingized' version. The starlings routinely rearranged and edited, and otherwise changed the sounds they heard so that a phrase such as "we'll see you later" eventually diverged into many phrases including "see ya later," "see you," and "we'll see, we'll see." Several of our starlings honored us by quoting the phrase, "basic research," but even that was vulnerable to interpretation, morphing into "sick research" and then simply "sick, sick."

Starlings seem to see any setting as potential vocal turf. The most obvious sign that they are in a music-making mode was when they were

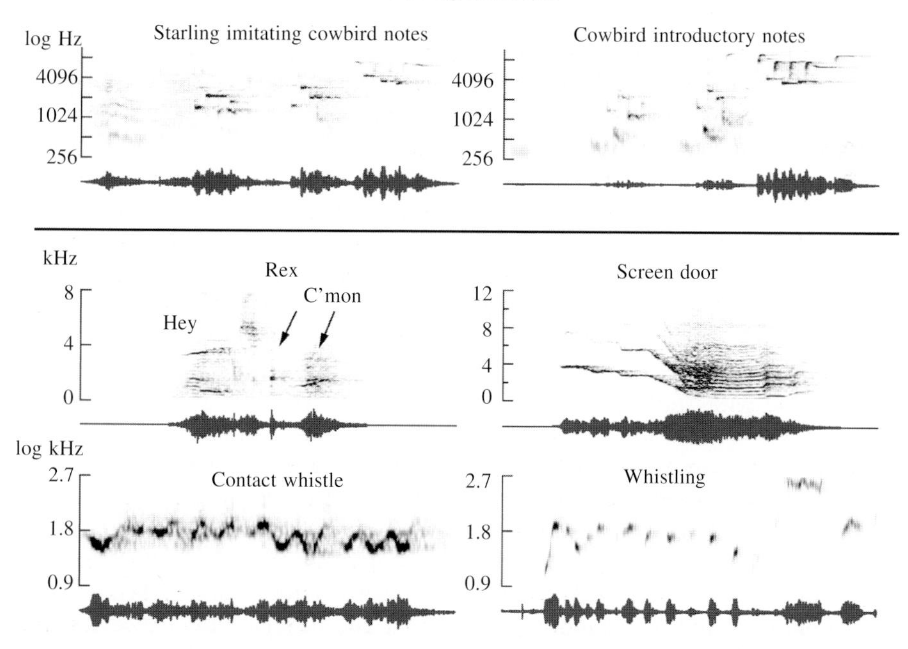

Figure 14.2 Examples of starling mimicry from a two-year-old male starling (**CD1 #97**). The upper panel shows mimicry of the introductory notes of a brown-headed cowbird song on the left with the same notes by a cowbird on the right. The bottom panels show starling mimicry of speech, a squeaky door, and two examples of whistling by the starling.

quiet, cocking their head to and fro listening to whistles, music, or may be the teapot. They did not vocalize at these times but later repeated parts of what occurred while in their listening posture, a posture found in zebra finches and perhaps other birds as well. In zebra finches, there is also a connection between assuming a listening posture (the bird flies near another bird, stretches his neck forward, and freezes in that position) and learning the songs of another finch species, at least in juveniles (ten Cate 1986).

Indeed, the best way to quiet a noisy starling is to feed him a new sound; he or she must stop vocalizing to digest the vocal bite. Whistles are probably most effective, or sounds from instruments such as recorders, which, of course, have been used for centuries to train birds to sing. But taken as a whole, the work in our lab suggests that starlings need new sounds (mimicked or created) as building blocks for vocal learning. In understanding why some sounds persist, we propose a form of social sonar, bouncing vocal bits off a social (or vocal) sounding board, and observing the sounds and sights returned to them. The social requirement seems strong as studies of starling development and mimicry suggest that learning from live tutors, compared to tape tutors, is much more effective (Chaiken et al. 1997).

Mozart's Starling

We are not the only people to have been befriended by starlings; the Internet now has many sites for starling 'owners.' And there are several books on life with starlings (Corbo & Barras 1983). The words others use to describe their relationship are congruent with our experiences: the birds are seen as comical, smart, and attentive (Suthers 1982). If we had not studied starlings, it would be easy to dismiss some of the anecdotes and the general emotional tone. But, we do have a musical collaborator of sorts to confirm our observations of starlings. Our expert witness to their charms is the quite well-known human mentioned earlier in another context, Wolfgang Amadeus Mozart.

Upon finding Mozart had owned a starling, we embarked on a time travel adventure back to the 1700s in search of an answer to questions about his pet. One of us (MJW) first discovered Mozart's starling looking through a biography of the composer and noting in the index to the book, an entry entitled "Mozart buries his starling" (Hildesheimer 1983). Further reading, however, revealed that historians and musicologists did not share our opinion that a starling would be an ideal companion to Mozart. They believed his burial for his starling bird (who lived three years), the poem he read at the occasion, and the funeral garb of his guests was further evidence of his immaturity, the genius that never grew up. Some also complained that his behavior might make more sense if the bird had been something else, a lark, perhaps, or a nightingale. It is important to understand that Mozart scholars study everything about this great man – his left ear lobe is the subject of a set of scholarly and medical papers (Davies 1989).

When he had bought the bird, it could whistle part of his Concerto in G major, K. 453, an accomplishment Mozart noted in his diary. Thus, the beginning of their relationship contains a mystery in and of itself, as the piece had not yet been performed in public. Throughout his life, plagiarists plagued Mozart, and we wonder if he bought the bird to remove a vocal copyist from the public scene. A major disappointment, however, was that the research we did revealed no musical tribute to a bird we were sure Mozart had loved.

The answer came when we turned from reading books and searching through documentary evidence about his life to listening to music composed at or about the time Mozart had owned his starling. We looked especially around the time of the bird's death, which occurred in May 1787, also near to the time of Mozart's father's death. Leopold was his son's first and foremost teacher and critic.

Among the musical works entered in Mozart's autograph scores for June 1787 was one that attracted our attention, K. 522, entitled in English, 'A musical joke'. Experts in the field saw it, at worst, as a backhanded gesture of frustration by Mozart towards his father, as his father had urged Mozart to compose music more popular to paying patrons. At its best, the piece was earmarked as a parody of bad composing, "a marvelous and malicious prank" concluded a recent biographer (Gutman 1999, p. 668). It may well be these things but it may be something more. Further research into the chronology of Mozart's compositions revealed that the musical joke was not written in June 1787, but over a three-year period beginning in 1784, the year of the starling's acquisition. It even contained part of K. 453, the music that had brought the two together. As much as any of Mozart's biographers know about his life or music, it is doubtful few have ever paid attention to a starling in a rococo mood, hence their inability to link K. 522 to the starling (Gutman 1999).

But, when we listen to K. 522 (especially the Presto, with a nine measure trill), we hear the autograph of a starling. There is the fractured phrasing of the entire serenade, tiresome repetitions and an eccentric ending, sounding as if the instruments had ceased to work. These features would all be well known to starling 'experts.' The repeated trill is also of considerable interest because it sounds so much like the contact whistles starlings owners used 200 or so years later. Lorenz first noted such whistles in jackdaws (Lorenz 1957). An intriguing feature to us was that they escaped improvisation in all cases we know of.

Another possible explanation for K. 522 of course is that Mozart was having a bad spell. After all, his father (and starling) had recently died. But an entry placed in his autograph scores soon thereafter argues against such a hypothesis: the piece is one of Mozart's most beloved: K. 525, "Eine kleine Nachtmusik."

CONCLUSIONS

There is no other way to describe the feelings we had when discovering Mozart's starling in his music other than in terms such as those in the prologue from the film and book, *Close Encounters.* To be able to reach across time and hear something Mozart heard is as indescribable as meeting members of a different planet and finding you share something fundamental and intimate. We believe this is the case for other music–animal sound connections that Luis Baptista spoke about with such fondness (Milius 2000). When we hear a familiar refrain from nature, even if we do not at once recognize the source, we make a connection. Moreover, the familiarity of half of the equation, either the music or the sound from nature, may facilitate our memory of both. Humans are better at remembering test words when heard as music rather than as speech, as long as the music has a repetitive quality. If the melody is not repeated or if a new melody accompanies new texts, recall is no better than when the text is heard as prose (Wallace 1994).

Perhaps the starlings were so good at mimicking domestic parts of life because domesticity contains the necessary repetition. Every day we eat breakfast, go to work, shower, and so on. Starling song, like that of many passerines, has much internal repetition, thus setting up its melody for recall. Perhaps the internal rhythm reflects that in the wild, hours and days are spent doing similar things. For a colonial species, some kind of modulatory vigilance system to broadcast the state of the group may be useful for all members, not so much as 'passwords' but as 'watchwords.'

We have argued that music is a social and cognitive means of involvement; a simple, acoustic means to allow two groups or two or more individuals to assess whether a social relationship exists and perhaps to keep it going. Music is special because social interaction is special, and any medium that brings individuals together can be mutually reinforcing. It is important to remember that in the real world, not just movies, music can have impressive effects on human behavior. Long time observers of Central Park in New York City say the biggest crowd ever assembled for a music event was in 1991 for a Paul Simon concert – over 750,000

BOX 48

BIRDSONG AS INSPIRATION TO MUSICIANS

Like Mozart, many composers and musicians, from folk and classical music, jazz, reggae to modern pop and rock music, have found a source of inspiration in singing birds. Probably the most famous classical compositions inspired by birdsong are those of the French composer Olivier Messiaen. Renditions of songs of various bird species are found throughout in works such as 'Reveil des Oiseaux' and 'Oiseaux Exotiques.' Messiaen regarded birds as "probably the greatest musicians to inhabit our planet" (Messiaen 1994). His transcriptions of birdsongs are sometimes quite problematic, sometimes rendering the bird that inspired it unrecognizable to an ornithologist. On the other hand, musicians are quite definite that they can relate Messiaen's renditions to avian counterparts, perhaps because he delved not just into the surface features of birdsong, but into its deeper musical structure as well. His version of chaffinch song for violin is unmistakable (**CD2 #89**).

Duke Ellington is said to have heard a common mynah singing in a hotel garden while on tour in Asia. Co-composer Billy Strayhorn used the sounds of the bird to compose 'Bluebird of Delhi (Mynah),' on the album 'The Far East Suite' (**CD2 #90**). The result is an arrangement of instruments that emit sound bursts in a way resembling the bird's song.

'Three Little Birds' of Bob Marley refers to "sweet songs of melodies pure and true" which say to the listener: "don't worry about a thing, cause every little thing is gonna be alright." Presumably, Marley wrote this song on the doorstep of his house in Jamaica, while smoking marijuana. The inspiration apparently came from some doves that were attracted by seeds that he tossed in the process of rolling a joint.

Imitations of birdsongs also find their way into modern music by way of the human voice. A great example is 'Cu cu ru cu cu paloma' from Harry Belafonte. In this song Belafonte mimics a dove coo (**CD2 #91**); according to him it expresses a sadness comparable to his feelings over losing a lover.

Popular bands use bird recordings as a backdrop for their compositions, to create a particular atmosphere or to place their music in a natural setting. A nice example is 'Cirrus Minor' on the Pink Floyd album 'More,' used as a soundtrack to the French hippy movie of the same name. In the opening number, recordings of song thrush and nightingale alternate independently and repeatedly between left and right speaker, while a European cuckoo echoes in the background before the instruments enter the composition. Ian Anderson from Jethro Tull introduces vocalizations of a series of birds in the title song from his solo album 'The Secret Language of Birds' (Fig.). He has house sparrows and a wood pigeon singing at the start, and crows and a winter wren at the end of the song, in which he invites the listener to "stay and learn the secret language of birds."

Most famous of all is probably the wonderful duet between Paul McCartney and a European blackbird on 'The White Album' of The Beatles (**CD2 #92**). Yet another way to use birdsong is as an additional instrument. 'The Patty Patty Sound' album from the Scottish group The Beta Band provides some examples. They use songs and calls of different species, including aviary favorites such as the canary and the zebra finch. In 'The House Song,' the band scratches away with samples from blackbird song used together with a mixture of other instruments (**CD2 #93**). In 'Bird,' the group Dead Can Dance generates a true 'ambient orchestra' on their album 'A Passage in Time' (**CD2 #94**). Calling birds, chirping cicadas, and whale, gibbon, and human songs, are all used in a virtual celebration of 'Nature's Music.'

Hans Slabbekoorn

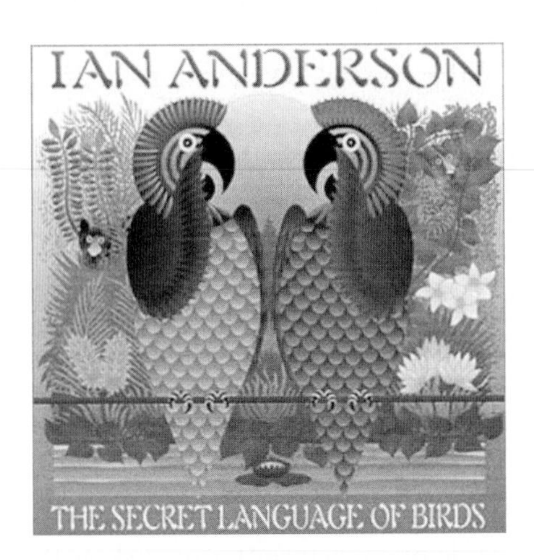

people (McFadden 1991). No national issue galvanized the crowd except the chance to hear and sing music probably heard and sung many times before. By comparison, the Reverend Billy Graham drew a third of the Simon's crowd. Three-quarters of a million people to hear one man and his band play popular folk music is formidable testimony to the effect of music.

So what can music do? It can bring together often unfamiliar individuals into a common umwelt. Such a coming together may be trivial when it is two friends singing in the car to the their favorite song, or deeply mysterious, when hearing a song one has not heard for 30 years and realizing the words and notes are all still there in a cognitive reservoir.

Some of the government probes the USA have sent into outer space have contained music – diverse selections from whales to the Beatles to Bach (Sagan 1978). Is the music a way to represent our cultures or to make our cultures inviting? We would like to think the latter is primary because nature's music shows animals at their best – we look, listen, and say, "there are no words for this." It is that shared emotion that creates the music effect.

Bibliography

Abs, M. 1980. Zur Bioakustik des Stimmbruchs bei Vögeln. *Zoölogische Jahrbucher-Abteilung für allgemeine Zoölogie und Physiologie der Tiere* **84**:289–382.

Adkins-Regan, E. & M. Ascenzi. 1990. Sexual differentiation of behavior in the zebra finch: Effect of early gonadectomy or androgen treatment. *Hormones & Behavior* **24**:114–127.

Adkisson, C. 1981. Geographic variation in vocalizations and evolution of North American Pine Grosbeaks. *Condor* **83**:277–288.

Adkisson, C.S. & R.N. Conner. 1978. Interspecific vocal imitation in White-eyed Vireos. *Auk* **95**:602–606.

Agate, R.J., W. Grisham, J. Wade, S. Mann, J. Wingfield, C. Schanen, A. Palotie & A.P. Arnold. 2003. Neural, not gonadal, origin of brain sex differences in a gynandromorphic finch. *Proceedings of the National Academy of Sciences USA* **100**:4873–4878.

Airey, D.C., K.L. Buchanan, T. Szekely, C.K. Catchpole & T.J. DeVoogd. 2000. Song, sexual selection, and a song control nucleus (HVc) in the brains of European sedge warblers. *Journal of Neurobiology* **44**:1–6.

Airey, D. & T. DeVoogd. 2000. Greater song complexity is associated with augmented song system anatomy in zebra finches. *Neuroreport* **11**:2339–2344.

Alatalo, R.V., C. Glynn & A. Lundberg. 1990. Singing rate and female attraction in the pied flycatcher: An experiment. *Animal Behaviour* **39**:601–603.

Alexander, M.P., M.A. Naeser & C.L. Palumbo. 1987. Correlations of subcortical CT lesion sites and aphasia profiles. *Brain* **110**:961–991.

Alexander, R.M. 1984. Walking and running: The biomechanics of traveling on foot. *American Scientist* **72**:348–354.

Allan, S.E. & R.A. Suthers. 1994. Lateralization and motor stereotypy of song production in the brown-headed cowbird. *Journal of Neurobiology* **25**:1154–1166.

Alstrom, P. & U. Olsson. 1995. A new species of *Phylloscopus* warbler from Sichuan Province China. *Ibis* **137**:459–468.

Altman, J. 1962. Are new neurons formed in the brain of adult animals? *Science* **135**:1127–1128.

Alvarez-Borda, B. & F. Nottebohm. 2002. Gonads and singing play separate, addititive roles in new neuron recruitment in adult canary brain. *Journal of Neuroscience* **22**:8684–8690.

Alvarez-Buylla, A. & J.R. Kirn. 1997. Birth, migration, incorporation, and death of vocal control neurons in adult songbirds. *Journal of Neurobiology* **33**:585–601.

Alvarez-Buylla, A., J.R. Kirn & F. Nottebohm. 1990. Birth of projection neurons in adult avian brain may be related to perceptual or motor learning. *Science* **249**:1444–1446.

Alvarez-Buylla, A., C.Y. Ling & F. Nottebohm. 1992. High vocal center growth and its relation to neurogenesis, neuronal replacement and song acquisition in juvenile canaries. *Journal of Neurobiology* **23**:396–406.

Alvarez-Buylla, A., M. Theelen & F. Nottebohm. 1988. Birth of projection neurons in the higher vocal center of the canary forebrain before, during, and after song learning. *Proceedings of the National Academy of Sciences* **85**:8722–8726.

Amagai, S., R.J. Dooling, S. Shamma, T.L. Kidd & B. Lohr. 1999. Detection of modulation in spectral envelopes and linear-rippled noises by budgerigars (*Melopsittacus undulatus*). *Journal of the Acoustical Society of America* **105**:2029–2035.

American Ornithologists' Union. 1998. *Check-list of North American Birds* (7th ed.). American Ornithologists' Union, Washington, D.C.

Ames, P.L. 1971. The morphology of the syrinx in passerine birds. *Bulletin Peabody Museum of Natural History* **37**:1–194.

Andelman, S.J. & W.F. Fagan. 2000. Umbrellas and flagships: Efficient conservation surrogates, or expensive mistakes? *Proceedings of the National Academy of Sciences* **97**:5954–5959.

Anderson, B.J., P. Vietze & P.R. Dokecki. 1977. Reciprocity in vocal interactions of mothers and infants. *Child Development* **48**:1676–1681.

Andersson, M. 1994. *Sexual Selection*. Princeton University Press, Princeton, NJ.

Andersson, S. 1989. Sexual selection and cues for female choice in leks of Jackson's widowbird, *Euplectes jacksoni*. *Behavioral Ecology-& Sociobiology* **25**:403–410.

Aoki, K. 1989. A sexual-selection model for the evolution of imitative learning of song in polygynous birds. *American Naturalist* **134**:599–612.

Appleby, B.M. 1995. *The behaviour and ecology of the tawny owl Strix aluco*. PhD thesis, Oxford University.

Appleby, B.M. & S.M. Redpath. 1997a. Variation in the male territorial hoot of the tawny owl *Strix aluco* in three English populations. *Ibis* **139**:152–158.

Appleby, B.M. & S.M. Redpath. 1997b. Indicators of male quality in the hoots of tawny owls (*Strix aluco*). *Journal of Raptor Research* **31**:65–70.

Appleby, B.M, N. Yamaguchi, P.J. Johnson & D.W. MacDonald. 1999. Sex specific territorial responses in tawny owls *Strix aluco*. *Ibis* **141**:91–99.

Appletants, D., P. Absil, J. Balthazart & G.F. Ball. 2000. Identification of the origin of catecholaminergic inputs to HVc in canaries by retrograde tract tracing combined with tyrosine hydroxylase immunocytochemistry. *Journal of Chemical Neuroanatomy* **18**:117–133.

Argyle, M. 1972. Non-verbal communication in human social interaction. In: *Non-verbal Communication*, R.A. Hinde (ed.), pp. 243–268. Cambridge University Press, Cambridge MA.

Armstrong, E.A. 1955. *The Wren*. Collins, London.

Arnold, A.P. 1975. The effects of castration and androgen replacement on song courtship, and aggression in zebra finches. *Journal of Experimental Zoology* **191**:309–326.

Arnold, A.P. 1997. Sexual differentiation of the zebra finch song system: Positive evidence, negative evidence, null hypotheses, and a paradigm shift. *Journal of Neurobiology* **33**:572–584.

Arnold, A., S. Bottjer, E. Brenowitz, E. Nordeen & K. Nordeen. 1986. Sexual dimorphism in the neural vocal control system in song birds: Ontogeny and phylogeny. *Brain, Behavior & Evolution* **295**:22–31.

Arnold, A. & G. Mathews. 1988. Sexual differentiation of brain and behavior in birds. In: *Handbook of Sexology*, J. Sitsen (ed.), pp. 122–144. Elsevier, Amsterdam.

Arnold, A.P., F. Nottebohm, & D.W. Pfaff. 1976 Hormone concentrating cells in vocal control and other areas of the brain of the zebra finch (*Poephila guttata*). Journal of Comparative Neurology **165**:487–512.

Arnold, M.L. & S.A. Hodges. 1995. Are natural hybrids fit or unfit relative to their parents? *Trends in Ecology & Evolution* **10**:67–71.

Aubin, T. & J.C. Bremond. 1992. Perception of distress call harmonic structure by the starling (*Sturnus vulgaris*). *Behaviour* **120**:151–163.

Aubin, T. & P. Jouventin. 1998. Cocktail-party effect in king penguin colonies. *Proceedings of the Royal Society of London, B* **265**:1665–1673.

Aubin, T. & P. Jouventin. 2002. How to vocally identify kin in a crowd: The penguin model. *Advances in the Study of Behavior* **31**:243–277.

Aubin, T., P. Jouventin & C. Hildebrand. 2000. Penguins use the two-voice system to recognise each other. *Proceedings of the Royal Society of London, B* **267**:1081–1087.

Awbrey, F.T., D. Hunsaker II & R. Church. 1995. Acoustical responses of California gnatcatchers to traffic noise. *Inter-Noise* **95**:971–974.

Aylor, D. 1971. Noise reduction by vegetation and ground. *Journal of the Acoustical Society of America* **51**:197–205.

Badyaev, A.V. & E.S. Leaf. 1997. Habitat associations of song characteristics in *Phylloscopus* and *Hippolais* warblers. *Auk* **114**:40–46.

Baker, M. C. 1994. Loss of function in territorial song: Comparison of island and mainland populations of the singing honeyeater (*Meliphaga virescens*). *Auk* **111**:178–184.

Baker, M.C. 1996. Depauperate meme pool of vocal signals in an island population of singing honeyeaters. *Animal Behaviour* **51**:853–858.

Baker, M.C. 2000. Cultural diversification in the flight call of the ringneck parrot in western Australia. *Condor* **102**:905–910.

Baker, M.C., E.M. Baker & M.S.A. Baker. 2001. Island and island-like effects on vocal repertoire of singing honeyeaters. *Animal Behaviour* **62**:767–774.

Baker, M.C., T.K. Bjerke, H. Lampe & Y. Espmark. 1986. Sexual response of female great tits to variation in size of males' song repertoires. *American Naturalist* **128**:491–498.

Baker, M.C. & J.T. Boylan. 1995. A catalog of song syllables of indigo and lazuli buntings. *Condor* **97**:1028–1040.

Baker, M.C. & J.T. Boylan. 1999. Singing behavior, mating associations and reproductive success in a population of hybridizing lazuli and indigo buntings. *Condor* **101**:493–504.

Baker, M.C. & M.A. Cunningham. 1985. The biology of bird-song dialects. *Behavioral & Brain Sciences* **8**:85–133.

Baker, M.C., T.M. Howard & P.W. Sweet. 2000. Microgeographic variation and sharing of the gargle vocalization and its component syllables in black-capped chickadee (Aves, Paridae, *Poecile atricapillus*) populations. *Ethology* **106**:819–838.

Baker, M.C. & D.M. Logue. 2003. Population differences in a complex bird sound: A comparison of three bioacoustical analysis procedures. *Ethology* **109**:223–242.

Baker, M.C., P.K. McGregor & J.R. Krebs. 1987a. Sexual response of female great tits to local and distant songs. *Ornis Scandinavica* **18**:186–188.

Baker, M.C. & L.R. Mewaldt. 1978. Song dialects as barriers to dispersal in white-crowned sparrows (*Zonotrichia leucophrys nuttalli*). *Evolution* **32**:712–722.

Baker, M.C., K.J. Spitler-Nabors & D.C. Bradley. 1981. Early experience determines song dialect responsiveness of female sparrows. *Science* **214**:819–821.

Baker, M.C., K.J. Spitler-Nabors, A.D.J. Thompson & M.A. Cunningham. 1987b. Reproductive behaviour of female white-crowned sparrows: Effect of dialects and synthetic hybrid songs. *Animal Behaviour* **35**:1766–1774.

Baker, M.C. & D.B. Thompson. 1985. Song dialects of white-crowned sparrows (*Zonotrichia leucophrys nuttalli*): Historical processes inferred from patterns of geographic variation. *Condor* **87**:127–141.

Balaban, E. 1988. Cultural and genetic variation in swamp sparrows (*Melospiza georgiana*). II. Behavioural salience of geographic song variants. *Behaviour* **105**:292–322.

Balda, R.P. & J.H. Balda. 1978. The care of young Piñon jays and their integration into the flock. *Journal für Ornithologie* **119**:146–171.

Baldwin, D.A. 1991. Infants' contributions to the achievement of joint reference. *Child Development* **62**:875–890.

Ball, G.F. & S.H. Hulse. 1998. Birdsong. *American Psychologist* **53**:37–58.

Ball, G. & J. Balthazart. 2002. Neuroendocrine mechanisms regulating reproductive cycles and reproductive behavior in birds. In: *Hormones, Brain and Behavior,* D. Pfaff, A. Arnold, A. Etgen, S. Fahrbach & R. Rubin (eds.), pp. 649–798. Elsevier Science (USA).

Ball, G. & S. Macdougall-Shackleton. 2001. Sex differences in songbirds 25 years later: What have we learned and where do we go? *Microscopy Research and Technique* **54**:327–334.

Ball, G., L. Riters & J. Balthazart. 2002. Neuroendocrinology of song behavior and avian brain plasticity: Multiple sites of action on sex steriod hormones. *Frontiers in Neuroendocrinology* **23**:137–178.

Ballentine, B., J. Hyman & S. Nowicki. 2004. Vocal performance influences female response to male bird song: an experimental test. Behavioral Ecology **15**:163–168.

Ballintijn, M.R. & C. ten Cate. 1997. Sex differences in the vocalizations and syrinx of the collared dove (*Streptopelia decaocto*). *Auk* **114**:22–39.

Ballintijn, M.R. & C. ten Cate. 1998. Sound production in the collared dove: A test of the 'whistle' hypothesis. *Journal of Experimental Biology* **201**:1637–1649.

Ballintijn, M.R. & C. ten Cate. 1999a. Acoustic differentiation in the coo vocalizations of the collared dove. *Bioacoustics* **10**:1–17.

Ballintijn, M.R. & C. ten Cate. 1999b. Variation in the number of elements in the perch-coo vocalization of the collared dove (*Streptopelia decaocto*) and what it may tell about the sender. *Behaviour* **136**:847–864.

Ballintijn, M.R., C. ten Cate, F.W. Nuijens & H. Berkhoudt. 1995. The syrinx of the collared dove (*Streptopelia decaocto*): Structure, interindividual variation and development. *Netherlands Journal of Zoology* **45**:455–479.

Balthazart, J., P. Absil, V. Fiasse & G. Ball. 1995. Effects of the brain aromatase inhibitor R76713 on sexual differentiation of brain and behavior in zebra finches. *Behaviour* **120**:225–260.

Balthazart, J. & E. Adkins-Regan. 2002. Sexual differentiation of brain and behavior in birds. In: *Hormones, Brain and Behavior,* D. Pfaff, A. Arnold, A. Etgen, S. Fahrbach & R. Rubin (eds.), pp. 223–301. Academic Press, San Diego, CA.

Balthazart, J., A. Foidart, E.M. Wilson & G.F. Ball. 1992. Immunocytochemical localization of androgen receptors in the male songbird and quail brain. *Journal of Comparative Neurology* **317**:407–420.

Bandura, A. 1971. Analysis of social modeling processes. In: *Psychological Modeling*, A. Bandura (ed.), pp. 1–62. Aldine-Atherton, Chicago, IL.

Banta Lavenex, P. 1999. Vocal production mechanisms in the budgerigar (*Melopsittacus undulatus*): The presence and implications of amplitude modulation. *Journal of the Acoustical Society of America* **106**:491–505.

Baptista, L.F. 1975. Song dialects and demes in sedentary populations of the white-crowned sparrow (*Zonotrichia leucophrys nuttalli*). *University of California Publications in Zoology* **105**:1–52.

Baptista, L.F. 1983. Song learning. In: *Perspectives in Ornithology*, A.H. Brush & G.A. Clark Jr. (eds.), pp. 500–506. Cambridge University Press, Cambridge, MA.

Baptista, L.F. 1985. Bird song dialects: Social adaptation of assortative mating? *Behavioural Brain Science* **8**:100–101.

Baptista, L.F. 1990. Dialectal variation in the raincall of the chaffinch (*Fringilla coelebs*). *Die Vogelwarte* **35**:249–256.

Baptista L.F. 1996. Nature and its nurturing in avian vocal development. In: *Ecology and Evolution of Acoustic Communication in Birds,* D.E. Kroodsma & E.H. Miller (eds.), pp. 39–60. Cornell University Press, Ithaca, NY.

Baptista, L.F. 1999. Field and laboratory studies of the biology of the white-crowned sparrow. *Poultry and Avian Biology Reviews* **10**:101–107.

Baptista, L.F. & M. Abs. 1983. Vocalizations. In: *Physiology and Behaviour of the Pigeon*, M. Abs (ed.), pp. 309–325. Academic Press, London.

Baptista, L.F., W.I. Boarman & P. Kandianidis. 1983. Behavior and taxonomic status of Grayson's dove. *Auk* **100**:907–919.

Baptista, L.F. & C.K. Catchpole. 1989. Vocal mimicry and interspecific aggression in songbirds: Experiments using white-crowned sparrow imitation of song sparrow song. *Behaviour* **109**:247–257.

Baptista, L.F. & S.L.L. Gaunt. 1997a. Social interaction and vocal development in birds. In: *Social Influences on Vocal Development*, C.T. Snowdon & M. Hausberger (eds.), pp. 23–40. Cambridge University Press, Cambridge, MA.

Baptista, L.F. & S.L.L. Gaunt. 1997b. Bioacoustics as a tool in conservation studies. In: *Behavioral Approaches to Conservation in the Wild*, R. Buchholz & J.R. Clemmons (eds.), pp. 212–242. Cambridge University Press, New York.

Baptista, L.F. & D.E. Kroodsma. 2001. Foreword: Avian bioacoustics: A tribute to Luis Felipe Baptista. In: *Handbook of the Birds of the World. Vol. 6*, J. del Hoyo, A. Elliott & J. Sargatal (eds.), pp. 11–52. Lynx Edicions, Barcelona.

Baptista, L.F. & M.L. Morton. 1981. Interspecific song acquisition by a white-crowned sparrow. *Auk* **98**:383–385.

Baptista, L.F. & M.L. Morton. 1982. Song dialects and mate selection in montane white-crowned sparrows. *Auk* **99**:537–547.

Baptista, L.F. & M.L. Morton. 1988. Song learning in montane white-crowned sparrows: From whom and when? *Animal Behaviour* **36**:1753–1764.

Baptista, L.F., M.L. Morton & M.E. Pereyra. 1981. Interspecific song mimesis by a Lincoln sparrow. *Wilson Bulletin* **93**:265–267.

Baptista, L.F. & L. Petrinovich. 1984. Social interaction, sensitive phases and the song template hypothesis in the white-crowned sparrow. *Animal Behaviour* **32**:172–181.

Baptista, L.F. & L. Petrinovich. 1986. Song development in the white-crowned sparrow: Social factors and sex differences. *Animal Behaviour* **34**:1359–1371.

Baptista, L.F. & L. Petrinovich. 1987. Song development in the white-crowned sparrow: Modification of learned song. *Animal Behaviour* **35**:964–974.

Baptista, L.F. & K.L. Schuchmann. 1990. Song learning in the Anna hummingbird (*Calypte anna*). *Ethology* **84**:15–26.

Baptista, L.F. & P.W. Trail. 1992. The role of song in the evolution of passerine diversity. *Systematic Biology* **41**:242–247.

Baptista, L.F., P.W. Trail, B.B. Dewolfe & M.L. Morton. 1993. Singing and its functions in female white-crowned sparrows. *Animal Behaviour* **46**:511–524.

Barrington, D. 1773. Experiments and observations on the singing of birds. *Philosophical Transactions of the Royal Society* **63**:249–291.

Bartlett, P. & P.J.B. Slater. 1999. The effect of new recruits on the flock specific call of budgerigars (*Melopsittacus undulatus*). *Ethology, Ecology and Evolution* **11**:139–147.

Baylis, J.R. 1982. Avian vocal mimicry: Its function and evolution. In: *Acoustic Communication in Birds Vol. 2 Song Learning and its Consequences*, D. Kroodsma &E. H. Miller (eds.). pp. 51–83. Academic Press, New York.

Beani, L. & F. Dessi-Fulghieri. 1995. Mate choice in the grey partridge, *Perdix perdix*: Role of physical and behavioural traits. *Animal Behaviour* **49**:347–356.

Beani, L., G. Panzica, F. Briganti, P. Persichella & F.

Dessifulgheri. 1995. Testosterone-induced changes of call structure, midbrain and syrinx anatomy in partridges. *Physiology and Behavior* **58**:1149–1157.

Becker, P.H. 1982. The coding of species-specific characteristics in bird sounds. In: *Acoustic Communication in Birds. Vol. I, Production, Perception, and Design Features of Sounds*, D.E. Kroodsma & E.H. Miller (eds.), pp. 214–252. Academic Press, New York.

Beckers, G.J.L., B.M.A. Goossens & C. ten Cate. 2003a. Perceptual salience of acoustic differences between conspecific and allospecific vocalizations in African collared-doves. *Animal Behaviour* **65**:605–614.

Beckers, G.J.L., R.A. Suthers & C. ten Cate. 2003b. Pure-tone birdsong by resonance filtering of harmonic overtones. *Proceedings of the National Academy of Sciences* **100**:7372–7376.

Beckers, G.J.L., R.A. Suthers & C. ten Cate. 2003c. Mechanisms of frequency and amplitude modulation in ring dove song. *Journal of Experimental Biology* **206**:1833–1843.

Beckers, G.J.L. & C. ten Cate. 2001. Perceptual relevance of species-specific differences in acoustic signal structure in *Streptopelia* doves. *Animal Behaviour* **62**:519–525.

Beckers, G.J.L. & C. ten Cate. In press. Nonlinear phenomena and song evolution in *Streptopelia* doves. *Acta Zoologica Sinica.*

Beddard, F.E. 1898. *The Structure and Classification of Birds.* Longman, Green, New York.

Beecher, M.D. 1990. The evolution of parent–offspring recognition in swallows. In: *Contemporary Issues in Comparative Psychology*, D.A. Dewsbury (ed.), pp. 360–380. Sinauer Associates, Sunderland, MA.

Beecher, M.D. 1996. Birdsong learning in the laboratory and field. In: *Ecology and Evolution of Acoustic Communication in Birds,* D.E. kroodsma & E.H. Miller (eds.), pp. 61–78. Cornell University Press, Ithaca, NY.

Beecher, M.D., I.M. Beecher & S. Hahn. 1981a. Parent–offspring recognition in bank swallows (*Riparia riparia*): II. Development and acoustic bias. *Animal Behaviour* **29**:95–101.

Beecher, M.D., I.M. Beecher & S. Lumpkin. 1981b. Parent–offspring recognition in bank swallows (*Riparia riparia*): I. Natural history. *Animal Behaviour* **29**:86–94.

Beecher, M.D., S.E. Campbell, J.M. Burt, C.E. Hill & J.C. Nordby. 2000a. Song-type matching between neighbouring song sparrows. *Animal Behaviour* **59**:29–37.

Beecher, M.D., S.E. Campbell & J.C. Nordby. 2000b. Territory tenure in song sparrows is related to song sharing with neighbours, but not to repertoire size. *Animal Behaviour* **59**:29–37.

Beecher, M.D., P.K. Stoddard, S.E. Campbell & C.L. Horning. 1996. Repertoire matching between neighbouring song sparrows. *Animal Behaviour* **51**:917–923.

Beeman, K. 2002. *Signal and Sigwin Users Guide.* Engineering Design, Belmont, MA.

Beer, C.G. 1969. Laughing gull chicks: Recognition of their parents' voices. *Science* **166**:1030–1032.

Beer, C.G. 1970. Individual recognition of voice in the social behavior of birds. In: *Advances in the Study of Behavior. Vol. 3,* D.S. Lehrman, R.A. Hinde & E. Shaw (eds.), pp. 27–74. Academic Press, New York.

Béguin, N., G. Leboucher & M. Kreutzer. 1998. Sexual preferences for mate song in female canaries. *Behaviour* **135**:1185–1196.

Beier, J., B. Leisler, & M. Wink. 1997. A great reed warbler x reed warbler (*Acrocephalus arundinaceus* x *A. scirpaceus*) hybrid and its parentage. *Journal für Ornithologie* **138**:51–60.

Beletsky, L.D., B.J. Higgins & G.H. Orians. 1986. Communication by changing signals: Call switching in red-winged blackbirds. *Behavioral Ecology & Sociobiology* **18**:221–229.

Bell, D.A., P.W. Trail & L.F. Baptista. 1998. Song learning and vocal tradition in Nuttall's white-crowned sparrows. *Animal Behaviour* **55**:939–956.

Bell, P.R. (ed.). 1959. *Darwin's Biological Work: Some Aspects Reconsidered.* Cambridge University Press, Cambridge, MA.

Bengtsson, H. & O. Rydén. 1983. Parental feeding rate in relation to begging behavior in asynchronously hatched broods of the great tit *Parus major*. *Behavioral Ecology & Sociobiology* **12**:243–251.

Benkman, C.W. 1993. Adaptation to single resources and the evolution of crossbill (*Loxia*) diversity. *Ecological Monographs* **63**:305–325.

Benney, K.S. & R.F. Braaten. 2000. Auditory scene analysis in estrildid finches (*Taenopygia guttata* and *Lonchura striata domestica*): A species advantage for detection of conspecific song. *Journal of Comparative Psychology* **114**:174–182.

Bensch, S., D. Hasselquist, B. Nielsen & B. Hasson. 1998. Higher fitness for philopatric than for immigrant males in a semi-isolated population of great reed warblers. *Evolution* **52**:877–883.

Benson, D.F. & A. Ardila. 1996. *Aphasia: A Clinical Perspective*.Oxford University Press, New York.

Benton, S., D.A. Nelson, P. Marler & T.J. DeVoogd. 1998. Anterior forebrain pathway is needed for stable song expression in adult male white-crowned sparrows (*Zonotrichia leucophrys*). *Behavioural Brain Research* **96**:135–150.

Berck, K.H. 1961. Beiträge zur Ethologie des Feldsperlings (*Passer montanus*) und dessen Beziehung zum Haussperling (*Passer domesticus*). *Vogelwelt* **82**:129–72; **83**:8–26.

Berglund, A., A Bisazza & A. Pilastro. 1996. Armaments and ornaments: An evolutionary explanation of traits of dual utility. *Biological Journal of the Linnean Society* **58**:385–399.

Bergmann, H.-H. 1978. Beziehungen zwischen Habitatstructur und Motivgesang bei europäischen Grasmücken (Gattung *Sylvia*). *Journal für Ornithologie* **119**:236–237.

Bergmann, H.-H. 1993. *Der Buchfink*. AULA-Verlag, Wiesbaden.

Bergmann, H.-H. & H.-W. Helb. 1982. *Stimmen der Vögel Europas*. BLV Verlagsgesellschaft, München.

Bernard, D.J., G.E. Bentley, J Balthazart, F.W. Turek & G.F. Ball. 1999. Androgen receptor, estrogen receptor a, and estrogen receptor b show distinct patterns of expression in forebrain song control nuclei of European starlings. *Endocrinology* **140**:4633–4643.

Binder, J.R. 1997. Neuroanatomy of language processing studied with functional MRI. *Clinical Neuroscience* **4**:87–94.

Birkhead, T.R., F. Fletcher & E.J. Pellatt. 1999. Nestling diet, secondary sexual traits and fitness in the zebra finch. *Proceedings of the Royal Society of London, B* **266**: 385–390.

Birmani, T. 1999. Birds in song contest. *Dawn*, the internet edition, 20 December 1999.

Bischof, H.J. 1997. Song learning, filial imprinting, and sexual imprinting: Three variations of a common theme? *Biomedical Research* Tokyo **18**:133–146.

Bjerke, T.K. & T.H. Bjerke. 1981. Song dialects in the redwing (*Turdus iliacus*). *Ornis Scandinavica* **12**:40–50.

Blair, R.B. 1999. Birds and butterflies along an urban gradient: Surrogate taxa for assessing biodiversity? *Ecological Applications* **9**:164–170.

Blanchard, B.D. 1941. The white-crowned sparrows (*Zonotrichia leucophrys*) of the Pacific seaboard: Environment and annual cycle. *University of California Publications in Zoology* **46**:1–178.

Blokpoel, H. & J. Neuman. 1997. Sound levels in 3 ring-billed gull colonies of different size. *Colonial Waterbirds* **20**:221–226.

Boag, P.T. & P.R. Grant. 1981. Intense natural selection in a population of Darwin's finches (*Geospizinae*) in the Galápagos. *Science* **214**:82–85.

Boco, T. & D. Margoliash. 2001. NIf is a major source of auditory and spontaneous drive to HVc. *Society for Neuroscience* (*Abstracts*) 27:841.

Böhner, J. 1983. Song learning in the zebra finch (*Taeniopygia guttata*): Selectivity in the choice of a tutor and accuracy of song copies. *Animal Behaviour* **31**:231–237.

Böhner, J. 1990. Early acquisition of song in the zebra finch, *Taeniopygia guttata*. *Animal Behaviour* **39**:369–374.

Böhner, J., M. Chaiken, G.F. Ball & P. Marler. 1990. Song acquisition in photo-sensitive and photo-refractory male European starlings. *Hormones & Behavior* **24**:582–594.

Böhner, J. & D. Todt. 1996. Influence of auditory stimulation on the development of syntactical and temporal features in European starling song. The *Auk* **113**:450–456.

Bolhuis, J.J., G.G.O. Zijlstra, A.M. den Boer-Visser & E.A. Van der Zee. 2000. Localized neuronal activation in the zebra finch brain is related to the strength of song learning. *Proceedings of the National Academy of Sciences USA* **97**:2282–2285.

Bolsinger, J.S. 2000. Use of two song categories by golden-cheeked warblers. *Condor* **102**:539–552.

Borg, E., S.A. Counter & J. Lannergren. 1982. Analysis of the avian middle ear muscle contraction by strain gauge and volume and impedance change measures. *Comparative Biochemistry and Physiology. A: Comparative Physiology* **71**:619–621.

Borgia, G. & S.W. Coleman. 2000. Co-option of male courtship signals from aggressive display in bowerbirds. *Proceedings of the Royal Society of London, B* **267**:1735–1740.

Borror, D.J. 1956. Variation in Carolina Wren songs. Auk **73**:211–229.

Borror, D.J. 1959. Songs of the chipping sparrow. *The Ohio Journal of Science* **59**:347–356.

Borror, D.J. 1960. The analysis of animal sounds. In: *Animal Sounds and Communication*, W.E.

Lanyon & W.N. Tavolga (eds.), American Institution of Biological Sciences, Washington, D.C.

Borror, D.J. 1964. Songs of the thrushes (*Turdidae*), wrens (*Troglodytidae*), and mockingbirds (*Mimidae*) of eastern North America. *Ohio Journal of Science* **64**:195–207.

Borror, D.J. & C.R. Reese. 1953. The analysis of bird song by means of a vibralyser. *Wilson Bulletin* **65**:271–276.

Borror, D.J. & C.R. Reese. 1956. Vocal gymnastics in wood thrush songs. *Ohio Journal of Science* **56**:177–182.

Bottjer, S.W., J.D. Brady & B. Cribbs. 2000. Connections of a motor cortical region in zebra finches: *Relation* to pathways for vocal learning. *Journal of Comparative Neurology* **420**:244–260.

Bottjer, S.W., K.A. Halsema, S.A. Brown & E.A. Miesner. 1989. Axonal connections of a forebrain nucleus involved with vocal learning in zebra finches. *Journal of Comparative Neurology* **279**:312–326.

Bottjer, S.W. & F. Johnson. 1997. Circuits, hormones, and learning: Vocal behavior in songbirds. *Journal of Neurobiology* **33**:602–618.

Bottjer, S.W., E.A. Miesner & A.P. Arnold. 1984. Forebrain lesions disrupt development but not maintenance of song in passerine birds. *Science* **224**:901–903.

Boudreau, G.W. 1968. Alarm sounds and responses of birds and their application in controlling problem species. *Living Bird* 7:27–46.

Boughey, M.J. & S. Nicholas. 1990. Song variation in the brown thrasher (*Toxostoma rufum*). *Zeitschrift fur Tierpsychologie* **56**:47–58.

Boughey, M.J. & N.S. Thompson. 1981. Song variety in the brown thrasher (*Toxostoma rufum*). *Zeitschrift für Tierpsychologie* **56**:47–58.

Boughman, J.W. 1997. Greater spear-nosed bats give group-distinctive calls. *Behavioral Ecology & Sociobiology* **40**:61–70.

Bower, G.H. 1970. Organizational factors in memory. *Cognitive Psychology* **1**:18–46.

Bower, J.L. 2000. *Acoustic interactions during naturally occurring territorial conflict in a song sparrow (Melospiza melodia) neighborhood*. Doctoral dissertation, Cornell University, Ithaca, NY.

Bowman, R.I. 1961. Morphological differentiation and adaptation in the Galápagos finches. *University of California Publications in Zoology* **58**:1–302.

Bowman, R.I. 1963. Evolutionary patterns in Darwin's finches. *Occasional Papers of the California Academy of Sciences* **44**:107–140.

Bowman, R.I. 1979. Adaptive morphology of song in Darwin's finches. *Journal für Ornithologie* **120**:353–389.

Bowman, R.I. 1983. The evolution of song in Darwin's finches. In: *Patterns of Evolution in Galápagos Organisms*, R.I. Bowman, M. Berson & A.E. Leviton (eds.), pp. 237–537. American Association for the Advancement of Science, Pacific Division, San Francisco CA.

Brackenbury, J.H. 1979. Power capabilities of the avian sound-producing system. *Journal of Experimental Biology* **78**:163–166.

Brackenbury, J.H. 1982. The structural basis of voice production and its relationship to sound characteristics. In: *Acoustic Communication in Birds, Vol. 1. Production, Perception, and Design Features of Sounds*, D.E. Kroodsma & E.H. Miller (eds.), pp. 53–73. Academic Press, New York.

Bradbury, J.W. 2003. Vocal communication in wild parrots. In: *Animal Social Complexity: Intelligence, Culture and Individualized Societies*, F.B.M. DeWaal & P.L. Tyack (eds.), pp. 293–316. Harvard University Press, Cambridge, MA.

Bradbury, J.W., K.A. Cortopassi & J.R. Clemmons. 2001. Geographical variation in the contact calls of orange-fronted parakeets. *Auk* **118**:958–972.

Bradbury, J.W. & S.L. Vehrencamp. 1994. SINGIT! A program for interactive playback on the Macintosh. *Bioacoustics* **5**:310.

Bradbury, J.W. & S.L. Vehrencamp. 1998. *Principles of Animal Communication*. Sinauer Associates, Sunderland, MA.

Bradshaw, J. 1989. Hemispheric specialization and psychological function. In: (eds), pp. 121–147, 179–191. John Wiley & Sons, Chichester, NY.

Brainard, M. & A. Doupe. 2000. Interruption of a basal ganglia–forebrain circuit prevents plasticity of learned vocalizations. *Nature* **404**:762–766.

Brainard, M.S. & A.J. Doupe. 2002. What songbirds teach us about learning. *Nature* **417**: 351–358.

Brauth, S.E., J.T. Heaton, S.E. Durand, W. Liang & W.S. Hall. 1994. Functional anatomy of forebrain auditory pathways in the budgerigar (*Melopsittacus undulatus*). *Brain, Behavior & Evolution* **44**:210–233.

Brauth, S.E., W. Liang & T.F. Roberts. 2001. Projections of the oval nucleus of the hyperstriatum ventrale in the budgerigar: relationships with the auditory system. *Journal of Comparative Neurology* **432**:481–511.

Brauth, S.E. & C.M. McHale. 1988. Auditory pathways in the budgerigar. II. Intratelencephalic pathways. *Brain, Behavior & Evolution* **32**:193–207.

Brauth, S.E., C.M. McHale, C.A. Brasher & R.J. Dooling. 1987. Auditory pathways in the budgerigar. I. Thalamo-telencephalic projections. *Brain, Behavior & Evolution* **30**:174–199.

Bregman, A.S. 1990. *Auditory Scene Analysis.* MIT Press, Cambridge, MA.

Bregman, A.S. & J.D. Campbell. 1971. Primary auditory stream segregation and perception of order in rapid sequences of tones. *Journal of Experimental Psychology* **89**:244–249.

Brémond, J.-C. 1968. Recherches sur la semantique et les elements vecteurs d'information dans les signaux acoustiques du rouge-gorge (*Erithacus rubecula L.*). *Terre Vie* **2**:109–220.

Brémond, J.-C. 1976. Specific recognition in the song of Bonelli's warbler (*Phylloscopus bonelli*). *Behaviour* **58**:99–116.

Brémond, J.-C. 1978. Acoustic competition between the song of the wren (*Troglodytes troglodytes*) and the songs of other species. *Behaviour* **65**:89–98.

Brémond, J.-C. & T. Aubin. 1990. Responses to distress calls by black-headed gulls, *Larus ridibundus*: The role of non-degraded features. *Animal Behaviour* **39**:503–511.

Brenowitz, E.A. 1982a. The active space of red-winged blackbird song. *Journal of Comparative Physiology A.* **147**:511–522.

Brenowitz, E.A. 1982b. Long-range communication of species identity by song in the red-winged blackbird. *Behavioural Ecology & Sociobiology* **10**:29–38.

Brenowitz, E.A. 1991. Evolution of the vocal control system in the avian brain. *The Neurosciences* **3**:399–407.

Brenowitz, E.A. 1997. Comparative approaches to the avian song system. *Journal of Neurobiology* **33**:517–531.

Brenowitz, E.A. & A.P. Arnold. 1986. Interspecific comparisons of the size of neural song control regions and song complexity in duetting birds: Evolutionary implications. *Journal of Neuroscience* **6**:2875–2879.

Brenowitz, E.A. & D.E. Kroodsma. 1996. The neuroethology of birdsong. In: *Ecology and Evolution of Acoustic Communication in Birds,* D.E. Kroodsma & E.H. Miller (eds), pp. 285–304. Cornell University Press, Ithaca, NY.

Brenowitz, E.A., K. Lent & D.E. Kroodsma. 1995. Brain space for learned song in birds develops independently of song learning. *Journal of Neuroscience* **15**:6281–6286.

Brindley, E.L. 1991. Response of European robins to playback of song: Neighbor recognition and overlapping. *Animal Behaviour* **41**:503–512.

Briskie, J.V., P.R. Martin & T.E. Martin. 1999. Nest predation and the evolution of nestling begging calls. *Proceedings of the Royal Society of London, B* **266**:2153–2159.

Brittan-Powell, E.F. & R.J. Dooling, In press. Development of auditory sensitivity in budgerigars (*Melopsittacus undulatus*). Journal of the Acoustical Society of America.

Brittan-Powell, E.F., R.J. Dooling & S.M. Farabaugh. 1997. Vocal development in budgerigars (*Melopsittacus undulatus*): Contact calls. *Journal of Comparative Psychology* **111**:226–241.

Brittan-Powell, E.F., R.J. Dooling, T. Wright, P.C. Mundinger & B.M. Ryals. 2002. *Development of auditory sensitivity in Belgian Waterslager (BWS) canaries.* Association for Research in Otolaryngology, St. Petersburg, FL.

Broca, P. 1861. Nouvelle observation d'aphemie produite par une lesion de la moitie posterierure des deuxieme et troisieme circonvolutions frontales. *Bulletin de la Societe Anatomy de Paris* **VI**:398–407.

Brooks, R.J. & J.B. Falls. 1975a. Individual recognition by song in white-throated sparrows I: Discrimination of songs of neighbors and strangers. *Canadian Journal of Zoology* **53**:879–888.

Brooks, R.J. & J.B. Falls. 1975b. Individual recognition by song in white-throated sparrows III: Song features used in individual recognition. *Canadian Journal of Zoology* **53**:1749–1761.

Brosset, A. & C. Erard. 1986. Les oiseaux des régions forestières du nord-est du Gabon. Volume I. *Rev. Ecol.* (Terre et Vie): Numéro Spéciale.

Brown, A.L. 1990. Measuring the effect of aircraft noise on sea birds. *Environment International* **16**:587–592.

Brown, C.H. 1989. The active space of the blue monkey and grey-cheeked mangabey vocalizations. *Animal Behaviour* **37**:1023–1034.

Brown, C.R., M.B. Brown & M.L. Shaffer. 1991. Food-sharing signals among socially foraging cliff swallows. *Animal Behaviour* **42**:551–564.

Brown, E.D. & S.M. Farabaugh. 1991. Song sharing in a group-living songbird, the Australian magpie, *Gymnorhina tibicen*: Part III. Sex specificity and individual specificity of vocal parts in communal chorus and duet songs. *Behaviour* **118**:244–274

Brown, E.D. & S.M. Farabaugh. 1997. What birds with complex social relationships can tell us about vocal learning: Vocal sharing in avian groups. In: *Social influences on vocal development*, C.T. Snowdon & M. Hausberger, (eds.), pp. 98–127. Cambridge University Press, Cambridge, MA.

Brown, J. 1965. Loss of vocalizations caused by lesions in the nucleus mesencephalicus lateralis of the redwinged blackbird. *American Zoology* **5**:693.

Brown, J. & R. Hunsperger. 1963. Neuro-ethology and the motivation of agonistic behavior. *Animal Behaviour* **11**:439–448.

Brown, J.L. 1964. The integration of agonistic behavior in the Steller's jay *Cyanocitta stelleri* (Gmelin). *University of California Publications in Zoology* **60**:223–328.

Brown, P. & C.D. Marsden. 1998. What do the basal ganglia do? *Lancet* **351**:1801–1804.

Brown, S.D., R.J. Dooling & K. O'Grady. 1988. Perceptual organization of acoustic stimuli by budgerigars (*Melopsittacus undulatus*): III. Contact calls. *Journal of Comparative Psychology* **102**:236–47.

Brown, T.J. & P. Handford. 1996. Acoustical signal amplitude patterns: a computer simulation investigation of the acoustic adaptation hypothesis. *Condor* **98**:608–623.

Brown, T.J. & P. Handford. 2000. Sound design for vocalizations: Quality in the woods, consistency in the fields. *Condor* **102**:81–92.

Brown, T.J. & P. Handford. 2003. Why birds sing at dawn: The role of consistent song transmission. *Ibis* **145**:120–129.

Brown, R. 1958. *Words and Things*. Free Press, New York.

Brown, R. 1973. *A First Language: The Early Stages*. Harvard University Press, Cambridge, MA.

Browne, P.W.P. 1953. Nocturnal migration of thrushes in Ireland. *British Birds* **46**:370–374.

Brumm, H. & H. Hultsch. 2001. Pattern amplitude is related to pattern imitation during the song development of nightingales. *Animal Behaviour* **61**:747–754.

Brumm, H. & D. Todt. 2002. Noise-dependent song amplitude regulation in a territorial songbird. *Animal Behaviour* **63**:891–897.

Buchanan, K.L. & C.K. Catchpole. 1997. Female choice in the sedge warbler, *Acrocephalus schoenobaenus*: Multiple cues from song and territory quality. *Proceedings of the Royal Society of London, B* **264**:521–526.

Buchanan, K.L. & C.K. Catchpole. 2000. Song as an indicator of male parental effort in the sedge warbler. *Proceedings of the Royal Society of London, B* **267**:321–326.

Buchanan, K.L., C.K. Catchpole, J.W. Lewis & A. Lodge. 1999. Song as an indicator of parasitism in the sedge warbler. *Animal Behaviour* **57**:307–314.

Buckland, S.T., D.R. Anderson, K.P. Burnham, J.L. Laake, D.L. Borchers & L. Thomas. 2001. *Introduction to Distance Sampling: Estimating Abundance of Biological Populations*. Oxford University Press, Oxford.

Buckley, P.A. & F.G. Buckley. 1972. Individual egg and chick recognition by adult royal terns (*Sterna maxima maxima*). *Animal Behaviour* **20**:457–462.

Budney, G.F. & R.W. Grotke. 1997. Techniques for audio recording vocalizations of tropical birds. In: *Studies in Neotropical Ornithology Honoring Ted Parker*, J.V. Remsen, Jr. (ed.), pp. 147–163. American Ornithologists' Union, Washington, D.C.

Bugnyar, T., M. Kijne & K. Kotrschal. 2001. Food calling in ravens: Are yells referential signals? *Animal Behaviour* **61**:949–958.

Burgoon, J.K. & T. Saine. 1978. *The Unspoken Dialogue*. Houghton Mifflin Co, Dallas, Palo Alto, London.

Burnett, S.C. 2001. *Individual Variation in the Echolocation Calls of Big Brown Bats (Eptesicus fuscus) and their Potential for Acoustic Identification and Censusing*. Doctoral dissertation, The Ohio State University, Columbus.

Burnett, S.C., K.A. Kazial & W.M. Masters. 2001. Discriminating individual big brown bat (*Eptesicus fuscus*) sonar vocalizations in different recording situations. *Bioacoustics* **11**:189–210.

Burnett, S.C. & W.M. Masters. 1999. The use of neural networks to classify echolocation in bats. *Journal of the Acoustical Society of America* **106**:2189.

Burt, J.M., S.E. Campbell & M.D. Beecher. 2001. Song type matching as threat: A test using interactive playback. *Animal Behaviour* **62**:1163–1170.

Burt, J.M., K.L. Lent, M.D. Beecher & E.A. Brenowitz. 2000. Lesions of the anterior forebrain song control pathway in female canaries affect song perception in an operant task. *Journal of Neurobiology* **42**:1–13.

Busnel, R.G. (ed.). 1963. *Acoustic Behaviour of Animals*. Elsevier, New York.

Busse, K. & K. Busse. 1977. Prägungsbedingte Bindung von Küstenseeschwalbenkücken (*Sterna paradisea* Pont.) an die Eltern und ihre Fähigkeit, sie an der Stimme zu erkennen. *Zeitschrift für Tierpsychologie* **43**:287–294.

Byers, B.E. 1995. Song types, repertoires and song variability in a population of chestnut-sided warblers. *Condor* **97**:390–401.

Byers, B.E. 1996a. Geographic variation of song form within and among chestnut-sided warbler populations. *Auk* **113**:288–299.

Byers, B.E. 1996b. Messages encoded in the songs of chestnut-sided warblers. *Animal Behaviour* **52**:691–705.

Byers, B.E. & D.E. Kroodsma. 1992. Development of two song categories by chestnut-sided warblers. *Animal Behaviour* **44**:799–810.

Byrkjedal, I. 1990. Song flight of the pintail snipe *Gallinago stenura* on the breeding grounds. *Ornis Scandinavica* **21**:239–247.

Calder, W.A. 1970. Respiration during song in the canary (*Serinus canaria*). *Comparative Biochemistry and Physiology* **32**:251–258.

Calder III, W.A. 1990. The scaling of sound output and territory size: Are they matched? *Ecology* **71**:1810–1816.

Caldwell, M.C. & D.K. Caldwell. 1972. Vocal mimicry in the whistle mode by an Atlantic bottlenosed dolphin. *Cetology* **9**:1–8.

Cameron, H., C. Wooley, B. McEwen & E. Gould. 1993. Differentiation of newly born neuron and glia in the dentate gyrus of the adult rat. *Neuroscience* **56**:337–344.

Capp, M.S. & W.A. Searcy. 1991. Acoustical communication of aggressive intentions by territorial male bobolinks. *Behavioural Ecology* **2**:319–326.

Capsius, B. & H. Leppelsack. 1999. Response patterns and their relationship to frequency analysis in auditory forebrain centers of a songbird. *Hearing Research* **136**:91–99.

Carlson, G. & C.H. Trost. 1992. Sex determination of the whooping crane by analysis of vocalizations. *Condor* **94**:532–536.

Carr, C. & R. Code. 2000. The central auditory system of reptiles and birds. In: *Comparative Hearing: Birds and Reptiles,* R. Dooling, R. Fay & A. Popper (eds.), pp. 197–248. Springer, New York.

Carson, R. 1962. *Silent Spring*. Houghton Mifflin Co., Boston.

Catchpole, C.K. 1976. Temporal and sequential organisation of song in the sedge warbler (*Acrocephalus schoenobaenus*). *Behaviour* **59**:226–246.

Catchpole, C.K. 1980 Sexual selection and the evolution of complex songs among European warblers of the genus *Acrocephalus*. *Behaviour* **74**:149–166.

Catchpole, C.K. 1983. Variation in the song of the great reed warblers *Acrocephalus arundinaceus* in relation to mate attraction and territorial defense. *Animal Behaviour* **31**:1217–1225.

Catchpole, C.K. 1986. Song repertoires and reproductive success in the great reed warbler, *Acrocephalus arundinaceus*. *Behavioural Ecology & Sociobiology* **19**:439–445.

Catchpole, C.K. 2000. Sexual selection and the evolution of song and brain structure in Acrocephalus warblers. *Advances in the Study of Behaviour* **29**:45–97.

Catchpole, C.K., J. Dittami & B. Leisler. 1984. Differential responses to male song repertoires in female songbirds implanted with oestradiol. *Nature* **312**:563–564.

Catchpole, C.K. & B. Leisler. 1996. Female aquatic warblers (*Acrocephalus paludicola*) are attracted by playback of longer and more complicated songs. *Behaviour* **133**:1153–1164.

Catchpole, C.K. & A. Rowell. 1993. Song sharing and local dialects in a population of the European wren *Troglodytes troglodytes*. *Behaviour* **125**:67–78.

Catchpole, C.K. & P.J.B. Slater. 1995 . *Bird Song: Biological Themes and Variations*. Cambridge University Press, Cambridge, MA.

Cerchio, S., J.K. Jacobsen & T.F. Norris. 2001. Temporal and geographical variation in songs of humpback whales, *Megaptera novaeangliae*: Synchronous change in Hawaiian and Mexican breeding assemblages. *Animal Behaviour* **62**:313–329.

Chaiken, M. 2000. Rehabilitation of isolate song in adult European starlings, *Sturnus vulgaris*. *Society for Neuroscience (Abstracts)* **26**:269.4.

Chaiken, M., J. Böhner & P. Marler. 1993. Song acquisition in European starlings, *Sturnus vulgaris*: A comparison of the songs of live-tutored, tape-tutored, untutored, and wild-caught males. *Animal Behaviour* **46**:1079–1090.

Chaiken, M., J. Böhner & P. Marler. 1994. Repertoire turnover and the timing of song acquisition in European starlings. *Behaviour* **128**:25–39.

Chaiken, M.L., T.Q. Gentner & S.H. Hulse. 1997. Effects of social interactions on the development of starling song and the perception of those effects by conspecifics. *Journal of Comparative Psychology* **111**:379–392.

Chamberlin, T.C. 1965. The Method of Multiple Working Hypotheses; With this method the dangers of parental affection for a favorite theory can be circumvented. *Science* **148**:754–759.

Chance, E.P. 1940. *The Truth About the Cuckoo.* Country Life, London.

Chance, M.R.A. 1962. An interpretation of some agonistic postures; the role of "cut-off" acts and postures. *Symposia of the Zoological Society of London* **8**:71–89.

Chapin, J.P. 1954. The birds of the Belgian Congo. *Bulletin of the American Museum of Natural History* **75B**.

Chappuis, C. 1971. Un example de l'influence du milieu sur les emissions vocales des oiseaux: l'evolution des chants en forêt equatoriale. *Terre et Vie* **118**:183–202.

Chappuis, C. 2001. *Oiseaux d'Afrique, Vol. 2* (11 CDs and book). SEOF. Alauda, Paris.

Chappuis, C. & C. Erard. 1993. Species limits in the genus *Bleda Bonaparte*, 1857 (Aves, *Pycnonotidae*). *Zeitschrift für Zoölogische Systematik und Evolutionsforschung* **31**:280–299.

Charrier, I., P. Jouventin, N. Mathevon & T. Aubin. 2001. Acoustic communication in a black-headed gull colony: How to identify their parents? *Ethology* **107**:961–974.

Cheng, M.F. 1992. For whom does the female dove coo? A case for the role of vocal self-stimulation. *Animal Behaviour* **43**:1035–1044.

Chew, S.J., C. Mello, F. Nottebohm, E. Jarvis & D.S. Vicario. 1995. Decrements in auditory responses to a repeated conspecific song are long-lasting and require two periods of protein synthesis in the songbird forebrain. *Proceedings of the National Academy of Sciences* **92**:3406–3410.

Chew, S.J., D.S. Vicario & F. Nottebohm. 1996. A large-capacity memory system that recognizes the calls and songs of individual birds. *Proceedings of the National Academy of Sciences* **93**:1950–1955.

Chilton, G. & M.R. Lein. 1996. Songs and sexual responses of female white-crowned sparrows (*Zonotrichia leucophrys*) from a mixed-dialect population. *Behaviour* **133**:173–198.

Chilton, G., M.R. Lein & L.F. Baptista. 1990. Mate choice by female white-crowned sparrows in a mixed dialect population. *Behavioural Ecology & Sociobiology* **27**:223–227.

Chisholm, A.H. 1946. Nature's Linguists: A Study of the Riddle of Vocal Mimicry. Brown, Prior, Anderson Pty. Ltd., Melbourne.

Chu, M. 2001a. Vocal mimicry in distress calls of phainopeplas. *Condor* **103**:389–395.

Chu, M. 2001b. Heterospecific responses to scream calls and vocal mimicry by phainopeplas (*Phainopepla nitens*) in distress. *Behaviour* **138**:775–787.

Cicero, C. 2000. Oak titmouse (*Baeolophus inornatus*) and juniper titmouse (*Baeolophus ridgwayi*). In: *The Birds of North America, No. 485*, A. Poole & F. Gill (eds.), pp. 1–28. The Birds of North America, Inc., Philadelphia, PA.

Clark, C.W., P. Marler & K. Beeman. 1987. Quantitative analysis of animal vocal phonology: An application to swamp sparrow song. *Ethology* **76**:101–115.

Clayton, D.F. 1997. Role of gene regulation in song circuit development and song learning. *Journal of Neurobiology* **33**:549–571.

Clayton, N.S. 1987. Song tutor choice in zebra finches. *Animal Behaviour* **35**:714–722.

Clayton, N.S. 1988. Song discrimination and learning in zebra finches. *Animal Behaviour* **36**:1016–1924.

Clayton, N.S. 1990a. Subspecies recognition and song learning in zebra finches. *Animal Behaviour* **40**:1009–1017.

Clayton, N.S. 1990b. Assortative mating in zebra finch subspecies, *Taeniopygia guttata guttata* and *T. g. castanotis*. *Philosophical Transactions of the Royal Society of London, B* **330**: 351–370.

Cody, M.L. & J.H. Brown. 1969. Song asynchrony in chapparal birds. *Nature* **222**:778–780.

Coles, R.B., M. Konishi & J.D. Pettigrew. 1987. Hearing and echolocation in the Australian grey swiftlet, *Collocalia spodiopygia*. *Journal of Experimental Biology* **129**:365–371.

Collar, N.J. 1999. Forward: Risk indicators and status assessment in birds. In: *Handbook of Birds of the World, Vol. 5, Barn-owls to Hummingbirds*, J. del Hoyo, A. Elliott & J. Sargatal (eds.), pp. 13–28. Lynx Edicions, Barcelona.

Collias, N.E. 1960. An ecological and functional

classification of animal sounds. In: *Animal Sounds and Communication*, W.E. Lanyon & W.N. Tavolga (eds.), pp. 368–391. American Institute of Biological Sciences, Washington, D.C.

Collias, N.E. 1963. A spectrographic analysis of the vocal repertoire of the African village weaverbird. *Condor* **65**:517–527.

Collias, N.E. 1987. The vocal repertoire of the red junglefowl: A spectrographic classification and the code of communication. *Condor* **89**:510–524.

Collias, N.E. & M. Joos. 1953. The spectrographic analysis of sound signals of the domestic fowl. *Behaviour* **5**:175–188.

Collins, S.A. 1999. Is female preference for male repertoires due to sensory bias? *Proceedings of the Royal Society of London, B.* **266**: 2309–2314.

Collins, S.A. & A.R. Goldsmith. 1998. Individual and species differences in quail calls (*Coturnix c. japonica, C. c. coturnix* and a hybrid). *Ethology* **104**:977–990.

Collins, S.A., C. Hubbard, & A.M. Houtman. 1994. Female mate choice in the zebra finch: The effect of male beak colour and male song. *Behavioural Ecology & Sociobiology* **35**:21–25.

Colver, K.J., D. Stokes & L. Stokes. 1999. *Stokes Field Guide to Bird Songs*. Time Warner Audio Books, New York.

Conover, M.R. 1994. Stimuli eliciting distress calls in adult passerines and response of predators and birds to their broadcast. *Behaviour* **131**:19–37.

Conover, M.R. & J.J. Perrito. 1981. Response of starlings to distress calls and predator models holding conspecific prey. *Zeitschrift für Tierpsychologie* **57**:163–172.

Conway, C.J., W.R. Eddleman & S.H. Anderson. 1994. Nesting success and survival of Virginia rails and soras. *Wilson Bulletin* **106**:466–473.

Conway, C.J. & J.P. Gibbs. 2001. Factors influencing detection probabilities and the benefits of call broadcast surveys for monitoring marsh birds. Final Report, USGS Patuxent Wildlife Research Center, Laurel, MD. 58 pp.

Cooke, B., C. Hegstrom, L. Villeneuve & S. Breedlove. 1998. Sexual differentiation of the vertebrate brain: Principles and mechanisms. *Frontiers in Neuroendocrinology* **19**:211–213.

Cooper, R.P. & R.N. Aslin. 1990. Preference for infant-directed speech in the first month after birth. *Child Development* **61**:1584–1595.

Cooney, R. & A. Cockburn. 1995. Territorial defence is the major function of female song in the superb fairy-wren *Malarus cyaneus*. *Animal Behaviour* **49**:1635–1647.

Corbo, M.S. & D.M. Barras. 1983. *Arnie, the Darling Starling*. Houghton Mifflin, Boston, MA.

Cornell Laboratory of Ornithology. 1992. *A Field Guide to Western Bird Song, Peterson Field Guide*. Houghton Mifflin Company, Boston, MA.

Cosens, S.E. & B. Falls. 1984. A comparison of sound propagation and song frequency in temperate marsh and grassland habitats. *Behavioural Ecology & Sociobiology* **15**:161–170.

Counter, S.A. & E. Borg. 1982. The avian stapedius muscle. Influence on auditory sensitivity and sound transmission. *Acta Otolaryngolica* **94**:267–274.

Cowan, N. 2001 The magical number 4 in short-term memory: A reconsideration of mental storage capacity. *Behavavioral and Brain Sciences* **24**:87–185.

Crawford, R.D. 1977. Polygynous breeding of short-billed marsh wrens. *Auk* **94**:359–362.

Cresswell, W. 1994. Song as a pursuit-deterrent signal, and its occurrence relative to other anti-predation behaviours of skylark (*Alauda arvensis*) on attack by merlins (*Falco columbarius*). *Behavioural Ecology & Sociobiology* **34**:217–223.

Croll, D., J.K. Nansen, M.E. Goebel, P.L. Boveng & J.L. Bengtson. 1996. Foraging behavior and reproductive success in chinstrap penguins: The effects of transmitter attachment. *Journal of Field Ornithology* **67**:1–9.

Crowder, R.G. 1976. *Principles of Learning and Memory*. Erlbaum, Hillsdale, New Jersey.

Cruickshank, A.J., J.-P. Gautier & C. Chappuis. 1993. Vocal mimicry in wild African grey parrots, *Psittacus erithacus*. *Ibis* **135**:293–299.

Culbertson, G.H. & L.R. Caporael. 1983. Baby talk speech to the elderly: Complexity and content of messages. *Personality and Social Psychology Bulletin* **9**:305–312.

Cunningham, M.A. & M.C. Baker. 1983. Vocal learning in white-crowned sparrows: Sensitive phase and song dialects. *Behavioural Ecology & Sociobiology* **13**:259–269.

Curio, E. 1971. Die akustische Wirkung von Feindalarmen auf einige Singvögel. *Journal für Ornithologie* **112**:365–372.

Cutler, B. 1970. *Anatomical studies of the syrinx of Darwin's finches*, pp. 272. San Francisco State University, San Francisco.

Cynx, J. 1990. Experimental determination of a unit

of song production in the zebra finch (*Taeniopygia guttata*). *Journal of Comparitive Psychology* **104**:3–10.

Cynx, J. 1993. Auditory frequency generalization and a failure to find octave generalization in a songbird, the European starling (*Sturnus vulgaris*). *Journal of Comparative Psychology* **107**:140–146.

Cynx, J. 1995. Similarities in absolute and relative pitch perception in song birds (starling and zebra finch) and a non-song bird (pigeon).*Journal of Comparative Psychology*, **109**: 261–267.

Cynx, J. 1999. Effects of social context on amplitude in zebra finch vocalizations. *Abstracts of the Midwinter Meeting*, Association for Research in Otolaryngology, St. Petersburg, FL.

Cynx, J., & Gell, C. 2004. Social mediation of vocal amplitude in a songbird, *Taeniopygia guttata*. *Animal Behaviour* **67**:451–455.

Cynx, J., S.H. Hulse & S. Polyzois. 1986. A psychophysical measure of pitch discrimination loss resulting from a frequency range constraint in European starlings (*Sturnus vulgaris*). *Journal of Experimental Psychology: Animal Behavior Processes* **12**:394–402.

Cynx, J., R. Lewis, B. Tavel & H. Tse. 1998. Amplitude regulation of vocalizations in noise by a songbird, *Taeniopygia guttata*. *Animal Behaviour* **56**:107–113.

Cynx, J. & U. von Rad. 2001. Immediate and transitory effects of delayed auditory feedback on birdsong production. *Animal Behaviour* **62**:305–312.

Cynx, J. & Shapiro, M. 1986. Perception of missing fundamental by a species of songbird (*Sturnus vulgaris*). *Journal of Comparative Psychology* **100**:356–360.

Cynx, J., H. Williams & F. Nottebohm. 1990. Timbre discrimination in zebra finch (*Taeniopygia guttata*) song syllables. *Journal of Comparative Psychology* **104**:303–308.

Cziko, G. 1995. *Without Miracles. Universal Selection Theory and the Second Darwinian Revolution*. MIT Press, Cambridge, MA.

Dabelsteen, T. 1984. An analysis of the full song of the blackbird *Turdus merula* with respect to message coding and adaptations for acoustic communication. *Ornis Scandinavica* **15**:227–239.

Dabelsteen, T., O.N. Larsen & S.B. Pedersen. 1993. Habitat-induced degradation of sound signals: Quantifying the effects of communication sounds and bird location on blur ratio, excess attenuation, and signal-to-noise ratio in blackbird song. *Journal of Acoustical Society of America* **93**:2206–2220.

Dabelsteen, T. & P.K. McGregor. 1996. Dynamic acoustic communication and interactive playback. In: *Ecology and Evolution of Acoustic Communication in Birds*, D.E. Kroodsma & E.H. Miller (eds.), pp. 398–408. Cornell University Press, Ithaca, NY.

Dabelsteen, T., P.K. McGregor, J. Holland, J.A. Tobias & S.B. Pedersen. 1997. The signal function of overlapping singing in male robins. *Animal Behaviour* **53**:249–256.

Dabelsteen, T., P.K. McGregor, M. Sheppard, X. Whittaker & S.B. Pedersen. 1996. Is the signal of overlapping different from that of alternating during matched singing in great tits? *Journal of Avian Biology* **27**:189–194.

Dabelsteen, T. & S.B. Pedersen. 1985. Correspondence between messages in the full song of the blackbird *Turdus merula* and meanings to territorial males, as inferred from responses to computerized modifications of natural song. *Zeitschrift für Tierpsychology* **69**:149–165.

Dabelsteen, T. & S.B. Pedersen. 1988a. Do female blackbirds, *Turdus merula*, decode song in the same way as males? *Animal Behaviour* **36**:1858–1960.

Dabelsteen, T. & S.B. Pedersen. 1988b. Song parts adapted to function both at long and short ranges may communicate information about the species to female blackbirds *Turdus merula*. *Ornis Scandinavica* **19**:195–1998.

Dabelsteen, T. & S.B. Pedersen. 1990. Song information about aggressive responses of blackbirds, *Turdus merula*: Evidence from interactive playback experiments with territory owners. *Animal Behaviour* **40**:1158–1168.

Dabelsteen, T. & S.B. Pedersen. 1992. Song features essential for species discrimination and behaviour assessment by male blackbirds (*Turdus merula*). *Behaviour* **121**:259–287.

Damasio, A.R., H. Damasio, M. Rizzo, N. Varney & F. Gersh. 1982. Aphasia with nonhemorrhagic lesions in the basal ganglia and internal capsule. *Archives of Neurology* **39**:15–24.

Dantzker, M.S., G.B. Deane & J.W. Bradbury. 1999. Directional acoustic radiation in the strut display of male sage grouse *Centrocercus urophasianus*. *Journal of Experimental Biology* **202**:2893–2909.

Darwin, C.R. 1860. *On the Origin of Species by Means of Natural Selection, or the Preservation of Favoured Races in the Struggle for Life*. John Murray, London.

Darwin, C. 1871. *The Descent of Man, and Selection in Relation to Sex.* John Murray, London.

Darwin, C.R. 1875. *The Variation of Animals and Plants Under Domestication. 2nd* ed. John Murray, London.

Dave, A. & D. Margoliash. 2000. Song replay during sleep and computational rules for sensorimotor vocal learning. *Science* **290**:812–816.

Dave, A.S., A.C. Yu & D. Margoliash. 1998. Behavioral state modulation of auditory activity in a vocal motor system. *Science* **282**:2250–2254.

Davies, N.B. 2000. *Cuckoos, Cowbirds and Other Cheats.* T. & A. D. Poyser, London.

Davies, N.B., R.M. Kilner & D.G. Noble. 1998. Nestling cuckoos, *Cuculus canorus,* exploit hosts with begging calls that mimic a brood. *Proceedings of the Royal Society of London, B* **265**:673–678.

Davies, P.J. 1989. *Mozart in person: His character and health.* Greenwood, New York.

Davies, S.J.J.F. 1970. Patterns of inheritance in the bowing display and associated behaviour of some hybrid *Streptopelia* doves. *Behaviour* **36**:187–214.

Davies, S.J.J.F. & R. Carrick. 1962. On the ability of crested terns, *Sterna burghi,* to recognize their own chicks. *Australian Journal of Zoology* **10**:171–177.

Davis, D.E. 1946. A seasonal analysis of mixed flocks of birds in Brazil. *Ecology* **27**:1–128.

Deckert, G. 1962. Zur Ethologie des Feldsperlings (*Passer m. montanus* L.) *Journal für Ornithologie* **103**:428–486.

DeClerq, N., A. DePrez, W. Godefroid, F. Santens & L. Van Acker. 1995. *Hinke de Vinke.* 400 jaar Vinkensport in Vlaanderen. 60 jaar Avibo. Vichte, Belgium.

Deecke, V.B., J.K.B. Ford & P. Spong. 1999. Quantifying complex patterns of bioacoustic variation: Use of a neural network to compare killer whale (*Orcinus orca*) dialects. *Journal of the Acoustical Society of America* **105**:2499–2507.

De Kort, S.R. 2002. *Pigeon pidgin: Evolution of vocal signals through interspecific interactions in Steptopelia* doves. Ph.D. thesis, Leiden University, Leiden, The Netherlands.

De Kort, S.R., P.M. den Hartog, & C. ten Cate. 2002. Diverge or merge? The effect of sympatric occurrence on the territorial vocalizations of the vinaceous dove *Streptopelia vinacea* and the ring-necked dove *S. capicola. Journal of Avian Biology* **33**:150–158.

De Kort, S.R. & C. ten Cate. 2001. Response to interspecific vocalisations is affected by degree of relatedness in *Streptopelia* doves. *Animal Behaviour* **61**:239–247.

De Kort, S.R. & C. ten Cate. 2004. Repeated decrease in vocal repertoire size in *Streptopelia* doves. Animal Behaviour **67**:549–557.

Delacour, J. & E. Mayr. 1945. The family *Anatidae. Wilson Bulletin* **57**:3–55.

Del Hoyo, J., A. Elliott & J. Sargatal. 2001. *Handbook of the Birds of the World: Mousebirds to Hornbills.* Lynx Edicions, Barcelona.

Del Negro, C., M. Kreutzer & M. Gahr. 2000. Sexually stimulating signals of canary (*Serinus canaria*) songs: Evidence for a female-specific auditory representation in the HVc nucleus during the breeding season. *Behavioral Neuroscience* **114**:526–542.

DeLong, M.R. & A.P. Georgopoulos. 1981. Motor functions of the basal ganglia. In: *Handbook of physiology – the nervous system,* V.B. Brooks (eds.), pp. 1017–1062. American Physiological Society, Bethesda.

Dent, M.L., E.F. Brittan-Powell, R.J. Dooling & A. Pierce. 1997a. Perception of synthetic /ba/-/wa/ speech continuum by budgerigars (*Melopsittacus undulatus*). *Journal of the Acoustical Society of America* **102**:1891–1897.

Dent, M.L., R.J. Dooling & A.S. Pierce. 2000. Frequency discrimination in budgerigars (*Melopsittacus undulatus*): effects of tone duration and tonal context, *J. Acoust. Soc. Am.* **107**:2657–64.

Dent, M.L., O.N. Larsen & R.J. Dooling. 1997b. Free-field binaural unmasking in budgerigars (*Melopsittacus undulatus*). *Behavioral Neuroscience* **111**:590–598.

Depraz, V., G. Leboucher & M. Kreutzer. 2000. Early tutoring and adult reproductive behaviour in female domestic canary. *Animal Cognition* **3**:45–51.

Deregnaucourt, S. & J.C. Guyomarc'h. 2003. Mating call discrimination in female European (*Coturnix c. coturnix*) and Japanese quail (*Coturnix c. japonica*). *Ethology* **109**:107–119.

Derrickson, K.C. 1987. Yearly and situational changes in the estimate of repertoire size in northern mockingbirds (*Mimus polyglottos*). *Auk* **104**:198–207.

Derrickson, K.C. 1988. Variation in repertoire presentation in northern mockingbirds. *Condor* 90:592–606.

Desante, D.F. 1986. A field test of the variable circular-plot censusing method in a Sierran (California, USA) subalpine forest habitat. *Condor* **88**:129–142.

Detert, H. & H.-H. Bergmann. 1984. Regenrufdialekte von Buchfinken (*Fringilla coelebs* L.): Untersuchungen an einer Population von Mischrufern. *Ökologie der Vögel* **6**:101–118.

DeVoogd, T.J. 1991. Endocrine modulation of the development and adult function of the avian song system. *Psychoneuroendocrinology* **16**:41–66.

DeVoogd, T.J., J.R. Krebs, S.D. Healy & A. Purvis. 1993. Relations between song repertoire size and the volume of brain nuclei related to song: Comparative evolutionary analyses amongst oscine birds. *Proceedings of the Royal Society of London, B* **254**:75–82.

DeVoogd, T.J. & F. Nottebohm. 1981. Gonadal hormones induce dendritic growth in the adult brain. *Science* **214**:202–204.

DeVoogd, T.J., D.J. Pyskaty & F. Nottebohm. 1991. Lateral asymmetries and testosterone-induced changes in the gross morphology of the hypoglossal nucleus in adult canaries. *Journal of Comparative Neurology* **307**:65–76.

DeWolfe, B.B., L.F. Baptista & L. Petrinovich. 1989. Song development and territory establishment in Nuttall's white-crowned sparrows. *Condor* **91**:397–407.

DeWolfe, B.B. D.D. Kaska & L.J. Peyton. 1974. Prominent variations in the songs of Gambel's white-crowned sparrows. *Bird-Banding* **45**:224–252.

Dittus, W.P.J. & R.E. Lemon. 1969. Effects of song tutoring and acoustic isolation on the song repertoires of cardinals. *Animal Behavior* **17**:523–533.

Doetsch, F., I. Caille, D. Lim, J. Garcia-Verdugo & A. Alvarez-Buylla. 1999. Subventricular zone astrocytes are neural stem cells in adult mammalian brain. *Cell* **97**:703–716.

Doetsch, F., J.M. Garcia-Verdugo & A. Alvarez-Buylla. 1997. Cellular composition and three-dimensional organization of the subventricular germinal zone in the adult mammalian brain. *Journal of Neuroscience* **17**:5046–5061.

Dooling, R.J. 1980. Behavior and psychophysics of hearing in birds. In: *Comparative studies of hearing in vertebrates*, A.N. Popper & R.R. Fay (eds.), pp. 261–288. Springer-Verlag, New York.

Dooling, R.J. 1982a. Auditory perception in birds. In: *Acoustic Communication in Birds, Vol. 1. Production, Perception, and Design Features of Sounds*, D.E. Kroodsma & E.H. Miller (eds.), pp. 95–130. Academic Press, New York.

Dooling, R.J. 1982b. Ontogeny of song recognition in birds. *American Zoologist* **22**:571–580.

Dooling, R.J. 1992. Perception of speech sounds by birds. In: *The 9th International Symposium on Hearing: Auditory Physiology and Perception*, Y. Cazals, L. Demany & K. Horner (eds.), pp. 407–413. Pergamon Press, Oxford.

Dooling, R.J., C.T. Best & S.D. Brown. 1995. Discrimination of synthetic full-formant and sinewave /ra-la/ continua by budgerigars (*Melopsittacus undulatus*) and zebra finch (*Taeniopygia guttata*). *Journal of the Acoustical Society of America* **97**:1839–1846.

Dooling, R.J. & S.D. Brown. 1990. Speech perception by budgerigars (*Melopsittacus undulatus*): Spoken vowels. *Perception and Psychophysics* **47**:568–574.

Dooling, R.J., S.D. Brown, G.M. Klump & K. Okanoya. 1992. Auditory perception of conspecific and heterospecific vocalizations in birds: Evidence for special processes. *Journal of Comparative Psychology* **106**:20–28.

Dooling, R.J., M.L. Dent, M.R. Leek & O. Gleich. 2001. Masking by harmonic complexes in birds: Behavioral thresholds and cochlear responses. *Hearing Research* **152**:159–172.

Dooling, R.J., B.F. Gephart, P.H. Price, C. McHale & S.E. Brauth. 1987a. Effects of deafening on the contact calls of the budgerigar (*Melopsittacus undulatus*). *Animal Behaviour* **35**: 1264–1266.

Dooling, R.J. & R.J. Haskell. 1978. Auditory duration discrimination in the parakeet (*Melopsittacus undulatus*). *Journal of the Acoustical Society of America* **63**:1640–1642.

Dooling, R.J., M.R. Leek, O. Gleich & M.L. Dent. 2002. Auditory temporal resolution in birds: Discrimination of harmonic complexes. *Journal of the Acoustical Society of America* **112**:748–759.

Dooling, R.J., B. Lohr & M.L. Dent. 2000. Hearing in birds and reptiles. In: *Comparative Hearing: Birds and Reptiles*, R.J. Dooling, A.N. Popper & R.R. Fay (eds.), pp. 308–359. Springer-Verlag, New York.

Dooling, R.J., K. Okanoya & S.D. Brown. 1989. Speech perception by budgerigars (*Melopsittacus undulatus*) – the voiced-voiceless distinction. *Perception & Psychophysics* **46**:65–71.

Dooling, R.J., T.J. Park, S.D. Brown, K. Okanoya & S.D. Soli. 1987b. Perceptual organization of acoustic stimuli by budgerigars (*Melopsittacus undulatus*): II. Vocal signals. *Journal of Comparative Psychology* **101**:367–381.

Dooling, R.J., B.M. Ryals & K. Manabe. 1997. Recovery of hearing and vocal behavior after hair-cell regeneration. *Proceedings of the National Academy of Sciences USA* **94**:14206–14210.

Dooling, R.J. & J.C. Saunders. 1975. Auditory intensity discrimination in the parakeet (*Melopsittacus undulatus*). *Journal of the Acoustical Society of America* **58**:1308–1310.

Dooling, R.J. & M.H. Searcy. 1980. Early perceptual selectivity in the swamp sparrow. *Developmental Psychobiology* **13**:499–506.

Doupe, A.J. 1993. A neural circuit specialized for vocal learning. *Current Opinion in Neurobiology* **3**:104–111.

Doupe, A.J. 1997. Song- and order-selective neurons in the songbird anterior forebrain and their emergence during vocal development. *Journal of Neuroscience* **17**:1147–1167.

Doupe, A.J. 1998. Development and learning in the birdsong system: Are there shared mechanisms? In: *Mechanistic Relationships Between Development and Learning*, T.J. Carew, R. Menzel & C.J. Shatz (eds.), pp. 29–52. John Wiley and Sons, New York.

Doupe, A.J. & M. Konishi. 1991. Song-selective auditory circuits in the vocal control system of the zebra finch. *Proceedings of the National Academy of Sciences* **88**:11339–11343.

Doupe, A.J. & P.K. Kuhl. 1999. Birdsong and human speech: common themes and mechanisms. *Annual Review of Neuroscience* **22**:567–631.

Doupe, A.J. & M.M. Solis. 1997. Song- and order-selective neurons develop in the songbird anterior forebrain during vocal learning. *Journal of Neurobiology* **33**:694–709.

Doutrelant, C. & M.M. Lambrechts. 2001. Macrogeographic variation in song–a test of competition and habitat effects in Blue Tits. *Ethology* **107**:533–544.

Dowsett-Lemaire, F. 1979a. The imitative range of the song of the marsh warbler *Acrocephalus palustris*, with special reference to imitations of African birds. *Ibis* **121**:453–468.

Dowsett-Lemaire, F. 1979b. Vocal behaviour of the marsh warbler, *Acrocephalus palustris*. *Gerfaut* **69**:475–502.

Dowsett-Lemaire, F. 1981a. The transition period from juvenile to adult song in the European marsh warbler (*Acrocephalus palustris*). *Ostrich* **52**: 253–255.

Dowsett-Lemaire, F. 1981b. Eco-ethological aspects of breeding in the marsh warbler, *Acrocephalus palustris*. *Revue d'Ecologie (Terre et Vie)* **35**:437–491.

Draganoiu T., L. Nagle & M. Kreutzer. 2002. Directional female preference for an exaggerated male trait in canary (*Serinus canaria*) song. *Proceedings of the Royal Society of London, B* **269**:2525–2531.

Dunn, A.M. & R.A. Zann. 1996a. Undirected song in wild zebra finch flocks: contexts and effects of mate removal. *Ethology* **102**:529–539.

Dunn, A.M. & R.A. Zann. 1996b. Undirected song encourages the breeding female zebra finch to remain in the nest. *Ethology* **102**:540–548.

Durand, S.E., J.T. Heaton, S.K. Amateau & S.E. Brauth. 1997. Vocal control pathways through the anterior forebrain of a parrot (*Melopsittacus undulatus*). *Journal of Comparative Neurology* **377**:179–206.

Durlach, N. & H.S. Colburn. 1978. Binaural phenomena. In: *Handbook of Perception, Vol. IV: Hearing*, M.P. Friedman (ed.), pp. 365–466. Academic Press, New York.

Eales, L.A. 1985. Song learning in zebra finches: Some effects of song model availability on what is learnt and when. *Animal Behaviour* **33**:1293–1300.

Eales, L.A. 1987. Do zebra finch males that have been raised by another species still tend to select a conspecific song tutor? *Animal Behaviour* **35**:1347–1355.

Eddleman, W.R., F.L. Knopf, B. Meanley, F.A. Reid & R. Zembal. 1988. Conservation of North American rallids. *Wilson Bulletin* **100**:458–475.

Edinger, L. 1908. Vorlesungen uber den bau der nervosen zentralorgane des menschen und der Tiere – fur Artze und strdierende. In: *Zweiter Band. Vergleichende Anatomie des Gerhirns Siebente, umgearbeitete und vermehrte Auflage*. Vogel Verlag, Leipzig.

Edinger, L., A. Wallenberg & G. Holmes. 1903. Untersuchungen über die vergleichende anatomie des gehirns. 5. das vorderhirn der vögel. Abhandl.d. *Senckenb. nat. Gesellsch., Frankfurt am Main,* **20**:343–426.

Eens, M. 1997. Understanding the complex song of the European starling: An integrated ethological approach. *Advances in the Study of Behavior* **26**:355–434.

Eens, M., R. Pinxten & R.F. Verheyen. 1991. Male song as a cue for mate choice in the European starling. *Behaviour* **116**:210–238.

Eens, M., R. Pinxten & R.F. Verheyen. 1992. Song learning in captive European starlings. *Animal Behaviour* **44**:1131–1143.

Eens, M., R. Pinxten & R.F. Verheyen. 1993. Function of the song and song repertoire in the European starling (*Sturnus vulgaris*): An aviary experiment. *Behaviour* **125**:51–66.

Eibl-Eibesfeldt, I. 1972. Similarities and differences between culture in expressive movements. In: *Non-verbal Communication*, R.A. Hinde (ed.), pp. 297–312. Cambridge University Press, Cambridge, MA.

Eisenberg, J.F. 1974. The function and motivational basis of hystricomorph vocalizations. In: *The Biology of Hystricomorph Rodents (Symposium no. 34)*, I.W. Rowlands & B. Weir (eds.), pp. 211–244. Zoological Society of London, London.

Ekman, P. (ed.). 1982. *Emotion in the Human Face* (2nd ed.). Cambridge University Press, New York.

Elgar, M.A. 1986. House sparrows establish foraging flocks by giving chirrup calls if the resource is divisible. *Animal Behaviour* **34**:169–174.

Ellers, J. & H. Slabbekoorn. 2003. Song divergence and male dispersal among bird populations: A spatially explicit model testing the role of vocal learning. *Animal Behaviour* **65**:671–681.

Elliot, L. 1999. *Music of the Birds*. Houghton Mifflin Co., New York.

Ellis, C.R. Jr. & A.W. Stokes. 1966. Vocalizations and behavior in captive Gambel quail. *Condor* **68**:72–80.

Ellis, D.H., S.R. Swengel, G.W. Archibald & C.B. Kepler. 1998. A sociogram for the cranes of the world. *Behavioural Processes* **43**:125–151.

Embleton, T.F.W. 1963. Sound propagation in homogeneous deciduous and evergreen woods. *Journal of the Acoustical Society of America* **35**:1119–1125.

Embleton, T.F.W. 1996. Tutorial on sound propagation outdoors. *Journal of Acoustical Society of America* **100**:31–48.

Emery, N.J. 2000. The eyes have it: the neuroethology, function and evolution of social gaze. *Neuroscience and Biobehavioral Reviews* **24**:581–604.

Emlen, S.T. 1971. Geographic variation in indigo bunting song (*Passerina cyanea*). *Animal Behaviour* **19**:407–408.

Emlen, S.T. 1972. An experimental analysis of the parameters of bird song eliciting species recognition. *Behaviour* **41**:130–171.

Emlen, S.T., J.D. Rising & W.L. Thompson. 1975. A behavioral and morphological study of sympatry in the indigo and lazuli buntings of the Great Plains. *Wilson Bulletin* **87**:145–179.

Endler, J.A. 1992. Signals, signal conditions, and the direction of evolution. *American Naturalist* (Suppl.) **139**:125–153.

Endler, J.A. & A.L. Basolo. 1998. Sensory ecology, receiver biases and sexual selection. *Trends In Ecology & Evolution* **13**:415–420.

Engle, M.S. 2000. *The importance of social and acoustic environments in the production of song and mimicry in the European starling*. Doctoral dissertation, Indiana University, Bloomington.

Eriksson, D. & L. Wallin. 1986. Male bird song attracts females: A field experiment. *Behavioural Ecology & Sociobiology* **19**:297–300.

Eriksson, P.S., E. Perfilieva, T. Bjork-Eriksson, A.-M. Alborn, C. Nordborg, D.A. Peterson & F.H. Gage. 1998. Neurogenesis in the adult human hippocampus. *Nature Medicine* **4**:1313–1317.

Espada, M.A.M. *A hard task, how to obtain good singers*. Internet publication.

Espmark, Y.O., H.M. Lampe & T.K. Bjerke. 1989. Song conformity and continuity in song dialects of redwings ,*Turdus iliacus*, and some ecological correlates. *Ornis Scandinavica* **20**:1–12.

Esser, K.H. 1994. Audio-vocal learning in a non-human mammal: The lesser spear-nosed bat *Phyllostomus discolor*. *Neuroreport* **5**:1718–1720.

Evans, C.S. 1997. Referential signals. *Perspectives in Ethology* **12**:99–143.

Evans, C.S. 2002. Cracking the code: Communication and cognition in birds. In: *The Cognitive Animal: Empirical and Theoretical Perspectives on Animal Cognition*, M. Bekoff, C. Allen & G.M. Burghardt (eds.), pp. 315–322. MIT Press, Cambridge, MA.

Evans, C.S., L. Evans & P. Marler. 1993. On the meaning of alarm calls: Functional reference in an avian vocal system. *Animal Behaviour* **46**:23–38.

Evans, C.S. & L. Evans. 1999. Chicken food calls are functionally referential. *Animal Behaviour* **58**:307–319.

Evans, C.S. & P. Marler. 1994. Food calling and audience effects in male chickens, *Gallus gallus*: Their relationships to food availability, courtship and social facilitation. *Animal Behaviour* **47**:1159–1170.

Evans, R.M. 1970. Parental recognition and the "mew call" in black-billed gulls (*Larus bulleri*). *Auk* **87**:503–513.

Evans, W.R. 1994. Nocturnal flight call of Bicknell's thrush. *Wilson Bulletin* **106**:55–61.

Evans, W.R. & D.K. Mellinger. 1999. Monitoring grassland birds in nocturnal migration. *Studies in Avian Biology* **19**:219–229.

Ewert, D.N. & D.E. Kroodsma. 1994. Song sharing and repertoires among migratory and resident rufous-sided towhees. *Condor* **96**:190–196.

Falls, J. 1963. Properties of bird song eliciting responses from territorial males. *Proceedings of the International Ornithological Congress* **13**: 359–371.

Falls, J.B. 1978. Bird song and territorial behavior. In: *Aggression, Dominance and Individual Spacing*, L. Krames, P. Pliner & T. Alloway (eds.), pp. 61–89. Academic Press, New York.

Falls, J.B. 1982. Individual recognition by sound in birds. In: *Acoustic Communication in Birds Vol. 2: Song Learning and its Consequences*, D.E. Kroodsma & E.H. Miller (eds.), pp. 237–278. Academic Press, New York.

Falls, J.B. 1985. Song matching in western meadowlarks, *Sturnella neglecta*. *Canadian Journal of Zoology* **63**:2520–2524.

Falls, J.B. 1988. Does song deter intruders in white-throated sparrows (*Zonotrichia albicolis*)? *Canadian Journal of Zoology* **66**:206–211.

Falls, J.B. 1992. Playback: A historical perspective. In: *Playback and Studies of Animal Communication*, P.K. McGregor (ed.), pp. 11–33. Plenum Press, New York.

Falls, J.B. & J.R. Brooks. 1975. Individual recognition by song in white-throated sparrows. II. Effects of location. *Canadian Journal of Zoology* **53**:412–420.

Falls, J.B. & L.G. D'Agincourt. 1982. Why do meadowlarks switch song types? *Canadian Journal of Zoology* **60**:3400–3408.

Falls, J.B. & J.R. Krebs. 1975. Sequence of songs in repertoires of western meadowlarks (*Sturnella neglecta*). *Canadian Journal of Zoology* **53**:1165–1178.

Falls, J.B., J.R. Krebs & P.K. McGregor. 1982. Song matching in the great tit (*Parus major*): The effect of similarity and familiarity. *Animal Behaviour* **30**:991–1009.

Fant, G. 1960. *Acoustic Theory of Speech Production*. Morton, The Hague.

Farabaugh, S. & R.J. Dooling. 1996. Acoustic communication in parrots: Laboratory and field studies of budgerigars, *Melopsittacus undulatus*. In: *Ecology and Evolution of Acoustic Communication in Birds*, D.E. Kroodsma & E.H. Miller (eds.), pp. 97–117. Cornell University Press, Ithaca, NY.

Farabaugh, S.M., A. Linzenbold & R.J. Dooling. 1994. Vocal plasticity in budgerigars (*Melopsittacus undulatus*): Evidence for social factors in the learning of contact calls. *Journal of Comparative Psychology* **108**:81–92.

Farabaugh, S.M. & J.M. Wild. 1997. Reciprocal connections between primary and secondary auditory pathways in the telencephalon of the budgerigar (*Melopsittacus undulatus*). *Brain Research* **747**:18–25.

Fay, R.R. 1988. *Hearing in Vertebrates: A Psychophysics Databook*. Hill-Fay Associates, Winnetka, IL.

Feduccia, A. 1995. Explosive evolution in tertiary birds and mammals. *Science* **267**:637–638.

Fee, M. & A. Leonardo. 2001. Miniature motorized microdrive and commutator system for chronic neural recording in small animals. *Journal of Neuroscience Methods* **15**:83–94.

Fee, M.S., B. Shraiman, B. Pesaran & P.P. Mitra. 1998. The role of nonlinear dynamics of the syrinx in the vocalizations of a songbird. *Nature* **395**:67–71.

Feekes, F. 1977. Colony-specific song in *Cacicus cela* (*Icteridae*, Aves): The pass-word hypothesis. *Ardea* **65**:197–202.

Fernald, A. 1985. Four-month-old infants prefer to listen to motherese. *Infant Behavior and Development* **8**:181–195.

Fernald, A. 1989. Intonation and communicative intent in mothers' speech to infants: Is the melody the message? *Child Development* **60**:1497–1510.

Fernald, A. 1992. Meaningful melodies in mothers' speech to infants. In: *Nonverbal Vocal Communication: Comparative and Developmental Approaches*, H. Papousek, U. Jergens & M. Papousek (eds.). Cambridge University Press, New York.

Fernald, A. & P. Kuhl. 1987. Acoustic determinants of infant preference for motherese speech. *Infant Behavior and Development* **10**:279–293.

Ficken, M.S. 2000. Call similarities among mixed species flock associates. *Southwestern Naturalist* **45**:154–158.

Ficken, M.S. & R.W. Ficken. 1962. The comparative ethology of wood warblers: A review. *Living Bird* **1**:103–122.

Ficken, M.S.,& R.W. Ficken. 1967. Singing behaviour of blue-winged and golden-winged warblers and their hybrids. *Behaviour* **28**:149–181.

Ficken, R.W., M.S. Ficken & J.P. Hailman. 1974. Temporal pattern shift to avoid acoustic interference in singing birds. *Science* **183**:762–763.

Ficken, M.S., R.W. Ficken & S.R. Witkin. 1978. Vocal repertoire of the black-capped Chickadee. *Auk* **95**:34–48.

Ficken, M.S. & J.W. Popp. 1992. Syntactical organization of the gargle vocalization of the black-capped chickadee, *Parus atricapillus*. *Ethology* **91**:156–168.

Ficken, M.S. & J.W. Popp. 1995. Long-term persistence of a culturally transmitted vocalization of the black-capped chickadee. *Animal Behaviour* **50**:683–693.

Ficken, M.S. & J.W. Popp. 1996. A comparative analysis of passerine mobbing calls. *Auk* **113**:370–380.

Ficken, R.W., J.W. Popp & P.E. Matthiae. 1985. Avoidance of acoustics interference by ovenbirds. *Wilson Bulletin* **97**:569–571.

Ficken, M.S., K.M. Rusch, S.J. Taylor & D.R. Powers. 2000. Blue throated hummingbird song: A pinnacle of nonoscine vocalizations. *Auk* **117**:120–128.

Ficken, M.S. & C.M. Weise. 1984. A complex call of the black-capped chickadee (*Parus atricapillus*). I. Microgeographic variation. *Auk* **101**:349–360.

Ficken, M.S., C.M. Weise & J.A. Reinartz. 1987. A complex vocalization of the black-capped chickadee. II. Repertoires, dominance and dialects. *Condor* **89**:500–509.

Fish, M.P. & W.H. Mowbray. 1970. *Sounds of Western North Atlantic Fishes: A Reference File of Biological Underwater Sounds*. Johns Hopkins Press, Baltimore.

Fisher, R.A. 1930. *The Genetical Theory of Natural Selection*. Oxford University Press, Oxford.

Fitch, W.T. 1999. Acoustic exaggeration of size in birds via tracheal elongation: Comparative and theoretical analyses. *Journal of Zoology, London* **248**:31–48.

Fitch, W.T. 2000a. The evolution of speech: A comparative review. *Trends in Cognitive Sciences* **4**:258–267.

Fitch, W.T. 2000b. Skull dimensions in relation to body size in nonhuman mammals: The causal bases for acoustic allometry. *Zoology* **103**:40–58.

Fitch, W.T. & M.D. Hauser. 2003. Unpacking "honesty": Vertebrate vocal production and the evolution of acoustic signals. In: *Acoustic Communication*, A.M. Simmons, A.N. Popper & R.R. Fay (eds.), pp. 65–137. Springer, New York.

Fitch, W.T. & J.P. Kelley. 2000. Perception of vocal tract resonances by whooping cranes, *Grus americana*. *Ethology* **106**:559–574.

Fitch, W.T., J. Neubauer & H.P. Herzel. 2002. Calls out of chaos: The adaptive significance of nonlinear phenomena in mammalian vocal production. *Animal Behaviour* **63**:407–418.

Fitzpatrick, J.W. 1988. Why so many passerine birds? A response to Raikow. *Systematic Zoology* **37**:71–76.

Fletcher, L.E. & D.G. Smith. 1978. Some parameters of song important in conspecific recognition by gray catbirds. *Auk* **95**:338–347.

Fletcher, N.H. 1988. Bird song – a quantitative acoustic model. *Journal of Theoretical Biology* **135**:455–481.

Fletcher, N.H. 1992. *Acoustic Systems in Biology*. Oxford University Press, Oxford.

Fleuster W. 1973. Versuche zur Reaktion freilebender Vögel auf Klangattrappen verschiedener Buchfinkenalarme. *Journal für Ornithologie* **114**:417–428.

Forman, R.T.T. & L.E. Alexander. 1998. Roads and their major ecological effects. *Annual Reviews in Ecology and Systematics* **29**:207–231.

Forrest, T.G. 1994. From sender to receiver: Propagation and environmental effects on acoustic signals. *American Zoologist* **34**:644–654.

Forshow, J.M. 1989. *The Parrots of the World*. Landsdowne Editions, Sydney.

Forstmeier, W. & T.J.S. Balsby. 2002. Why mated dusky warblers sing so much: Territory guarding and male quality announcement. *Behaviour* **139**:89–111.

Fortune, E.S. & D. Margoliash. 1995. Parallel pathways converge onto HVc and adjacent neostriatum of adult male zebra finches (*Taeniopygia guttata*). *Journal of Comparative Neurology* **360**:413–441.

Foster, E.F. & S.W. Bottjer. 1998. Axonal connections of the high vocal center and surrounding cortical regions in juvenile and adult male zebra finches. *Journal of Comparative Neurology* **397**:118–138.

Foster, E.F. & S.W. Bottjer. 2001. Lesions of a telencephalic nucleus in male zebra finches: Influences on vocal behavior in juveniles and adults. *Journal of Neurobiology* **46**:142–165.

Foster, E.F., R.P. Mehta & S.W. Bottjer. 1997. Axonal connections of the medial magnocellular nucleus of the anterior neostriatum in zebra finches. *Journal of Comparative Neurology* **382**:364–381.

Franz, M. & F. Goller. 2002. Respiratory units of motor production and song imitation in the zebra finch. *Journal of Neurobiology* **51**:129–141.

Freeberg, T.M., S.D. Duncan, T.L. Kast & D.A. Enstrom. 1999. Cultural influences on female mate choice; an experimental test in cowbirds, *Molorus ater*. *Animal Behaviour* **57**:421–426.

Freeberg, T.M. & J.R. Lucas. 2002. Receivers respond differently to chick-a-dee calls varying in note composition in Carolina chickadees, *Poecile cardinensis*. *Animal Behaviour* **63**:837–845.

Fridlund, A.J. 1994. *Human Facial Expression: An Evolutionary View*. Academic Press, San Diego, CA.

Friedmann, H. 1929. *The Cowbirds: A Study in the Biology of Social Parasitism*. C.C. Thomas, Springfield, IL.

Frith, C.B. 1994. Adaptive significance of tracheal elongation in manucodes (*Paradisaeidae*). *Condor* **96**:552–555.

Fry, C.H. 1974. Vocal mimesis in nestling greater honey-guides. *Bulletin of the British Ornithologists' Club* **94**:58–59.

Fusani, L., L. Beani & F. Dessi-Fulghieri. 1994. Testosterone affects the acoustic structure of male call in the grey partridge, *Perdix perdix*. *Behaviour* **128**:301–310.

Gaddis, P. 1980. Mixed flocks, accipiters, and antipredator behavior. *Condor* **82**:348–349.

Gaddis, P.K. 1987. Social interactions and habitat overlap between plain and bridled titmice. *Southwestern Naturalist* **32**:197–202.

Gage, F. 2000. Mammalian neural stem cells. *Science* **287**:1433–1438.

Gahr, M. 2000. Neural song control system of hummingbirds: Comparison to swifts, vocal learning (songbirds) and nonlearning (suboscines) passerines, and vocal learning (budgerigars) and nonlearning (dove, owl, gull, quail, chicken) nonpasserines. *Journal of Comparative Neurology* **426**:182–196.

Gahr, M. & H.-R. Güttinger. 1986. Functional aspects of singing in male and female *Uraeginthus bengalus* (*Estrildidae*). *Ethology* **72**:123–131.

Gahr, M. & M. Konishi. 1988. Developmental changes in estrogen-sensative neurons in the forebrain of the zebra finch. *Proceedings of the National Academy of Sciences USA* **85**:7380–7383.

Gahr, M. & E. Kosar. 1996. Identification, distribution, and developmental changes of a melatonin binding site in the song control system of the zebra finch. *Journal of Comparative Neurology* **367**:308–318.

Gahr, M. & R. Metzdorf. 1997. Distribution and dynamics in the expression of androgen and estrogen receptors in vocal control systems of songbirds. *Brain Research Bulletin* **44**:509–517.

Gaioni, S.J. & C.S. Evans. 1986. Perception of distress calls in mallard *ducklings (Anas platyrhynchos)*. *Behaviour* **99**:250–274.

Galef, B.G. 1988. Imitation in animals: History, definitions and interpretation of data from the psychological laboratory. In: *Social Learning*, T. Zentall & B.G. Galef (eds.), pp. 3–28. Erlbaum, Hillsdale, NJ.

Galeotti, P. 1998. Correlates of hoot rate and structure in male tawny owls, *Strix aluco*: Implications for male rivalry and female mate choice. *Journal of Avian Biology* **29**:25–33.

Galeotti, P. & G. Pavan. 1991. Individual recognition of male tawny owls (*Strix aluco*) using spectrograms of their territorial calls. *Ethology, Ecology and Evolution* 3: 113–126.

Galeotti, P. & G. Pavan. 1993. Differential responses of territorial tawny owls, *Strix aluco*, to the hooting of neighbours and strangers. *Ibis* **135**:300–304.

Galeotti, P., N. Saino, R. Sacchi & A.P. Mørller. 1997. Song correlates with social context, testosterone and body condition in male barn swallows. *Animal Behaviour* **53**:687–700.

Gall, F. & G. Spurzheim. 1810–1819. *Anatomie et Physiologie du Systeme nerveux en general et du Cerveau en particuleir avec des Observations sur la Possibilite de Reconnoitre Plusiers Dispositions intellectuelles et morales de l'Homme et des Animaux par la Configuration de leurs Tetes*. 4 Volumes F. Schoell, Paris.

Gardner, T., G. Cecchi, M. Magnasco, R. Laje & G.B. Mindlin. 2001. Simple motor gestures for birdsongs. *Physical Review Letters* **87**:208101.

Garland, T.J. & J.B. Losos. 1994. Ecological morphology of locomotor performance in squamate reptiles. In: *Ecological Morphology: Integrative*

Organismal Biology, P.C. Wainwright & S.M. Reilly (eds.), pp. 240–302. The University of Chicago Press, Chicago.

Garrod, A. H. 1873. On some points in the anatomy of *Steatornis*. *Proceedings of the Zoological Society of London*, **1873**:457–472.

Garstang, W. 1922. *Songs of the Birds*. Witherby & Sons, London.

Gaunt, A.S. 1983. An hypothesis concerning the relationship of syringeal structure to vocal abilities. *Auk* **100**:853–862.

Gaunt, A.S. 1987. Phonation. In: *Bird Respiration, Vol. 1*. T.J. Seller (ed.), pp. 71–94. Academic Press, New York.

Gaunt, A.S. 1988. Interaction of syringeal structure and airflow in avian phonation. In: *Acta XIX Congressus Internationalis Ornithologici, Vol. I*, H. Ouellet (ed.), pp. 915–924. University of Ottawa Press, Ottawa.

Gaunt, S.L.L., L.F. Baptista, J.E. Sánchez & D. Hernandez. 1994. Song learning as evidenced from song sharing in two species of hummingbird species (*Colibri coruscans* and *C. thalassinus*). *The Auk* **111**:87–193.

Gaunt, A.S. & S.L.L. Gaunt. 1977. Mechanics of the syrinx in *Gallus gallus*. II. Electromyographic studies of *ad libitum* vocalizations. *Journal of Morphology* **152**:1–20.

Gaunt, A.S., S.L.L. Gaunt & R.M. Casey. 1982. Syringeal mechanics reassessed: Evidence from *Streptopelia*. *The Auk* **99**:474–494.

Gaunt, A.S., S.L.L. Gaunt, H.D. Prange & J.S. Wasser. 1987. The effects of tracheal coiling on the vocalizations of cranes (Aves: *Gruidae*). *Journal of Comparative Physiology A* **161**:43–58.

Gaunt, A.S. & S. Nowicki. 1998. Sound production in birds: Acoustics and physiology revisited. In: *Animal Acoustic Communication: Sound Analysis and Research Methods*, S.L. Hopp, M.J. Owren & C.S. Evans (eds.), pp. 291–322. Springer-Verlag, Berlin.

Gaunt, S.L.L., L.F. Baptista, J.E. Sánchez & D. Hernandez. 1994. Song learning as evidenced from song sharing in two hummingbird species (*Colibri coruscans* and *C. thalassinus*). *Auk* **111**:87–103.

Geberzahn, N. & H. Hultsch. 2003. Long-time storage of song types in birds: Evidence from interactive playbacks. *Proceedings of the Royal Society of London, B* **270**:1085–1090.

Geberzahn, N., H. Hultsch & D. Todt. 2002. Latent song-type memories are accessible through auditory stimulation in a hand-reared songbird. *Animal Behaviour* **64**:783–790.

Gelter, H.P. 1987. Song differences between the pied flycatcher, *Ficedula hypoleuca*, the collared flycatcher, *F. albicollis*, and their hybrids. *Ornis Scandinavica* **18**:205–215.

Geschwind, N. 1979. Specializations of the human brain. *Scientific American* **241**:180–199.

Getting, P.A. 1989. Emerging principles governing the operation of neural networks. *Annual Review of Neuroscience* **12**:185–204.

Gibbs, H.L. & P.R. Grant. 1987. Oscillating selection on Darwin's finches. *Nature* **327**:511–513.

Gibbs, J.P. & S.M. Melvin. 1993. Call-response surveys for monitoring breeding waterbirds. *Journal of Wildlife Management* **57**:27–34.

Gibson, R.M. 1996. Female choice in sage grouse: the roles of attraction and active comparison. *Behavioural Ecology & Sociobiology* **39**:55–59.

Gil, D., J.L.S. Cobb & P.J.B Slater. 2001. Song characteristics are age dependent in the willow warbler, *Phylloscopus trochilus*. *Animal Behaviour* **62**:689–694.

Gil, D. & M. Gahr. 2002. The honesty of bird song: Multiple constraints for multiple traits. *Trends in Ecology & Evolution* **17**:133–141.

Gill, F.B. 1997. Local cytonuclear extinction of the golden-winged warbler. *Evolution* **51**:519–525.

Gish, S.L. & E.S. Morton. 1981. Structural adaptations to local habitat acoustics in Carolina wren songs. *Zeitschrift für Tierpsychologie* **56**:74–81.

Gittleman, J.L., C.G. Anderson, M. Kot & H.K. Luh. 1996. Phylogenetic lability and rates of evolution: A comparison of behavioral, morphological and life history traits. In: *Phylogenies and the Comparative Method in Animal Behavior*, E.P. Martins (ed.), pp. 166–205. Oxford University Press, Oxford.

Gleich, O., R.J. Dooling & G.A. Manley. 1994a. Inner-ear abnormalities and their functional consequences in Belgian Waterslager canaries (*Serinus canarius*). *Hearing Research* **79**:123–136.

Gleich, O. & G.M. Klump. 1994. Hereditary sensorineural hearing loss in a bird. *Naturwissenschaften* **81**:320–323.

Gleich, O. & G.A. Manley. 2000. The hearing organ of birds and *Crocodilia*. In: *Comparative Hearing: Birds and Reptiles*, R.J. Dooling, A.N. Popper & R.R. Fay (eds.), pp. 70–138. Springer-Verlag Publishers, New York.

Gleich, O., G.A. Manley, A. Mandl & R.J. Dooling. 1994b. Basilar papilla of the canary and zebra finch: A quantitative scanning electron microscopical description. *J. Morphology* **22**:1–24.

Gnam, R. 1988. Preliminary results on the breeding biology of Bahama Amazon. *Parrot Letter* **1**:23–26.

Godfray, H.C.J. 1991. Signalling of need by offspring to their parents. *Nature* **352**:328–330.

Godfray, H.C.J. 1995. Signaling of need between parents and young: Parent–offspring conflict and sibling rivalry. *American Naturalist* **146**:1–24.

Goelet, P., V.F. Castellucci, S. Schacher & E.R. Kandel. 1986. The long and the short of long-term memory–a molecular framework. *Nature* **322**:419–422.

Goldman, S. & F. Nottebohm. 1983. Neuronal production, migration and differentiation in a vocal control nucleus of the adult female canary brain. *Proceedings of the National Academy of Sciences USA* **80**:2390–2394.

Goller, F. & M.A. Daley. 2001. Novel motor gestures for phonation during inspiration enhance the acoustic complexity of birdsong. *Proceedings of the Royal Society of London, B* **268**:2301–2305.

Goller, F. & O.N. Larsen. 1997a. *In situ* biomechanics of the syrinx and sound generation in pigeons. *Journal of Experimental Biology* **200**:2165–2176.

Goller, F. & O.N. Larsen. 1997b. A new mechanism of sound generation in songbirds. *Proceedings of the National Academy of Sciences USA* **94**:14787–14791.

Goller, F. & O.N. Larsen. 2002. New perspectives on mechanisms of sound generation in songbirds. *Journal of Comparative Physiology* A **188**:841–850.

Goller, F. & R.A. Suthers. 1995a. Contributions of expiratory muscles to song production in brown thrashers. In: *Nervous Systems and Behaviour. Proceedings of the 4th International Congress of Neuroethology*, M. Burrows, T. Matheson, P. Newland & H. Schuppe (eds.), pp. 334. Georg Thieme Verlag, Stuttgart.

Goller, F. & R.A. Suthers. 1995b. Implications for lateralization of bird song from unilateral gating of bilateral motor patterns. *Nature* **373**:63–66.

Goller, F. & R.A. Suthers. 1996a. Role of syringeal muscles in controlling the phonology of bird song. *Journal of Neurophysiology* **76**:287–300.

Goller, F. & R.A. Suthers. 1996b. Role of syringeal muscles in gating airflow and sound production in singing brown thrashers. *Journal of Neurophysiology* **75**:867–876.

Goller, F. & R.A. Suthers. 1999. Bilaterally symmetrical respiratory activity during lateralized birdsong. *Journal of Neurobiology* **41**:513–523.

Gompertz, T. 1961. The vocabulary of the great tit. *British Birds* **54**:369–418.

Gompertz, T. 1967. The hiss-display of the great tit (*Parus major*). *Vogelwelt* **88**:165–169.

Goodwin, D. 1982. *Estrildid Finches of the World.* British Museum of National History, London.

Goodwin, D. 1983. *Pigeons and Doves of the World.* 3rd ed. British Museum of Natural History, London.

Göransson, G., G. Högstedt, J. Karlsson, H. Källander & S. Ulfstrand 1974. Sångens roll för revirhållandet hos näktergal *Luscinia luscinia* – några experiment med playback–teknik. *Vår Fågelvärld*, **33**:201–209.

Goth, A., U. Vogel & E. Curio. 1999. The acoustic communication of the Polynesian megapode, *Megapodius pritchardi G. R. Gray. Zoologische Verhandelingen* **327**:37–51.

Gottfried, G.M. & J.M. Tonks. 1996. Specifying the relation between novel and known: Input affects the acquisition of novel color terms. *Child Development* **67**:850–866.

Gottlieb, G. 1971. *Development of Species Identification in Birds: An Inquiry into the Prenatal Determinants of Perception.* University of Chicago Press, Chicago.

Gottlieb, G. 1976. The roles of experience in the development of behavior and the nervous system. In: *Neural and Behavioral Specificity*, G. Gottlieb (ed.), pp. 25–54. Academic Press, New York.

Gottlieb, G. 1992. *Individual Development and Evolution.* Oxford University Press, New York.

Gould, E., A. Beylin, P. Tanapat, A. Reeves & T. Shors. 1999a. Learning enhances adult neurogenesis in the hippocampal formation. *Nature Neuroscience* **2**:209–211.

Gould, E. & B. McEwen. 1993. Neuronal birth and death. *Current Opinion in Neurobiology* **3**:676–682.

Gould, E., A. Reeves, M. Graziano & C. Gross. 1999b. Neurogenesis in the neocortex of adult primates. *Science* **286**:548–552.

Gould, E. & P. Tanapat. 1999. Stress and hippocampal neurogenesis in the hippocampal formation. *Biological Psychiatry* **46**:1472–1479.

Gould, S.J. 1980. The evolutionary biology of constraint. *Daedalus* **109**:39–52.

Gottlander, K. 1987. Variation in the song rate of the male pied flycatcher (*Ficedula hypoleuca*): Causes and consequences. *Animal Behaviour* **35**:1037–1043.

Graber, R.R. & W.W. Cochran. 1959. An audio technique for the study of nocturnal migration of birds. *Wilson Bulletin* **71**:220–236.

Graber, R.R. & W.W. Cochran. 1960. Evaluation of an aural record of nocturnal bird migration. *Wilson Bulletin* **72**:253–273.

Grafen, A. 1991. Biological signals as handicaps. *Journal of Theoretical Biology* **144**:517–546.

Graff-Radford, N.R., H. Damasio, T. Yamada, P.J. Eslinger & A.R. Damasio. 1985. Nonhaemorrhagic thalamic infarction. Clinical, neuropsychological and electrophysiological findings in four anatomical groups defined by computerized tomography. *Brain* **108**:485–516.

Grant, B.R. & P.R. Grant. 1996a. Cultural inheritance of song and its role in the evolution of Darwin's finches. *Evolution* **50**:2471–2487.

Grant, B.R. & P.R. Grant. 1998. Hybridization and speciation in Darwin's finches: The role of sexual imprinting on a culturally transmitted trait. In: *Endless Forms: Species and Speciation*, D.J. Howard & S.H. Berlocher (eds.), pp. 404–422. Oxford University Press, Oxford.

Grant, B.R. & P.R. Grant. 2002a. Lack of premating isolation at the base of a phylogenetic tree. *American Naturalist* **160**:1–19.

Grant, P.R. 1999. *Ecology and Evolution of Darwin's Finches*. Princeton University Press, Princeton NJ.

Grant, P.R. & B.R. Grant. 1992. Hybridization of bird species. *Science* **256**:193–197.

Grant, P.R. & B.R. Grant. 1996b. Speciation and hybridization in island birds. *Philosophical Transactions of the Royal Society of London,* **351**:765–772.

Grant, P.R. & B.R. Grant. 1997a. Genetics and the origin of bird species. *Proceedings of the National Academy of Sciences USA* **94**:7768–7775.

Grant, P.R. & B.R. Grant. 1997b. Hybridization, sexual imprinting, and mate choice. *American Naturalist* **149**:1–28.

Grant, P.R. & B.R. Grant. 2002b. Adaptive radiation of Darwin's finches. *American Scientist* **90**:130–139.

Gray, D.A. & J.C. Hagelin. 1996. Song repertoires and sensory exploitation: Reconsidering the case of the common grackle. *Animal Behaviour* **52**:795–800.

Greene, E. 1989. A diet-induced developmental polymorphism in a caterpillar. *Science* **243**:643–646.

Greenewalt, C. 1968. *Bird Song: Acoustics and Physiology*. Smithsonian Institution Press, Washington, D.C.

Greig-Smith, P.W. 1978. The formation, structure and function of mixed-species insectivorous bird flocks in West African savanna woodland. *Ibis* **120**:284–297.

Greig-Smith, P.W. 1982a. Song-rates and parental care by individual male stonechats (*Saxicola torquata*). *Animal Behaviour* **30**:245–252.

Greig-Smith, P.W. 1982b. Distress calling by woodland birds. *Animal Behaviour* **30**:299–301.

Griffin, D.R. 1953. Acoustic orientation in the oilbird, *Steatornis. Proceedings of the National Academy of Sciences USA* **39**:884–893.

Griffin, D.R. & R.A. Suthers. 1970. Sensitivity of echolocation in cave swiftlets. *Biological Bulletin* **139**:495–501.

Griffin, D.R. & D. Thompson. 1982. Echolocation by cave swiftlets. *Behavioral Ecology & Sociobiology* **10**:119–123.

Grimes, L.G. 1974. Dialects and geographical variation in the song of the splendid sunbird *Nectarinia coccinigaster. Ibis* **116**:314–329.

Griscom, L. 1937. A monographic study of the red crossbill. *Proceedings of the Boston Society of Natural History* **41**:77–210.

Grisham, J., J. Lee, M.E. McCormick, K. Yang-Stayner & A.P. Arnold. 2002. Antiandrogen blocks estrogen-induced masculinization of the song system in female zebra finches. *Journal of Neurobiology* **51**:1–8.

Grosslight, J.H. & W.C. Zaynor. 1967. Vocal behavior of the mynah bird. In: *Research in Verbal Behavior and Some Neurophysiological Implications*, K. Salzinger & S. Salzinger (eds.), pp. 5–9. Academic Press, New York.

Groth, J.G. 1988. Resolution of cryptic species in Appalachian red crossbills. *Condor* **90**:745–760.

Groth, J.G. 1993a. Evolutionary differentiation in morphology, vocalizations, and allozymes among nomadic sibling species in the North American red crossbill (*Loxia curvirostra)* complex. *University of California Publications in Zoology* **127**:1–143.

Groth, J.G. 1993b. Call matching and positive assortative mating in Red Crossbills. *Auk* **110**:398–401.

Groth, J.G. 2001. Finches and allies. In: *The Sibley Guide to Bird Life & Behavior*, C. Elphick, J.B. Dunning Jr. & D.A. Sibley (eds.), pp. 552–560. Alfred A. Knopf, New York.

Guilford, T. & M.S. Dawkins. 1995. What are

conventional signals? *Animal Behaviour* **49**:1689–1695.

Guinee, L.H. & K.B. Payne. 1988. Rhyme-like repititions in songs of humpback whales. *Ethology* **79**:295–306.

Gurney, M.E.& M. Konishi. 1980. Hormone-induced sexual differentiation of brain and behavior in zebra finches. *Science* **208**:1380–1383.

Gutman, R.W. 1999. *Mozart: A Cultural Biography*, P. 668, Harcourt Brace & Co., New York.

Güttinger, H.-R. 1979. The integration of learnt and genetically programmed behavior: A study of hierarchal organization in songs of canaries, green finches, and their hybrids. *Zeitschrift für Tierpsychologie* **49**:285–303.

Güttinger, H.-R. 1985. Consequences of domestication on the song structures in the canary. *Behaviour* **94**:255–278.

Güttinger, H.-R. & J. Nicolai 1973. Struktur und Funktion der Rufe bei Prachtfinken (*Estrildidae*). *Zeitschrift für Tierpsychologie* **33**:319–334.

Güttinger, H.-R., T. Turner, S. Dobmeyer & J. Nicolai. 2002. Melodiewarnehmung und Wiedergabe beim Gimpel: Untersuchungen an liederpfeifenden und Kanariengesang imitierenden Gimpeln (*Pyrrhula pyrrhula*). *Journal für Ornithologie* **143**:303–318.

Guyomarc'h, J.C. & C. Guyomarc'h. 1996. Vocal communication in European quail; comparison with Japanese quail. *Comptes Rendus de l'Acamie des Sciences Serie III–Sciences de la Vie* **319**:827–834.

Gyger, M., S.J. Karakashian, A.M. Dufty Jr. & P. Marler. 1988. Alarm signals in birds: The role of testosterone. *Hormones and Behavior* **22**:305–314.

Gyger, M., S. Karakashian & P. Marler. 1986. Avian alarm calling: Is there an audience effect? *Animal Behaviour* **34**:1570–1572.

Gyger, M. & P. Marler. 1988. Food calling in the domestic fowl, *Gallus gallus*: The role of external referents and deception. *Animal Behaviour* **36**:358–365.

Gyger, M., P. Marler & R. Pickert. 1987. Semantics of an avian alarm call system: The male domestic fowl, *Gallus domesticus*. *Behaviour* **102**:15–40.

Haftorn, S. 1999. Calls by willow tits (*Parus montanus*) during ringing and after release. *Journal für Ornithologie* **140**:51–56.

Hahnloser, R.H.R., A.A. Kozhevnikov & M.S. Fee. 2002. An ultrasparse code underlies the generation of neural sequences in a songbird. *Nature* **419**:65–70.

Hailman, J.P. 1977. *Optical Signals: Animal Communication and Light*. Indiana University Press, Bloomington.

Hailman, J.P. & M.S. Ficken. 1986. Combinatorial animal communication with computable syntax: Chick-a-dee calling qualifies as 'language' by structural linguistics. *Animal Behaviour* **34**:1899–1901.

Hailman, J.P. & M.S. Ficken. 1996. Comparative analysis of vocal repertoires, with reference to chickadees. In: *Ecology and Evolution of Acoustic Communication* in Birds, D.E. Kroodsma & E.H. Miller (eds.), pp. 136–159, Cornell University Press, Ithaca, NY.

Hailman, J.P. & C.K. Griswold. 1996. Syntax of black-capped chickadee (*Parus atricapillus*) gargles sorts many types into few groups: Implications for geographic variation, dialect drift, and vocal learning. *Bird Behavior* **11**:39–57.

Haldane, J.B.S. 1946. The interaction of nature and nurture. *Annals of Eugenics* **13**:197–205.

Halkin, S.L. & S.U. Linville. 1999. Northern Cardinal (*Cardinalis cardinalis*). In: *The Birds of North America, No. 440*, A. Poole & F. Gill (eds.). pp. 1–32. The Birds of North America Inc., Philadelphia, PA.

Hallé, F., M. Gahr & M. Kreutzer. Effects of unilateral lesions of HVc on song patterns of male domesticated canaries. *Journal of Neurobiology*.

Halpern, D. 1992. *Sex Differences in Cognitive Abilities*, 2nd ed. Chapter 3: Empirical evidence for cognitive sex differences, pp. 59–96; Chapter 5: Biological hypothesis part III: brains and brain–behavior interactions, pp. 137–170. Lawrence Erlbaum Associates, Hillsdale, NJ.

Hamann, I., G.M. Klump, C. Fichtel & U. Langemann. 1999. *CMR in a songbird studied with narrow-band maskers*. Association for Research in Otolaryngology, St. Petersburg, FL.

Hamilton, K.S., A.P. King, D.R. Sengelaub & M.J. West. 1997. A brain of her own: A neural correlate of song assessment in a female songbird. *Neurobiology of Learning & Memory* **68**:325–332.

Hamilton, W.D. 1964. The genetical evolution of social behaviour. *Journal of Theoretical Biology* **7**:1–52.

Hamilton, W.D. & M. Zuk. 1982. Heritable true fitness and bright birds: A role for parasites? *Science* **218**:384–387.

Hamilton, W.J. III. 1962. Evidence concerning the function of nocturnal call notes of migratory birds. *Condor* **64**:390–401.

Handford, P. 1981 Vegetational correlates of variation

in the song of *Zonotrichia capensis* . Behav. Ecol. Sociobiol. **8**:203–206.

Handford, P. 1988. Trill rate dialects in the rufous-collared sparrow, *Zonotrichia capensis*, in northwestern Argentina, Canadian Journal Zoology **66:** 2658–2670.

Handford, P. & S.C. Lougheed. 1991. Variation in duration and frequency characters in the song of the rufous-collared sparrow, *Zonotrichia capensis*, with respect to habitat, trill dialects and body size. *Condor* **93**:644–658.

Handford, P. & F. Nottebohm 1976. Allozymic and morphological variation in population samples of rufous-collared sparrows, *Zonotrichia capensis*, in relation to vocal dialects. *Evolution* **30:** 802–817.

Handley, H.G. & D.A. Nelson. In review. Ecological and phylogenetic influences on vocal dialect formation in song birds. *Ethology*.

Hansen, P. 1979. Vocal learning: Its role in adapting sound structures to long-distance propagation, and a hypothesis on its evolution. *Animal Behaviour* 27:1270–1271.

Hansen, P. 1981. Coordinated singing in neighboring yellowhammers (*Emberiza citrinella*). *Natura Jutlandica* **19**:121–138.

Harasty, J., K.L. Double, G.M. Halliday, J.J. Kril & D.A. McRitchie. 1997. Language-associated cortical regions are proportionally larger in the female brain. *Archives of Neurology* **54**:171–176.

Harcus, J.L. 1977. The functions of mimicry in the vocal behaviour of the Chorister Robin. *Zeitschrift für Tierpsychologie* **44**:178–193.

Hardy, J.W. 1967. The puzzling vocal repertoire of the South American collared jay, *Cyanolyca viridicyana merida*. *Condor* **69**:513–521.

Hardy, J.W. 1984. Depositing sound specimens. *Auk* **101**:623–624.

Hardy, J.W. & T.A. Parker, III. 1997. The nature and probable function of vocal copying in Lawrence's thrush, *Turdus lawrencii*, pp. 307–320 In: *Studies in Neotropical Ornithology Honoring ted Parker*, J.V. Remsen, Jr. (Ed.). American Ornithologists' Union, Washington, D. C.

Harper, D.G.C. 1991. Communication. In: *Behavioural Ecology: An Evolutionary Approach*, J.R. Krebs & N.B. Davies (eds.), pp. 374–397. Blackwell Scientific Publications, London.

Harrison, K.G. 1977. Perch height selection of grassland birds. *Wilson Bulletin* **89**:486–487.

Hartley, R.S. 1990. Expiratory muscle activity during song production in the canary. *Respiration Physiology* **81**:177–188.

Hartley, R.S., M.S. Chinn & N.F.E. Ullrich. 1997. Left syringeal dominance in testosterone-treated female canaries. *Neurobiology of Learning and Memory* **67**:248–253.

Hartley, R.S. & R.A. Suthers. 1989. Airflow and pressure during canary song: Direct evidence for mini-breaths. *Journal of Comparative Physiology A* **165**:15–26.

Hartmann, W.M., S. Mcadams & B.K. Smith. 1990. Hearing a mistuned harmonic in an otherwise periodic complex tone. *Journal of the Acoustical Society of America* **88**:1712–1724.

Hartshorne, C. 1956. The monotony-threshold in singing birds. *Auk* **73**:176–192.

Hartshorne, C. 1973. *Born to Sing. An Interpretation and World Survey of Bird song.* Indiana University Press, Bloomington.

Haselmayer, J. & J.S. Quinn. 2000. A comparison of point counts and sound recording as bird survey methods in Amazonian southeast Peru. *Condor* **102**:887–893.

Hashino, E. & M. Sokabe. 1989. Hearing loss in the budgerigar (*Melopsittacus undulatus*). *Journal of the Acoustical Society of America* **85**:289–294.

Hashino, E., M. Sokabe & K. Miyamoto. 1988. Frequency specific susceptibility to acoustic trauma in the budgerigar. *Journal of the Acoustical Society of America* **83**:2450–2452.

Haskell, D. 1994. Experimental evidence that nestling begging behaviour incurs a cost due to nest predation. *Proceedings of the Royal Society of London, B* **257**:161–164.

Hasselquist, D., S. Bensch & T. Von Schantz. 1996. Correlation between male song repertoire, extra-pair paternity and offspring survival in the great reed warbler. *Nature* **381**:229–232.

Hasson, O. 1991. Pursuit-deterrent signals: Communication between prey and predator. *Trends in Ecology and Evolution* **6**:325–329.

Hausberger, M. 1997. Social influences on song acquisition and sharing in the European starling (*Sturnus vulgaris*). In: *Social Influences on Vocal Development*, C.T. Snowdon & M. Hausberger (eds.), pp. 128–156. Cambridge University Press, New York.

Hausberger, M., P.F. Jenkins & J. Keene 1991. Species-specificity and mimicry in bird song: Are they paradoxes? *Behaviour* **117:** 53–81.

Hauser, M.D. 1993. The evolution of nonhuman primate vocalizations: Effects of phylogeny, body weight, and social context. *American Naturalist* **142**:528–542.

Hauser, M.D. 1996. *The Evolution of Communication*. MIT Press, Cambridge, MA.

Hauser, M.D., N. Chomsky & W.T. Fitch. 1999. The faculty of language: What is it, who has it, and how did it evolve? *Science* **283**:1–11.

Hayes, K. & C. Hayes. 1951. The intellectual development of a home-raised chimpanzee. *Proceedings of the American Philosophical Society* **95**:105–109.

Heaton, J.T. & S.E. Brauth. 1999. Effects of deafening on the development of nestling and juvenile vocalizations in budgerigars (*Melopsittacus undulatus*). *Journal of Comparative Psychology* **113**: 314–320.

Heaton, J.T. & S.E. Brauth. 2000a. Effects of lesions of the central nucleus of the anterior archistriatum on contact call and warble song production in the budgerigar (*Melopsittacus undulatus*). *Neurobiology of Learning & Memory* **73**:207–242.

Heaton, J.T. & S.E. Brauth. 2000b. Telencephalic nuclei control late but not early nestling calls in the budgerigar. *Behavioural Brain Research* **109**:129–135.

Heaton, J.T., R.J. Dooling & S.M. Farabaugh. 1999. Effects of deafening on the calls and warble song of adult budgerigars (*Melopsittacus undulatus*). *Journal of the Acoustical Society of America* **105**:2010–2019.

Hebb, D.O. 1949. *The Organization of Behavior: A Neuropsychological Theory*. John Wiley & Sons, New York.

Hebb, D.O. 1953. Heredity and environment in mammalian behavior. *British Journal of Animal Behaviour* **1**:43–47.

Hebb, D.O. 1958. *A Textbook of Psychology*. WB Saunders Co., Philadelphia.

Hedenström, A. 1995. Song flight performance in the skylark *Alauda arvensis*. *Journal of Avian Biology* **26**:337–342.

Heinrich, B. 1988. Winter foraging at carcasses by three sympatric corvids, with emphasis on recruitment by the raven, *Corvus corax*. *Behavioral Ecology &Sociobiology* **23**: 141–156.

Heinrich, B. 1989. *Ravens in Winter*. Simon & Schuster, New York.

Heinrich, B., J.M. Marzluff & C.S. Marzluff. 1993. Common ravens are attracted by appeasement calls of food discoverers when attacked. *Auk* **110**:247–254.

Helb, H.W., F. Dowsett-Lemaire, H.H. Bergmann & K. Condrads. 1985. Mixed singing in the European songbirds-–a review. *Zeitschrift für Tierpsychologie* **69**:27–41.

Helekar, S.A., S. Marsh, N.S. Viswanath & D.B. Rosenfield. 2000. Acoustic pattern variations in the female-directed bird songs of a colony of laboratory-bred zebra finches. *Behavioural Processes* **49**:99–110.

Henwood, K. & A. Fabrick. 1979. A quantitative analysis of the dawn chorus: Temporal selection for communicatory optimization. *American Naturalist* **114**:260–274.

Herdegen, T. & J.D. Leah. 1998. Inducible and constitutive transcription factors in the mammalian nervous system: Control of gene expression by Jun, Fos and Krox, and CREB/ATF proteins. *Brain Research. Brain Research Reviews* **28**:370–490.

Hérissant, M. 1753. Recherches sur les organes de la voix de quadrupèdes et de celle des oiseaux. *Academie Royal Memoires Scientifiques*: 279–295.

Herrick, C.J. 1924. *Neurological Foundations of Behaviour*. Holt, New York.

Herrick, C.J. 1956. *The Evolution of Human Nature*. University of Texas Press, Austin.

Herrmann, K. & A. Arnold. 1991. The development of afferent projections to the robust archistriatal nucleus in male zebra finches: A quantitative electron microscope study. *Journal of Neuroscience* **11**:2063–2074.

Herting, B.L. & J.R. Belthoff. 2001. Bounce and double trill songs of male and female western screech-owls: Characterization and usefulness for classification of sex. *Auk* **118**:1095–1101.

Hessler, N.A. & A.J. Doupe. 1999a. Singing-related neural activity in a dorsal forebrain–basal ganglia circuit of adult zebra finches. *Journal of Neuroscience* **19**:10461–10481.

Hessler, N.A. & A.J. Doupe. 1999b. Social context modulates singing-related neural activity in the songbird forebrain. *Nature Neuroscience* **2**:209–211.

Heuwinkel, H. 1982. Schalldruckpegel und Frequenzspektren der Gesänge von *Acrocephalus arundinaceus*, *A. scirpeus*, *A. schoenobaenus* und *A. palustris* und ihre Beziehung zur Biotopakustik. *Ökologie der Vögel* **4**:85–174.

Heyer, W.R., R.W. Donnelly, R.W. McDiarmid, L.C. Hayec & M.S. Foster. 1993. *Measuring and Monitoring Biological Diversity: Standard Methods for Amphibians*. Smithsonian Institution Press, Washington, D.C.

Hiebert, S.M., P.K. Stoddard & P. Races. 1989.

Repertoire size, territory acquisition and reproductive success in the song sparrow. *Animal Behaviour* **37**:266–273.

Hienz, R.D., S.E. Lukas & J.V. Brady. 1981. The effects of pentobarbital upon auditory and visual thresholds in the baboon. *Pharmacology Biochemistry and Behavior* **15**:799–805.

Hienz, R.D., J.M. Sinnott & M.B. Sachs. 1980. Auditory intensity discrimination in blackbirds and pigeons. *Journal of Comparative and Physiological Psychology* **94**:993–1002.

Higgins, P.J., J.M. Peter & W.K. Steele. 2001. *Handbook of Australian, New Zealand and Antarctic Birds*. Oxford University Press, Melbourne.

Hildesheimer, W. 1983. *Mozart* (M. Faber, Trans.). Vintage, New York.

Hile, A.G., T.K. Plummer & G.F. Striedter. 2000. Male vocal imitation produces call convergence during pair bonding in budgerigars, *Melopsittacus undulatus*. *Animal Behaviour* **59**:1209–1218.

Hile, A.G. & G.F. Striedter. 2000. Call convergence within groups of female budgerigars (*Melopsittacus undulatus*). *Ethology* **106**:1105–1114.

Hill, G.E. 1990. Female house finches prefer colorful males: sexual selection for a condition-dependent trait. *Animal Behaviour* **40**:563–572.

Hill, G.E. 1991. Plumage coloration is a sexually selected indicator of male quality. *Nature* **350**:337–339.

Hill, B.G. & M.R. Lein. 1985. The non-song vocal repertoire of the white-crowned sparrow. *Condor* **87**:327–335.

Hinde, R.A. 1952. The behaviour of the great tit (*Parus major*) and some other related species. *Behaviour Supplement* **II**:1–201.

Hinde, R.A. 1956a. The biological significance of the territories of birds. *Ibis* **98**:340–369.

Hinde, R.A. 1956b. The behaviour of certain cardueline F1 inter-species hybrids. *Behaviour* **9**:202–213.

Hinde, R.A. 1958. Alternative motor patterns in chaffinch song. *Animal Behaviour* **6**:211–218.

Hinds, D.S. & W.A. Calder. 1971. Tracheal dead space in the respiration of birds. *Evolution* **25**:429–440.

Hinojosa-Huerta, O., S. DeStefano & W.W. Shaw. 2002. Evaluation of call-response surveys for monitoring breeding yuma clapper rails (*Rallus longirostris yumanensis*). *Journal of Field Ornithology* **73**:151–155.

Hirsh, I.J. 1971. Masking of speech and auditory localization. *Audiology* **10**:110–114.

Hixon, T.J. 1973. Respiratory function in speech. In: *Normal Aspects of Speech, Hearing and Language*, F.D. Minifie, T.J. Hixon & F. Williams (eds.), pp. 73–126. Prentice-Hall, Englewood Cliffs, NJ.

Hoese, W.J., J. Podos, N.C. Boetticher & S. Nowicki. 2000. Vocal tract function in birdsong production: Experimental manipulation of beak movements. *Journal of Experimental Biology* **203**:1845–1855.

Högstedt, G. 1983. Adaptation unto death: Function of fear screams. *American Naturalist* **121**:562–570.

Holland, J., T. Dabelsteen & S.B. Pedersen. 1998. Degradation of wren *Troglodytes troglodytes* song: Implications for Information transfer and ranging. *Journal of Acoustical Society of America* **103**:2154–2166.

Hölldobler, B. & E.O. Wilson. 1990. *The Ants*. Harvard University Press, Cambridge, MA.

Hollich, G.J., K. Hirsh-Pasek & R.M. Golinkoff. 2000. Breaking the language barrier: An emergentist coalition model for the origins of word learning. *SRCD Monographs* **262**:1–138.

Hollien, H. 1975. Neural control of the speech mechanism. In: *The Nervous System: Human and Communication and Its Disorders,* D.B. Tower (ed.), pp. 483–491. Raven Press, New York.

Holloway, C. & D. Clayton. 2001. Estrogen synthesis in the male brain triggers development of the avian song control pathway *in vitro*. *Nature Neuroscience* **4**:170–175.

Honda, E. & K. Okanoya. 1999. Acoustical and syntactical comparisons between songs of the white-backed munia (*Lonchura striata*) and its domesticated strain, the Bengalese finch (*Lonchura striata* var. *domestica*) *Zoological Science (Tokyo)* **16**:319–326.

Hope, S. 1980. Call form in relation to function in the Steller's jay. *American Naturalist* **116**:788–820.

Hopkins, C.D., M. Rosetto & A. Lutjen. 1974. A continuous sound spectrum analyzer for animal sounds. *Zeitschrift für Tierpsychologie* **34**:313–320.

Hopp, S.L., P. Jablonski & J.L. Brown. 2001. Recognition of group membership by voice in Mexican jays, *Aphelocoma ultramarina*. *Animal Behaviour* **62**:297–303.

Horn, A.G., M.L. Leonard, L. Ratcliffe, S.A. Shackleton & R.G. Weisman. 1992. Frequency variation in the songs of black-capped chickadees (*Parus atricapillus*). *Auk* **109**:847–852.

Horn, A.G., M.L. Leonard & D.M. Weary. 1995. Oxygen consumption during crowing by roosters:

Talk is cheap. *Animal Behaviour* **50:** 1171–1175.
Hosino, T. & K. Okanoya. 2000. Lesion of a higher-order song nucleus disrupts phrase level complexity in Bengalese finches. *Neuroreport* **11**:2091–2095.
Hough, G.E., D.A. Nelson & S.F. Volman. 2000. Re-expression of songs deleted during vocal development in white-crowned sparrows, *Zonotrichia leucophrys. Animal Behaviour* **60**:279–287.
Houtman, A.M. 1990. *Sexual selection in the zebra finch, Poephila guttata.* PhD thesis, University of Oxford, Oxford.
Howard, E. 1920. *Territory in Bird Life.* John Murray, London.
Hudspeth, A.J. 1997. How hearing happens. *Neuron* **19**:947–950.
Hughes, M., H. Hultsch & D. Todt. 2002. Imitation and invention in song learning in nightingales (*Luscinia megarhynchos* B., *Turdidae*). *Ethology* **108**:97–113.
Hughes, M.S., S. Nowicki & B. Lohr. 1998. Call learning in black-capped chickadees (*Parus atricapillus*): The role of experience in the development of 'chick-a-dee' calls. *Ethology* **104**:232–249.
Hulse, S.H. 2002. Auditory scene analysis in animal communication. *Advances in the Study of Behavior* **31**:163–200.
Hulse, S.H. & J. Cynx. 1985. Relative pitch perception is constrained by absolute pitch in songbirds (*Mimus, Molothrus, Sturnus*). *Journal of Comparative Psychology* **99**:176–196.
Hulse, S.H. & J. Cynx. 1986. Interval and contour in serial pitch perception by a passerine bird, the European starling (*Sturnus vulgaris*). *Journal of Comparative Psychology* **100**:215–228.
Hulse, S.H., J. Cynx, & J. Humpal 1984. Absolute and relative discrimination in serial pitch perception by birds. *Journal of Experimental Psychology: General,* **113**:38–54.
Hulse, S.H., S.A. MacDougall-Shackleton & A.B. Wisniewski. 1997. Auditory scene analysis by songbirds: Stream segregation of birdsong by European starlings (*Sturnus vulgaris*). *Journal of Comparative Psychology* **111**:3–13.
Hultsch, H. 1980. *Beziehungen zwischen Struktur, zeitlicher Variabilität und sozialem Einsatz des Gesangs der Nachtigall (Luscinia megarhynchos).* Doctoral dissertation, Freie Universität, Berlin.
Hultsch, H. 1985. Sub-Repertoire-Bildung: Ein Organisationsprinzip bei Erwerb und Einsatz großer vokaler Repertoires. *Verhandlungen der Deutschen Zoologen Gesellschaft* **78**:229.
Hultsch, H. 1991a. Song ontogeny in birds: Closed or open developmental programs? In: *Synapse-transmission, Modulation,* N. Elsner & H. Penzlin (eds.), p. 576. Thieme Verlag, Stuttgart.
Hultsch, H. 1993a. Tracing the memory mechanisms in the song acquisition of birds. *Netherlands Journal of Zoology* **43**:155–171.
Hultsch, H. 1993b. Ecological versus psychobiological aspects of song learning in birds. *Etologia* **3**:309–323.
Hultsch, H., N. Geberzahn & F. Schleuss. 1998. Song invention in nightingales – cues from song development. *Ostrich* **69**:254.
Hultsch, H. & M.L. Kopp. 1989. Early auditory learning and song improvisation in nightingales, *Luscinia megarhynchos. Animal Behaviour* **37**:510–512.
Hultsch, H., R. Lange & D. Todt. 1984. Mustertyp-markierte Lernangebote: Eine Methode zur Untersuchung auditorisch/vokaler Strophentypgedächtnisse. *Verhandlungen der Deutschen Zoologen Gesellschaft* 77:249.
Hultsch, H., R. Mundry & D. Todt. 1999a. Learning, representation and retrieval of rule-related knowledge in the song system of birds. In: *Learning: Rule Extraction and Representation,* A.D. Friederici & R. Menzel (eds.), pp. 89–115. De Gruyter, Berlin.
Hultsch, H., F. Schleuss & D. Todt. 1999b. Auditory–visual stimulus pairing enhances perceptual learning in a songbird. *Animal Behaviour* **58**:143–149.
Hultsch, H. & D. Todt. 1982. Temporal performance roles during vocal interactions in nightingales (*Luscinia megarhynchos* B.). *Behavioral Ecology & Sociobiology* **11**:253–260.
Hultsch, H. & D. Todt. 1986. Signal matching. *Z. Semiotik* **8**:584–586.
Hultsch, H & D. Todt. 1989a. Song acquisition and acquisition constraints in the nightingale (*Luscinia megarhynchos*). *Naturwissenschaften* **76**:83–86.
Hultsch, H & D. Todt. 1989b. Context memorization in the song learning of birds. *Naturwissenschaften* **76**:584–586.
Hultsch, H & D. Todt. 1989c. Memorization and reproduction of songs in nightingales (*Luscinia megarhynchos*): Evidence for package formation. *Journal of Comparative Physiology* A **165**:197–203.
Hultsch, H & D. Todt. 1992. The serial order effect in the song acquisition of birds. *Animal Behaviour* **44**:590–592.
Hultsch, H. & D. Todt. 1996. Discontinuous and

incremental processes in the song learning of birds: Evidence for a primer effect. *Journal of Comparative Physiology* A **179**:291–299.

Hunter, M.L. 1980. Microhabitat selection for singing and other behaviour in great tits, *Parus major*: Some visual and acoustical considerations. *Animal Behaviour* **28**:468–475.

Hurley, T.A., L.M. Ratcliffe, D.M. Weary, & R.G. Weisman 1992. White-throated sparrows (*Zonotrichia albicollis*) can perceive pitch change in conspecific song by using the frequency ratio independent of the frequency difference. *Journal of Comparative Psychology* **106**:388–391.

Hurley, T.A., L.M. Ratcliffe & R.G. Weisman. 1992. Relative pitch recognition in white-throated sparrows. *Animal Behaviour*, **40**:176–181.

Hutto, R.L. 1987. A description of mixed-species insectivorous bird flocks in western Mexico. *Condor* **89**:282–292.

Huxley, J.S. 1942. *Evolution: The Modern Synthesis*. George Allen and Unwin Ltd, London.

Huxley, J.S. 1947. Song variants in the yellowhammer. *British Birds* **XL**:162–164.

Huxley, J.S. 1963. Lorenzian ethology. *Zeitschrift für Tierpsychologie* **20**:402–409.

Hygonstrom, S.E., R.M. Timm & G.E. Larson. 1994. *Wildlife Damage Handbook 202: Prevention and Control of Wildlife Damage*. University of Nebraska & US Dept. of Agriculture, Lincoln.

Ikebuchi, M., M. Futamatsu & K. Okanoya. 2003. Sex differences in song perception in Bengalese finches measured by the cardiac response. *Animal Behaviour* **65**:123–130.

Ikebuchi, M. & K. Okanoya. 2000. The site of hearing-induced gene expression in the avian telencephalon (NCM) is responsible for behaviorally mediated auditory memory retrieval. *Society for Neuroscience (Abstracts)* **26**:725.

Il'ichev, V.D., I.I. Kamenskii & O.L. Silaeva. 1995. Ecological and technogenous factors of noise pollution of natural habitats of birds. *Russian Journal of Ecology* **26**:345–348.

Immelmann, K. 1969. Song development in the zebra finch and other estrildid finches. In: *Bird Vocalizations, Their Relation to Current Problems in Biology and Psychology*, R.A. Hinde (ed.), pp. 61–74. Cambridge University Press, Cambridge.

Immelmann, K. & C. Beer. 1989. *A Dictionary of Ethology*. Harvard University Press, Cambridge, MA.

Immelmann, K. & S.J. Suomi. 1981. Sensitive phases in development. In: *Behavioral Development*, K. Immelmann, G.W. Barlow, L. Petrinovich & M. Main (eds.), pp. 395–431. Cambridge University Press, Cambridge.

Ince, S.A., P.J.B. Slater & C. Weismann. 1980. Changes with time in the songs of a population of chaffinches. *Condor* **82**:285–290.

Inglis, I.R., M.R. Fletcher, C.J. Feare, P.W. Greig-Smith & S. Land. 1982. The incidence of distress calling among British birds. *Ibis* **124**:351–355.

Ingold, P. 1973. Zur lautlichen Beziehung des Alters zu seinem Kücken bei Tordalken (*Alca torda*). *Behaviour* **45**:154–190.

Irwin, D.E. 2000. Song variation in an avian ring species. *Evolution* **54**:998–1010.

Irwin, D.E., S. Bensch & T.D. Price. 2001a. Speciation in a ring. *Nature* **409**:333–337.

Irwin, D.E., J.H. Irwin & T.D. Price. 2001b. Ring species as bridges between microevolution and speciation. *Genetica* **112–113**:223–243.

Irwin, D.E. & T. Price. 1999. Sexual imprinting, learning and speciation. *Heredity* **82**:347–354.

Irwin, R.E. 1988. The evolutionary importance of behavioural development: The ontogeny and phylogeny of bird song. *Animal Behaviour* **36**:814–824.

Isler, M.L., P.R. Isler & B.M. Whitney. 1997. Biogeography and systematics of the *Thamnophilus punctatus* (*Thamnophilidae*) complex. In: *Studies in Neotropical Ornithology Honoring Ted Parker*, (J. Remsen, Jr. (ed.) pp. 355–382. American Ornithologists' Union, Washington, D. C.

IUCN. 2000. *Threatened Birds of the World*. A.J. Stattersfiled & D.R. Cooper (eds.). Birdlife International, Cambridge, UK.

Iyengar, S., S.S. Viswanathan & S.W. Bottjer. 1999. Development of topography within song control circuitry of zebra finches during the sensitive period for song learning. *Journal of Neuroscience* **19**:6037–6057.

Jacobs, E., A. Arnold & A. Campbell. 1999. Developmental regulation of the distribution of aromatase- and estrogen-receptor-mRNA-expressing cells in the zebra finch brain. *Journal of Neurobiology* **27**:513–519.

Jaffe, J., B. Beebe, S. Feldstein, C.L. Crown & M.D. Jasnow. 2001. Rhythms of dialogue in infancy. *Monographs of the Society for Research in Child Development* **66**:vi–131.

Janik, V.M. & P.J.B. Slater. 1997. Vocal learning in mammals. *Advances in the Study of Behavior* **26**:59–99.

Jarvi, T., T. Rædsater & S. Jakobsson. 1980. The

song of the willow warbler *Phylloscopus trochilus* with special reference to singing behaviour in agonistic situations. *Ornis Scandinavica* **11**:236–242.

Jarvis, E.D. 2001. Insights from vocal learning birds into the neurobiology of human language. *Society for Neuroscience (Abstracts)* **27**:843.

Jarvis, E.D. 2004. Evolution of vocal learning brain structures in birds: a synopsis. *Acta Zoologica* in press.

Jarvis, E.D., O. Güntürkün, L. Bruce, A. Csillag, H. Karten, W. Kuenzel, L. Medina, G. Paxinos, D.J. Perkel, T. Shimizu, G. Striedter, M. Wild, G.F. Ball, J. Dugas-Ford, S. Durand, G. Hough, S. Husband, L. Kubikova, D. Lee, C.V. Mello, A. Powers, C. Siang, T.V. Smulders, K. Wada, S.A. White, K. Yamamoto, J. Yu, A. Reiner & A. Butler. In review. A paradigm shift in understanding the organization, evolution, and function of the avian brain. *Public Library of Science* in review.

Jarvis, E.D., V. Smith, K. Wada, M. Rivas, M. McElroy, T. Smulders, P. Carninci, Y. Hayashisaki, F. Dietrich, X. Wu, J. Yu, P. McConnell, P. Wang, A. Hartemink & S. Lin. 2002. A framework for integrating the songbird brain. *Journal of Comparative Physiology A* **188**:961–980.

Jarvis, E.D. & C.V. Mello. 2000. Molecular mapping of brain areas involved in parrot vocal communication. *Journal of Comparative Neurology* **419**:1–31.

Jarvis, E.D., C.V. Mello & F. Nottebohm. 1995. Associative learning and stimulus novelty influence the song-induced expression of an immediate early gene in the canary forebrain. *Learning & Memory* **2**:62–80.

Jarvis, E.D. & F. Nottebohm. 1997. Motor-driven gene expression. *Proceedings of the National Academy of Sciences USA* **94**:4097–4102.

Jarvis, E.D., S. Ribeiro, J. Vielliard, M.L. Da Silva, D. Ventura & C.V. Mello. 2000. Behaviorally driven gene expression reveals song nuclei in hummingbird brain. *Nature* **406**:628–632.

Jarvis, E.D., C. Scharff, M.R. Grossman, J.A. Ramos & F. Nottebohm. 1998. For whom the bird sings: Context-dependent gene expression. *Neuron* **21**:775–788.

Jenkins, P.F. 1978. Cultural transmission of song patterns and dialect development in a free-living bird population. *Animal Behaviour* **26**:50–78.

Jilka, A. & B. Leisler. 1974. Die Einpassung dreier Rohrsängerarten *Acrocephalus schoenobaenus, A. scirpaceus, A. arundinaceus* in ihre Lebensräume in bezug auf das Frequenzspectrum ihrer Reviergesänge. *Journal für Ornithologie* **115**:192–212.

Jin, H. & D.F. Clayton. 1997. Localized changes in immediate-early gene regulation during sensory and motor learning in zebra finches. *Neuron* **19**:1049–1059.

Johnsgard, P.A. 1973. *Grouse and Quails of North America.* University of Nebraska Press , Lincoln.

Johnsgard, P.A. 1981. *The Plovers, Sandpipers, and Snipes of the World.* University of Nebraska Press, Lincoln.

Johnson, F., M.M. Sablan & S.W. Bottjer. 1995. Topographic organization of a forebrain pathway involved with vocal learning in zebra finches. *Journal of Comparative Neurology* **358**:260–278.

Johnson, K.P., S. de Kort, K. Dinwoodey, A.C. Mateman, C. ten Cate, C.M. Lessells & D.H. Clayton. 2001. A molecular phylogeny of the dove genera *Streptopelia* and *Columba. Auk* **118**:874–887.

Johnson, L.S. & W.A. Searcy. 1996. Female attraction to male song in house wrens (*Troglodytes aedon*). *Behaviour* **133**:357–366.

Johnson, N.K. & R.E. Jones. 2001. A new species of Tody-tyrant (*Tyrannidae*: *Poecilotriccus*) from northern Peru. *Auk* **118**:334–341.

Johnson, R.R., B.T. Brown, L.T. Haight & J.M. Simpson. 1981. Playback recordings as a special avian censusing technique. In: *Estimating Numbers of Terrestrial Birds*, C.J. Ralph & J.M. Scott (eds.), pp. 68–75. Allen Press, Lawrence, KS.

Johnson, R.R. & J.J. Dinsmore. 1986. The use of tape-recorded calls to count Virginia rails and soras. *Wilson Bulletin* **98**:303–306.

Johnson, R.S. 1975. *Messiaen.* University of California Press, Berkeley, CA.

Johnston, T.D. 1988. Developmental explanation and the ontogeny of birdsong: Nature/nurture redux. *Behavioural Brain Science* **11**:617–663.

Jouventin, P. 1982. Visual and vocal signals in penguins, their evolution and adaptive characters. *Advances in Ethology* **24**:1–149.

Jouventin, P., T. Aubin & T. Lengagne. 1999. Finding a parent in a king penguin colony: The acoustic system of individual recognition. *Animal Behaviour* **57**:1175–1183.

Jung, R.E., E.S. Morton & R.C. Fleischer. 1994. Behavior and parentage of a white-throated sparrow x dark-eyed junco hybrid. *Wilson Bulletin* **106**:189–202.

Jürgens, U. 1995. Neuronal control of vocal production in non-human and human primates. In: *Current Topics in Primate Vocal Communication,* E. Zimmermann, J.D. Newman & U. Jürgens (eds.), pp. 199–206. Plenum Press, New York.

Jürgens, U. 1998. Neuronal control of mammalian vocalization, with special reference to the squirrel monkey. *Naturwissenschaften* **85**:376–388.

Jurisevic, M.A. & K.J. Sanderson. 1994. Alarm vocalisations in Australian birds: Convergent characteristics and phylogenetic differences. *Emu* **94**:69–77.

Jurisevic, M.A. & K.J. Sanderson. 1998. A comparative analysis of distress call structure in Australian passerine and non-passerine species: Influence of size and phylogeny. *Journal of Avian Biology* **29**:61–71.

Jusczyk, P.W., K. Hirsh-Pasek, D.B. Kemler Nelson, L.J. Kennedy, A. Woodward & J. Piwoz. 1992. Perception of acoustic correlates of major phrasal units by young infants. *Cognitive Psychology* **24**:252–293.

Kalischer, O. 1905. Das grosshirn der Papageien in anatomischer und physiologischer Bezeihung. *Abhandlungen der Königlich Preussischenn Akademie der Wissenschaften* **IV**:1–105.

Kaplan, P.S., J. Bachorowski, M.J. Smoski & M. Zinser. 2001. Role of clinical diagnosis and medication use in effects of maternal depression on infant-directed speech. *Infancy* **2**:537–548.

Kaplan, P.S., J. Bachorowski & P. Zarlengo-Strouse. 1999. Child-directed speech produced by mothers with symptoms of depression fails to promote associative learning in four-month-old infants. *Child Development* **70**:560–570.

Kaplan, P.S., J. Bachorowski & P. Zarlengo-Strouse. 2002. Infants of depressed mothers, although competent learners, fail to learn in response to their own mothers' infant-directed speech. *Psychological Science* **13**:268–271.

Kaplan, P.S. & K.B. Fox. 1991. Cross-modal associative transfer of response sensitization in infants. *Developmental Psychobiology* **24**:265–276.

Kaplan, P.S., M.H. Goldstein, E.R. Huckeby & R.P. Cooper. 1995a. Habituation, sensitization, and infants' responses to motherese speech. *Developmental Psychobiology* **28**:45–57.

Kaplan, P.S., M.H. Goldstein, E.R. Huckeby, M.J. Owren & R.P. Cooper. 1995b. Dishabituation of visual attention by infant—versus adult-directed speech: Effects of frequency modulation and spectral composition. *Infant Behavior and Development* **18**:209–223.

Kaplan, P.S., P.C. Jung, J.S. Ryther & P. Zarlengo-Strouse. 1996. Infant-directed versus adult-directed speech as signals for faces. *Developmental Psychobiology*, **32**:880–891.

Karakashian, S.J., M. Gyger & P. Marler. 1988. Audience effects on alarm calling in chickens (*Gallus gallus*). *Journal of Comparative Psychology* **102**:129–135.

Karr, J.R. 1981. Surveying birds in the tropics. In: *Estimating Numbers of Terrestrial Birds*, C.J. Ralph & J.M. Scott (eds.), pp. 548–553. Allen Press, Lawrence, KS.

Katz, L.C. & M.E. Gurney. 1981. Auditory responses in the zebra finch's motor system for song. *Brain Research* **221**:192–197.

Keefe, K.A. & C.R. Gerfen. 1999. Local infusion of the (+/–)-alpha-amino-3-hydroxy-5-methylisoxazole-4-propionate/kainate receptor antagonist 6-cyano-7-nitroquinoxaline-2,3-dione does not block D1 dopamine receptor-mediated increases in immediate early gene expression in the dopamine-depleted striatum. *Neuroscience* **89**:491–504.

Kelley, D.B. & F. Nottebohm. 1979. Projections of a telencephalic auditory nucleus–field L–in the canary. *Journal of Comparative Neurology* **183**:455–469.

Kent, R.D. 2000. Research on speech motor control and its disorders: A review and perspective. *Journal of Communication Disorders* **33**:391–427.

Kerlinger, P. 1995. *How Birds Migrate* Stackpole Books, Pennsylvania.

Kerr, J.T., A. Sugar & L. Packer. 2000. Indicator taxa, rapid biodiversity assessment, and nestedness in an endangered ecosystem. *Conservation Biology* **14**:1726–1734.

Kimpo, R.R. & A.J. Doupe. 1997. FOS is induced by singing in distinct neuronal populations in a motor network. *Neuron* **18**:315–325.

Kimura, D. 1996. Sex, sexual orientation and sex hormones influence human cognitive function. *Current Opinion in Neurobiology* **6**:259–263.

King, A.P. & M.J. West. 1977. Species identification in the North American cowbird: Appropriate responses to abnormal song. *Science* **195**:1002–1004.

King, A.P. & M.J. West. 1983. Dissecting cowbird

song potency: Assessing a song's geographic identity and relative appeal. *Zeitschrift für Tierpsychologie* **63**:37–50.

King, A.P. & M.J. West. 1990. Variation in species-typical behavior: A contemporary theme for comparative psychology. In: *Contemporary Issues in Comparative Psychology*, D.A. Dewsbury (ed.), pp. 331–339. Sinauer, Sunderland, MA.

King, A.P., M.J. West & D.H. Eastzer. 1980. Song structure and song development as potential contributors to reproductive isolation in cowbirds (*Molothrus ater*). *Journal of Comparative and Physiological Psychology* **94**:1028–1036.

King, A.P., M.J. West, D.H. Eastzer & J.E.R. Staddon. 1981. An experimental investigation of the bioacoustics of cowbird song. *Behavioural Ecology and Sociobiology* **9**:211–217.

King, A.P., M.J. West & M.H. Goldstein. In review. Cross modal signaling; a new form of plasticity in bird song.

King, A.S. 1989. Functional anatomy of the syrinx. In: *Form and Function in Birds*, A.S. King & J. McClelland (eds.), pp. 105–191. Academic Press, London.

Kirn, J.R., Y. Fishman, K. Sasportas, A. Alvarez-Buylla & F. Nottebohm. 1999. Fate of new neurons in adult canary high vocal center during the first 30 days after their formation. *Journal of Comparative Neurology* **411**:487–494.

Kirn, J., B. O'Loughlin, S. Kasparian & F. Nottebohm. 1994. Cell death and neuronal recruitment in the high vocal center of adult male canaries are temporally related to changes in song. *Proceedings of the National Academy of Sciences* **91**:7844–7848.

Kirn, J.R., A. Alvarez-Buylla & F. Nottebohm. 1991. Production and survival of projection neurons in a forebrain vocal center of adult male canaries. *Journal of Neuroscience* **11**:1756–1762.

Kleunder, K.R., R.L. Diehl & P.R. Killeen. 1987. Japanese quail can learn phonetic categories. *Science* **237**:1195–1197.

Klump, G.M. 1996. Bird communication in the noisy world. In: *Ecology and Evolution of Acoustic Communication in Birds*, D.E. Kroodsma & E.H. Miller (eds.), pp. 321–338. Cornell University Press, Ithaca, NY.

Klump, G.M. & E. Curio 1983. Reactions of blue tits *Parus caeruleus* to hawk models of different sizes. *Bird Behaviour* **4**:78–81.

Klump, G.M., R.J. Dooling & R.R. Fay. 1995. *Methods in Comparative Psychoacoustics*. Birkhauser-Verlag, Basel.

Klump, G.M., E. Kretzschmar & E. Curio. 1986. The hearing of an avian predator and its avian prey. *Behavioral Ecology & Sociobiology* **18**:317–323.

Klump, G.M. & U. Langemann. 1995. Comodulation masking release in a songbird. *Hearing Research* **87**:157–164.

Klump, G.M. & O.N. Larsen. 1992. Azimuth sound localization in the European starling (*Sturnus vulgaris*): Physical binaural cues. *Journal of Comparative Physiology A* **170**:243–251.

Klump, G.M. & E.H. Maier. 1990. Temporal summation in the European starling (*Sturnus vulgaris*). *Journal of Comparative Psychology* **104**:94–100.

Klump, G.M. & M.D. Shalter. 1984. Acoustic behaviour of birds and mammals in the predator context. I. Factors affecting the structure of alarm signals. II. The functional significance and evolution of alarm signals. *Zeitschrift für Tierpsychologie* **66**:189–226.

Kobayasi, K., H. Uno & K. Okanoya. 2001. Partial lesions in the anterior forebrain pathway affect song production in adult Bengalese finches. *Neuroreport* **12**:353–358.

Koenig, W.D., M.T. Stanback, P.N. Hooge & R.L. Mumme. 1991. Distress calls in the acorn woodpecker. *Condor* **93**:637–643.

Konishi, M. 1963. The role of auditory feedback in the vocal behavior of the domestic fowl. *Zeitschrift für Tierpsychologie* **20**:349–367.

Konishi, M. 1964. Effects of deafening on song development in two species of juncos. *Condor* **66**:85–102.

Konishi, M. 1965a. The role of auditory feedback in the control of vocalization in the white-crowned sparrow. *Zeitschrift für Tierpsychologie* **22**:770–783.

Konishi, M. 1965b. Effects of deafening on song development in American robins and black-headed grosbeaks. *Zeitschrift für Tierpsychologie* **22**:584–99.

Konishi, M. 1969. Time resolution by single auditory neurones in birds. *Nature* **222**:566–567.

Konishi, M. 1985. Birdsong: From behavior to neuron. *Annual Review of Neuroscience* **8**:125–170.

Konishi, M. 1989. Birdsong for neurobiologist. *Neuron* **3**:541–549.

Konishi, M. 1994. An outline of recent advances in birdsong neurobiology. *Brain Behavior and Evolution* **44**:279–285.

Konishi, M. & E.I. Knudsen. 1979. The oilbird: Hearing and echolocation. *Science* **204**:425–427.

Konishi, M. & F. Nottebohm. 1969. Experimental studies on the ontogeny of avian vocalizations. In: *Bird Vocalizations*, R.A. Hinde (ed.), pp. 29–48. Cambridge University Press, Cambridge.

Kornack, D. & P. Rakic. 2001. Cell proliferation without neurogenesis in adult primate neocortex. *Science* **294**:2127–2130.

Korsia, S. & S. Bottjer. 1991. Chronic testosterone treatment impairs vocal learning in male zebra finches during a restricted period of development. *Journal of Neuroscience* **11**:2362–2371.

Kowalski, M.P. 1983. Factors affecting the performance of flight songs and perch songs in the common yellowthroat. *Wilson Bulletin* **95**:140–142.

Kramer, H.G. & R.E. Lemon. 1983. Dynamics of territorial singing between neighbouring song sparrows (*Melospiza melodia*). *Behaviour* **85**:198–223.

Kramer, H.G., R.E. Lemon & M.J. Morris. 1985. Song switching and agonistic stimulation in the song sparrow (*Melospiza melodia*): Five tests. *Animal Behaviour* **33**:135–149

Krams, I. 2001. Communication in crested tits and the risk of predation. *Animal Behaviour* **61**:1065–1068.

Krams, I. & T. Krams. 2002. Interspecific reciprocity explains mobbing behaviour of the breeding chaffinches, *Fringilla coelebs*. *Proceedings of the Royal Society of London, B* **269**: 2345–2350.

Krebs, J.R. 1976. Habituation and song repertoires in the great tit. *Behavioural Ecology & Sociobiology* **1**:215–227

Krebs, J.R. 1977a. The significance of song repertoires: The Beau Geste hypothesis. *Animal Behaviour* **25**:428–478

Krebs, J. R. 1977b. Song and territory in the great tit. In: *Evolutionary Ecology*, B Stonehouse & C.M. Perrins, (ed.), pp. 47–62. MacMillan and Co, London.

Krebs, J.R., R. Ashcroft & K. van Orsdol. 1981. Song matching in the great tit, *Parus major*. *Animal Behaviour* **29**:918–923.

Krebs, J.R., R. Ashcroft & M. Webber. 1978. Song repertoires and territory defence in the great tit. *Nature* **271**:539–542.

Krebs, J.R.& D.E. Kroodsma. 1980. Repertoires and geographical variation in bird song. In *Advances in the Study of Behavior*, J.S. Rosenblatt, R.A. Hinde, C. Beer & M.C. Busnel (eds.), pp. 143–177. Academic Press, New York.

Kress, S.W. 1995. The use of decoys, sound recordings, and gull control for re-establishing a tern colony in Maine. *Colonial Waterbirds* **6**:185–196.

Kress, S.W. & D.N. Nettleship. 1988. Re-establishment of Atlantic puffins (*Fratercula arctic*) at a former breeding site in the Gulf of Mexico. *Journal of Field Ornithology* **59**:161–170.

Kreutzer, M., I. Beme, E. Vallet & L. Kiosseva. 1999. Social stimulation modulates the use of the 'A' phrase in male canary songs. *Behaviour* **11**:1–10.

Kreutzer, M.L., E. Vallet & L. Nagle. 1996. Female canaries display to songs of early isolated males. *Experientia* **52**: 277–280.

Kroodsma, D.E. 1973. Coexistence of Bewick's wrens and house wrens in Oregon. *Auk* **90**:341–352.

Kroodsma, D.E. 1974. Song learning, dialects, and dispersal in the Bewick's wren. *Zeitschrift für Tierpsychologie* **35**:352–380.

Kroodsma, D.E. 1976. Reproductive development in a female songbird: Differential stimulation by quality of male song. *Science* **192**:574–575.

Kroodsma, D.E. 1977. A re-evaluation of song development in the song sparrow. *Animal Behaviour* **25**:390–399.

Kroodsma, D.E. 1978. Aspects of learning in the ontogeny of bird song: Where, from whom, when, how many, which, and how accurately? In: *Development of behavior*, G. Burghardt & M. Bekoff, (eds.) pp. 215–230. Garland Publishing Co., New York.

Kroodsma, D.E. 1979. Vocal dueling among male marsh wrens: Evidence for ritualized expression of dominance/subordinance. *Auk* **96**:506–515.

Kroodsma, D.E. 1981. Geographical variation and functions of song types in warblers (*Parulidae*). *Auk* **98**:743–751.

Kroodsma, D.E. 1984. Songs of the alder flycatcher (*Empidonax alnorum*) and willow flycatcher (*Empidonax traillii*) are innate. *Auk* **101**:13–24.

Kroodsma, D.E. 1985. Development and use of two song forms by the eastern phoebe. *Wilson Bulletin* **97**:21–29.

Kroodsma, D.E. 1986. Song development by castrated marsh wrens. *Animal Behaviour* **34**:1572–1575.

Kroodsma, D.E. 1988. Song types and their use: Developmental flexibility of the male blue-winged warbler. *Ethology* **79**:235–247.

Kroodsma, D.E. 1989. Male eastern phoebes (*Tyrannidae, Passeriformes*) fail to imitate songs. *Journal of Comparative Psychology* **103**: 227–232.

Kroodsma, D.E. 1990. Using appropriate experimental

designs for intended hypotheses in song playbacks, with examples for testing effects of song repertoire sizes. *Animal Behaviour* **40**:1138–1150.

Kroodsma, D.E. 1996. Ecology of passerine song development. In: *Ecology and Evolution of Acoustic Communication in Birds*, D.E. Kroodsma & E.H. Miller (eds.), pp. 3–19. Cornell University Press, Ithaca, NY.

Kroodsma, D.E. 1999. Making ecological sense of song development by songbirds. In: *The Design of Animal Communication*, M.D. Hauser & M. Konishi, (eds.), pp. 319–342. Massachusetts Institute of Technology, Cambridge, MA.

Kroodsma, D.E. 2000. Vocal behavior. Cornell Laboratory of Ornithology Home Study Course in Bird Biology, Chapter 7 pp. 7.1–7.18.

Kroodsma, D.E., M.C. Baker, L.F. Baptista & L. Petrinovich. 1985. Vocal "dialects" in Nuttall's White-crowned sparrow. *Current Ornithology* **2**:103–133.

Kroodsma, D.E. & J.R. Baylis. 1982. Appendix: A world survey of evidence for vocal learning in birds. In: *Acoustic Communication in Birds. Vol. 2. Song Learning and its Consequences*, D.E. Kroodsma & E.H. Miller (eds.), pp. 311–337. Academic Press, New York.

Kroodsma, D.E., R.E. Bereson, B.E. Byers & E. Minear. 1989. Use of song types by the chestnut-sided warbler: Evidence for both intra- and inter-sexual functions. *Canadian Journal of Zoology* **67**:447–456.

Kroodsma, D.E., G.F. Budney, R.W. Grotke, J. Vielliard, R. Ranft, O.D. Veprintsev & S.L.L. Gaunt. 1996. Natural sound archives. In: *Ecology and Evolution of Acoustic Communication among Birds*, D.E. Kroodsma & E.J. Miller (eds.), pp. 474–486. Cornell University Press, Ithaca, NY.

Kroodsma, D.E., & B.E. Byers. 1991. The function(s) of bird song. *American Zoologist* **31**:318–328.

Kroodsma, D.E. & B.E. Byers. 1998. Songbird song repertoires: An ethological approach to studying cognition. In: *Animal Cognition in Nature*, R. Balda, I.M. Pepperberg & A. Kamil (eds.) pp. 305–336. *Academic Press, London.*

Kroodsma, D.E., B.E. Byers, E. Goodale, S. Johnson & W.-C. Liu. 2001a. Pseudoreplication in playback experiments, revisited a decade later. *Animal Behaviour* **61**:1029–1033.

Kroodsma, D.E., B.E. Byers, S.L. Halkin, C. Hill, D. Minis, J.R. Bolsinger, J.-A. Dawson, E. Donelan, J. Farrington, F. Gill, P. Houlihan, D. Innes, G. Keller, L. Macaulay, C.A. Marantz, J. Ortiz, P.K. Stoddard & K. Wilda. 1999a. Geographic variation in black-capped chickadee songs and singing behavior. *Auk* **116**:387–402.

Kroodsma, D.E. & R.A. Canady. 1985. Differences in repertoire size, singing behavior, and associated neuroanatomy among marsh wren populations have a genetic basis. *Auk* **102**:439–446.

Kroodsma, D.E., P.W. Houlihan, P.A. Fallon, & J.A. Wells. 1997. Song development by grey catbirds. *Animal Behaviour* **54**:457–464.

Kroodsma, D.E., & M. Konishi. 1991. A suboscine bird (eastern phoebe, *Sayornis phoebe*) develops normal song without auditory feedback. *Animal Behaviour* **42**:477–488.

Kroodsma, D.E., W.-C. Liu, E. Goodwin & P.A. Bedell. 1999b. The ecology of song improvisation as illustrated by North American sedge wrens. *Auk* **116**:373–386.

Kroodsma, D.E. & E.H. Miller. 1982. *Acoustic Communication in Birds 2 volumes*. Academic Press, New York.

Kroodsma, D.E. & E.H. Miller. 1996. *Ecology and Evolution of Acoustic Communication in Birds*. Cornell University Press, Ithaca, NY.

Kroodsma, D.E. & H. Momose. 1991. Songs of the Japanese population of the winter wren, *Troglodytes troglodytes*. *Condor* **93**:424–432.

Kroodsma, D.E. & L.D. Parker. 1977. Vocal virtuosity in the brown thrasher. *Auk* **94**:783–785.

Kroodsma, D.E. & R. Pickert. 1980. Environmentally dependent sensitive periods for avian vocal learning. *Nature* **288**:477–479.

Kroodsma, D.E., J. Sánchez, D.W. Stemple, E. Goodwin, M.L. da Silva & J.M.E. Vielliard. 1999c. Sedentary lifestyle of neotropical sedge wrens promotes song imitation. *Animal Behaviour* **57**:855–863.

Kroodsma, D.E., & J. Verner. 1978. Complex singing behaviors among *Cistothorus* wrens. *Auk* **95**:703–716.

Kroodsma, D.E., & J. Verner. 1997. Marsh wren (*Cistothorus palustris*). In: *The Birds of North America, No. 308*, A. Poole & F. Gill, (eds.), pp. 1–32. The Academy of Natural Sciences, Philadelphia, PA; The American Ornithologists' Union, Washington, D.C.

Kroodsma, D.E., M.E. Vielliard & F.G. Stiles. 1996. Study of bird sounds in the neotropics: Urgency and opportunity. In: *Ecology and Evolution of Acoustic Communication in Birds,* D.E. Kroodsma

& E.H. Miller (eds.), pp. 269–281. Cornell University Press, Ithaca, NY.

Kroodsma, D.E., K. Wilda, V. Salas & R. Muradian. 2001b. Song variation among *Cistothorus* wrens, with a focus on the Merida wren. *Condor* **103**:855–861.

Kroodsma, D.E., R.W. Woods & E.A. Goodwin. 2002. Falkland Island sedge wrens (*Cistothorus platensis*) imitate rather than improvise large song repertoires. *Auk* **119**:523–528.

Krumhansl, C.L. 1990. *Cognitive Foundations of Musical Pitch*, Oxford University Press, New York.

Kuczaj, S.A. 1983. *Crib Speech and Language Play*. Springer-Verlag, New York.

Kuczaj, S.A. 1998. Is an evolutionary theory of language play possible? *Cahiers Psychologie Cognitive* **17**:135–154.

Kuhl, P.K. 1989. On babies, birds, modules, and mechanisms: A comparative approach to the acquisition of vocal communication. In: *The Comparative Psychology of Audition: Perceiving Complex Sounds*, R.J. Dooling & S.H. Hulse (eds.), pp. 379–419. Lawrence Erlbaum, Hillsdale, NJ.

Kuhl, P., J.E. Andruski, I.A. Chistovich & L.A. Chistovich. 1997. Cross-language analysis of phonetic units in language addressed to infants. *Science* **277**:684–686.

Kumar, A. & D. Bhatt. 2000. Vocal signals in a tropical avian species, the redvented bulbul *Pychonotus cafer*: Their characteristics and importance. *Journal of Biosciences* **25**:387–396.

Kumar, A. & D. Bhatt. 2001. Characteristics and significance of calls in oriental magpie robin. *Current Science* **80**:77–82.

Kuypers, H.G.J.M. 1958a. Corticobulbar connexions to the pons and lower brain-stem in man. *Brain* **81**:364–388.

Kuypers, H.G.J.M. 1958b. Some projections from the peri-central cortex to the pons and lower brain stem in monkey and chinpanzee. *Journal of Comparative Neurology* **100**:221–255.

Lack, D. 1947. *Darwin's Finches*. Cambridge University Press, Cambridge.

Lack, D. 1965. *The Life of the Robin*. Witherby, London.

Lade, B.I. & W.H. Thorpe. 1964. Dove songs as innately coded patterns of specific behaviour. *Nature* **202**:366–368.

Ladefoged, P. 1982. *A Course in Phonetics*. Harcourt Brace Javonovitch, San Diego, CA.

Laje, R., T.J. Gardner & G.B. Mindlin. 2002. Neuromuscular control of vocalizations in birdsong: A model. *Physical Review* E **65**:051921.

Lambrechts, M.M. 1996. Organization of birdsong and constraints on performance. In: *Ecology and Evolution of Acoustic Communication in Birds*, D.E. Kroodsma & E.H. Miller (eds.), pp. 305–320. Cornell University Press, Ithaca, NY.

Lambrechts, M.M. 1997. Song frequency plasticity and composition of phrase versions in great tits *Parus major*. *Ardea* **85**:99–109.

Lambrechts, M. & A.A. Dhondt. 1986. Male quality, reproduction, and survival in the great tit (*Parus major*). *Behavioral Ecology & Sociobiology* **19**:57–63.

Lambrechts, M.M. & A.A. Dhondt. 1987. Differences in singing performance between male great tits. *Ardea* **75**:43–52.

Lambrechts, M.M. & A.A. Dhondt. 1988. The anti-exhaustion hypothesis: A new hypothesis to explain song performance and song switching in the great tit. *Animal Behaviour* **36**:327–334.

Lambrechts, M. & A.A. Dhondt. 1995. Individual voice discrimination in birds. *Current Ornithology* **12**:115–139. D.M. Power (ed), Plenum Press, New York.

Lampe, H.M. & Y.O. Epsmark. 1994. Song structure reflects male quality in pied flycatchers, *Ficedula hypoleuca*. *Animal Behaviour* **47**:869–876.

Lampe, H.M. & G.-P. Sætre. 1995. Female pied flycatchers prefer males with larger song repertoires. *Proceedings of the Royal Society of London, B* **262**:163–167.

Landers, P.B., J. Verner & J.W. Thomas. 1988. Ecological uses of vertebrate indicator species: a critique. *Conservation Biology* **2**:316–329.

Lane, H. & B. Tranel. 1971. Lombard sign and role of hearing in speech. *Journal of Speech and Hearing Research* **14**:677–709.

Langmore, N.E. 1997. Song switching in monandrous and polyandrous dunnocks, *Prunella modularis*. *Animal Behaviour* **53**:757–766.

Langmore, N.E. 1998. Functions of duet and solo songs of female birds. *Trends in Ecology & Evolution* **13**:136–140.

Langmore, N.E. 2000. Female bird song. In: *Signalling and Signal Design in Animal Communication,* Y. Espmark, T. Amundsen & G. Rosenqvist (eds.), pp. 319–327. Tapir Academic Press, Trondheim, Norway.

Langmore, N.E., N.B. Davies, B.J. Hatchwell & I.R. Hartley. 1996. Female dunnocks use vocalisations to compete for males. *Animal Behaviour* **53**:881–890.

Langemann, U., B. Gauger & G.M. Klump. 1998. Auditory sensitivity in the great tit: Perception of signals in the presence and absence of noise. *Animal Behaviour* **56**:763–769.

Lanyon, W.E. 1957. The comparative biology of the meadowlarks in Wisconsin. *Publications of the Nuttall Ornithological Club* **1**:1–67.

Lanyon, W.E. 1960. The ontogeny of vocalizations in birds. In: *Animal Sounds and Communication*, W.E. Lanyon & W.N. Tavolga (eds.), pp. 321–347. American Institute of Biological Sciences, Washington, D.C.

Lanyon, W.E. 1978. Revision of the *Myiarchus* flycatchers of South America. *Bulletin of the American Museum of Natural History* **161**:427–628.

Lanyon, W.E. 1979. Development of song in the wood thrush (*Hylocichla mustelina*) with notes on a technique for hand-rearing passerines from the egg. *American Museum Novitates*. **2666**:1–27.

Larom, D., M. Garstang, M. Lindeque, R. Raspet, M. Zunckel, Y. Hong, K. Brassel, S. O'Beirne, & F. Sokolic. 1997a. Meteorology and elephant infrasound at Etosha National Park, Namibia. *Journal of the Acoustical Society of America* **101**:1710–1717.

Larom, D., M. Garstang, K. Payne, R. Raspet & M. Lindeque. 1997b. The influence of surface atmospheric conditions on the range and area reached by animal vocalizations. *Journal of Experimental Biology* **200**:421–431.

Larsen, O.N. & T. Dabelsteen. 1990. Directionality of blackbird vocalization. Implications for vocal communication and its further study. *Ornis Scandinavica* **21**:37–45.

Larsen, O.N., M.L. Dent & R.J. Dooling. 1994. Free-field release from masking in the budgerigar (*Melopsittacus undulatus*). In: *Sensory transduction, Volume 2*. H. Breer & N. Elsner (eds.), p. 370, Thieme Verlag, Stuttgart.

Larsen, O.N., R.J. Dooling & B.M. Ryals. 1997. Roles of intracranial air pressure on hearing in birds. In: *Diversity in Auditory Mechanics*, E.R. Lewis (ed.), pp. 253–259. World Scientific Publishers, Singapore.

Larsen, O.N. & F. Goller. 1999. Role of syringeal vibrations in bird vocalizations. *Proceedings of the Royal Society of London, B* **266**:1609–1615.

Larsen, O.N. & Goller, F. 2002. Direct observation of syringeal muscle function in songbirds and a parrot. *Journal of Experimental Biology* **205**:25–35.

Lashley, K.S. 1915. Notes on the nesting activity of the noddy and sotty terns. *Carnegie Institution of Washington Publications* 7:61–83.

Latimer, W. 1977. A comparative study of the songs and alarm calls of some *Parus* species. *Zeitschrift für Tierpsychologie* **45**:414–433.

Laurance, W.F. 1991. Ecological correlates of extinction proneness in Australian tropical rain forest mammals. *Conservation Biology* **5**:79–89.

Lavenex, P.B. 2000. Lesions in the budgerigar vocal control nucleus NLc affect production, but not memory, of English words and natural vocalizations. *Journal of Comparative Neurology* **421**:437–460.

Leboucher G., V. Depraz, M. Kreutzer & L. Nagle. 1998. Male song stimulation of female reproduction in canaries: Features relevant to sexual displays are not relevant to nest-building or egg laying. *Ethology* **104**:613–624.

Leck, C.F. 1979. Avian extinctions in an isolated tropical wet forest preserve, Ecuador. *Auk* **96**:343–352.

Legare, M.L., W.R. Eddleman, P.A. Buckley & C. Kelly. 1999. The effectiveness of tape playback in estimating black rail density. *Journal of Wildlife Management* **63**:116–125.

Lehrman, D.S. 1953. A critique of Konrad Lorenz's theory of instinctive behavior. *Quarterly Review of Biology* **28**:337–363.

Lehrman, D.S. 1970. Semantic and conceptual issues in the nature–nurture problem. In: *Development and Evolution of Behavior; Essays in Memory of T.C. Schneirla*, L.R. Aronson, E. Tobach, D.S. Lehrman & J. Rosenblatt (eds.), pp. 17–52. W.H. Freeman, San Francisco.

Lein, M.R. 1978. Song variation in a population of chestnut-sided warblers (*Dendroica pensylvanica*): Its nature and suggested significance. *Canadian Journal of Zoology* **56**:1266–1283.

Lein, M.R. 1980. Display behaviors of ovenbirds (*Seiurus aurocapillus*). I. Non-song vocalizations. *Wilson Bulletin* **92**:312–329.

Leitão, A. & K. Riebel. 2003. Are good ornaments bad armaments? Male chaffinch perception of songs with varying flourish length. *Animal Behaviour* **66**:161–167.

Leitner, S., J. Nicholson, B. Leisler, T.J. DeVoogd & C.K. Catchpole. 2002. Song and the song control pathway in the brain can develop independently of exposure to song in the sedge warbler. *Proceedings of the Royal Society of London, B* **269**:2519–2524.

Leitner, S., C. Voigt & M. Gahr. 2001. Seasonal changes in the song pattern of the non-domesticated island canary (*Serinus canaria*), a field study. *Behaviour* **138**:885–904.

Lemon, R.E. 1968. The relation between organization and function of song in cardinals. *Behaviour* **32**:158–178.

Lemon, R.E. 1974. Song dialects, song matching and species recognition by cardinals *Richmondena cardinalis*. *Ibis* **116**:545–548.

Lemon, R.E. & C. Chatfield. 1971. Organization of song in cardinals. *Animal Behaviour* **19**:1–17.

Lemon, R.E., R. Cotter, R.C. Macnally, & S. Monette. 1985. Song repertoires and song sharing by American redstarts. *Condor* **87**:457–470.

Lemon, R.E., & M. Harris. 1974. The question of dialects in the songs of white-throated sparrows. *Canadian Journal of Zoology* **52**:83–98.

Lemon, R.E., M.J. Lechowicz & R.F. Norman. 1981. Song features and singing heights of American warblers: Maximization or optimization of distance. *Journal of Acoustical Society of America* **69**:1169–1176.

Lemon, R.E., S. Perreault, & D.M. Weary. 1994. Dual strategies of song development in American redstarts, *Setophaga ruticilla*. *Animal Behaviour* **47**:317–329.

Lemon, R.E. & D.M. Scott. 1966. On the development of song in young cardinals. *Canadian Journal of Zoology* **44**:191–197.

Lengagne, T., T. Aubin, J. Lauga & P. Jouventin. 1999. How do king penguins *Aptenodytes patagonicus* apply the mathematical theory of information to communicate in windy conditions? *Proceedings of the Royal Society of London, B* **266**:1623–1628.

Lengagne, T., J. Lauga & T. Aubin. 2001. Intra-syllabic acoustic signatures used by the king penguin in parent–chick recognition: An experimental approach. *Journal of Experimental Biology* **204**:663–672.

Lengagne, T. & P.J.B. Slater. 2002. The effects of rain on acoustic communication: Tawny owls have good reason for calling less in wet weather. *Proceedings of the Royal Society of London, B* **269**:2121–2125.

Leonard, M.L., N. Fernandez & G. Brown. 1997. Parental calls and nestling behavior in tree swallows. *Auk* **114**:668–672.

Leonard, M.L. & A.G. Horn. 1995. Crowing in relation to status in roosters. *Animal Behaviour* **48**:1283–1290.

Leonard, M.L. & J. Picman. 1988. Mate choice by marsh wrens: The influence of male and territory quality. *Animal Behaviour* **36**:517–528.

Leonardo, A. & M. Konishi. 1999. Decrystallization of adult birdsong by perturbation of auditory feedback. *Nature* **399**:466–470.

Leppelsack, H.-J. 1978. Unit responses to species-specific sounds in the auditory forebrain center of birds. *Federation Proceedings* **37**:2336–2341.

Levin, R.N. 1996a. Song behaviour and reproductive strategies in a duetting wren, *Thryothorus nigricapillus*: I. Removal experiments. *Animal Behaviour* **52**:1093–1106 .

Levin, R.N. 1996b. Song behaviour and reproductive strategies in a duetting wren, *Thryothorus nigricapillus*: II. Playback experiments. *Animal Behaviour* **52**: 1107–1117.

Levinson, S.T. 1980. The social behavior of the white-fronted Amazon (*Amazona albifrons*). In: *Conservation of New World Parrots*, R.F. Pasquier (ed.), pp. 403–417. ICBP Technical Publication No. 1.

Lewald, J. 1990. The directionality of the ear of the pigeon (*Columba livia*). *Journal of Comparative Physiology A* **167**:533–543.

Li, X., E. Jarvis, B. Alvarez-Borda, D. Lim & F. Nottebohm. 2000. A relationship between behavior, neurotrophin expression, and new neuron survival. *Proceedings of the National Academy of Sciences USA* **97**:8584–8589.

Li, X.-C. & E. Jarvis. 2001. Sensory- and motor-driven BDNF expression in a vocal communication system. *Society for Neuroscience (Abstracts)* **27**:1425.

Liberman, A., F. Cooper, D. Shankweiler & M. Studdert-Kennedy. 1967. Perception of the speech code. *Psychology Reviews* **74**:431–461.

Liberman, A. & I. Mattingly. 1985. The motor theory of speech perception revised. *Cognition* **21**:1–36.

Lieberburg, I. & F. Nottebohm. 1979. High-affinity androgen binding proteins in syringeal tissues of songbirds. *General and Comparative Endocrinology* **37**: 286–293.

Liittschwager, J.C. & E.M. Markman. 1994. Sixteen- and 24-month olds' use of mutual exclusivity as a default assumption in second-label learning. *Developmental Psychology* **30**:955–968.

Lincoln, L., G. Boxshall & P. Clark. 1998. *A Dictionary of Ecology, Evolution and Systematics* (2nd ed.). Cambridge University Press, Cambridge.

Ling, C.-Y., M. Zuo, A. Alvarez-Buylla & M. Cheng. 1997. Neurogenesis in juvenile and adult ring doves. *Journal of Comparative Neurology* **379**:300–312.

Liu, W.-C. 2001. Song development and singing behavior of the chipping sparrow (*Spizella passerina*) in western Massachusetts. Doctoral dissertation, University of Massachusetts, Amherst.

Liu, W.-C. & D.E. Kroodsma. 1999. Song development by field sparrows (*Spizella pusilla*) and chipping sparrows (*Spizella passerina*). *Animal Behaviour* **57**:1275–1286.

Livingston, F.S. & R. Mooney. 1997. Development of intrinsic and synaptic properties in a forebrain nucleus essential to avian song learning. *Journal of Neuroscience* **17**:8997–9009.

Loesche, P., P.K. Stoddard, B.J. Higgins & M.D. Beecher. 1991. Signature vs. perceptual adaptations for individual vocal recognition in swallows. *Behaviour* **118**:15–25.

Lohr, B., S. Bartone & R.J. Dooling. 2000. The discrimination of fine-scale temporal changes in call-like harmonic stimuli by birds. Association for Research in Otolaryngology, St. Petersburg, FL.

Lohr, B. & R.J. Dooling. 1998. Detection of changes in timbre and harmonicity in complex sounds by zebra finches (*Taeniopygia guttata*) and budgerigars (*Melopsittacus undulatus*). *Journal of Comparative Psychology* **112**:36–47.

Lohr, B., T.F. Wright & R.J. Dooling. 2003. Detection and discrimination of natural calls in masking noise by birds: Estimating the active space signal. *Animal Behaviour* **65**:763–777.

Lois, C. & A. Alvarez-Buylla. 1993. Proliferating subventricular zone cells in the adult mammalian forebrain can differentiate into neurons and glia. *Proceedings of the National Academy of Sciences USA* **90**:2074–2077.

Lois, C., J.M. Garcia-Verdugo & A. Alvarez-Buylla. 1996. Chain migration of neuronal precursors. *Science* **271**:978–981.

Lombardino, A.J. & F. Nottebohm. 2000. Age at deafening affects the stability of learned song in adult male zebra finches. *Journal of Neuroscience* **20**:5054–5064.

Lorenz. K. 1941. Vergleichende Bewegungsstudien an Anatinen. *Journal für Ornithologie* **89**:194–293.

Lorenz, K. 1957. Companionship in bird life. In: *Instinctive Behavior: The Development of a Modern Concept,* C.H. Schiller (ed.), pp. 82–128. International Universities Press, New York.

Lorenz, K. 1965. *Evolution and Modification of Behavior*. University of Chicago Press, Chicago.

Lott, D.F. 1991. *Intraspecific Variation in Social Systems of Wild Vertebrates.* Cambridge University Press, New York.

Lougheed, S.C., & P. Handford. 1992. Vocal dialects and the structure of geographic variation in morphological and allozymic characters in the rufous-collared sparrow, *Zonotrichia capensis. Evolution* **46**:1443–1456.

Luo, M., L. Ding & D.J. Perkel. 2001. An avian basal ganglia pathway essential for vocal learning forms a closed topographic loop. *Journal of Neuroscience* **21**:6836–6845.

Luo, M. & D.J. Perkel. 1999a. A GABAergic, strongly inhibitory projection to a thalamic nucleus in the zebra finch song system. *Journal of Neuroscience* **19**:6700–6711.

Luo, M. & D.J. Perkel. 1999b. Long-range gabaergic projection in a circuit essential for vocal learning. *Journal of Comparative Neurology* **403**:68–84.

Lynch, A. 1996. The population mimetics of birdsong. In: *Ecology and Evolution of Acoustic Communication in Birds*, D.E. Kroodsma & E.H. Miller (eds.), pp. 181–197. Cornell University Press, Ithaca, NY.

Lynch, A., G.M. Plunkett, A.J. Baker & P.F. Jenkins. 1989. A model of cultural evolution of chaffinch song derived with the meme concept. *American Naturalist* **135**:634-653.

Lynch, J.F. 1995. Effects of point count duration, time-of-day, and aural stimuli on detectability of migratory and resident bird species in Quintana Roo, Mexico. In: *Monitoring Bird Populations by Point Count,* C.J. Ralph, J.R. Sauer & S. Droege (eds.), pp. 1–6. Forest Service USDA PSW-GTR-149, Albany, CA.

MacDonald, D.W. & D.G. Henderson. 1977. Aspects of the behaviour and ecology of mixed-species bird flocks in Kashmir. *Ibis* **119**:481–493.

MacDougall-Shackleton, S.A. & S.H. Hulse. 1996. Concurrent absolute and relative pitch processing by European starlings (*Sturnus vulgaris*). *Journal of Comparative Psychology* **110**:139–146.

MacDougall-Shackleton, S.A., S.H. Hulse & G.F. Ball. 1998. Neural correlates of singing behavior in male zebra finches (*Taeniopygia guttata*). *Journal of Neurobiology* **36**:421–430.

MacDougall-Shackleton, E.A., & S.A. MacDougall-Shackleton. 2001. Cultural and genetic evolution in mountain white-crowned sparrows: Song dialects are associated with population structure. *evolution* **55**:2568–2575.

Mac Nally, R., A.F. Bennett, G.W. Brown, L.F. Lumsden, A. Yen, S. Hinkley, P. Lillywhite & D.A. Ward. 2002. How well do ecosystem-based planning units represent different components of biodiversity? *Ecological Applications* **12**:900–912.

Mac Nally, R. & E. Fleishman. 2002. Using "indicator" species to model species richness: Model development and predictions. *Ecological Applications* **12**:79–92.

Mace, R. 1987. Why do birds sing at dawn? *Ardea* **75**:123–132.

Macnamara, J. 1982. *Names for Things: A Study of Human Learning*. MIT Press, Cambridge, MA.

Mahler, B. & P.L. Tubaro. 2001. Relationship between song characters and morphology in New World pigeons. *Biological Journal of the Linnean Society* **74**:533–539.

Maier, N.R.F. & T.C. Schneirla. 1935. *Principles of Animal Psychology*. McGraw-Hill Book Company, New York.

Maier, V. 1982. Acoustic communication in the guinea fowl (*Numida meleagris*): Structure and use of vocalizations, and the principles of message coding. *Zeitschrift für Tierpsychologie* **59**:29–83.

Maier, V., O.A.E. Rasa & H. Scheich. 1983. Call-system similarity in a ground-living social bird and a mammal in the bush habitat. *Behavioral Ecology & Sociobiology* **12**:5–9.

Mammen, D.L. & S. Nowicki. 1981. Individual differences and within-flock convergence in chickadee calls. *Behavioral Ecology & Sociobiology* **9**:179–186.

Manabe, K., E.I. Sadr & R.J. Dooling. 1998. Control of vocal intensity in budgerigars (*Melopsittacus undulatus*): Differential reinforcement of vocal intensity and the Lombard Effect. *Journal of the Acoustical Society of America* **103**:1190–1198.

Maney, D.L., D.J. Bernard & G.F. Ball. 2001. Gonadal steroid receptor mRNA in catecholaminergic nuclei of the canary brainstem. *Neuroscience Letters* **31**:189–192.

Mann, V., S. Sasanuma, N. Sakuma & S. Masaki. 1990. Sex differences in cognitive abilities: A cross cultural perspective. *Neuropsychologia* **28**:1063–1077.

Marean, G.C., J.M. Burt, M.D. Beecher & E.W. Rubel. 1993. Hair cell regeneration in the European starling (*Sturnus vulgaris*) – Recovery of pure-tone detection thresholds. *Hearing Research* **71**:125–136.

Margoliash, D. 1983. Acoustic parameters underlying the responses of song-specific neurons in the white-crowned sparrow. *Journal of Neuroscience* **3**:1039–1057.

Margoliash, D. 1986. Preference for autogenous song by auditory neurons in a song system nucleus of the white-crowned sparrow. *Journal of Neuroscience* **6**:1643–1661.

Margoliash, D. 1997. Functional organization of forebrain pathways for song production and perception. *Journal of Neurobiology* **33**:671–693.

Margoliash, D. & E.S. Fortune. 1992. Temporal and harmonic combination-sensitive neurons in the zebra finch's HVc. *Journal of Neuroscience* **12**:4309–4326.

Margoliash, D., D.A. Staicer & S.A. Inove. 1991. Stereotyped and plastic song in adult indigo buntings, *Passerina cyanea*. *Animal Behaviour* **42**:367–388.

Marin, O., W.J. Smeets & A. Gonzalez. 1998. Evolution of the basal ganglia in tetrapods: A new perspective based on recent studies in amphibians. *Trends in Neurosciences* **21**:487–494.

Marion, W.R., T.E. O'Meara & D.S. Maehr. 1981. Use of playback recordings in sampling elusive or secretive birds. In: *Estimating Numbers of Terrestrial Birds*, C.J. Ralph & J.M. Scott (eds.), pp. 81–85. Allen Press, Lawrence, KS.

Markman, E.M. 1990. Constraints children place on word meaning. *Cognitive Science* **14**:57–77.

Marler, P. 1952. Variation in the song of the chaffinch, *Fringilla coelebs*. *Ibis* **94**:458–472.

Marler, P. 1955. Characteristics of some animal calls. *Nature* **176**:6–8.

Marler, P. 1956. The voice of the chaffinch and its function as a language. *Ibis* **98**:231–261.

Marler, P. 1957. Specific distinctiveness in the communication signals of birds. *Behaviour* **11**:13–39.

Marler, P. 1959. Developments in the study of animal communication. In: *Darwin's Biological Work*, P.R. Bell (ed.), pp. 150–206 & 329–335. Cambridge University Press, Cambridge.

Marler, P. 1963. Inheritance and learning in the development of animal vocalizations. In: *Acoustic Behavior of Animals*, R.G. Busnel (ed.), pp. 228–243. Elsevier, Amsterdam.

Marler, P. 1967. Comparative study of song development in sparrows. *Proceedings of XIV*

International Ornithological Congress, pp. 231–244. Blackwell Scientific Publications, Oxford.

Marler, P. 1969. Tonal quality of bird sounds. In: *Bird Vocalizations*, R.A. Hinde (ed.), pp. 5–18. Cambridge University Press, Cambridge.

Marler, P. 1970a. A comparative approach to vocal learning: Song development in white-crowned sparrows. *The Journal of Comparative and Physiological Psychology* **71**:1–25.

Marler, P. 1970b. Birdsong and speech development: Could there be parallels? *American Scientist* **58**:669–673.

Marler, P. 1976. Sensory templates in species-specific behavior. In: *Simpler Networks and Behavior*, J.C. Fentress (ed.), pp. 314–329. Sinauer, Sunderland, MA.

Marler, P. 1984. Song learning: Innate species differences in the learning process. In: *The Biology of Learning*, P. Marler & H.S. Terrace (eds.), pp. 289–309. Springer-Verlag, Berlin.

Marler, P. 1987. Sensitive periods and the role of specific and general sensory stimulation in birdsong learning. In: *Imprinting and Cortical Plasticity*, J.P. Rauschecker & P. Marler (eds.), pp. 99–135. John Wiley & Sons, New York.

Marler, P. 1990a. Innate learning preferences: Signals for communication. *Developmental Psychobiology* **23**:557–568.

Marler, P. 1990b. Song learning: The interface between behavior and neuroethology. *Philosophical Transactions of the Royal Society of London, B* **329**: 109–114.

Marler, P. 1991. Differences in behavioural development in closely related species: Birdsong. In: *The Development and Integration of Behaviour: Essays in Honour of Robert Hinde*, P. Bateson (ed.), pp. 41–70. Cambridge University Press, Cambridge.

Marler, P. 1997. Three models of song learning: Evidence from behavior. *Journal of Neurobiology* **33**:501–516.

Marler, P. 1998. Animal communication and human language. In: *The Origin and Diversification of Language*, N.G. Jablonski & L.C. Aiello (eds.), pp.1–19. California Academy of Sciences, San Francisco.

Marler, P. 1999. How much does a human environment humanize a chimp? *American Anthropologist* **101**:432–435.

Marler, P. 2004. Innateness and the instinct to learn. In: *Annals of the XIX International Bioacoustics Congress*, and *Anais da Academia Brasileira de Ciências – Annals of the Brazilian Academy of Sciences* **76**: 189–200.

Marler, P. & D.J. Boatman. 1951. Observations on the birds of Pico, Azores. *Ibis* **93**:90–99.

Marler, P. & A.J. Doupe. 2000. Singing in the brain. *Proceedings of the National Academy of Sciences USA* **97**:2965–2967.

Marler, P., A. Dufty & R. Pickert. 1986a. Vocal communication in the domestic chicken: I. Does a sender communicate information about the quality of a food referent to a receiver? *Animal Behaviour* **34**:188–193.

Marler, P., A. Dufty & R. Pickert. 1986b. Vocal communication in the domestic chicken: II. Is a sender sensitive to the presence and nature of a receiver? *Animal Behaviour* **34**:194–198.

Marler, P. & C.S. Evans. 1997. Communication signals of animals: Contributions of emotion and reference. In: *Nonverbal Communication: Where Nature Meets Culture*, V. Segerstrale & P. Molnar (eds.), pp. 151–170. Lawrence Erlbaum, Mahwah, NJ.

Marler, P., C.S. Evans & M.D. Hauser. 1992. Animal signals: Motivational, referential, or both? In: *Nonverbal Vocal Communication: Comparative and Developmental Approaches*, H. Papousek, U. Jürgens & M. Papousek (eds.), pp. 66–68. Cambridge University Press, Cambridge, MA.

Marler, P., S. Karakashian & M. Gyger. 1991. Do animals have the option of withholding signals when communication is inappropriate? The audience effect. In: *Cognitive Ethology: The Minds of Other Animals*, C. Ristau (ed.), pp. 187–208. Lawrence Erlbaum, Hillsdale, NJ.

Marler, P., M. Konishi, A. Lutjen & M.S. Waser. 1973. Effects of continuous noise on avian hearing and vocal development. Proceedings of the National Academy of Sciences USA **70**:1393–1396.

Marler, P., M. Kreith & M. Tamura. 1962. Song development in hand-raised Oregon juncos. *Auk* **79**:12–30.

Marler, P. & P.C. Mundinger. 1975. Vocalizations, social organization and breeding biology of the twite, *Acanthus flavirostris*. *Ibis* **117**:1–17.

Marler, P. & D.A. Nelson. 1993. Action-based learning: A new form of developmental plasticity in birdsong. *Netherlands Journal of Zoology* **43**:91–103.

Marler, P. & S. Peters. 1977. Selective vocal learning in a sparrow. *Science* **198**:519–521.

Marler, P. & S. Peters. 1980. Birdsong and speech:

Evidence for special processing. In: *Perspectives on the Study of Speech*, P. Eimas & J. Miller (eds.) pp. 75–112. Lawrence Erlbaum, Hillsdale, NJ.

Marler, P. & S. Peters. 1981. Sparrows learn adult song and more from memory. *Science* **213**:780–782.

Marler, P. & S. Peters. 1982a. Subsong and plastic song: Their role in the vocal learning process. In: *Acoustic Communication in Birds, Vol. 2. Song Learning and its Consequences*, D.E. Kroodsma & E.H. Miller (eds.), pp. 25–50. Academic Press, New York.

Marler, P. & S. Peters. 1982b. Developmental overproduction and selective attrition: New processes in the epigenesis of birdsong. *Developmental Psychobiology* **15**:369–378.

Marler, P. & S. Peters. 1982c. Structural changes in song ontogeny in the swamp sparrow *Melospiza georgiana*. *Auk* **99**:446–458.

Marler, P. & S. Peters. 1987. A sensitive period for song acquisition in the song sparrow, *Melospiza melodia*, a case of age-limited learning. *Ethology* **76**:89–100.

Marler, P. & S. Peters. 1988a. Sensitive periods for song acquisition from tape recordings and live tutors in the swamp sparrow, *Melospiza georgiana*. *Ethology* 77:76–84.

Marler, P. & S. Peters. 1988b. The role of song phonology and syntax in vocal learning preferences in the song sparrow *Melospiza melodia*. *Ethology* 77:125–149.

Marler, P. & S. Peters. 1989. Species differences in auditory responsiveness in early vocal learning. In: *The Comparative Psychology of Audition: Perceiving Complex Sounds*, R.J. Dooling & S. Hulse (eds.), pp. 243–273. Lawrence Erlbaum, Hillsdale, NJ.

Marler, P., S. Peters, G. Ball, A. Duffy Jr. & J. Wingfield. 1988. The role of sex steriods in the acquisition and production of birdsong. *Nature* **336**:770–772.

Marler, P. & R. Pickert. 1984. Species-universal microstructure in the learned song of the swamp sparrow, *Melospiza georgiana*. *Animal Behaviour* **32**:673–689.

Marler, P. & V. Sherman. 1983. Song structure without auditory feedback: Emendations of the auditory template hypothesis. *Journal of Neuroscience* **3**:517–531.

Marler, P. & V. Sherman. 1985. Innate differences in singing behavior of sparrows reared in isolation from adult conspecific song. *Animal Behaviour* **33**:57–71.

Marler, P., & M. Tamura. 1962. Song dialects in three populations of the white-crowned sparrow. *Condor* **64**:368–377.

Marler, P. & M. Tamura. 1964. Culturally transmitted patterns of vocal behavior in sparrows. *Science* **146**:1483–1486.

Marler, P. & M.S. Waser. 1977. Role of auditory feedback in canary song development. *Journal of Comparative and Physiological Psychology* **91**:8–16.

Marshall, A.J. 1954. *Bower-Birds: Their Displays and Breeding Cycles*. Clarendon Press, Oxford.

Marshall, J.T. 1988. Birds lost from a giant sequoia forest during fifty years. *Condor* **90**:359–372.

Marten, K. & P. Marler. 1977. Sound transmission and its significance for animal vocalization. I. Temperate habitats. *Behavioral Ecology & Sociobiology* **2**:271–290.

Marten, K., D. Quine & P. Marler. 1977. Sound transmission and its significance for animal vocalization. II. Tropical forest habitats. *Behavioral Ecology & Sociobiology* **2**:291–302.

Martens, J. 1996. Vocalizations and speciation of palearctic birds. In: *Ecology and Evolution of Acoustic Communication in Birds*, D.E. Kroodsma & E.H. Miller (eds.), pp. 221–240. Cornell University Press, Ithaca, NY.

Martens, M.J.M. 1980. Foliage as a low-pass filter: Experiments with model forests in an anechoic chamber. *Journal of the Acoustical Society of America* **67**:66–72.

Martins, E.P. (ed.). 1996. *Phylogenies and the comparative method in animal behavior*. Oxford University Press, Oxford.

Martin-Vivaldi, M., J.J. Palomino & M. Soler 2000. Attraction of hoopoe *Upupa epops* females and males by means of song playback in the field: Influence of strophe length. *Journal of Avian Biology* **31**: 351–359.

Masataka, N. & K. Fujita. 1989. Vocal learning of Japanese and rhesus monkeys. *Behaviour* **109**:191–199.

Matessi, G., A. Pilastro & G. Marin. 2000. Variation in quantitative properties of song among European populations of reed bunting (*Emberiza schoeniclus*) with respect to bill morphology. *Canadian Journal of Zoology* **78**:428–437.

Mathevon, N. 1997. Individuality of contact calls in the greater flamingo *Phoenicopterus ruber* and the problem of background noise in a colony. *Ibis* **139**:513–517.

Mathevon, N. & T. Aubin. 1996. Reaction to conspecific degraded song by the wren *Troglodytes Troglodytes*: Territorial response and choice of song post. *Behavioural Processes* **39**:77–84.

Mathevon, N. & T. Aubin. 2001. Sound-based species-specific recognition in the blackcap *Sylvia atricapilla* shows high tolerance to signal modifications. *Behaviour* **138**:511–524.

Mathevon, N., T. Aubin & J-C. Brémond. 1997. Propagation of bird acoustic signals: Comparative study of starling and blackbird distress calls. *Life Sciences* **320**:869–876.

Mathevon, N., T. Aubin, & T. Dabelsteen. 1996. Song degradation during propagation: Importance of song post for the wren *Troglodytes troglodytes*. *Ethology* **102**:397–412.

May, L. 1998. Individually distinctive corncrake *Crex crex* calls: A further study. *Bioacoustics* **9**:135–148.

Mayr, E. 1942. *Systematics and the Origin of Species.* Columbia University Press, New York.

Mayr, E. 1974. Behavior programs and evolutionary strategies. *American Scientist* **62**:650–659.

McArthur, P.D. 1982. Mechanisms and development of parent–young vocal recognition in the piñon jay (*Gymnorhinus cyanocephalus*). *Animal Behaviour* **30**:62–74.

McBride, G., I.P. Parer & F. Foenander. 1969. The social organization and behaviour of the feral domestic fowl. *Animal Behaviour Monographs* **2**:125–181.

McCasland, J.S. 1987. Neuronal control of bird song production. *Journal of Neuroscience* **7**:23–39.

McCasland, J.S. & M. Konishi. 1981. Interaction between auditory and motor activities in an avian song control nucleus. *Proceedings of the National Academy of Sciences USA* **78**:7815–7819.

McClure, H.E. 1967. The composition of mixed species flocks in lowlands and submontane forests of Malaya. *Wilson Bulletin* **79**:131–154.

McConnell, P.B. 1991. Lessons from animal trainers: The effect of acoustic structure on an animal's response. In: *Perspectives in Ethology, Volume 9: Human Understanding and Animal Awareness,* P. Bateson & P. Klopfer (eds.), pp. 165–187. Plenum Press, New York.

McCowan, B. & D. Reiss. 1995. Whistle contour development in captive-born infant bottlenose dolphins (*Tursiops truncatus*): Role of learning. *Journal of Comparative Psychology* **109**:242–260.

McCowan, B. & D. Reiss. 1997. Vocal learning in captive bottlenose dolphins: A comparison to humans and nonhuman animals. In: *Social Influences on Vocal Development,* C.T. Snowdon & M. Hausberger (eds.), pp. 178–207. Cambridge University Press, Cambridge.

McCracken, K.G. 1999. *Systematics, ecology, and social biology of the musk duck (Biziura lobata) of Australia.* Ph.D. dissertation, Louisiana State University, Baton Rouge, Louisiana.

McCracken, K.G. & F.H. Sheldon. 1997. Avian vocalizations and phylogenetic signal. *Proceedings of the National Academy of Sciences USA* **94**:3833–3836.

McEwen, B. 1994. Steriod hormone actions on the brain: When is the genome involved? *Hormones & Behavior* **28**:396–405.

McFadden, R.D. 1991. 25,000 in Central Park hear word of hope from Graham. *New York Times,* September 23, 1991, p. A16.

McGregor, P.K. 1980. Song dialects in the corn bunting (*Emberiza calandra*). *Zeitschrift für Tierpsychologie* **54**:285–297.

McGregor, P.K. 1983. The response of corn buntings to playback of dialects. *Zeitschrift für Tierpsychologie* **62**:256–260.

McGregor, P.K. 1992. *Playback and Studies of Animal Communication.* Plenum Press, New York.

McGregor, P.K. & P. Byle. 1992. Individually distinctive bittern booms: Potential as a census tool. *Bioacoustics* **4**:93–109.

McGregor, P.K., C.K. Catchpole, T. Dabelsteen, J.B. Falls, L. Fusani, H.C. Gerhardt, F. Gilbert, A.G. Horn, G.M. Klump, D.E. Kroodsma, M.M. Lambrechts, K.E. McComb, D.A. Nelson, I.M. Pepperberg, L. Ratcliffe, W.A. Searcy & D.M. Weary. 1992. Design of playback experiments: The Thornbridge Hall NATO ARW Consensus. In: *Playback and Studies of Animal Communication,* P.K. McGregor (ed.), pp. 1–9. Plenum Press, New York.

McGregor, P.K. & T. Dabelsteen. 1996. Communication networks. In: *Ecology and Evolution of Acoustic Communication in Birds,* D.E. Kroodsma & E.H. Miller (eds.), pp. 409–425. Cornell University Press, Ithaca, NY.

McGregor, P.K., T. Dabelsteen, M. Shepherd & S.B. Pedersen. 1992. The singing value of matched singing in great tits: Evidence from interactive playback experiments. *Animal Behaviour* **43**:987–998.

McGregor, P.K. & J.B. Falls. 1984. The response of western meadowlarks (*Sturnella neglecta*) to the playback of degraded and undegraded songs. *Canadian Journal of Zoology* **62**:2125–2128.

McGregor, P.K. & J.R. Krebs. 1984. Sound degradation as distance cue in great tit (*Parus major*). *Behavioral Ecology Sociobiology* **16**:49–56.

McGregor, P.K., J.R. Krebs & C.M. Perrins. 1981. Song repertoires and lifetime reproductive success in the great tit (*Parus major*). *Behavioral Ecology & Sociobiology* **19**:57–63.

McGregor, P.K. & T.M. Peake. 1998. The role of individual identification in conservation biology. In: *Behavior Ecology and Conservation Biology*, T.M. Caro (ed.), pp. 31–55. Oxford University Press, New York.

McGregor, P.K., T.M. Peake & G. Gilbert. 2000. Communication behaviour and conservation. In: *Behaviour and Conservation*, L.M. Gosling & W.J. Sutherland (eds.), pp. 261–280. Cambridge University Press, Cambridge.

McGregor, P.K. & R.D. Ranft. 1994. Equipment for sound analysis and playback: A survey. *Bioacoustics* **6**:83–86.

McGregor, P.K. & D.B.A. Thompson. 1988. Constancy and change in local dialects of the corn bunting. *Ornis Scandinavica* **19**:153–159.

McLean, I.G. & J.R. Waas. 1987. Do cuckoo chicks mimic the begging calls of their hosts? *Animal Behaviour* **35**:1896–1898.

Medellín, R.A., M. Equihua & M.A. Amin. 2000. Bat diversity and abundance as indicators of disturbance in neotropical rainforests. *Conservation Biology* **14**:1666–1675.

Medway, L. 1967. The function of echonavigation among swiftlets. *Animal Behaviour* **15**:416–420.

Mello, C.V. & D.F. Clayton. 1994. Song-induced ZENK gene expression in auditory pathways of songbird brain and its relation to the song control system. *Journal of Neuroscience* **14**:6652–6666.

Mello, C.V., F. Nottebohm & D. Clayton. 1995. Repeated exposure to one song leads to a rapid and persistent decline in an immediate early gene's response to that song in zebra finch telencephalon. *Journal of Neuroscience* **15**:6919–6925.

Mello, C.V., G.E. Vates, S. Okuhata & F. Nottebohm. 1998. Descending auditory pathways in the adult male zebra finch (*Taeniopygia guttata*). *Journal of Comparative Neurology* **395**:137–160.

Mello, C.V., D.S. Vicario & D.F. Clayton. 1992. Song presentation induces gene expression in the songbird forebrain. *Proceedings of the National Academy of Sciences USA* **89**:6818–6822.

Mennill, D.J. & L. Ratcliffe. 2001. A field test of syrinx sound analysis software in interactive playback. *Bioacoustics* **11**:77–86.

Merila, J. & J. Sorjonen. 1994. Seasonal and diurnal patterns of singing and song-flight activity in bluethroats (*Luscinia svecica*). *Auk* **111**:556–562.

Merriman, W.E. 1991. The mutual exclusivity bias in children's word learning: A reply to Woodward and Markman. *Developmental Review* **11**:164–191.

Messiaen, O. 1994. *Music and Color: Conversations with Claude Samuel,* E. Thomas Glasow (trans.). Amadeus Press, Portland, OR.

Metzdorf, R., M. Gahr & L. Fusani. 1999. Distribution of aromatase, estrogen receptor, and androgen receptor mRNA in the forebrain of songbirds and nonsongbirds. *Journal of Comparative Neurology* **407**:115–129.

Milius, S. 2000. Music without borders. *Science News* **157**:252–254.

Miller, D.E. & J.T. Emlen Jr. 1975. Individual chick recognition and family integrity in the ring-billed gull. *Behaviour* **52**:124–144.

Miller, E.H. 1996. Acoustic differentiation and speciation in shorebirds. In: *Ecology and Evolution of Acoustic Communication in Birds*, D.E. Kroodsma & E.H. Miller (eds.), pp. 241–257. Cornell University Press, Ithaca, NY.

Miller, R.C. 1921. The flock behavior of the coast bush-tit. *Condor* **23**:121–127.

Miskimen, M. 1951. Sound production in passerine birds. *Auk* **68**:493–504.

Miyasato, L.E. & M.C. Baker. 1999. Black-capped chickadee call dialects along a continuous habitat corridor. *Animal Behaviour* **57**:1311–1318.

Møller, A.P. 1988. False alarm calls as a mean of resource usurpation in the great tit *Parus major*. *Ethology* **79**:25–30.

Møller, A.P. 1991. Parasite load reduces song output in a passerine bird. *Animal Behaviour* **41**:723–730.

Møller, A.P., P.Y. Henry & J. Erritzoe. 2000. The evolution of song repertoires and immune defence in birds. *Proceedings of the Royal Society of London, B* **267**:165–169.

Møller, A.P., N. Saino, G. Taramino, P. Galeotti & S. Ferrario. 1998. Paternity and multiple signaling: Effects of a secondary sexual character and song on paternity in the barn swallow. *American Naturalist* **151**:236–242.

Mondloch, C.J. 1995. Chick hunger and begging affect parental allocation of feeding in pigeons. *Animal Behaviour* **49**:601–613.

Monson, G. & A.R. Phillips. 1981. The races of red crossbill, *Loxia curvirostra*, in Arizona. In: *Checklist of Birds of Arizona*, G. Monson & A.R. Phillips, (eds.) pp.223–230. University of Arizona Press, Tucson.

Mooney, H.A., R. Lubchenco & O.E. Sala. 1995. Biodiversity and ecosystem functioning: Basic principles. In: *Biodiversity Assessment. United Nations Environment Programme*, V.H. Heywood (ed.), pp. 275–325. Global, Cambridge University Press, UK.

Mooney, R. 1999. Sensitive periods and circuits for learned birdsong. *Current Opinion in Neurobiology* **9**:121–127.

Mooney, R. 2000. Different subthreshold mechanisms underlie song selectivity in identified HVc neurons of the zebra finch. *Journal of Neuroscience* **20**:5420–5436.

Mooney, R. & M. Konishi 1991. Two distinct inputs to an avian song nucleus activate different glutamate receptor subtypes on individual neurons. *Proceedings of the National Academy of Sciences USA* **88**:4075–4079.

Mooney, R. & M. Rao. 1994. Waiting periods versus early innervation: The development of axonal connections in the zebra finch song system. *Journal of Neuroscience* **14**:6532–6543.

Mooney, R, M.J. Rosen & C.B. Sturdy. 2002. Respiratory and telencephalic modulation of vocal motor neurons in the zebra finch. *Journal of Neuroscience* **23**:1072–1086.

Moore, B.C.J., B.R. Glasberg & R.W. Peters. 1986. Thresholds for hearing mistuned partials as separate tones in harmonic complexes. *Journal of the Acoustical Society of America* **80**:479–483.

Moore, B.C.J., R.W. Peters & B.R. Glasberg. 1985a. Thresholds for the detection of inharmonicity in complex tones. *Journal of the Acoustical Society of America* 77:1861–1867.

Moore, B.C.J., B.R. Glasberg & R.W. Peters. 1985b. Relative dominance of individual partials in determining the pitch of complex tones. *Journal of the Acoustical Society of America* 77: 1853–1860.

Moreau, R.E. & P. Wayre. 1968. On the palearctic quails. *Ardea* **56**:209–227.

Moriyama, K. & K. Okanoya. 1996. Effect of beak movement in singing Bengalese finches. *Abstracts. Acoustical Society of America and Acoustical Society of Japan, Third Joint Meeting. Honolulu, 2–6 Dec* **1996**:129–130.

Morse, D.H. 1970a. Territorial and courtship songs of birds. *Nature* **226**:659–661.

Morse, D.H. 1970b. Ecological aspects of mixed-species foraging flocks of birds. *Ecological Monographs* **40**:119–168.

Morse, D.H. 1980. *Behavioral Mechanisms in Ecology*. Harvard University Press, Cambridge.

Morton, E.S. 1970. *Ecological sources of selection on avian sounds*. Dissertation, Yale University, New Haven, Connecticut.

Morton, E.S. 1975. Ecological sources of selection on avian sounds. *American Naturalist* **109**:17–34.

Morton, E.S. 1976. Vocal mimicry in the thick-billed euphonia. *Wilson Bulletin* **88**:485–487.

Morton, E.S. 1977. On the occurrence and significance of motivation-structural rules in some bird and mammal sounds. *American Naturalist* **111**:855–869.

Morton, E.S. 1982. Grading, discreteness, redundancy, and motivation-structural rules. In: *Acoustic Communication in Birds, Vol. 1. Production, Perception, and Design Features of Sounds*, D.E. Kroodsma & E.H. Miller (eds.), pp.183–212. Academic Press, New York.

Morton, E.S. 1996a. A comparison of vocal behavior among tropical and temperate passerine birds. In: *Ecology and Evolution of Acoustic Communication in Birds*, D.E. Kroodsma & E.H. Miller (eds.), pp. 258–268. Cornell University Press, Ithaca, NY

Morton, E.S. 1996b. Why songbirds learn songs: An arms race over ranging? *Poultry and Avian Biology Reviews* 7:65–71.

Morton, E.S., S.L. Gish, & M. van der Voort. 1986. On the learning of degraded and undegraded songs in the Carolina wren (*Thryothorus ludovicianus*). *Animal Behaviour* **34**:815–820.

Morton, E.S., M. Naguib, & R.H. Wiley. 1998. Degradation and signal ranging in birds: Memory matters (and replies). *Behavioral Ecology & Sociobiology* **42**:135–148.

Mountjoy, J. & D.W. Leger. 2001. Vireo song repertoires and migratory distance: Three sexual selection hypotheses fail to explain the correlation. *Behavioral Ecology* **12**:98–102.

Mountjoy, D.J. & R.E. Lemon. 1991. Song as an attractant for male and female European starlings and the influence of song complexity on their response. *Behavioral Ecology & Sociobiology* **28**:97–100.

Mountjoy, D.J. & R.E. Lemon. 1996. Female choice for complex song in the European starling: A field experiment. *Behavioral Ecology & Sociobiology* **38**:65–71.

Mowbrey, T.B. 1997. Swamp sparrow (*Melospiza georgiana*). In: *Birds of North America*, A. Poole & F.Gill (eds.), pp. 1–24. American Ornithologists' Union, Washington, D.C.

Mowrer, O.H. 1950. *Learning Theory and Personality Dynamics*. Ronald, New York.

Moynihan, M. 1962. The organization and probable evolution of some mixed species flocks of neotropical birds. *Smithsonian Miscellaneous Collections* **143**:1–140.

Müller, C.M. & H.-J. Leppelsack. 1985. Feature extraction and tonotopic organization in the avian auditory forebrain. *Experimental Brain Research* **59**:587–599.

Müller, J. 1841. Über die Anatomie des *Steatornis caripensis. Akademie der Wissenschaften Bericht, Berlin*:172–179.

Müller, S.C. & H. Scheich. 1985. Functional organization of the avian auditory field L.A. comparative 2 deoxyglucose study. *Journal of Comparative Physiology A* **156**:1–12.

Müller-Bröse, M. & D. Todt 1991. Lokomotorische Aktivität von Nachtigallen (*Luscinia megarhynchos*) während auditorischer Stimulation in ihrer lernsensiblen Altersphase. *Verhandlungen der Deutschen Zoologen Gesellschaft* **84**:476–477.

Mulligan, J.A. 1963. A description of song sparrow song based on instrumental analysis. In: *Proceedings of the XIII International Ornithological Congress. Vol. 1*, C.G. Sibley (ed.), pp. 272–284. The American Ornithologists' Union, Museum of Zoology, Louisiana State University.

Mulligan, J.A. 1966. Singing behaviour and its development in the song sparrow, *Melospiza melodia. University of California Publications in Zoology* **81**:1–76.

Mundinger, P.C. 1970. Vocal imitation and individual recognition of finch calls. *Science* **168**:480–482.

Mundinger, P. 1975. Song dialects and colonization in the house finch, *Carpodacus mexicanus*, on the east coast. *Condor* 77:407–422.

Mundinger, P.C. 1979. Call learning in the *Carduelinae*: Ethological and systematic considerations. *Systematic Zoology* **28**:270–283.

Mundinger, P.C. 1982. Microgeographic and macrogeographic variation in the acquired vocalizations of birds. In: *Acoustic Communication in Birds Vol. 2 Song Learning and its Consequences*, D.E. Kroodsma & E.H. Miller (eds.), pp. 147–208. Academic Press, New York.

Mundinger, P.C. 1999. Genetics of canary song learning: Innate mechanisms and other neurobiological considerations. In: *The Design of Animal Communication*, M.D. Hauser & M. Konishi (eds.), pp. 369–389. MIT Press, Cambridge, MA.

Mundry, R. 2000. *Struktur und Einsatz des Gesanges bei Sprosser-Mischsängern* (*Luscinia luscinia* L.). Doctoral dissertation, Freie Universität, Berlin.

Mundy, P.J. 1973. Vocal mimicry of their hosts by nestlings of the great spotted cuckoo and striped crested cuckoo. *Ibis* **115**:602–604.

Munn, C.A. 1984. Birds of different feather also flock together. *Natural History* **11**:34–42.

Munn, C.A. 1986. Birds that 'cry wolf'. *Nature* **319**:143–145.

Munn, C.A. & J.W. Terbourgh. 1979. Multi-species territoriality in neotropical foraging flocks. *Condor* **81**:338–347.

Myers, N. 1987. The extinction spasm impending: Synergisms at work. *Conservation Biology* **1**:14–21.

Myers, N. 1988. Threatened biotas: Hot spots in tropical forests. *Environmentalist* **8**:187–208.

Naeser, M.A., M.P. Alexander, N. Helm-Estabrooks, H.L. Levine, S.A. Laughlin & N. Geshwind. 1982. Aphasia with predominantly subcortical lesion sites: Description of three capsular/putaminal aphasia syndromes. *Archives of Neurology* **39**:2–14.

Nagle, L. & M.L. Kreutzer. 1997. Song tutoring influences female song preferences in domesticated canaries. *Behaviour* **134**:89–104.

Naguib, M. 1995. Auditory distance assessments in Carolina wrens (*Thryothorus ludovicianus*): The role of reverberation and frequency dependent attenuation of conspecific song. *Animal Behaviour* **50**:1297–1307.

Naguib, M. 1996. Ranging by song in Carolina wrens *Thryothorus ludovicianus*: Effects of environmental acoustics and strength of song degradation. *Behaviour* **133**:541–559.

Naguib, M. 1999. Effects of song overlapping and alternating on nocturnally singing nightingales. *Animal Behaviour* **58**:1061–1067.

Naguib, M. & D. Todt. 1997. Effects of dyadic vocal interactions on other conspecific receivers in nightingales. *Animal Behaviour* **54**:1535–1543.

Naguib, M. & R.H. Wiley. 2001. Estimating the

distance to a source of sound: Mechanisms and adaptations for long-range communication. *Animal Behaviour* **62**:825–837.

Nantais, K.M. & E.G. Schellenberg. 1999. The Mozart effect: An artifact of preference. *Psychological Science*, **10**:370–373.

Neal, K.B. 1996. *The development of a vocalization in an African grey parrot* (*Psittacus erithacus*). Senior thesis, University of Arizona, Tucson.

Nealen, P. & D. Perkel. 2000. Sexual dimorphism in the song system of the Carolina wren *Thryothorus ludovicianus*. *Journal of Comparative Neurology* **418**:346–360.

Nealen, P.M. & M.F. Schmidt. 2002. Comparative approaches to avian song system function: Insights into auditory and motor processing. *Journal of Comparative Physiology A* **188**:929–941.

Nelson, D.A. 1988. Feature weighting in species song recognition by the field sparrow (*Spizella pusilla*). *Behaviour* **106**:158–182.

Nelson, D.A. 1989a. The importance of invariant and distinctive features in species recognition of bird song. *Condor* **91**:120–130.

Nelson, D.A. 1989b. Song frequency as a cue for recognition of species and individuals in the field sparrow (*Spizella pusilla*). *Journal of Comparative Psychology* **103**:171–176.

Nelson, D.A. 1992. Song overproduction and selective attrition lead to song sharing in the field sparrow (*Spizella pusilla*). *Behavioral Ecology & Sociobiology* **30**:415–424.

Nelson, D.A. 1997. Social interaction and sensitive phases for song learning: A critical review. In: *Social Influences on Vocal Development*, C.T. Snowdon & M. Hausberger (eds.), pp. 7–22. Cambridge University Press, Cambridge.

Nelson, D.A. 1999. Ecological influences on vocal development in the white-crowned sparrow. *Animal Behaviour* **58**:21–36.

Nelson, D.A. 2000a. Ecological influences on vocal development in the white-crowned sparrow. *Animal Behaviour* **58**:21–36.

Nelson, D.A. 2000b. A preference for own-subspecies' song guides vocal learning in a song bird. *Proceedings of the National Academy of Sciences USA* **97**:13348–13353.

Nelson, D.A. 2000c. Song overproduction, selective attrition, and vocal dialects in the white-crowned sparrow. *Animal Behaviour* **60**:887–898.

Nelson, D.A., H. Khanna & P. Marler. 2001. Learning by instruction or selection: Implications for patterns of geographic variation in bird song. *Behaviour* **138**:1137–1160.

Nelson, D.A. & P. Marler. 1989. Categorical perception of a natural stimulus continuum: Birdsong. *Science* **244**:976–978.

Nelson, D.A. & P. Marler. 1990. The perception of birdsong and an ecological concept of signal space. In: *Comparative Perception Volume II: Complex Signals*, M.A. Berkley & W.C. Stebbins (eds.), pp. 443–478. John Wiley & Sons, New York.

Nelson, D.A. & P. Marler. 1993. Innate recognition of song in white-crowned sparrows: A role in selective vocal learning? *Animal Behaviour* **46**:806–808.

Nelson, D.A. & P. Marler. 1994. Selection-based learning in bird song development. *Proceedings of the National Academy of Sciences USA* **91**:10498–10501.

Nelson, D.A., P. Marler & M.L. Morton. 1996. Overproduction in song development: An evolutionary correlate with migration. *Animal Behaviour* **51**:1127–1140.

Nelson, D.A., P. Marler & A. Palleroni. 1995. A comparative approach to vocal learning: Intraspecific variation in the learning process. *Animal Behaviour* **50**:83–97.

Nelson, D.A., P. Marler, J.A. Soha & A.L. Fullerton. 1997. The timing of song memorization differs in males and females: A new assay for avian vocal learning. *Animal Behaviour* **54**: 587–97.

Nelson, K. 1989. Monologues in the crib. In: *Narratives from the Crib,* K. Nelson (ed.), pp. 1–23. Harvard University Press, Cambridge, MA.

Nemeth, E. & H. Winkler. 2001. Differential degradation of antbird songs in a neotropical rainforest: Adaptation to perch height? *Journal of the Acoustical Society of America* **110**:3263–3274.

Newton, I. 1989. *Lifetime Reproduction in Birds.* Academic Press, London.

Newton, I. 1995. The contribution of some recent research on birds to ecological understanding. *Journal of Animal Ecology* **64**:675–696.

Nice, M.M. 1943. Studies in the life history of the song sparrow. II. The behavior of the song sparrow and other passerines. *Transactions of the Linnean Society of New York* **6**:1–328.

Nicholson, E.M. & L. Koch. 1936. *Songs of Wild Birds.* H.F. & G. Witherby, London.

Nick, T.A. & M. Konishi. 2001. Dynamic control of auditory acitivity during sleep: Correlation between song response and EEG. *Proceedings of the National*

Academy of Sciences USA **20**:14012–14016.

Nicolai, J. 1959. Familientradition in der Gesangsentwicklung des Gimpels *(Pyrrhula pyrrhula)*. *Journal für Ornithologie* **100**:39–46.

Nicolai, J. 1964. Der Brutparasitismus der Viduinae als ethologisches Problem. Prägungsphaenomene als Faktoren der Rassen- und Artbildung. *Zeitschrift für Tierpsychologie* **21**:129–204.

Nielsen, B.M.B. & S.L. Vehrencamp. 1995. Responses of song sparrows to song-type matching via interactive playback. *Behavioural Ecology & Sociobiology* **37**:109–117.

Niemi, G.N., J.M. Hanowski, A.R. Lima, T. Nicholls & N. Weiland. 1997. A critical analysis on the use of indicator species in management. *Journal of Wildlife Management* **61**:1240–1252.

Niemiec, A.J., Y. Raphael & D.B. Moody. 1994. Return of auditory function following structural regeneration after acoustic trauma: Behavioral measures from quail. *Hearing Research* **75**: 209–24.

Nijhout, H.F. 1991. *The Development and Evolution of Butterfly Wing Patterns*. Smithsonian Institution Press, Washington, D.C.

Nixdorf-Bergweiler, B. 2001. Lateral magnocellular nucleus of the anterior neostriatum (LMAN) in the zebra finch: Neuronal connectivity and the emergence of sex differences in cell morphology. *Microscopy Research Techniques* **54**:335–353.

Nixdorf-Bergweiler, B.E., M.B. Lips & U. Heinemann. 1995. Electrophysiological and morphological evidence for a new projection of LMAN-neurones towards area X. *Neuroreport* **6**:1729–1732.

Noad, M.J., D.H. Cato, M.M. Bryden, M.-N. Jenner & K.C.S. Jenner. 2000. Cultural revolution in whale songs. *Nature* **408**:537.

Nordby, J.C., S.E. Campbell & M.D. Beecher. 1999. Ecological correlates of song learning in song sparrows. *Behavioral Ecology* **10**:287–297.

Nordby, J.C., S.E. Campbell & M.D. Beecher 2001. Late song learning in song sparrows. *Animal Behaviour* **61**:835–846.

Nordeen, K. & E. Nordeen. 1992. Auditory feedback is necessary for the maintenance of stereotyped song in adult zebra finches. *Behavioral & Neural Biology* **57**:58–66.

Nordeen, K.W. & E.J. Nordeen. 1993. Long-term maintenance of song in adult zebra finches is not affected by lesions of a forebrain region involved in song learning. *Behavioral & Neural Biology* **59**:79–82.

Norris, J.C. & W.E. Evans. 1995. Sources of variance in acoustic censusing of cetaceans. *Conference on Biology of Marine Mammals* **11**:83.

Noss, R.F. 1990. Indicators for monitoring biodiversity: A hierarchical approach. *Conservation Biology* **4**:355–364.

Nottebohm, F. 1966. *The role of sensory feedback in the development of avian vocalizations.* Doctoral dissertation, University of California, Berkeley.

Nottebohm, F. 1968. Auditory experience and song development in the chaffinch, *Fringilla coelebs*. *Ibis* **110**:549–568.

Nottebohm, F. 1969. The "critical" period for song learning. *Ibis* **111**:386–387.

Nottebohm, F. 1969. The song of the chingolo, *Zonotrichia capensis*, in Argentina: Description and evaluation of a system of dialects. *Condor* **71**:299–315.

Nottebohm, F. 1971a. Neural laterization of vocal control in a passerine bird. I. Song. *Journal of Experimental Zoology* **177**:229–262.

Nottebohm, F. 1971b. Neural lateralization of vocal control in a passerine bird. II. Subsong, calls, and a theory of vocal learning. *Journal of Experimental Zoology* **179**:35–50.

Nottebohm, F. 1972. The origins of vocal learning. *American Naturalist* **106**:116–140.

Nottebohm, F. 1977. Asymmetries in neural control of vocalization in the canary. In: *Lateralization in the nervous system.*, S. Harnad, R.W. Doty, L. Goldstein, J. Jaynes & G. Krauthamer (ed.), pp. 23–44. Academic Press, New York.

Nottebohm, F. 1980. Testosterone triggers growth of brain vocal control nuclei in adult female canaries. *Brain Research* **189**:429–436.

Nottebohm, F. 1984. Vocal learning and its possible relation to replaceable synapses and neurons. In: *Biological Perspectives on Language,* D. Caplan (ed.), pp. 65–95. MIT Press, Cambridge MA.

Nottebohm, F. 1985. Neuronal replacement in adulthood. *Annals of the New York Academy of Sciences* **457**:143–161.

Nottebohm, F. 1989. From bird song to neurogenesis. *Scientific American* **260**:74–79.

Nottebohm, F. 1991. Reassessing the mechanisms and origins of vocal learning in birds. *Trends in Neuroscience* **14**:206–210.

Nottebohm, F. 1993. The search for neural mechanisms that define the sensitive period for song learning in birds. *Netherlands Journal of Zoology* **43**:193–234.

Nottebohm, F. 1999. The anatomy and timing of

vocal learning in birds. In: *The Design of Animal Communication*, (M.D. Hauser & M. Konishi (eds.) pp. 63–110. MIT Press, Cambridge, MA.

Nottebohm, F. 2002. Why are some neurons replaced in adult brain? *Journal of Neuroscience* **22**:624–628.

Nottebohm, F. & A. Alvarez-Buylla. 1993. Neurogenesis and neuronal replacement in adult birds. In: *Neuronal Cell Death and Repair*, A.C. Cuello (ed.), pp. 227–236. Elsevier, Amsterdam.

Nottebohm, F., A. Alvarez-Buylla, J. Cynx, J. Kirn, C.Y. Ling, M. Nottebohm, R. Suter, A. Tolles & H. Williams. 1990. Song learning in birds: The relation between perception and production. *Philosophical Transactions of the Royal Society of London, B* **329**:115–124.

Nottebohm, F. & A.P. Arnold. 1976. Sexual dimorphism in vocal control areas of the songbird brain. *Science* **194**:211–213.

Nottebohm, F., S. Kasparian & C. Patton. 1981. Brain space for a learned task. *Brain Research* **213**:99–109.

Nottebohm, F., D.B. Kelley & J.A. Paton. 1982. Connections of vocal control nuclei in the canary telencephalon. *Journal of Comparative Neurology* **207**:344–357.

Nottebohm, F. & M.E. Nottebohm. 1976. Left hypoglossal dominance in the control of canary and white-crowned sparrow song. *Journal of Comparative Physiology* **108**:171–192.

Nottebohm, F. & M.E. Nottebohm. 1978. Relationship between song repertoire and age in the canary *Serinus canaria*. *Zeitschrift für Tierpsychologie* **46**:298–305.

Nottebohm, F., M.E. Nottebohm & L. Crane. 1986. Developmental and seasonal changes in canary song and their relation to changes in the anatomy of song control nuclei. *Behavioral & Neural Biology* **46**:445–471.

Nottebohm, F., T.M. Stokes & C.M. Leonard. 1976. Central control of song in the canary, *Serinus canarius*. *Journal of Comparative Neurology* **165**:457–486.

Novacek, M.J. 1992. Mammalian phylogeny: Shaking the tree. *Nature* **356**:121–125.

Nowicki, S. 1983. Flock-specific recognition of chickadee calls. *Behavioral Ecology & Sociobiology* **12**:317–320.

Nowicki, S. 1987. Vocal tract resonances in oscine bird sound production: Evidence from birdsongs in a helium atmosphere. *Nature* **325**:53–55.

Nowicki, S. 1989. Vocal plasticity in captive black-capped chickadees: The acoustic basis and rate of call convergence. *Animal Behaviour* **37**:64–73.

Nowicki, S. & R.R. Capranica. 1986a. Bilateral syringeal coupling during phonation of a songbird. *Journal of Neuroscience* **6**:3595–3610.

Nowicki, S. & R.R. Capranica. 1986b. Bilateral syringeal interaction in vocal production of an oscine bird sound. *Science* **231**:1297–1299.

Nowicki, S., D. Hasselquist, S. Bensch & S. Peters. 2000. Nestling growth and song repertoire sire in great reed warblers: Evidence for song learning as an indicator mechanism in mate choice. *Proceedings of the Royal Society of London, B* **267**:2419–2424.

Nowicki, S, M. Hughes & P. Marler. 1991. Flight songs of swamp sparrows: Alternative phonology of an alternative song category. *Condor* **1**:1–11.

Nowicki, S. & P. Marler. 1988. How do birds sing? *Music Perception* **5**:391–426.

Nowicki, S., J.C. Mitani, D.A. Nelson & P. Marler. 1989. The communicative significance of tonality in birdsong: Responses to songs produced in helium. *Bioacoustics* **2**:35–46.

Nowicki, S., S. Peters & J. Podos. 1998a. Song learning, early nutrition and sexual selection in songbirds. *American Zoologist* **38**:179–190.

Nowicki, S., W.A. Searcy & M. Hughes 1998b. The territory defense function of song in song sparrows: A test with the speaker occupation design. *Behaviour* **135**: 615–628.

Nowicki, S., W.A. Searcy, M. Hughes & J. Podos. 2001. The evolution of bird song: Male and female response to song innovation in swamp sparrows. *Animal Behaviour* **62**:1189–1195.

Nowicki, S., W.A. Searcy & S. Peters. 2002a. Quality of song learning affects female response to male bird song. *Proceedings of the Royal Society of London, B* **269**:1949–1954.

Nowicki, S., W.A. Searcy & S. Peters. 2002b. Brain development, song learning and mate choice in birds: A review and experimental test of the "nutritional stress hypothesis". *Journal of Comparative Physiology A* **188**:1003–1014.

Nowicki, S., M.W. Westneat & W.J. Hoese. 1992. Birdsong: Motor function and the evolution of communication. *Seminars in Neuroscience* **4**:385–390.

Nuechterlein, G.L. 1981. Courtship behavior and reproductive isolation between western grebe color morphs. *Auk* **98**: 335–349.

Nuechterlein, G.L. & D. Buitron. 1998. Interspecific

mate choice by late-courting male western grebes. *Behavioral Ecology* **9**:313–321.

Obwerger, K. & F. Goller. 2001. The metabolic costs of birdsong production. *Journal of Experimental Biology* **204**:3379–3388.

O'Connell-Rodwell, C., N. Rojek, T. Rodwell & P. Shannon. 2002. *Social attraction techniques induce group displays and nesting behavior in Caribbean flamingos on Guana Island, BVI.* Waterbird Society, Nov. (abstract).

Ohala, J.J. 1983. The origin of sound patterns in vocal tract constraints. In: *The Production of Speech*, P.F. MacNeilage (ed.), pp. 189–216. Springer-Verlag, New York.

Okanoya, K. 2002. Sexual display as a syntactical vehicle: The evolution of syntax in birdsong and human language through sexual selection. In: *The Transition to Language*, A. Wray (ed.), pp. 46–63. Oxford University Press, Oxford.

Okanoya, K. & R. Dooling. 1985. Colony differences in auditory thresholds in the canary (*Serinus canarius*). *Journal of the Acoustical Society of America* **78**:1170–1176.

Okanoya, K. & A. Yamaguchi. 1997. Adult bengalese finches (*Lonchura striata* var. *domestica*) require real-time auditory feedback to produce normal song syntax. *Journal of Neurobiology* **33**:343–356.

Okuhata, S. & N. Saito. 1987. Synaptic connection of thalamocerebral vocal nuclei of the canary. *Brain Research Bulletin* **18**:35–44.

Oller, D.K. & R.E. Eilers. 1988. The role of audition in infant babbling. *Child development* **59**: 441–449.

O'Loghlen, A.L. & M.D. Beecher. 1999. Mate, neighbour and stranger songs: A female song sparrow perspective. *Animal Behaviour* **58**:12–20.

O'Loghlen, A.L. & S.I. Rothstein. 1995. Culturally correct song dialects are correlated with male age and female song preferences in wild populations of brown-headed cowbirds. *Behavioral Ecology & Sociobiology* **36**:251–259.

Olson, S.L. 2001. Why so many kinds of passerine birds? *BioScience* **51**:268–269.

Orians, G.H. & G.M. Christman. 1968. A comparative study of the behavior of red-winged, tricolored, and yellow-headed blackbirds. *University of California Publications in Zoology* **84**:1–81.

Orr, R.T. 1945. A study of captive Galápagos finches of the genus *Geospiza*. *Condor* **47**:177–201.

Otter, K., L. Ratcliffe, D. Michaud, & P.T. Boag. 1998. Do female black-capped chickadees prefer high-ranking males as extra-pair partners? *Behavioral Ecology & Sociobiology* **43**:25–36.

Owings, D.H. & E.S. Morton. 1998. *Animal Vocal Communication: A New Approach.* Cambridge University Press, Cambridge.

Owren, M.J., J.A. Dieter, R.M. Seyfarth & D.L. Cheney. 1993. Vocalizations of rhesus (*Macaca mulatta*) and Japanese (*M. fuscata*) macaques cross-fostered between species show evidence of only limited modification. *Developmental Psychobiology* **26**:389–406.

Page, S.C., S.H. Hulse & J. Cynx. 1989. Relative pitch perception in the European starling (*Sturnus–vulgaris*) – Further evidence for an elusive phenomenon. *Journal of Experimental Psychology–Animal Behaviour Processes* **15**:137–146.

Palacios, M.G. & P.L. Tubaro. 2000. Does beak size affect acoustic frequencies in woodcreepers? *Condor* **102**:553–560.

Panov, E.N., A.S. Roubtsov & D.G. Monzikov. 2003. Hybridization between yellowhammer and pine bunting in Russia. *Dutch Birding* **25**:17–31.

Papousek, M. 1992. Early ontogeny of vocal communication in parent–infant interactions. In: *Nonverbal Vocal Communication: Comparative and Developmental Approaches,* H. Papousek, U. Jergens & M. Papousek (eds.), pp. 230–261. Cambridge University Press, New York.

Papousek, M., H. Papousek & M.H. Bornstein. 1985. The naturalistic vocal environment of young infants: On the significance of homogeneity and variability in parental speech. In: *Social Perception in Infants,* T. Field & N. Fox (eds.). Ablex, Norwood, NJ.

Parent, A. & L. Hazrati. 1995. Functional anatomy of the basal ganglia. I. The cortico-basal ganglia-thalamo-cortical loop. *Brain Research. Brain Research Reviews* **20**:91–127.

Parker, M., J. Boyce, R. Young, N. Rojek, C. Hamilton, V. Slowik, H. Gellerman, S.W. Kress, H. Carter, G. Moore & J.J. Cohen. 2000. Restoration of common murre colonies in central coastal California. In: *San Francisco Bay National Wildlife Refuge Complex Annual Report 1999.* U.S. Fish and Wildlife Service, Newark, CA.

Parker, T.A. 1991. On the use of tape recorders in avifaunal surveys. *Auk* **108**:443–444.

Paton, J.A. & F. Nottebohm. 1984. Neurons generated in the adult brain are recruited into functional circuits. *Science* **225**:1046–1048.

Paton, J.A., K.R. Manogue & F. Nottebohm. 1981.

Bilateral organization of the vocal control pathway in the budgerigar, *Melopsittacus undulatus*. *Journal of Neuroscience* **1**:1279–1288.

Patterson, D.K. & I.M. Pepperberg. 1994. A comparative study of human and grey parrot phonation: Acoustic and articulatory correlates of vowels. *Journal of the Acoustical Society of America* **96**:634–648.

Patterson, D.K. & I.M. Pepperberg. 1998. A comparative study of human and grey parrot phonation: Acoustic and articulatory correlates of stop consonants. *Journal of the Acoustical Society of America* **103**:2197–2213.

Payne, K. 2001. *Year-end Technical Report: Elephant Listening Project 2000*, pp. 1–17. U.S. Fish and Wildlife Service, Washington, D.C.

Payne, R.B. 1973a. Vocal mimicry of the paradise whydahs (*Vidua*) and response of female whydahs to the songs of their hosts (*Pytillia*) and their mimics. *Animal Behaviour* **21**:762–771.

Payne, R.B. 1973b. Behavior, mimetic songs and song dialects, and species relationships of the parasitic indigobirds (*Vidua*) of Africa. *Ornithological Monographs* **11**:1–333.

Payne, R.B. 1978. Microgeographic variation in songs of splendid sunbirds *Nectarinia coccinigaster*: Population phenetics, habitats, and song dialects. *Behaviour* **65**:282–308.

Payne, R.B. 1980. Behavior and songs of hybrid parasitic finches. *Auk* **97**:118–134.

Payne, R.B. 1981. Song learning and social interaction in indigo buntings. *Animal Behaviour* **29**:688–697.

Payne, R.B. 1982. Ecological consequences of song matching: Breeding success and intraspecific song mimicry in indigo buntings. *Ecology* **61**:401–411.

Payne, R.B. 1986. Bird songs and avian systematics. In: *Current Ornithology* (*Vol. 3*), R.F. Johnston (ed.), pp. 88–123. Plenum Press, New York.

Payne, R.B. 1996. Song traditions in indigo buntings: Origin, improvisation, dispersal, and extinction in cultural evolution. In: *Ecology and Evolution of Acoustic Communication in Birds*, D.E. Kroodsma & E.H. Miller (eds.), pp. 198–220. Cornell University Press, Ithaca, NY.

Payne, R.B. & K. Payne. 1977. Social organization and mating success in local song populations of village indigobirds, *Vidua chalybeata*. *Zietschrift für Tierpsychologie* **45**:113–173.

Payne, R.B. & L.L. Payne. 1994. Song mimicry and species associations of west African indigobirds *Vidua* with quail-finch *Ortygospiza atricollis*, goldbreast *Amandava subflava*, and brown twinspot *Clytospiza monteiri*. *Ibis* **136**:291–304.

Payne, R.B. & L.L. Payne. 1997. Field observations, experimental design, and the time and place of learning bird songs. In: Social influences on vocal development (C.T. Snowdon & M. Hausberger, (eds.), pp. 57–84 Cambridge University Press, Cambridge, MA.

Payne, R.B., L.L. Payne & S.M. Doehlert. 1984. Interspecific song learning in a wild chestnut-sided warbler. *Wilson Bulletin* **96**:292–294.

Payne, R.B., L.L. Payne & J.L. Woods. 1998. Song learning in brood-parasitic indigobirds *Vidua chalybeata*: Song mimicry of the host species. *Animal Behaviour* **55**:1537–1553.

Payne, R.B., L.L. Payne, J.L. Woods & M.D. Sorenson. 2000. Imprinting and the origin of parasite–host species associations in brood-parasitic indigobirds, *Vidua chalybeata*. *Animal Behaviour* **59**:69–81.

Payne, R.B., W.L. Thompson, K.L. Fiala & L.L. Sweany. 1981. Local song traditions in indigo buntings: Cultural transmission of behavior patterns across generations. *Behaviour* 77:199–221.

Payne, R.S. & S. McVay. 1971. Songs of humpback whales. *Science* **173**:585–597.

Peake, T.M. 1997. *Variation in the vocal behaviour of the corncrake Crex crex: Potential for conservation.* Doctoral dissertation, University of Nottingham, UK.

Peake, T.M. & P.K. McGregor. 1999. Geographical variation in the vocalization of the corncrake *Crex crex*. *Ethology, Ecology and Evolution* **11**:123–137.

Peake, T.M., P.K. McGregor, K.W. Smith, G. Tyler, G. Gilbert & R.E. Green. 1998. Individuality in corncrake *Crex crex* vocalizations. *Ibis* **140**:120–127.

Pearson, K.G. 1993. Common principles of motor control in vertebrates and invertebrates. *Annual Review of Neuroscience* **16**:265–297.

Pepperberg, I.M. 1981. Functional vocalizations by an African grey parrot *(Psittacus erithacus)*. *Zeitschrift für Tierpsychologie* **55**:139–160.

Pepperberg, I.M. 1985. Social modeling theory: A possible framework for understanding avian vocal learning. *Auk* **102**:854–868.

Pepperberg, I.M. 1987a. Acquisition of the same/different concept by an African grey parrot (*Psittacus erithacus*): Learning with respect to categories of color, shape, and material. *Animal Learning & Behavior* **15**:423–432.

Pepperberg, I.M. 1987b. Interspecies communication: A tool for assessing conceptual abilities in the African grey parrot (*Psittacus erithacus*). In: *Language, Cognition, and Consciousness: Integrative Levels*, G. Greenberg & E. Tobach (eds.), pp. 31–56. Lawrence Erlbaum, Hillsdale, NJ.

Pepperberg, I.M. 1988. An interactive modeling technique for acquisition of communication skills: Separation of 'labeling' and 'requesting' in a psittacine subject. *Applied Psycholinguistics* **9**:59–76.

Pepperberg, I.M. 1990a. An investigation into the cognitive capacities of an African grey Parrot. In: *Advances in the Study of Behavior*, P.J.B. Slater, J.S. Rosenblatt & C. Beer (eds.), pp. 357–409. Academic Press, New York.

Pepperberg, I.M. 1990b. Cognition in an African grey parrot (*Psittacus erithacus*): Further evidence for comprehension of categories and labels. *Journal of Comparative Psychology* **104**:41–52.

Pepperberg, I.M. 1990c. Referential mapping: Attaching functional significance to the innovative utterances of an African grey parrot. *Applied Psycholinguistics* **11**:23–44.

Pepperberg, I.M. 1992. Proficient performance of a conjunctive, recursive task by an African grey parrot (*Psittacus erithacus*). *Journal of Comparative Psychology* **106**:295–305.

Pepperberg, I.M. 1993. A review of the effects of social interaction on vocal learning in African grey parrots (*Psittacus erithacus*). *Netherlands Journal of Zoology* **43**:104–124.

Pepperberg, I.M. 1994a. Evidence for numerical competence in an African grey parrot (*Psittacus erithacus*). *Journal of Comparative Psychology* **108**:36–44.

Pepperberg, I.M. 1994b. Vocal learning in African grey parrots: Effects of social interaction. *Auk* **111**:300–313.

Pepperberg, I.M. 1996. Categorical class formation by an African grey parrot (*Psittacus erithacus*). In: *Stimulus Class Formation in Humans and Animals*, T.R. Zentall & P.R. Smeets (eds.), pp. 71–90. Elsevier, Amsterdam.

Pepperberg, I.M. 1997. Social influences on the acquisition of human-based codes in parrots and nonhuman primates. In: *Social Influences on Vocal Development*, C.T. Snowdon & M. Hausberger (eds.), pp. 157–177. Cambridge University Press, Cambridge.

Pepperberg, I.M. 1999. *The Alex Studies: Cognitive and Communicative Abilities of Grey Parrots*. Harvard University Press, Cambridge, MA.

Pepperberg, I.M. 2001. Lessons from cognitive ethology: Animal models for ethological computing. *Proceedings of the First Conference on Epigenetic Robotics*, Lund, Sweden.

Pepperberg, I.M. 2002. Allospecific referential speech acquisition in grey parrots (*Psittacus erithacus*): Evidence for multiple levels of avian vocal imitation. In: *Imitation in Animals and Artifacts*, K. Dautenhahn & C. Nehaniv (eds.), pp. 109–131. MIT Press, Cambridge, MA.

Pepperberg I.M., K.J. Brese & B.J. Harris. 1991. Solitary sound play during acquisition of English vocalizations by an African grey parrot (*Psittacus erithacus*): Possible parallels with children's monologue speech. *Applied Psycholinguistics* **12**:151–177.

Pepperberg, I.M. & M.V. Brezinsky. 1991. Relational learning by an African grey parrot: Discriminations based on relative size. *Journal of Comparative Psychology* **105**:286–294.

Pepperberg, I.M. & M.S. Funk. 1990. Object permanence in four species of psittacine birds: An African grey parrot (*Psittacus erithacus*), an Illiger macaw (*Ara maracana*), a parakeet (*Melopsittacus undulatus*), and a cockatiel (*Nymphus hollandicus*). *Animal Learning & Behavior* **18**:97–108.

Pepperberg, I.M., S.E. Garcia, E.C. Jackson & S. Marconi. 1995. Mirror use by African grey parrots (*Psittacus erithacus*). *Journal of Comparative Psychology* **109**:182–195.

Pepperberg I.M., L.I. Gardiner, & L.J. Luttrell. 1999. Limited contextual vocal learning in the grey parrot: The effect of co-viewers on videotaped instruction. *Journal of Comparative Psychology* **113**:158–172.

Pepperberg, I.M. & F.A. Kozak. 1986. Object permanence in the African grey parrot (*Psittacus erithacus*). *Animal Learning & Behavior* **14**:322–330.

Pepperberg I.M. & M.A. McLaughlin. 1996. Effect of avian–human joint attention on allospecific vocal learning by grey parrots. *Journal of Comparative Psychology* **110**:286–297.

Pepperberg, I.M., J.R. Naughton & P.A. Banta. 1998. Allospecific vocal learning by grey parrots (*Psittacus erithacus*): A failure of videotaped instruction under certain conditions. *Behavioural Processes* **42**:139–158.

Pepperberg I.M., R.M. Sandefer, D. Noel, & C.P. Ellsworth. 2000. Vocal learning in the grey parrot:

Effect of species identity and number of trainers. *Journal of Comparative Psychology* **114**:371–380.

Pepperberg I.M.& D.V. Sherman. 2000. Proposed use of two-part interactive modeling as a means to increase functional skills in children with a variety of disabilities. *Teaching and Learning in Medicine* **12**:213–220.

Pepperberg I.M. & H.A. Shive. 2001. Hierarchical combinations by a grey parrot: Bottle caps, lids, and labels. *Journal of Comparative Psychology* **115**:376–384.

Pepperberg I.M. & S.E. Wilcox. 2000. Evidence for a form of mutual exclusivity during label acquisition by grey parrots? *Journal of Comparative Psychology* **114**:219–231.

Pepperberg, I.M., M.R. Willner & L.B. Gravitz. 1997. Development of Piagetian object permanence in a grey parrot (*Psittacus erithacus*). *Journal of Comparative Psychology* **111**:63–75.

Pereyra, M.E. 1998. *Effects of environment and endocrine function on control of reproduction in a high elevation tyrannid.* Doctoral dissertation, Northern Arizona University.

Perkel, D. & M. Farries. 2000. Complementary 'bottom-up' and 'top-down' approaches to basal ganglia function. *Current Opinion in Neurobiology* **10**:725–731.

Perrett, D.I. & N.J. Emery. 1994. Understanding the intentions of others from visual signals: Neurophysiological evidence. *Current Psychology of Cognition* **13**:683–694.

Perrett, D.I., P.A.J. Smith, D.D. Potter, A.J. Mistlin, A.S.M. Head & M.A. Jeeves. 1984. Visual cells in the temporal cortex sensitive to face view and gaze direction. *Proceedings of the Royal Society of London, B* **223**:293–317.

Perrin, N. & J. Travis. 1992. On the use of constraints in evolutionary biology and some allergic reactions to them. *Functional Ecology* **6**:361–363.

Perrone, M., Jr. 1980. Factors affecting the incidence of distress calls in passerines. *Wilson Bulletin* **92**:404–408.

Perry, E.A. & J.M. Terhune. 1999. Variation of harp seal (*Pagophilus groenlandicus*) underwater vocalizations among three breeding locations. *Journal of Zoology London* **249**:181–186.

Peters, S. & S. Nowicki. 1996. Development of tonal quality in birdsong: Further evidence from song sparrows. *Ethology* **102**:323–335.

Peterson, R.T. 1947. *A Field Guide to the Birds. Giving Field Marks of all Species Found East of the Rockies.* Houghton Mifflin Co., Cambridge MA.

Petren, K., B.R. Grant & P.R. Grant. 1999. A phylogeny of Darwin's finches based on microsatellite DNA length variation. *Proceedings of the Royal Society of London, B* **266**:321–329.

Petrinovich, L. & L.F. Baptista. 1984. Song dialects, mate selection, and breeding success in white-crowned sparrows. *Animal Behaviour* **32**:1078–1088.

Petrinovich, L. & L.F. Baptista. 1987. Song development in the white-crowned sparrow: Modification of learned song. *Animal Behaviour* **35**:961–974.

Petrinovich, L., T. Patterson & L.F. Baptista. 1981. Song dialects as barriers to dispersal: A re-evaluation. *Evolution* **35**:180–188.

Pierce, J.R. 1983. *The Science of Musical Sounds.* W.H. Freeman, New York.

Podolsky, R.H. 1990. Effectiveness of social stimuli in attracting Laysan albatross to new potential nesting sites. *Auk* **107**:119–125.

Podolsky, R.H. & S.W. Kress. 1989. Factors affecting colony formation in Leache's storm-petrel. *Auk* **106**:332–336.

Podolsky, R.H. & S.W. Kress. 1992. Attraction of the endangered dark-rumped petrel to recorded vocalizations in the Galapagos Islands. *Condor* **94**:448–453.

Podos, J. 1996. Motor constraints on vocal development in a songbird. *Animal Behaviour* **51**:1061–1070.

Podos, J. 1997. A performance constraint on the evolution of trilled vocalizations in a songbird family (Passeriformes: *Emberizidae*). *Evolution* **51**:537–551.

Podos, J. 2001. Correlated evolution of morphology and vocal signal structure in Darwin's finches. *Nature* **409**:185–188.

Podos, J., S. Nowicki & S. Peters. 1999. Permissiveness in the learning and development of song syntax in swamp sparrows. *Animal Behaviour* **58**:93–103.

Podos, J., J.K. Sherer, S. Peters & S. Nowicki. 1995. Ontogeny of vocal-tract movements during song production in song sparrows. *Animal Behaviour* **50**:1287–1296.

Poeppel, D. 1996. A critical review of PET studies of phonological processing. *Brain & Language* **55**:317–385.

Pohl-Apel, G. & R. Sossinka. 1984. Hormonal determination of song capacity in females of the zebra finch: Critical phase of treatment. *Zeitschrift für Tierpsychologie* **64**:330–336.

Popp, J. & M.S. Ficken. 1991. Comparative analysis

of acoustic structure of passerine and woodpecker nestling calls. *Bioacoustics* **3**:255–274.

Poole, A. & F. Gill. 2002. *The Birds of North America*. The Birds of North America, Inc., Philadelphia, PA.

Poot, M., F. Engelen & J. van der Winden. 1999. A mixed breeding pair of Blyth's reed warbler *Acrocephalus dumetorum* and marsh warbler *A. palustris* near Utrecht in spring 1998. *Limosa* **72**:151–157.

Potash, L.M. 1972. Noise-induced changes in calls of the Japanese quail. *Psychonomic Science* **26**:252–254.

Potter, R.K. 1945. Visible patterns of sound. *Science* **102**:463–470.

Potter, R.K., G.A. Kopp & H.C. Green. 1947. *Visible Speech*. Dover Publications Inc., New York.

Poulsen, H. 1951. Inheritance and learning in the song of the chaffinch (*Fringilla coelebs*). *Behaviour* **3**:216–228.

Poulsen, H. 1958. The calls of the chaffinch (*Fringilla coelebs* L.) in Denmark. *Dansk Ornithologisk Forening Tidsskrift* **52**:89–105.

Powell, G.V.N. 1980. Migrant participation in neotropical mixed species flocks. In: *Migrant Birds in the Neotropics: Ecology, Behavior, Distribution, and Conservation*, A. Keast & E.S. Morton (eds.), pp. 477–483. Smithsonian Institution Press, Washington, D.C.

Prendergast, J.R., R.M. Quinn, J.H. Lawton, B.C. Eversham & D.W. Gibbons. 1993. Rare species, the coincidence of diversity hotspots and conservation strategies. *Nature* **365**:335–337.

Price, J.J. & S.M. Lanyon. 2002a. A robust molecular phylogeny of the oropendolas: Polyphyly revealed by mitochondrial sequence data. *Auk* **119**:335–348.

Price, J.J. & S.M. Lanyon. 2002b. Reconstructing the evolution of complex bird song in the oropendolas. *Evolution* **56**:1514–1529.

Prum, R.O. 1992. Syringeal morphology, phylogeny, and evolution of the neotropical manakins (Aves: *Pipridae*). *American Museum Novitates* **3043**:1–65.

Prum, R.O. 1998. Sexual selection and the evolution of mechanical sound production in manakins (Aves: *Pipridae*). *Animal Behaviour* **55**:977–994.

Prum, R.O. 2002. Why ornithologists should care about the theropod origin of birds. *Auk* **119**:1–17.

Puelles, L., E. Kuwana, E. Puelles & J.L.R. Rubenstein. 1999. Comparison of the mammalian and avian telencephalon from the perspective of gene expression data. *European Journal of Morphology* **37**:139–150.

Pumphrey, R.J. 1961. Sensory organs: Hearing. In: *Biology and Comparative Physiology of Birds*, A.J. Marshall (ed.), pp. 69–86. Academic Press, New York.

Pyle, P. 1997. *Identification Guide to North American Birds, Part I*. Slate Creek Press, Bolinas, CA.

Pytte, C.L., K.M. Rusch & M.S. Ficken. 2003. Regulation of vocal amplitude by the blue-throated hummingbird (*Lampornis clemenciae*). *Animal Behaviour* **1111**: in press.

Pytte, C.L. & R.A. Suthers. 2000. Sensitive period for sensorimotor integration during vocal motor learning. *Journal of Neurobiology* **42**:172–189.

Rædesater, T., S. Jakobsson, N. Andbjer, A. Bylin & K. Nystrom. 1987. Song rate and pair formation in the willow warbler, *Phylloscopus trochilus*. *Animal Behaviour* **35**:1645–1651.

Raff, R.A. & T.C. Kaufman. 1983. *Embryos, Genes and Evolution*. Macmillan, New York.

Raikow, R.J. 1986. Why are there so many kinds of passerine birds? *Systematic Zoology* **35**:255–259.

Raisman, G. & P. Field. 1971. Sexual dimorphism in the preoptic area of the rat. *Science* **173**:731–733.

Rakic, P. 1985. Limits of neurogenesis in primates. *Science* **227**:1054–1056.

Rakic, P. 2002a. Adult neurogenesis in mammals: An identity crises. *Journal of Neuroscience* **22**:614–618.

Rakic, P. 2002b. Neurogenesis in adult primate neocortex: An evaluation of the evidence. *Nature Reviews Neuroscience* **3**:65–71.

Ralls, K., P. Fiorelli & S. Gish. 1985. Vocalizations and vocal mimicry in captive harbor seals, *Phoca vitulina*. *Canadian Journal of Zoology* **63**:1050–1056.

Ramenofsky, M., R. Agatsuma, M. Barga, R. Cameron, J. Harm, M. Landys & T. Ramfar. 2003. Migratory behavior: New insights from captive studies. In: *Avian Migration*, P. Berthold, E. Gwinner & E. Sonnenschein (eds.), pp. 97–112. Springer-Verlag, Berlin.

Ramirez, V., J. Zheng & K. Siddique. 1996. Membrane receptors for estrogen, progesterone, and testosterone in the rat brain: Fantasy or reality. *Cellular & Molecular Neurobiology* **16**:175–198.

Ramon y Cajal, S. 1913. *Degeneration and regeneration*

of the nervous system. (1928 translation from Spanish edition, reprinted 1991). Oxford University Press.

Rasika, S., F. Nottebohm & A. Alvarez-Buylla. 1994. Testosterone increases the recruitment and/or survival of new high vocal center neurons in adult female canaries. *Proceedings of the National Academy of Sciences USA* **91**:7854–7858.

Ratcliffe, L.M. 1981. Species recognition in Darwin's ground finches (*Geospiza* Gould). McGill University, Montreal.

Ratcliffe, L.M. & P.R. Grant. 1983a. Species recognition in Darwin's finches (*Geospiza* Gould). I. Discrimination by morphological cues. *Animal Behaviour* **31**:1139–1153.

Ratcliffe, L.M. & P.R. Grant. 1983b. Species recognition in Darwin's finches (*Geospiza* Gould). II. Geographic variation in mate preference. *Animal Behaviour* **31**:1154–1165.

Ratcliffe, L.M. & P.R. Grant. 1985. Species recognition in Darwin's finches (*Geospiza* Gould). III. Male responses to playback of different song types, dialects, and heterospecific songs. *Animal Behaviour* **33**:290–307.

Ratcliffe, L. & K. Otter. 1996. Sex differences in song recognition. In: *Ecology and Evolution of Acoustic Communication in Birds*, D.E. Kroodsma & E.H. Miller (eds.), pp. 339–355. Cornell University Press, Ithaca, NY.

Ratcliffe, L. & R.G. Weisman. 1985. Frequency shift in the fee-bee song of the black-capped chickadee. *Condor* **87**:555–556.

Rattenborg, N.C., C.J. Amlaner & D. Phil. 2002. Phylogeny of sleep. In: *Sleep Medicine*, T.L. Lee-Chiong Jr., M.J. Sateia & M.A. Carskadon (eds.), pp. 7–22. Hanley & Belfus, Inc., Philadelphia, PA.

Rauch, N. 1978. Struktur der Lautäusserungen eines sprach imitierenden Graupapageis (*Psittacus erithacus* L.) *Behaviour* **66**:56–105.

Raup, D.M. 1966. Geometric analysis of shell coiling: General problems. *Journal of Paleontology* **40**:1178–1190.

Rauschecker, J.P. & P. Marler. 1987. *Imprinting and Cortical Plasticity: Comparative Aspects of Sensitive Periods.* Wiley, New York.

Rauscher, F.H., G.L. Shaw & K.N. Kye. 1993. Music and spatial task performance. *Nature* **365**:611.

Rauscher, F.H., G.L. Shaw & K.N. Kye. 1995. Listening to Mozart enhances spatial–temporal reasoning: Towards a neurophysiological basis. *Neuroscience Letters* **185**:44–87.

Rauscher, F.H., G.L. Shaw, L.J. Levine, E.L. Wright, W.R. Dennis & R.L. Newcomb. 1997. Music training causes long-term enhancement of preschool children's spatial–temporal reasoning. *Neurological Research* **19**:2–8.

Read, A.F. & D.M. Weary. 1992. The evolution of bird song: Comparative analyses. *Philosophical Transactions of the Royal Society of London, B* **338**:165–187.

Reby, D., S. Lek, I. Dimopoulos, J. Joachim, J. Lauga & S. Aulagnier. 1997. Artificial neural networks as a classification method in the behavioural sciences. *Behavioural Processes* **40**:35–43.

Redondo, T. & L. Arias de Reyna. 1988a. Locatability of begging calls in nestling altricial birds. *Animal Behaviour* **36**:653–661.

Redondo, T. & L. Arias de Reyna. 1988b. Vocal mimicry of hosts by great spotted cuckoo *Clamator glandarius*: Further evidence. *Ibis* **130**:540–544.

Redondo, T. & F. Castro. 1992. Signalling of nutritional need by magpie nestlings. *Ethology* **92**:193–204.

Redpath, S.M., B.M. Appleby & S.J. Petty. 2001. Do male hoots betray parasite loads in tawny owls? *Journal of Avian Biology* **31**: 457–462.

Reed, J.M. & A.P. Dobson. 1993. Behavioral constraints and conservation biology: Conspecific attraction and recruitment. *Trends in Ecology and Evolution* **8**:253–257.

Reed, R.A. 1968. Studies of the diederik cuckoo *Chrysococcyx caprius* in the Transvaal. *Ibis* **110**:321–331.

Rehkaemper, G., K.-L. Schuchmann, A. Schleicher & K. Zilles. 1991. Encephalization in hummingbirds (*Trochilidae*). *Brain, Behavior and Evolution* **37**:85–91.

Rehsteiner, U., H. Geisser & H.-U. Reyer. 1998. Singing and mating success in water pipits: One specific song element makes all the difference. *Animal Behaviour* **55**:1471–1481.

Reid, M.L. & P.J. Weatherhead. 1990. Mate choice criteria of Ipswich sparrows – the importance of variability. *Animal Behaviour* **40**:538–544.

Reijnen, R., R. Foppen, C. Ter Braak & J. Thissen. 1995. The effects of car traffic on breeding bird populations in woodland: III. Reduction of density in relation to the proximity of main roads. *Journal of Applied Ecology* **32**:187–202.

Reijnen, R., R. Foppen & G. Veenbaas. 1997. Disturbance by traffic of breeding birds: Evaluation of the effect and considerations in planning and managing road corridors. *Biodiversity and Conservation* **6**:567–581.

Reiner, A., D.J. Perkel, L. Bruce, A.B. Butler, A. Csillag, W. Kuenzel, L. Medina, G. Paxinos, T. Shimizu, G.F. Striedter, M. Wild, G.F. Ball, S. Durand, O. Güntürkün, D.W. Lee, C.V. Mello, A. Powers, S.A. White, G. Hough, L. Kubikova, T.V. Smulders, K. Wada, J. Dugas-Ford, S. Husband, K. Yamamoto, J. Yu, C. Siang & E.D. Jarvis. 2004a. Revised nomenclature for avian telencephalon and some related brainstem nuclei. *Journal of Comparative Neurology* **473**:377–414.

Reiner, A., L. Bruce, A. Butler, A. Csillag, W. Kuenzel, L. Medina, G. Paxinos, D. Perkel, T. Shimizu, G. Striedter, M. Wild, G. Ball, S. Durand, O. Güntürkün, D. Lee, C.V. Mello, A. Powers, S. White, G. Hough, L. Kubikova, T.V. Smulders, K. Wada, J. Dugas-Ford, S. Husband, K. Yamamoto, J. Yu, C. Siang & E.D. Jarvis. In press B. The avian brain nomenclature forum: a new century in comparative neuroanatomy. *Journal of Comparative Neurology* conditional acceptance upon review.

Reiner, A., L. Medina & C.L. Veenman. 1998. Structural and functional evolution of the basal ganglia in vertebrates. *Brain Research. Brain Research Reviews* **28**:235–285.

Remsen, J.V. 1976. Observations of vocal mimicry in the thick-billed Euphonia. *Wilson Bulletin* **88**:487–488.

Remsen, J.V., Jr. 2001. True winter range of the veery (*Catharus fuscescens*): Lessons for determining winter ranges of species that winter in the tropics. *Auk* **118**:838–848.

Remsen, J.V., K. Garrett & R.A. Erickson. 1982. Vocal copying in Lawrence's and lesser goldfinches. *Western Birds* **13**:29–33.

Ribeiro, S. 2000. Habit and learning. In: *Song, sleep, and the slow evolution of thoughts: Gene expression studies on brain representation,* (eds), pp. 54–60. Doctoral dissertation, Rockefeller University, New York.

Ribeiro, S., G.A. Cecchi, M.O. Magnasco & C.V. Mello. 1998. Toward a song code: Evidence for a syllabic representation in the canary brain. *Neuron* **21**:359–371.

Ribic, C.A., S. Lewis, S. Melvin, J. Bart & B. Peterjohn. 1999. *Proceedings of the Marsh Bird Monitoring Workshop*. USFWS Region 3 Administrative Report, Fort Snelling MN.

Richards, D.G. 1981a. Alerting and message components in songs of the rufous-sided towhees. *Behaviour* **76**:223–249.

Richards, D.G. 1981b. Estimation of distance of singing conspecifics by the Carolina wren. *Auk* **98**:127–133.

Riebel, K. 2000. Early exposure leads to repeatable preferences for male song in female zebra finches. *Proceedings of the Royal Society of London, B* **267**: 2553–2558.

Riebel, K. & P.J.B. Slater. 1998a. Testing female chaffinch preferences by operant conditioning. *Animal Behaviour* **56**:1443–1453.

Riebel, K. & P.J.B. Slater. 1998b. Male chaffinches (*Fringilla coelebs*) can copy calls from a tape tutor. *Journal für Ornithologie* **139**:353–355.

Riebel, K., I.M. Smallegange, N.J. Terpstra & J.J. Bolhuis. 2002. Sexual equality in zebra finch song preference: Evidence for a dissociation between song recognition and production learning. *Proceedings of the Royal Society of London, B* **269**:729–733.

Riebel, K., & D. Todt 1997. Light flash stimulation alters the nightingale's singing style: Implications for song control mechanisms. *Behaviour,* **134**:9–10.

Ritchison, G. 1991. The flight songs of common yellowthroats: Description and causation. *Condor* **93**:12–18.

Riters, L.V. & G.F. Ball. 1999. Lesions to the medial preoptic area affect singing in the male European starling (*Sturnus vulgaris*). *Hormones and Behavior* **36**:276–286.

Robbins, C.S., B. Bruun & H.S. Zimm. 1966. *A Guide to Field Identification: Birds of North America.* Golden Press, New York.

Robbins, M.B. & S.N.G. Howell. 1995. A new species of pygmy-owl (*Strigidae*: *Glaucidium*) from the eastern Andes. *Wilson Bulletin* **107**:1–6.

Robbins, M.B. & F.G. Stiles. 1999. A new species of pygmy-owl (*Strigidae*: *Glaucidium*) from the Pacific slope of the northern Andes. *Auk* **116**:305–315.

Roberts, T.S. 1880. The convolution of the trachea in the sandhill and whooping cranes. *American Naturalist* **14**:108–114.

Robinson, F.N. 1974. The function of vocal mimicry in some avian displays. *Emu* **74**:9–10.

Robinson, F.N. 1975. Vocal mimicry and the evolution of bird song. *Emu* **75**:23–37.

Robinson, F.N. & H.S. Curtis. 1996. The vocal displays of the lyrebirds (*Menuridae*). *Emu* **96**:258–275.

Robisson, P. 1987. L'adaptation des règles de décodage des signaux acoustiques des oiseaux au canal de transmission. Etude appliquée aux cris de détresse du vanneau huppé *Vanellus vanellus*. *C.R. Acad. Sci. Paris* **304**:275–278.

Roby, D.D., K. Collis, D.E. Lyons, D.P. Craig, J.Y. Adkins, A.M. Myers & R.M. Suryan. 2002. Effects of colony relocation on diet and productivity of Caspian terns. *Journal of Wildlife Management* **66**:662–673.

Röll, K. 2002. *Zum gesanglichen Erfindungsreichtum von Nachtigallen*. Master thesis, Freie Universität, Berlin.

Rooke, I.J. & T.A. Knight. 1977. Alarm calls of honeyeaters with reference to locating sources of sound. *Emu* 77:193–198.

Rothstein, S.I. & R.C. Fleischer. 1987. Vocal dialects and their possible relation to honest status signalling in the brown-headed cowbird. *Condor* **89**:1–23.

Rothstein, S.I., D.A. Yokel & R.C. Fleischer. 1986. Social dominance, mating and spacing systems, female fecundity, and vocal dialects in captive and free-ranging brown-headed cowbirds. In: *Current Ornithology Vol. 3*, R.F. Johnston (ed.), pp.127–185. Plenum Press, New York.

Rowley, I. & G. Chapman. 1986. Cross-fostering, imprinting and learning in two sympatric species of cockatoo. *Behaviour* **96**:1–16.

Rowley, I. & E. Russell. 1997. *Fairy-wrens and Grasswrens: Maluridae.* Oxford University Press, Oxford.

Rubsamen, R. & M. Schafer. 1990. Audiovocal interactions during development? Vocalisation in deafened young horseshoe bats vs. audition in vocalisation-impaired bats. *Journal of Comparative Physiology A Sensory Neural and Behavioral Physiology* **167**:771–784.

Rusch, K., Pytte, C. & Ficken, M. 1996. Organization of agonistic vocalizations in black-chinned hummingbirds. *Condo*r, 557–566.

Ryals, B.M., R.J. Dooling, E. Westbrook, M.L. Dent, A. MacKenzie & O.N. Larsen. 1999. Avian species differences in susceptibility to noise exposure. *Hearing Research* **131**:71–88.

Ryals, B.M. & E.W. Rubel. 1985a. Differential susceptibility of avian hair cells to acoustic trauma. *Hearing Research* **19**:73–84.

Ryals, B.M. & E.W. Rubel. 1985b. Ontogenetic changes in the position of hair cell loss after acoustic overstimulation in avian basilar papilla. *Hearing Research* **19**:135–142.

Ryals, B.M. & E.W. Rubel. 1988. Hair cell regeneration after acoustic trauma in adult Coturnix quail. *Science* **240**:1774–1776.

Ryan, M.J. 1986. Neuroanatomy influences speciation rates among anurans. *Proceedings of the National Academy of Sciences USA* **83**: 1379–1382.

Ryan, M.J. & E.A. Brenowitz. 1985. The role of body size, phylogeny, and ambient noise in the evolution of bird song. *American Naturalist* **126**:87–100.

Ryan, M.J.& A.S. Rand. 1993. Sexual selection and signal evolution: The ghost of biases past. *Proceedings of the Royal Society of London, B* **340**:187–195.

Rydén, O. & H. Bengtsson. 1980. Differential begging and locomotory behaviour by early and late hatched nestlings affecting the distribution of food in asynchronously hatched broods of altricial birds. *Zeitschrift für Tierpsychologie* **53**:209–224.

Sætre, G.-P., T. Fossnes. & T. Slagsvold. 1995. Food provisioning in the pied flycatcher: Do females gain direct benefits from choosing bright coloured males. *Journal of Animal Ecology* **64**:21–30.

Sagan, C. 1978. *Murmurs of Earth: The Voyager Interstellar Record.* Random House, New York.

Saldanha, C., M. Tuerk, Y.-H. Kim, A. Fernandes, A. Arnold & B. Schlinger. 2000. Distribution and regulation of telencephalic aromatase expression in the zebra finch revealed with a specific antibody. *Journal of Comparative Neurology* **423**:619–630.

Samson, F.B. 1978. Vocalizations of Cassin's finch in northern Utah. *Condor* **80**:203–210.

Sartor, J.J. and G.F. Ball. 2001. Song output following manipulation of social factors and effects on song control nuclei in European starlings. Society for Neuroscience Abstracts. 27(2). 1709.

Sartor, J.J., T. Charlier, C. Pytte, J. Balthazart & G. Ball. 2002. Converging evidence that song performance modulates seasonal changes in the avian song control system. *Society for Neuroscience (Abstract Viewer and Itinerary Planner)* No. 781.10.

Saunders, A.A. 1935. *A Guide to Birdsongs.* Appleton-Century Co., New York.

Saunders, D.A. 1983. Vocal repertoire and individual vocal recognition in the short-billed white-tailed black cockatoo, *Calyptorhynchus funereus latirostris* Carnaby. *Australian Wildlife Research* **10**:527–536.

Saunders, J.C., R.K. Duncan, D.E. Doan & Y.L. Werner. 2000. The middle ear of reptiles and birds. In: *Comparative Hearing: Birds and Reptiles*, A.N. Popper (ed.), pp. 13–69. Springer-Verlag, New York.

Saunders, S.S. & R.J. Salvi. 1995. Pure tone masking patterns in adult chickens before and after recovery from acoustic trauma. *Journal of the Acoustical Society of America* **98**:1365–71.

Scharff, C., J. Kirn, M. Grossman, J. Macklis & F. Nottebohm. 2000. Targeted neuronal death affects neuronal replacement and vocal behavior in adult songbirds. *Neuron* **25**:481–492.

Scharff, C. & F. Nottebohm. 1991. A comparative study of the behavioral deficits following lesions of various parts of the zebra finch song system: Implications for vocal learning. *Journal of Neuroscience* **11**:2896–2913.

Schlinger, B.A. & A.P. Arnold. 1991. Brain is the major site of estrogen synthesis in a male songbird. *Proceedings of the National Academy of Sciences, USA* **88**:4191–4194.

Schlinger, B.A. & A.P. Arnold. 1992. Circulating estrogens in a male songbird originate in the brain. *Proceedings of the National Academy of Sciences USA* **89**:7650–7653.

Schlinger, B.A., K.K. Soma & S.E. London. 2001. Neurosteriods and brain sexual differentiation. *Trends in Neurosciences* **24**:429–431.

Schluter, D. 2000. *The Ecology of Adaptive Radiation.* Oxford University Press, Oxford.

Schluter, D., T.D. Price & P.R. Grant. 1985. Ecological character displacement in Darwin's finches. *Science* **227**:1056–1059.

Schmalhausen, I.I. 1949. *Factors of Evolution.* University of Chicago Press, Chicago.

Schmidt, M.F. & M. Konishi. 1998. Gating of auditory responses in the vocal control system of awake songbirds. *Nature Neuroscience* **1**:513–518.

Schmidt, V., H.M. Schaefer & B. Leisler. 1999. Song behaviour and range use in the polygamous aquatic warbler *Acrocephalus paludicola. Acta Ornithologica* **34**:209–213.

Schmidt-Nielsen, K. 1972. *How Animals Work.* Cambridge University Press, New York.

Schroeder, D.J. & R.H. Wiley. 1983. Communication with shared song themes in tufted titmice. *Auk* **100**:414–424.

Schroeder, M.R. 1970. Synthesis of low-peak-factor signals and binary sequences with lowl Autocorrelation. *IEEE Transactions on Information Theory* **IT16**: 85–95.

Schuchmann, K.L. 1999. Order *Apodiformes*, family *Trochelidae* (hummingbirds). In: *Handbook of the Birds of the World: Barn Owls to Hummingbirds*, J. del Hoyo, A. Elliott & J. Sargatal (eds.), pp. 468–535. Lynx Ediciones, Barcelona.

Scherer, K.R. 1985. Vocal affect signaling: A comparative approach. *Advances in the Study of Behavior* **15**:189–244.

Schuchmann, K.-L. 1999. Family *Trochilidae* (hummingbirds). In: *Handbook of the Birds of the World: Barn-owls to Hummingbirds*, J. del Hoyo, A. Elliott & J. Sargatal (eds.), pp. 468–680. Lynx Ediciones, Barcelona.

Schwartzkopff, J. 1968. Structure and function of the ear and the auditory brain areas in birds. In: *Hearing Mechanisms in Vertebrates*, A.V.S. DeReuck & J. Knight (eds.), pp. 41–59. Little, Brown, Boston, MA.

Schwartzkopff, J. 1973. Mechanoreception. In: *Avian Biology, Vol 3.* D.S. Farner, J.R. King & K.C. Parks (eds.), pp. 417–477. Academic Press, New York.

Searcy, W.A. 1983. Response to multiple song types in the male song sparrows and field sparrows. *Animal Behaviour* **31**:948–949.

Searcy, W.A. 1988. Dual intersexual and intrasexual functions of song in red-winged blackbirds. *Proceedings XIX International Congress of Ornithology* **1**: 1373–1381.

Searcy, W.A. 1990. Species recognition of song by female red-winged blackbirds. *Animal Behaviour* **40**:1119–1127.

Searcy, W.A. 1992a. Measuring responses of female birds to male song. In: *Playback and Studies of Animal Communication*, P.K. McGregor (ed.), pp. 175–189. Plenum, New York.

Searcy, W. 1992b. Song repertoire and mate choice in birds. *American Zoologist* **32**:71–80.

Searcy, W.A. & M. Andersson. 1986. Sexual selection and the evolution of song. *Annual Review of Ecology and Systematics* **17**:507–533.

Searcy, W.A., E. Balaban, R.A. Canady, S.J. Clark, S. Runfeldt & H. Williams. 1981a. Responsiveness of male swamp sparrows to temporal organization of song. *Auk* **98**:613–615.

Searcy, W.A., S. Coffman & D.F. Raikow. 1994. Habituation, recovery and the similarity of song types with repertoires in red-winged blackbirds (*Agelaius phoeniceus*). *Ethology* **98**:38–49.

Searcy, W.A. & P. Marler. 1981. A test for responsiveness to song structure and programming in female sparrows. *Science* **213**:926–928.

Searcy, W.A. & P. Marler. 1984. Interspecific differences in the response of female birds to song repertoires. *Zeitschrift für Tierpsychologie* **66**:128–142.

Searcy, W.A. & P. Marler. 1987. Response of sparrows to songs of deaf and isolation-reared males: Further evidence for innate auditory templates. *Developmental Psychobiology* **20**:509–520.

Searcy, W.A., P. Marler & S.S. Peters. 1981b. Species song discrimination in adult female song and swamp sparrows. *Animal Behaviour* **29**:997–1003.

Searcy, W.A., P. Marler & S.S. Peters. 1985. Songs of isolation-reared sparrows function in communication, but are significantly less effective than learned songs. *Behavioral Ecology & Sociobiology* **17**:223–229.

Searcy, W.A., P.D. McArthur, S.S. Peters & P. Marler. 1981c. Response of male song and swamp sparrows to neighbour, stranger, and self songs. *Behaviour* **77**:152–166.

Searcy, W.A. & S. Nowicki. 1999. Functions of song variation in song sparrows. In: *The Design of Animal Communication*, M.D. Hauser & M. Konishi (eds.), pp. 577–595. The MIT Press, Cambridge, MA.

Searcy, W.A. & S. Nowicki. 2000. Male–male competition and female choice in the evolution of vocal signaling. In: *Animal Signals: Signalling and Signal Design in Animal Communication*, Y. Espmark, T. Amundsen & G. Rosenqvist (eds.), pp. 301–315. Tapir Academic Press, Trondheim, Norway.

Searcy, W.A., S. Nowicki & C. Hogan. 2000. Song type variants and aggressive context. *Behavioral Ecology & Sociobiology* **48**:358–363.

Searcy, W.A., S. Nowicki & M. Hughes. 1997. The response of male and female song sparrows to geographic variation in song. *Condor* **99**:651–657.

Searcy, W.A., S. Nowicki, M. Hughes & S. Peters. 2002. Geographic song discrimination in relation to dispersal distances in song sparrows. *American Naturalist* **159**:221–230.

Searcy, W.A., S. Nowicki & S. Peters. 1999. Song types as fundamental units in vocal repertoires. *Animal Behaviour* **58**:37–44.

Searcy, W.A., J. Podos, S. Peters & S. Nowicki. 1995. Discrimination of song types and variants in song sparrows. *Animal Behaviour* **49**:1219–1226.

Searcy, W.A., M.H. Searcy & P. Marler. 1982. The response of song sparrows to acoustically distinct song types. *Behaviour* **80**:70–83.

Searcy, W.A. & K. Yasukawa. 1990. Use of the song repertoire in intersexual and intrasexual contexts by male red-winged blackbirds. *Behavioral Ecology & Sociobiology* **27**:123–128.

Searcy, W.A. & K. Yasukawa. 1995. *Polygyny and Sexual Selection in Red-Winged Blackbirds*. Princeton University Press, Princeton, NJ.

Searcy, W.A. & K. Yasukawa. 1996. Song and female choice. In: *Ecology and Evolution of Acoustic Communication in Birds*, D.E. Kroodsma & E.H. Miller (eds.), pp. 454–473. Cornell University Press, Ithaca, NY.

Seddon, N., J.A. Tobias & A. Alvarez. 2002. Vocal communication in the pale-winged trumpeter (*Psophia leucoptera*): Repertoire, context and functional reference. *Behaviour* **139**:1331–1359.

Seddon, P.J. & Y. van Heezik. 1993. Parent–offspring recognition in the jackass penguin. *Journal of Field Ornithology* **64**:27–31.

Sedgwick, J.A. 2001. Geographic variation in the song of willow flycatchers: Differentiation between *Empidonax traillii adastus* and *E. t. extimus*. *Auk* **118**:366–379.

Seller, T. 1981. Midbrain vocalization centers in birds. *Trends in Neurosciences* **12**:301–303.

Sen, K., F. Theunissen & A. Doupe. 2000. Feature analysis of natutral sounds in the songbird auditory forebrain. *Journal of Neurophysiology* **86**:1445–1458.

Sewall, K., R. Kelsey & T.P. Hahn. 2004. Discrete variants of Evening Grosbeak flight calls. *Condor* **106**:161–165.

Shackelton, S.A. & L. Ratcliffe. 1994. Matched counter-singing signals escalation of aggression in black-capped chickadees (*Parus atricapillus*). *Ethology* **97**:310–316.

Shapiro, A.M. 1976. Seasonal polyphenism. *Evolutionary Biology* **9**:259–333.

Shaywitz, B.A., S.E. Shaywitz, K.R. Pugh, R.T. Constable, P. Skudlarski, R.K. Fulbright, R.A. Bronen, J.M. Fletcher, D.P. Shankweiler, L. Katz & et al. 1995. Sex differences in the functional organization of the brain for language. *Nature* **373**:607–609.

Shiovitz, K.A. 1975. The process of species-specific song recognition in the indigo bunting (*Passerina cyanea*) and its relationship to the organization of avian acoustic behaviour. *Behaviour* **55**:128–179.

Shiovitz, K.A. & W.L. Thompson. 1970. Geographical variation in song composition of the indigo bunting, *Passerina cyanea*. *Animal Behaviour* **18**:151–158.

Short, L.L., Jr. 1961. Interspecies flocking of birds of montane forest in Oaxaca, Mexico. *Wilson Bulletin* **73**:341–347.

Sibley, C.G. & J.E. Ahlquist. 1990. *Phylogeny and*

Classification of Birds: A Study in Molecular Evolution. Yale University Press, New Haven.

Sibley, C.G. & B.L. Monroe, Jr. 1990. *Distribution and Taxonomy of Birds of the World*. Yale University Press, New Haven, Connecticut.

Sick, H. 1939. Über die Dialektbildung beim Regenruf des Buchfinken. *Journal für Ornithologie* **87**:568–592.

Siegel, J.M. 1995. Phylogeny and the function of REM sleep. *Behavioural Brain Research* **69**:29–34.

Silverstone, J.L. 1989. *Numerical Abilities in the African Grey Parrot: Sequential Numerical Tags*. Senior Honors Thesis, Northwestern University.

Simon, H.A. 1974. How big is a chunk ? *Science* **183**:482–488.

Simpson, H.B. & D.S. Vicario. 1990. Brain pathways for learned and unlearned vocalizations differ in zebra finches. *Journal of Neuroscience* **10**:1541–1556.

Singh, T., M. Basham, E. Nordeen & K. Nordeen. 2000. Early sensory and harmonal experience modulate age-related changes in NR2B mRNA within a forebrain region controlling avian vocal learning. *Journal of Neurobiology* **44**:82–94.

Sinnott, J.M., M.B. Sachs & R.D. Hienz. 1980. Aspects of frequency discrimination in passerine birds and pigeons. *Journal of Comparative and Physiological Psychology* **94**:401–415.

Slabbekoorn, H., S. de Kort & C. ten Cate. 1999. Comparative analysis of perch-coo vocalizations in *Streptopelia* doves. *Auk* **116**:737–748.

Slabbekoorn, H., J. Ellers & T.B. Smith. 2002. Bird song and sound transmission: The benefits of reverberations. *Condor* **104**:564–573.

Slabbekoorn, H.A. Jesse & D.A. Bell. 2003. Microgeographic song variation in island populations of white-crowned sparrow (*Zonotrichia leucophrys nutalli*): innovation through recombination. *Behaviour* **140**:947–963.

Slabbekoorn, H. & M. Peet. 2003. Birds sing at a higher pitch in urban noise. *Nature* **424**:267.

Slabbekoorn, H. & T.B. Smith. 2000. Does bill size polymorphism affect courtship song characteristics in the African finch *Pyrenestes ostrinus* ? *Biological Journal of the Linnean Society* **71**:737–753.

Slabbekoorn, H. & T.B. Smith. 2002a. Habitat-dependent song divergence in the little greenbul: An analysis of environmental selection pressures on acoustic signals. *Evolution* **56**:1849–1858.

Slabbekoorn, H. & T.B. Smith. 2002b. Bird song, ecology, and speciation. *Philosophical Transactions of the Royal Society of London, B* **357**:493–503.

Slabbekoorn, H. & C. ten Cate. 1997. Stronger territorial responses to frequency modulated coos in collared doves. *Animal Behaviour* **54**:955–965.

Slabbekoorn, H. & C. ten Cate. 1998a. Multiple parameters in the territorial coo of the collared dove: Interactions and meaning. *Behaviour* **135**:879–895.

Slabbekoorn, H. & C. ten Cate. 1998b. Perceptual tuning to frequency characteristics of territorial signals in collared doves. *Animal Behaviour* **56**:847–857.

Slabbekoorn, H. & C. ten Cate. 1999. Collared doves responses to playback: Slaves to the rhythm. *Ethology* **105**:377–391.

Slater, P.J.B. 1981. Chaffinch song repertoires: Observations, experiments and a discussion of their significance. *Ethology* **56**:1–24.

Slater, P.J.B. 1986. The cultural transmission of bird song. *Trends in Ecology & Evolution* **1**:94–97.

Slater, P.J.B. 1989. Bird song learning: Causes and consequences. *Ethology Ecology & Evolution* **1**:19–46.

Slater, P.J.B., F.A. Clements & D.J. Goodfellow. 1984. Local and regional variations in chaffinch song and the question of dialects. *Behaviour* 88:76–97.

Slater, P.J.B., L.A. Eales & N.S. Clayton. 1988. Song learning in zebra finches (*Taeniopygia guttata*): Progress and prospects. *Advances in the Study of Behavior* **18**:1–34.

Slater, P.J.B & S.A. Ince. 1982. Song development in chaffinches: What is learnt and when? *Ibis* **124**:21–26.

Slater, P.J.B., A. Jones & C. ten Cate. 1993. Can lack of experience delay the end of the sensitive phase for song learning? *Netherlands Journal of Zoology* **43**:80–90.

Smith, A.T. & M.M. Peacock. 1990. Conspecific attraction and the determination of metapopulation colonization rates. *Conservation Biology* **4**:320–323.

Smith, H.G. & R. Montgomerie. 1991. Nestling American robins compete with siblings by begging. *Behavioral Ecology & Sociobiology* **29**:307–312.

Smith, S.T. 1972. Communication and other social behavior in *Parus carolinensis. Nuttall Ornithological Club* **11**:1–125.

Smith, T.B., R.K. Wayne, D.J. Girman & M.W. Bruford. 1997. A role for ecotones in generating rainforest biodiversity. *Science* **276**:1855–1857.

Smith, V.A., A.P. King & M.J. West. 2000. A role of her own: Female cowbirds (*Molothrus ater*) influence male song development and the outcome of song learning. *Animal Behaviour* **60**:599–609.

Smith, W.J. & A.M. Smith. 1996. Information about behaviour provided by Louisiana waterthrush, *Seiurus motacilla* (*Parulinae*), songs. *Animal Behaviour* **51**:785–799.

Snow, B.K. 1973. The behavior and ecology of hermit hummingbirds in the Kanaku Mountains, Guyana. *Wilson Bulletin* **85**:163–177.

Snow, B.K. 1977. Territorial behavior and courtship of the male three-wattled bellbird. *Auk* **94**:623–645.

Snow, D.W. 1958. *A Study of Blackbirds*. Allen and Unwin, London.

Snow, D. 1968. The singing assemblies of little hermits. *The Living Bird* 7:47–55.

Soha, J.A. & P. Marler. 2000. A species-specific acoustic cue for selective song learning in the white-crowned sparrow. *Animal Behaviour* **60**:297–306.

Soha, J.A. & P. Marler. 2001. Cues for early discrimination of conspecific song in the white-crowned sparrow (*Zonotrichia leucophyrys*). *Ethology* **107**:813–826.

Soha, J.A., T. Shimizu & A.J. Doupe. 1996. Development of the catecholaminergic innervation of the song system of the male zebra finch. *Journal of Neurobiology* **29**:473–489.

Sohrabji, F., E.J. Nordeen & K.W. Nordeen. 1990. Selective impairment of song learning following lesions of a forebrain nucleus in the juvenile zebra finch. *Behavioral & Neural Biology* **53**:51–63.

Soma, K., R. Bindra, J. Gee, J. Wingfield & B. Schlinger. 1999. Androgen-metabolizing enzymes show region-specific changes across the breeding season in the brain of a wild songbird. *Journal of Neurobiology* **41**:176–188.

Sonnenschein, E. & H.-U. Reyer. 1983. Mate-guarding and other functions of antiphonal duets in the slate coloured boubou (*Laniarius funebris*). *Zeitschrift für Tierpsychologie* **63**:112–140.

Sorensen, M.D. & R.B. Payne. 2001. A single ancient origin of brood parasitism in African finches: Implications for host–parasite coevolution. *Evolution* **55**:2550–2567.

Sorjonen, J. 1986. Factors affecting the structure of song and the singing behavior of some northern European passerine birds. *Behaviour* **98**:286–304.

Sorjonen, J. & J. Merila. 2000. Response of male bluethroats *Luscinia svecica* to song playback: Evidence of territorial function of song and song flights. *Ornis Fennica* 77:43–47.

Sossinka, R. & J. Bohner. 1980. Song types in the zebra finch (*Poephila guttata castanotis*). *Zeitschrift für Tierpsychologie* **53**:123–132.

Sparling, D.W. 1979. Evidence for vocal learning in prairie grouse. *Wilson Bulletin* **91**:618–621.

Spector, D.A. 1992. Wood-warbler song systems: A review of paruline singing behaviors. *Current Ornithology* **9**:199–238.

Spector, D.A., L.K. McKim & D.E. Kroodsma. 1989. Yellow warblers are able to learn songs and situations in which to use them. *Animal Behaviour* **38**:723–725.

Specter, M. 2001. Rethinking the brain. *The New Yorker* **July 23**:42–53.

Spielberg, S. 1977. *Close Encounters of the Third Kind*, pp. 279. Dell, New York.

Staicer, C.A. 1996a. Acoustical features of song categories of the Adelaide's warbler (*Dendroica adelaidae*). *Auk* **113**:771–783.

Staicer, C.A. 1996b. Honest advertisement of pairing status: Evidence from a tropical resident wood-warbler. *Animal Behaviour* **51**:375–390.

Staicer, C.A., D.A. Spector & A.G. Horn. 1996. The dawn chorus and other diel patterns in acoustic signals. In: *Ecology and Evolution of Acoustic Communication in Bird*, D.E. Kroodsma & E.H. Miller (eds.), pp. 426–453. Cornell University Press, Ithaca, NY.

Stamps, J., A. Clark, P. Arrowood & B. Kus. 1989. Begging behavior in budgerigars. *Ethology* **81**:177–192.

Stebbins, W.C. 1970. Studies of hearing and hearing loss in the monkey. In: *Animal Psychophysics: The Design and Conduct of Sensory Experiments*, W.C. Stebbins (ed.), pp. 41–66. Appleton, New York.

Steele, K.M. 2000. Arousal and mood factors in the "Mozart effect". *Perceptual and Motor Skills* **91**:188–190.

Steele, K.M., K.E. Bass & M.D. Crook. 1999. The mystery of the Mozart effect: Failure to replicate. *Psychological Science* **10**:366–369.

Stefanski, R.A. & J.B. Falls. 1972a. A study of distress calls of song, swamp, and white-throated sparrows (Aves: *Fringillidae*). I. Intraspecific responses and functions. *Canadian Journal of Zoology* **50**:1501–1512.

Stefanski, R.A. & J.B. Falls. 1972b. A study of distress calls of song, swamp, and white-throated sparrows (Aves: *Fringillidae*). II. Interspecific responses and properties used in recognition. *Canadian Journal of Zoology* **50**:1513–1525.

Stein, R.C. 1963. Isolating mechanisms between populations of Traill's flycatchers. *Proceedings of the American Philosophical Society* **107**:22–50.

Stein, R.C. 1968. Modulation in bird sound. *Auk* **94**:229–243.

Stephens, J. 2000. No place for cheap trills. Inside the costly world of songbird breeding. *Asiaweek.com Magazine.* Volume 29, no. 24.

Stevens, E.F. 1991. Flamingo breeding: The role of group displays. *Zoo Biology* **10**:53–64.

Stevens, E.F. & C. Pickett. 1994. Managing the social environments of flamingos for reproductive success. *Zoo Biology* **13**:501–507.

Stevens, K.N. 1972. The quantal nature of speech: Evidence from articulatory–acoustic data. In: *Human Communication: A Unified View*, E.E. David & P.B. Denes (eds.), pp. 51–66. McGraw-Hill, New York.

Stiles, G.F. 1982. Aggressive and courtship displays of the male Anna's hummingbird. *Condor* **84**:208–225.

Stocker, R.C. 1998. The role of artificial neural networks in the analysis of ultrasonic bat calls: A case study. *Complexity International* **5**:1–13.

Stoddard, P.K., M.D. Beecher, S.E. Campbell & C.L. Horning. 1992. Song type matching in the Song Sparrow. *Canadian Journal of Zoology* **70**:1440–1444.

Stoddard, P.K., M.D. Beecher & M.S. Willis. 1988. Response of territorial male song sparrows to song types and variations. *Behavioural Ecology & Sociobiology* **22**:125–130.

Stokes, A.W. 1961. Voice and social behavior of the chukar partridge. *Condor* **63**:111–127.

Stokes, A.W. 1967. Behavior of the bobwhite, *Colinus virginianus*. *Auk* **84**:1–33.

Stokes, A.W. 1971. Parental and courtship feeding in red jungle fowl. *Auk* **88**:21–29.

Stokes, A.W. & H.W. Williams. 1971. Courtship feeding in gallinaceous birds. *Auk* **88**:543–559.

Stokes, A.W. & H.W. Williams. 1972. Courtship feeding in gallinaceous birds. *Auk* **89**:177–180.

Stone, E. 2000. Separating the noise from the noise: A finding in support of the "niche hypothesis," that birds are influenced by human-induced noise in natural habitats. *Anthrozoös* **13**:225–231.

Storer, R.W. 1965. The color phases of the western grebe. *Living Bird* **4**:59–63.

Striedter, G.F. 1994. The vocal control pathways in budgerigars differ from those in songbirds. *Journal of Comparative Neurology* **343**:35–56.

Striedter, G.F. & E.T. Vu. 1997. Bilateral feedback projections to the forebrain in the premotor network for singing in zebra finches. *Journal of Neurobiology* **34**:27–40.

Stripling, R., A. Kruse & D. Clayton. 2001. Development of song responses in the zebra finch caudomedial neostriatum: Role of genomic and electrophysiological activities. *Journal of Neurobiology* **48**:163–180.

Stripling, R., S.F. Volman & D.F. Clayton. 1997. Response modulation in the zebra finch neostriatum: Relationship to nuclear gene regulation. *Journal of Neuroscience* **17**:3883–3893.

Sullivan, K.A. 1984. Information exploitation by downy woodpeckers in mixed-species flocks. *Behaviour* **91**:294–311.

Sumner, E.L., Jr. 1935. A life history study of the California quail with recommendations for its conservation and management. *California Fish and Game* **21**:167–342.

Suter, R., A. Tolles, M. Nottebohm & F. Nottebohm. 1990. Bilateral LMAN lesions in adult male canaries affect song in different ways with different latencies. *Society for Neuroscience (Abstracts)* **16**:1249.

Suthers, H.B. 1982. Starling mimics human speech. *Birdwatcher's Digest* **2**:37–39.

Suthers, R.A. 1990. Contributions to birdsong from the left and right sides of the intact syrinx. *Nature* **347**:473–477.

Suthers, R.A. 1992. Lateralization of sound production and motor action on the left and right sides of the syrinx during bird song. *14th International Congress on Acoustics*:I1–5.

Suthers, R.A. 1994. Variable asymmetry and resonance in the avian vocal tract: A structural basis for individually distinct vocalizations. *Journal of Comparative Physiology* **175**:457–466.

Suthers, R.A. 1997. Peripheral control and lateralization of birdsong. *Journal of Neurobiology* **33**:632–652.

Suthers, R.A. 1999a. The motor basis of vocal performance in songbirds. In: *The Design of Animal Communication*, M.D. Hauser & M. Konishi (eds.). pp. 37–62. MIT Press, Cambridge, MA.

Suthers, R.A. 1999b. Peripheral mechanisms for singing: Motor strategies for vocal diversity. In: *Proceedings of 22nd International Ornithological Congress, Durban*, N.J. Adams &R.H. Slotow (eds.), pp. 491–508. BirdLife South Africa, Johannesburg.

Suthers, R.A. 2001. Peripheral vocal mechanisms in birds: Are songbirds special? *Netherlands Journal of Zoology* **51**:217–242.

Suthers, R.A. & F. Goller. 1996. Respiratory and syringeal dynamics of song production in northern cardinals. In: *Nervous Systems and Behaviour. Proceedings of the 4th International Congress of Neuroethology*, M. Burrows, T. Matheson, P. Newland & H. Schuppe (eds.), p. 333. Georg Thieme Verlag, Stuttgart.

Suthers, R.A. & F. Goller. 1997. Motor correlates of vocal diversity in songbirds. In: *Current Ornithology 14*, V. Nolan Jr, E. Ketterson & C.F. Thompson (eds.), pp. 235–288. Plenum Press, New York.

Suthers, R.A. & F. Goller. 1998a. Ontogeny of song lateralization in juvenile northern cardinals. *Society for Neuroscience (Abstracts)* **24**:1187.

Suthers, R.A. & F. Goller. 1998b. Respiratory-syringeal motor coordination during song learning in northern cardinals. *Fifth International Congress of Neuroethology (Abstracts)*No. 298.

Suthers, R.A., F. Goller & R.S. Hartley. 1994. Motor dynamics of song production by mimic thrushes. *Journal of Neurobiology* **25**:917–936.

Suthers, R.A., F. Goller & R.S. Hartley. 1996. Motor stereotypy and diversity in songs of mimic thrushes. *Journal of Neurobiology* **30**:231–245.

Suthers, R.A., F. Goller & C. Pytte. 1999. The neuromuscular control of birdsong. *Philosophical Transactions of the Royal Society of London, B* **354**:927–939.

Suthers, R.A., F. Goller & J.M. Wild. 2002. Somatosensory feedback modulates the respiratory motor program of crystallized birdsong. *Proceedings of the National Academy of Sciences USA* **99**:5680–5685.

Suthers, R.A. & D.H. Hector. 1982. Mechanism for the production of echolocating clicks by the grey swiftlet, *Collocalia spodiopygia*. *Journal of Comparative Physiology A* **148**:457–470.

Suthers, R.A. & D.H. Hector. 1985. The physiology of vocalization by the echolocating oilbird, *Steatornis caripensis*. *Journal of Comparative Physiology A* **156**:243–266.

Suthers, R.A. & D.H. Hector. 1988. Individual variation in vocal tract resonance may assist oilbirds in recognizing echoes of their own sonar clicks. In: *Animal Sonar: Processes and Performance*, P.E. Nachtigall & P.W.B. Moore (eds.), pp. 87–91. Plenum Press, New York.

Suthers, R.A. & D. Margoliash. 2002. Motor control of birdsong. *Current Opinion in Neurobiology* **12**:684–690.

Suthers, R.A., E.M. Vallet & M. Kreutzer. 2001. Bilateral song production in domestic canaries. *6th International Congress of Neuroethology (Abstracts)*, 125. Bonn, Germany.

Suthers, R.A. & J.M. Wild. 2000. Real-time modulation of the syringeal motor program in response to externally imposed respiratory perturbations in adult songbirds. *Society for Neuroscience (Abstracts)* **26**:269.8.

Sutter, M.L. & D. Margoliash. 1994. Global synchronous response to autogenous song in zebra finch HVc. *Journal of Neurophysiology* **72**:2105–2123.

Svensson, B.W. 1987. Structure and vocalizations of display flights in the broad-billed sandpiper, *Limicola falcinellus*. *Ornis Scandinavica* **18**:47–52.

Swanson, L. 2000a. Cerebral hemisphere regulation of motivated behavior. *Brain Research* **886**:113–164.

Swanson, L. 2000b. What is the brain? *Trends in Neurosciences* **23**:519–527.

Swarth, H.S. 1934. The bird fauna of the Galápagos Islands in relation to species formation. *Biological Reviews* **9**:213–234.

Székely, T., C.K. Catchpole, A. DeVoogd, Z. Marchl & T.J. DeVoogd. 1996. Evolutionary changes in a song control area of the brain (HVC) are associated with evolutionary changes in song repertoire among European warblers (*Sylviidae*). *Proceedings of the Royal Society of London, B* **263**:607–610.

Szymczak, J.T., H.-W. Helb & W. Kaiser. 1993. Electrophysiological and behavioral correlates of sleep in the blackbird (*Turdus merula*). *Physiology and Behavior* **53**:1201–1210.

Takasaka, T. & C.A. Smith. 1971. The structure and innervation of the pigeon's basilar papilla. *J Ultrastruct Res* **35**:20–65.

Taoka, M., T. Sato, T. Kamada & H. Okumura. 1989. Sexual dimorphism of chatter-calls and vocal sex recognition in Leach's storm-petrels (*Oceanodroma leucorhoa*). *Auk* **106**:498–501.

Tardif, B. & J.L. Des Granges. 1998. Correspondence between bird and plant hotspots of the St. Lawrence River and influence of scale on their location. *Biological Conservation* **84**:53–63.

Tauber, M.J., C.A. Tauber & S. Masaki. 1986. *Seasonal Adaptations of Insects*. Oxford University Press, New York.

Tchernichovski, O., F. Nottebohm, C.E. Ho, B. Pesaran & P.P. Mitra. 2000. A procedure for an automated measurement of song similarity. *Animal Behaviour* **59**:1167–1176.

Tchernichovski, O., P.P. Mitra, T. Lints & F. Nottebohm. 2001. Dynamics of the vocal imitation process: How a zebra finch learns its song. *Science* **291**:2564–2569.

ten Cate, C. 1986. Listening behavior and song learning in zebra finches. *Animal Behaviour* **34**:1267–1268.

ten Cate, C. 1989. Behavioural development: Toward understanding processes. In: *Perspectives in Ethology, Vol. 8. Whither Ethology?* P.P.G. Bateson & P.H. Klopfer (eds.), pp. 243–269. Plenum Press, New York.

ten Cate, C. 1992. Coo-types in the collared dove *Streptopelia decaocto*: One theme, distinctive variations. *Bioacoustics* **4**:161–183.

ten Cate, C. 2000. How learning mechanisms might affect evolutionary processes. *Trends in Ecology & Evolution* **15**:179–181.

ten Cate, C. & M.R. Ballintijn. 1996. Dove coos and flashed lights: Interruptibility of "song" in a nonsongbird. *Journal of Comparative Psychology* **110**:267–275.

ten Cate, C., H. Slabbekoorn & M.R. Ballintijn. 2002. Birdsong and male-male competition: Causes and consequences of vocal variability in the collared dove (*Streptopelia decaocto*). *Advances in the Study of Behavior* **31**:31–75.

ten Cate, C. & D.R. Vos. 1999. Sexual imprinting and evolutionary processes in birds: A reassessment. *Advances in the Study of Behaviour* **28**:1–31.

ten Cate, C., D.R. Vos & N. Mann. 1993. Sexual imprinting and song learning: Two of one kind. *Netherlands Journal of Zoology* **43**:34–45.

Terry, A.M.R., P.K. McGregor & T.M. Peake. 2001. A comparison of some techniques used to assess vocal individuality. *Bioacoustics* **11**:169–188.

Theunissen, F.E. & A.J. Doupe. 1998. Temporal and spectral sensitivity of complex auditory neurons in the nucleus HVc of male zebra finches. *Journal of Neuroscience* **18**:3786–3802.

Thielcke, G. 1965. Die Ontogenese der Bettellaute von Garten- und Waldbaum-läufer (*Certhia brachydactyla* Brehm und *C. familiaris* L.). *Zoologischer Anzeiger* **174**:237–241.

Thielcke, G. 1970a. Lernen von Gesang als möglicher Schrittmacher der Evolution. *Zeitschrift für Zoologische Systematik und Evolutionsforschung* **8**:309–320.

Thielcke, G. 1970b. Die sozialen Funktionen der Vogelstimmen. *Vogelwarte* **25**:204–229.

Thielcke, G. 1976. *Bird Sounds.* University of Michigan Press, Ann Arbor, MI.

Thomas, J.W., E.E. Forsman, J.B. Lint, E.C. Meslow, B.N. Noon & J.A. Verner. 1990. *Conservation Strategy for the Northern Spotted Owl.* Department of Agriculture and US Department of Interior, Portland, OR.

Thompson, D.H. & J.T. Jr. Emlen. 1968. Parent–chick individual recognition in the Adélie penguin. *Antarctic Journal of the United States* **3**:132.

Thompson, N.S. 1982. A comparison of cawing in the European carrion crow (*Corvus corone*) and the American common crow (*Corvus brachyrhynchos*). *Behaviour* **80**:106–117.

Thompson, W.F., E.G. Schellenberg & G. Husain. 2001. Arousal, mood, and the Mozart effect. *Psychological Science* **12**:248–251.

Thompson, W.L. 1970. Song variation in a population of indigo buntings. *Auk* **87**:58–71.

Thompson, W.L. 1976. Vocalizations of the lazuli bunting. *Condor* **78**:195–207.

Thompson, W.L. & J.O. Rice. 1970. Calls of the indigo bunting, *Passerina cyanea. Zeitschrift für Tierpsychologie* **27**:35–46.

Thorpe, W.H. 1950. The concepts of learning and their relation to those of instinct. *Symposium of the Society for Experimental Biology on Physiological Mechanisms in Animal Behaviour,* **4**:387–408.

Thorpe, W.H. 1951. The learning of abilities of birds. *Ibis* **93**:1–52 & 252–296.

Thorpe, W.H. 1954. The process of song-learning in the chaffinch as studied by means of the sound spectrograph. *Nature* **173**:465.

Thorpe, W.H. 1955. Comments on '*The Bird Fancyer's Delight* together with notes on imitation in the sub-song of the chaffinch. *Ibis* **97**:247–251.

Thorpe, W.H. 1958. The learning of song patterns by birds, with especial reference to the song of the chaffinch *Fringilla coelebs. Ibis* **100**:535–570.

Thorpe, W.H. 1959. Talking birds and the mode of action of the vocal apparatus of birds. *Proceedings of the Zoological Society of London* **132**:441–455.

Thorpe, W.H. 1961. *Bird-Song: The Biology of Vocal Communication and Expression in Birds.* Cambridge University Press, Cambridge.

Thorpe, W.H. & P.M. Pilcher. 1958. The nature and characteristics of subsong. *British Birds* **51**:509–514.

Ticehurst, C.B. 1938. *A systematic review of the genus*

Phylloscopus. Trustees of the British Museum, London.

Tinbergen, N. 1951. *The Study of Instinct*. Clarendon Press, Oxford.

Tinbergen, N. 1963. [On aims and methods of ethology.] *Zeitschrift für Tierpsychologie* **20**:410–433.

Tintle, R.F. 1982. Relationship of multiple nest building to female mate choice in long-billed marsh wrens, *Cistothorus palustris*. Unpublished M.A. Thesis, State University of New York at Stony Brook, Stony Brook.

Titze, I.R. (ed.). 1994. *Principles of Voice Production*. Prentice Hall, Englewood Cliffs NJ.

Todt, D. 1970. Gesang und gesangliche Korrespondenz der Amsel. *Naturwissenschaften* **57**:61–66.

Todt, D. 1971. Äquivalente und konvalente gesangliche Reaktion einer extrem regelmäig singenden Nachtigall (*Luscinia megarhynchos* L.). *Zeitschrift für vergleichende Physiologie* **71**:262–285.

Todt, D. 1974. Zur Bedeutung der 'richtigen' Syntax auditiver Muster für deren vokale Beantwortung durch Amseln. *Zeitschrift für Naturforschung* **29**:157–160.

Todt, D. 1975. [Social learning of vocal patterns and models of their applications in grey parrots.] *Zeitschrift für Tierpsychologie* **39**:178–188.

Todt, D. 1981. On functions of vocal matching: Effect of counter-replies on song post choice and singing. *Zeitschrift für Tierpsychologie* **57**:73–93.

Todt, D. & J. Böhner. 1994. Former experience can modify social selectivity during song learning in the nightingale (*Luscinia megarhynchos*). *Ethology* **97**:169–176.

Todt, D., J. Cirillo, N. Geberzahn & F. Schleuss. 2001. The role of hierarchy levels in vocal imitations of birds. *Cybernetics and Systems* **32**:257–283.

Todt, D. & N. Geberzahn. 2003. Age dependent effects of song exposure: Song crystallization sets a boundary between fast and delayed vocal imitation. *Animal Behaviour*. In press.

Todt, D. & H. Hultsch. 1996. Acquisition and performance of song repertoires: Ways of coping with diversity and versatility. In: *Ecology and Evolution of Acoustic Communication in Birds*, D.E. Kroodsma & E.H. Miller (eds.), pp. 79–96. Cornell Univerity Press, Ithaca, NY.

Todt, D. & H. Hultsch. 1998. Hierarchical learning, development and representation of song. In: *Animal Cognition in Nature: The Convergence of Psychology and Biology in Laboratory and Field*, R.P. Balda, I.M. Pepperberg & A.C. Kamil (eds.), pp. 275–303. Academic Press, San Diego, CA.

Todt, D. & H. Hultsch. 1999. How nightingales develop their vocal competence. In: *Proceedings of the 22nd International Ornithological Congress, Durban*, N.J. Adams & R.H. Slotow (eds.), pp. 193–215. BirdLife South Africa, Johannesburg.

Todt, D., H. Hultsch & D. Heike. 1979. [Conditions affecting song acquisition in nightingales (*Luscinia megarhynchos* L.).] *Zeitschrift für Tierpsychologie* **51**:23–35.

Todt, D. & M. Naguib. 2000. Vocal interactions in birds: The use of song as a model in communication. *Advances in the Study of Behavior* **29**:247–296.

Tomback, D.F. & M.C. Baker. 1984. Assortative mating by white-crowned sparrows at song dialect boundaries. *Animal Behaviour* **32**:465–469.

Trainer, J.M. 1989. Cultural evolution in song dialects of yellow-rumped caciques in Panama. *Ethology* **80**:190–204.

Trainer, J.M. & D.B. McDonald. 1993. Vocal repertoire of the long-tailed manakin and its relation to male–male cooperation. *Condor* **95**:769–781.

Trainer, J.M., D.B. McDonald & W.A. Learn. 2002. The development of coordinated singing in cooperatively displaying long-tailed manakins. *Behavioral Ecology* **13**:65–69.

Tramontin, A.D. & E.A. Brenowitz. 1999. A field study of seasonal neuronal incorporation into the song control system of a songbird that lacks adult song learning. *Journal of Neurobiology* **40**:316–326.

Tramontin, A.D. & E.A. Brenowitz. 2000. Seasonal plasticity in the adult brain. *Trends in Neurosciences* **23**:251–258.

Tramontin, A.D., V.N. Hartman & E.A. Brenowitz. 2000. Breeding conditions induce rapid and sequential growth in adult avian song control circuits: A model of seasonal plasticity in the brain. *Journal of Neuroscience* **20**:854–861.

Transportation Canada. 2002. *Wildlife Control Procedures Manual; Active Management Using Dispersal techniques (TP11500)*. Canada Civil Aviation, pp. 1–55.

Trehub, S.E. & L.J. Trainor. 1998. Singing to infants: Lullabies and play songs. *Advances in Infancy Research* **12**:43–77.

Tretzel, E. 1965. Imitation und Variation von

Schäterpfiffen durch Haubenlerchen (*Galerida c. cristata* [L.]). Ein Beispiel für spezielle Spottmotiv-Prädisposition. *Zeitschrift für Tierpsychologie* **22**:784–809.

Tretzel, E. 1967. Imitation und Transposition menschlicher Pfiffe durch Amseln (*Turdus m. merula* L.). Ein weiterer Nachweis relativen Lernens und akustischer Abstraktion bei Vögeln. *Zeitschrift für Tierpsychologie* **24**:37–161.

Trivers, R.L. 1974. Parent–offspring conflict. *American Zoologist* **14**:249–264.

Troyer, T. & A. Doupe. 2000. An associational model of birdsong sensorimotor learning I. Efference copy and the learning of song syllables. *Journal of Neurophysiology* **84**:1204–1223.

Tschanz, B. 1965. Beobachtungen und Experimente zur Entstehung der 'persönlichen' Beziehung zwischen Jungvogel und Eltern bei Trottellummen. *Verhandlungen Schweizerische Naturforschende Gesellschaft* **1964**:211–216.

Tschanz, B. 1968. Trottellummen. *Zeitschrift für Tierpsychologie Supplement* **4**:1–103.

Tubaro, P.L. & B. Mahler. 1998. Acoustic frequencies and body mass in New World doves. *Condor* **100**:54–61.

Tuttel, J. 2000. Hanenkraaiwedstrijd "De Schellekraaiers"; Het fenomeen Hanenkraaiwedstrijd nader verklaard. *Volkskunde en Traditie*, November 5, 2000.

Tyack, P. 1983. Male competition in large groups of wintering humpback whales. *Behaviour* **83**:132–154.

Tyler, W.M. 1916. The call-notes of some nocturnal migrating birds. *Auk* **33**:132–141.

Vallet, E., I. Beme & M. Kreutzer. 1998. Two-note syllables in canary songs elicit high levels of sexual display. *Animal Behaviour* **55**:291–297.

Vallet, E. & M. Kreutzer. 1995. Female canaries are sexually responsive to special song phrases. *Animal Behaviour* **49**:1603–1610.

VanBuskirk, J. 1997. Independent evolution of song structure and note structure in American wood warblers. *Proceedings of the Royal Society of London, B* **264**:755–761.

Van Essen, D. 1997. A tension-based theory of morphogenesis and compact wiring in the central nervous system. *Nature* **385**:313–318.

Van Pragg, H., G. Kempermann & F. Gage. 1999. Running increases cell proliferation and neurogenesis in the adult mouse dentate gyrus. *Nature Neuroscience* **2**:266–270.

Vates, G.E., B.M. Broome, C.V. Mello & F. Nottebohm. 1996. Auditory pathways of caudal telencephalon and their relation to the song system of adult male zebra finches. *Journal of Comparative Neurology* **366**:613–642.

Vates, G.E. & F. Nottebohm. 1995. Feedback circuitry within a song-learning pathway. *Proceedings of the National Academy of Sciences USA* **92**:5139–5143.

Vates, G.E., D.S. Vicario & F. Nottebohm. 1997. Reafferent thalamo "cortical" loops in the song system of oscine songbirds. *Journal of Comparative Neurology* **380**:275–290.

Veen, T., T. Borge, S.C. Griffith, G.-P. Sætre, S. Bures, L. Gustafsson & B.C. Sheldon. 2001. Hybridization and adaptive mate choice in flycatchers. *Nature* **411**:45–50.

Veenman, C.L., J.M. Wild & A. Reiner. 1995. Organization of the avian "corticostriatal" projection system: A retrograde and anterograde pathway tracing study in pigeons. *Journal of Comparative Neurology* **354**:87–126.

Vehrencamp, S.L. 2000. Handicap, index and conventional signal elements of bird song. In: *Signalling and Signal Design in Animal Communication*, Y. Epsmark, T. Amundsen & G. Rosenqvist (eds.), pp. 277–300. Tapir Academic Press, Trondheim, Norway.

Vehrencamp, S.L. 2001. Is song-type matching a conventional signal of aggressive intentions? *Proceedings of the Royal Society of London, B* **268**:1637–1642.

Vehrencamp, S.L., J.W. Bradbury & R.M. Gibson. 1989. The energetic cost of display in male sage grouse. *Animal Behaviour* **38**:885–898.

Vehrencamp, S.L., A.F.Ritter, M. Keever & J.W. Bradbury. 2003. Responses to playback of local vs. distant contact calls in the orange-fronted conure, *Aratinga canicularis*. *Ethology* **109**:37–54.

Ventura, D.F. & E. Takase1994. Ultraviolet color discrimination in the hummingbird. *Investigative Ophthalmology and Visual Science* **35**:2168.

Venuto, V., V. Ferraiuolo, L. Bottoni & R. Massa. 2001. Distress call in six species of African *Poicephalus* parrots. *Ethology, Ecology & Evolution* **13**:49–68.

Vermeij, G.J. 1988. The evolutionary success of passerines: A question of semantics? *Systematic Zoology* **37**:69–71.

Verner, J. 1976. Complex song repertoire of male

long-billed marsh wrens in eastern Washington. *Living Bird* **14**:263–300.

Verner, J. 1976. Complex song repertoire of male long-billed marsh wrens in eastern Washington. *Living Bird* **14**:263–300.

Vicario, D.S. 1991. Neural mechanisms of vocal production in songbirds. *Current Opinion in Neurobiology* **1**:595–600.

Vicario, D.S. 1994. Motor mechanisms relevant to auditory–vocal interactions in songbirds. *Brain, Behavior & Evolution* **44**:265–278.

Vicario, D.S. & F. Nottebohm. 1988. Organization of the zebra finch song control system: I. Representation of syringeal muscles in the hypoglossal nucleus. *Journal of Comparative Neurology* **271**:346–354.

Vielliard, J. 1994. *Catalogo dos troquilideos do Museu de Biologia Mello Leitao* MBML, Santa Teresa.

Viemeister, N.F. & C.J. Plack. 1993. Time analysis. In: *Human Psychophysics*, W.A. Yost, A.N. Popper & R.R. Fay (eds.), pp. 116–154. Springer-Verlag, New York.

Vince, M.A. 1964. Social facilitation of hatching in the bobwhite quail. *Animal Behaviour* **12**:531–534.

Vogel, S. 1988. *Life's Devices.* Princeton University Press, Princeton.

von Haartman, L. 1953. Was reizt den Trauerfliegenschnäpper (*Muscicapa hypoleuca*) zu füttern? *Vogelwelt* **16**:157–164.

von Haartmann, L. 1957. Adaptations in hole-nesting birds. *Evolution* **11**:284–347.

von Haartmann, L. & H. Löhrl. 1950. Die Lautäusserungen des Trauer- und Halsband-fliegenschnäppers, *Muscicapa h. hypoleuca* (Pall.) und *M. a. albicollis* Temminck. *Ornis Fennica* **27**:85–97.

Vu, E.T., M.E. Mazurek & Y.-C. Kuo. 1994. Identification of a forebrain motor programming network for the learned song of zebra finches. *Journal of Neuroscience* **14**:6924–6934.

Wada, K., H. Sakaguchi, E.D. Jarvis and M. Hagiwara 2004. Differential expression of glutamate receptors in avian neural pathways for learned voralization. Journal of Comparative Neurology in press.

Waddington, C.H. 1957. *The Strategy of the Genes.* Allen and Unwin, London.

Wade, J. 2001. Zebra finch sexual differentiation: The aromatization hypothesis revisited. *Microscopy Research and Technique* **54**:354–363.

Wade J., D.A. Swender, T.L. McElhinny 1999. Sexual differentiation of the zebra finch song system parallels genetic, not gonadal, sex. Hormones and Behavior **36**:141–152.

Wake, D. & A. Larson. 1987. Multidimensional analysis of an evolving lineage. *Science* **238**:42–48.

Wallace, W.T. 1994. Memory for music. *Journal of Experimental Psychology: Learning, Cognition, and Memory* **20**:1471–1485.

Wallschläger, D. 1980. Correlation of song frequency and body weight in passerine birds. *Experientia* **36**:69–94.

Wang, X., M.M. Merzenich, R. Beitel & C.E. Schreiner. 1995. Representation of a species-specific vocalization in the primary auditory cortex of the common marmoset: Temporal and spectral characteristics. *Journal of Neurophysiology* **74**:2685–2706.

Wanker, R. & J. Fischer. 2001. Intra- and interindividual variation in the contact calls of spectacled parrotlets (*Forpus conspicillatus*). *Behaviour* **138**:709–726.

Ward, B., E. Nordeen & K. Nordeen. 1998. Individual variation in neuron number predicts differences in the propensity for avian vocal imitation. *Proceedings of the National Academy of Sciences USA* **95**:1277–1282.

Ward, B., E. Nordeen & K. Nordeen. 2001. Anatomical and ontogenetic factors producing variation in HVc neuron number in zebra finches. *Brain Research* **904**:318–326.

Ward, P. & A. Zahavi. 1973. The importance of certain assemblages of birds as "information-centres" for food-finding. *Ibis* **115**:517–534.

Ward, W.D. & E.M. Burns 1982. Absolute pitch. In: *The Psychology of Music* D. Deutsch (ed.). pp. 431–451. Academic Press, New York.

Warren, D.K., D.K. Patterson & I.M. Pepperberg. 1996. Mechanisms of American English vowel production in a grey parrot (*Psittacus erithacus*). *Auk* **113**:41–58.

Warring, R.H. 1972. *Handbook of Noise and Vibration Control* (2nd ed.). Trade and Technical Press Ltd., Morden, England.

Waser, M.S. & P. Marler. 1977. Song learning in canaries. *Journal of Comparative and Physiological Psychology* **91**:1–7.

Waser, P.M. & C.H. Brown. 1986. Habitat acoustics and primate communication. *American Journal of Primatology* **10**:135–154.

Waser, P.M. & M.S. Waser. 1977. [Experimental

studies of primate vocalization – specializations for long-distance propagation.] *Zeitschrift Fur Tierpsychologie* **43**:239–263.

Wasserman, F.E. & J.A. Cigliano. 1991. Song output and stimulation of the female in white-throated sparrows. *Behavioural Ecology & Sociobiology* **29**:55–59.

Weary, D.M., J.R. Krebs, R. Eddyshaw, P.K. McGregor & A. Horn. 1988. Decline in song output by great tits: Exhaustion or motivation? *Animal Behaviour* **36**:1241–1244.

Weary, D.M., R.E. Lemon & S. Perreault. 1994. Different responses to different song types in American redstarts. *Auk* **111**:730–734.

Weary, D.M. & R.G. Weisman 1991. Operant discrimination of frequency and frequency ratios in the black-capped chickadee (*Parus atricapillus*). *Journal of Comparative Psychology* **105**: 253–259.

Weir. R. 1962. *Language in the Crib.* Mouton, The Hague.

Wells, K.D. & T.L. Taigen. 1986. The effect of social interactions on calling energetics in the gray treefrog (*Hyla versicolor*). *Behavioral Ecology & Sociobiology* **19**:9–18.

Welty, J.C. & L. Baptista. 1990. *The Life of Birds* (4th ed.). Harcourt Brace Jovanovich College Publishers, New York.

Wenzel, J.W. 1993. Application of the biogenetic law to behavioral ontogeny: A test using nest architecture in paper wasps. *Journal of Evolutionary Biology* **6**:229–247.

Werker, J.F. & P.J. McLeod. 1989. Infant preference for both male and female infant-directed talk: A developmental study of attentional and affective responsiveness. *Canadian Journal of Psychology* **43**:230–246.

Wertheim, N. 1969. The amusias. In: *Handbook of Clinical Neurology, Vol. 4*, P.J. Vinken & G.W. Bruyn (eds.), pp. 195–206. North Holland, Amsterdam.

West, M.J. & A.P. King. 1988b. Female visual displays affect the development of male song in the cowbird. *Nature* **334**:244–246.

West, M.J. & A.P. King 1988a. Vocalizations of juvenile cowbirds (Molothrus aterater) evoke copulatory responses from females. Developmental Psychobiology **21**:543–552.

West, M.J. & A.P. King. 1990. Mozart's starling. *American Scientist* **78**:106–114.

West, M.J., A.P. King & D.H. Eastzer. 1981. Validating the female bioassay of cowbird song: Relating differences in song potency to mating success. *Animal Behaviour* **29**:490–501.

West, M.J., A.P. King, D.H. Eastzer & J.E.R. Staddon. 1979. A bioassay of isolate cowbird song. *Journal of Comparative and Physiological Psychology* **93**:124–133.

West, M.J., A.N. Stroud & A.P. King. 1983. Mimicry of the human voice by European starlings: The role of social interaction. *Wilson Bulletin* **95**:635–640.

West-Eberhard, M.J. 1989. Phenotypic plasticity and the origins of diversity. *Annual Reviews in Ecology and Systematics* **20**:249–278.

West-Eberhard, M.J. 2003. *Developmental Plasticity and Evolution.* Oxford University Press, Oxford.

Westneat, M.W., J.H. Long, W. Hoese & S. Nowicki. 1993. Kinematics of birdsong: Functional correlation of cranial movements and acoustic features in sparrows. *Journal of Experimental Biology* **182**:147–171.

Whaling, C.S., D.A. Nelson & P. Marler. 1995. Testosterone-induced shortening of the storage phase of song development in birds interferes with vocal learning. *Developmental Psychobiology* **28**:367–376.

White, S.A., F.S. Livingston & R. Mooney. 1999. Androgens modulate NMDA receptor-mediated EPSCs in the zebra finch song system. *Journal of Neurophysiology* **82**:2221–2234.

White, S.J. & R.E.C. White. 1970. Individual voice production in gannets. *Behaviour* **37**:40–54.

White, S.J., R.E.C. White & W.H. Thorpe. 1970. Acoustic basis for individual recognition in the gannet. *Nature* **225**:1156–1158.

Whitney, B.M., J.F. Pacheco, D.R.C. Buzetti & R. Parrini. 2000. Systematic revision and biogeography of the *Herpsilochmus pileatus* complex, with description of a new species from northeastern Brazil. *Auk* **117**:869–891.

Whitney, C.L. 1989. Geographical variation in wood thrush song: A comparison of samples recorded in New York and South Carolina, USA. *Behaviour* **111**:49–60.

Whitney, C.L. & J. Miller. 1983. Song matching in the wood thrush (*Hylocichla mustelina*): A function of song dissimilarity. *Animal Behaviour* **31**:457–461.

Whitney, C.L. & J. Miller. 1987. Distribution and variability of song types in the wood thrush. *Behaviour* **103**:49–67.

Whitney, O., K. Soderstrom & F. Johnson. 2000. Posttranscriptional regulation of zenk expression associated with zebra finch vocal development. *Brain Research. Molecular Brain Research* **80**:279–290.

Wickler, W. & E. Sonnenschein. 1989. Ontogeny of song in captive duet-singing slate-colored boubous (*Laniarius funebris*): A study in birdsong epigenesis. *Behaviour* **111**:220–233.

Wier, C.C., W. Jesteadt & D.M. Green. 1977. Frequency discrimination as a function of frequency and sensation level. *Journal of the Acoustical Society of America* **61**:177–184.

Wild, J.M. 1994. Visual and somatosensory inputs to the avian song system via nucleus uvaeformis (Uva) and a comparison with the projections of a similar thalamic nucleus in a nonsongbird, *Columba livia*. *Journal of Comparative Neurology* **349**:512–535.

Wild, J.M. 1997. Neural pathways for the control of birdsong production. *Journal of Neurobiology* **33**:653–670.

Wild, J.M., F. Goller & R.A. Suthers. 1998. Inspiratory muscle activity during birdsong. *Journal of Neurobiology* **36**:441–453.

Wild, J.M., H.J. Karten & B.J. Frost. 1993. Connections of the auditory forebrain in the pigeon (*Columba livia*). *Journal of Comparative Neurology* **337**:32–62.

Wild, J.M., D. Li & C. Eagleton. 1997a. Projections of the dorsomedial nucleus of the intercollicular complex (DM) in relation to respiratory–vocal nuclei in the brainstem of pigeon (*Columba livia*) and zebra finch (*Taeniopygia guttata*). *Journal of Comparative Neurology* **377**:392–413.

Wild, J.M., H. Reinke & S.M. Farabaugh. 1997b. A non-thalamic pathway contributes to a whole body map in the brain of the budgerigar. *Brain Research* **755**:137–141.

Wild, J.M., M.N. Williams & R.A. Suthers. 2000. Neural pathways for bilateral vocal control in songbirds. *Journal of Comparative Neurology* **423**:413–426.

Wilden, I., H. Herzel, G. Peters & G. Tembrock. 1998. Subharmonics, biphonation, and deterministic chaos in mammal vocalisation. *Bioacoustics* **9**:171–196.

Wiley, R.H. 1971. Song groups in a singing assembly of little hermits. *Condor* **73**:28–35.

Wiley, R.H. 1980. Multispecies antbird societies in lowland forests of Suriname and Ecuador: Stable membership and foraging differences. *Journal of Zoology (London)* **191**:127–145.

Wiley, R.H. 1991. Associations of song properties with habitats for territorial oscine birds of eastern North America. *American Naturalist* **138**:973–993.

Wiley, R.H. 1994. Errors, exaggeration, and deception in animal communication. In: *Behavioral Mechanisms in Evolutionary Ecology*, L.A. Real (ed.), pp. 157–189. University of Chicago Press, Chicago.

Wiley, R.H. & D.G. Richards. 1978. Physical constraints on acoustic communication in the atmosphere: Implications for the evolution of animal vocalizations. *Behavioral Ecology Sociobiology* **3**:69–94.

Wiley, R.H. & D.G. Richards. 1982. Adaptations for acoustic communication in birds Sound transmission and signal detection. In: *Acoustic Communication in Birds, Vol. 1. Production, Perception, and Design Features of Sounds*, D.E. Kroodsma & E.H. Miller (eds.), pp. 132–181. Academic Press, New York.

Wilkinson, R. 1980. Calls of nestling chaffinches, *Fringilla coelebs*: The use of two sound sources. *Zeitschrift für Tierpsychologie* **54**:346–356.

Williams, H. 1989. Multiple representations and auditory–motor interactions in the avian song system. *Annals of the New York Academy of Sciences* **563**:148–164.

Williams, H. 1990. Models for song learning in the zebra finch: Fathers or others? *Animal Behaviour* **39**:745–757.

Williams, H. 2001. Choreography of song, dance and beak movements in the zebra finch (*Taeniopygia guttata*). *Journal of Experimental Biology* **204**:3497–3506.

Williams, H. & J.R. McKibben, 1992. Changes in stereotyped central motor patterns controlling vocalization are induced by peripheral nerve injury. *Behavioral & Neural Biology* **57**:67–78.

Williams, H. & N. Mehta. 1999. Changes in adult zebra finch song require a forebrain nucleus that is not necessary for song production. *Journal of Neurobiology* **39**:14–28.

Williams, H. & F. Nottebohm. 1985. Auditory responses in avian vocal motor neurons: A motor theory for song perception in birds. *Science* **229**:279–282.

Williams, H. & K. Staples. 1992. Syllable chunking in zebra finch (*Taeniopygia guttata*) song. *Journal of Compartive Psychology* **106**:278–286.

Williams, H. & D. Vicario. 1993. Temporal patterning of song production: Participation of nucleus uvaeformis of the thalamus. *Journal of Neurobiology* **39**:14–28.

Williams, H.W. 1969. Vocal behavior of adult California quail. *Auk* **86**:631–659.

Williams, H.W., A.W. Stokes & J.C. Wallen. 1968. The food call and display of the bobwhite quail (*Colinus virginianus*). *Auk* **85**:464–476.

Wilson, M.A., B.L. McNaughton. 1994. Reactivation of hippocampal ensemble memories during sleep. *Science* **265**:676–679.

Wilson, N.S.H. 2001. Song structure and syllable repertoires in the European sedge warbler, *Acrocephalus schoenobaenus*. Master of Science, Zoology and Entomology, University of Pretoria, Pretoria.

Wilson, P.L., M.C. Towner & S.L. Vehrencamp. 2000. Survival and song-type sharing in a sedentary subspecies of the song sparrow. *Condor* **102**:355–363.

Wilson, P.L. & S.L. Vehrencamp. 2001. A test of the deceptive mimicry hypothesis in song-sharing song sparrows. *Animal Behaviour* **62**:1197–1205.

Wilson, S.A.K. 1914. An experimental research into the anatomy and physiology of the corpus striatum. *Brain* **36**:427–492.

Wimberger, P.H. & A. de Queiroz. 1996. Comparing behavioral and morphological characters as indicators of phylogeny. In: *Phylogenies and the Comparative Method in Animal Behavior,* E.P. Martins (ed.), pp. 166–205. Oxford University Press, Oxford.

Wise, K.K., M.R. Conover & F.F. Knowlton. 1999. Responses of coyotes to avian distress calls: Testing the startle-predator and predator-attraction hypotheses. *Behaviour* **136**:935–949.

Wisniewski, A.B. & S.H. Hulse. 1997. Auditory scene analysis in European starlings (*Sturnus vulgaris*): Discrimination of song segments, their segregation from multiple and reversed conspecific songs, and evidence for conspecific categorization. *Journal of Comparative Psychology* **111**:337–350.

Witelson, S., I. Glezer & D. Kigar. 1995. Women have greater density of neurons in posterior temporal cortex. *Journal of Neuroscience* **15**:3418–3428.

Witkin, S.R. 1977. The importance of directional sound radiation in avian vocalization. *Condor* **79**:490–493.

Witmer, M.C., D.J. Mountjoy & L. Elliot. 1997. Cedar waxwing (*Bombycilla cedrorum*). In: *The Birds of North America, No. 309*, A. Poole & F. Gill (eds.) pp 1–28. The Academy of Natural Sciences, Philadelphia, Pennsylvania; The American Ornithologists' Union, Washington, D. C.

Wolffgramm, J. & D. Todt. 1982. Pattern and time specificity in vocal responses of blackbirds *Turdus merula* L. *Behaviour* **81**:264–286.

Wolnik, G. 1994. Song canaries in the U.S. *Finch and Canary World* Volume 1, no. 1.

Woolley, S.M.N. & E.W. Rubel. 1997. Bengalese finches, *Lonchura striata-domestica,* depend upon auditory feedback for the maintenance of adult song. *Journal of Neuroscience* **17**: 6380–6390.

Wooley, S.M.N. & E.W. Rubel. 2002. Vocal memory and learning in adult bengalese finches with regenerated hair cells. *Journal of Neuroscience* **22**: 7774–7787.

Worley, P.F., B.A. Christy, Y. Nakabeppu, R.V. Bhat, A.J. Cole & J.M. Baraban. 1991. Constitutive expression of zif268 in neocortex is regulated by synaptic activity. *Proceedings of the National Academy of Sciences USA* **88**:5106–5110.

Wright, S. 1931. Evolution in Mendelian populations. *Genetics* **16**:97–159.

Wright, T.F. 1996. Regional dialects in the contact calls of a parrot. *Proceedings of the Royal Society London, B* **263**:867–872.

Wright, T.F. 1997. *Vocal communication in the yellow-naped Amazon* (*Amazona auropalliata*). Doctoral dissertation, University of California, San Diego.

Wright, T.F. & M. Dorin. 2001. Pair duets in the yellow-naped Amazon (Psittaciformes: *Amazona auropalliata*): Responses to playbacks of different dialects. *Ethology* **107**:111–124.

Wright, T.F., & G.S. Wilkinson. 2001. Population genetic structure and vocal dialects in an Amazon parrot. *Proceedings of the Royal Society of London, B* **268**:609–616.

Wyndham, E. 1980. Diurnal cycle, behaviour, and social organization of the budgerigar (*Melopsittacus undulatus*). *Emu* **80**:25–33.

Yamaguchi, A. 1998. A sexually dimorphic learned birdsong in the northern cardinal. *Condor* **100**:504–511.

Yamashita, C. 1987. Field observations and comments on the indigo macaw (*Anodorhynchus leari*), a highly endangered species from northeastern Brazil. *Wilson Bulletin* **99**:280–282.

Yasukawa, K. 1981a. Song and territory defense in the red-winged blackbird (*Agelaius phoeniceus*). *Auk* **98**:185–187.

Yasukawa, K. 1981b. Song repertoires in the red-winged blackbird (*Agelaius phoeniceus*): A test of the Beau Geste hypothesis. *Animal Behaviour* **29**: 114–125.

Yasukawa, K., J.L. Blank, & C.B. Patterson. 1980.

Song repertoires and sexual selection in the red-winged blackbird. *Behavioral Ecology & Sociobiology* **7**:233–238.

Yasukawa, K. & W.A. Searcy. 1982. Aggression in female red-winged blackbirds: a strategy to ensure male parental investment. *Behavioural Ecology & Sociobiology* **11**:12–17.

Yoneda, T. & K. Okanoya. 1991. Ontogeny of sexually dimorphic distance calls in Bengalese finches (*Lonchura domestica*). *Journal of Ethology* **9**:41–46.

Yu, A.C. & D. Margoliash. 1996. Temporal hierarchical control of singing in birds. *Science* **273**:1871–1875.

Zahavi, A. 1975. Mate selection: a selection for A handicap. *Journal of Theoretical Biology* **53**:205–214.

Zahavi, A. 1977. The cost of honesty (further remarks on the handicap principle). *Journal of Theoretical Biology* **67**:603–605.

Zahavi, A. 1982. The theory of signal selection and some of its implications. In: *Proceedings of the International Symposium of Biological Evolution*, V.P. Delfino (ed.), pp. 305–327. Adriatica Edtrice, in Ban.

Zahavi, A. & A. Zahavi. 1997. *The Handicap Principle*. Oxford University Press, Oxford.

Zann, R. 1975. Inter- and intraspecific variation in the calls of three species of grassfinches of the subgenus *Poephila* (Gould) (*Estrildidae*). *Zeitschrift für Tierpsychologie* **39**:85–125.

Zann, R. 1990. Song and call learning in wild zebra finches in south-east Australia. *Animal Behaviour* **40**:818–828.

Zann, R. 1997. Vocal learning in wild and domesticated zebra finches: Signature cues for kin recognition of epiphenomena?. In: *Social Influences on Vocal Development*, C.T. Snowdon & M. Hausberger (eds.), pp. 85–97. Cambridge University Press, Cambridge.

Zeier, H. & H.J. Karten. 1971. The archistriatum of the pigeon: Organization of afferent and efferent connections. *Brain Research* **31**:313–326.

Zink, R.M. & R.C. Blackwell. 1996. Patterns of allozyme, mitochondrial DNA, and morphometric variation in four sparrow genera. *Auk* **113**:59–67.

Zink, R.M. & M.C. McKitrick. 1995. The debate over species concepts and its implications for ornithology. *Auk* **112**:701–719.

Zollinger, S.A. & R.A. Suthers. 2004. Motor mechanisms of a vocal mimic: implications for birdsong production. *Proceedings of the Royal Society of London, B* **271**:483–491.

Glossary

Acoustic structure – the total set of temporal, spectral, and amplitude characteristics of a sound.

Acoustic space – a multi-dimensional space available for sound signals in a given community. An ecological concept used in analyzing the distinctiveness of songs and evaluating the acoustic overlap among bird species in a multi-variate way.

Action-based learning – learning during the sensorimotor phase in which decisions to keep or discard learned songs or song elements in the adult repertoire are influenced by social interactions with conspecifics.

Active space – the distance from a singing bird over which its voice is detectable and recognizable by conspecifics. The active space is influenced by the song amplitude, receiver sensitivity, attenuation and degradation during transmission, and interference by ambient noise.

Adaptation – a trait that has evolved in a particular way because of a selective advantage in the local environment.

Adaptive radiation – diversification followed by a series of speciation events within a clade during a relatively short period of evolutionary time, often hypothesized to be associated with sudden range expansion and invasion of new niches.

Advertisement song – a stereotyped complex vocalization that advertises an individual's presence and such qualities as species, sex, strength, and motivation.

Age-limited learner – a bird species in which song development is affected by learning up to a certain age after which no new songs can be acquired. Equivalent to 'closed-ended' learner.

Air sac – thin membranous structures that act as a bellows to ventilate the lungs without participating directly in gas exchange. Most bird species have nine air sacs. The lungs themselves are relatively small, located dorsally, and do not change volume during breathing.

Allopatry – when the geographic distribution of two species or other taxonomic units is non-overlapping.

Altruistic behavior – behavior benefiting others at the expense of the actor.

Ambient noise – the sum of all sounds present in a natural environment.

Amniotes – taxonomic group comprising all vertebrates that posses an amnion, including birds, reptiles and mammals. The amnion is a thin membranous sac enclosing the developing embryo.

Amusia – inability to recognize or play music.

Aphasia – inability to use or understand language due to brain damage.

Area X – a brain area involved in song production, part of the anterior vocal pathway, so-named because its function was not clear at the time of discovery.

Artificial selection – biased reproductive success during domestication intended by breeders to make certain traits less or more common or exaggerated.

Assessment signal – a signal that permits reliable assessment by a receiver of some quality of the sender (see honest signal).

Assortative mating – non-random mate choice.

Audience effect – a difference in signal production in response to a particular stimulus dependent on whether or not another individual is present during the interaction.

Auditory scene analysis – the detection, recognition, and spatial perception of relevant acoustic signals in the continuous and complex mixtures of sounds occurring in natural environments.

Auditory template – a representation of a song pattern present in the brain, acquired through learning after hearing a song, used by a bird to shape its own singing by matching the pattern.

Autoradiography – a photographic procedure for identifying cells and tissues that have taken up radioactive material.

BOS – abbreviation for 'bird's own song', used in the context of a neural representation or behavioral response specific to playback of a bird's own song.

Brain stem cells – populations of cells that retain the embryonic ability to divide and differentiate as specialized brain cells.

Bronchi – paired airways connecting the trachea with the lungs.

Call – vocalizations, *usually* short, simple one-note

sounds, used in a wide variety of contexts, including gathering at food sources, predator alarm, aggression, and courtship.

Carrier frequency – in frequency- or amplitude-modulated tones, the frequency that is modulated to produce amplitude and spectral variation in the form of side-bands above and below the carrier (see overtone and harmonic).

Catecholamines – a class of neurotransmitters including dopamine and epinephrine.

Chorus – a relatively uncoordinated vocal performance by two or more birds.

Clade – a set of species descended from the same ancestral species.

Cladistic – based on the branching patterns by which species have diverged from common ancestors.

Closed-ended learner – a song learning species in which acquisition of a new song is restricted to a period early in life.

Cochlea – fluid-filled cavity of the inner ear where sound vibrations are converted into neuronal activity via the sensory epithelium of the basilar papilla.

Cocktail-party effect – the ability to hear and understand acoustic signals under noisy conditions such as at a cocktail party.

Cognition – collective term for the mental processes by which animals acquire, process, store, recover, and integrate information about the environment and plan future actions.

Columella – the ossicle in the bird's middle ear cavity conducting sound vibrations from the ear drum to the oval window of the cochlea.

Communication – immediate or eventual modification of the behavior or thinking of another on receiving signals from others, such as songs or calls.

Complex tone – sound with energy in a broad band of multiple frequencies.

Conspecific – of the same species.

Conventional signal – a signal with arbitrary acoustic structure not obviously and immediately related to sender characteristics such as strength or condition; the meaning is thus based on 'convention'.

Convergent evolution – evolution of similar trait characteristics in distantly- or un-related taxa.

Cranial – related to the bird's skull

Cretaceous-Tertiary boundary – historical time zone from the end of the Cretaceous to the Tertiary period, during which many vertebrate taxa went extinct and after which new taxonomic radiations arose.

Critical periods – limited time windows critical to development, often with heightened sensitivity to particular types of stimulation. Hence, also known as sensitive periods.

Cross-correlation – statistical procedure to calculate similarity between two sounds by digitally overlaying sonograms in a way that maximizes the overlap in the sound energy distribution pattern. Correlation coefficients varying between 0 for no similarity and 1 for complete similarity. Cross-correlations can also be performed on the amplitude wave patterns or the power spectrograms of two sounds.

Crystallized song – a stabilized stage of song development, preceded in the case of learned song by the more unstable stages of subsong and plastic song.

Dawn chorus – a period of high singing activity by many bird species, typically starting about an hour before sunrise.

Dialect – songs of a set of individuals at one location that sing similarly, and differently from others nearby.

Dissonant – strident non-harmonious sounds in birdsongs and music.

Divergent evolution – when subgroups of one species develop different trait characteristics over evolutionary time.

DLM – lateral portion of the dorsal thalamus, a brain area involved in song learning and part of the so-called anterior vocal pathway.

Duet – a coordinated vocal performance by pairs of birds of the same species, vocalizing synchronously, alternately or in sequence.

Echo – a reflection of sound that arrives at a receiver later than the primary sound.

Ecological niche (see niche)

Electromyogram – graphic representation of electrical currents in muscle fibers that reflect activation through innervating motor neurons.

Epigenetic effects – genetic effects that are not directly attributable to a gene, or when offspring development is affected by traits of the mother.

Event detector – computer software programmed to scan a digital sound file and identify acoustic events meeting specific criteria of frequency, amplitude, and phrasing.

Eventual variety – when a bird repeats successive songs that are the same, before switching to a new series with a different song type; in contrast to 'immediate variety'.

Evolutionary constraint – limitations on change in traits within or across generations due to genotypic characteristics inherent in a species' phylogenetic history.

Evolutionary stasis – period on an evolutionary time scale in which a species or trait remains unchanged.

Fitness – relative genetic contribution to subsequent generations: an individual that has many relatives in the next generation as through a long life or a high rate of reproduction has a high fitness.

Formant – spectral peaks (concentrations of energy in the sound spectrum, as with vowels in speech) resulting from vocal tract resonances, distinct from harmonics.

Fourier transform – a mathematical integration technique that characterizes phase and amplitude at the frequencies present over time in a sound recording, used to calculate power spectrograms and sonograms.

Frequency – a sound property expressed as the rate of cycles per second, or hertz, roughly equivalent to perceived pitch.

Frequency bandwidth – the range between the lowest and highest pitch of a bird's song or call.

Fundamental frequency – the lowest frequency at which the underlying sound-producing structure vibrates, often accompanied by integer-multiple harmonics. The term fundamental frequency is applicable only to sounds with a periodic spectrum. Also referred to as the first harmonic.

Gene – a DNA segment coding for RNA and protein synthesis.

Gene flow – genetic transfer from one organism or population to another through migration and subsequent interbreeding.

Genetic predisposition – a genetically-preordained preference for acting in certain ways, and responding more strongly to certain stimuli.

Genotype – the heritable information encoded in the DNA of a living organism, which operates as a set of instructions for building and maintaining the phenotype in interaction with environmental factors.

Glial cells – neural tissue cells interspersed between signal-transmitting cells (neurons) that provide insulation and support.

Habituation – a stimulus-specific decrease in the strength of response to repeated presentations of the same stimulus.

Harmonic – in complex tones, the spectral sound energy that is produced simultaneously with, and at *integer multiples* of, the fundamental frequency. Harmonics are a subset of possible complex tone frequencies (or overtones) above the fundamental frequency (see fundamental frequency and overtone).

Hemispheric dominance – a predominant focus of certain neurobiological activity in one side of the brain.

Heterodyne – having alternating currents of two different frequencies that combine to produce two new frequencies.

Heterospecific – of another species.

Honest signal – a signal directly reflecting certain qualities of the sender, which cannot be produced by individuals lacking these qualities.

HVC – a songbird brain area involved in song production as part of the posterior vocal pathway, originally an abbreviation for hyperstriatum ventrale pars caudale, now an acronym for 'high vocal center'.

Hypoglossal nerve – the twelfth cranial which contains the axons of motor neurons from the hindbrain to the syrinx, controlling the activity of the syringeal muscles.

Imitation – an acquired feature of song or other behavior copied from another individual, in contrast to an invention.

Immediate-early genes – genes stimulated early in a cascade of gene activation, sometimes initiated through neuron activity, to produce proteins which regulate mRNA synthesis.

Immediate variety – singing with successive songs that are different, in contrast to 'eventual variety'.

Imprinting – a form of rapid, relatively irreversible learning early in life when parents or their surrogates are taken later as a standard for choosing social partners (social imprinting) or mating preferences (sexual imprinting).

Improvisation – an imitated song feature that is transformed by the singer into a self-generated variant; in contrast to invention.

Infrasound – very low frequency sound that falls outside the audible range for humans. Biologically significant infrasound ranges from 0.1 Hz to about 20 Hz.

Invention – a novel, individually-distinctive, self-generated song feature that is not copied from another bird in contrast with imitations and improvisations.

Ipsilateral – on the same side, as opposed to 'contralateral,' on the other side.
Isolate song – song sung by an individual raised in isolation without adult conspecifics.
Key adaptation – a novel trait variant that after emerging in the evolution of a species allows range expansion through invasion of new niches, which in turn may lead to adaptive radiation.
Larynx – the sound producing organ of non-avian tetrapods (anurans, reptiles, and mammals), a cartilaginous and muscular hollow structure connecting the trachea to the oral cavity and containing the vocal cords.
Lateralization – when a brain function is localized on either the left or the right side.
Lek – location used for 'lekking,' where males aggregate and compete for access to females.
lMAN – abbreviation for the lateral part of the magnocellular nucleus of the anterior nidopallium, a brain area involved in song production and part of the so-called anterior vocal pathway.
Lombard effect – a louder voice in response to high noise levels.
Magnetic resonance imaging (see MRI)
Mimicry – restricted here to imitation of the song of another species
Mini-breath – brief, shallow inhalation between successive notes or syllables within a song that is produced while exhaling
Mobbing – conspicuous individual or group harassment of a predator.
Monogamy – when an individual has only one mate.
Monologue speech – active practice of vocal communication without a partner
Motivation-structural rules – hypothetical rules that relate the motivation of an individual to certain acoustic features of its vocalizations.
Motor constraint – mechanistic restrictions on vocal production imposed by the syrinx, the supra-syringeal vocal tract, the musculature or their innervation.
Motor control – neural input to muscles that governs how they act to determine bodily functions, activity and behavior.
MRI – magnetic resonance imaging based on nuclear magnetic resonance. A magnetic field temporarily alters the electrical status of molecules within the living body, making it possible to trace them in three-dimensional images.
mRNA – messenger RNA is a form of ribonucleic acid that serves as a template for protein synthesis, as a transcript of DNA.
Mutual exclusivity – an assumption in early acquisition and comprehension of language, that each object has only one label, sometimes resulting in confusion, as when an object is given multiple labels.
Natural selection – differential survival and/or reproduction among individuals or traits, that differ genetically, leading to genetic change in subsequent generations.
Neural predisposition (see genetic predisposition)
Neurotransmitter – chemicals released at a synapse facilitating transmission of impulses to muscles and other neurons.
Niche – the functional position or role of a species in an ecosystem.
Nidicolous – when chicks remain for a period within the nest after hatching
Nidifugous – when newly-hatched chicks quickly leave the nest following the mother.
Noisy sound – sound of variable acoustic structure with energy continuously distributed or simultaneously present at multiple irregularly spaced frequencies.
Note complex – a coherent cluster of notes making up a song phrase that is not repeated consecutively, as occurs in a trill.
Octave – interval between two sound frequencies, where the second one is twice the first.
Open-ended learner – a bird species in which the ability to acquire new songs is not restricted to one early period in life, but is retained beyond the first or second breeding season.
Operant conditioning – the experimental shaping of behavior by reinforcement of responses without the stimulus that would elicit the behavior under natural conditions.
Oscines – 'true' songbirds, a suborder of the order of Passeriformes or 'perching birds' (see Passerines).
Overlapping – when songs of rivals overlap in time during vocal interactions.
Overtone – bands of spectral sound energy produced simultaneously with the fundamental (F_0). Overtones placed as integer multiples of the fundamental are harmonics (F_1, F_2, etc.). (The fundamental frequency is sometimes incorrectly referred to as the first harmonic).
Passerines – 'perching birds' including the suborders of the oscines, the 'true' songbirds and sub-oscines, such as the new world flycatchers.
Perceptual bias – a sensitivity beyond the range of values used in current signaling that can drive signal evolution.

Performance constraint (see motor constraint)

PET – positron emission tomography, an imaging technique tracing emissions from an injected radioactive substance, used for measuring blood flow and metabolism of internal body tissues.

Phenetic – based on phenotypic similarity

Phenotype – the outward physical manifestation of a living organism, which is the outcome of the interplay between the heritable information in the genotype and environmental factors during ontogeny.

Phenotypic plasticity – the potential of a given genotype to react to different environmental stimuli with changes in form, physiological state, or behavior.

Philopatry – tendency for offspring to breed in their natal home range

Phonation – production of sound during respiration.

Phoneme – short, distinctively-different, syllable-like speech sounds, meaningless in themselves, used to compose words in a language.

Phrase – a subdivision within a song, either a repetitive trill or a non-repetitive sequence of different notes, known as a note complex.

Phylogenetic – based on evolutionary relationships.

Phylogeny – the reconstructed genealogy of a species group, usually depicted as a branching pattern reflecting the hypothesized evolutionary relationships between ancestral and extant species.

Plastic song – an intermediate stage in development of learned birdsongs after subsong, where unstable imitations and species typical song patterns first emerge.

Polyandry – when a female has several mates.

Polygamy – when an individual mates with multiple mates.

Polygyny – when a male has several mates.

Positron emission tomography (see PET)

Potential acoustic variation – the range of song characteristics that a species or group could theoretically produce.

Power spectrogram – graphic representation of all or part of a sound depicting the sum of frequencies present on the x-axis and amplitude on the y-axis (also named 'power spectrum').

Preoptic region – a brain area concerned with reproductive behavior in the hypothalamus.

Pure tone – sound with a single or narrow band of frequencies at any one moment.

RA – the robust nucleus of the arcuopallium, a songbird brain area involved in song production, part of the posterior vocal pathway.

Race – a set of individuals or populations with a distinctive phenotype, naturally or through domestication. It is poorly defined and often equivalent to subspecies.

Radiation – the occurrence of multiple speciation events within a clade.

Ranging – assessment of a singer's distance based on signal attenuation and degradation.

Realized acoustic variation – the range of song characteristics actually realized by members of a particular group of birds.

Real time – analysis mode in computer software that generates sound spectrograms in real time, which allows viewing and listening to sounds at the same time at the moment of receiving the analogue or digital sound signal.

Referent – that which a signal stands for or denotes.

Referential signal – a vocal label for an object or event understood as such by others.

REM sleep – rapid eye-movement sleep, associated with brain activity and dreaming.

Repertoire – a set of distinct signals or signal elements used by one individual or group.

Repertoire matching – preferential use of the part of one bird's repertoire shared with another's during counter-singing.

Reproductive isolation – absence of interbreeding between individuals from different taxonomic units.

Reproductive success – the number of offspring in a year or a lifetime.

Resonance – vibration of a structure or cavity at the frequency at which it is most strongly energized.

Respiratory cycle – the rhythmic pattern of inhalation and exhalation with its associated gas exchange in the lungs during breathing.

Respiratory system – the set of muscles and organs used in respiration, including lungs, air sacs, and the upper airways from the air sacs through the branching pattern of bronchi and the trachea to the nares and mouth of a bird.

Respiratory tidal volume – the amount of air that moves through the respiratory system in one cycle of inhalation and exhalation.

Reverberation – the lingering after-effect of the passage of a sound resulting from multiple reflections and refractions, often visible as a smear or a tail on a sonogram.

Roost – a place where birds aggregate to rest or sleep.

Rostral – anterior, or pertaining to the bird's bill.

Selection-based learning (see action-based learning)

Selective attrition – a learning mechanism in which part of the learned song repertoire, accumulated early in life, is discarded, and not represented in the adult repertoire of crystallized songs.

Sensitive phase – one or more periods in life with a heightened readiness to learn (see also critical period).

Sensory bias (see perceptual bias)

Sensory feedback – self-perception of the sensory consequences of an action, as when a bird hears or otherwise senses its own song production.

Sensory-motor phase – a period of vocal practice following song memorization that includes subsong, plastic and crystallized song.

Sexual selection – natural selection related to competition among individuals of the same sex for access to mates of the other sex.

Side-bands – multiple zones of sound energy present above and below a carrier frequency when it is subject to amplitude or frequency modulation. The interval between side bands corresponds to the modulation rate: an increase in rate leads to larger frequency intervals.

Signal – a stimulus emitted by a sender that can provide a receiver with information it can subsequently use in decision-making.

Signal degradation – changes in spectral, temporal and structural characteristics of a signal during transmission through the environment.

Signal redundancy – excess information produced by repetition of all or part of a signal or multiple encoding of its message.

Signal space – a multi-dimensional space available for signals. This is an ecological concept used in analyzing the similarity among different signals and evaluating the overlap within a bird community in a multi-variate way.

Signal-to-noise ratio – signal amplitude relative to the ambient noise level.

Sister species – species in a clade that are phylogenetically closely related.

Sonogram – graphic representation of sound depicting time on the x-axis, frequency on the y-axis, and relative amplitude as a grey-scale.

Song – stereotyped polysyllabic vocalizations, usually relatively long and complex, typically used in territory defense and/or mate attraction.

Song acquisition – memorization of songs heard as a first step in vocal learning.

Songbirds – oscine birds that are members of the order Passeriformes or 'perching birds,' that also includes sub-oscines, which are mostly new world species, like the tyrranid flycatchers.

Song control system – the set of brain nuclei and their neuronal connections associated with the ability to learn new vocalizations.

Song crystallization – a stage when developing song changes from variable rambling sounds to a loud and stereotyped song. It is the culmination of the stages of subsong and plastic song.

Song learning – development of new vocalizations by imitation of others, by improvisation or by invention.

Song matching – when birds with a repertoire produce the same or shared song types in response to a singing neighbor.

Song note – the smallest sound element in the acoustic structure of birdsong. It appears in a sonogram as a continuous sound trace.

Song overproduction – singing more song types early in life than will be retained in the adult repertoire.

Song phrase (see phrase)

Song sharing – when two individuals have one or more similar song types in their repertoire.

Song syllable – a cluster of two or more song notes produced as a coherent unit.

Song template (see auditory template)

Song type – category of songs with very similar acoustic structure, distinct from that of songs of another category.

Speciation – the emergence of a new species.

Species – according to the biological species concept, this is a group of organisms whose members do or can interbreed under natural conditions.

Storage phase – the period between memorization of a song and the first production of an imitation of it.

Sub-oscines – a suborder of the order of Passeriformes or 'perching birds' that includes new world flycatchers and their relatives.

Subsong – the earliest stage in development of learned birdsongs, consisting of soft and highly variable sounds, followed by plastic song and crystallized song.

Subspecies – a taxonomically labeled race.

Sympatry – when geographic distributions overlap and individuals of two species or other taxonomic units live in the same area.

Synapse – the point of contact between neurons where information is transmitted between them by chemical or electrical means.

Syntax – the temporal arrangement of song notes, syllables, and phrases in birds or the construction of words and sentences in human language.

Syrinx – the sound producing organ of birds, located in songbirds where the two bronchi merge with the trachea, composed of ossified rings, membranes, and muscles.

Taxonomy – the naming and assignment of organisms to taxa and study of their relationships.

Temporal summation – a perceptual phenomenon that makes signals of longer duration easier to hear.

Territory – an area defended by one or more individuals against others, usually of the same species.

Testosterone implant – a small hormone-filled plastic tube placed beneath the skin that gradually releases testosterone into the blood.

Tonal – a sound consisting mainly or entirely of pure tones.

Tonotopic – spatial arrangement in the auditory pathway, such that tones are transmitted at different localities or by different cells depending on their frequencies.

Trachea – the tubular structure of bony or cartilaginous rings, connective and muscle tissue that forms an airway connecting the bronchi to the mouth, nose, and larynx. The trachea is separated from the larynx by the slit-like opening of the glottis.

Trill – a phrase type in birdsong consisting of a repetitive sequnce of identical or very similar syllables.

Two-voicing – the capability of songbirds to produce two independent and harmonically unrelated sounds on right and left sides of the synrinx.

Ultrasound – high frequency sound beyond the audible range for humans above about 15 kHz.

Vocalization – a sound produced by a bird's ssyrinx and shaped by the trachea and other parts of the vocal tract.

Vocal learning (see song learning)

Vocal tract – the part of the respiratory system along which sound is transmitted; in birds from the syrinx to the beak.

Warble – tonal sound with periodic fluctuations in pitch.

White-noise – 'Noisy' sound with a frequency spectrum that is continuous and random over a specified frequency band; 'white' noise usually covers the range of human hearing.

Species list

Common name	Scientific name
acorn woodpecker	*Melanerpes tormicivorus*
Acrocephalus warblers	Sylviidae
Aerodramus Asian swiftlets	*Collocalia brevirostris*
African cuckoo	*Cuculus gularis*
African grey parrot	*Psittacus erithacus*
auks	*Alcidae*
alpine accentor	*Prunella collaris*
American crow	*Corvus brachyrhynchos*
American goldfinch	*Carduelis tristis*
American mockingbird	*Mimus polyglottos*
American parulid warblers	(see northern parula)
American redstart	*Setophaga ruticilla*
American robin	*Turdus migratorius*
Andean pygmy-owl	*Glaucidium jardinii*
ant wrens	*Thamnophilidae*
Apolinar's wren	*Cistothorus apolinarae*
aquatic warbler	*Acrocephalus paludicola*
Australian lyrebird	(see superb lyrebird)
Australian magpie	*Gymnorhina tibicen*
Bachman's sparrow	*Aimophila aestivalis*
Bahia antwren	*Herpsilochmus pileatus*
Baltimore oriole	*Icterus galbula*
banded wren	*Campylorhynchus zonatus*
bank swallow	*Riparia riparia*
bare-throated bellbird	*Procnias nudicollis*
barn swallow	*Hirundo rustica*
bay wren	*Thryothorus nigricapillus*
bee-eaters	*Meropidae*
Bengalese finch	*Lonchura striata*
Bewick's wren	*Thryomanes bewickii*
Bicknell's thrush	*Catharus bicknelli*
bitterns	*Ardeidae*
blackbird	(see European blackbird)
blackcap	*Sylvia atricapilla*
black cuckoo	*Cuculus clamosus*
black partridge	*Francolinus francolinus*
black rail	*Laterallus jamaicencis*
black redstart	*Phoenicurus ochruros*
black-bellied seedcracker	*Pyrenestis ostrinus*
black-capped chickadee	*Poecile atricapillus*
black-faced solitaire	*Myadestes melanops*
black-headed grosbeak	*Pheucticus melanocephalus*
black-throated sparrow	*Amphispiza bilineata*
blue cuckoo-shrike	*Coracina azurea*
bluethroat	*Luscinia svecica*
blue-breasted kingfisher	*Halcyon malimbica*
blue-breasted waxbill	*Uraeginthus angolensis*
blue-headed crested flycatcher	*Trochocercus nitens*
blue-winged warbler	*Vermivora pinus*
blue-throated hummingbird	*Lampornis clemenciae*
bluish-slate antshrike	*Thamnomanes schistogynus*
Blyth's leaf warbler	*Phylloscopus reguloides*
bobolink	*Dolichonyx oryzivorus*
bobwhite	(see northern bobwhite)
Bonin petrel	*Pterodroma hypoleuca*
Boran cisticola	*Cisticola bodessa*
boreal chickadee	*Parus hudsonicus*
Brewer's blackbird	*Euphagus cyanocephalus*
bridled tit	*Parus wollweberi*
bristle-nosed barbet	*Gymnobucco peli*
broad-billed sandpiper	*Limicola falcinellus*
brown creeper	*Certhia americana*
brown-backed solitaire	*Myadestes occidentalis*
brown-headed cowbird	*Molothrus ater*
brown thrasher	*Toxostoma rufum*
brubru shrike	*Nilaus afer*
budgerigar	*Melopsittacus undulatus*
bulbuls	*Picnonotidae*
bullfinch	*Pyrrhula pyrrhula*
Bullock's oriole	*Icterus bullockii*
bush tit	*Psaltriparus minimus*
Cabanis' bunting	*Emberiza cabanisi*
cactus finch	*Geospiza scandens*
California condor	*Gymnogyps californianus*
California quail	*Callipepla californica*
canary	(see common canary)
cape sparrow	*Passer melanurus*
cardinal	(see northern cardinal)
Carolina chickadee	*Parus carolinenis*
Carolina wren	*Thryothorus ludovicianus*
Caribbean flamingo	*Phoenicopterus ruber ruber*
Caspian tern	*Sterna caspia*
cedar waxwing	*Bombycilla cedrorum*
chaffinch	*Fringilla coelebs*
chestnut wattle-eye	*Platysteira castanea*
chestnut-sided warbler	*Dendroica pensylvanica*
chickadee	(see black-capped chickadee)
chicken	*Gallus gallus domesticus*

chingolo	(see rufous-collared sparrow)
chipping sparrow	*Spizella passerina*
chocolate-backed kingfisher	*Halcyon badia*
Chopi blackbird	*Gnorimopsar chopi*
chukar partridge	*Alectoris chukar*
cirl bunting	*Emberiza cirlus*
clapper rail	*Rallus longirostris*
Clark's grebe	*Aechmophorus clarkii*
cliff swallow	*Hirundo pyrrhonota*
collared dove	*Streptopelia decaocto*
collared flycatcher	*Muscicapa albicollis*
collared jay	*Cyanolyca viridicyana merida*
common canary	*Serinus canaria*
common cuckoo	*Cuculus canorus*
common grackle	*Quiscalus quiscula*
common murre	*Uria aalge*
common mynal	*Acridotheres tristis*
common nightingale	*Luscinia megarhynchos*
common yellowthroat	*Geothlypis trichas*
Cooper's hawk	*Accipiter cooperii*
cordon-bleu finches	*Uraeginthus bengalus, U. cyanocephalus, U. angolensis*
corn bunting	*Miliaria calandra*
crested pigeon	*Geophaps lophotes*
crested lark	*Galerida cristata*
crested tit	*Parus cristatus*
crow	*Corvus corone*
crowned crane	*Balearica regulorum*
dark-eyed junco	*Junco hyemalis*
dark-rumped or Galapagos petrel	*Pterodroma phaeopygia*
Darwin's finches	*Geospizinae*
day-herons	*Ardeidae*
Dendroica	(see New World wood warblers)
Diederik cuckoo	*Chrysococcyx caprius*
domestic chicken	(see chicken)
domestic pigeon	(see pigeon)
doves	(see p. 000)
downy woodpecker	*Picoides pubescens*
dunlin	*Calidris alpina*
dunnock	(see hedge accentor)
dusky antbird	*Cercomacra tyrannina*
dusky flycatcher	*Empidonax oberholseri*
dusky seaside-sparrow	*Ammodramus maritimus nigrescens*
dusky warbler	*Phylloscopus fuscatus*
eastern phoebe	*Sayornis phoebe*
eastern towhee	*Pipilo erythrophthalmus*
Emei leaf warbler	*Phylloscopus emeiensis*
emerald cuckoo	*Chrysococcyx cupreus*
emu	*Dromaius novaehollandiae*
Eurasian collared dove	*Streptopelia decaocto*
Eurasian jay	*Garrulus glandarius*
Eurasian tree-creeper	*Certhia familiaris*
European bee-eater	*Merops apiaster*
European blackbird	*Turdus merula*
European cuckoo	*Cuculus canorus*
European goldfinch	*Carduelis carduelis*
European nightingale	*Luscinia megarhynchos*
European quail	*Coturnix coturnix coturnix*
European robin	*Erithacus rubecula*
European siskin	*Carduelis spinus*
European starling	*Sturnus vulgaris*
European tree-creeper	*Certhia brachdactyla*
European wren	*Troglodytes troglodytes*
fairy-wrens	*Malurinae*
fawn-breasted bowerbird	*Chlamydera cerviniventris*
field sparrow	*Spizella pusilla*
finches	*Fringillidae*
firecrest	*Regulus ignicapillus*
firefinch	*Lagnostica senegala*
fox sparrow	*Passerella iliaca*
galah	*Eolophus roseicapilla*
Gambel's white-crowned sparrow	*Zonotrichia leucophrys gambelii*
gannet	*Sula bassana*
glossy ibis	*Plegadis falcinellus*
goldcrest	*Regulus regulus*
golden-crowned kinglet	*Regulus satrapa*
golden-crowned sparrow	*Zonotrichia atricapilla*
golden-winged warbler	*Vermivora chrysoptera*
grassfinches	*Spermestidae*
grasshopper sparrow	*Ammodramus savannarum*
grass wrens	*Maluridae*
gray catbird	*Dumetella carolinensis*
great bittern	*Botaurus stellaris*
great reed warbler	*Acrocephalus arundinaceus*
great tit	*Parus major*
greater roadrunner	*Geococcyx californianus*
green hylia	*Hylia prasina*
green violet-ear	*Colibri thalassinus*
greenfinch	*Chloris chloris*
greenish warbler	*Phylloscopus trochiloides*
green-tailed towhee	*Pipilo chlorurus*
grey catbird	*Dumetella carolinensis*
grey warbler	*Gerygone igata*
grouse	*Tetraonidae*
guinea fowl	*Numida meleagris*
gulls	*Laridae*
Harris' sparrow	*Zonotrichia querula*
Hawaiian goose or nene	*Branta sandvicensis*
hawfinch	*Coccothraustes coccothraustes*
hedge accentor	*Prunella modularis*
hen	*Gallus domesticus*

Henslow's sparrow	*Passerherbulus henslowi*
hill mynah	*Gracula religiosa*
house finch	*Carpodacus mexicanus*
house sparrow	*Passer domesticus*
house wren	*Troglodytes troglodytes*
hummingbirds	*Trochilidae*
Indian painted snipe	*Rostratula benghalensis*
indigo bunting	*Passerina cyanea*
Ipswich sparrow	*Passerculus sandwichensis*
ivory-billed woodpecker	*Campephilus principalis*
Japanese quail	*Coturnix cortunix japonica*
junco	(see dark-eyed junco)
junglefowl	*Gallus gallus*
juniper titmouse	*Baeolophus ridgwayi*
king penguin	*Aptenodytes patagonicus*
king rail	*Rallus elegans*
large cactus finch	*Geospiza conirostris*
large ground finch	*Geospiza magnirostris*
large tree finch	*Camarhynchus psittacula*
lark bunting	*Calamospiza melanocorys*
lark sparrow	*Chondestes grammacus*
laugher pigeon	(see pigeon)
Lawrence's goldfinch	*Carduelis lawrencii*
Lawrence's thrush	*Turdus lawrencii*
lazuli bunting	*Passerina amoena*
Leach's storm-petrel	*Oceanodroma leucorhoa*
leadbeater cockatoo	*Cacatua leadbeateri*
least flycatcher	*Empidonax minimus*
lesser bristlebill	*Bleda notata*
lesser goldfinch	*Carduelis psaltria*
little greenbul	*Andropadus virens*
Lincoln's sparrow	*Melospiza lincolnii*
linnet	*Carduelis cannabina*
long-billed cuckoo	*Rhamphomantis megarhynchus*
long-tailed cuckoo (koel)	*Eudynamys taitensis*
long-tailed manakin	*Chiroxiphia linearis*
Louisiana waterthrush	*Seiurus motacilla*
Lühders bush shrike	*Laniarius luchderi*
Lulu's tody-tyrant	*Poecilotriccus luluae*
lyrebird	(see superb lyrebird)
manakins	*Pipridae*
marsh warbler	*Acrocephalus palustris*
marsh wren	*Cistothorus palustris*
medium ground finch	*Geospiza fortis*
Merida wren	*Cistothorus meridae*
merlins	*Falco columbarius*
Mexican chickadee	*Parus sclateri*
Mexican jay	*Aphelocoma ultramarina*
Mimidae	mockingbird family
mistle thrush	*Turdus viscivorus*
mockingbird	(see American mockingbird)
mourning dove	*Zenaida macroura*
murre	(see common murre)
musician wren	*Cyphorhinus arada*
musk duck	*Biziura lobata*
neotropical hummingbirds	*Trochilidae*
neotropical oilbird	*Steatornis caripensis*
New World vultures	*Cathartidae*
New World wood warblers	*Parulinae*
night-herons	*Ardeidae*
nightingale	(see European nightingale)
North American wood warblers	*Parulinae*
northern black flycatcher	*Melaenornis edolioides*
northern bobwhite	*Colinus virginianus*
northern cardinal	*Cardinalis cardinalis*
northern mockingbird	*Mimus polyglottos*
northern parula	*Parula americana*
oak titmouse	*Baeolophus inornatus*
olive-green cameroptera	*Cameroptera chloronota*
olive-sided flycatcher	*Contopus borealis*
orange-fronted conure	*Aratinga canicularis*
orange-chinned parakeet	*Brotogeris jugularis*
oropendolas	(see fig. **10.1**, p. 302)
ovenbird	*Seiurus aurocapillus*
owls	*Strigidae*
pale-winged trumpeter	*Psophia leucoptera*
partridge	*Perdix perdix*
penguins	*Spheniscidae*
phainopepla	*Phainopepla nitens*
pied flycatcher	*Muscicapa hypoleuca*
pigeon	*Columba livia*
pintail snipe	*Gallinago stenura*
piñon jay	*Gymnorhinus cyanocephalus*
Polynesian megapode	*Megapodius pritchardi*
prairie warbler	*Dendroica discolor*
prothonotary warbler	*Protonotaria citrea*
purple sandpiper	*Calidris maritima*
raven	*Corvus corax*
razorbill	*Alca torda*
red avadavat	*Amandava amandava*
red bishop	*Euplectes orix*
red crossbill	*Loxia curvirostra*
redwing	*Turdus iliacus*
red-eyed vireo	*Vireo olivaceus*
red-winged blackbird	*Agelaius phoeniceus*
reed bunting	*Emberiza schoeniclus*
reed warbler	*Acrocephalus scirpaceus*
ringneck parrot	*Platycercus zonarius*
ringdove	*Streptopelia decaocto risoria*
ring-necked pheasant	*Phasianus colchicus*
robin	(see American robin and European robin)
robin-chat	*Cossypha spp.*

rock doves	*Columba livia*
rock ptarmigan	*Lagopus mutus*
rock sandpiper	*Calidris ptilocnemis*
rook	*Corvus frugilegus*
rooster	(see chicken)
ruby-crowned kinglet	*Regulus calendula*
rufous-breasted hermit	*Gluacis hirsuta*
rufous-collared sparrow	*Zonotrichia capensis*
rufous-crowned tody-tyrant	*Poecilotriccus ruficeps*
rufous-sided towhee	*Pipilio erythrophthalmus*
saddleback	*Philesturnus carunculatus*
sage grouse	*Centrocercus urophasianus*
sedge warbler	*Acrocephalus schoenobaenus*
sedge wren	*Cistothorus platensis*
shama	*Copsychus malabaricus*
shining cuckoo	*Chrysococcyx lucidus*
short-toed tree-creeper	*Certhia brachydactyla*
singing honeyeater	*Lichenostomus virescens*
skylark	*Alauda arvensis*
slate-colored boubou	*Laniarius funebris*
small ground finch	*Geospiza fuliginosa*
small tree finch	*Camarhynchus parvulus*
Socorro dove	*Zenaida graysoni*
sombre hummingbird	*Aphanotochroa cirrochloris*
song sparrow	*Melospiza melodia*
song thrush	*Turdus philomelos*
sora (rail)	*Porzana carolina*
spotted bowerbird	*Chlamydera maculata*
spotted owl	*Strix occidentalis*
spotted towhee	*Pipilo maculatus*
starling	(see European starling)
Steller's jay	*Cyanocitta stellari*
stonecha	*Saxicola torquata*
strawberry finch	(see red avadavat)
sub-oscine antbirds	*Thamnophilidae*
sub-oscine flycatchers	*Tyrannidae*
subtropical pygmy-owl	*Glaucidium parkeri*
superb lyrebird	*Menura novaehollandiae*
superb fairy wren	*Malarus cyaneus*
Swainson's thrush	*Catharus ustulatus*
swamp sparrow	*Melospiza georgiana*
Sylvia warblers	*Sylviidae*
tawny owl	*Strix aluco*
terns	*Sternidae*
thick-billed euphonia	*Euphonia laniirostris*
three-wattled bellbird	*Procnias tricarunculata*
thrushes	*Turdidae*
thrush nightingale	*Luscinia luscinia*
tree-pipit	*Anthus trivialis*
tree sparrow	*Passer montanus*
tree swallow	*Tachycineta bicolor*
tree-creeper	(see Eurasian and short-toed tree-creeper)
trumpeter pigeon	(see pigeon)
tufted titmouse	*Parus bicolor*
tui	*Prosthemadera novaeseelandiae*
turkey	*Meleagris gallopavo*
turkey vulture	*Cathartes aura*
turtle doves	(see fig. **10.2**, p. 305)
twite	*Carduelis flavirostris*
vesper sparrow	*Pooecetis gramineus*
viduine finches	*Vidua spp.*
village indigobird	*Vidua chalybeata*
village weaverbird	*Ploceus cucullatus*
vinaceous dove	*Streptopelia vinacea*
Virginia rail	*Rallus limicola*
warbler finch	*Certhidea olivacea*
water pipit	*Anthus spinoletta*
western grebe	*Aechmophorus occidentalis*
western meadowlark	*Sturnella neglecta*
western screech-owl	*Otus kenniottii*
whistling eagle	*Haliastur sphenurus*
whitehead	*Mohoua albicilla*
white-backed munia	*Lonchura striata*
white-browed tit	*Poecile superciliosa*
white-crowned sparrow	*Zonotrichia leucophrys*
white-eyed vireo	*Vireo griseus*
white-shouldered black tit	*Parus leucomelas*
white-spotted flufftail	*Sarothrura pulchra*
white-tailed leaf warbler	*Phylloscopus davisoni disturbans*
white-throated sparrow	*Zonotrichia albicollis*
white-winged shrike tanager	*Lanio versicolor*
whooping crane	*Grus americana*
willow tit	*Parus montanus*
willow warbler	*Phylloscopus trochilus*
Wilson's warbler	*Wilsonia pulsilla*
winter wren	*Troglodytes troglodytes*
wood pigeon	*Columba palumbus*
wood thrush	*Hylocichla mustelina*
wood warbler	*Phylloscopus sibilatrix*
woodcock	*Scolopax rusticola*
woodpecker finch	*Cactospiza pallida*
wrens	*Troglodytidae*
wrentit	*Chamaea fuscata*
yellow longbill	*Macrosphenus flavicans*
yellow warbler	*Dendroica petechia*
yellowhammer	*Emberiza citrinella*
yellowhead	*Mohoua ochrocephala*
yellow-eyed junco	*Junco phaeonotus*
yellow-headed blackbird	*Xanthocephalus xanthocephalus*
yellow-naped Amazon	*Amazona auropalliata*
yellow-rumped cacique	*Cacicus cela*
zebra dove	*Geopelia striata*
zebra finch	*Taenopygia guttata*

Index

B after a page number indicated material in a box
Figures and tables are in Italic
P after a roman numeral indicates a plate number

Introduction to the CD's

Recordings on the two CD's that accompany Nature's Music are all related in some way to the science of birdsong. We are deeply indebted to the many authors who provided them. We are especially grateful to the following: to Albery Albuquerque Jr., for permission to use excerpts from his "Timbres da Natureza Amazonica II Fontes de Clavenários": to Jack Bradbury for allowing access to recordings from "The Diversity of Animal Sounds," published by the Macaulay Library of Natural Sounds at the Cornell Laboratory of Ornithology, and to Don Kroodsma for recordings from the CD of the Home Study Course in Bird Biology on Vocal Behavior (Chapter 7) published by the Cornell Laboratory of Ornithology. Richard Ranft at the British Library of Wildlife Sounds at the National Sound Archive provided several recordings. Others who were most helpful included Allan Baker, Hans-Heiner Bergmann, Jack Bradbury, Ian Burrows, Lang Elliot, Bruce Falls, Robin Jung, Albertine Leitao, Gene Morton, Jürgen Nicolai, Susan Peters, Katharina Riebel, and Sandy Vehrencamp.

Nature's Music. The science of Music – CD1

1 Sound track from "Tímbres da Natureza Amazônica' by Albery Albuquerque Júnior
2 This is CD 1 of 2 belonging to Nature's Music, a book on the current state of the art in the science of birdsong. The sounds on this CD are associated with the text in the fourteen chapters.

		Recorded by	Date	Location
Chapter 1				
3 chaffinch recordings by Thorpe from 1954 and 1956	a normal Cambridge chaffinch	W.H. Thorpe	4/20/54	Cambridge, England
	another normal Cambridge chaffinch	W.H. Thorpe	6/15/54	Cambridge, England
	an isolate 'Kaspar Hauser' song	W.H. Thorpe	5/7/56	Cambridge, England
	an abnormal chaffinch tutored with pipit song	W.H. Thorpe	5/30/56	Cambridge, England
4 English chaffinch from Sussex		A.J. Baker	5/2/94	Medstead, England
New Zealand chaffinch		A.J. Baker	5/2/94	Karioi, New Zealand
5 indigo bunting		W.L. Hershberger	7/25/96	West Virginia, USA
6 winter wren	two winter wren songs then one at half speed	G.F. Budney	5/14/94	Franklin Co, Bay Pond Park, New York, US
7 fawn-breasted bowerbird	imitating a carpenter at work	I. Burrows	8/23/93	University campus, Port Moresby, Papua New Guinea
Indian mynah bird (hill mynah)	mimicry of human voice	C. Greenwalt		
8 common nightingale		A.B. van den Berg	5/22/92	Kennemerduinen, Bloemendaal, The Netherlands
wood thrush		D.J. Borror	7/2/54	Woods, Franklin, County, Ohio
brown-backed solitaire		W.A. Thurber	3/4/72	El Salvador, Santa Ana, Cerro Verde
superb lyrebird		F.N. Robinson		Australia
tui		L.B. McPherson	1956	Kapiti Island, New Zealand
Australian magpie		M.J. Widdowson	1/11/03	near Carlsbrook, Victoria, Australia
9 musician wren	from Peru	T.A. Parker III	8/27/82	Cocha Cashu, Manu N.P., Madre de Dios, Peru
10 bullfinch	imitation of German folk tones	J. Nicolai		Germany
Chapter 2				
11 barn swallow	two songs both ending with a rattle	P. Galeotti	2001	Northern Italy

	water pipit	song including snarr notes	U. Rehsteiner	6/13/91	Dischma Valley, Davos, Switzerland
12	European blackbird	high intensity song	T. Dabelsteen		Danmark
		low intensity song	T. Dabelsteen		Danmark
13	collared dove	coo bout with abrupt frequency changes within coos	H. Slabbekoorn	5/9/94	Moordrecht, The Netherlands
14	great reed warbler	syllable repertoire	D. Hasselquist	1997	Sweden
	great tit	song type repertoire, six song types of one bird	M. Peet	5/8/02	Leiderdorp, The Netherlands
15	European starling	song with heterospecific mimicry including a chicken and a goat	M. Eens	1984	Antwerp, Belgium
16	zebra finch	artificially constructed repertoire	S. Collins	1997	Leiden, The Netherlands
		artificially constructed repetition	S. Collins	1997	Leiden, The Netherlands
17	black cap	male song	S. Collins	2002	Sobreda, Portugal
18	superb fairy wren	type I song	C. Blackmore		Australia
		type II song	C. Blackmore		Australia
19	red-cheeked cordon bleu	male song	S. Collins	2001	Nottingham, UK
		female song	S. Collins	2001	Nottingham, UK
20	blue-breasted cordon bleu	male song	S. Collins	2001	Nottingham, UK
		female song	S. Collins	2001	Nottingham, UK
Chapter 3					
21	swamp sparrow	subsong		~1980–1982	Milbrook, New York: laboratory
	song thrush	subsong	Inst. Anim. Behav. FU	1994	Berlin, Germany: laboratory
	nightingale	subsong	Inst. Anim. Behav. FU	1998	Berlin, Germany: laboratory
22	nightingale	early plastic song	Inst. Anim. Behav. FU	1998	Berlin, Germany: laboratory
		plastic song	Inst. Anim. Behav. FU	1998	Berlin, Germany: laboratory
		late plastic song	Inst. Anim. Behav. FU	1998	Berlin, Germany: laboratory
		crystallized song	Inst. Anim. Behav. FU	1998	Berlin, Germany: laboratory
23	song thrush	early plastic song	Inst. Anim. Behav. FU	1994	Berlin, Germany: laboratory
		plastic song	Inst. Anim. Behav. FU	1994	Berlin, Germany: laboratory
		late plastic song	Inst. Anim. Behav. FU	1994	Berlin, Germany: laboratory
		crystallized song	Inst. Anim. Behav. FU	1994	Berlin, Germany: laboratory
24	swamp sparrow	subsong	S. Peters	~1980–1982	Milbrook, New York: laboratory
		early plastic song	S. Peters	~1980–1982	Milbrook, New York: laboratory
		late plastic song	S. Peters	~1980–1982	Milbrook, New York: laboratory
		crystallized song	S. Peters	~1980–1982	Milbrook, New York: laboratory
25	nightingale	solo-singing	H. Hultsch	1987	Berlin, Germany: Gatow
		counter-singing	D. Todt	1987	Berlin, Germany: Gatow

			Recorded by	Date	Location
Chapter 4					
26	eastern phoebe	fee-bee and fee-b-bee song			
27	three-wattled bellbird	Monteverde dialect	D. Kroodsma	7/1/01	Monteverde, Costa Rica
		Panama dialect	D. Kroodsma	July 2000	Monteverde, Costa Rica
28	sedge wren	song sequence	G.A. Keller	10/2/94	Coos County, Oregon, US
29	chipping sparrow	male 1	D.S. Herr		Washington, US
		male 2	R.S. Little	5/28/98	Van Etten, New York, US
		male 3	G.A. Keller	6/17/00	Klamath, Oregon, US
30	chestnut-sided warbler	'accented ending' song, to attract females	W.W.H. Gunn	6/11/56	Sinclair Twp. Limberlost Rd, Ontario, Canada
	chestnut-sided warbler	'unaccented ending' song, to repel males	D. Kroodsma	June 1990	Berkshire Mts., Massachusetts, US
31	Bewick's wren	subsong	D. Kroodsma	1971	Corvallis, Oregon, US
		adult song	D. Kroodsma	1971	Corvallis, Oregon, US
32	example of simple song	Henslow's sparrow	G.A. Keller	5/22/95	Johnson County, Indiana, US
	example of complex song	brown thrasher	G.A. Keller	5/1194	Florida, Marion County, Florida, US
33	marsh wren		J. Brazie	3/21/96	Wild Gustin, California, US
34	black-capped chickadee		D. Kroodsma	May 1996	Montague, Massachusetts
35	superb lyrebird	mimicry of various species	F.N. Robinson		Australia
36	northern mockingbird	mimicry of various species	G.F. Budney	June 1994	Luray, Groton Plantation, South Carolina, US
Chapter 5					
37	chaffinch	'chink' separation/alarm calls	K. Riebel	6/8/96	Kippo Plantation, near St. Andrews, UK
		'tupe' flight calls		10/17/01	Ijmuiden, Noord-Holland, Netherlands
		'huit' rain calls	K. Riebel	5/17/96	Magus Muir, near St. Andrews, UK
38	chicken	food calls	C. Evans	~1992	Davis, California, US
		ground alarm call	C. Evans	~1992	Davis, California, US
		aerial alarm calls	C. Evans	~1997	Sydney, Australia
39	orange-fronted conure	loud contact calls	J. Bradbury	May 2001	Finca Centeno, Guancaste Province, Costa Rica
		pre-flight calls	J. Bradbury	May 2001	Finca Centeno, Guancaste Province, Costa Rica
		soft contact calls	J. Bradbury	May 2001	Finca Centeno, Guancaste Province, Costa Rica
		begging call	J. Bradbury	May 2001	Finca Centeno, Guancaste Province, Costa Rica

		agonistic protest calls	J. Bradbury	May 2001	Finca Centeno, Guancaste Province, Costa Rica
		warble notes	J. Bradbury	May 2001	Finca Centeno, Guancaste Province, Costa Rica
40	seet alarm calls by different species	American robin	L. Elliot	8/3/90	Deer Park, Upper Peninsula of Michigan, US
		European blackbird	H. Slabbekoorn	5/30/04	De Pijp, Amsterdam, the Netherlands
		tufted titmouse	G.A. Keller	4/8/87	Laguna Atascosa NWR, Texas, US
		black-capped chickadee	W.W.H. Gunn	8/13/57	Algonquin Park, Ontario, Canada
41	mobbing calls by different species	blue jay	H. McIsaac	9/23/79	Ithaca, New York, US
		house wren	W.W.H. Gunn	July 1974	Parkland, Alberta, Canada
		tufted titmouse	G.A. Keller	4/8/87	Laguna Atascosa NWR, Texas, US
		Mexican chickadee, western tanager & red-breasted nuthatch	R.S. Little	5/27/77	Herb martyr, Basin Trail, Arizona, US
42	distress calls by different species	white-throated sparrow	B. Falls & D. Robinson	October 1963	Long Point Bird Observatory, Lake Erie, Canada
		swamp sparrow	B. Falls & D. Robinson	October 1963	Long Point Bird Observatory, Lake Erie, Canada
		song sparrow	B. Falls & D. Robinson	October 1963	Long Point Bird Observatory, Lake Erie, Canada
43	acorn woodpecker	distress calls	W. Koenig	3/30/02	Hastings, California, US
44	California quail	contact calling	J. Gee	March 1998	Mountain Center, Palm Desert, California, US
45	crested tit	variety of calls	I. Krams	10/28/01	Southeastern Latvia near Kraslava
46	nocturnal flight calls of different species	American redstart	W.R. Evans	May 1989	Eastern Florida, US
		veery	W.R. Evans	Aug 1988	Near Ithaca, New York, US
		bobolink	W.R. Evans	May 1989	Eastern Florida, US
		upland sandpiper	W.R. Evans	Aug 1989	West/Central New York, US
47	raven	yells	B. Heinrich	winter 1991	western Maine, US
48	barn swallow	begging calls, 4 recording	M. Beecher, P. Stoddard & M. Medvin	~1982–1983	Okanogan & Kittas Counties, Washington, US
49	cliff swallow	begging calls, 4 recordings	M. Beecher, P. Stoddard & M. Medvin	~1982–1983	Okanogan & Kittas Counties, Washington, US
50	cuckoo chick, in a reed warbler nest, 9 days old	begging calls	N. Davies	7/5/02	Wicken Fen, Cambridgeshire, UK
51	brown-headed cowbird	call dialect 1, Convict	S. Rothstein, A. O'Loghlen & R. Fleischer	April–July 1980's	Convict, California, US

Track	Description	Recorded by	Date	Location
	call dialect 2, Lee Vining	S. Rothstein, A. O'Loghlen & R. Fleischer	April–July 1980's	LeeVining, California, US
	call dialect 3, Mammoth	S. Rothstein, A. O'Loghlen & R. Fleischer	April–July 1980's	Mammoth, California, US
Chapter 6				
52 white-crowned sparrow	clean song	H. Slabbekoorn	6/18/01	Sierra Nevada, California, US
	song after high-frequency filtering	H. Slabbekoorn	6/18/01	Sierra Nevada, California, US
	song with artificial echoes	H. Slabbekoorn	6/18/01	Sierra Nevada, California, US
53 green hylia	song with natural ambient noise	H. Slabbekoorn	4/24/99	Eboum-Etoum, Cameroon
	same song without ambient noise	H. Slabbekoorn	4/24/99	Eboum-Etoum, Cameroon
54 ambient noise examples	riparian scrubs	H. Slabbekoorn	5/28/01	Sierra Nevada, California, US
	montane meadow	H. Slabbekoorn	6/23/00	Rocky Mountains, Colorado, US
	street noise	H. Slabbekoorn	12/14/01	San Francisco, California, US
	rainforest	H. Slabbekoorn	4/15/99	Dja Reserve, Cameroon
	littoral forest	H. Slabbekoorn	3/22/99	Nguti, Cameroon
	gallery forest	H. Slabbekoorn	2/25/99	Wakwa, Cameroon
55 king penguin	colony noise	T. Aubin	11/3/99	Possession Island, Crozet Archipelago
	calls of one individual	T. Aubin	11/10/99	Possession Island, Crozet Archipelago
56 song examples from closed habitat	yellow longbill	H. Slabbekoorn	4/15/99	Dja Reserve, Cameroon
	olive-green cameroptera	H. Slabbekoorn	4/12/99	Dja Reserve, Cameroon
	lesser bristlebill	H. Slabbekoorn	4/16/99	Dja Reserve, Cameroon
57 song examples from open habitat	northern black flycatcher	H. Slabbekoorn	2/27/99	Ranch de Ngaundaba, Cameroon
	Cabanis' bunting	H. Slabbekoorn	2/27/99	Ranch de Ngaundaba, Cameroon
	white-shouldered black tit	H. Slabbekoorn	2/15/99	Betare Oya, Cameroon
58 example of song convergence	white-spotted flufftail	H. Slabbekoorn	4/24/99	Eboum-Etoum, Cameroon
	blue-headed crested flycatcher	H. Slabbekoorn	2/15/99	Betare Oya, Cameroon
	same song without ambient noise	H. Slabbekoorn	2/15/99	Betare Oya, Cameroon
example of song divergence	blue-breasted kingfisher	H. Slabbekoorn	3/22/99	Nguti, Cameroon
		H. Slabbekoorn	4/12/99	Dja Reserve, Cameroon
59 little greenbul	song sequence 2-1-3-4	H. Slabbekoorn	3/10/99	Yoko, Cameroon
	song sequence 1-4-2-3	H. Slabbekoorn	4/13/99	Dja Reserve, Cameroon
Chapter 7				
60 four calls of three different species	budgerigar	B. Dooling		College Park, Maryland, US, laboratory

	zebra finch	B. Dooling		College Park, Maryland, US, laboratory
	canary	B. Dooling		College Park, Maryland, US, laboratory
61 two times the same standard tone series of the baseline 'chirp,' as tested on birds		H. Slabbekoor	3/19/03	Leiden, The Netherlands, laboratory
change in second chirp in tone 2		H. Slabbekoorn	3/19/03	Leiden, The Netherlands, laboratory
change in second chirp in tone 1		H. Slabbekoorn	3/19/03	Leiden, The Netherlands, laboratory
change in second chirp in tone 7		H. Slabbekoorn	3/19/03	Leiden, The Netherlands, laboratory
62 two times the same standard tone series of a slow 'chirp,' slowed down for the human ear		H. Slabbekoorn	3/19/03	Leiden, The Netherlands, laboratory
change in second slow chirp in tone 2		H. Slabbekoorn	3/19/03	Leiden, The Netherlands, laboratory
change in second slow chirp in tone 4		H. Slabbekoorn	3/19/03	Leiden, The Netherlands, laboratory
63 two of the same harmonic tones		R. Dooling	~1997	College Park, Maryland, US, laboratory
two harmonic tones that differ: the second has a mistuned baseline of 10 Hz		R. Dooling	~1997	College Park, Maryland, US, laboratory
64 two harmonic tones that differ with respect to the Schroeder complex with		R. Dooling	~2001	College Park, Maryland, US, laboratory
relatively high frequency and of relatively low frequency		R. Dooling	~2001	College Park, Maryland, US, laboratory
Chapter 8				
65 canary	before HVC lesion	M. & F. Nottebohm	6/6/73	Rockefeller University Field Center
	after HVC lesion	M. & F. Nottebohm	3/20/74	Rockefeller University Field Center
66 zebra finch	juvenile, post lAreaX lesion	C. Scharff	5/24/90	Rockefeller University Laboratory of Anim. Behav.
	juvenile, post lMan lesion	C. Scharff	5/20/90	Rockefeller University Laboratory of Anim. Behav.

		Recorded by	Date	Location
	adult, pre lAreaX lesion	C. Scharff	10/17/89	Rockefeller University Laboratory of Anim. Behav.
	adult, post lAreaX lesion	C. Scharff	11/27/89	Rockefeller University Laboratory of Anim. Behav.
Chapter 9				
67 northern cardinal	adult song	R. Suthers	1996	Bloomington, Indiana, US, laboratory
	late subsong	R. Suthers	1996	Bloomington, Indiana, US, laboratory
	plastic song	R. Suthers	1996	Bloomington, Indiana, US, laboratory
68 brown thrasher	adult song	R. Suthers	1994	Bloomington, Indiana, US, laboratory
69 brown-headed cowbird	song at normal speed	R. Suthers	1995	Bloomington, Indiana, US, laboratory
	song at 1/8 speed with spectral adjustment	R. Suthers	1995	Bloomington, Indiana, US, laboratory
70 northern cardinal experiment	synthesized cardinal tutor song	R. Suthers	2000	Bloomington, Indiana, US, laboratory
	mockingbird imitation of cardinal song	R. Suthers	2000	Bloomington, Indiana, US, laboratory
71 domestic canary	song including 'sexy' phrases	R. Suthers	1999	University of Paris X, Nanterre, France
	'sexy' phrases only at 1/8 speed	R. Suthers	1999	University of Paris X, Nanterre, France
Chapter 10				
72 Eurasian collared dove	perch-coo	M. Ballintijn	1994	captivity, Leiden, the Netherlands
	bow-coo	M. Ballintijn	1994	captivity, Leiden, the Netherlands
	nest-coo	M. Ballintijn	1994	captivity, Leiden, the Netherlands
73 African collared dove	perch-coo	S. de Kort	1998	Waza National Park, Cameroon
	bow-coo	S. de Kort	1998	Waza National Park, Cameroon
	nest-coo	S. de Kort	1998	Waza National Park, Cameroon
74 African mourning dove	perch-coo	S. de Kort	1998	Waza National Park, Cameroon
	bow-coo	S. de Kort	1998	Waza National Park, Cameroon
	nest-coo	S. de Kort	1998	Waza National Park, Cameroon
75 red-eyed dove	perch-coo	S. de Kort	1998	Waza National Park, Cameroon
	bow-coo	S. de Kort	1999	Adamawa plateau, Cameroon
	nest-coo	S. de Kort	1999	Matopo National park, Zimbabwe
76 ring-necked dove	perch-coo	S. de Kort	1999	Nylsvlei, South Africa
	bow-coo	S. de Kort	1999	Hwange National park, Zimbabwe
	nest-coo	S. de Kort	1999	Hwange National park, Zimbabwe
77 vinaceous dove	perch-coo	S. de Kort	1997	Waza National Park, Cameroon
	bow-coo	S. de Kort	1997	Waza National Park, Cameroon
	nest-coo	S. de Kort	1997	Waza National Park, Cameroon

No.	Track	Call/Song	Recordist	Date	Location/Source
78	Adamawa turtle dove	perch-coo	S. de Kort	1999	Benoue National park, Cameroon
		bow-coo	S. de Kort	1999	Benoue National park, Cameroon
		nest-coo	S. de Kort	1999	Benoue National park, Cameroon
79	European turtle dove	perch-coo	S. de Kort	1996	Meyendel, the Netherlands
		bow-coo	S. de Kort	1995	captivity, the Netherlands
		nest-coo	S. de Kort	1995	captivity, the Netherlands
80	Oriental turtle dove	perch-coo	S. de Kort	1995	captivity, the Netherlands
		bow-coo	S. de Kort	1995	captivity, the Netherlands
		nest-coo	S. de Kort	1995	captivity, the Netherlands
81	island turtle dove	perch-coo			Java, Indonesia (National sound archive)
		bow-coo	S. de Kort	2000	Luzon, Philippines
		nest-coo			Java, Indonesia (National sound archive)
82	red turtle dove	perch-coo	S. de Kort	1995	captivity, the Netherlands
		bow-coo	S. de Kort	1995	captivity, the Netherlands
		nest-coo	S. de Kort	1995	captivity, the Netherlands
83	pink pigeon	perch-coo	M. Griekspoor		captivity, the Netherlands
		nest-coo	M. Griekspoor		captivity, the Netherlands
84	Madagascar turtle dove	perch-coo	S. de Kort	1995	captivity, the Netherlands
		bow-coo	S. de Kort	1995	captivity, the Netherlands
85	spotted dove	perch-coo	S. de Kort	1999	Hongkong
		bow-coo	S. de Kort	1995	captivity, the Netherlands
		nest-coo	S. de Kort	1995	captivity, the Netherlands
86	laughing dove	perch-coo	S. de Kort	1998	Waza National Park, Cameroon
		bow-coo	S. de Kort	1998	Waza National Park, Cameroon
		nest-coo	S. de Kort	1997	Adamawa plateau, Cameroon

Chapter 11

No.	Track	Call/Song	Recordist	Date	Location/Source
87	examples of sparrow songs	Bachman's sparrow	AAA, PPK	5/16/54	LNS 14976
		black-throated sparrow, song 1	G. Keller	5/17/90	LNS 50164
		song 2	G. Keller	5/17/90	LNS 50164
		lark bunting	RCS, MCM	6/12/58	LNS 15174
		lark sparrow	K. Colver	6/11/90	LNS 49778
		yellow-eyed junco	G. Keller	5/31/87	Portal, Arizona, US
		Lincoln's sparrow	L.V. Peyton	6/11/72	LNS 16706
		vesper sparrow	R.S. Little	4/2/64	LNS 15373
		chipping sparrow	G. Keller	6/12/88	LNS 42245

No.	Species	Song	Recorded by	Date	Location
88	natural and manipulated swamp sparrow songs	song A	S. Nowicki	5/1/88	Duchess County, NY
		manipulated song B	J. Podos & S. Nowicki	6/1/93	laboratory recordings
		song C	S. Nowicki	5/1/88	Duchess County, NY
		manipulated song D	J. Podos & S. Nowicki	6/1/93	laboratory recordings
		song E	S. Nowicki	5/1/88	Duchess County, NY
		manipulated song F	J. Podos & S. Nowicki	6/1/93	laboratory recordings
89	swamp sparrow song copies	copy A	J. Podos & S. Nowicki	5/3/94	laboratory recordings
		copy B	J. Podos & S. Nowicki	5/17/94	laboratory recordings
		copy C	J. Podos & S. Nowicki	5/5/94	laboratory recordings
		copy D	J. Podos & S. Nowicki	4/29/94	laboratory recordings
		copy E	J. Podos & S. Nowicki	5/25/94	laboratory recordings
90	songs of the Darwin's finches	large ground finch	J. Podos	2/16/99	Santa Cruz Island, Galapagos
		medium ground finch	J. Podos	2/22/99	Santa Cruz Island, Galapagos
		small ground finch	J. Podos	3/11/99	Santa Cruz Island, Galapagos
		cactus finch	J. Podos	2/15/99	Santa Cruz Island, Galapagos
		small tree finch	J. Podos	2/23/99	Santa Cruz Island, Galapagos
		large tree finch	J. Podos	3/7/99	Santa Cruz Island, Galapagos
		woodpecker finch	J. Podos	3/11/99	Santa Cruz Island, Galapagos
		warbler finch	J. Podos	3/10/99	Santa Cruz Island, Galapagos
Chapter 12					
91	subtropical pygmy-owl		F. Sornoza Molina	7/22/92	MLNS 63426
	Andian pygmy-owl		P. Swartz	10/18/70	Tachira, Venezuela (ARA 16, MLNS)
	Emei leaf warbler		P. Alstrom		Emei Shan, Sichuan Province, China (BLNSA Kina94 II#01)
	Blyth's leaf warbler		P. Alstrom		China (BLNS Kina94 I#83)
	Lulu's tody-tyrant		B.M. Whitney	8/25/89	Florida de Pomacochas, Depto. Amazonas, Peru
	rufous-crowned tody-tyrant		P. Donahue	4/23/90	Prov. Zamora-Chinchipe, Ecuador (MLNS 55778)
	Socorro dove		L.F. Baptista		captive, Calif. Academy (ARA Record 14)
	mourning dove		S.L.L. Gaunt	5/25/96	Franklin Co., OH (BLB 21146)
92	Galapagos petrel	colony calls	R.J. Tompkins	3/31/79	MLNS 20301
93	black-faced solitaire		L.F. Baptista	2/15/92	Monteverde, Costa Rica (BLB 21615)
	white-browed tit		S.L.L. Gaunt	5/19/91	San Jose Prov. Costa Rica (BLB 18073)

Track	Species	Description	Recordist	Date	Location
	dusky seaside-sparrow		W. Post	6/2/79	Bervard Co, FL (FSMNH 724 cut2)
	ivory-billed woodpecker	calls, drumming	A. Allen, P. Kellogg	4/9/35	MLNS 6784
	white-tailed leaf warbler		P. Alstrom	4/29/96	China (BLNS Kina94 I#82)
94	western screech-owl	call of male and female	B.L. Herting	6/1/93	Owyhee Co, ID
	whooping crane	call of male and female	Crane Foundation	6/1/93	Baraboo Wisconsin, US
	Leach's storm-petrel	call of female	L. Elliot	6/6/02	Kent Island, NB, Canada (BLB 28010)
		call of male	L. Elliot	3/5/83	Kent Island, NB, Canada (BLB 28010)
Chapter 13					
95	African grey parrot	training session with Irene and Alex	I. Pepperberg		Tucson, Arizona, US
Chapter 14					
96	brown-headed cowbird	four different songs	A. King	6/6/83	Bloomington, Indiana, US
97	European starling, mimicry examples	cowbird notes	A. King	3/5/83	Mebane, North Carolina, US
		the real cowbird	A. King	3/5/83	Mebane, North Carolina, US
		"hey Rex c'mon"	A. King	3/5/83	Mebane, North Carolina, US
		"screen door"	M. West	3/5/83	Mebane, North Carolina, US
		contact whistle	M. West	3/5/83	Mebane, North Carolina, US
		human whistling	M. West	3/5/83	Mebane, North Carolina, US
98	adult-directed speech			11/21/02	Franklin and Marshall College, Lancaster, PA
	infant-directed speech			11/21/02	Franklin and Marshall College, Lancaster, PA

Nature's Music. The science of Music – CD2

1 Sound track from 'Tímbres da Natureza Amazônica' by Albery Albuquerque Júnior

2 This is CD 2 of 2 belonging to Nature's Music, a book on the current state of the art in the science of birdsong. The sounds on this CD are associated with the text in 28 out of the 48 boxes that are distributed throughout the book.

Box 2 (Chapter 1)			
3 sonogram A, three artificial tones of different frequency	H. Slabbekoorn	11/4/02	Leiden, the Netherlands
4 sonogram B, African cuckoo	H. Slabbekoorn	3/5/99	Djohong, Cameroon
5 sonogram C, introductory note of a white-crowned sparrow	H. Slabbekoorn	6/18/01	Sierra Nevada, California, US
6 sonogram D, alarm call of a great tit	H. Slabbekoorn	4/8/02	De Vuursche, the Netherlands
7 sonogram E, three artificial tones with changing frequency	H. Slabbekoorn	11/4/02	Leiden, the Netherlands
8 sonogram F, fox sparrow song	H. Slabbekoorn	6/23/00	Gothic, Colorado, US
9 same fox sparrow song, at 1/8 speed	H. Slabbekoorn	6/23/00	Gothic, Colorado, US
10 sonogram G + H, artificial tones plus yellow warbler song	H. Slabbekoorn	6/7/00	Gothic, Colorado, US
11 sonogram I, two songs of a Diederik cuckoo	H. Slabbekoorn	6/28/99	Lendi, Cameroon
12 sonogram J, artificially-created harmonic tones	H. Slabbekoorn	11/4/02	Leiden, the Netherlands
13 sonogram K, zebra finch song	N. Terpstra	9/20/02	captivity, Leiden, the Netherlands
14 sonogram L, calls of the Eurasian jay	H. Slabbekoorn	4/10/94	Soestduinen, the Netherlands
15 sonogram M, black-bellied seedcracker, two songs of different individuals	T.B. Smith	5/22/86	Ndibi, Cameroon
Box 4 (Chapter 2)			
16 domestic canary, song containing sexy syllables	A. Tanvez		
Box 5 (Chapter 2)			
17 song sparrow, normal song	R. Anderson	5/18/02	Geneva, Crawford County, PA
isolate song	S. Peters	1981	Millbrook, New York, US
song of a deaf bird	S. Peters	1979	Millbrook, New York, US
Box 6 (Chapter 2)			
18 tawny owl, male 1, hoot 1	S. Redpath	3/1/94	Cambridgeshire, UK
male 1, hoot 2	S. Redpath	3/1/94	Cambridgeshire, UK

	male 2	S. Redpath	3/1/94	Cambridgeshire, UK
	male 3	S. Redpath	3/1/94	Cambridgeshire, UK
Box 8 (Chapter 2)				
19	Clark's grebe, female advertising calls	G. Nuechterlein	4/1/79	Klamath Falls, Oregon, US
	male advertising calls	G. Nuechterlein	4/1/79	Klamath Falls, Oregon, US
20	western grebe, female advertising calls	G. Nuechterlein	4/1/79	Klamath Falls, Oregon, US
	male advertising calls	G. Nuechterlein	4/1/79	Klamath Falls, Oregon, US
Box 9 (Chapter 3)				
21	swamp sparrow	D. Borror	7/16/78	Duckpuddle Pond, Lincoln Co, ME
22	song sparrow	J. Soha	5/17/00	Pacific City, Tillamook Co, OR
23	white-crowned sparrow	J. Orjuela/M. Morton	6/23/70	Onion Valley, Inyo Co, CA
24	white-crowned sparrow, song with whistle removed	L.F. Baptista	6/25/72	Steens Mountain, Harney Co, OR
	heterospecific song with whistle, created with Belding's ground squirrel call	J. Soha	7/1/94	Tioga Pass, Mono Co, CA
	same heterospecific song without whistle	J. Soha/R. Lein	7/1/94 & 1970's	Tioga Pass, Mono Co, CA + Turner Valley, Alberta, Canada
25	white-crowned sparrow pupil's learned song	J. Soha	6/14/95	Animal Communication Lab at UCDavis, Yolo Co, CA
Box 12 (Chapter 3)				
26	eastern white-crowned sparrow	D.A. Nelson	6/11/95	Churchill (20 km east of), Manitoba, Canada
	2nd individual	D.A. Nelson	6/7/02	Kuujjuarapik, Nunavik, Quebec, Canada
27	Gambel's white-crowned sparrow	K. Colver	6/12/97	Cantwell (16 km east of), Maunuska-Susitna, Alaska, US
	2nd individual	G.D. McNett	6/24/97	Churchill (20 km east of), Manitoba, Canada
28	mountain white-crowned sparrow	D.A. Nelson	7/9/93	Bridgeport (37 km west of), Mono and Tuolumne, California, US
	2nd individual	D.A. Nelson	7/5/95	Markleeville (18 km west of), Alpine, California, US
29	Nuttall's white-crowned sparrow	D.A. Nelson	3/25/98	Inverness, Marin, California, US
	2nd individual	D.A. Nelson	3/26/97	Pajaro, Monterey, California, US
30	Puget Sound white-crowned sparrow	D.J. Borror	7/15/65	Corvallis (12 mi south of), Benton, Oregon, US
	2nd individual	D.J. Borror	6/23/65	Lewisburg (6 mi northwest of), Benton, Oregon, US
Box 17 (Chapter 4)				
31	sedge warbler	L. Macaulay	4/3/97	Oued Masso south of Agadir, Marocco

No.	Recording	Recordist	Date	Location
Box 18 (Chapter 4)				
32	Boran cisticola	C. Chappuis	1974	West-Africa
	mimicry by marsh warbler	F. Dowsett-Lemaire	1975	Belgium
33	vinaceous dove	C. Chappuis	1974	West-Africa
	mimicry by marsh warbler	F. Dowsett-Lemaire	1974-1975	Belgium
34	brubru shrike	C.H. Haagner	1961	Kruger Park, South Africa
	mimicry by marsh warbler	F. Dowsett-Lemaire	1974-1975	Belgium
Box 20 (Chapter 5)				
35	red crossbill flight calls, type2, variant1	T. Hahn	11/19/88	Devils Table, WA
	type2, variant2	T. Hahn	8/21/98	Old Station, CA
36	type4, variant1	T. Hahn	7/6/88	Cedar Rock, Shaw Island, WA
	type4, variant4	T. Hahn	2/17/89	Swank Pass, WA
37	type2, two birds of a pair	T. Hahn	8/21/98	Old Station, CA
Box 22 (Chapter 6)				
38	common yellowthroat, perched song	L. Elliot	5/3/90	Tompkins Co., NY
	flight song	L. Elliot	5/28/90	Tompkins Co., NY
Box 23 (Chapter 6)				
39	chipping sparrow	D. Borror	6/24/76	Worcester County, MD
40	field sparrow	D. Borror	5/10/87	Strafford County, NH
41	field sparrow, control song used for playback	D.A. Nelson	~1985–1988	New York, US
	song increased in frequency with 1 JMD	D.A. Nelson	~1985–1988	New York, US
	song decreased in pause duration with 1 JMD	D.A. Nelson	~1985–1988	New York, US
Box 24 (Chapter 6)				
42	rufous-collared sparrow, habitat-related dialects, grassland dialect	P. Handford	1984	Tucuman province, Argentina
43	desert scrub dialect	P. Handford	1999	Mendoza province, Argentina
44	puna scrub dialect	P. Handford	1985	Jujuy province, Argentina
45	chaco dialect	P. Handford	1984	Tucuman province, Argentina
46	alder dialect	P. Handford	1984	Tucuman province, Argentina
47	transition dialect	P. Handford	1984	Tucuman province, Argentina
Box 25 (Chapter 6)				
48	greenish warbler, songs from 8 populations, population TL	D. Irwin	5/29/95	Teletsk Lake, Altai Mountains, Russia
49	population AA	D. Irwin	6/22/97	Ala Archa N.P., Kyrgyzstan

#	Recording	Recordist	Date	Location
50	population PK	S. Gross/T. Price	1999	Pakistan
51	population MN	Z. Benowitz-Fredericks/K. Marchetti	1997	Manali, Himachal Pradesh, India
52	population LN	D. Irwin	5/21/97	Langtang N.P., Nepal
53	population EM	J. Irwin	5/17/02	Emeishan, Schezwan Province, China
54	population XN	J. Irwin/D. Irwin	6/4/00	Laoye Shan, Qinghai Province, China
55	population ST	D. Irwin	7/5/96	Stolbi N.P., Krasnoyarsk Region, Russia
Box 26 (Chapter 7)				
56	Bengalese finch, song before deafening	K. Okanoya	6/1/92	National Agriculture Research Center
	song 5 days after deafening	K. Okanoya	6/9/92	National Agriculture Research Center
	song 30 days after deafening	K. Okanoya	7/3/92	National Agriculture Research Center
Box 27 (Chapter 7)				
57	four series of artificial tones, each repeated three times, falling in pitch	J. Cynx	3/1/03	Poughkeepsie, NY
58	rising in pitch	J. Cynx	3/1/03	Poughkeepsie, NY
59	falling in pitch and shifted down	J. Cynx	3/1/03	Poughkeepsie, NY
60	rising in pitch and shifted down	J. Cynx	3/1/03	Poughkeepsie, NY
Box 31 (Chapter 8)				
61	Bengalese finch, song pre- Area X lesion	K. Kobayasi	9/11/96	Chiba University, Japan
	song post- Area X lesion	K. Kobayasi	9/19/96	Chiba University, Japan
Box 35 (Chapter 8)				
62	sombre hummingbird	A. Ferreira	September 1999	Altantic Forest reserve, Espirito Santo, Brazil
63	rufous-breasted hermit hummingbird	A. Ferreira	September 1999	Altantic Forest reserve, Espirito Santo, Brazil
Box 36 (Chapter 9)				
64	swiftlet, echolocating clicks	R. Suthers	~1980	Lab recording, Queensland, Autralia
	oilbird, echolocating clicks in a dark cave	R. Suthers	~1963	Oropouche Cave, Trinidad, WI
65	oilbird, social vocalizations	R. Suthers	~1963	Oropouche Cave, Trinidad, WI
Box 38 (Chapter 9)				
66	zebra finch, normal song	A. Leonardo	1998	Caltech; Pasadena, CA
	decrystallized song	A. Leonardo	1998	Caltech; Pasadena, CA
Box 39 (Chapter 10)				
67	rock dove or feral pigeon, part of a coo bout	H. Slabbekoorn	1/20/03	De Pijp, Amsterdam, the Netherlands

68	'trumpeter' rock dove, one completed cooing sequence	F. Dorritie	1/23/03	"The Lort of John de Carlo", Gilroy, California, USA
69	Spanish timbrado canary, song sequence by 4 males as they enter song contests	E. Parra	9/1/00	Seville, Spain
70	wild canary, two songs recorded in captivity	H. Slabbekoorn	10/18/02	captivity, Leidschenveen, the Netherlands
Box 40 (Chapter 10)				
71	European quail, one growl and three calls	S. Collins	1986	captivity, Bristol, England
72	Japanese quail, two calls	S. Collins	1986	captivity, Bristol, England
73	European x Japanese quail hybrid, two calls each for two different individuals	S. Collins	1986	captivity, Bristol, England
74	Oregon junco	G.A. Keller	7/1/90	
75	white-throated sparrow	L. Ellliott	6/9/95	Franklin County, New York, USA
76	junco x white-throated sparrow hybrid	R. Jung	1/8/92	Washington, DC, USA
Box 41 (Chapter 10)				
77	firefinch	R.B. Payne	1993	captivity, Ann Arbor, MI
	mimicry by village indigobird	R.B. Payne	1993	captivity, Ann Arbor, MI
78	Bengalese finch	R.B. Payne	1992	captivity, Ann Arbor, MI
	mimicry by village indigobird	R.B. Payne	1992	captivity, Ann Arbor, MI
Box 43 (Chapter 11)				
79	crowned crane	M.E.W. North	4/13/62	Tanzania
	whooping crane	A.A. Allen	1/29/54	Texas, US
Box 45 (Chapter 12)				
80	green violet-ear, recording 1	L. F. Baptista	2/21/95	San Jose Prov., Costa Rica
	recording 2	S.L.L. Gaunt	2/21/92	San Jose Prov., Costa Rica
Box 46 (Chapter 13)				
81	Lühder's bush shrike	C. Chappuis	February	Gabon
	mimicry by African grey parrot	J.P. Gautier	8/21/1991	Zaïre
82	bristle-nosed barbet	C. Chappuis	July	Ivory Coast
	mimicry by African grey parrot	J.P. Gautier	8/21/1991	Zaïre
83	blue cuckoo-shrike	C. Chappuis	April	Ivory Coast
	mimicry by African grey parrot	C. Chappuis	1968	Ivory Coast
84	emerald cuckoo	J. Brunel		Ivory Coast
	mimicry by African grey parrot	J.P. Gautier	8/21/1991	Zaïre

85	black cuckoo	C. Chappuis	July	Ivory Coast
	mimicry by African grey parrot	J.P. Gautier	8/21/1991	Zaïre
86	chestnut wattle-eye	C. Chappuis	December	Cameroon
	mimicry by African grey parrot	J.P. Gautier	8/21/1991	Zaïre
Box 47 (Chapter 13)				
87	yellow-naped Amazon, Northern dialect	T. Wright	March-June 1994	Costa Rica
	Southern dialect	T. Wright	March-June 1994	Costa Rica
88	ringneck parrot, Perth dialect	M. Baker	4/28/99	Perth, Western Australia
	Margaret River dialect	M. Baker	4/29/99	Margaret River, Western Australia
Box 48 (Chapter 14)				
89	Composition by Messiaen inspired by chaffinch song			
	real chaffinch song	A. Leitao	5/3/99	Warmond, The Netherlands
90	Duke Ellington inspired by song of the common mynah			
91	Harry Belafonte inspired by mourning dove cooing			
92	The Beatles using song of a European blackbird			
93	The Beta Band using song of a European blackbird			
94	Dead Can Dance using a variety of sounds from birds and other animals			

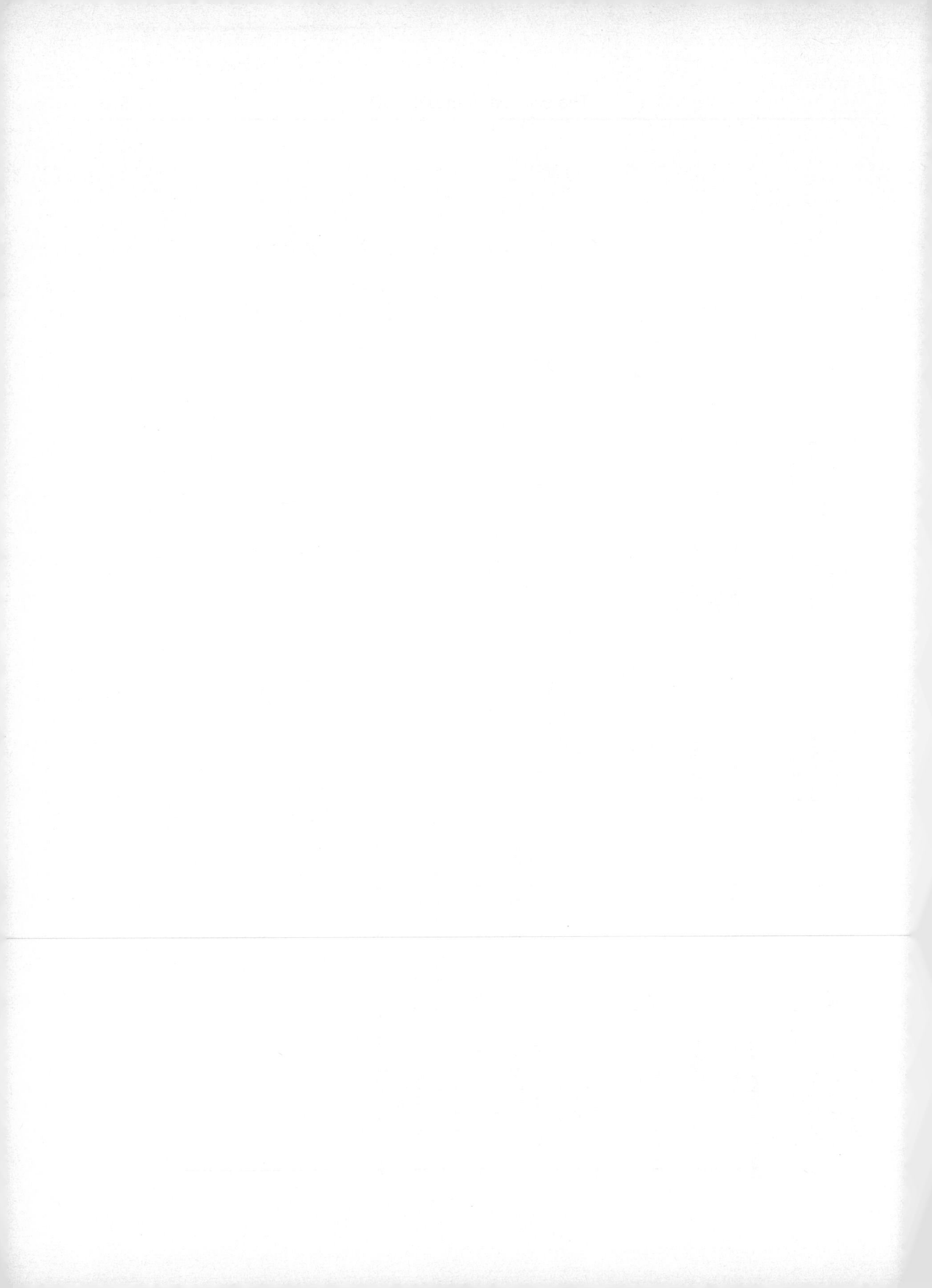

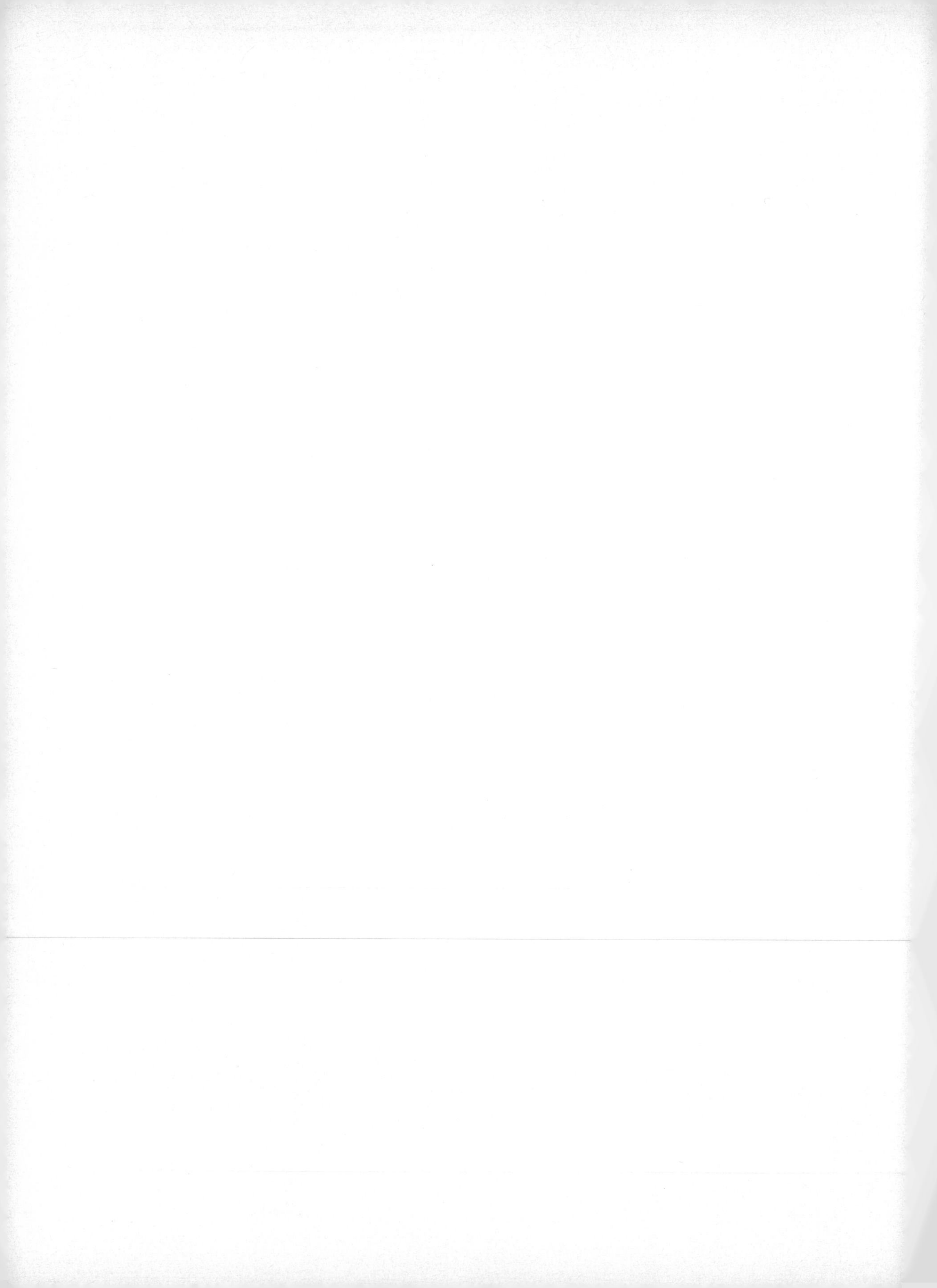

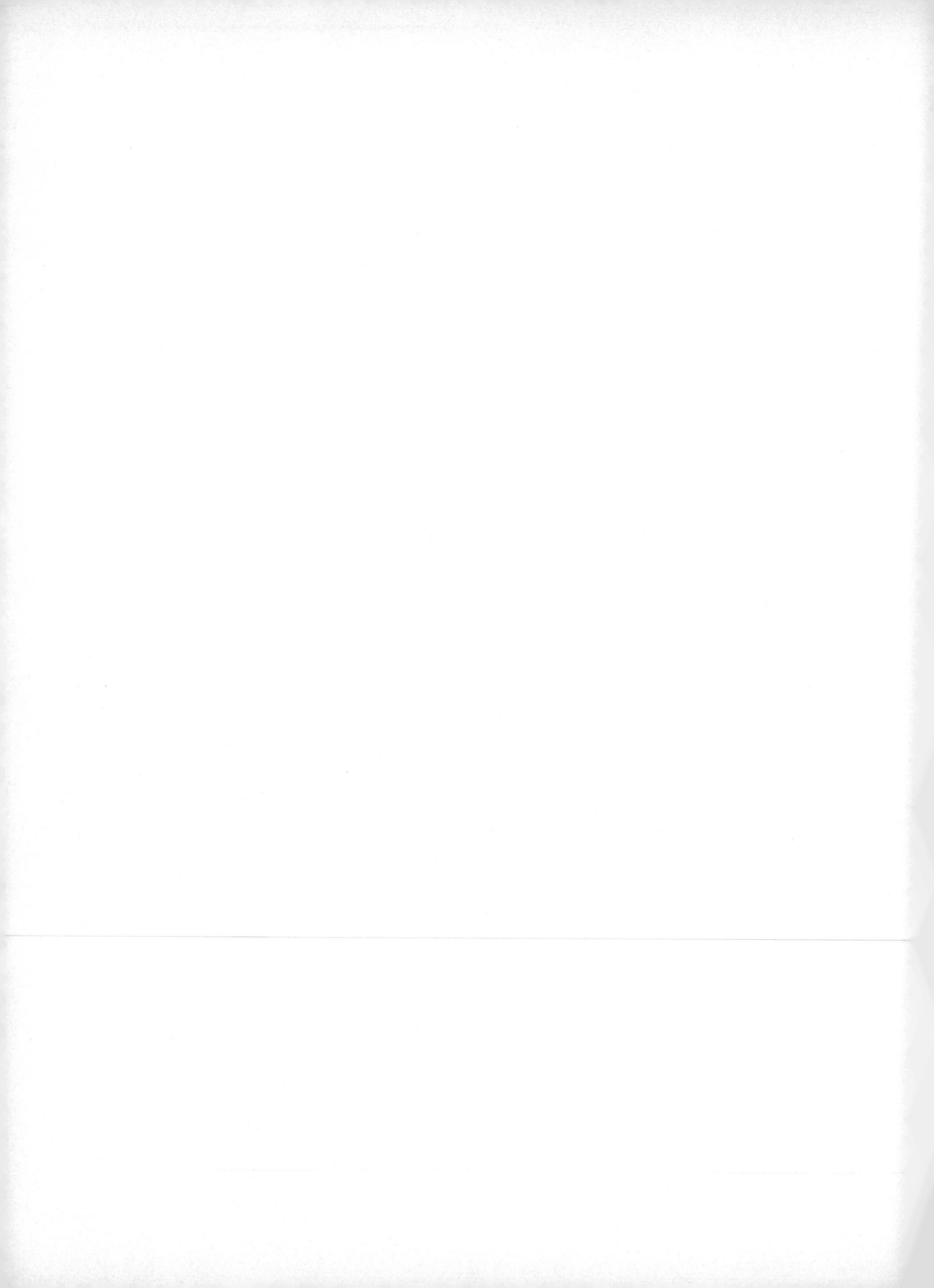

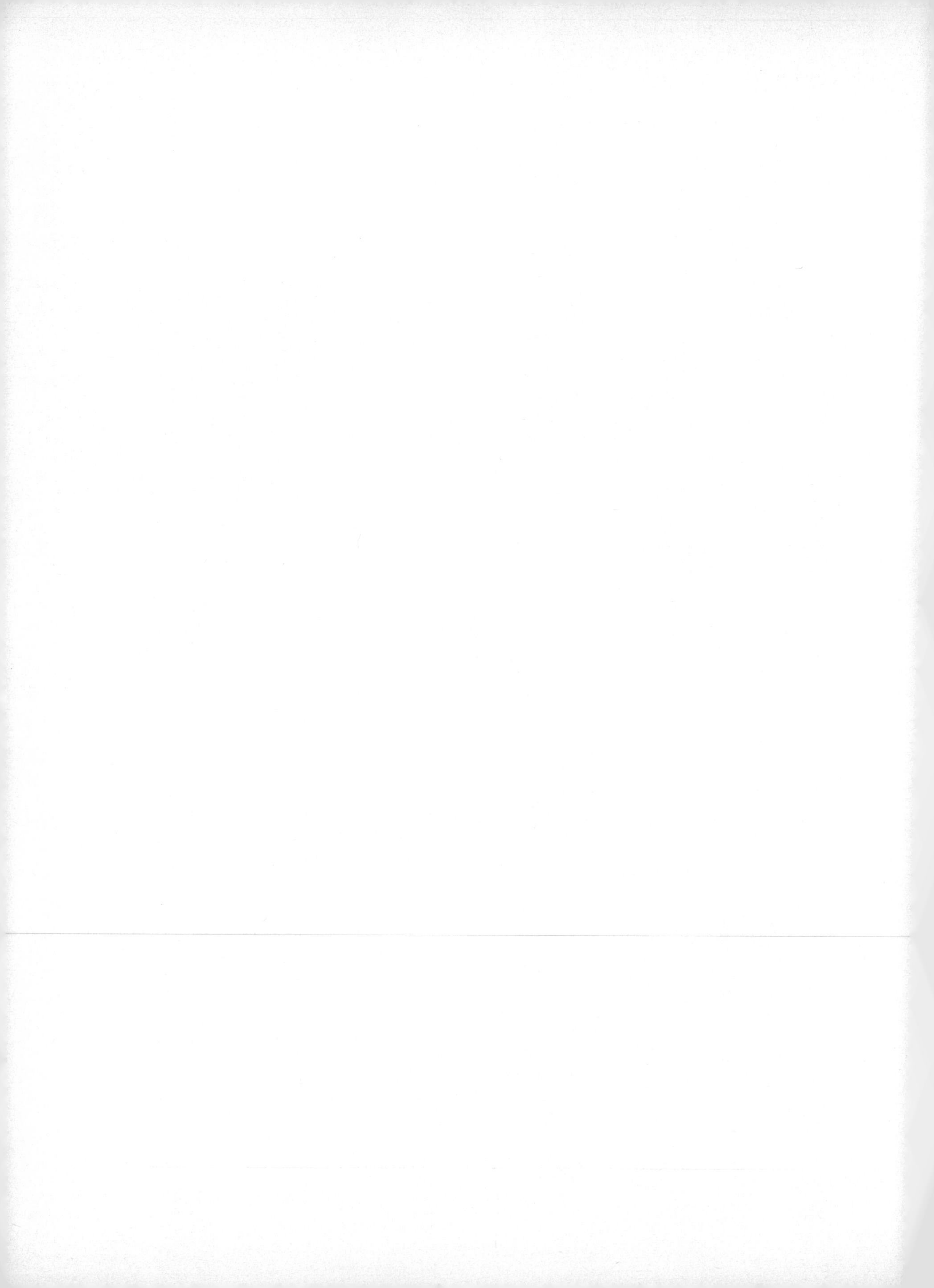